CHEMISTRY

WILLIAM L. MASTERTON

University of Connecticut, Storrs, Connecticut

EMIL J. SLOWINSKI

Macalester College, St. Paul, Minnesota

EDWARD T. WALFORD

Cheyenne Mountain High School
Colorado Springs, Colorado

HOLT, RINEHART AND WINSTON, PUBLISHERS

New York • Toronto • London • Sydney

Front cover photograph is a photomicrograph of a crystal of a cyclic monomer grown in a gaseous environment. This monomer is a building block for the production of a solid-state polymer. (Photo courtesy of National Bureau of Standards, and originally published in *Chemistry* magazine.)

Printed in the United States of America
ISBN:0-7216-6164-5
3456 032 111098765

PREFACE

This book is designed for the mainstream course in high school chemistry. We believe it is suitable for students who will not pursue the subject further as well as those who will go on to take college chemistry. With this in mind, the principles covered are restricted to those which we feel are basic to a beginning course in chemistry. These principles are developed in considerable detail from a very elementary point of view. More advanced ideas are treated in optional sections set off from the rest of the text. These may be covered at the discretion of the instructor, but are not built upon in subsequent chapters.

To make sure that students become thoroughly familiar with basic principles, we have referred to them repeatedly throughout the text. Chemical formulas, equations, and stoichiometric calculations are introduced early and reviewed frequently. Most students do not grasp all the implications of a quantitative concept the first time it is presented. Only by frequent review does an idea such as the mole become a useful tool to the student.

Descriptive chemistry, both inorganic and organic, is emphasized throughout the book. Of the 27 chapters, 12 are primarily descriptive. These are distributed throughout the book and are designed to apply and illustrate chemical principles. We believe that this approach is necessary if the student is to get an accurate picture of what chemistry is all about. Too often, he or she emerges from a first course in the subject convinced that chemistry is a sterile, intellectual exercise with little application to the real world.

One unusual feature of the book is the large number of solved examples, about 200 in all. In the solutions, we have emphasized the reasoning involved rather than which number goes where. There are also a large number of questions (500) and problems (600) at the ends of chapters. The problems are graded in difficulty. The first several problems in a set are keyed directly to the solved examples and are answered in an appendix. These are followed by similar, unanswered problems which cover all the basic ideas presented in the chapter. Finally, there are a few more advanced problems, marked with asterisks, for which answers are provided.

Most students, upon their first exposure to chemistry, have difficulties with the vocabulary. Realizing this, we have listed at the end of each chapter the new terms introduced. All of these are defined in a glossary at the back of the book. We have also tried to make sure that the reading level is appropriate for the average high school student.

There is a laboratory manual available for use with this text. The sequence of experiments follows the chapter order. We have also prepared a teacher's guide. This contains answers to the questions in the text and detailed solutions to all the problems. It also includes learning goals (basic skills) for each chapter, suggestions as to coverage of material, and appropriate demonstrations.

A great many people have contributed, directly or indirectly, to this book. It is always a pleasure to acknowledge the assistance of our editor, John Vondeling. He and Kay Dowgun have made our writing task enjoyable, insofar as that is possible. We would also like to acknowledge the copy-editorial skills of David Milley. Amy Shapiro went to a great deal of trouble to obtain appropriate photographs, many of which we used. The art work was in the capable hands of George Kelvin. Our colleagues, Ray Boyington and Ruven Smith provided us with slides for the color plates in the center of the book.

REVIEWING BOARD

We are grateful to the teachers listed below for critical reviews of the original manuscript:

Floyd Sturtevant
Ames Senior High School, Ames, Iowa

James Gardner
Millburn High School, Millburn, NJ

Selma Koury Wunderlich
Academy of Notre Dame de Namur, Villanova, PA

Jeanne Maguire
Windham High School, Willimantic, CT

Peter Dahl
Lowell High School, San Francisco, CA

Mithal Patel
Oak Park High School, Winnipeg, Manitoba, Canada

Jon Thompson
Mitchell High School, Colorado Springs, CO

Conrad Stanitski
Randolph-Macon College, Ashland, VA

Ray Boyington
University of Connecticut, Storrs, CT

Eugene Rochow
Harvard University, Boston, MA

Lowell A. King
Air Force Academy, Colorado Springs, CO

Chemistry students at Cheyenne Mountain High School
Colorado Springs, CO

CONTENTS

Chapter 15

Chapter 16

Chapter 17

One of the few elements discovered by the alchemists was phosphorus. Hennig Brand, searching for the philosopher's stone in urine, a most unlikely source, isolated phosphorus instead. Bettman Archive.

1 AN INTRODUCTION TO CHEMISTRY

You are just beginning to read this book and probably just beginning a course in high school chemistry. You are in a position similar to that of a person beginning to climb a mountain. Both you and the climber should be asking yourself two questions: (1) Why am I here? and (2) What is the nature of this course over which I intend to spend considerable time and effort traveling?

In this chapter you will be introduced to some basic ideas of chemistry. This way, you will know better what to expect in your journey. You will also learn of some of the skills that will be required. With these skills, your experience in chemistry can be both enjoyable and rewarding.

1.1 THE NATURE OF CHEMISTRY

At the root of chemistry (and all other sciences) is curiosity, the desire to know. Ancient peoples were curious for practical reasons. They wanted to find answers to their needs for food, water, and shelter. Later, as civilization developed, men and women became curious for the sake of knowledge itself. We can only guess some of their first questions. What is everything made of? Why do things change? What happens when wood burns? when water freezes? when iron rusts?

For centuries, chemists have sought answers to questions such as these. They have used many different approaches, some more rewarding than others. In this section, we will trace briefly the evolution of modern chemistry. It began as abstract theory in ancient Greece. Later, during the Middle Ages, chemistry became an art, fascinating but obscure. Only within the past 200 years has chemistry become the useful science that we know today.

Chemistry as an Exercise in Logic: the Greeks

Early Greek philosophers developed curiosity into a system of reasoning. Answers to questions about the nature of things were stated on the basis of abstract logic. No attempt was made to do experiments. Apparently, Greek intellectuals felt it was demeaning to work with their hands. Aristotle (384–322 B.C.) was the most famous of this group. He had an answer to the question: what are things made of? According to Aristotle, there were four "elements" of matter: fire, air, water, and earth. All matter was made up of these elements in varying amounts. An object that was "hot and dry" was mostly fire. "Hot and moist" was associated with air, "cold and moist" with water, "cold and dry" with earth.

Aristotle was actually a better philosopher than he was a scientist.

Today we can smile at these ideas or dismiss them as utter nonsense. Yet they persisted for nearly 2000 years. In part, this was due to the remarkable success of the Greeks in certain areas of science, notably geometry. Perhaps more important, there was no way to test these ideas in the laboratory. Any attempt to combine fire with water is doomed to failure. Indeed, it would never have occurred to Aristotle to try such an experiment. The moral is clear:

1. *A theory or explanation which cannot be tested by experiment is worthless.* It may satisfy curiosity for a time, but it leads only to a dead end.

Chemistry as Fun and Games: the Alchemists

The pseudo-science of alchemy began around 300 A.D. It flourished in Europe as late as 1700 A.D. A major goal of the alchemists was to find the "philosopher's stone" which would convert base metals to gold. To achieve this they carried out a great many experiments. In this respect, the alchemists progressed beyond the Greek philosophers.

The laboratory work of the alchemists left a great deal to be desired. Their procedures and descriptions were, to say the least, obscure. Consider, for example, the following recipe for obtaining the philosopher's stone.

"Take all the mineral salts there are, also all salts of animal and vegetable origin. Add all the metals and minerals, omitting none. Take two parts of the salts and grate in one part of the metals and minerals. Melt this in a crucible, forming a mass that reflects the essence of the world in all its colors. Pulverize this and pour vinegar over it. Pour off the red liquid into English wine bottles, filling them half-full. Seal them with the bladder of an ox (*not* that of a pig). Punch a hole in the top with a coarse needle. Put the bottles in hot sand for three months. Vapor will escape through the hole in the top, leaving a red powder, . . ."

This experiment, as you can imagine, is not easy to duplicate. If you wanted to repeat it, you might be able to locate English wine bottles. Perhaps you could even find the bladder of an ox. However, you could spend a lifetime collecting *all* the salts, metals, and minerals in the world. From this story we conclude that:

A scientific experiment usually isn't done just once.

2. *An experiment which cannot be repeated is worthless.* It must be described in such a way that others can perform it, confirming or refuting the results.

To their credit, some of the later alchemists carried out rational experiments. Consider, for example, Jon Baptista van Helmont (1577–1644). He planted a young willow tree, weighing 5 lb, in a tub filled with 200 lb of dry earth (Figure 1.1). For five years, he added only water to the tub. At that point, the tree weighed 169 lb. Virtually all of the 200 lb of earth remained. Van Helmont concluded that:

"164 lb of wood, bark, and roots arose out of water only"

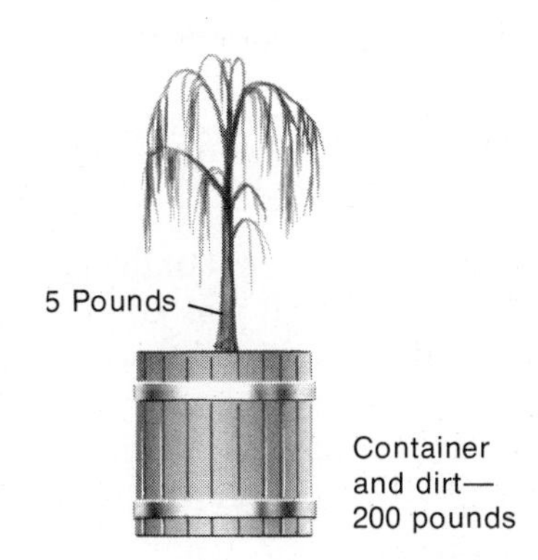

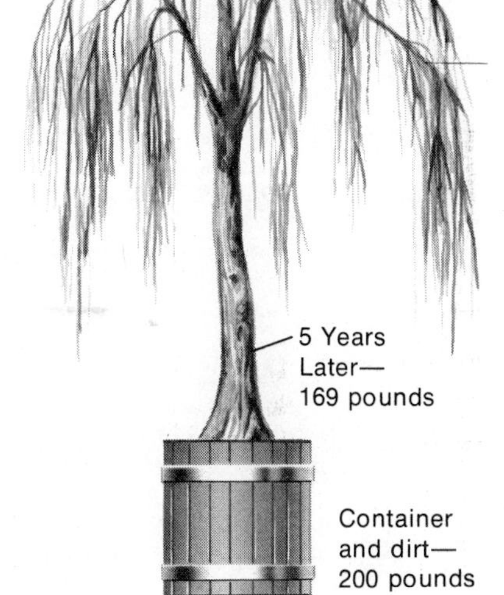

FIGURE 1.1

Van Helmont found that the mass of his tree went from 5 to 169 lbs in five years. Since he had added only water to the tub, he figured that the tree had gained its mass from the water added. He didn't realize that the added mass could have, and did, come from the air around the tree as well as from the water that was added.

This observation overlooked a possible source of the increase in mass of the willow tree. This is the air over the tree. Today we know that growing plants take in carbon dioxide from the air as well as water. In fact, more than half of the increase in mass came from the carbon dioxide. Without being too critical of van Helmont, we can say that:

3. *An experiment must be carried out under carefully controlled conditions.* Any factor which could affect the results must be taken into account.

The Dawn of Modern Chemistry: Lavoisier

One man more than any other transformed chemistry from an art to a science. Antoine Lavoisier was born in Paris in 1743. He died in

TABLE 1.1 QUANTITATIVE EXPERIMENT ON THE FERMENTATION OF WINE (LAVOISIER)

Reactants	Mass (Relative)	Products	Mass (Relative)
water	400	carbon dioxide	35.3
sugar	100	alcohol	57.7
yeast	10	acetic acid	2.5
		water	409.0
		sugar (unreacted)	4.1
		yeast (unreacted)	1.4

1794 on the guillotine during the French revolution. Above all else, Lavoisier recognized the importance of carefully controlled experiments. These were described clearly in his book, "Elements of Chemistry." Published in 1789, it is illustrated with diagrams, drawn by his wife. One such diagram is given in Figure 1.2. It shows the equipment Lavoisier used to study the fermentation of wine.

All of Lavoisier's experiments involved accurate measurements. They were *quantitative.* That is, Lavoisier carefully measured the amounts of reactants and products. The results of one such experiment are shown in Table 1.1. The data are taken directly from Lavoisier. They relate to the fermentation of wine.

If you add up the masses of products and reactants in Table 1.1, you will find them to be the same, 510. As Lavoisier put it:

> "In all of the operations of man and nature, nothing is created. An equal quantity of matter exists before and after the experiment."

The Law is true, but was not obvious, since many reactions produce gases.

This was the first statement of the Law of Conservation of Mass, a fundamental law of nature. It was confirmed by Lavoisier in many more experiments. This Law was the cornerstone for the growth of chemistry and physics in the 19th century. You can readily see why Lavoisier is called the father of modern chemistry. His work illustrates an important principle:

4. *Quantitative experiments, involving accurate measurements, can reveal the laws of nature.* Such experiments are essential to all sciences, including chemistry.

Chemistry Today

In 200 years, chemistry has progressed far beyond Lavoisier. The equipment you use in the laboratory is more sophisticated than that shown in Figure 1.2. The laws and principles discussed in this text are more advanced than those discovered by Lavoisier. Yet, there are similarities. Today's chemists, like Lavoisier, blend theory with experiment. In this way, they reveal the nature of matter and chemical change.

One of the goals of modern chemistry is to improve the quality of life. Chemists do this in many ways. They change natural products into

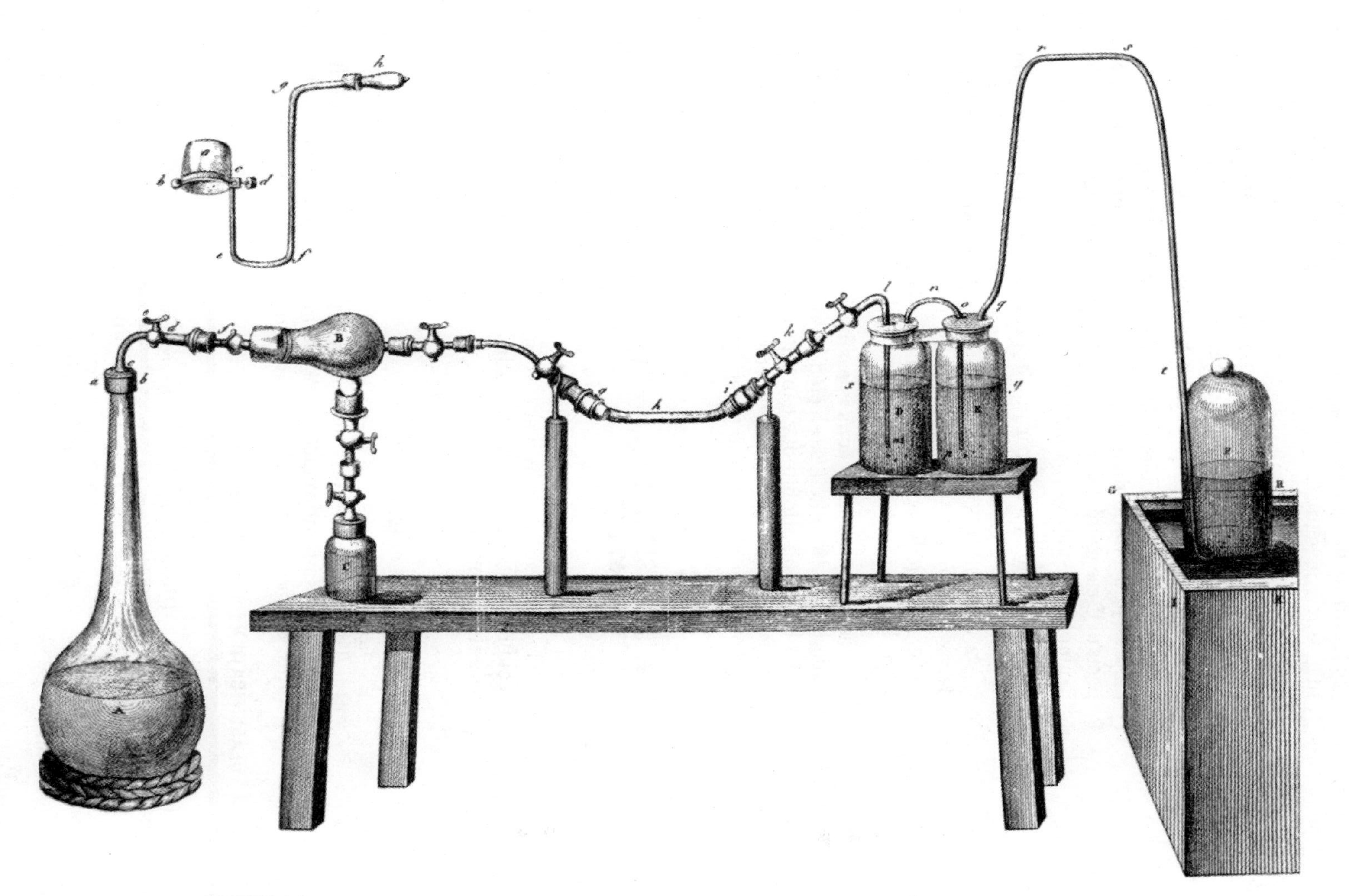

FIGURE 1.2

Apparatus used by Lavoisier to study the fermentation of wine, which led to the Law of Conservation of Mass. Antoine Lavoisier: *Elements of Chemistry*, 1789. Facsimile reprint by Dover Publications, 1965.

materials which better meet our needs. Products such as nylon and penicillin are made by chemists. At the same time, chemists are engaged in research on many of the problems that concern modern society. These include:

1. depletion of our natural resources, including petroleum.
2. pollution of the environment.
3. the nature of the aging process, and indeed of life itself.

Each of these problems has a chemical basis. Hence, chemistry can help us understand their nature and find solutions to them. Many aspects of these problems will require individual or community decisions. Examples of such decisions are:

1. Should our nuclear power program be expanded despite possible risks to the environment?

2. Should DDT and other insecticides be used to control insects at the expense of other living species?

How would you answer these questions?

3. Should laboratory experiments with DNA which create new life forms be allowed?

None of these decisions is easily reached. In the end, they will be made by public officials, elected by ordinary men and women at the ballot box. No nation today can afford a citizenry that is not informed. In today's world, an informed person must have an elementary knowledge of chemistry and the other sciences.

1.2 CLASSIFICATION AND PROPERTIES OF MATTER

The science of chemistry is the study of matter. Matter is the stuff of which the universe is made. It includes anything that has mass and occupies space. The air we breathe, the water we drink, and the food we eat all consist of matter.

Classification of Matter

The most useful way of classifying matter is based upon its purity (Table 1.2). Pure substances have a definite chemical composition. Mixtures, on the other hand, contain two or more substances in no fixed

TABLE 1.2 CLASSIFICATION OF MATTER WITH EXAMPLES

Pure Substances		Mixtures (Homogeneous)	Mixtures (Heterogeneous)
ELEMENTS	COMPOUNDS	SOLUTIONS	CRUDE MIXTURES
oxygen	water	air	granite
chlorine	ethyl alcohol	seawater	concrete
gold	carbon dioxide	Coca-Cola	pizza

ratio. Approximately one hundred pure substances are unique in one important way. They cannot, by any ordinary chemical process, be broken down into two or more simpler substances. We refer to such substances as **elements.**

Compounds contain two or more elements in chemical combination. In a compound, the percentages by mass of the elements are fixed. The most common compound on earth is water, where the two elements hydrogen and oxygen are combined (11.2% hydrogen, 88.8% oxygen). As you might guess, there are many more compounds than elements. Indeed, a single element, carbon, is found combined with other elements in literally thousands of different compounds.

Pure substances rarely occur as such in nature. Instead, they are usually found in mixtures with other substances. Air is one such mixture. Its principal components are the two elements nitrogen and oxygen. There are smaller amounts of the element argon and the two compounds water and carbon dioxide. Many other substances, including such pollutants as carbon monoxide and sulfur dioxide, may also be present in trace amounts. Another familiar mixture is seawater. This contains a large number of different salts, including sodium chloride, dissolved in water.

Ordinary samples of matter are usually chemically complex.

Air and seawater are examples of *homogeneous* mixtures or *solutions*. In a solution, the various components are distributed uniformly. If you were to analyze several samples of air from different parts of the room you are in, you would find the same percentage of water vapor in each case. The percentage of water vapor in a shower room would be quite different. However, it would not vary from one part of the room to another.

The components of a *heterogeneous* or *crude* mixture are not distributed uniformly. Instead, different components are concentrated in different regions of the sample. A piece of granite shows differences in color from one area to another. The various colors are those of the several substances making up the mixture.

Separation of Mixtures

An important task of a chemist is to extract pure substances from mixtures. Many different approaches are possible. Suppose, for example, you wanted to separate pure sand (silicon dioxide) from a mixture with sugar. In principle, you could use a magnifying glass and a pair of tweezers to remove the crystals of sugar. A more effective method would be to add water to the mixture, stir to bring the sugar into solution, and filter (Figure 1.3A, p. 8). This separates the sand, which is insoluble.

Some solutions to problems are better than others.

Solids contaminated by small amounts of impurities are often purified by *recrystallization* (Figure 1.3B). The sample is first heated with a solvent, often water, to bring all or nearly all of it into solution. Insoluble materials can be removed at this stage by filtering the hot solution. This solution is then cooled to room temperature or below. The desired product is ordinarily less soluble at low temperatures. Hence, most of it separates on cooling as finely divided crystals. Soluble impurities, if present in small amounts, remain in the solution.

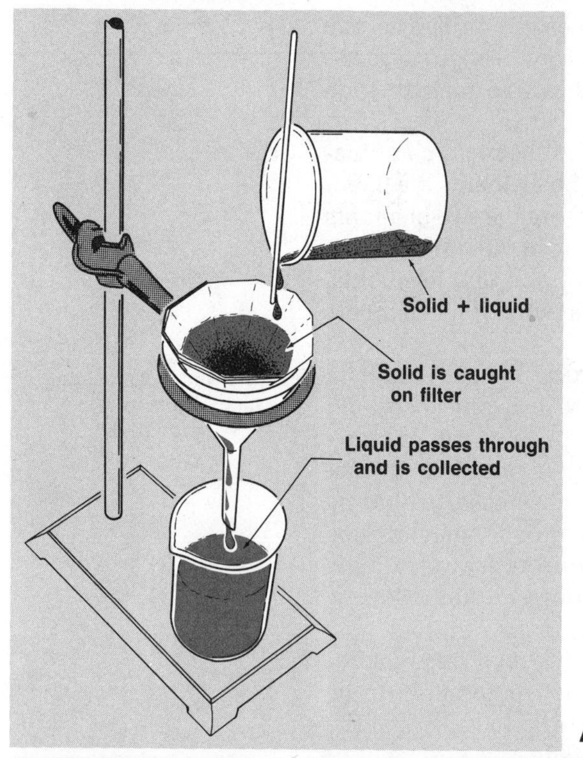

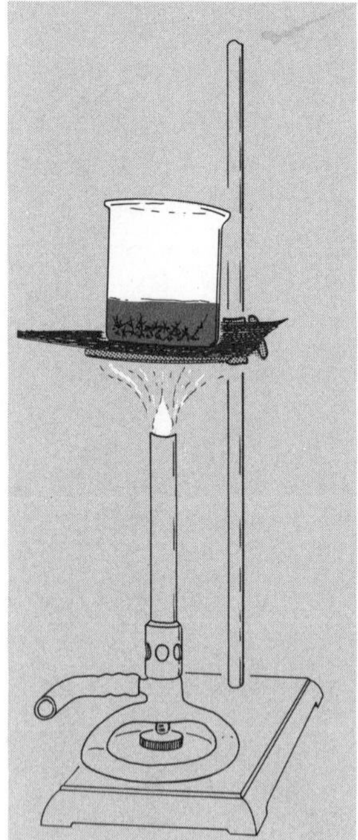

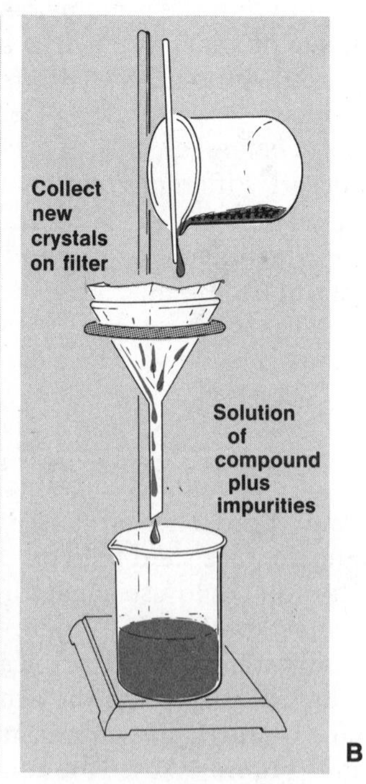

FIGURE 1.3

Filtration, shown in (A) at the top, is one of the steps involved in recrystallization (B).

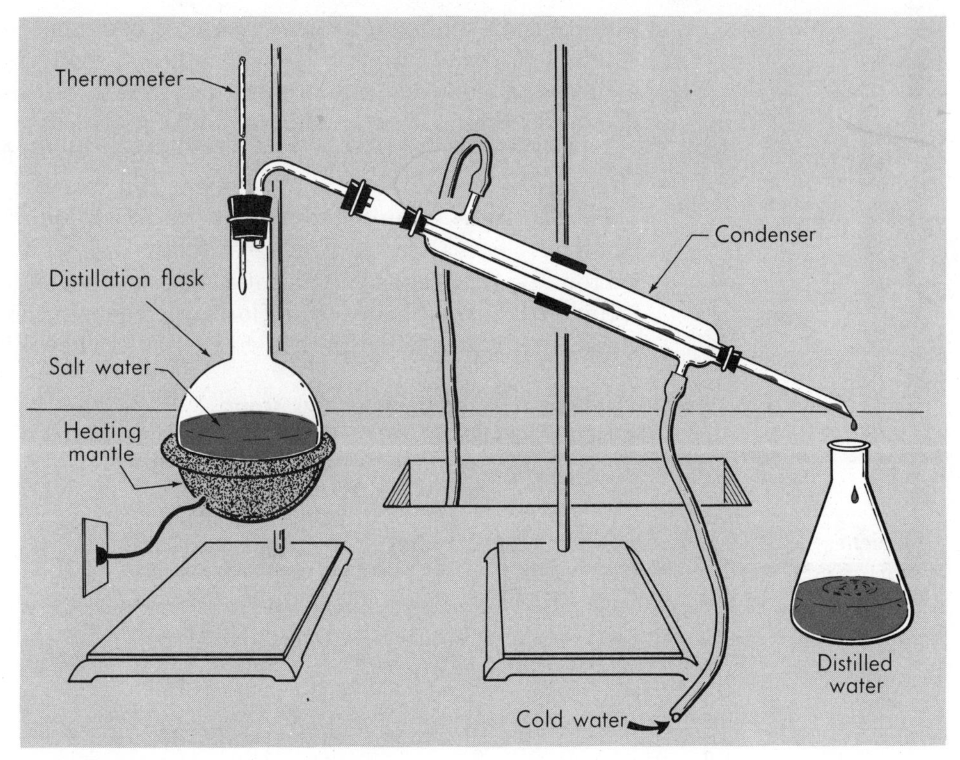

FIGURE 1.4

When salt water is boiled, the water leaves the solution as a vapor and can be condensed as pure water. The salt remains in the flask.

A solution of a solid in a liquid, such as seawater, can be separated by *distillation* (Figure 1.4). The solution is heated to boiling. This converts the water to vapor. The water vapor is then cooled by passing through a glass "condenser" through which tap water is flowing. The vapor condenses as drops of pure water which are collected. The dissolved salts do not vaporize at the boiling temperature. They remain behind in the distilling flask.

There are many ways to resolve mixtures. The approaches we have discussed (solution, recrystallization, distillation) are among the simpler. In recent years, chemists have developed more powerful methods to separate complex mixtures. One of these methods, called chromatography, is discussed on p. 11 at the end of this section.

Properties of Matter

Every pure substance has a characteristic set of properties. Certain of these, known as **physical properties,** can be measured without a change in chemical composition. A simple example is color. Another is *density,* which gives the mass-to-volume ratio of a sample of a substance.

$$\text{density} = \frac{\text{mass}}{\text{volume}} \qquad (1.1)$$

Density is most often expressed in grams per cubic centimeter (g/cm^3). Another physical property is *melting point*, the temperature at which a substance changes from a solid to a liquid. Still another is *boiling point*, the temperature at which a liquid boils.

If all the physical properties of two pure samples are the same, both samples contain the same substance.

Physical properties are often used to identify a substance. Suppose you were given a sample of an element and asked to identify it, using Table 1.3. If the sample happened to be a red liquid, your task would be an easy one. Color and physical state alone would establish that it is bromine. If it were a yellow solid, its identity would be less obvious. It could be either gold or sulfur. However, a density measurement would easily distinguish between the two elements. They differ widely in this

TABLE 1.3 PHYSICAL PROPERTIES OF SOME OF THE MORE FAMILIAR ELEMENTS

Element	State (25°C, 1 atm)	Color	Density, g/cm^3 (25°C, 1 atm)	Melting Point (°C)	Boiling Point (°C at 1 atm)
Aluminum	solid	–	2.7	660	2330
Argon	gas	–	0.00163	−189	−186
Bismuth	solid	–	9.8	271	1560
Bromine	liquid	red	3.1	−7	59
Cadmium	solid	–	8.6	321	767
Chlorine	gas	yellow	0.00290	−101	−34
Chromium	solid	–	7.2	1860	2670
Copper	solid	red	8.9	1080	2570
Gold	solid	yellow	19.3	1060	2970
Helium	gas	–	0.000164	−272	−269
Hydrogen	gas	–	0.000082	−259	−253
Iodine	solid	violet	4.9	114	184
Iron	solid	–	7.9	1540	2750
Lead	solid	–	11.3	328	1750
Magnesium	solid	–	1.7	650	1120
Mercury	liquid	–	13.5	−39	357
Neon	gas	–	0.000826	−249	−246
Nickel	solid	–	9.9	1450	2730
Nitrogen	gas	–	0.00115	−210	−196
Oxygen	gas	–	0.00131	−218	−183
Silver	solid	–	10.5	961	2210
Sodium	solid	–	0.97	98	889
Sulfur	solid	yellow	2.1	119	444
Tin	solid	–	7.3	232	2270
Zinc	solid	–	7.1	420	907

physical property (19.3 g/cm^3 *vs.* 2.1 g/cm^3). Suppose, though, that you were given a shiny, metallic solid. Here, density alone might not be enough to identify the element. Notice, for example, that chromium, tin, and zinc have nearly equal densities. You should check another property, perhaps melting point, before making a decision.

Certain properties cannot be measured without changing chemical composition. These are referred to as **chemical properties.** Among the chemical properties of the element hydrogen are the following:

In chemical reactions, substances are converted to other substances.

1. It reacts explosively with oxygen to form water.
2. It combines with chlorine upon exposure to ultraviolet light to form gaseous hydrogen chloride.
3. When passed over hot copper oxide, it reacts to form copper metal and water vapor.

These properties serve to distinguish hydrogen from all of the other gaseous elements listed in Table 1.3. Helium, for example, has chemical properties quite different from those of hydrogen. Indeed, helium is completely unreactive towards all other substances.

Chromatography

One of the most useful separation techniques in modern chemistry is called chromatography. To show how it works, consider the apparatus shown in Color Plate 1A, center of book. The column is packed with a finely divided solid, such as aluminum oxide. A green solution containing two dissolved solids, A and B, is poured into the column. The solid mixture is adsorbed on the surfaces of the aluminum oxide. It forms a green band at the top of the column.

At this point, a liquid in which both A and B are soluble is poured into the column. As the liquid moves down the column, it carries A and B with it. However, these two substances move to different extents. As you can see from the color plate, A, which has a blue color, moves farther in a given time. Solid B, which has a yellow color, lags behind. As time passes, two distinct bands appear in the column. Near the bottom is a blue band consisting of A adsorbed on the aluminum oxide. Further up the column, adsorbed B forms a yellow band.

By passing more liquid down the column, it is possible to flush out first A and then B. In this way, the two solids are separated from each other. At the end of the experiment we have two solutions. One, containing only A, is blue. The other, collected last, contains B and is colored yellow.

The principle behind this method is a simple one. To separate a mixture by chromatography, the components must differ in one or both of the following ways.

1. *Solubility in the liquid.* The more soluble component, in this case A, is carried farther down the column. The less soluble component, B, stays behind.
2. *Extent of adsorption.* The component most strongly adsorbed on the aluminum oxide, B, resists being moved by the liquid. A component such as A which is only weakly adsorbed is swept along by the liquid.

Separations of this type are most easily carried out when the components have different colors. Indeed, the word chromatography comes from the Greek *chroma*, meaning color. However, it is possible to separate components which are colorless by chromatography. Here it is necessary to use some other property of the components to locate them on the column.

1.3 MAKING MEASUREMENTS

Measurements are at the heart of every science, chemistry included. Most physical properties require quantitative measurement. They are ordinarily described with a number and a unit (aluminum: density = 2.7 g/cm^3, melting point = 660°C). In this section, we will describe some simple devices used to measure volume, mass, and temperature. We will also consider the units in which these quantities are commonly expressed.

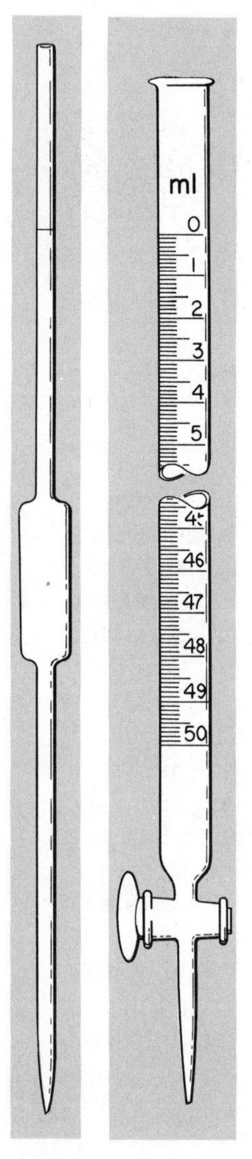

FIGURE 1.5

Volumes of liquids are often measured with a pipet or a buret. A pipet is calibrated to deliver a specific volume when filled to the mark and allowed to drain. With a buret one reads the level of liquid before and after using. The liquid volume delivered is the difference between the final and initial readings.

Measuring Devices

Perhaps the simplest measuring device is the meter stick, used to measure the *length* of an object. The stick is exactly one meter long. It is ordinarily divided into 100 equal parts, each one centimeter (1 cm) long. Using a meter stick or a rule graduated in centimeters, you should find that this page is about 27 cm long.

At least one device used to measure *volume*, a graduated cylinder, is familiar to you. Volumes of liquids can be measured more accurately with the other two pieces of glassware shown in Figure 1.5. A pipet, filled to the mark on its narrow stem and allowed to drain normally, delivers a fixed volume. This volume can be expressed in cubic centimeters (cm^3). Depending on the size of the pipet, it may deliver 5 cm^3, 10 cm^3, 25 cm^3, and so on. A buret, marked from 0 to 50 cm^3 in 0.1 cm^3 divisions, can be used to measure out any desired volume. Thus we might use a buret to obtain 6.2 cm^3, 21.7 cm^3, or a similar volume of water or other liquid.

Mass, which is a measure of the amount of matter in a sample, is commonly measured with a balance. Three types of balances found in the chemistry laboratory are shown in Figure 1.6. Of these, the double pan balance is the simplest in principle (and the most tedious to use). The sample whose mass is to be determined is placed on the left pan. Metal pieces having a known mass in grams (50 g, 5 g, 0.1 g, and so on) are added to the right pan to "balance" the sample. That is, they are added until the two pans are at exactly the same height. At this point, the mass of the sample is equal to that of the pieces of metal.

The density of a substance can be determined by measuring the mass and volume of a sample of that substance. A method frequently used with liquids is illustrated in Example 1.1.

EXAMPLE 1.1

To determine the density of ethyl alcohol, a student runs out a sample from a buret into a previously weighed flask. The data are as follows:

mass empty flask	= 31.4 g	initial buret reading	= 0.2 cm^3
mass flask + alcohol	= 47.7 g	final buret reading	= 20.6 cm^3

Calculate the density of ethyl alcohol.

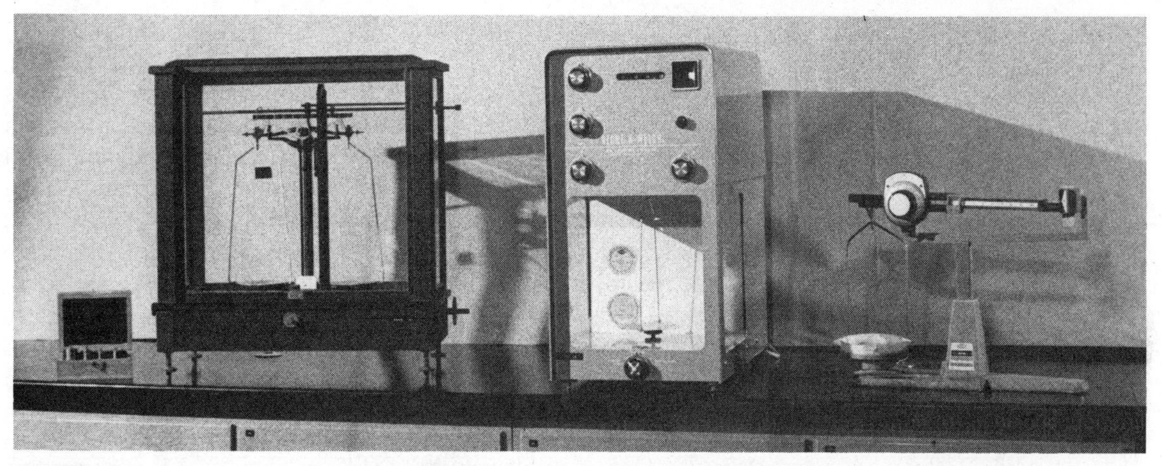

FIGURE 1.6

Three types of laboratory balances. The two-pan balance pan on the left was in common use until about 1960. It has been largely replaced by the single pan balance in the middle. In this balance the weights are inside and are manipulated by turning knobs on the front. Both balances can weigh to ±0.0001 g. The balance on the right is a triple beam balance, used for rough weighing to about ±0.01 g.

SOLUTION

The mass of the alcohol is:

$$47.7\ \text{g} - 31.4\ \text{g} = 16.3\ \text{g}$$

The volume is obtained by subtracting the initial from the final buret reading:

$$20.6\ \text{cm}^3 - 0.2\ \text{cm}^3 = 20.4\ \text{cm}^3$$

Since density is defined as mass per unit volume:

$$\text{density} = \frac{\text{mass}}{\text{volume}} = \frac{16.3\ \text{g}}{20.4\ \text{cm}^3} = 0.799\ \text{g/cm}^3$$

Densities of solids are difficult to determine accurately for several reasons. A major problem is measuring the volume of a solid sample. If the solid has a regular shape, such as a cube, its volume can be calculated from its measured dimensions. More frequently, the solid has an irregular shape or is finely divided. In this case, its volume is obtained indirectly. One approach is shown in Figure 1.7. The volume of the solid is taken to be that of the liquid it displaces when added to a graduated cylinder.

All of us are familiar with the concept of temperature. Quite simply, temperature is an indicator of the direction of heat flow. When you drink a cup of hot coffee, you feel warm because heat is transferred from the coffee to your body. With a glass of iced tea, which is below body temperature, heat flows in the opposite direction. In general, heat flows

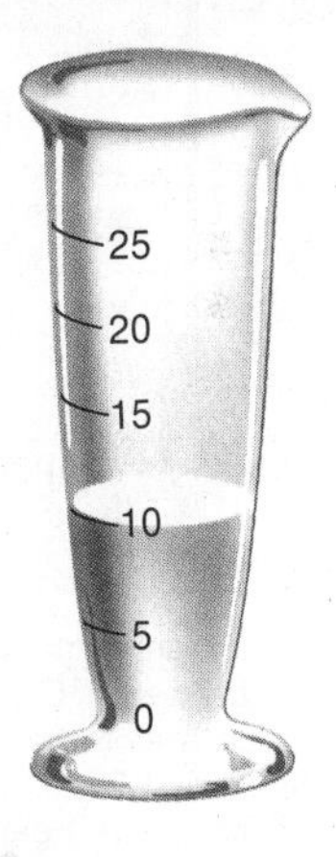

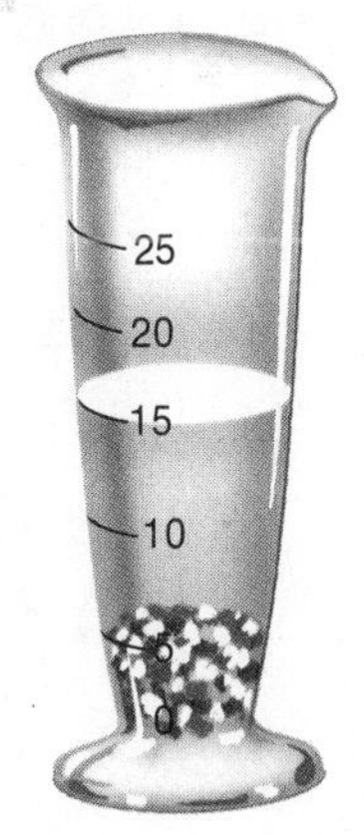

FIGURE 1.7

Measuring the density of a solid. A solid sample weighing 12.5 g was added to 10.0 cm^3 of water in a cylinder. The water level went up to 15.0 cm^3. This indicates that the volume of the solid is 5.0 cm^3. The density of the solid, mass/volume, equals 12.5 g/ 5.0 cm^3 = 2.5 g/cm^3.

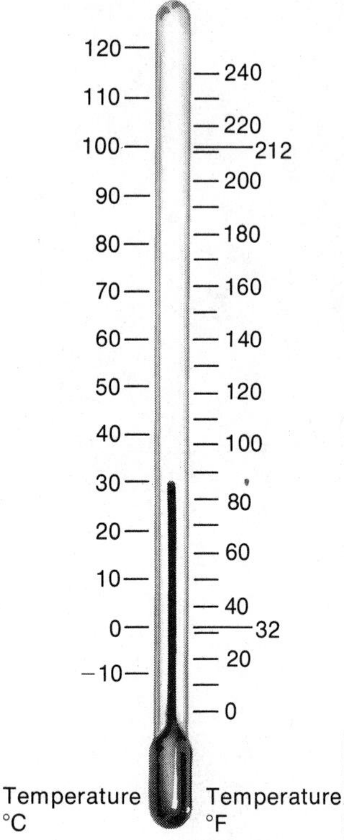

FIGURE 1.8

A thermometer showing the relationship between °C and °F. The scales are engraved so as to fit Eqn. 1.2. A temperature of 30°C is the same as a temperature of 86°F.

from a "hot" (higher temperature) to a "cold" (lower temperature) object.

To measure temperature, we determine its effect on the properties of substances. Density ordinarily decreases as temperature rises. Hence, a sample expands to occupy a greater volume when it is heated and contracts when it is cooled. The most common temperature-measuring device in the laboratory is the mercury thermometer (Figure 1.8). It contains liquid mercury sealed into a very fine tube which widens into a bulb at the lower end. The height of the mercury depends upon the temperature around it. If the temperature increases, the mercury expands and the level rises.

To express temperature as a number, a thermometer must be marked according to some scale. On the Celsius (centigrade) scale, 0°C is taken to be the freezing point of pure water and 100°C its boiling point at one atmosphere pressure. Another temperature scale commonly used in the United States is the Fahrenheit scale (°F). Here, these two points are taken to be 32°F and 212°F in that order. Conversions between the two scales can be made using the relation:

$$°F = 1.8(°C) + 32° \qquad (1.2)$$

EXAMPLE 1.2

A TV weatherman in Miami gives the temperature as 30°C; another, in Boston, reports a temperature of 68°F. Convert

a. the Miami temperature to °F b. the Boston temperature to °C

SOLUTION

a. Direct substitution into Equation 1.2 gives:

$$°F = 1.8(30°) + 32° = 54° + 32° = 86°$$

b. Substituting for °F, we have: 68° = 1.8(°C) + 32°
To solve this equation for °C, we subtract 32° from both sides and then divide by 1.8

$$36° = 1.8(°C);\ 36/1.8 = °C = 20$$

Clearly, however you look at it, Boston is cooler than Miami!

Systems of Units: the Metric System

To express a quantity such as length, volume, or mass, we need a unit of measurement. There are two systems of units currently in use in the United States. These are the English system and the metric system. Most other nations, and all of the world of science, use the metric system. The United States is gradually converting to the metric system. However, until this is complete, we must work with both systems.

In the metric system, the units used to express a quantity are related to each other by powers of 10. To illustrate, consider length, where the basic unit is the *meter*. A larger unit, often used to express highway distances, is the *kilometer* (1000 m). Smaller units of length include the *centimeter* (1/100 meter) and the *millimeter* (1/1000 m). A very small unit which we will use in later chapters is the *nanometer* (10^{-9} m).

$$1\ \text{km} = 1000\ \text{m} = 10^3\ \text{m}$$
$$1\ \text{cm} = 0.01\ \text{m} = 10^{-2}\ \text{m}$$
$$1\ \text{mm} = 0.001\ \text{m} = 10^{-3}\ \text{m}$$
$$1\ \text{nm} = 0.000\ 000\ 001\ \text{m} = 10^{-9}\ \text{m}$$

Some things in chemistry are very small indeed.

(The use of exponents is described in Appendix I, p. A.1)

These prefixes retain their meaning when used with other units. In dealing with mass, we refer to the *kilogram* (1000 g) and the *milligram* (0.001 g). The most common unit of volume in the metric system is the *liter*. A liter is 1000 milliliters. A milliliter, in turn, has a volume equal to that of a cube one centimeter on an edge.

$$1\ \ell = 1000\ \text{ml} = 1000\ \text{cm}^3 \qquad (1.3)$$

$1\ \text{ml} = 1\ \text{cm}^3$

All scientific measurements are made in the metric system. In making comparisons with the English system, the following rough approximations may be helpful:

—the meter is about 10% longer than the yard;
—the liter (1000 cm^3) is about 5% larger than the U.S. liquid quart;
—the kilogram is a little more than twice as large as a pound.

Further comparisons are shown in Figure 1.9. A series of more exact relationships is given in Table 1.4, p. 17.

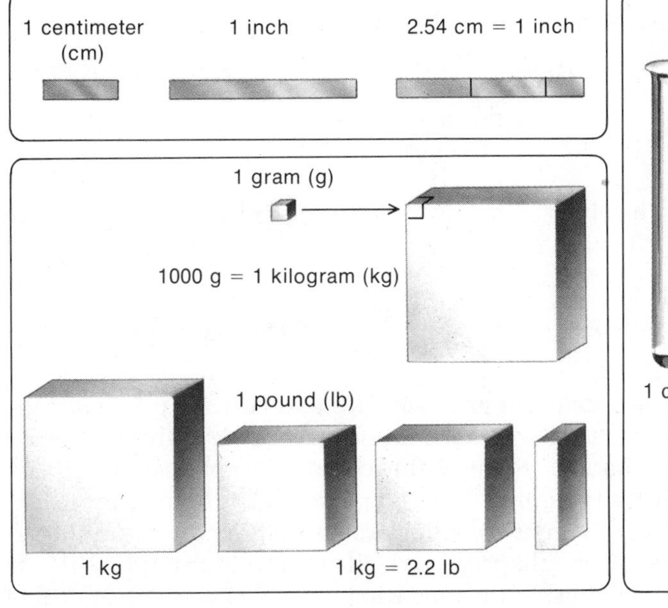

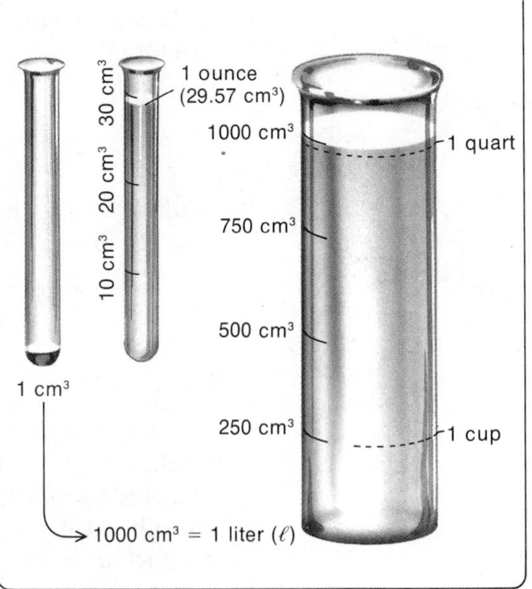

FIGURE 1.9

Some relationships between the metric and English systems of measurement. American scientists are familiar with both systems, but almost always use the metric system in their work.

Conversion of Units

Frequently we need to convert a quantity expressed in one unit to a different unit. Conversions within the metric system are easy to carry out because of the decimal relationship between units (1 km = 10^3 m, for example). However, conversions within the English system, where the units are not simply related (1 mile = 5280 ft) or between the two systems (1 km = 0.621 mile) are more difficult.

Unit conversions of all types can be handled by a process known as *unit analysis* which involves the use of **conversion factors.** To illustrate the approach, let us suppose you want to find the number of eggs in 5 dozen eggs. Here, the relationship is:

$$12 \text{ eggs} = 1 \text{ dozen eggs}$$

If we divide both sides of this equation by the quantity "1 dozen eggs" we obtain a ratio equal to unity. This ratio is called a conversion factor:

$$\frac{12 \text{ eggs}}{1 \text{ dozen eggs}} = 1$$

Multiplying the quantity "5 dozen eggs" by this ratio does not change its value. However, it does accomplish the desired conversion of units.

$$5 \text{ \sout{dozen eggs}} \times \frac{12 \text{ eggs}}{1 \text{ \sout{dozen eggs}}} = 60 \text{ eggs}$$

The same approach could be used to make a conversion in the opposite direction. Suppose, for example, you wanted to know how many dozen eggs there are in 108 eggs. Again we start with the equation:

$$12 \text{ eggs} = 1 \text{ dozen eggs}$$

When you multiply something by 1, you don't change its magnitude, but you can change its units.

This time we divide both sides of the equation by "12 eggs" to obtain the ratio:

$$\frac{1 \text{ dozen eggs}}{12 \text{ eggs}} = 1$$

Multiplying by the given quantity, 108 eggs, gives us the answer we are looking for:

$$108 \text{ \sout{eggs}} \times \frac{1 \text{ dozen eggs}}{12 \text{ \sout{eggs}}} = 9 \text{ dozen eggs}$$

We have illustrated unit analysis with two very simple examples. Don't be fooled into thinking that a formal procedure of this sort is not worth the effort. Granted, you probably could have obtained the correct answers by "multiplying by 12" in one case and "dividing by 12" in the other. However, with units which are less familiar, it may not be at all obvious how you should proceed. Unit analysis is a simple, fail-safe approach to conversions. In the long run it will save you a great deal of

TABLE 1.4 COMMONLY USED METRIC UNITS AND ENGLISH EQUIVALENTS

Quantity	Metric Unit	English Equivalent
length	meter (m)	39.4 in (1.09 yd)
	centimeter (cm)	0.394 in
	kilometer (km)	0.621 mile (1090 yd)
volume	liter (ℓ)	1.06 U.S. liq qt
	cubic centimeter (cm^3) milliliter (ml)	0.0610 in^3
mass	kilogram (kg)	2.20 lb
	gram (g)	0.0353 oz

time and effort. It will be used again and again throughout this text to solve a wide variety of chemical problems.

It may be helpful to list the steps that we went through in converting between numbers and dozens of eggs. They are common to all types of unit analysis.

1. **Write down the equation relating the two units** (12 eggs = 1 dozen eggs).

2. **Use this equation to obtain a ratio equal to unity, a conversion factor.** Two such conversion factors are available (12 eggs/1 dozen eggs or 1 dozen eggs/12 eggs). *Choose the conversion factor which will allow you to cancel the unit you started with, leaving the desired unit.*

Conversion factors are easy to set up, and are very helpful in many chemical problems. Learn to use them properly.

3. **Multiply the given quantity by this conversion factor,** cancelling units so as to obtain the answer.

Clearly, in following this process you should retain units in all your calculations. This way, you can be sure that they cancel properly in the end.

The use of this approach is further illustrated in Examples 1.3 and 1.4. Conversion factors between the metric and English systems are listed in Table 1.4.

EXAMPLE 1.3

The distance from Boston to New York is given on a highway sign as 188 miles. Express this distance in kilometers.

SOLUTION

From Table 1.4, we see that the relation required is:

$$1 \text{ km} = 0.621 \text{ mile}$$

This gives us two conversion factors:

$$\frac{1 \text{ km}}{0.621 \text{ mile}} \quad \text{and} \quad \frac{0.621 \text{ mile}}{1 \text{ km}}$$

We want to convert from miles to kilometers. Hence we use the first conversion factor, which has kilometers in the numerator and miles in the denominator.

$$\text{distance (km)} = 188 \cancel{\text{miles}} \times \frac{1 \text{ km}}{0.621 \cancel{\text{mile}}} = 303 \text{ km}$$

EXAMPLE 1.4

The density of water is 1.00 g/cm^3. Express this in kilograms per liter.

SOLUTION

Here two conversions are required. Grams must be changed to kilograms and cubic centimeters to liters.

$$1 \text{ kg} = 1000 \text{ g}$$
$$1 \ell = 1000 \text{ cm}^3$$

In the given quantity, 1.00 g/cm^3, "grams" are in the numerator. We need a conversion factor with kilograms in the numerator and grams in the denominator:

$$(1) \quad \frac{1 \text{ kg}}{1000 \text{ g}}$$

To obtain the second conversion factor, we note that "cm^3" is in the denominator of 1.00 g/cm^3. To cancel units, we need a conversion factor with cubic centimeters in the numerator:

$$(2) \quad \frac{1000 \text{ cm}^3}{1 \ell}$$

The two conversions are carried out, one after the other, in a single calculation:

$$\text{density } (\text{kg}/\ell) = 1.00 \frac{\cancel{\text{g}}}{\cancel{\text{cm}^3}} \times \frac{1 \text{ kg}}{1000 \cancel{\text{g}}} \times \frac{1000 \cancel{\text{cm}^3}}{1 \ell} = 1.00 \frac{\text{kg}}{\ell}$$

1.4 MEASUREMENT ERRORS

There are two kinds of numbers, exact and approximate. We say that there are twelve inches in one foot:

$$12 \text{ in} = 1 \text{ ft}$$

Here, the number 12 is exact. This equation defines the relation between inches and feet. There are exactly twelve inches in one foot, no more, no less. Similarly, in the relation:

$$1000 \text{ g} = 1 \text{ kg}$$

the number 1000 is exact.

The number of students in a classroom can be found exactly. "Counting" is not the same as "measuring."

The situation is quite different with numbers based on measurements. When you say that you weigh "135 lb," the number 135 is approximate. It is based on a scale which is perhaps accurate to the nearest pound. The true value might be, for all you know, 134 lb or 136 lb. Putting it another way there is an *uncertainty* of one pound in this value.

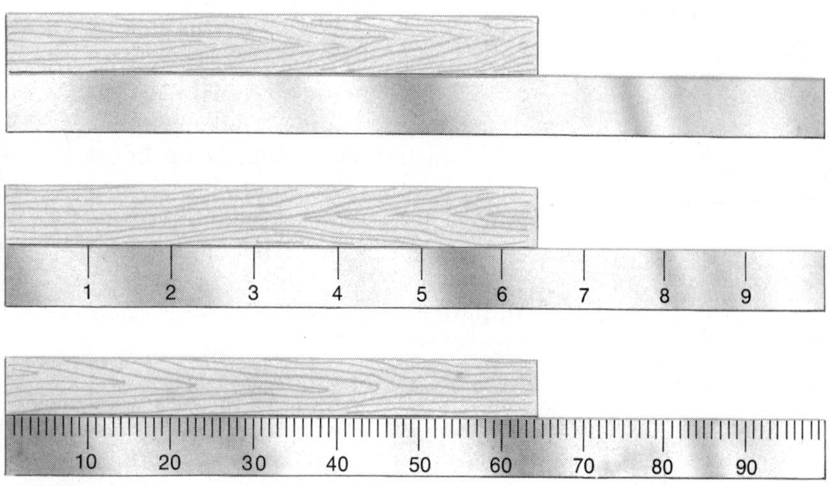

FIGURE 1.10

Measuring the length of a board with a meter stick. The uncertainty of the measurement will depend on the spacing of the markings on the stick. With a stick calibrated in cm one can estimate the length to ±0.001 m. No matter how close one puts the markings on the meter stick, there will always be some uncertainty in the measurement of the length of the board.

You might better say that you weigh:

$$135 \pm 1 \text{ lb}$$

where the symbol ± means "plus or minus."

Uncertainties in Measurements

No measured quantity is exact. Associated with every measurement is an error or uncertainty of the type just discussed. The amount of uncertainty depends upon the nature of the measuring device. Suppose you were asked to measure the length of a board, using the three meter sticks shown in Figure 1.10. The first meter stick contains no graduations. With it, you would do well to estimate the length to the nearest tenth of a meter. You might report that the board is 0.6 m long. Using a meter stick marked in tenths of a meter, you could do better. You might obtain a length of 0.64 m, with an uncertainty of 0.01 m. Finally, consider the meter stick marked in centimeters. With this, you could find the length to the nearest 0.001 m, perhaps 0.643 m.

Table 1.5 summarizes the results of these three measurements. The uncertainties listed (0.1 m, 0.01 m, 0.001 m) indicate the *precision* of the measurement. Other students, using the same meter sticks, should obtain the same values. They might differ by one digit in the last number quoted (for example, 0.644 m instead of 0.643 m). The precision of a measurement depends upon the sensitivity of the instrument you use. It also depends upon the care you take in using it. However, you can never achieve "absolute" precision. No matter how sensitive the instrument, there will always be some uncertainty in your measurement.

Seems unjust somehow, but that's the way it is.

TABLE 1.5 UNCERTAINTIES IN MEASURED QUANTITIES

Measuring Device	Length of Board	Uncertainty	
Unmarked meter stick	0.6 m	0.1 m	increasing precision ↓
Meter stick marked in 0.1 m	0.64 m	0.01 m	
Meter stick marked in 0.01 m	0.643 m	0.001 m	

Significant Figures in Measurements

As we have seen, the precision of a measurement can be described by giving the uncertainty. Another way is to cite the number of *significant figures*. A significant figure is exactly what it sounds like. It is a digit that has experimental meaning. In the measured length 0.6 m, the digit "6" was measured. We say that there is *one* significant figure in 0.6 m. Similarly, there are *two* significant figures in 0.64 m and *three* in 0.643 m. In another case, we might weigh a series of objects with the following results:

6.142 g	4 significant figures
2.3 g	2 significant figures
65.1 g	3 significant figures

Notice that the location of the decimal point has nothing to do with the number of significant figures. A digit is "significant" if it has been measured.

Counting significant figures is often a simple task, as in the above examples. However, if one of the digits is a zero you may be in doubt as to whether it is significant. Here, common sense is the best guide. Consider the quantities 12.40 g and 0.004 g. In the first case, the zero must be significant. It was placed there to indicate that the object was weighed to 0.01 g. Its mass is closer to 12.40 g than to 12.39 or 12.41 g. On the other hand, the zeros in 0.004 g are not significant. They serve only to locate the decimal point. The last digit, 4, is the only measured number.

If you express a number in exponential notation (see Appendix), the digits in the power of 10 are all significant: $0.018 = \underline{1.8} \times 10^{-2}$ (2 significant figures)

12.40 g	4 significant figures (zero is significant)
0.004 g	1 significant figure (zeros are not significant)

EXAMPLE 1.5

Give the number of significant figures in the quantities

a. 642 cm^3 b. 1.717 cm^3 c. 2.0 cm^3 d. 0.018 g

SOLUTION

In (a), (b), and (c), all the numbers are meaningful; there are 3, 4 and 2 significant figures in that order. In (d), the zeros serve only to fix the position of the decimal point; there are only two significant figures.

Errors in Calculated Results

Most of the measurements we make in the laboratory are not end results in themselves. Instead, they are combined to calculate the quantity we are interested in. We might, for example, measure the mass and volume of a liquid sample to determine its density. Clearly the precision of the quantity we calculate will be limited by those of the ones we measure. Suppose we measured a mass of 18.50 g and a volume of 4.2 cm^3. It would be absurd to report a density of 4.404761905 g/cm^3, the value obtained by dividing on a 10-place calculator. This would imply a precision in the density far greater than those of the measured quantities, mass and volume. To decide how many digits to retain in this calculation, we use the general rule:

In multiplication or division, the number of significant figures in the answer is the same as that in the measured quantity with the fewest significant figures.

Applying this rule to obtain the density of the liquid, we note that there are:

4 significant figures in the mass (18.50 g)
2 significant figures in the volume (4.2 cm^3)

We should then retain only two significant figures in the density:

$$\text{density} = \frac{\text{mass}}{\text{volume}} = \frac{18.50 \text{ g}}{4.2 \text{ cm}^3} = 4.4 \text{ g/cm}^3$$

Notice that the extra significant figures in the mass are "wasted". The precision of the calculated density is limited by that of the volume measurement.

In experiments, you are usually most efficient when you measure all quantities with the same precision.

EXAMPLE 1.6

The density of aluminum is 2.701 g/cm^3. Consider a cube of aluminum 1.24 cm on an edge. Calculate, to the correct number of significant figures:

a. the volume of the cube b. the mass of the cube

SOLUTION

a. $V = (1.24 \text{ cm})^3 = 1.91 \text{ cm}^3$ (3 significant figures)

b. To obtain the mass, m, we start with the defining equation for density, d:

$$d = m/V$$

Multiplying both sides of this equation by V, we obtain:

$$m = d \times V$$

$$= 2.701 \frac{\text{g}}{\text{cm}^3} \times 1.91 \text{ cm}^3 = 5.16 \text{ g}$$

We retain 3 significant figures in the mass, since we know the volume to 3 figures; knowing the density to 4 doesn't help!

Often we need to add or subtract measured quantities. Here, as with multiplication or division, the precision of the result depends upon that of the measured quantities. However, the rules for addition and subtraction are different. They deal with the estimated error, or *uncertainty*, rather than the number of significant figures. To illustrate the approach used, suppose you were to add 5.12 cm^3 of water to a sample with a volume of 62 cm^3. To obtain the total volume, we note that the quantities "62 cm^3" and "5.12 cm^3" each have an uncertainty of 1 in the last digit:

Volume	*Uncertainty*
62 cm^3	1 cm^3
5.12 cm^3	0.01 cm^3
67 cm^3	1 cm^3

Since 62 cm^3 is known to only 1 cm^3, the total volume can not be known more precisely than that. The fact that 5.12 cm^3 has an uncertainty of 0.01 cm^3 does us no good. Here, as before, it is the least precise measurement which limits the precision of the result.

This calculation illustrates the general rule:

In addition or subtraction, the uncertainty in the result is the same as that in the measured quantity which has the greatest uncertainty.

EXAMPLE 1.7

Calculate the mass remaining when 5.12 g is withdrawn from a sample weighing

a. 7.1 g b. 5.18 g

SOLUTION

a. 7.1 g − 5.12 g = 2.0 g (uncertainty = 0.1 g)

b. 5.18 g − 5.12 g = 0.06 g (uncertainty = 0.01 g)

Notice that in (b), the answer is known to only one significant figure, even though both measured quantities are known to three. This situation is quite common; we frequently "lose" significant figures when we subtract.

1.5 SUMMARY

In this chapter, you were introduced to some basic ideas which will be used over and over again throughout this text. It is important that you thoroughly understand:

1. The distinction between elements, compounds and mixtures (homogeneous and heterogeneous).

2. How substances can be separated and identified on the basis of their physical and chemical properties.
3. How volume, mass, density and temperature are measured in the laboratory.
4. The use of the metric system.
5. The use of conversion factors.
6. The rules governing the use of significant figures.

NEW TERMS

These terms are all defined in the Glossary at the end of the book. Chemistry is a lot easier if you know what the terms mean.

boiling point
buret
Celsius scale
centi
chemical property
compound
conversion factor
crude mixture
density
distillation
element
Fahrenheit scale
filtration
kilo
liter
mass
melting point
milli
mixture
nano
physical property
pipet
pure substance
recrystallization
significant figure
solution
temperature
uncertainty

QUESTIONS

1. How did Lavoisier's contributions to chemistry differ from those of:

   a. the Greek philosophers? b. the alchemists?

2. A student places a penny in a crucible and pours nitric acid over it. He notices that a brown gas forms and the solution turns blue. He makes no further observations. Would you say that, in his approach, he most closely resembles:

   a. the Greeks b. the alchemists c. Lavoisier

3. A student makes the statement that, "Hot water freezes faster than cold water". Describe an experiment that would prove or disprove this statement.

4. Classify each of the following as a pure substance or a mixture. For each pure substance, indicate whether it is an element or a compound. Which of the mixtures are solutions?

   a. air b. uranium c. wood d. table salt e. sulfur
   f. tea g. carbon dioxide

5. Name three compounds and three solutions commonly found in the kitchen.

6. List all of the elements that you have seen (outside the chem lab) in pure form.

7. What kind of experiment would you carry out to demonstrate that:

   a. water is a compound rather than an element.
   b. seawater is a solution rather than a pure substance.
   c. granite is a heterogeneous mixture.

8. Suggest a method of separating the following mixtures.

   a. charcoal and sugar
   b. sugar and water
   c. quartz and gold
   d. iron filings and powdered aluminum
   e. sheep and goats

9. You are given a beaker containing a water solution of a solid A, contaminated with a small amount of another solid, B. At the bottom of the beaker is a layer of finely divided sand. How would you obtain, from this mixture:

   a. pure water? b. pure solid A? c. pure sand?

10. The water solubility of sodium chloride is virtually independent of temperature. Could you purify sodium chloride by recrystallization? Explain.

11. List as many physical properties of water as you can think of.

12. You are given two samples of pure aluminum, one weighing 10 g, the other 100 g. How would these two samples compare in:

    a. melting point? b. boiling point? c. volume? d. density?

13. Consider the following paragraph:

    Chlorine is a greenish-yellow gas with a density about 2.4 times that of air (density air = 1.2 g/ℓ at 25°C, 1 atm). It condenses to a liquid at −34°C and to a solid at −101°C. When calcium is exposed to chlorine gas, it becomes coated with a white, brittle solid. With hydrogen, chlorine forms a gaseous compound, hydrogen chloride.

    Based on this information, list several physical and chemical properties of chlorine. Be as specific as possible.

14. You are given a solid and told that it is one of the elements listed in Table 1.3. You attempt to determine its melting point and find that it does not melt upon heating to 600°C. Which of the elements listed in the table could it be? What experiment(s) would you carry out to decide upon its identity?

15. Describe how you would determine the density of

    a. ethyl alcohol b. aluminum

16. To determine the density of a solid, a student first weighs it. He then adds it to a graduated cylinder containing water to determine its volume. Will this approach work if the solid is wood? table salt? Suggest how you might determine the densities of these solids.

17. Cite several examples, other than those mentioned in the text, where heat flows from a hot to a cold object.

18. What is meant by each of the following prefixes?

    a. kilo b. milli c. nano d. centi

19. One store sells milk by the quart, another sells it in liter containers. If the prices are the same, which is the better buy? (See Table 1.4.)

20. A metric ton is defined as 1000 kg. Which has the greater mass, a metric ton of coal or an English ton (2000 lb)? (See Table 1.4.)

21. Which of the following quantities could be determined exactly? Which would have at least a small uncertainty?

    a. attendance at a football game
    b. distance between goalposts at a football stadium
    c. number of grams in a sample of sugar
    d. number of inches in one yard
    e. your grade in this course

22. What exactly does the phrase "significant figure" mean?

PROBLEMS

1. *Density (Example 1.1)* A student runs out 5.00 cm^3 of benzene from a pipet and finds that it weighs 4.40 g. What is the density of benzene?

2. *Density (Example 1.1)* Using the density calculated in Problem 1, determine the volume of a sample of benzene weighing 1.00 g.

3. *Temperature (Example 1.2)* Convert:

   a. 112°C to °F b. 68°F to °C

4. *Conversion of Units (Examples 1.3, 1.4)* The distance from Conway, New Hampshire to Storrs, Connecticut is 202 miles. Express this in kilometers (1 km = 0.621 mile).

5. *Conversion of Units (Examples 1.3, 1.4)* A sample of water has a volume of 25.2 cm^3. Express this in cubic inches (1 cm = 0.394 in).

6. *Conversion of Units (Examples 1.3, 1.4)* The density of ethyl alcohol is 0.799 g/cm^3. Express this in:

   a. g/ℓ b. lb/qt

7. *Counting Significant Figures (Example 1.5)* How many significant figures are there in:

   a. 12.26 cm^3 b. 6.40 g c. 0.023 g

8. *Significant Figures in Calculations (Example 1.6)* A piece of paper is 11.2 cm wide and 51 cm long. What is its area in square centimeters?

9. *Uncertainty in Calculations (Example 1.7)* Using the following data:

$$\text{mass crucible + sample} = 12.612 \text{ g}$$
$$\text{mass crucible} = 11.5 \text{ g}$$

   calculate the mass of the sample.

* * * * *

10. A certain solid has a mass of 27.5 g. When added to a graduated cylinder, it displaces 15.1 cm^3 of water. Calculate its density.

11. The density of mercury is 13.5 g/cm^3. Calculate the:

    a. mass of 25.2 cm^3 of mercury.
    b. volume occupied by 1.00 g of mercury.

12. Mercury freezes at $-40°C$. What is this temperature in °F?

13. A garage fills your radiator with an antifreeze which is guaranteed not to freeze at $-20°F$. On a certain night the temperature drops to $-20°C$. The radiator freezes. What is the temperature in °F? Can you collect on the guarantee?

14. Make the following conversions, using Table 1.4 where necessary.

    a. 16.2 g to milligrams; to pounds.
    b. 248 cm^3 to liters; to quarts
    c. 1.52 miles to feet; to kilometers

15. The density of mahogany is 0.031 lb/in^3. What is its density in g/cm^3?

16. Given that: 1 nerd = 2.60 troll; 4 troll = 8.92 curd, how many nerds are there in 62.7 curd?

17. If a first down in football were 10 m instead of 10 yards, how many extra feet would be required?

18. Convert your height to meters and your mass to kilograms.

19. Give the number of significant figures in each of the following masses:

    a. 1.69 g b. 3.004 g c. 0.000164 g d. 1.6×10^{-4} kg
    e. 2.0 g

20. Perform each of the following operations involving measured quantities. In each case express your answer in the proper units with the proper number of significant figures.

    a. 2.60 cm × 1.8 cm
    b. 3.9025 g/0.75 cm^3
    c. $(3.1\ cm)^3$
    d. 61 g/$(2.91\ cm)^3$
    e. 52.0 $\ell \times$ 1.927 g/ℓ

21. Perform the following operations, giving answers with the proper uncertainties.

    a. 16.0 g + 3.106 g + 0.8 g
    b. 12.45 cm^3 + 0.001 cm^3 + 68 cm^3
    c. 9.02 m − 3.1 m
    d. 9.61 m − 52 cm
    e. 12 in − 2 cm (answer in inches)

22. You start with a sample weighing 9.81 g and carry out the following operations in order. State the number of grams remaining after each addition or subtraction.

    a. add 0.18 g b. remove 9.220 g c. add 1.83 g d. add 6 g

*23. Oil spreads on water to form a film about 1×10^{-6} cm thick. Its density is 0.85 g/cm^3. What will be the area of the slick formed when one kilogram of oil is spilled on water? Express this in square centimeters; in square miles.

*24. At what point is the temperature in °F twice that in °C?

*25. A certain element not listed in Table 1.3 is a very reactive metal which melts at 367°F and has a density less than that of water. Its compounds are used in treating mental disorders. The metal itself has been suggested as a substitute for lead in storage batteries. Using any source of information you can find, identify the element.

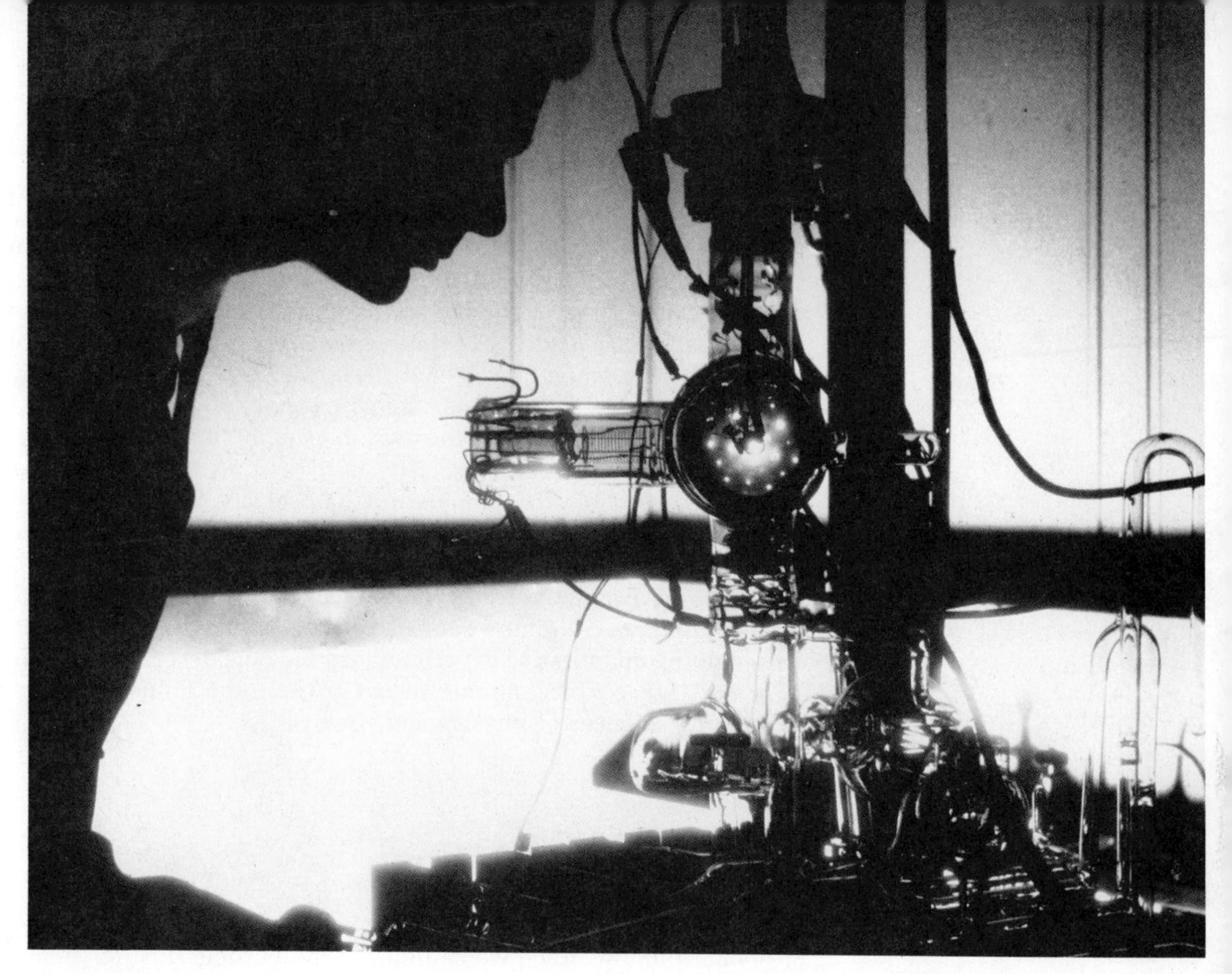

2

ATOMS, MOLECULES, AND IONS

Chemistry, as we have seen, deals with the properties of pure substances, both elements and compounds. These properties can be studied by experiments of the type described in Chapter 1. We can, for example, determine that the compound methane is a colorless gas at room temperature and atmospheric pressure with a density of 0.000654 g/cm^3. It condenses to a liquid at −162°C and solidifies at −183°C. Sodium chloride, on the other hand, is a white solid (density = 2.16 g/cm^3). This compound must be heated to 800°C to melt it and to 1400°C to convert it to a vapor.

We are naturally curious as to why methane and sodium chloride have such very different properties. To interpret properties of substances, we need to examine their particle structures. That is, we need to consider the tiny "building blocks" of matter called atoms, molecules, and ions. In this chapter we will study the nature of these particles. We will also consider how they are assembled to form the visible matter that we work with in the laboratory.

The idea of the atomic nature of matter is an old one. It was suggested by Leucippus in about 450 B.C. He challenged the prevailing idea of Greek philosophers that matter could be subdivided forever. His disciple, Democritus, expanded on the ideas of Leucippus. He believed that all matter was made up of tiny particles which could not be broken down further. Democritus named these particles "atomos," meaning indivisible. He speculated that atoms of elements differed from one another in size and shape. In this way, he sought to explain differences in properties among elements.

Today we know that Democritus had remarkable insight into the nature of matter. However, neither he nor anyone else suggested experiments to prove or disprove the existence of atoms. As a result, atoms remained an abstract concept, of interest only to philosophers. This situation remained for more than 2000 years. It changed when scientists started doing quantitative experiments of the type described in Chapter 1. In the 19th century, atomic theory became the foundation for the developing sciences of physics and chemistry.

2.1 ATOMIC THEORY

Any theory about an ultimate particle of matter must use a model, or mental picture, since we cannot see the particle (it's too small). A reliable model is consistent with experiment. Frequently, the model is changed as more and better observations are made. Improved models become more and more like the systems which the scientist is trying to describe. The growth of atomic theory over the past two centuries is an excellent example of model-building.

Chemistry, as a science, is only about 200 years old.

Our present model of the atom is based on the ideas of John Dalton, an English schoolteacher and chemist. His publication in 1808 of a *New System of Chemical Philosophy* was the beginning of modern atomic theory. Dalton arrived at his conclusions by using the results of others. Robert Boyle (1662) had previously concluded that gases must consist of particles. Dalton expanded on this to include all states of matter.

Some of Dalton's ideas have not stood the test of time. The theory has been modified as new facts became known. The major features of modern atomic theory are summarized in the following statements.

1. **An element is composed of extremely small, invisible particles called atoms.** *Example:* The element hydrogen is made up of atoms which have an average mass of 1.67×10^{-24} g. These atoms may be regarded as spheres with a diameter of about 0.74×10^{-10} m (0.074 nm). To give you some idea of just how small atoms are, consider that:

—more than two billion (2×10^{9}) hydrogen atoms would have to be laid end-to-end to form a line stretching across this page.

—nearly a billion, billion (10^{18}) hydrogen atoms would be required to give a sample that could be weighed on the most sensitive balance.

2. **Atoms of a particular element have a unique set of chemical**

properties which distinguish them from atoms of all other elements. *Example:* All atoms of the element hydrogen behave chemically in the same way. Their behavior is different from oxygen atoms or those of any other element.

3. **In the course of an ordinary chemical reaction, atoms do not disappear or change into atoms of another element.** *Example:* In any reaction involving hydrogen, there is no change in the number of hydrogen atoms. There are exactly as many hydrogen atoms in the products as in the reactants.

4. **Compounds are formed when atoms of two or more elements combine with each other.** *Example:* The compound water is formed when hydrogen and oxygen atoms combine chemically. Ammonia is formed by the combination of hydrogen with nitrogen atoms.

5. **In a particular compound, the ratio of different kinds of atoms is fixed. Ordinarily, this ratio can be expressed in terms of simple whole numbers.** *Example:* In water, there are always two hydrogen atoms for every oxygen atom present. In ammonia, the ratio of hydrogen to nitrogen atoms is 3:1.

The atomic theory offers a simple explanation of certain basic laws of science. The third statement, for example, leads directly to the **Law of Conservation of Mass.** This Law says that there is no detectable change in mass in an ordinary chemical reaction. If the products of the reaction contain the same atoms as the starting materials, they must have the same mass. In other words, if atoms are "conserved" in a reaction, mass must also be conserved.

Most scientific laws have simple explanations, once we know what's going on.

The fifth statement of the atomic theory explains the **Law of Constant Composition.** This Law says that a compound always contains the same elements in the same percentages by mass. For example, all samples of pure water contain 11.2% by mass of hydrogen and 88.8% by mass of oxygen. Clearly, if the atom ratio of hydrogen to oxygen is fixed (2 hydrogen atoms for every oxygen atom), the mass ratio must also be fixed. The same reasoning applies to any other compound. So long as the elements are present in a constant atom ratio, their mass percentages must be constant.

2.2 COMPOSITION OF THE ATOM

Dalton believed that atoms were the ultimate particles of matter. He felt that they could not be broken down into still smaller particles. This model of the "indestructible" atom lasted for nearly 100 years. Then, as experimental techniques advanced, several different subatomic particles were discovered. We will consider three such particles that are of particular interest in chemistry.

Electrons, Protons, and Neutrons

J. J. Thomson (England) in 1897 is credited with the discovery of the first subatomic particle. He used a glass tube, with most of the air evacuated, connected to a spark coil (Figure 2.1). As the voltage across the tube was increased, rays of light became visible. These are referred to as cathode rays. Thomson found that they were deflected by both electric and magnetic fields. From the nature of the deflection, he concluded that the cathode rays were made up of very small, negatively charged particles. Thomson named the new particle the **electron.** It has a unit negative charge (−1).

In further experiments, Thomson showed that the same particle was produced regardless of what gas was used in the tube. This meant that electrons must be present in all atoms. In 1909, Robert Millikan (United States) determined the mass of the electron: 9.11×10^{-28} g. This is only a little more than 1/2000 of the mass of a hydrogen atom.

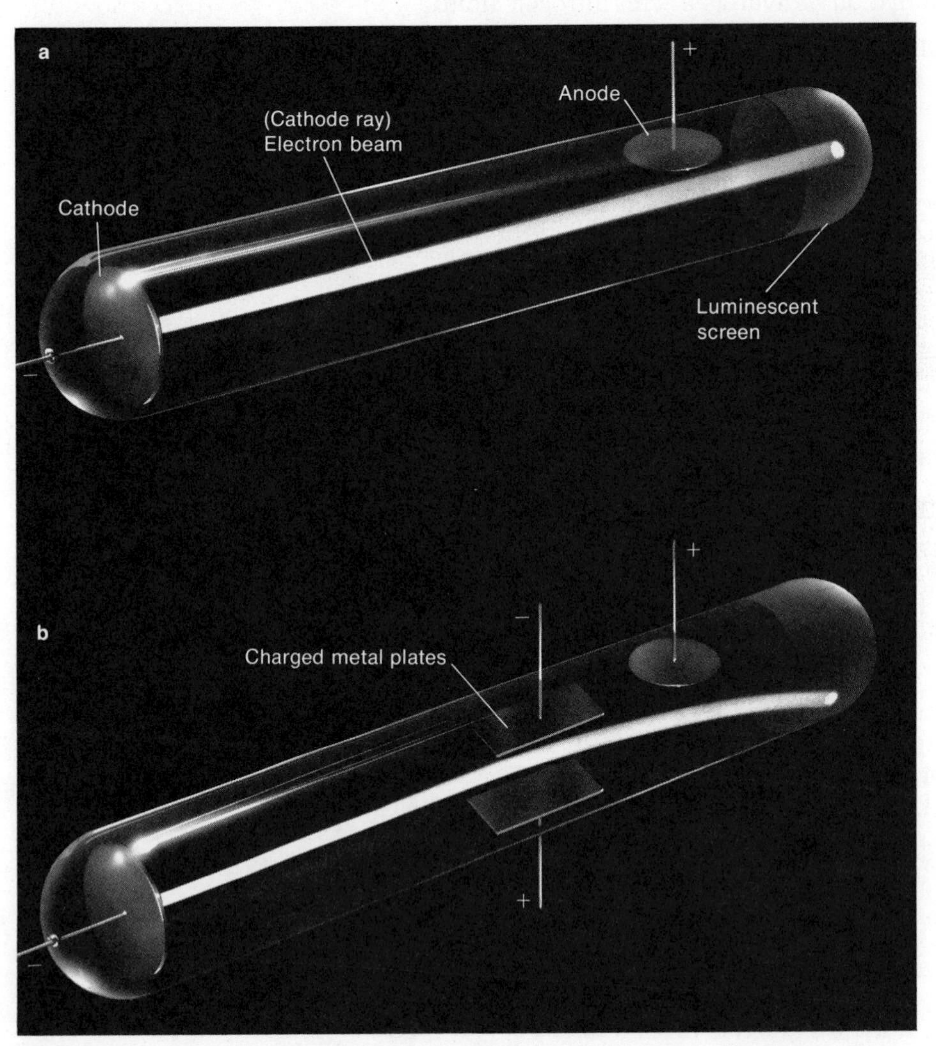

FIGURE 2.1

(a) A cathode ray tube containing two metal electrodes (anode and cathode) and a gas at low pressure. If several thousand volts of electrical potential are applied to the electrodes a visible discharge (cathode ray) is observed. (b) If charged metal plates are placed in the tube, the cathode ray is attracted to the positive plate, showing that the particles in the ray have a negative charge. Opposite charges attract.

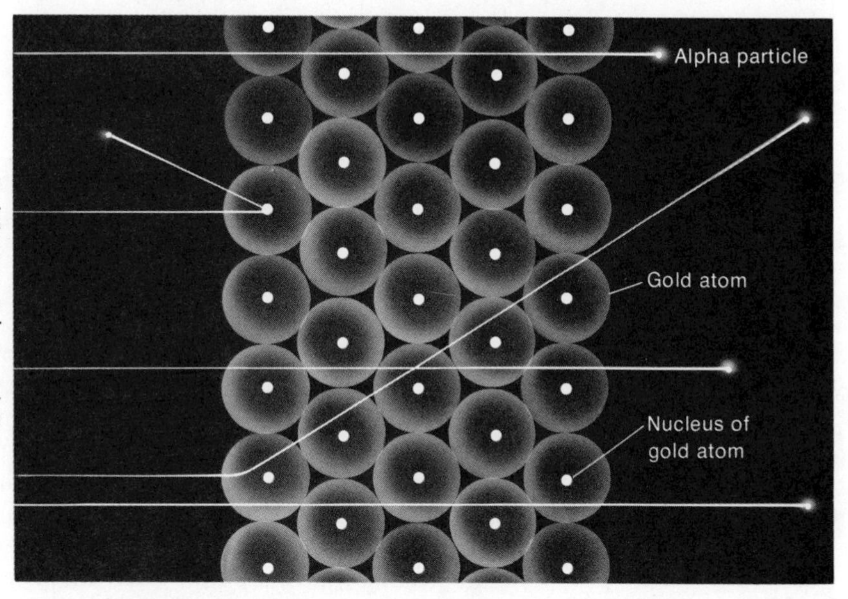

FIGURE 2.2

Rutherford's experiment. Alpha particles from radium were allowed to hit a gold foil. Most of the particles were almost undeflected by the foil, but a few were reflected back toward the radium source. For this to occur with alpha particles (helium nuclei) there must be very small, heavy, positively charged particles in the gold foil (gold nuclei).

In 1911, Ernest Rutherford (England) performed an experiment which had a far-reaching result. As shown in Figure 2.2, he bombarded a very thin gold foil with alpha particles (helium atoms stripped of their electrons). Using a fluorescent screen, he was able to measure the effect of the gold atoms on the path of the alpha particles. Most of them went through the foil almost undeflected. However, about one of every 8,000 bounced back at an angle of more than 90°. This was unexpected. At that time atoms were thought to be rather soft and mushy. In Rutherford's words, "It was about as credible as if you had fired a 15-inch shell at a piece of tissue paper and it came back and hit you."

Rutherford repeated his experiments with foils of different metals. The effects were similar to those observed with gold. Most of the alpha particles were undeflected. This suggests that atoms are mostly empty space. A few alpha particles were deflected sharply. This means that there must be a very small region within the atom where most of its mass is located.

From his experiments, Rutherford proposed a model of the atom which we use today. According to the model, atoms have a very small center, called the **nucleus.** The nucleus is positively charged and contains almost all of the mass of the atom. Outside the nucleus is a large region which defines the volume of the atom. Within that region, there are enough electrons to make the atom electrically neutral.

Rutherford showed that the diameter of an atomic nucleus is about 10^{-14} m. In contrast, the diameter of an average atom is about 10^{-10} m. This means that the diameter of the nucleus is only 1/10,000 of that of the atom. The space assigned to the nucleus within the atom is comparable to that occupied by a fly in the middle of Yankee Stadium.

No relation to a pop fly.

Further experiments showed that the nucleus, small as it is, consists of still smaller particles. One of these is a **proton,** which has a mass about equal to that of a hydrogen atom. It has a positive charge which is equal in magnitude to the negative charge of the electron (+1 *vs.* −1). The existence of protons within the nucleus explains its positive charge.

TABLE 2.1 COMPONENTS OF THE ATOM

Particle	Charge	Mass*	Location
Electron	−1	1/2000	outside nucleus
Proton	+1	1	in the nucleus
Neutron	0	1	in the nucleus

*relative to a hydrogen atom (approximate)

Rutherford realized that protons, by themselves, could not account for the entire mass of the nucleus. He predicted the existence of a new nuclear particle which would be neutral and would account for the missing mass. In 1932, James Chadwick (England) discovered this particle, the **neutron.** The neutron has zero charge. Its mass is about the same as that of the proton.

For our purposes in chemistry, we can stop our discussion of subatomic particles at this point. Their important properties are listed in Table 2.1. Dalton's atom was certainly not indestructible! Keep in mind, however, that the atom is the smallest particle which retains the properties of an element. Electrons, protons and neutrons are found in atoms of all elements.

Atoms aren't indestructible, but they hold together in chemical reactions.

Atoms differ from one another in the number of protons and neutrons in the nucleus. They also differ in the number of electrons found outside the nucleus. We will now consider how these differences come about and how they are expressed.

Atomic Number

The simple model of the atom that we have developed has direct application to chemistry. We can use it to explain why all atoms of a given element behave chemically in the same way. Moreover, it explains why atoms of different elements show different chemical properties. The principle is a simple one.

The chemical behavior of an atom is determined by the number of protons in its nucleus. This number is characteristic of a particular element. All atoms of a given element contain the same number of protons. An atom of a different element has a different number of protons in its nucleus.

As an example, consider the four elements hydrogen, helium, oxygen and uranium.

All hydrogen atoms have 1 proton in the nucleus.
All helium atoms have 2 protons in the nucleus.
All oxygen atoms have 8 protons in the nucleus.
All uranium atoms have 92 protons in the nucleus.

The number of protons in an atom of an element is referred to as its **atomic number.**

$$\text{atomic number} = \text{number of protons in nucleus} \qquad (2.1)$$

Thus we would say that:

Hydrogen has an atomic number of 1.
Helium has an atomic number of 2.
Oxygen has an atomic number of 8.
Uranium has an atomic number of 92.

Isolated atoms are electrically neutral. This means that the number of electrons (−1 charge) outside the nucleus must equal the number of protons (+1 charge) in the nucleus. In other words, the atomic number represents not only the number of protons but also the number of electrons in a neutral atom.

Atom	Atomic Number	Number of Protons	Number of Electrons
Hydrogen	1	1	1
Helium	2	2	2
Oxygen	8	8	8
Uranium	92	92	92

It takes one electron to balance the charge on each proton, and so produce a net charge of zero on each atom.

Mass Number. Isotopes

You will recall (Table 2.1) that the mass of an electron is very much smaller than that of a proton or neutron. Hence the mass of an atom is determined mainly by the total number of protons and neutrons in the nucleus. This quantity is known as the **mass number.**

$$\text{mass number} = \text{number of protons} + \text{number of neutrons} \quad (2.2)$$

Each proton and each neutron have a relative mass equal to 1.

All atoms of a given element contain the same number of protons. However, they need not contain the same number of neutrons. Consider, for example, the element hydrogen. All hydrogen atoms have one proton in the nucleus. Most of them (99.985%) have no neutrons in the nucleus. A few (0.015%) have one neutron. These two different kinds of atoms are often referred to as "light" hydrogen and "heavy" hydrogen (or deuterium).

"light" hydrogen: 1 proton, 0 neutrons
"heavy" hydrogen: 1 proton, 1 neutron

The situation we have described with hydrogen is typical of most elements. They exist in nature as a mixture of **isotopes.** *Isotopes are atoms which have the same number of protons but a different number of neutrons.* A familiar example is the element uranium which consists of three different isotopes:

"uranium-234": 92 protons, 142 neutrons
"uranium-235": 92 protons, 143 neutrons
"uranium-238": 92 protons, 146 neutrons

Since isotopes of an element contain different numbers of neutrons, they differ from one another in mass number.

Isotope	Number of Protons	Number of Neutrons	Mass Number
"light hydrogen"	1	0	1
"heavy hydrogen"	1	1	2
"uranium-234"	92	142	234
"uranium-235"	92	143	235
"uranium-238"	92	146	238

Atomic and Nuclear Symbols

Elements are assigned symbols derived from their names. A symbol is a kind of shorthand notation for the element. The symbols are international. Therefore, they are not always based on the English name of the element. The symbol may be one or two letters. If two letters are used, the second is always in lower-case. Examples include:

Sooner or later, chemists learn the names and symbols of the elements. Better sooner than later.

H hydrogen
P phosphorus
K potassium (from the Latin, kalium)
Si silicon
Ra radium
Au gold (from the Latin, aurum)
U uranium

Frequently we use the symbol of an element to refer to an atom of that element. Thus we may take the symbol P to represent an atom of phosphorus. In another case, a silicon atom is represented by the symbol Si. This interpretation of symbols is particularly useful in writing chemical equations (Chapter 3).

To identify a particular isotope, we show the atomic number and mass number in addition to the symbol of the element. The atomic number appears as a subscript at the lower left. The mass number is written as a superscript at the upper left. Thus, for the isotopes of hydrogen (1 proton; 0 or 1 neutron) and uranium (92 protons; 142, 143, or 146 neutrons), we write:

$$^{1}_{1}H,\ ^{2}_{1}H;\ ^{234}_{92}U,\ ^{235}_{92}U,\ ^{238}_{92}U$$

In general, for an isotope of atom X, the nuclear symbol is:

$$^{\text{Mass no.}}_{\text{At. no.}}X$$

EXAMPLE 2.1

a. Write the nuclear symbol for the isotope of oxygen (atomic number = 8) which has 10 neutrons in the nucleus.

b. How many protons, electrons, and neutrons are there in the neutral atom $^{37}_{17}Cl$?

SOLUTION

a. mass number = no. protons + no. neutrons = 8 + 10 = 18; $^{18}_{8}O$

b. no. of protons = no. of electrons = atomic number = 17
no. of neutrons + no. of protons = mass number = 37
hence, no. of neutrons = 37 − 17 = 20

All the isotopes of an element have the same chemical properties.

Stable and Radioactive Isotopes

About 20 elements exist in nature as a single isotope. Included among these are some of the most familiar elements: fluorine ($^{19}_{9}F$), sodium ($^{23}_{11}Na$), aluminum ($^{27}_{13}Al$), and gold ($^{197}_{79}Au$). More commonly, elements occur as a mixture of two or more isotopes. Tin (at. no. = 50) is an extreme case. It has ten stable isotopes, ranging in mass number from 112 to 124.

Ordinarily the different isotopes of an element occur in a fixed ratio to one another. Consider, for example, the element copper. We find that 68.94% of all copper atoms have a mass number of 63 ($^{63}_{29}Cu$). A smaller fraction, 31.06%, have a mass number of 65 ($^{65}_{29}Cu$). These percentages remain constant regardless of the source of the copper. They are often referred to as "abundances." We say that the natural abundances of $^{63}_{29}Cu$ and $^{65}_{29}Cu$ are 68.94% and 31.06% respectively.

In the stable isotopes of the elements of low atomic number, the ratio of neutrons to protons is close to one. Thus we have

$^{4}_{2}He$	2 protons, 2 neutrons	n/p = 2/2 = 1
$^{12}_{6}C$	6 protons, 6 neutrons	n/p = 6/6 = 1
$^{16}_{8}O$	8 protons, 8 neutrons	n/p = 8/8 = 1

With elements of higher atomic number, stable isotopes ordinarily contain more neutrons than protons. The neutron-to-proton ratio in the heavier elements approaches a value of about 1.5.

$^{59}_{27}Co$	27 protons, 32 neutrons	n/p = 32/27 = 1.2
$^{133}_{55}Cs$	55 protons, 78 neutrons	n/p = 78/55 = 1.4
$^{209}_{83}Bi$	83 protons, 126 neutrons	n/p = 126/83 = 1.5

This trend is shown graphically in Fig. 2.3, p. 36. As the number of protons in the nucleus increases, repulsive forces between these positively charged particles become greater. More neutrons are required to insulate the protons from each other (or so it seems).

Isotopes which fall outside the stability region shown in Figure 2.3 are radioactive. They decompose to more stable isotopes. In this process, described in more detail in Chapter 26, high-energy radiation is given off. A few radioactive isotopes decompose slowly enough to be found in nature. One of these is the most abundant isotope of uranium, $^{238}_{92}U$. It decomposes to give an alpha particle, $^{4}_{2}He$, and an isotope of thorium, $^{234}_{90}Th$. The alpha particle can be detected by means of a Geiger counter (Figure 2.4).

Radioactive isotopes can be made in the laboratory by nuclear reactions. At least one such isotope is available for all of the common elements. They have a variety of uses. "Cobalt-60" ($^{60}_{27}Co$) is effective in treating cancer. The radiation given off when it decomposes destroys malignant tissue. Very small samples are effective. The radiation is directed in a narrow beam so that damage to healthy tissue is held to a minimum.

There are many uses for radioactive isotopes in industry. They are well suited for finding small leaks in buried pipelines. A small amount of a radioactive isotope is added to the gas or liquid moving through the pipe. By traveling along the line with a Geiger counter, it is possible to pinpoint the location of a leak.

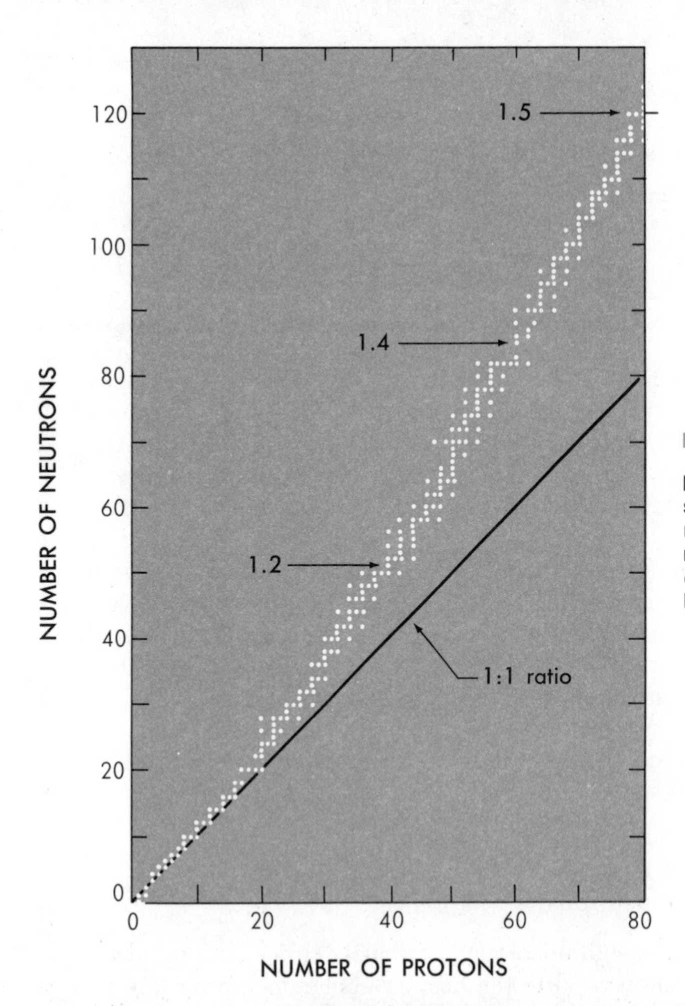

FIGURE 2.3

Diagram showing number of neutrons in the stable isotopes of the elements. The neutron-proton ratio is about 1:1 for the light elements (low atomic numbers). For the heavier elements the ratio becomes larger, becoming about 1.5 for elements like lead and mercury.

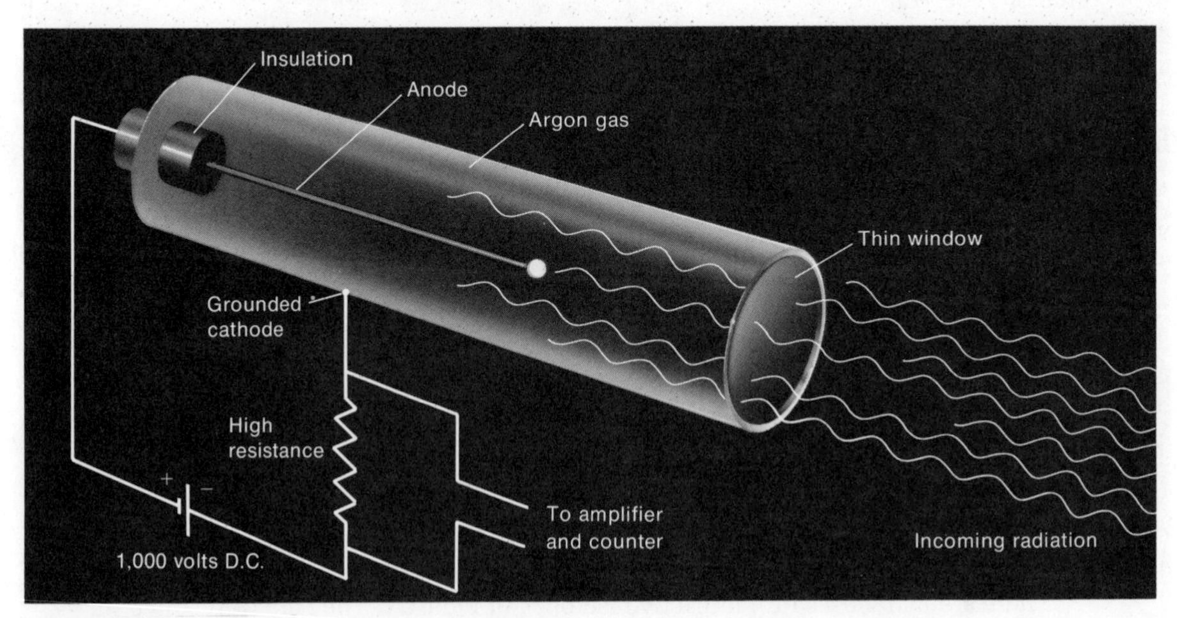

FIGURE 2.4

A Geiger counter. If an alpha particle enters the counter it knocks electrons off some of the argon atoms. The electrons and charged atoms are attracted to the + and − electrodes in the counter, causing an electric pulse, which can be detected and so serve to "count" the alpha particles that enter.

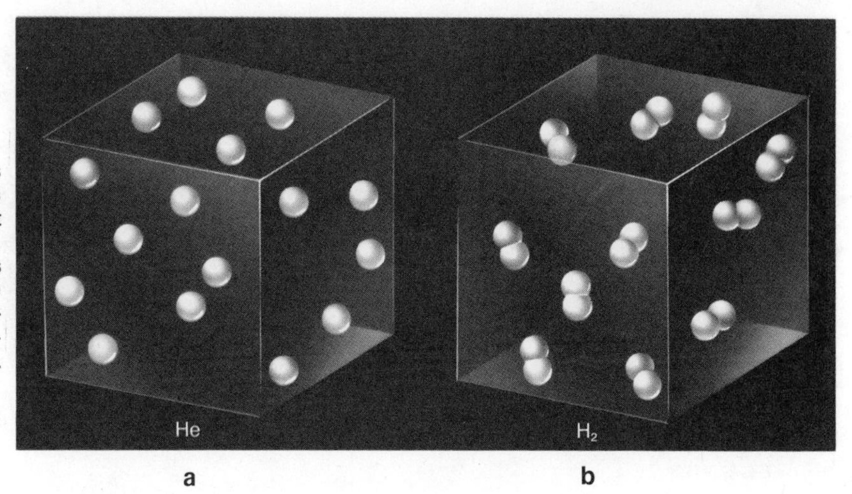

FIGURE 2.5

(a) In helium gas the particles are single He atoms. He atoms are inert chemically and do not combine with any other atoms. (b) In hydrogen gas the particles contain two H atoms which are combined chemically in a hydrogen molecule. There are no single H atoms in hydrogen gas.

2.3 MOLECULES

Isolated atoms are rarely found in nature, since they are very reactive chemically. An exception to this rule is gaseous helium, where individual He atoms are the basic building block (Figure 2.5a). Atoms of most elements act quite differently. Consider, for example, hydrogen. When two hydrogen atoms approach one another closely, they react to form a *molecule* (Figure 2.5b). This molecule consists of two atoms held together by a strong force called a chemical *bond*. This force is extremely strong. Temperatures as high as 2000°C are required to split hydrogen molecules into atoms.

"Super glue" is weak compared to chemical bonds.

In general, we can describe a molecule as **a group of two or more atoms held together by chemical bonds.** Molecules are the basic building blocks of many elements. These include:

—the gases fluorine, chlorine, nitrogen and oxygen. All of these elements, like hydrogen, form molecules containing two atoms.

—liquid bromine (two atoms per molecule).

—solid iodine (two atoms per molecule) and white phosphorus (four atoms per molecule). The structure of white phosphorus and its molecules is shown in Color Plate 1B, center of book.

Molecules are very common among compounds. Molecules of a few simple compounds are shown in Figure 2.6, p. 38. These correspond to the compounds hydrogen fluoride, water, ammonia, and methane. Notice that in each molecule, the chemical bonds join *unlike* atoms. In methane, for example, each hydrogen atom is bonded to the central carbon atom. This arrangement is typical of the molecules of most compounds. In a compound, we rarely find two like atoms (e.g., H) bonded to each other.

Molecular Formulas

The composition of a molecule, or that of the corresponding substance, is indicated by writing its molecular formula. Here, the number

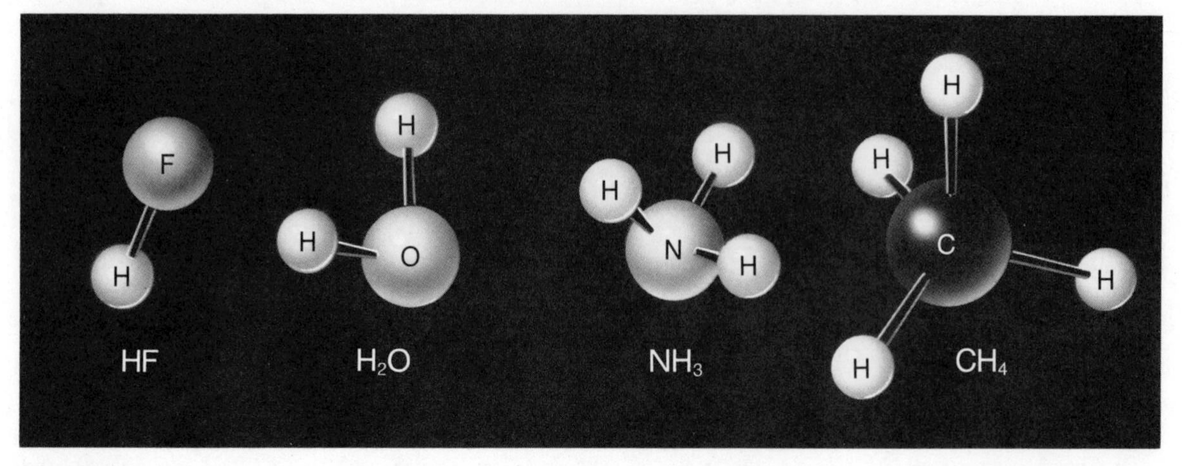

FIGURE 2.6

Structures of some molecules. In these particles the atoms are bound chemically as shown. The bonds between atoms are strong and not easily broken.

of atoms of each element in the molecule is shown as a subscript immediately after the symbol of the element. Thus, for the diatomic molecules of hydrogen, chlorine, nitrogen, and oxygen, we have:

$$H_2, Cl_2, N_2, O_2$$

The formulas of the molecules shown in Figure 2.6 are written as:

HF	(1 H atom, 1 F atom per molecule)
H_2O	(2 H atoms, 1 O atom per molecule)
NH_3	(1 N atom, 3 H atoms per molecule)
CH_4	(1 C atom, 4 H atoms per molecule)

Notice that the subscript 1 is not used. When no subscript appears after a symbol, it is understood that there is one atom of that element per molecule.

The molecules we have considered so far are relatively simple. They contain, at most, 5 atoms of two different elements (CH_4). Many familiar substances are built up of more complex molecules. Their formulas are handled in exactly the same way (Example 2.2).

EXAMPLE 2.2

Consider the compounds benzene and sucrose (sugar). In the benzene molecule there are 6 C atoms and 6 H atoms. Sucrose has the molecular formula $C_{12}H_{22}O_{11}$. Give:

a. the molecular formula of benzene.

b. the total number of atoms in a molecule of sucrose.

SOLUTION

a. C_6H_6

b. 12 C atoms, 22 H atoms, 11 O atoms
total no. atoms = 12 + 22 + 11 = 45

2.4 IONS

As we have seen, many compounds have a molecular structure. The basic structural unit in water is the H_2O molecule. In methane, it is the CH_4 molecule. However, certain familiar compounds have a very different structure. They do not contain molecules. Instead, they consist of charged particles called *ions*.

Ordinary table salt, sodium chloride, is a simple ionic compound. In a crystal of sodium chloride, there are two different structural units.

If you remove a negative charge from a neutral particle, the particle becomes positively charged.

1. A positive ion with a +1 charge, Na^+. This ion is formed when a sodium atom loses an electron:

$$\text{Na atom (11 protons, 11 electrons)} \rightarrow \text{Na}^+ \text{ ion (11 protons, 10 electrons)} + e^-$$

2. A negative ion with a −1 charge, Cl^-. This ion is formed when a chlorine atom gains an electron:

$$\text{Cl atom (17 protons, 17 electrons)} + e^- \rightarrow \text{Cl}^- \text{ ion (17 protons, 18 electrons)}$$

Figure 2.7, p. 40 shows the structure of a tiny portion of a crystal of sodium chloride. Notice that each Na^+ ion is surrounded by Cl^- ions. The oppositely charged ions are held together by strong electrical attractive forces. These forces, called ionic *bonds*, extend throughout the entire crystal. There are no small, discrete molecules, as there would be in water or methane.

Many compounds in addition to sodium chloride are ionic. Like sodium chloride, they contain both positive and negative ions. Examples include:

1. Calcium chloride. Here, the positive ion is Ca^{2+}. This is derived from a Ca atom by the loss of two electrons:

$$\text{Ca atom (20 protons, 20 electrons)} \rightarrow \text{Ca}^{2+} \text{ ion (20 protons, 18 electrons)} + 2\,e^-$$

The negative ion in this compound is the chloride ion, Cl^-. This is the same ion that is present in sodium chloride.

2. Calcium oxide. The positive ion in this compound is the calcium ion, Ca^{2+}. The negative ion is the oxide ion, O^{2-}. This ion is formed when an oxygen atom gains two electrons:

$$\text{O atom (8 protons, 8 electrons)} + 2\,e^- \rightarrow \text{O}^{2-} \text{ ion (8 protons, 10 electrons)}$$

These ionic compounds, and many others that we might mention, have a structure similar to sodium chloride. They contain two different building blocks. One is a positive ion, which contains fewer electrons than the corresponding atom. It may have a charge of +1 (Na^+), +2 (Ca^{2+}), or even +3 (Al^{3+}). The other structural unit is a negative ion. It contains more electrons than the corresponding atom. Ordinarily it has

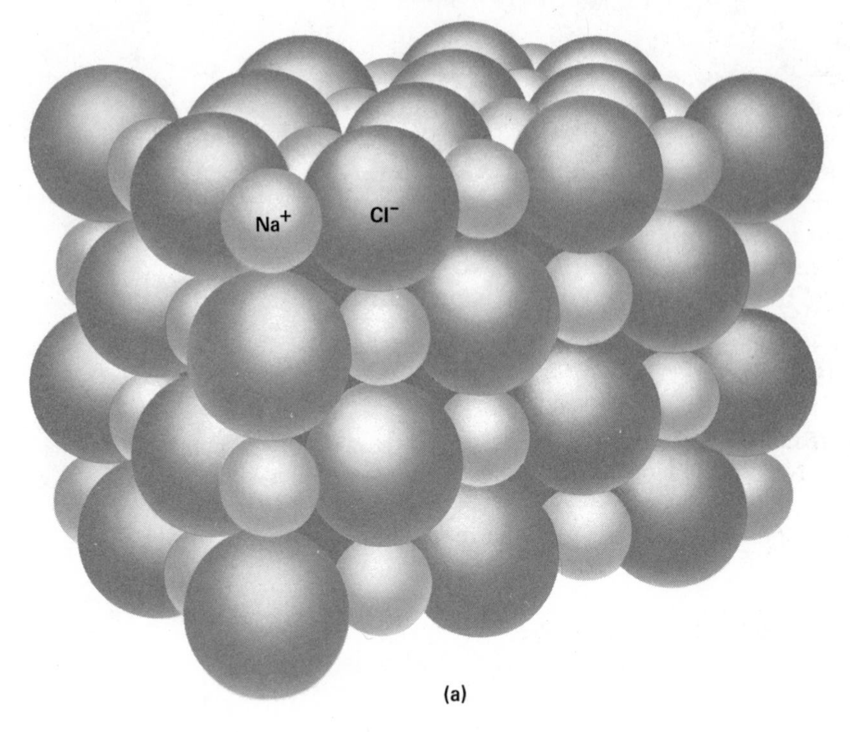

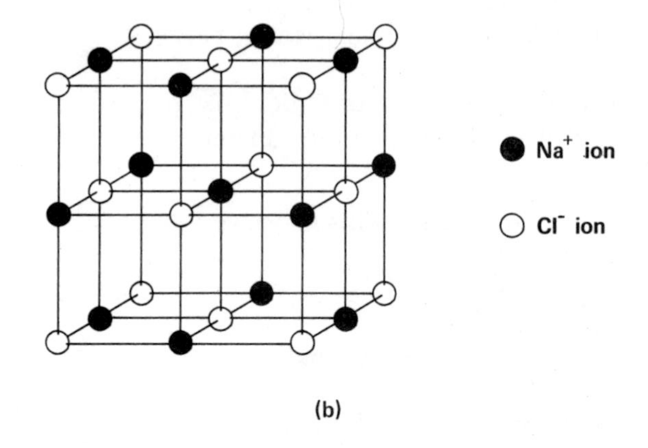

FIGURE 2.7

Structure of the sodium chloride crystal. (a) In this model the relative sizes of the Na^+ and Cl^- ions are shown along with the manner in which the ions are packed in the crystal. (b) In this ball and stick model the cubic arrangement of the ions is shown. Each Na^+ ion is surrounded by 6 Cl^- ions; each Cl^- ion is surrounded similarly by 6 Na^+ ions. The cubic structure will extend perfectly in three dimensions to include many thousands of ions.

a charge of -1 (Cl^-) or -2 (O^{2-}). In the solid ionic compound, positive ions are held to negative ions by strong electrical attractive forces. These forces extend throughout the solid. There are no molecules present in ionic compounds.

Formulas of Ionic Compounds

Ionic compounds are electrically neutral. This condition applies, for example, to sodium chloride. In this case the ions (Na^+, Cl^-) have equal but opposite charges. Hence, there must be one Na^+ ion for every

Cl^- ion. To indicate this, we write the formula of sodium chloride as NaCl. A similar situation applies with calcium oxide, which consists of Ca^{2+} and O^{2-} ions. Its formula is CaO. This indicates that in any sample of calcium oxide there are equal numbers of +2 and −2 ions.

The principle of electrical neutrality can be used to predict the formulas of ionic compounds in which the charges of the two ions differ. Consider, for example, calcium chloride. As we have seen, this is made up of Ca^{2+} and Cl^- ions. Since a crystal of calcium chloride is electrically neutral, there must be two Cl^- ions for every Ca^{2+} ion. To indicate this, we write the formula as $CaCl_2$. Notice that *in all ionic compounds, the positive ion appears first in the formula as it does in the name.*

EXAMPLE 2.3

Give the formulas of

a. potassium oxide (K^+, O^{2-} ions)

b. aluminum oxide (Al^{3+}, O^{2-} ions)

SOLUTION

a. Two K^+ ions are required to balance one O^{2-} ion: K_2O

b. The simplest ratio of ions which will give electrical neutrality is 2 Al^{3+}:3 O^{2-} (total positive charge = +6; total negative charge = −6). The formula is Al_2O_3.

Formulas such as NaCl and $CaCl_2$ are often referred to as **simplest** or empirical formulas. The simplest formula of a compound tells us only the simplest ratio of atoms or ions present. Thus, in NaCl there is one Na^+ ion for every Cl^- ion; in $CaCl_2$, the ion ratio of Ca^{2+} to Cl^- is 1:2.

The only formula an ionic compound has is its simplest formula.

2.5 RELATIVE MASSES OF ATOMS AND MOLECULES

Atoms and molecules are very tiny. Typically, their masses range from 10^{-24} to 10^{-19} g. Later in this chapter, we will see how actual masses of atoms and molecules can be calculated. Here we consider a somewhat different, more basic problem. We are concerned with the *relative* masses of atoms and molecules. That is, we want to know how one such particle compares to another in mass. This will allow us to answer questions such as the following:

Which is heavier, a hydrogen atom or an oxygen atom?

How heavy is a sugar molecule compared to a water molecule?

Atomic Masses

Formerly, atomic masses were called atomic weights, so you may see that term in other books on chemistry.

To compare the masses of different atoms, we use a table of *atomic masses* (abbreviated AM). Such a table is inside the back cover of this book. As you can see, each element is assigned a number called its atomic mass. These numbers are proportional to the masses of atoms of the elements. Stated another way:

The atomic mass (AM) of an element is a number which tells us how heavy an atom of that element is compared to an atom of another element.

To show what this statement means, consider the three elements hydrogen, oxygen, and sulfur. Their atomic masses are approximately 1, 16, and 32 in that order. This means that:

A hydrogen atom weighs $1/16$ as much as an oxygen atom.
A hydrogen atom weighs $1/32$ as much as a sulfur atom.
An oxygen atom weighs $1/2$ (i.e., $16/32$) as much as a sulfur atom.

The meaning of atomic masses is further illustrated in Example 2.4.

EXAMPLE 2.4

The atomic mass of fluorine is 19.00.

a. Is a fluorine atom heavier or lighter than a helium atom (AM = 4.00)? What is the ratio of the mass of a F atom to a He atom?

b. What is the atomic mass of an element whose atoms are 2.11 times as heavy as those of fluorine?

SOLUTION

a. Since the atomic mass of fluorine is greater than that of helium, a fluorine atom must be heavier than a helium atom. To obtain the mass ratio:

$$\frac{\text{mass F atom}}{\text{mass He atom}} = \frac{\text{AM F}}{\text{AM He}} = \frac{19.00}{4.00} = 4.75$$

In other words, a fluorine atom weighs about 4.75 times as much as a helium atom.

b. If the atom is 2.11 times as heavy as a fluorine atom, the atomic mass must be 2.11 times as great as that of fluorine.

$$\text{AM} = 2.11 \times 19.00 = 40.1 \text{ (3 significant figures)}$$

Looking at the table of atomic masses on the inside back cover, can you identify the element?

Atomic masses are based on the so-called *carbon-12 scale*. That is, the most common isotope of carbon, ${}^{12}_{6}C$, is assigned an atomic mass of exactly twelve. Atomic masses of other elements are quoted relative to that of the ${}^{12}_{6}C$ isotope. For many of the lighter elements, the atomic mass is very nearly a whole number. Fluorine, for example, has an atomic mass of almost exactly 19. This reflects the fact that fluorine consists of a single isotope of mass number 19 ($^{19}_{9}F$).

Protons and neutrons have atomic masses that are almost exactly equal to 1 on the $^{12}_{6}C$ relative mass scale.

Some elements have atomic masses that are not close to a whole number. One such element is boron. Its atomic mass is 10.81. As you might guess, boron consists of a mixture of two isotopes, $^{10}_{5}B$ and $^{11}_{5}B$. The atomic mass of the element is intermediate between those of the two isotopes. The $^{11}_{5}B$ isotope is more abundant than $^{10}_{5}B$. Hence, the atomic mass of boron is closer to 11 than to 10.

With some elements, this effect is less obvious. Consider, for example, the case of carbon. Carbon consists mainly of the $^{12}_{6}C$ isotope upon which the atomic mass scale is based. However the element, as it occurs in nature, contains a small amount of the heavier isotope, $^{13}_{6}C$. About one out of every hundred carbon atoms has a mass number of 13. This explains why the atomic mass of carbon is slightly greater than 12, about 12.01.

We can readily calculate the atomic mass of an element which is a mixture of isotopes. To do this, we must know the atomic masses of the individual isotopes. We also need the percentages *(abundances)* of these isotopes. The principle behind the calculation is a simple one. Suppose, for example, that an element X contains two isotopes of atomic masses 50 and 55. These have abundances of 40% and 60% in that order. For every 100 atoms of X, there are 40 atoms of the light isotope and 60 atoms of the heavy isotope. The 100 atoms have a total mass, on the atomic mass scale, of:

$$50(40) + 55(60) = 5300$$

The "average mass" of an atom of X is found by dividing the total mass by the total number of atoms, 100. This gives a value of 53:

The average atomic mass is the one listed in tables of atomic masses.

$$\frac{50(40) + 55(60)}{100} = \frac{5300}{100} = 53$$

More generally, we can say that for an element X consisting of several isotopes A, B, . . .

$$\text{AM of X} = \text{mass A}\,\frac{(\%\ \text{A})}{100} + \text{mass B}\,\frac{(\%\ \text{B})}{100} + \ldots \qquad (2.3)$$

where "mass A," "mass B," . . . refer to the masses of these isotopes on the atomic mass scale. The use of this relationship is shown in Example 2.5.

EXAMPLE 2.5

The element chlorine is made up of two isotopes. One of these has a mass of 34.97, the other a mass of 36.97. The light isotope has an abundance of 75.5%, the heavy isotope an abundance of 24.5%. Calculate the atomic mass of chlorine.

SOLUTION

$$\text{AM of Cl} = 34.97 \times \frac{75.5}{100} + 36.97 \times \frac{24.5}{100} = 35.5$$

Notice that the atomic mass of chlorine, 35.5, is closer to that of the light isotope (35) than it is to that of the heavy isotope (37). This is reasonable, since the light isotope is more abundant.

Chemists studying an element ordinarily work with the naturally occurring mixture of isotopes. Hence, they use average atomic masses such as those calculated above. These are the ones listed on the inside back cover of this book. Only when a particular isotope is studied do we need individual isotopic masses.

Molecular Masses

Relative masses of molecules are expressed in terms of *molecular masses* (MM). The molecular mass, like the atomic mass, is a number based on the carbon-12 scale. It is interpreted in the same way as an atomic mass. When we say that a substance has a molecular mass of 20, we mean that its molecules are:

$^{20}/_{12}$ as heavy as a carbon-12 atom
$^{20}/_{16}$ as heavy as an oxygen atom
$^{20}/_{18}$ as heavy as another molecule of molecular mass 18

Molecular masses are readily calculated if the molecular formula is known. All we need do is to add up the atomic masses of all the atoms in the molecule. Thus we have:

Finding the molecular mass of a substance is no problem if we know its molecular formula.

$$\text{MM } H_2 = 2(\text{AM H}) = 2(1.008) = 2.016$$
$$\text{MM } H_2O = 2(\text{AM H}) + (\text{AM O}) = 2(1.008) + 16.00 = 18.02$$
$$\text{MM } CCl_4 = (\text{AM C}) + 4(\text{AM Cl}) = 12.01 + 4(35.45) = 153.81$$

EXAMPLE 2.6

The molecular formula of sucrose (sugar) is $C_{12}H_{22}O_{11}$.

a. What is the molecular mass of sucrose?

b. Compare the relative masses of sucrose and water molecules.

SOLUTION

a. $\text{MM } C_{12}H_{22}O_{11} = 12(\text{AM C}) + 22(\text{AM H}) + 11(\text{AM O})$
$= 12(12.01) + 22(1.008) + 11(16.00) = 342.30$

b. Since the molecular mass of water is 18.02, the ratio of masses must be:

$$\frac{342.30}{18.02} = 19.00$$

In other words, a sugar molecule is almost exactly 19 times as heavy as a water molecule.

2.6 THE MOLE

As we said earlier, individual atoms and molecules are far too small to be weighed. The chunks of matter that we work with in the laboratory contain huge numbers of these particles. It is convenient when dealing with small items to use a larger counting unit. Eggs are sold by the dozen (12), paper clips by the gross (144). Neither of these units would be appropriate for tiny particles like atoms and molecules. They are far, far smaller than eggs or paper clips. A completely new unit is needed.

The counting unit used by chemists for atoms, molecules, and other very small particles is the *mole* (abbr. *mol*). A mole, like a dozen or a gross, is a particular number of items. However, a mole contains a very large number of items. To be exact:

A mole refers to 6.022×10^{23} items.

Thus we say that:

$$\text{a mole of H atoms} = 6.022 \times 10^{23}\ \text{H atoms}$$
$$\text{a mole of O atoms} = 6.022 \times 10^{23}\ \text{O atoms}$$
$$\text{a mole of } H_2O \text{ molecules} = 6.022 \times 10^{23}\ H_2O \text{ molecules}$$
$$\text{a mole of electrons} = 6.022 \times 10^{23}\ \text{electrons}$$

As you can imagine, 6.022×10^{23} is a very large number. Written out it is:

602,200,000,000,000,000,000,000

The vastness of this number is almost impossible to grasp. If you could collect 6×10^{23} eggs, they would cover the earth to a depth of about six kilometers (four miles). You will never in your life possess anything like 6×10^{23} visible objects, whether they be eggs, pennies, or skateboards. Yet, as we will see in a moment, 6×10^{23} atoms or molecules add up to a relatively small amount of matter. If you pick up a piece of charcoal weighing only 12 g, you hold in your hand 6×10^{23} carbon atoms. Again, 6×10^{23} water molecules have a volume of only 18 cm^3, easily swallowed in one gulp. All of this simply points up the fact that, by ordinary standards, molecules and atoms are very, very small. It takes a huge number of them to furnish a sample large enough to work with in the laboratory.

That would sure take a lot of chickens.

Mass of a Mole

As we have seen, a mole always represents a definite number of items, 6.022×10^{23}. However, its mass will vary depending upon the type of item involved. A mole of hydrogen atoms weighs considerably less than a mole of oxygen atoms and very, very, very, . . . much less than a mole of eggs.

The mass of a mole of atoms is found by applying the rule:

A mole of atoms has a mass in grams which is numerically equal to the atomic mass of the corresponding element. Thus we have:

	AM	
H	1.0	1 mol of H atoms weighs 1.0 g
O	16.0	1 mol of O atoms weighs 16.0 g
S	32.1	1 mol of S atoms weighs 32.1 g

Similarly, for molecules:

A mole of molecules has a mass in grams which is numerically equal to the molecular mass of the corresponding substance. Thus we have:

	MM	
H_2	2.0	1 mol of H_2 molecules weighs 2.0 g
CH_4	16.0	1 mol of CH_4 molecules weighs 16.0 g
H_2O	18.0	1 mol of H_2O molecules weighs 18.0 g

More generally, we can say that for any substance:

A mole has a mass in grams which is numerically equal to the formula mass of the substance. The formula mass is the sum of the atomic masses of all the atoms in the formula. This rule can be applied to any substance provided its formula is known. It makes no difference whether the substance consists of atoms, molecules, or ions (Example 2.7).

The size of the mole was picked so as to make this statement true.

EXAMPLE 2.7

Determine the mass in grams of one mole of

a. Zn b. C_6H_6 c. $CaCl_2$

SOLUTION

a. Here the "formula mass" is simply the atomic mass of zinc, 65.38.

1 mol Zn weighs 65.38 g

A mole of Zn^{2+} ions also weighs 65.38 grams.

b. To find the formula mass of C_6H_6 (benzene), we add up the atomic masses of the atoms in the formula.

$$\text{formula mass } C_6H_6 = 6(\text{AM C}) + 6(\text{AM H}) = 6(12.01) + 6(1.01) = 78.12$$

1 mol C_6H_6 weighs 78.12 g

c. Here we proceed exactly as in (b):

$$\text{formula mass } CaCl_2 = \text{AM Ca} + 2(\text{AM Cl})$$
$$= 40.08 + 2(35.45) = 110.98$$
$$1 \text{ mol } CaCl_2 \text{ weighs } 110.98 \text{ g}$$

Color Plate 1C, center of book shows the amount of matter represented by one mole of various substances.

Mole-Gram Conversions

The mole is a quantity which is used constantly in chemistry. We will refer to it, in one way or another, in almost every chapter of this book. Frequently you will need to convert a given mass in grams to a number of moles. At other times, you will find the mass in grams of a given number of moles. Conversions of this type are readily made, knowing:

—the formula of the substance involved.

—the atomic mass of each atom in the formula.

Example 2.8 illustrates both conversions.

The mole is the main chemical unit for any substance.

EXAMPLE 2.8

Find the

a. mass in grams of 1.42 mol of H_2O.

b. number of moles in 31.6 g of K_2O.

SOLUTION

a. Formula mass $H_2O = 2(\text{AM H}) + \text{AM O}$

$$= 2(1.01) + 16.00 = 18.02$$
$$1 \text{ mol } H_2O = 18.02 \text{ g } H_2O$$

This relationship gives us the conversion factor we need. We are going from moles to grams, so we put 18.02 g in the numerator and 1 mol in the denominator.

$$\text{mass } H_2O = 1.42 \text{ mol } H_2O \times \frac{18.02 \text{ g } H_2O}{1 \text{ mol } H_2O} = 25.6 \text{ g } H_2O$$

This kind of conversion gets us from the chemical unit to the mass unit.

b. As in (a), we first find the mass in grams of one mole.

$$\text{Formula mass } K_2O = 2\ (\text{AM K}) + \text{AM O}$$
$$= 2(39.10) + 16.00 = 94.20$$
$$1 \text{ mol } K_2O = 94.20 \text{ g } K_2O$$

This gives us the conversion factor we need to go from grams to moles.

$$\text{moles } K_2O = 31.6 \text{ g } K_2O \times \frac{1 \text{ mol } K_2O}{94.20 \text{ g } K_2O} = 0.335 \text{ mol } K_2O$$

Avogadro's Number

The number of items in a mole is often referred to as Avogadro's number. This honors the Italian scientist Amadeo Avogadro (1776–1856). Contrary to legend, Avogadro did not measure this number. Indeed, it seems unlikely that he realized that such a number existed. At this point, you might well ask the following questions.

1. Why *should* 1.0 g of H, 16 g of O, 32 g of S, and so on, contain the same number of atoms?
2. How do we know that this number is, to four significant figures, 6.022×10^{23}?
3. What practical use can we make of this number?

The first question is readily answered. Since an O atom is 16 times as heavy as an H atom, 16 g of oxygen and 1.0 g of hydrogen must contain the same number of atoms. The same argument can be extended to any other element, such as sulfur. It can also be used with molecular substances. Consider, for example, H_2O. We know that the molecular mass of water is 18. This means that an H_2O molecule is 18 times as heavy as an H atom. Hence, there must be the same number of H_2O molecules in 18 g of water as there are H atoms in 1.0 g of hydrogen.

A variety of methods can be used to determine Avogadro's number in the laboratory. One of the most direct was first applied by Rutherford and Geiger. They studied the radioactive decomposition of radium, which gives off alpha particles (^4_2He). Using a primitive type of Geiger counter (Figure 2.4), they found that 3.4×10^{10} alpha particles were emitted by a gram of radium in one second. This corresponds to about 1.07×10^{18} He atoms per year.

$$3.4 \times 10^{10}\,\frac{\text{atoms}}{\text{s}} \times \frac{60\text{ s}}{1\text{ min}} \times \frac{60\text{ min}}{1\text{ hr}} \times \frac{24\text{ hr}}{1\text{ d}} \times \frac{365\text{ d}}{1\text{ yr}} = 1.07 \times 10^{18}\,\frac{\text{atoms}}{\text{yr}}$$

Later, another English scientist, Dewar, measured the amount of helium produced by a gram of radium. He found that in one year 1.74×10^{-6} mol of He was formed. Using these numbers, we obtain 6.15×10^{23} for Avogadro's number:

$$\frac{1.07 \times 10^{18}\text{ atoms}}{1.74 \times 10^{-6}\text{ mol}} = 6.15 \times 10^{23}\text{ atoms/mol}$$

These experiments were done between 1905 and 1912. As time passed, other, more accurate methods were developed. They gave somewhat lower values. The currently accepted value of Avogadro's number is $(6.02209 \pm 0.00001) \times 10^{23}$.

Avogadro's number can be used for many different purposes. One thing we can do with it is to calculate the masses of atoms and molecules. Knowing that the atomic mass of silver is 107.87, it follows that:

$$6.022 \times 10^{23}\text{ Ag atoms weigh }107.87\text{ g}$$

Hence a silver atom must weigh:

$$\frac{107.87\text{ g}}{6.022 \times 10^{23}} = 1.791 \times 10^{-22}\text{ g}$$

Again, since the molecular mass of water is 18.02:

$$\text{mass of } H_2O \text{ molecule} = \frac{18.02 \text{ g}}{6.022 \times 10^{23}} = 2.992 \times 10^{-23} \text{ g}$$

These calculations emphasize a point we have made throughout this chapter. Atoms, molecules, and ions are very tiny particles. They are far too small to be weighed individually on any balance.

2.7 SUMMARY

In this chapter, you have been introduced to three structural units of matter. These are:

atoms, which consist of a small nucleus containing protons and neutrons surrounded by a "cloud" of electrons. The number of protons in the nucleus is called the atomic number. It is characteristic of all atoms of a particular element. Atoms of the same element may differ in the number of neutrons in the nucleus and hence in mass number (no. of protons + no. of neutrons).

molecules, which are small, discrete groups of atoms held together by strong forces. The composition of a molecular substance is indicated by giving its molecular formula (H_2, H_2O, and so on).

ions, which are derived from atoms by the loss of electrons (positive ions) or gain of electrons (negative ions). In ionic compounds, + and − ions are bound to one another by strong electrical forces. The composition of an ionic compound is indicated by its simplest formula (NaCl, $CaCl_2$, and so on).

Relative masses of atoms of different elements are expressed in terms of their atomic masses. Relative masses of molecules are expressed on the same scale by giving their molecular masses. These are found by adding the atomic masses of the atoms in the molecular formula (MM H_2 = 2.016, MM H_2O = 18.02).

A mole may refer to a furry animal who digs holes in the ground. More frequently in chemistry it means Avogadro's number (6.022×10^{23}) of items. These items are most commonly atoms or molecules. The mass of a mole in grams is found from the symbol or formula of the substance involved. For example, 1 mol H atoms = 1.008 g; 1 mol H_2 = 2.016 g; 1 mol NaCl = 58.44 g.

NEW TERMS

abundance (isotopes)
atom
atomic mass
atomic number
atomic theory
Avogadro's number
carbon-12 scale
electron
ion
isotope
Law of Conservation of Mass
Law of Constant Composition

mass number
mole
molecular formula
molecular mass
molecule
neutron
nucleus
proton
simplest formula
symbol

QUESTIONS

1. Which of the following statements are true?

   a. A H atom is too small to be seen by the naked eye.
   b. A H atom can be weighed on a grocer's scale.
   c. All H atoms have the same mass.
   d. All H atoms behave chemically in the same way.

2. State the Law of Constant Composition. Suggest an explanation in terms of atomic theory.

3. Describe briefly the contribution made by each of the following scientists to determining the structure of the atom.

   a. Thomson b. Millikan c. Rutherford

4. Consider the three particles: electron, proton, neutron.

   a. Which one has the smallest mass?
   b. Which one is uncharged?
   c. Which one is found outside the nucleus?
   d. Which two have nearly the same mass?

5. Which of the following are isotopes?

   a. ${}^{2}_{1}H$ and ${}^{1}_{1}H$ b. ${}^{4}_{2}He$ and ${}^{4}_{3}Li$ c. ${}^{16}_{8}O$ and ${}^{16}_{8}O^{2-}$
   d. ${}^{14}_{7}N$ and ${}^{15}_{7}N$ e. ${}^{1}_{1}H^{+}$ and ${}^{1}_{1}H^{-}$

6. Look up the symbols of the following elements.

   a. argon b. boron c. strontium d. tin e. tungsten

7. Give the names of the elements with the following symbols:

   a. Cu b. Sn c. K d. P e. As

8. A certain element contains 75% of an isotope of mass number X and 25% of an isotope of mass number Y. Is the atomic mass of the element closer to X or to Y?

9. Explain, in your own words, what is meant by a molecule. How does a molecule differ from an atom?

10. How does a Cl^- ion differ from a Cl atom? from a Cl_2 molecule?

11. Referring to the table of atomic masses, arrange the following atoms in order of decreasing mass: Al, B, Be.

12. Arrange the following particles in order of decreasing mass.

    a. H atom b. C atom c. CH_4 molecule d. C_2H_6 molecule
    e. C_2H_4 molecule

13. Which of the following statements concerning the $^{12}_{6}C$ atom are true?

    a. It is the basis for the atomic mass scale.
    b. It weighs 12.0 g.
    c. It is more abundant in nature than $^{13}_{6}C$.
    d. It has an atomic number of 12.

14. Which of the following statements are true?

    a. A mole of Zn contains the same number of atoms as a mole of Cl.
    b. A mole of Zn has the same mass as a mole of Cl.
    c. A mole of H_2O weighs more than a mole of H_2.

PROBLEMS

1. *Composition of the Atom (Example 2.1)* Complete the following table on a separate sheet of paper. Use the table on the inside back cover to find the missing symbols.

	Nuclear Symbol	Number of Protons	Atomic Number	Number of Neutrons	Mass Number
a.	$^{12}_{6}C$	–	–	–	–
b.	–	5	–	6	–
c.	–	–	12	13	–
d.	–	–	4	–	9

2. *Molecules (Example 2.2)* Complete the following table on a separate sheet of paper. All of the compounds involved are hydrocarbons, containing only the elements carbon and hydrogen.

	Molecular Formula	Number of C Atoms per Molecule	Number of H Atoms per Molecule	Total No. of Atoms per Molecule
a.	CH_4	1	4	5
b.	C_2H_2	–	–	–
c.	–	2	6	–
d.	–	4	–	12

3. *Ions (Example 2.3)* Give the simplest formula of the compound containing:

    a. Li^+ and F^- ions b. Li^+ and S^{2-} ions c. Mg^{2+} and F^- ions
    d. Mg^{2+} and S^{2-} ions

4. *Atomic Masses (Example 2.4)* Zinc has an atomic mass of 65.38. The atomic mass of Cl is 35.45. A Zn atom is how many times as heavy as a Cl atom?

5. *Atomic Masses (Example 2.4)* A certain atom weighs 0.460 times as much as a nickel atom (AM = 58.70). What is its atomic mass? Can you find the element in a table of atomic masses?

6. *Atomic Masses (Example 2.5)* A certain element consists of two isotopes which have atomic masses of 10.0 and 11.0. Their abundances are 18.8% and 81.2% in that order. What is the atomic mass of the element?

7. *Molecular Masses (Example 2.6)* Determine the molecular mass of

   a. C_2H_2 b. CO_2 c. H_2S

8. *Molecular Masses (Example 2.6)* An H_2O molecule is how many times as heavy as an O atom? as an H_2S molecule?

9. *The Mole (Example 2.7)* Determine the mass in grams of one mole of

   a. C_2H_2 b. CO_2 c. K_2S d. $KMnO_4$

10. *The Mole (Example 2.8)* Determine the number of grams in 2.50 mol of

    a. C_2H_2 b. H_2S c. KNO_3

11. *The Mole (Example 2.8)* How many moles are there in 1.00 g of

    a. N_2 b. HF c. $COCl_2$

* * * * *

12. An oxygen atom has a diameter of 1.32×10^{-10} m. How many oxygen atoms would have to be placed end-to-end to stretch from top to bottom of this page? Use a metric ruler to measure the page length.

13. The atomic numbers of Li. F, Mg, and S are 3, 9, 12, and 16 in that order. Give the number of protons and electrons in the following ions.

    a. Li^+ b. F^- c. Mg^{2+} d. S^{2-}

14. Suppose we were to define Walford's number as 1.00×10^{23}. How many of these units would there be in one mole?

15. Complete the following table on a separate sheet of paper.

	Nuclear Symbol	Number of Protons	Number of Electrons	Number of Neutrons	Charge
a.	${}^{16}_{8}O^{2-}$	8	10	8	−2
b.	${}^{27}_{13}Al^{3+}$	–	–	–	–
c.	–	12	–	12	+2
d.	–	–	18	19	−1

16. Glucose has the molecular formula $C_6H_{12}O_6$.

    a. What is the molecular mass of glucose?
    b. What is the mass in grams of one mole of glucose?
    c. How many molecules are there in a mole of glucose?
    d. How many atoms are there in a mole of glucose?

17. In a molecule of the explosive "TNT" there are 7 carbon atoms, 5 hydrogen atoms, 3 nitrogen atoms, and 6 oxygen atoms.

    a. What is the molecular formula of TNT?
    b. What is the molecular mass of TNT?
    c. Convert 1.20 mol of TNT to grams.

18. Limestone has the simplest formula $CaCO_3$.

    a. What is the mass in grams of one mole of limestone?
    b. How many grams are there in 0.239 mol $CaCO_3$?
    c. Convert 24.5 g of $CaCO_3$ to moles.

19. Calculate the mass in grams of 1.42 mol of:

    a. KNO_3 b. C_6H_6 c. NH_4Cl

20. Calculate the number of moles in 12.3 g of:

    a. KNO_3 b. C_6H_6 c. NH_4Cl

*21. How many atoms are there in the smallest sample of copper (AM = 63.55) that can be detected on a microgram balance (1×10^{-6} g)?

*22. Copper (AM = 63.55) consists of two isotopes with masses of 62.93 and 64.93. What are the percentages of these isotopes?

*23. What is the volume of a mole of ice cubes 2.50 cm on an edge? What is their mass ($d = 0.92$ g/cm^3)? How does this compare to the mass of the earth (6×10^{24} kg)?

*24. Estimate the mass in grams of a proton (AM = 1.0). Taking it to be spherical with a diameter of 1×10^{-12} m, estimate its density in g/cm^3. The volume of a sphere is $4\pi r^3/3$, where r is the radius.

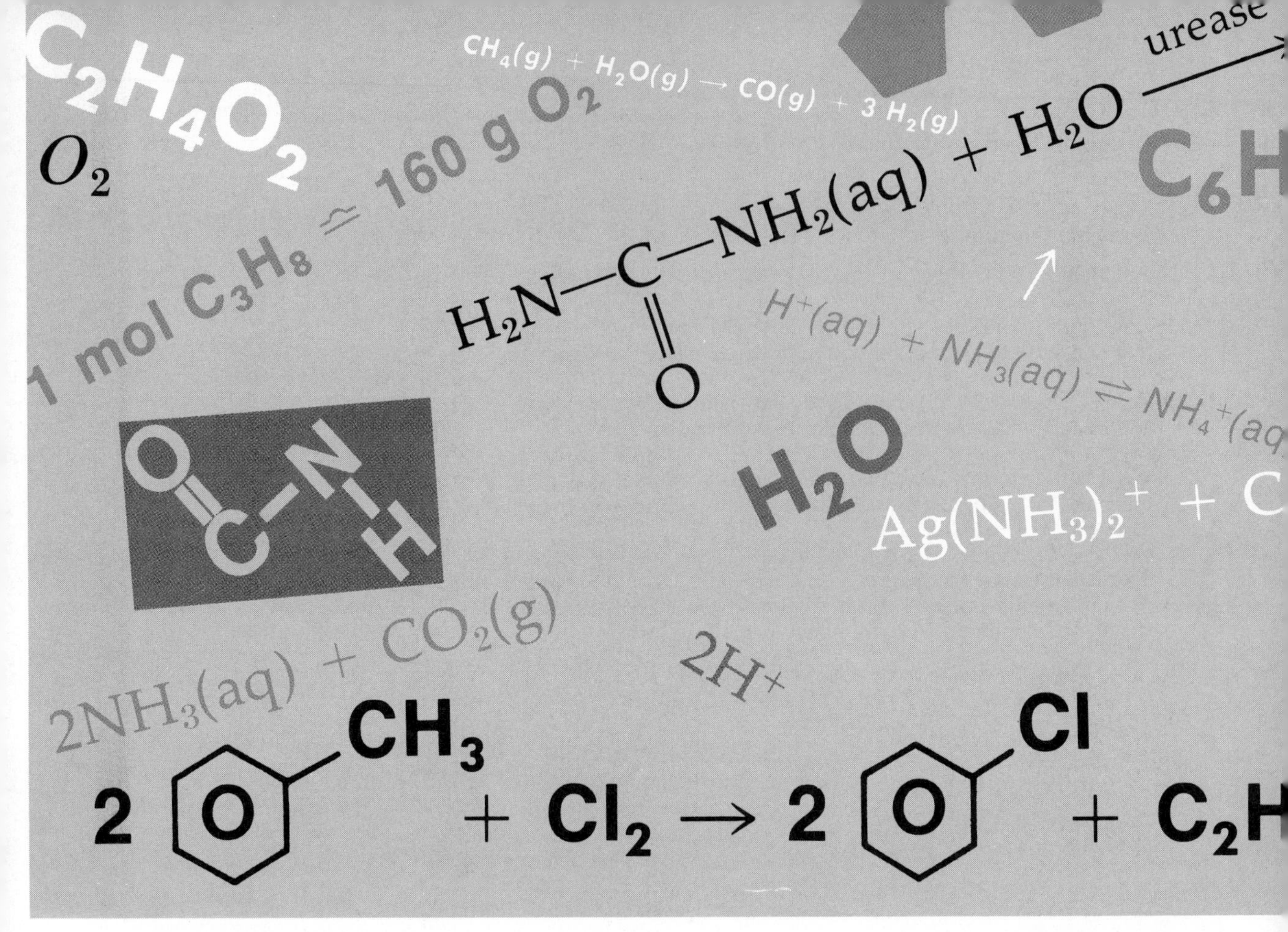

3

CHEMICAL FORMULAS AND EQUATIONS

They live, and love, and sometimes make merry.

The chemist is commonly pictured in a white laboratory coat holding up a test tube containing a wonderful new substance. There is some truth in this image, but not very much. The fact is that chemists are quite ordinary human beings. We might say, very simply, that chemists are people who are curious as to what happens when a chemical reaction takes place. Their scientific curiosity leads them into all sorts of activities. Most of these are of practical value to society.

Some chemical reactions occur naturally, such as photosynthesis. Others are carried out by men and women in laboratories. All of these reactions can be described by a special chemical language. As we saw in Chapter 2, the language has an alphabet, the chemical symbols, and words, the chemical formulas. These can be put together to make a statement, the chemical equation. In this chapter, we will look closely at the language of chemistry. We will consider how formulas of substances are determined by experiment. Going one step further, we will see how reactions can be expressed in terms of balanced equations. The principles developed in this chapter are part of the basic working knowledge of every chemist. They will be used repeatedly throughout the rest of this book.

3.1 CHEMICAL FORMULAS AND PERCENTAGE COMPOSITION

The chemical composition of a compound can be specified in either of two ways. We may cite its:

—**formula,** which tells us, at a minimum, the atom ratio in which the elements are combined in the compound.

This is the easy, although indirect, way to describe composition.

—**percentage composition,** which gives the percentages by mass of the elements in the compound.

Both of these quantities have been referred to before. Here we will examine them more closely. We will see how they can be determined experimentally and how they are related to each other.

Meaning of Percentage Composition

The term "percent" means parts per hundred. If a sample weighs 100 g and contains 20 g of element A, the mass percent of A is 20. Most samples, of course, don't weigh 100 g. However, we can still get the mass percent of A by taking the fraction of the sample that consists of A and multiplying by 100.

$$\text{mass \% A} = \frac{\text{mass of A}}{\text{mass of sample}} \times 100 \tag{3.1}$$

In many cases, we know the mass percent of an element and want to find how much of that element there is in a sample. To do this, we solve Equation 3.1 for the mass of A, obtaining:

$$\text{mass of A} = \frac{\text{mass \% A}}{100} \times \text{mass of sample} \tag{3.2}$$

Can you see how equation 3.2 comes from equation 3.1? If you don't, try to work it out step by step.

Let us apply these equations to the compound water. We find by experiment that a 1.000 g sample contains 0.112 g of the element hydrogen. It follows from Equation 3.1 that:

$$\text{mass \% H} = \frac{0.112\text{ g}}{1.000\text{ g}} \times 100 = 11.2$$

Suppose now that we wanted to find the mass of hydrogen in a sample of water weighing 30.0 g. Applying Equation 3.2:

$$\text{mass of H} = \frac{11.2}{100} \times 30.0 \text{ g} = 3.36 \text{ g}$$

This line of reasoning can be applied to any compound (Example 3.1).

EXAMPLE 3.1

A sample of nickel oxide weighing 8.144 g contains 6.400 g of nickel and 1.744 g of oxygen.

a. What are the mass percentages of nickel and oxygen in nickel oxide?

b. What are the masses of nickel and oxygen in a sample of nickel oxide weighing 3.112 g?

SOLUTION

In problems like this, rather than worry overmuch about *what* to do, try to see *why* we do it.

a. Using Equation 3.1:

$$\text{mass \% Ni} = \frac{\text{mass nickel}}{\text{mass nickel oxide}} \times 100$$

$$\text{mass \% O} = \frac{\text{mass oxygen}}{\text{mass nickel oxide}} \times 100$$

Substituting numbers:

$$\text{mass \% Ni} = \frac{6.400 \text{ g}}{8.144 \text{ g}} \times 100 = 78.59$$

$$\text{mass \% O} = \frac{1.744 \text{ g}}{8.144 \text{ g}} \times 100 = 21.41$$

b. Using Equation 3.2:

$$\text{mass Ni} = \frac{\text{mass \% Ni}}{100} \times \text{mass of sample}$$

$$= 0.7859 \times 3.112 \text{ g} = 2.446 \text{ g}$$

The mass of oxygen could be obtained in the same way. We could obtain it more simply, however, by subtracting the mass of nickel from the total mass of the sample.

$$\text{mass O} = \text{mass nickel oxide} - \text{mass Ni}$$
$$= 3.112 \text{ g} - 2.446 \text{ g} = 0.666 \text{ g}$$

Two points about the calculation in Example 3.1 are worth commenting upon.

1. The masses of nickel and oxygen in any sample of nickel oxide must add to the total mass of the sample.

$$\text{mass Ni} + \text{mass O} = \text{mass nickel oxide}$$

More generally, for a compound containing elements A, B, C, . . .

$$\text{mass A} + \text{mass B} + \text{mass C} + \ldots = \text{mass compound} \qquad (3.3)$$

This is really just the Law of Conservation of Mass.

2. The mass percentages of nickel and oxygen in nickel oxide must add to 100.

$$\text{mass \% Ni} + \text{mass \% O} = 100$$

Thus in Example 3.1a, we could have obtained the mass percent of oxygen by subtracting from 100:

$$\text{mass \% O} = 100.00 - \text{mass \% Ni} = 100.00 - 78.59 = 21.41$$

More generally, for a compound containing elements A, B, C, . . .

$$\text{mass \% A} + \text{mass \% B} + \text{mass \% C} + \ldots = 100 \qquad (3.4)$$

If we know the percentages of all but one of the elements in a compound, we can obtain that percentage by subtracting from 100.

Percentage Composition from Formula

If the formula of a compound is known, the percentages by mass of the elements present are readily determined. The approach used is illustrated in Example 3.2.

A chemical formula serves not only to identify a substance, but also to describe its composition.

EXAMPLE 3.2

The molecular formula of methane is CH_4. The atomic masses of carbon and hydrogen are 12.01 and 1.008 in that order. Calculate the mass percentages of carbon and hydrogen in methane.

SOLUTION

It is convenient here to work through the mole, which was introduced in Chapter 2. We start by noting that:

1 mol of CH_4 contains 1 mol of C atoms and 4 mol of H atoms

To find the mass percentage of carbon, we compare the mass of one mole of carbon to that of one mole of methane.

1 mol C weighs 12.01 g

1 mol CH_4 weighs 12.01 g + 4(1.008 g) = 16.04 g

Hence: $$\text{mass \% C} = \frac{\text{mass C}}{\text{mass CH}_4} \times 100 = \frac{12.01\text{ g}}{16.04\text{ g}} \times 100 = 74.88$$

$$\text{mass \% H} = 100.00 - 74.88 = 25.12$$

The calculations in Example 3.2 illustrate an important point:

The subscripts in a chemical formula represent not only the atom ratio but also the mole ratio in which the elements are combined.

Thus, given the formulas NaCl and $KClO_3$ we could say that:

NaCl: 1 atom of Na is combined with 1 atom of Cl
or: 1 mol Na (22.99 g) is combined with 1 mol Cl (35.45 g)
$KClO_3$: 1 atom of K is combined with 1 atom of Cl and 3 atoms of O
or: 1 mol K (39.10 g) is combined with 1 mol Cl (35.45 g) and 3 mol O (48.00 g)

Example 3.3 further illustrates the usefulness of this principle.

EXAMPLE 3.3

The formula of sugar (sucrose) is $C_{12}H_{22}O_{11}$. Determine:

a. the mass percentages of carbon, hydrogen, and oxygen in sugar.

b. the mass of carbon in a teaspoon of sugar, which weighs about 10 g (2 sig. fig.).

SOLUTION

Here, using one mole as the sample size is very useful.

a. Consider a sample consisting of one mole of sugar. In this sample there are 12 mol of C, 22 mol of H, and 11 mol of O. Taking the atomic masses of these three elements to be 12.01, 1.008, and 16.00, we have:

$$\begin{aligned} 12(12.01 \text{ g C}) &= 144.1 \text{ g C} \\ 22(1.008 \text{ g H}) &= 22.2 \text{ g H} \\ 11(16.00 \text{ g O}) &= \underline{176.0 \text{ g O}} \\ & \quad 342.3 \text{ g sugar} \end{aligned}$$

Hence: $\% \text{ C} = \frac{144.1 \text{ g}}{342.3 \text{ g}} \times 100 = 42.10$; $\% \text{ H} = \frac{22.2 \text{ g}}{342.3 \text{ g}} \times 100 = 6.49$

$$\% \text{ O} = \frac{176.0 \text{ g}}{342.3 \text{ g}} \times 100 = 51.42$$

b. $\text{mass carbon} = \text{mass sugar} \times \frac{\% \text{ C}}{100} = 10 \text{ g} \times 0.4210 = 4.2 \text{ g}$

Simplest Formula from Percentage Composition

We have seen that, given the formula of a substance, we can obtain the mass percentages of the elements present. It is also possible to go in the reverse direction. Knowing the percentage composition of a compound, its formula can be found. There is, however, one restriction on this calculation. *The formula obtained from percentage composition is the simplest formula.* This gives only the simplest atom ratio of the elements present, not necessarily the molecular formula. Example 3.4

illustrates the reasoning involved in arriving at the simplest formula from percentage composition.

EXAMPLE 3.4

Calcium fluoride occurs as the mineral fluorite. It contains 51.3% by mass of calcium and 48.7% by mass of fluorine. Determine its simplest formula.

SOLUTION

Let us start with a fixed mass of the compound, 100 g for convenience. In a 100 g sample of calcium fluoride, there are:

Can you see why using 100 grams as a sample size is convenient in this problem?

$$51.3 \text{ g Ca, } 48.7 \text{ g F}$$

We now calculate the number of moles of each element in the 100 g sample. Since the atomic masses of calcium and fluorine are 40.1 and 19.0:

$$1 \text{ mol Ca} = 40.1 \text{ g Ca; } 1 \text{ mol F} = 19.0 \text{ g F}$$

Using the conversion factor approach, we convert 51.3 g of calcium to moles:

$$\text{moles Ca} = 51.3 \text{ g Ca} \times \frac{1 \text{ mol Ca}}{40.1 \text{ g Ca}} = 1.28 \text{ mol Ca}$$

Similarly, for 48.7 g of fluorine:

$$\text{moles F} = 48.7 \text{ g F} \times \frac{1 \text{ mol F}}{19.0 \text{ g F}} = 2.56 \text{ mol F}$$

Looking at the two numbers just calculated, we see that the mole ratio of F to Ca in this compound must be:

$$\frac{2.56 \text{ mol F}}{1.28 \text{ mol Ca}} = 2.00 \text{ mol F/mol Ca}$$

But, as pointed out on p. 58, *the mole ratio must also be the atom ratio.* In other words, the atom ratio of F to Ca must be 2:1. The simplest formula is CaF_2.

It may be helpful to review the calculations in Example 3.4 to make sure you understand them. Three steps are involved.

1. *Using the percentage composition* (51.3% Ca, 48.7% F), *determine the masses of the elements in 100 g of the compound* (51.3 g Ca, 48.7 g F).
2. *Convert these masses in grams to moles* (1.28 mol Ca, 2.56

mol F). To do this, divide in each case by the number of grams per mole (51.3/40.1 = 1.28; 48.7/19.0 = 2.56).

3. *Determine the simplest mole ratio between the elements and set that equal to the atom ratio in the simplest formula.* This involves finding the simplest ratio between two numbers. To do this, *divide by the smallest number.* Often, this gives a simple, whole number ratio, as in Example 3.4:

$$\frac{2.56 \text{ mol F}}{1.28 \text{ mol Ca}} = 2.00 \frac{\text{mol F}}{\text{mol Ca}}\text{; simplest formula } CaF_2$$

Sometimes it isn't so simple. Suppose, for example, that we found the numbers of moles of X and Y in 100 g of a compound to be 1.96 and 2.94 in that order. Dividing by 1.96 we find that:

Since the data will have some experimental error, the number might come out to be 1.52 or 1.47. We would still take it as 1.5.

$$\frac{2.94 \text{ mol Y}}{1.96 \text{ mol X}} = 1.50 \frac{\text{mol Y}}{\text{mol X}}$$

Here we do not have a whole number ratio. To obtain one, we multiply numerator and denominator by 2:

$$\frac{3 \text{ mol Y}}{2 \text{ mol X}}$$

The simplest formula must be X_2Y_3.

The approach we have used can be applied to any compound, no matter how many elements it contains (Example 3.5).

EXAMPLE 3.5

Acetic acid, the compound which gives vinegar (and some home-made wine) its sour taste, contains the three elements carbon, hydrogen, and oxygen. The mass percentages of these elements are: 40.0% C, 6.73% H, 53.3% O. Determine the simplest formula of acetic acid (AM C = 12.01, H = 1.01, O = 16.00).

SOLUTION

In 100 g of acetic acid there are:

$$40.0 \text{ g C, } 6.73 \text{ g H, } 53.3 \text{ g O}$$

Converting to moles in each case:

$$\text{moles C} = 40.0 \text{ g C} \times \frac{1 \text{ mol C}}{12.01 \text{ g C}} = 3.33 \text{ mol C}$$

$$\text{moles H} = 6.73 \text{ g H} \times \frac{1 \text{ mol H}}{1.01 \text{ g H}} = 6.66 \text{ mol H}$$

$$\text{moles O} = 53.3 \text{ g O} \times \frac{1 \text{ mol O}}{16.00 \text{ g O}} = 3.33 \text{ mol O}$$

To find the simplest mole ratio, we might compare each element to carbon:

$$\frac{6.66 \text{ mol H}}{3.33 \text{ mol C}} = 2.00 \text{ mol H/mol C}$$

$$\frac{3.33 \text{ mol O}}{3.33 \text{ mol C}} = 1.00 \text{ mol O/mol C}$$

In other words, for every mole of carbon we have two moles of hydrogen and one mole of oxygen. The mole ratio or *the atom ratio* is:

1 C : 2 H : 1 O

Simplest formula = CH_2O

Molecular Formula from Simplest Formula

The molecular formula describes the molecule of the substance which is the basic particle of that substance. The simplest formula tells us only the atom ratio in the substance.

From the percentage composition of a compound, we obtain only the simplest atom ratio between the elements present. This gives us, as we have seen, the simplest formula. In some cases, as with H_2O, the simplest formula is also the molecular formula. The water molecule contains two hydrogen atoms and one oxygen atom. In other cases, the molecular formula differs from the simplest formula. An example is the compound hydrogen peroxide. Its simplest formula is HO (one hydrogen atom for every oxygen atom). The molecular formula of hydrogen peroxide is H_2O_2. A molecule of this compound contains two hydrogen and two oxygen atoms.

It is possible to obtain the molecular formula of a compound if two pieces of information are available:

—the simplest formula, and
—the molecular mass

The way in which these quantities are used is shown in Example 3.6.

EXAMPLE 3.6

The simplest formula of acetic acid, as determined in Example 3.5, is CH_2O. The molecular mass of acetic acid is known to be 60.0. What is the molecular formula?

SOLUTION

The formula mass corresponding to CH_2O is

$$12.0 + 2(1.01) + 16.0 = 30.0$$

Since the molecular mass, 60.0, is twice the formula mass of CH_2O, 30.0, the molecular formula must contain twice as many atoms of each type as the simplest formula. To obtain the molecular formula, we double each subscript in the simplest formula. The molecular formula is $C_2H_4O_2$.

TABLE 3.1 SIMPLEST AND MOLECULAR FORMULAS OF COMPOUNDS

Compound	Simplest Formula	Mass of Simplest Formula	Molecular Mass	Molecular Formula
Water	H_2O	18.0	18.0	H_2O
Hydrogen Peroxide	HO	17.0	34.0	H_2O_2
Acetic Acid	CH_2O	30.0	60.0	$C_2H_4O_2$
Propylene	CH_2	14.0	42.0	C_3H_6
Benzene	CH	13.0	78.0	C_6H_6

The method outlined in Example 3.6 can be used with any molecular compound. By dividing the molecular mass by the mass of the simplest formula, we obtain a number which may be 1, 2, 3, . . . The molecular formula is then found by multiplying the simplest formula by that number. If the number is 1, the two formulas are the same (water in Table 3.1). If it is 2, the molecular formula is twice the simplest formula, and so on.

In dealing with molecular substances, we always cite the molecular formula if it is known. It gives more information about the structure of the substance. Thus in writing equations (Section 3.2) we represent

The molecular formula tells us directly the number and kind of each atom in the molecule.

hydrogen as H_2 rather than H
oxygen as O_2 rather than O
hydrogen peroxide as H_2O_2 rather than HO
acetic acid as $C_2H_4O_2$ rather than CH_2O

and so on. For ionic compounds, however, simplest formulas (NaCl, $CaCl_2$, Al_2O_3, . . .) are used, since these substances contain no molecules.

Formulas from Analysis

We have seen that the simplest formula of a compound can be determined if its percentage composition is known. At this stage you might well ask, "Where do these percentages come from?" Not out of thin air, certainly. Instead, they are obtained in the laboratory by a procedure called *chemical analysis.* The chemist starts with a weighed sample of a pure compound. He or she subjects it to one or more reactions to convert it to other substances. These reactions may be rather simple or quite complex. In any case, they are designed to give the relative masses of the elements in the compound.

As an example of a simple chemical analysis, consider the apparatus shown in Figure 3.1. Using this set-up, it is possible to determine the mass percentages of the elements in certain metal oxides. For example, it can be used to obtain the data quoted in Example 3.1. We start by weighing out a sample of pure nickel oxide, a green powder, into the test tube shown at the right. Hydrogen gas is generated by reacting zinc with hydrochloric acid. This takes place in the Erlenmeyer flask at the left. This gas passes over the nickel oxide, which is strongly heated with a burner. A chemical reaction occurs. The hydrogen combines with the oxygen in the nickel oxide, forming water vapor. The water passes off with the excess hydrogen. A dull grey powder remains; this is pure nickel. After cooling, the mass of this residue is determined. Knowing the mass of nickel oxide and that of the nickel obtained from it, the percentage composition of the compound is readily calculated (recall Example 3.1). Finally, the formula is obtained from the percentage composition. It happens to be NiO, as you can prove for yourself by working with the data in the Example.

The experiment just described is one that you could carry out in the laboratory. (Don't try it though unless your instructor supervises your work. There are some safety hazards involved). Ordinarily, chemists do not carry out their own analyses. Instead, they send samples to a commercial analyst. Upon payment of a small fee ($50–$100 nowadays), they get back a card listing the mass percentages of the elements present.

3.2 CHEMICAL EQUATIONS

The chemical reactions that occur in the laboratory or in the outside world can be described in many different ways. Consider, for example, what happens when a test tube filled with hydrogen gas is brought up to a burner (Fig. 3.2, p. 64). A small explosion occurs; if we look carefully, we see a few drops of water forming in the test tube. The reaction that takes place might be described as follows:

"Elementary hydrogen, which consists of diatomic molecules (H_2), reacts with oxygen molecules (O_2) in the air to form molecules of liquid water (H_2O)."

In another case, we notice that a shiny piece of aluminum exposed to the air becomes covered with a dull film. Chemical analysis shows that the surface film is an oxide of aluminum with the formula Al_2O_3.

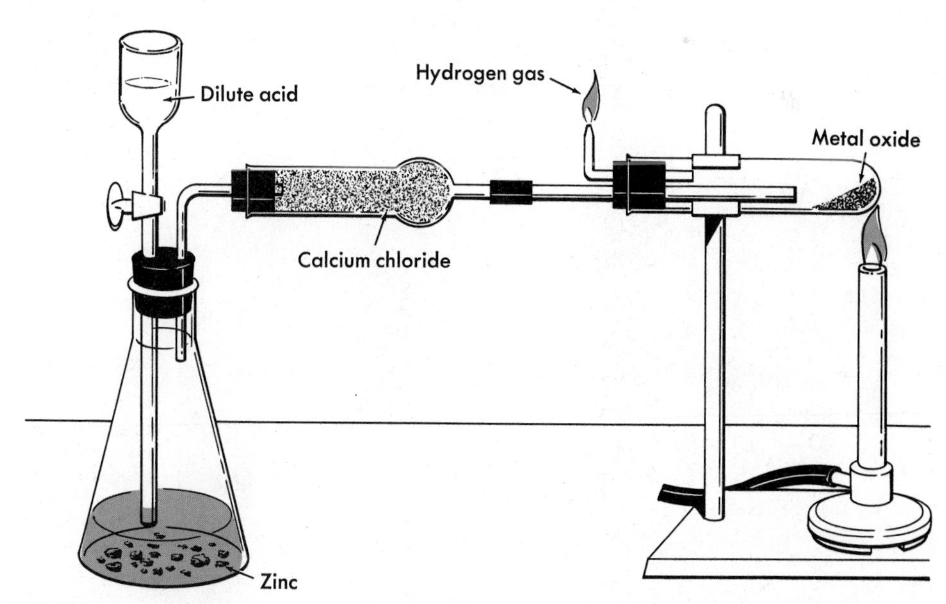

FIGURE 3.1

It is possible to convert many metal oxides to metals by heating the oxide in a stream of hydrogen gas. The hydrogen is made by reacting zinc metal with an acid. The products of the reaction are water and the pure metal. By taking the difference between the mass of the oxide and the mass of the metal obtained from it, one can find the mass of oxygen in the oxide.

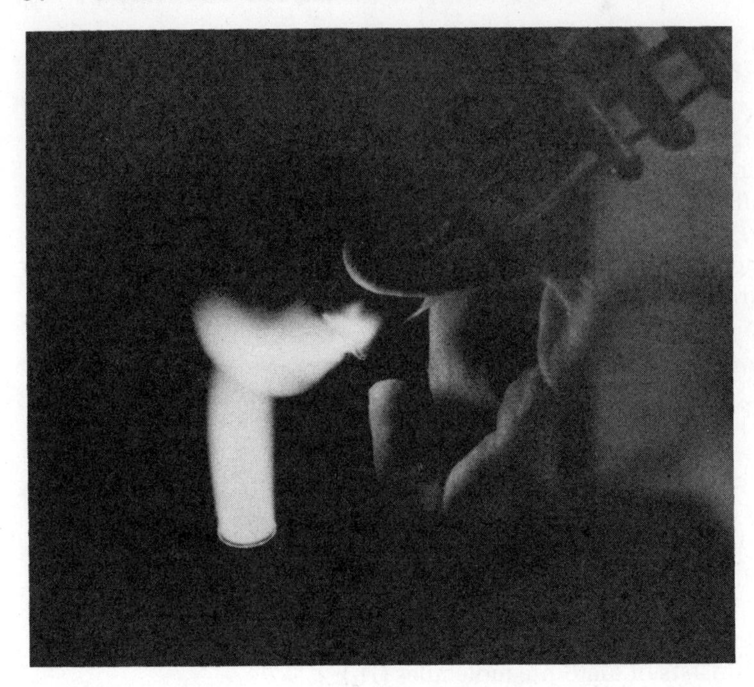

FIGURE 3.2

If pure hydrogen gas is ignited in air it burns with a hot, nearly nonluminous flame. Since hydrogen is much lighter than air, the tube is kept inverted to avoid losing the hydrogen before it is ignited. The gas ignites with a "pop" and the flame moves rapidly up the tube.

This reaction might be described by saying that:

"Aluminum atoms (Al) combine with oxygen molecules (O_2) to give the compound aluminum oxide, simplest formula Al_2O_3".

Word descriptions such as these help us to understand what happens in a chemical reaction. However, they have one important limitation. They do not allow us to say anything quantitative about the reaction. For example, they do not tell us how much product can be obtained from a given amount of reactant. For this purpose, it is helpful to represent a reaction by a chemical equation.

FIGURE 3.3

A sheet of aluminum metal exposed to air will quickly be covered by a thin layer of oxide. If aluminum powder is ignited in air it will burn rapidly, giving off a bright light and spectacular sparks. The product of the reaction is a white powder, Al_2O_3.

The Nature of a Chemical Equation

To show how a word description of a reaction is translated into an equation, consider the reactions referred to above. These reactions can be represented by the equations:

$$2\ H_2(g) + O_2(g) \rightarrow 2\ H_2O(l)$$
$$4\ Al(s) + 3\ O_2(g) \rightarrow 2\ Al_2O_3(s)$$

Looking carefully at these equations, we note that:

1. The *reactants* (hydrogen and oxygen; aluminum and oxygen) are written on the left. They are separated by an arrow from the *products* (water; aluminum oxide), written on the right.

Chemical equations tell us a good deal about the chemical reactions they describe.

2. Each substance taking part in the reaction is represented by its formula. If the substance is molecular, its molecular formula is used (H_2, O_2, H_2O). Elementary substances which are not molecular are indicated by their symbols (Al). Compounds which are not molecular, such as aluminum oxide, are represented by their simplest formulas (Al_2O_3).

3. The **coefficients** in the equation represent the relative numbers of particles that take part in the reaction. We interpret the equation:

$$2\ H_2(g) + O_2(g) \rightarrow 2\ H_2O(l)$$

to mean that *two* molecules of hydrogen react with *one* molecule of oxygen to give *two* molecules of water. (Note that a coefficient of 1 is understood for O_2.)

The coefficients in the equation:

$$4\ Al(s) + 3\ O_2(g) \rightarrow 2\ Al_2O_3(s)$$

tell us that *four* atoms of aluminum react with *three* molecules of oxygen to give two "formula units" of aluminum oxide. Here, a formula unit consists of a combination of two aluminum and three oxygen atoms.

4. There are the same number of atoms of each element on both sides of the equation. Looking at the equation for the reaction of hydrogen with oxygen, we see that there are:

In a chemical reaction, mass is conserved because atoms are conserved.

4 H atoms on the left (2 H_2 molecules); 4 H atoms on the right (2 H_2O molecules)
2 O atoms on the left (O_2 molecule); 2 O atoms on the right (2 H_2O molecules)

In the reaction between aluminum and oxygen there are:

4 Al atoms on the left; $2 \times 2 = 4$ Al atoms on the right
$3 \times 2 = 6$ O atoms on the left; $2 \times 3 = 6$ O atoms on the right

In both cases, atoms are "conserved". No atom of any element is created or destroyed in the reaction.

5. The physical states of reactants and products are indicated by the abbreviations:

(g)	gas	H_2, O_2
(l)	liquid	H_2O
(s)	solid	Al, Al_2O_3

Later, we will use another symbol, (aq), to represent a species in water (aqueous) solution.

Writing Chemical Equations

It is one thing to interpret a chemical equation that someone else has written. It is quite a different matter to write one yourself. To translate a chemical reaction into an equation, you must:

Students often have trouble writing chemical equations properly. The reason is, they have not learned these rules.

1. *Know what happens in the reaction.* That is, you must be able to identify the substances that react (the reactants) and those that are formed (the products). There is really only one way to obtain this information. Someone must carry out the reaction in the laboratory and make careful observations as to what happens. As you gain more knowledge of chemistry, you will be able to predict the products of certain types of reactions. For the time being, you will not be expected to do this. You will be told the identity of both products and reactants.

2. *Know the formulas and physical states of all products and reactants.* As you become more familiar with chemical substances, you will gradually acquire knowledge of this sort. At this stage you know that water has the molecular formula H_2O and is a liquid at room temperature and atmospheric pressure. A little while ago, if you didn't know already, you learned that aluminum oxide is a solid with the formula Al_2O_3. You may or may not know that carbon dioxide consists of CO_2 molecules and is a gas under ordinary conditions. (At any rate, you do now).

At the moment, you are not expected to know the formulas and physical states of more than a few, common compounds. However, you should be able to come up with this information for elementary substances. The majority of elements are

—solids at room temperature and atmospheric pressure.

—shown simply as individual atoms in chemical equations.

There are exceptions to these general rules, which you should be aware of (Table 3.2).

TABLE 3.2 FORMULAS AND PHYSICAL STATES OF ELEMENTS TO BE USED IN WRITING EQUATIONS

Formulas

MOLECULES	ATOMS
F_2, Cl_2, Br_2, I_2	all others (98 in all)
H_2, N_2, O_2	
P_4	

Physical States

GASES	LIQUIDS	SOLIDS
He, Ne, Ar, Kr, Xe, Rn	Br_2, Hg	all others (93 in all)
F_2, Cl_2		
H_2, N_2, O_2		

3. *Balance the equation.* You must, by adjusting the coefficients, make the number of atoms of each element the same on both sides of the equation. In doing this, you simply take account of the fact that atoms cannot be created or destroyed in a reaction.

Fortunately, it is possible to balance most chemical equations by a simple method. You need know very little chemistry and no mathematics beyond addition and multiplication. To illustrate the process involved, consider the reaction of lithium with oxygen to form lithium oxide. To write the equation, note from Table 3.2 that the element lithium is represented as Li(s) and oxygen as $O_2(g)$. We'll tell you that the simplest formula of lithium oxide, an ionic solid, is Li_2O. With this information, you should be able to write the "unbalanced" equation:

The formulas of all the metals are the same as their symbols.

$$Li(s) + O_2(g) \rightarrow Li_2O(s)$$

Clearly this equation is not balanced. There are too many oxygen atoms on the left (2 *vs.* 1 on the right) and too few lithium atoms (1 *vs.* 2 on the right).

We might start by balancing oxygen atoms. To do this, we write a coefficient of 2 before the formula for $Li_2O(s)$:

$$Li(s) + O_2(g) \rightarrow 2\ Li_2O(s)$$

so as to get two oxygen atoms on both sides. Now we turn our attention to lithium. We have $2 \times 2 = 4$ Li atoms on the right. To obtain an equal number on the left, we put a coefficient of 4 in front of Li(s):

$$4\ Li(s) + O_2(g) \rightarrow 2\ Li_2O(s)$$

This is the final, balanced equation for the reaction of lithium with oxygen. It tells us, among other things, that four lithium atoms are required to react with one oxygen molecule.

EXAMPLE 3.7

Bottled gas, which consists mostly of propane, C_3H_8, burns in pure oxygen (or in air) to form gaseous carbon dioxide and liquid water. Write a balanced equation for this reaction.

SOLUTION

We first write the formulas of the reactants, $C_3H_8(g)$ and $O_2(g)$, on the left and those of the products, $CO_2(g)$ and $H_2O(l)$, on the right:

$$C_3H_8(g) + O_2(g) \rightarrow CO_2(g) + H_2O(l)$$

Three elements are involved: carbon, hydrogen, and oxygen. In principle, we could start with any one of the three. In practice, it would be best to choose either carbon or hydrogen as a starting point. Unlike oxygen, they appear in only one species on both sides of the equation.

To balance the three carbon atoms in the C_3H_8 molecule, we write a coefficient of 3 in front of $CO_2(g)$:

$$C_3H_8(g) + O_2(g) \rightarrow 3\ CO_2(g) + H_2O(l)$$

Now we go on to hydrogen. The eight hydrogen atoms in C_3H_8 can be balanced by 4 H_2O molecules ($4 \times 2 = 8$):

$$C_3H_8(g) + O_2(g) \rightarrow 3\ CO_2(g) + 4\ H_2O(l)$$

Finally, we come to oxygen. On the right hand side of the equation we have:

$$3 \times 2 + 4 = 10 \text{ oxygen atoms}$$

To get 10 oxygen atoms on the left, we need a coefficient of 5 in front of $O_2(g)$. The final, balanced equation is:

$$C_3H_8(g) + 5\ O_2(g) \rightarrow 3\ CO_2(g) + 4\ H_2O(l)$$

It is well to keep two points in mind in trying to balance equations.

1. *Always adjust coefficients,* never subscripts. If asked to balance the equation for the reaction of hydrogen with oxygen to give water:

$$H_2(g) + O_2(g) \rightarrow H_2O(l)$$

you should adjust coefficients to arrive at:

$$2\ H_2(g) + O_2(g) \rightarrow 2\ H_2O(l)$$

It would *not* be correct to write:

$$H_2(g) + O(g) \rightarrow H_2O(l)$$

or:

$$H_2(g) + O_2(g) \rightarrow H_2O_2(l)$$

because neither of these equations corresponds to reality. Oxygen gas consists of O_2 molecules, not individual oxygen atoms. Water has the formula H_2O. Hydrogen peroxide, H_2O_2, is a different compound with properties quite different from those of water.

H_2O_2 is actually a very reactive chemical.

2. *The coefficients in the balanced equation are ordinarily the smallest possible whole numbers.* Thus for the reaction of hydrogen with oxygen we would write

$$2\ H_2(g) + O_2(g) \rightarrow 2\ H_2O(l)$$

rather than

$$4\ H_2(g) + 2\ O_2(g) \rightarrow 4\ H_2O(l)$$

The equation just written is legitimate in that there are the same number of atoms of each element on both sides. However, it is somewhat less convenient to work with than the first equation so is ordinarily not used.

Mole Relationships in Balanced Equations

As we have seen, the equation:

$$2\ H_2(g) + O_2(g) \rightarrow 2\ H_2O(l)$$

can be interpreted to mean that:

"2 H_2 molecules react with 1 O_2 molecule to form 2 H_2O molecules."

The chemical reaction actually occurs between the molecules.

However, the equation remains valid if each of the coefficients is multiplied by the same factor, such as 10, 100, and so on.

"20 H_2 molecules react with 10 O_2 molecules to form 20 H_2O molecules."

"200 H_2 molecules react with 100 O_2 molecules to form 200 H_2O molecules."

In particular, we can multiply each coefficient by Avogadro's number, N ($N = 6.022 \times 10^{23}$):

"2N H_2 molecules react with N O_2 molecules to form 2N H_2O molecules."

But since N molecules of any molecular substance represents one mole:

$$2 \text{ mol } H_2 + 1 \text{ mol } O_2 \rightarrow 2 \text{ mol } H_2O$$

The reasoning that we have just gone through illustrates a general principle that applies to all reactions. **The coefficients in a balanced equation can be interpreted in terms of moles of reactants and products.** Thus for the equations:

$$C_3H_8(g) + 5\ O_2(g) \rightarrow 3\ CO_2(g) + 4\ H_2O(l)$$
$$4\ Li(s) + O_2(g) \rightarrow 2\ Li_2O(s)$$

we can write:

$$1 \text{ mol } C_3H_8 + 5 \text{ mol } O_2 \rightarrow 3 \text{ mol } CO_2 + 4 \text{ mol } H_2O$$
$$4 \text{ mol Li} + 1 \text{ mol } O_2 \rightarrow 2 \text{ mol } Li_2O$$

From a slightly different point of view we can say that in these reactions:

$$1 \text{ mol } C_3H_8 \simeq 5 \text{ mol } O_2 \simeq 3 \text{ mol } CO_2 \simeq 4 \text{ mol } H_2O$$
$$4 \text{ mol Li} \simeq 1 \text{ mol } O_2 \simeq 2 \text{ mol } Li_2O$$

The symbol $\simeq$ means "is chemically equivalent to." 1 mol of C_3H_8 reacts exactly with 5 mol of O_2. In that sense, these two quantities are equivalent to each other. Again, 4 mol of Li is chemically equivalent to 2 mol of Li_2O; 4 mol of Li yields exactly 2 mol of Li_2O. We use the $\simeq$ symbol rather than $=$. It would sound a bit strange to say, for example, that 1 mol of C_3H_8 "equals" 5 mol of O_2. However, in calculations, an equivalence sign can be handled as if it were an equals sign.

Read this section until you understand it completely.

The relations we have just written lead directly to conversion factors that allow us to relate moles of different substances involved in a reaction. The reasoning involved is indicated in Example 3.8.

EXAMPLE 3.8

Consider the balanced equation for the combustion of propane gas:

$$C_3H_8(g) + 5\ O_2(g) \rightarrow 3\ CO_2(g) + 4\ H_2O(l)$$

If we start with 1.43 mol of propane, C_3H_8

a. how many moles of oxygen, O_2, are required to react with it?

b. how many moles of carbon dioxide, CO_2, are formed?

SOLUTION

a. We need a conversion factor relating moles of C_3H_8 (given) to moles of O_2 (required). This factor can be obtained from the relation:

$$1\ \text{mol}\ C_3H_8 \simeq 5\ \text{mol}\ O_2$$

Treating the equivalence sign, $\simeq$, as an equals sign, we obtain the conversion factor:

$$\frac{5\ \text{mol}\ O_2}{1\ \text{mol}\ C_3H_8}$$

Using moles as units, it is easy to relate amounts of reactants and products in chemical reactions.

We can now convert moles of C_3H_8 to moles of O_2:

$$\text{moles}\ O_2 = 1.43\ \text{mol}\ C_3H_8 \times \frac{5\ \text{mol}\ O_2}{1\ \text{mol}\ C_3H_8} = 7.15\ \text{mol}\ O_2$$

b. The appropriate relation, from the coefficients of the balanced equation, is:

$$1\ \text{mol}\ C_3H_8 \simeq 3\ \text{mol}\ CO_2$$

which leads to the conversion factor we need: $\frac{3\ \text{mol}\ CO_2}{1\ \text{mol}\ C_3H_8}$

To find the number of moles of CO_2 produced, we multiply the number of moles of propane given by this factor:

$$\text{moles}\ CO_2 = 1.43\ \text{mol}\ C_3H_8 \times \frac{3\ \text{mol}\ CO_2}{1\ \text{mol}\ C_3H_8} = 4.29\ \text{mol}\ CO_2$$

Mass Relations in Balanced Equations

As we saw in Chapter 2, a mole of a given substance has a fixed mass in grams. Hence, the mole relations just discussed can be converted to mass relations in grams. Consider, for example, the reaction:

$$C_3H_8(g) + 5\ O_2(g) \rightarrow 3\ CO_2(g) + 4\ H_2O(l)$$

In Example 3.8a, we saw how to calculate the number of moles of O_2 required to react with 1.43 mol of C_3H_8. To do that, we used the relation:

$$1 \text{ mol } C_3H_8 \simeq 5 \text{ mol } O_2 \tag{3.5}$$

Suppose now that we were asked to obtain the number of *grams* of O_2 required to react with 1.43 mol of C_3H_8. To do that, we need a relation between moles of C_3H_8 and grams of O_2. This is easily obtained. We know that one mole of O_2 weighs 32.0 g. Substituting in Equation 3.5:

$$1 \text{ mol } C_3H_8 \simeq 5(32.0 \text{ g}) \; O_2$$

or:

$$1 \text{ mol } C_3H_8 \simeq 160 \text{ g } O_2$$

This relation gives the conversion factor we need to obtain the mass of O_2 that reacts with 1.43 mol of C_3H_8:

$$\text{mass } O_2 = 1.43 \text{ mol } C_3H_8 \times \frac{160 \text{ g } O_2}{1 \text{ mol } C_3H_8} = 229 \text{ g } O_2$$

We conclude that 229 g of O_2 is required to react with 1.43 mol C_3H_8. Example 3.9 applies the same reasoning to a different reaction.

EXAMPLE 3.9

For the reaction: $4 \text{ Li}(s) + O_2(g) \rightarrow 2 \text{ Li}_2O(s)$, calculate the mass in grams of lithium oxide formed from 0.283 mol of lithium.

SOLUTION

From the coefficients of the balanced equation, we see that:

$$4 \text{ mol Li} \simeq 2 \text{ mol Li}_2O$$

or, simplifying:

$$2 \text{ mol Li} \simeq 1 \text{ mol Li}_2O$$

We need a relation between moles of Li and grams of Li_2O. To obtain this, we must find the mass of a mole of Li_2O. The atomic masses of Li and O are 6.94 and 16.00 in that order. The formula mass of Li_2O is: 2(6.94) + 16.00 = 29.88. So:

Try to understand the logic we use in attacking this problem.

$$1 \text{ mol Li}_2O = 29.88 \text{ g Li}_2O$$

Hence the relation we need is:

$$2 \text{ mol Li} \simeq 29.88 \text{ g Li}_2O$$

Now we can readily solve the problem, using the conversion factor approach.

$$\text{mass Li}_2O = 0.283 \text{ mol Li} \times \frac{29.88 \text{ g Li}_2O}{2 \text{ mol Li}} = 4.23 \text{ g Li}_2O$$

Very often, we need to know the mass of one substance which reacts with a given mass of another substance. We can solve problems of this type by a simple extension of the approach we have described. Consider again the equation:

$$C_3H_8(g) + 5\ O_2(g) \rightarrow 3\ CO_2(g) + 4\ H_2O(l)$$

Suppose now that we want to calculate the number of grams of O_2 required to react with 12.0 g of C_3H_8. As always, we start with the relation given by the coefficients of the balanced equation:

$$1 \text{ mol } C_3H_8 \simeq 5 \text{ mol } O_2 \quad (3.6)$$

Here we need a relation between grams of C_3H_8 and grams of O_2. The molecular masses are:

$$\text{MM } C_3H_8 = 3(12.0) + 8(1.0) = 44.0$$

$$\text{MM } O_2 = 2(16.0) = 32.0$$

The method used here shows how we go from a chemical reaction to the masses of reactants and products which are involved in a chemical reaction.

So, one mole of C_3H_8 weighs 44.0 g; one mole of O_2 weighs 32.0 g. Substituting in Equation 3.6:

$$44.0 \text{ g } C_3H_8 \simeq 5(32.0 \text{ g}) \ O_2$$

or:

$$44.0 \text{ g } C_3H_8 \simeq 160 \text{ g } O_2$$

This relation gives the conversion factor we need to obtain the mass of O_2 that reacts with 12.0 g of C_3H_8:

$$\text{mass } O_2 = 12.0 \text{ g } C_3H_8 \times \frac{160 \text{ g } O_2}{44.0 \text{ g } C_3H_8} = 43.6 \text{ g } O_2$$

We conclude that 43.6 g of O_2 is required to react with 12.0 g of C_3H_8.

EXAMPLE 3.10

For the reaction referred to in Example 3.9, calculate the mass in grams of Li_2O formed from 3.16 g of Li.

SOLUTION

As in Example 3.9, the basic relation is:

$$2 \text{ mol Li} \simeq 1 \text{ mol } Li_2O$$

We need a relation between grams of Li and grams of Li_2O. As pointed out in Example 3.9, one mole of Li_2O weighs 29.88 g. Since the atomic mass of Li is 6.94, one mole of Li weighs 6.94 g. Hence:

$$2(6.94 \text{ g}) \text{ Li} \simeq 29.88 \text{ g } Li_2O$$

or: $13.88 \text{ g Li} \triangleq 29.88 \text{ g } Li_2O$

This gives us the conversion factor we need to find the mass of Li_2O formed from 3.16 g of Li:

$$\text{mass } Li_2O = 3.16 \text{ g Li} \times \frac{29.88 \text{ g } Li_2O}{13.88 \text{ g Li}} = 6.80 \text{ g } Li_2O$$

3.3 SUMMARY

In this chapter we have considered several very important principles related to chemical formulas and equations. We have illustrated:

1. What is meant by percentage composition, and how it can be determined experimentally.
2. How percentage composition can be obtained from the formula of a substance, and vice versa.
3. How a molecular formula is obtained from a simplest formula, given the molecular mass.
4. What a chemical equation means, and how it can be balanced.
5. How the coefficients of a balanced equation are used to relate moles or grams of substances taking part in the reaction.

These relationships will provide you with much of the background that you will need in succeeding chapters. Take some time with this material so that you can build on a solid foundation. Work with your teacher and other students until you have mastered it. The pay-off will come as the tools learned in this chapter are used to open more doors in chemistry.

NEW TERMS

(aq) = species in water solution
balanced equation
chemical equation
chemical equivalence, $\triangleq$
coefficient
(g) = gaseous species
(l) = liquid species
mass percentage
percentage composition
(s) = solid species

QUESTIONS

1. The percentage of calcium in calcium oxide is 71.5%. What is the percentage of oxygen, the only other element present?
2. Describe, in your own words, how the percentage of carbon in CO_2 could be calculated. Do not carry out the calculation.
3. Describe, in your own words, the three steps involved in obtaining the simplest formula of a compound from its percentage composition.

4. A student is told that methane contains 74.9% C and 25.1% H. He is asked to determine the simplest formula. For some reason, he decides to base his calculation on a 200 g sample rather than 100 g. Will this make any difference? Explain.

5. In determining the simplest formula of a compound, the final step involves finding the mole ratio of the elements present. This is set equal to the atom ratio. Why must these two quantities be the same?

6. In vanadium oxide, the mole ratio is calculated to be: 2.50 mol O/mol V. What is the simplest formula of vanadium oxide?

7. A certain ore of iron called magnetite has the simplest formula Fe_3O_4. What is the mole ratio of oxygen to iron? (The answer should be given in terms of mol O/mol Fe.)

8. Which of the following must represent molecular formulas as opposed to simplest formulas? Which could be simplest formulas?

   a. C_6H_6 b. C_4H_{10} c. PCl_5 d. C_4H_8O e. N_2H_4

9. Consider the following balanced equation: $2\ SO_2(g) + O_2(g) \rightarrow 2\ SO_3(g)$. Interpret this equation, first in terms of number of molecules and then in terms of number of moles.

10. Which of the following statements are generally true for any chemical reaction?

    a. The number of grams of products equals the number of grams of reactants.
    b. The number of molecules of products equals the number of molecules of reactants.
    c. The number of moles of products equals the number of moles of reactants.
    d. There are the same number of atoms of each type in the products as in the reactants.

11. Consider each of the following elements. Indicate, on a sheet of notebook paper as in parts (a) and (b), how the element should be represented in a chemical equation.

    a. iron: Fe(s)
    b. chlorine: $Cl_2(g)$
    c. mercury:
    d. hydrogen:
    e. magnesium:
    f. nitrogen:
    g. xenon:
    h. copper:
    i. fluorine:
    j. bromine:

    Use Table 3.2 if you have to.

12. Give an example of an element which is, at room temperature and atmospheric pressure:

    a. a gas that consists of individual atoms.
    b. a liquid that consists of individual atoms.
    c. a solid that contains diatomic molecules.
    d. a liquid that contains diatomic molecules.

13. What is wrong with each of the following equations?

    a. $2\ H(g) + O(g) \rightarrow H_2O(l)$
    b. $Fe_3O_4(s) + H_2(g) \rightarrow 3\ Fe(s) + 4\ H_2O(l)$
    c. $2\ Fe_3O_4(s) + 8\ H_2(g) \rightarrow 3\ Fe(s) + 8\ H_2O(l)$
    d. $4\ P(s) + 5\ O_2(g) \rightarrow 2\ P_2O_5(s)$
    e. $2\ H_2(s) + O_2(g) \rightarrow 2\ H_2O(l)$

PROBLEMS

1. *Mass Percentages (Example 3.1)* The mass percentages of nitrogen and hydrogen in ammonia are: 82.3% N, 17.7% H. Determine the masses of the two elements in a sample of ammonia weighing 2.56 g.

2. *Mass Percentages (Example 3.1)* A piece of magnesium weighing 1.123 g burns in oxygen to form 1.862 g of magnesium oxide. What are the percentages of Mg and O in magnesium oxide?

3. *Percentage Composition from Formula (Example 3.2)* Give the mass percentages of the elements in:

   a. C_3H_8 b. C_2H_6O c. UF_6

4. *Percentage Composition from Formula (Example 3.3)* How many grams of iron can be obtained from a kilogram of hematite ore, Fe_2O_3?

5. *Simplest Formula from Percentage Composition (Example 3.4)* A certain compound contains 47.3% Cu and 52.7% Cl. What is its simplest formula?

6. *Simplest Formula from Percentage Composition (Example 3.5)* Ethyl alcohol contains the three elements carbon, hydrogen, and oxygen. Their mass percentages are 52.14, 13.13, and 34.73 in that order. What is the simplest formula of ethyl alcohol?

7. *Molecular Formula (Example 3.6)* A certain hydrocarbon has the simplest formula C_2H_5. Its molecular mass is 58. What is its molecular formula?

8. *Writing Equations (Example 3.7)* Translate the following word descriptions of reactions into balanced equations.

   a. Ethylene gas, C_2H_4, burns in oxygen to give carbon dioxide gas, CO_2, and liquid water.
   b. When ethylene is burned in a limited amount of air, the products are gaseous carbon monoxide, CO, and liquid water.
   c. Iron reacts with chlorine to give a solid ionic compound of simplest formula $FeCl_3$.
   d. Lithium, when heated in air from which oxygen has been removed, is converted to a solid of simplest formula Li_3N.

9. *Mole Relations in Equations (Example 3.8)* Consider the following equation:

$$N_2(g) + O_2(g) \rightarrow 2\ NO(g)$$

   a. How many moles of NO are produced from 2.26 mol of N_2?
   b. How many moles of N_2 are required to react with 0.128 mol of O_2?
   c. How many moles of O_2 are required to produce 2.34×10^6 mol of NO?

10. *Mole–Mass Relations in Equations (Example 3.9)* For the reaction in Problem 9, calculate:

    a. the mass in grams of NO produced from 8.19 mol of N_2.
    b. the number of moles of NO produced from 8.19 g of N_2.

11. *Mass Relations in Equations (Example 3.10)* For the reaction in Problem 9, calculate:

    a. the mass in grams of NO produced from 1.00 g of O_2.
    b. the mass in grams of O_2 required to form 1.00 g of NO.

* * * * *

12. Hydrogen peroxide contains two elements, hydrogen and oxygen. The mass percentage of hydrogen is 5.93. Calculate:
    a. the mass percentage of oxygen.
    b. the mass of oxygen in a sample of hydrogen peroxide weighing 6.46 g.
    c. the mass of hydrogen peroxide which contains 1.00 g of hydrogen.

13. Oxygen gas is often obtained in the laboratory by heating potassium chlorate, $KClO_3$.
    a. What are the mass percentages of K, Cl, and O in $KClO_3$?
    b. How many grams of oxygen can be obtained from a sample of $KClO_3$ weighing 1.00 g?
    c. How many grams of $KClO_3$ must be used to obtain 1.00 g of oxygen?

14. The oxide of boron contains 31.0% B; the remainder is oxygen. What is the simplest formula of this compound?

15. A sample of an oxide of tin is heated in a stream of hydrogen. The oxide is converted to pure tin. The following data are obtained:

    mass empty test tube = 15.254 g
    mass test tube + oxide = 15.565 g
    mass test tube + tin = 15.499 g

    Determine:

    a. the percentages of tin and oxygen in the oxide.
    b. the simplest formula of the oxide.

16. A certain compound has the simplest formula CH_2. Which of the following represent possible molecular masses for the compound?

    a. 14 b. 21 c. 28 d. 38 e. 42

    Write the molecular formula corresponding to each of your answers.

17. On a separate sheet of paper, balance the following equations:
    a. $? \ Fe_2O_3(s) + ? \ H_2(g) \rightarrow ? \ Fe(s) + ? \ H_2O(l)$
    b. $? \ K(s) + ? \ Br_2(l) \rightarrow ? \ KBr(s)$
    c. $? \ C_2H_2(g) + ? \ O_2(g) \rightarrow ? \ CO_2(g) + ? \ H_2O(l)$

18. Assuming the coefficients in the following equations are correct, determine the subscripts, x and y, in the incomplete formulas.
    a. $N_2(g) + 3 \ H_2(g) \rightarrow 2 \ N_xH_y(g)$
    b. $N_2(g) + 2 \ O_2(g) \rightarrow 2 \ N_xO_y(g)$
    c. $C_xH_y(l) + 8 \ O_2(g) \rightarrow 5 \ CO_2(g) + 6 \ H_2O(l)$

19. The mass percentages of H and O in water are 11.2 and 88.8 in that order. Calculate:
    a. the number of grams of hydrogen in 120 g of water.
    b. the number of grams of oxygen combined with 6.24 g of hydrogen in water.
    c. the number of grams of water that contains 10.0 g of oxygen.

20. In a certain reaction, 2.0 g of a certain element X combines with 8.0 g of another element Y to form 10.0 g of a compound Z.
    a. What are the mass percentages of X and Y in compound Z?
    b. How many grams of X would be required to form 6.0 g of Z?
    c. How many grams of Z could be formed from 1.2 g of Y?
    d. What additional information would you need to obtain the simplest formula of compound Z?

21. A 0.98 g sample of sulfur was burned in oxygen to produce 2.45 g of a gas. The gas has a molecular mass of 80. Find the simplest formula and then the molecular formula of the gas.

22. A sample of calcium metal weighing 1.296 g is heated in oxygen to form 1.813 g of calcium oxide.

    a. Determine the mass percentages of calcium and oxygen in calcium oxide.
    b. What is the simplest formula of calcium oxide?
    c. Write a balanced equation for the reaction of calcium with oxygen.

23. Consider the reaction: $2\ H_2(g) + O_2(g) \rightarrow 2\ H_2O(l)$. Calculate:

    a. the number of moles of H_2 required to react with 1.69 mol of O_2.
    b. the number of moles of H_2O formed from 0.918 mol of H_2.
    c. the number of moles of H_2O formed from 1.62 g of H_2.
    d. the number of grams of O_2 required to react with 12.2 g of H_2.

24. For the reaction: $N_2(g) + 3\ H_2(g) \rightarrow 2\ NH_3(g)$, complete the following table on a separate piece of paper.

$$0.50 \text{ mol } N_2 + ? \text{ mol } H_2 \rightarrow ? \text{ mol } NH_3$$
$$? \text{ mol } N_2 + ? \text{ mol } H_2 \rightarrow 1.28 \text{ mol } NH_3$$
$$? \text{ g } N_2 + ? \text{ g } H_2 \rightarrow 1.28 \text{ mol } NH_3$$
$$? \text{ g } N_2 + 16.0 \text{ g } H_2 \rightarrow ? \text{ g } NH_3$$

25. When sodium metal reacts with chlorine gas, table salt, NaCl, is produced.

    a. Write the equation for the reaction.
    b. Calculate the mass in grams of NaCl formed from 3.00 mol of sodium.

26. Aluminum reacts with hydrogen chloride, HCl(g), to produce solid aluminum chloride, $AlCl_3$, and hydrogen gas.

    a. Write the equation for the reaction.
    b. How many grams of aluminum would be required to form 5.00 mol of H_2?

27. When heated, potassium chlorate, $KClO_3(s)$, decomposes to potassium chloride, KCl(s), and oxygen gas.

    a. Write the equation for the reaction.
    b. How many grams of each product will be produced by the decomposition of 1.00 g of $KClO_3$?

*28. A certain compound contains the four elements Cr, N, H, and O. The mass percentages of Cr, N, and H are 15.3, 37.1, and 5.33 in that order. What is the simplest formula of the compound?

*29. A certain organic compound contains the three elements C, H, and O. Combustion of a sample weighing 1.000 g gives 1.998 g of CO_2 and 0.818 g of H_2O. What is the simplest formula of the compound?

*30. The reaction of potassium with bromine can be represented by the equation:

$$2\ K(s) + Br_2(l) \rightarrow 2\ KBr(s)$$

If 4.0 g of potassium is mixed with 10.0 g of Br_2, what is the maximum amount of KBr that can be obtained?

*31. Photosynthesis can be described by the equation:

$$6\ CO_2(g) + 6\ H_2O(l) \rightarrow C_6H_{12}O_6(s) + 6\ O_2(g)$$

The solid product, $C_6H_{12}O_6$, is a carbohydrate. It has been estimated that land plants alone use 25 billion metric tons of CO_2 in this reaction each year. How many grams of $C_6H_{12}O_6$ and O_2 are produced annually? (1 metric ton = 10^6 g).

This is one way to study the heat flow in an exothermic reaction. See page 97 for a better approach. *Courtesy of Eastman Kodak.*

4

ENERGY CHANGES

Both physical and chemical changes are accompanied by a flow of energy. Sometimes energy is absorbed from the surroundings. This happens when we supply heat from a Bunsen burner to boil water. In chemical reactions, energy is often evolved. Such is the case when natural gas burns. This process, taking place within a Bunsen burner, gives off energy in the form of heat to the surroundings.

In many reactions, it is the energy change which is of greatest interest. When you burn gasoline in your car, you are not much concerned with the products of the reaction. Instead, you want to know how much energy can be obtained from the gasoline. It is this energy which propels the automobile from one point to another.

We will look at the nature of energy changes in this chapter. For the most part, we will concentrate upon one type of energy: heat. This is the form which energy changes usually take in the laboratory. However, we will start with a general survey of all kinds of energy changes. Here, we will be interested in how one form of energy can be converted to another.

4.1 TYPES OF ENERGY

The energy produced in physical or chemical changes may appear in any of several forms. Most commonly, energy is evolved as **heat.** This happens when we burn fuels to heat our homes. In some cases, however, energy is given off as:

—**mechanical energy.** When food is metabolized in your body, part of the energy is released as heat. However, the major portion is in the form of mechanical energy. You use this type of energy to lift weights, climb mountains, or even open your chemistry textbook.

—**electrical energy.** The chemical reaction that occurs in the storage battery of an automobile produces electrical energy. This is used to start the car and operate the headlights, radio, and other electrical equipment.

—**light.** When a fuel burns, it usually gives off visible light. Years ago, people used kerosene lamps to provide light. Fireflies and some chemical mixtures give off "cold light".

Solar energy comes to the earth mostly as light.

To a certain extent, different forms of energy can be converted from one to another. The heat evolved when coal burns can be used to vaporize water to steam. The energy in the steam can turn a turbine, producing mechanical energy. The turbine runs a generator which produces electrical energy. Finally, the electrical energy is converted back to heat and mechanical energy by passing it through a toaster. This both burns the toast and pops it up afterwards.

Kinetic and Potential Energy

The energy which a system has may be either **potential energy** or **kinetic energy.** Potential energy is due to *position;* kinetic energy is energy of *motion.* To illustrate the difference, consider a boulder sitting on the side of a mountain. It has a high potential energy due to its position above the valley floor. If the boulder falls down the mountain, its potential energy is converted to kinetic energy. Very soon, the boulder comes to rest at the bottom of the valley. There, its potential energy is considerably less than it was originally (Figure 4.1, p. 80).

The concepts of kinetic and potential energy can also be applied at a molecular level. Consider, for example, a sample of the element hydrogen. The H_2 molecules in this sample possess three different kinds of kinetic energy (Figure 4.2, p. 80). In the first place, they are moving from one end of the container to the other. The energy associated with this motion is called *translational kinetic energy.* As they move, the molecules are flipping end over end, giving them *rotational kinetic energy.* Finally, within the H_2 molecule, the hydrogen atoms are vibrating with

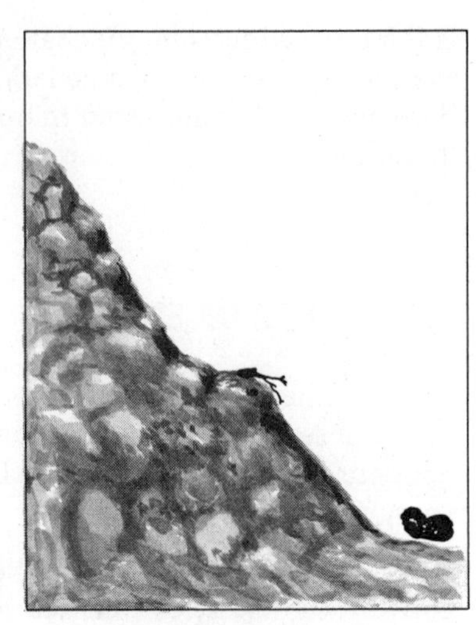

FIGURE 4.1

When a boulder is up on the side of a mountain, it has high potential energy. As it rolls down the mountain, its potential energy is converted into kinetic energy. As it slows down and comes to rest, its kinetic energy is converted to heat and is given to the earth. The potential energy when the boulder is at the bottom of the mountain is lower than when it is at the top.

respect to one another. The energy involved in this back and forth motion is called *vibrational kinetic energy.*

The H_2 molecules in Figure 4.2 also possess *potential energy* because of the chemical bond within the molecule. To break this bond, forming isolated atoms, the potential energy has to be increased. In this sense, the H_2 molecule is similar to a boulder at the bottom of a mountain. Both systems have a low potential energy. They are stable in that they tend to stay where they are. To break the molecule into atoms or to raise the boulder up the mountain, energy must be supplied. When

But not in any other sense!

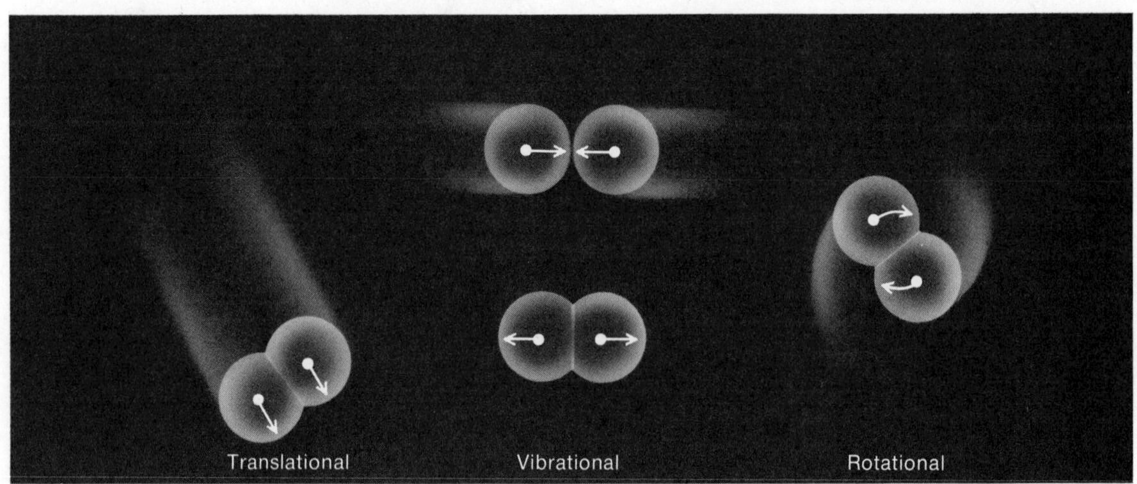

FIGURE 4.2

An H_2 molecule has three different kinds of kinetic energy. The energy of its motion through space is called translational. Its spinning and tumbling involve rotational energy. The molecule can also vibrate, changing the distance between the H nuclei.

that happens, both systems have a higher potential energy than they did to begin with. This means that they are unstable. The boulder has a tendency to fall back down the mountain. The hydrogen atoms tend to re-combine, forming an H_2 molecule.

Conservation of Energy

Experiment shows that energy is neither gained nor lost in physical or chemical changes. This principle is known as the Law of Conservation of Energy. It is often stated as follows:

Energy is neither created nor destroyed in ordinary physical and chemical changes.

All the energy that goes into one substance has to come out of another.

This Law applies to all the processes we have discussed. For example, the energy spent in carrying a boulder to the top of a mountain is exactly equal to the increase in its potential energy. To dissociate an H_2 molecule into atoms:

$$H_2(g) \rightarrow 2\ H(g)$$

we have to supply energy. The amount of energy required is equal to the potential energy difference between two H atoms and an H_2 molecule.

In using the Law of Conservation of Energy, we must take account of all the forms in which energy appears. Suppose, for example, you

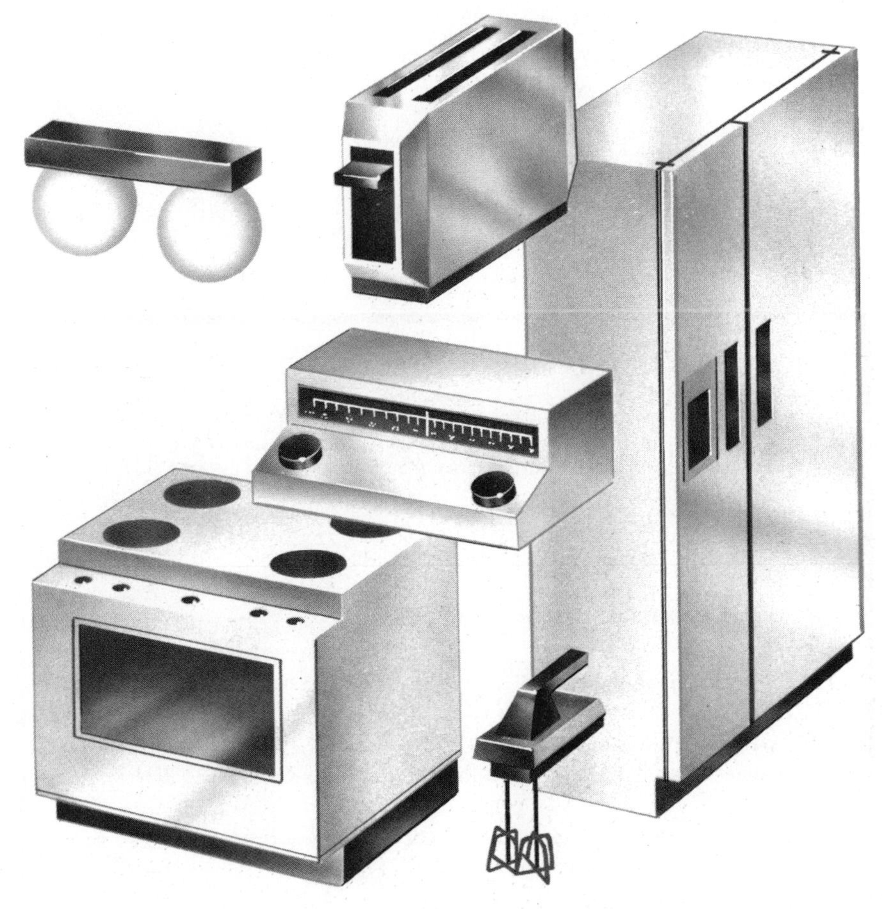

FIGURE 4.3

Electrical energy is used in the kitchen in many ways. A kitchen in 1900 would have had none of the items shown in the drawing. Life in those days required a lot more human energy.

wanted to account for the electrical energy that your family was charged for last month. You might find that:

—50% of it was in the form of heat. Some of this was used directly in cooking and in heating water. The rest of it was waste heat from light bulbs and so on.

—45% was converted to mechanical energy. This was used to run such machines as water pumps or refrigerators.

—5% was used as light (light bulbs, TV picture tubes, and so on). Only if each of these forms of energy is included will the sum add to 100%.

Energy Units

To make calculations involving energy, we need a unit to measure energy. A common unit is the **calorie.** The calorie is defined as the amount of energy in the form of heat required to raise the temperature of one gram of water by one degree Celsius (more exactly, from 14.5 to 15.5°C). Frequently, we use a larger unit, the **kilocalorie:**

$$1 \text{ kcal} = 10^3 \text{ cal}$$

When you're on a diet, you don't have much energy.

(The "calorie" referred to in nutrition is really a kilocalorie. When you are on a diet of "1000 calories" per day, you are eating food that produces 1000 kcal or 10^6 ordinary calories of energy).

Many other energy units are in common use. Electrical energy is often measured in kilowatt hours:

$$1 \text{ kwh} = 860 \text{ kcal} = 8.60 \times 10^5 \text{ cal} \qquad (4.1)$$

Another unit of energy which we will use occasionally is the joule:

$$1 \text{ cal} = 4.184 \text{ J}; \ 1 \text{ kcal} = 4.184 \text{ kJ} \qquad (4.2)$$

EXAMPLE 4.1

According to Mr. Hopper's electric bill, he used 212 kwh of electrical energy last month. How much energy did he use in kilocalories? in kilojoules?

SOLUTION

Here, as usual, we use unit analysis:

$$\text{energy (kcal)} = 212 \text{ kwh} \times \frac{860 \text{ kcal}}{1 \text{ kwh}} = 182{,}000 \text{ kcal} = 1.82 \times 10^5 \text{ kcal}$$

To obtain the amount of energy in kilojoules, we start with the answer just obtained and use the conversion factor: 4.184 kJ/1 kcal

$$\text{energy (kJ)} = 1.82 \times 10^5 \text{ kcal} \times \frac{4.184 \text{ kJ}}{1 \text{ kcal}} = 7.61 \times 10^5 \text{ kJ}$$

4.2 HEAT FLOW AND HEAT CONTENT

Throughout the rest of this chapter, we will deal with one particular type of energy: heat. We will look at the heat flow that accompanies physical or chemical changes. Here, we will be interested in two factors:

1. The direction in which heat is flowing.
2. The amount of heat involved.

In this section, we will concentrate upon the first of these factors. Later in the chapter, we will discuss the amount of heat flow.

Endothermic and Exothermic Processes

To make certain processes take place, it is necessary to supply heat. That is, heat must be absorbed if the process is to occur. Such processes are said to be *endothermic*. Other processes give off heat. When they occur, heat is evolved to the surroundings. We say that such processes are *exothermic*.

The direction of heat flow (exo or endo) always refers to the sample under study.

The most familiar endothermic processes involve changes in physical state. An example is vaporization, the conversion of a liquid to its vapor. You know from experience that heat has to be supplied to boil water. A "watched" pot may not boil, but an unheated one certainly won't. Heat has to flow into the water, perhaps from a Bunsen burner or an electric range. Only if this happens can we convert water to steam.

Another endothermic change in state is fusion (melting). Heat has to be supplied to melt ice. A tray of ice cubes remains frozen so long as it stays in the refrigerator. If you take it out and put it on the kitchen table, the ice melts. Heat flows from the warm air of the room to the ice, converting it to liquid water.

Many chemical reactions are endothermic. Such reactions often involve the decomposition of a compound. The products (Table 4.1) may be simpler compounds (CaO, CO_2 from $CaCO_3$). In other cases, they are elements (H_2, O_2 from H_2O). In these, as in all other endothermic processes, heat must be absorbed from the surroundings.

TABLE 4.1 EXAMPLES OF ENDOTHERMIC AND EXOTHERMIC PROCESSES

Changes of State		
1. Vaporization (boiling)	$H_2O(l) \rightarrow H_2O(g)$	ENDOTHERMIC
2. Fusion (melting)	$H_2O(s) \rightarrow H_2O(l)$	ENDOTHERMIC
3. Condensation	$H_2O(g) \rightarrow H_2O(l)$	EXOTHERMIC
4. Freezing	$H_2O(l) \rightarrow H_2O(s)$	EXOTHERMIC
Chemical Reactions		
1. Decomposition of limestone	$CaCO_3(s) \rightarrow CaO(s) + CO_2(g)$	ENDOTHERMIC
2. Decomposition of water	$2\ H_2O(l) \rightarrow 2\ H_2(g) + O_2(g)$	ENDOTHERMIC
3. Formation of water	$2\ H_2(g) + O_2(g) \rightarrow 2\ H_2O(l)$	EXOTHERMIC
4. Combustion	$C(s) + O_2(g) \rightarrow CO_2(g)$	EXOTHERMIC

As you can see from Table 4.1, changes in state can be exothermic. An example is the condensation of a vapor to a liquid. Steam gives off heat when it condenses to water. The heat given off in this process is the main reason that steam will give you a bad burn. Heat is also evolved when a liquid such as water freezes.

Exothermic chemical reactions are very common. Indeed, most reactions that take place at room temperature are of this type. Often (Table 4.1) they involve:

—the formation of a compound from the elements (H_2O from H_2 + O_2).

—combustion (C in this case). Combustion reactions are our major source of heat and other forms of energy. They are involved in the burning of all fossil fuels (natural gas, petroleum, coal).

Looking at Table 4.1, you will notice that:

If a given process is endothermic, the reverse process (taking place in the opposite direction) is exothermic.

For example, melting absorbs heat; the reverse process, freezing, evolves heat. The formation of water from the elements is exothermic. The decomposition of H_2O to H_2 and O_2 is endothermic.

Change in Heat Content, ΔH

We have seen that certain processes give off heat; they are exothermic. You might well ask: what is the source of this heat? Where does it come from? In an endothermic reaction, heat is absorbed. What happens to this heat? According to the Law of Conservation of Energy, it can't simply disappear.

If 10 cal goes into a sample, its heat content increases by 10 cal.

These questions can be answered in terms of a quantity known as *heat content* (H) sometimes referred to as *enthalpy*. Every system, whatever it may be, has a certain heat content. In that sense, it contains a certain amount of heat. The heat content of a system changes in the course of a physical or chemical change. The change in heat content, ΔH, is the difference between that of the products and reactants. That is:

$$\Delta H = H_{products} - H_{reactants} \qquad (4.3)$$

From Equation 4.3, we see that:

—if the heat content of the products is greater than that of the reactants, ΔH is a positive quantity ($\Delta H > 0$).

—if the heat content of the products is less than that of the reactants, ΔH is a negative quantity ($\Delta H < 0$).

These relations are shown graphically in Figure 4.4.

With this background, we can explain the source of heat flow in a process. There are two possibilities.

1. In an **endothermic** process, the heat absorbed goes to increase the heat content of the system. The products have a higher heat content

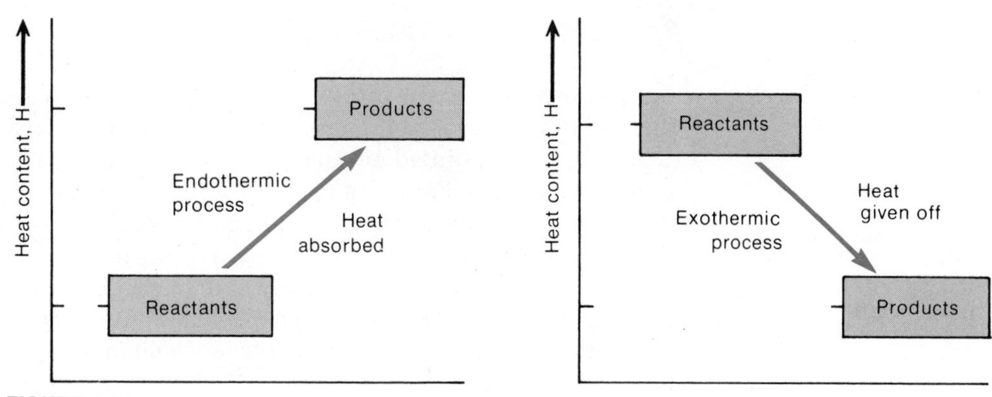

FIGURE 4.4

In an endothermic process the heat content of the products is higher than that of the reactants. The increase in heat content comes from the heat absorbed from the surroundings. In an exothermic process the heat content of the products is less than that of the reactants. The loss in heat content occurs as heat is given off to the surroundings.

("contain more heat") than the reactants. Hence, ΔH is a positive quantity.

2. In an **exothermic** process, heat is evolved at the expense of the heat content of the system. The products have a lower heat content ("contain less heat") than the reactants. Hence, ΔH is a negative quantity.

These statements apply to a change in state, a chemical reaction, or any other process you can think of. In summary:

$$\text{ENDOTHERMIC PROCESS: } \Delta H > 0\text{: heat is absorbed} \quad (4.4)$$

$$\text{EXOTHERMIC PROCESS: } \Delta H < 0\text{: heat is evolved} \quad (4.5)$$

EXAMPLE 4.2

The reaction: $CaCO_3(s) \rightarrow CaO(s) + CO_2(g)$ is endothermic.

a. How does the heat content of the products (1 mol CaO + 1 mol CO_2) compare to that of the reactant (1 mol $CaCO_3$)?

b. What is the sign of ΔH for this reaction?

c. What is the sign of ΔH for the reverse reaction?

SOLUTION

a. Since the reaction is endothermic, H of the products must be greater than that of the reactant.

b. ΔH must be a positive quantity (>0).

c. The reverse reaction must be exothermic. ΔH must be a negative quantity (<0).

4.3 ΔH FOR CHANGES IN STATE

In Section 4.2, we discussed the heat flows involved in certain changes in physical state. There, we were concerned with the direction of heat flow. We pointed out that heat is absorbed when a solid melts or a liquid vaporizes. For these processes, ΔH is a positive quantity. Conversely, when a liquid freezes or a gas condenses, heat is evolved. In these, as in all exothermic processes, ΔH is a negative quantity.

Here, we will look more closely at the heat effects in these changes of state. Our emphasis will be upon the amount of heat that is evolved or absorbed when one mole of a substance changes its state.

Heat of Fusion (ΔH_{fus})

It takes 80 times as much heat to melt a gram of ice as it does to raise the temperature of a gram of water 1°C.

To melt a solid we must supply heat. The heat absorbed is used to break down the rigid structure of the solid. It is called the *heat of fusion* and given the symbol ΔH_{fus}. The amount of heat absorbed depends upon the nature and amount of the solid that melts. In the case of ice, we find that 80.0 cal of heat are absorbed per gram of ice melted. For one mole of ice (MM = 18.0)

$$\Delta H_{fus} = 80.0\,\frac{\text{cal}}{\text{g}} \times 18.0\,\frac{\text{g}}{\text{mol}}$$

$$= 1440 \text{ cal/mol} = 1.44 \text{ kcal/mol}$$

In other words, the heat of fusion of ice is 1.44 kcal/mol. It takes 1.44 kcal of heat to melt a mole of ice.

The information just given can be summarized in a simple way. To do this, we use what is called a **thermochemical equation.** This is similar to an ordinary chemical equation except that the value of ΔH is included. Knowing that the heat of fusion of ice is 1.44 kcal/mol, we write:

$$H_2O(s) \rightarrow H_2O(l); \qquad \Delta H = +1.44 \text{ kcal} \qquad (4.6)$$

In this and all other thermochemical equations that we will write:

a. The sign of ΔH is given (+ or −).
b. The magnitude of ΔH is given in *kilocalories*.
c. The coefficients in the equation refer to numbers of moles.

In Equation 4.6, it is understood that ΔH = +1.44 kcal when one *mole* (18.0 g) of ice melts to form one *mole* of liquid water.

The heat of fusion is a characteristic property of a substance. In that sense, it resembles melting point or any other physical property. Consider, for example, benzene, C_6H_6. When one mole of benzene (78.0 g) melts, 2.35 kcal of heat is absorbed. The heat of fusion of benzene is 2.35 kcal/mol. The thermochemical equation for the melting of benzene is:

$$C_6H_6(s) \rightarrow C_6H_6(l); \qquad \Delta H = +2.35 \text{ kcal} \qquad (4.7)$$

Equations such as those just written can be interpreted in terms of heat content. To do this, we draw a **heat content diagram.** Figure 4.5

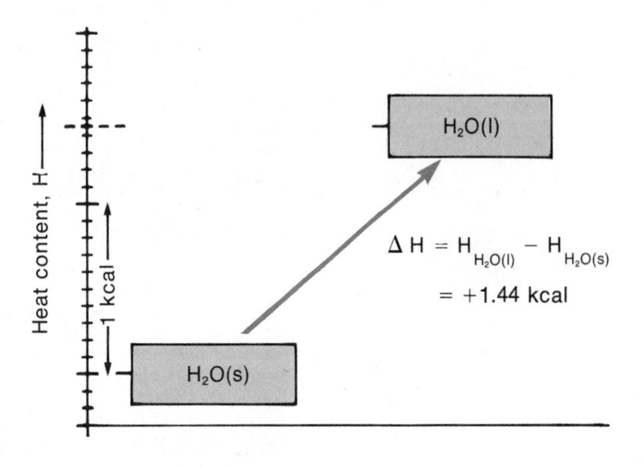

FIGURE 4.5

When a mole of ice is melted, it must absorb 1.44 kcal of heat from its surroundings. This heat goes to increase the heat content of the water produced. The process is endothermic and ΔH = +1.44 kcal.

is such a diagram for the melting of ice. Heat content, H, is plotted on the vertical axis. Reactants are located by an open bar at the left of the diagram. The box for the products is drawn at the right of the diagram. In this case, the product is at a higher level than the reactant. The heat content of one mole of liquid water is 1.44 kcal greater than that of one mole of ice. The diagram for benzene would look very much like this except that the difference would be 2.35 kcal.

You will notice from Figure 4.5 that there are no numbers along the vertical axis. This is the case with all heat content diagrams. It reflects the fact that we do not know the heat contents of individual substances. We cannot assign a heat content for one mole of ice. All we know is that it is 1.44 kcal *less* than the heat content of one mole of liquid water.

We measure changes in heat content, not heat content itself.

In drawing heat content diagrams, it doesn't matter where you start. The reactant line can be drawn at any convenient place. What does matter is the distance between the two lines. Since we know that ΔH for Reaction 4.6 is +1.44 kcal, the product must be located 1.44 kcal above the reactant. In Figure 4.5, each large space represents one kilocalorie. So, liquid water is located a bit less than 1.5 spaces above ice.

Heat of Vaporization (ΔH_{vap})

To vaporize a liquid, we must supply heat. The heat absorbed is used to separate the molecules from each other. It is called the *heat of vaporization* and given the symbol $\mathbf{\Delta H_{vap}}$. As in fusion, the heat of vaporization depends upon the nature and amount of the substance involved. For water, 540 cal of heat is absorbed per gram of liquid vaporized at 100°C. For one mole of water (MM = 18.0):

$$\Delta H_{vap} = 540 \frac{cal}{g} \times 18.0 \frac{g}{mol}$$

$$= 9720 \text{ cal/mol} = 9.72 \text{ kcal/mol}$$

We can represent this process by a thermochemical equation similar to 4.6. Since the heat of vaporization of water is 9.72 kcal/mol:

$$H_2O(l) \rightarrow H_2O(g); \qquad \Delta H = +9.72 \text{ kcal} \qquad (4.8)$$

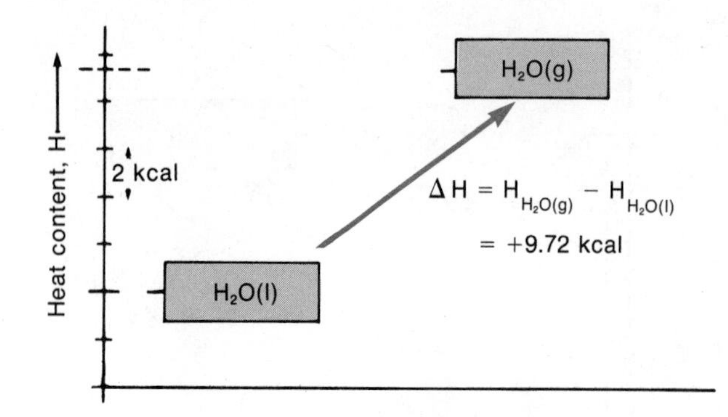

FIGURE 4.6

When a mole of water is boiled and converted to steam, its absorbs 9.72 kcal from the surroundings. This heat goes to increase the heat content of the steam by 9.72 kcal over that of the water from which it was formed.

This equation can be represented by a heat content diagram (Figure 4.6). From the figure, we see that the heat content of one mole of steam is greater, by 9.72 kcal, than that of one mole of water. To vaporize a mole of water, we must supply 9.72 kcal of heat.

EXAMPLE 4.3

The heat of vaporization of benzene is 7.36 kcal/mol.

a. Write a thermochemical equation for the vaporization of benzene.

b. Draw a heat content diagram corresponding to this equation.

SOLUTION

a. Remember that heat is absorbed in vaporization, so ΔH is a positive quantity.

$$C_6H_6(l) \rightarrow C_6H_6(g); \qquad \Delta H = +7.36 \text{ kcal}$$

b. On the vertical scale, we plot heat content, H. We might put the marks on the scale 5 mm apart. The space between two marks represents 1 kcal. We draw a line for the reactant (1 mol of liquid benzene) at the left of the diagram, near the bottom. Then we draw another line for the product (1 mol of benzene vapor) at the right. This line should be 7.36 kcal above that for the liquid. This means that the two lines are separated by a little more than 7 spaces on the vertical scale.

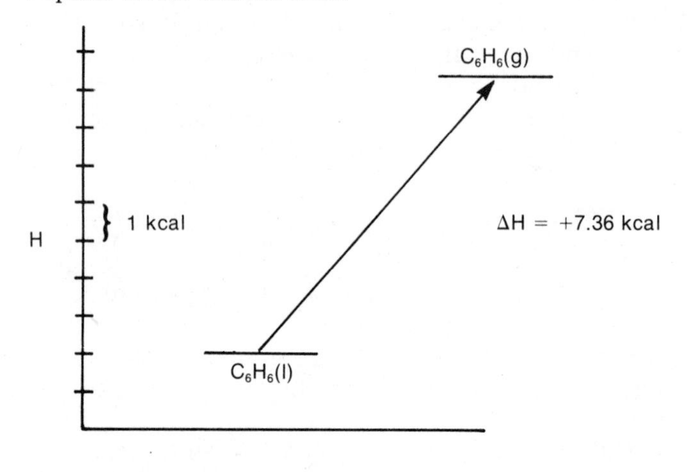

This drawing illustrates what the equation in (a) tells us.

4.4 ΔH FOR CHEMICAL REACTIONS

As pointed out earlier, certain chemical reactions are endothermic. Heat must be absorbed for such a reaction to occur. The change in heat content, ΔH, is positive. An example of such a reaction is the decomposition of limestone, $CaCO_3$, to give CaO and CO_2. In this case, 42.5 kcal of heat is absorbed to decompose one mole of calcium carbonate. The thermochemical equation is:

$$CaCO_3(s) \rightarrow CaO(s) + CO_2(g); \qquad \Delta H = +42.5 \text{ kcal} \qquad (4.9)$$

The heat content of the products (1 mol CaO + 1 mol CO_2) is 42.5 kcal greater than that of the reactant (1 mol $CaCO_3$).

An example of an exothermic reaction is that which occurs when carbon reacts with oxygen. Heat is given off as the product, CO_2, is formed. The change in heat content, ΔH, is negative. When one mole of carbon burns, ΔH is −94.1 kcal. The thermochemical equation is:

$$C(s) + O_2(g) \rightarrow CO_2(g); \qquad \Delta H = -94.1 \text{ kcal} \qquad (4.10)$$

This equation tells us that the heat content of the product (1 mol CO_2) is 94.1 kcal less than that of the reactants (1 mol C + 1 mol O_2). When a mole of carbon burns, 94.1 kcal of heat is given off.

Heats of Formation

A general type of reaction involves the formation of a compound from the elements. The change in heat content, **ΔH, when one mole of a compound is formed from the elements is called the heat of formation.** Table 4.2, p. 90 lists heats of formation for several common compounds. Notice that:

1. Fractional coefficients are used where necessary so that one mole of the compound will be formed.
2. Heats of formation are usually negative. The formation of a compound from the elements is usually an exothermic reaction. For example, when a mole of CaO is formed from the elements, 151.9 kcal of heat is given off.
3. The heat of formation of a compound depends upon its physical state. Compare, for example, the values given for $H_2O(l)$ and $H_2O(g)$. That for liquid water is more negative because $H_2O(l)$ has a lower heat content than $H_2O(g)$.

EXAMPLE 4.4

From Table 4.2 we see that for the reaction:

$$2\ Fe(s) + \tfrac{3}{2}\ O_2(g) \rightarrow Fe_2O_3(s); \qquad \Delta H = -196.5 \text{ kcal}$$

a. What is the heat of formation of Fe_2O_3, in kcal/mol?

b. Is this reaction exothermic or endothermic?

c. In this reaction, how many grams of iron react? how many grams of O_2?

SOLUTION

a. -196.5 kcal/mol

b. Exothermic; 196.5 kcal of heat is given off when one mole of Fe_2O_3 is formed from the elements

c. 2 mol Fe = 2(55.85 g Fe) = 111.7 g Fe
$\frac{3}{2}$ mol O_2 = 1.5(32.0 g O_2) = 48.0 g O_2

Try to state in words what the first line of Table 4.2 really means.

TABLE 4.2 HEATS OF FORMATION OF COMPOUNDS (kcal/mol at 25°C)

Compound	Reaction	ΔH (kcal)
$Ag_2O(s)$	$2\ Ag(s) + 0.5\ O_2(g) \rightarrow Ag_2O(s)$	−7.3
$C_6H_6(l)$	$6\ C(s) + 3\ H_2(g) \rightarrow C_6H_6(l)$	+11.6
$CaO(s)$	$Ca(s) + 0.5\ O_2(g) \rightarrow CaO(s)$	−151.9
$Ca(OH)_2(s)$	$Ca(s) + O_2(g) + H_2(g) \rightarrow Ca(OH)_2(s)$	−235.8
$CO(g)$	$C(s) + 0.5\ O_2(g) \rightarrow CO(g)$	−26.4
$CO_2(g)$	$C(s) + O_2(g) \rightarrow CO_2(g)$	−94.1
$CuO(s)$	$Cu(s) + 0.5\ O_2(g) \rightarrow CuO(s)$	−37.1
$Cu_2O(s)$	$2\ Cu(s) + 0.5\ O_2(g) \rightarrow Cu_2O(s)$	−39.8
$Fe_2O_3(s)$	$2\ Fe(s) + 1.5\ O_2(g) \rightarrow Fe_2O_3(s)$	−196.5
$HBr(g)$	$0.5\ H_2(g) + 0.5\ Br_2(l) \rightarrow HBr(g)$	−8.7
$HCl(g)$	$0.5\ H_2(g) + 0.5\ Cl_2(g) \rightarrow HCl(g)$	−22.1
$HF(g)$	$0.5\ H_2(g) + 0.5\ F_2(g) \rightarrow HF(g)$	−64.2
$HI(g)$	$0.5\ H_2(g) + 0.5\ I_2(s) \rightarrow HI(g)$	+6.2
$H_2O(g)$	$H_2(g) + 0.5\ O_2(g) \rightarrow H_2O(g)$	−57.8
$H_2O(l)$	$H_2(g) + 0.5\ O_2(g) \rightarrow H_2O(l)$	−68.3
$H_2S(g)$	$H_2(g) + S(s) \rightarrow H_2S(g)$	−4.8
$HgO(s)$	$Hg(l) + 0.5\ O_2(g) \rightarrow HgO(s)$	−21.7
$MgO(s)$	$Mg(s) + 0.5\ O_2(g) \rightarrow MgO(s)$	−143.8
$NaCl(s)$	$Na(s) + 0.5\ Cl_2(g) \rightarrow NaCl(s)$	−98.2
$NH_3(g)$	$0.5\ N_2(g) + 1.5\ H_2(g) \rightarrow NH_3(g)$	−11.0
$NO(g)$	$0.5\ N_2(g) + 0.5\ O_2(g) \rightarrow NO(g)$	+21.6
$NO_2(g)$	$0.5\ N_2(g) + O_2(g) \rightarrow NO_2(g)$	+8.1
$SO_2(g)$	$S(s) + O_2(g) \rightarrow SO_2(g)$	−71.0
$SO_3(g)$	$S(s) + 1.5\ O_2(g) \rightarrow SO_3(g)$	−94.5

TABLE 4.3 HEATS OF COMBUSTION (kcal/mol at 25°C)

Substance	Reaction	ΔH (kcal)
$H_2(g)$	$H_2(g) + 0.5\ O_2(g) \rightarrow H_2O(l)$	−68.3
$C(s)$	$C(s) + O_2(g) \rightarrow CO_2(g)$	−94.1
$CO(g)$	$CO(g) + 0.5\ O_2(g) \rightarrow CO_2(g)$	−67.7
$CH_4(g)$	$CH_4(g) + 2\ O_2(g) \rightarrow CO_2(g) + 2\ H_2O(l)$	−212.8
$CH_3OH(l)$	$CH_3OH(l) + 1.5\ O_2(g) \rightarrow CO_2(g) + 2\ H_2O(l)$	−173.7
$C_2H_2(g)$	$C_2H_2(g) + 2.5\ O_2(g) \rightarrow 2\ CO_2(g) + H_2O(l)$	−310.6
$C_2H_4(g)$	$C_2H_4(g) + 3\ O_2(g) \rightarrow 2\ CO_2(g) + 2\ H_2O(l)$	−337.2
$C_2H_6(g)$	$C_2H_6(g) + 3.5\ O_2(g) \rightarrow 2\ CO_2(g) + 3\ H_2O(l)$	−372.8
$C_2H_5OH(l)$	$C_2H_5OH(l) + 3\ O_2(g) \rightarrow 2\ CO_2(g) + 3\ H_2O(l)$	−326.7
$C_3H_8(g)$	$C_3H_8(g) + 5\ O_2(g) \rightarrow 3\ CO_2(g) + 4\ H_2O(l)$	−530.6
$C_4H_{10}(g)$	$C_4H_{10}(g) + 6.5\ O_2(g) \rightarrow 4\ CO_2(g) + 5\ H_2O(l)$	−688.0
$C_5H_{12}(l)$	$C_5H_{12}(l) + 8\ O_2(g) \rightarrow 5\ CO_2(g) + 6\ H_2O(l)$	−838.9
$C_6H_6(l)$	$C_6H_6(l) + 7.5\ O_2(g) \rightarrow 6\ CO_2(g) + 3\ H_2O(l)$	−781.1
$C_6H_{14}(l)$	$C_6H_{14}(l) + 9.5\ O_2(g) \rightarrow 6\ CO_2(g) + 7\ H_2O(l)$	−995.0
$C_7H_{16}(l)$	$C_7H_{16}(l) + 11\ O_2(g) \rightarrow 7\ CO_2(g) + 8\ H_2O(l)$	−1151.3
$C_8H_{18}(l)$	$C_8H_{18}(l) + 12.5\ O_2(g) \rightarrow 8\ CO_2(g) + 9\ H_2O(l)$	−1307.5
$C_{10}H_8(s)$	$C_{10}H_8(s) + 12\ O_2(g) \rightarrow 10\ CO_2(g) + 4\ H_2O(l)$	−1232.5

Heats of Combustion

Another general type of reaction is that involved when a substance burns. The other reactant is oxygen, which usually comes from the air. The change in heat content, **ΔH, when one mole of a substance burns is called the heat of combustion.** Table 4.3 lists heats of combustion for several elements and compounds.

Heats of combustion and formation are all determined experimentally.

Notice that heats of combustion are generally negative. Heat is evolved when a substance burns. For example, 212.8 kcal of heat is evolved when one mole of methane, CH_4, burns. The products are those obtained with excess oxygen, $CO_2(g)$ and $H_2O(l)$.

Fuels; Heating Values

Many of the substances listed in Table 4.3 are components of fuels. For example, natural gas is mostly methane, CH_4. Petroleum is a mixture of liquid hydrocarbons of which octane, C_8H_{18}, is typical. Coal contains solid hydrocarbons of high molecular mass. Their composition is similar to that of naphthalene, $C_{10}H_8$.

Perhaps the most important property of a fuel is its heating value. The heating value of a fuel is the amount of heat evolved when one gram of the fuel is burned. For pure substances, heating values can be calculated from heats of combustion. As an example, consider methane, CH_4. From Table 4.3, we see that 212.8 kcal is evolved when one mole of methane burns. The molecular mass of CH_4 is 12.0 + 4(1.0) = 16.0. One mole of methane weighs 16.0 g. So, the heating value of CH_4 is:

$$\frac{212.8 \text{ kcal}}{16.0 \text{ g}} = 13.3 \text{ kcal/g}$$

Table 4.4 lists the heating values of several fuels. Clearly, on this basis, natural gas and petroleum are the most effective fuels. Unfortunately, the United States is running out of these materials. We may well have to burn more coal, which has a heating value only about 60% of that of natural gas. Another possibility is to substitute ethyl alcohol or methyl alcohol for petroleum. These liquid fuels are being promoted to supplement or even replace gasoline. On the basis of heating value, ethyl alcohol would seem the better bet. The poorest fuel listed in Table 4.4 is wood, with a heating value of only 4.5 kcal/g. This is something to consider before you try heating your house with wood stoves. You have to burn a lot of wood to keep warm on a cold winter day. On the other hand, if the wood is free, your heating bill will be cut down considerably.

As a fuel, wood is said to warm you twice: once when you chop it and again when you burn it.

TABLE 4.4 HEATING VALUES OF FUELS

Fuel	Heating value
Natural gas	11.6 kcal/g
Petroleum, including gasoline	11.3 kcal/g
Hard coal (anthracite)	7.3 kcal/g
Soft coal (bituminous)	7.0 kcal/g
Ethyl alcohol (C_2H_5OH)	7.1 kcal/g
Methyl alcohol (CH_3OH)	5.4 kcal/g
Wood	4.5 kcal/g

4.5 USES OF ΔH

In this chapter, we have written a large number of thermochemical equations. We have tried to explain what these equations mean. For example, consider the equation:

$$2\ HgO(s) \rightarrow 2\ Hg(l) + O_2(g); \qquad \Delta H = +43.4 \text{ kcal} \qquad (4.11)$$

This tells us that 43.4 kcal of heat is absorbed when 2 mol of HgO decomposes to form 2 mol of Hg and 1 mol of O_2. Again, the equation:

$$2\ CO(g) + O_2(g) \rightarrow 2\ CO_2(g); \qquad \Delta H = -135.4 \text{ kcal} \qquad (4.12)$$

means that 135.4 kcal of heat is evolved when 2 mol of CO reacts with 1 mol of O_2 to form 2 mol of CO_2.

In this section, we will consider how thermochemical equations are used in practical calculations. To do this, we make use of certain rules that relate to the change in heat content, ΔH. We will look at three such rules. Each of them applies equally well to a change in state or to a chemical reaction.

Relation Between ΔH's of Forward and Reverse Processes

If we know the value of ΔH for a certain process, we can easily obtain ΔH for the reverse process. The rule is a simple one:

ΔH for a process is equal in magnitude but opposite in sign to ΔH for the reverse process.

To illustrate this rule, consider the equation:

The heat content, like the volume, of 1 mol of ice at 0°C and 1 atm has a fixed value.

$$H_2O(s) \rightarrow H_2O(l); \qquad \Delta H = +1.44 \text{ kcal}$$

This tells us that ΔH for the fusion of one mole of ice is +1.44 kcal. For the reverse process, ΔH must be −1.44 kcal:

$$H_2O(l) \rightarrow H_2O(s); \qquad \Delta H = -1.44 \text{ kcal}$$

In other words, 1.44 kcal of heat is evolved when a mole of liquid water freezes.

Glancing back at Figure 4.5 (p. 87), we can see why this rule works. The heat content of 1 mol of $H_2O(l)$ is 1.44 kcal above that of 1 mol of $H_2O(s)$. So, when ice melts, we move up the scale and ΔH = +1.44 kcal. If we move in the opposite direction, the magnitude of ΔH remains the same. However, the sign changes because we are now moving down instead of up. So, ΔH for the freezing of a mole of water must be −1.44 kcal.

This rule can be applied to chemical reactions as well as changes in state. Suppose, for example, we know that:

$$CaCO_3(s) \rightarrow CaO(s) + CO_2(g); \qquad \Delta H = +42.5 \text{ kcal}$$

It follows that:

$$CaO(s) + CO_2(g) \rightarrow CaCO_3(s); \qquad \Delta H = -42.5 \text{ kcal}$$

Relation Between ΔH and Amount of Reactant or Product

Thermochemical equations can be used to calculate the heat flow for any given amount of reactant or product. The rule used is:

ΔH is directly proportional to the amount of substance that reacts or is produced in a process.

To illustrate this rule, consider the change in state:

$$C_6H_6(l) \rightarrow C_6H_6(g); \qquad \Delta H = +7.36 \text{ kcal}$$

This equation tells us that 7.36 kcal of heat must be absorbed to vaporize a mole of benzene. If two moles were vaporized, ΔH would be twice as great:

$$2\ C_6H_6(l) \rightarrow 2\ C_6H_6(g); \qquad \Delta H = 2(+7.36 \text{ kcal}) = +14.72 \text{ kcal}$$

Similarly, if half a mole of benzene were vaporized we would have:

$$\tfrac{1}{2}\ C_6H_6(l) \rightarrow \tfrac{1}{2}\ C_6H_6(g); \qquad \Delta H = \tfrac{1}{2}(+7.36\ \text{kcal}) = +3.68\ \text{kcal}$$

Suppose now that we wished to calculate ΔH for the vaporization of one gram of benzene. The molecular mass of C_6H_6 is 6(12.0) + 6(1.0) = 78.0. One mole of benzene weighs 78.0 g. So, ΔH for one gram would be 1/78.0 of the value for one mole, or:

$$\frac{7.36\ \text{kcal}}{78.0} = 0.0944\ \text{kcal}$$

In other words, 0.0944 kcal (94.4 cal) of heat would have to be supplied to vaporize a gram of liquid benzene.

Example 4.5 illustrates the application of this rule to a chemical reaction.

EXAMPLE 4.5

Given the thermochemical equation:

$$C(s) + O_2(g) \rightarrow CO_2(g); \qquad \Delta H = -94.1\ \text{kcal}$$

calculate:

a. ΔH when one gram of carbon burns

b. ΔH when 1.80 g of carbon dioxide is formed.

SOLUTION

a. The equation tells us that when one *mole* of carbon burns, ΔH is −94.1 kcal. We might then say that:

$$1\ \text{mol C} \simeq -94.1\ \text{kcal}$$

This problem summarizes much of the content of this chapter. Be sure you understand each step.

Since the atomic mass of carbon is 12.0, one mole of C weighs 12.0 g. So:

$$12.0\ \text{g C} \simeq -94.1\ \text{kcal}$$

This relation gives us the conversion factor we need to find ΔH for the combustion of one gram of carbon:

$$\Delta H = 1.00\ \text{g C} \times \frac{-94.1\ \text{kcal}}{12.0\ \text{g C}} = -7.84\ \text{kcal}$$

We conclude that 7.84 kcal of heat is given off per gram of carbon burned.

b. Here we need a conversion factor relating grams of CO_2 to ΔH in kilocalories. From the thermochemical equation:

$$1\ \text{mol}\ CO_2 \simeq -94.1\ \text{kcal}$$

The molecular mass of CO_2 is 12.0 + 2(16.0) = 44.0. Hence:

$$44.0 \text{ g } CO_2 \mathrel{\widehat{=}} -94.1 \text{ kcal}$$

$$\Delta H = 1.80 \text{ g } CO_2 \times \frac{-94.1 \text{ kcal}}{44.0 \text{ g } CO_2} = -3.85 \text{ kcal}$$

That is, 3.85 kcal of heat is evolved when 1.80 g of CO_2 is formed in the reaction.

You will note from Example 4.5 that thermochemical equations are handled in much the same way as ordinary chemical equations. In particular, we use conversion factors to obtain ΔH in much the same way as in Chapter 3. This same approach is used again in Example 4.6.

EXAMPLE 4.6

Consider the thermochemical equation:

$$C_8H_{18}(l) + 12.5\ O_2(g) \rightarrow 8\ CO_2(g) + 9\ H_2O(l); \qquad \Delta H = -1308 \text{ kcal}$$

How many grams of octane, C_8H_{18}, have to be burned to evolve one kilocalorie of heat?

SOLUTION

From the equation we see that:

$$1 \text{ mol } C_8H_{18} \mathrel{\widehat{=}} -1308 \text{ kcal}$$

We want to know how many grams of C_8H_{18} must react when ΔH = −1.00 kcal. One mole of C_8H_{18} weighs:

$$8(12.0 \text{ g}) + 18(1.0 \text{ g}) = 114 \text{ g}$$

So:

$$114 \text{ g } C_8H_{18} \mathrel{\widehat{=}} -1308 \text{ kcal}$$

This relation gives us the conversion factor we need to find the mass of octane required.

$$\text{g } C_8H_{18} = -1.00 \text{ kcal} \times \frac{114 \text{ g } C_8H_{18}}{-1308 \text{ kcal}} = 0.0872 \text{ g } C_8H_{18}$$

We conclude that somewhat less than a tenth of a gram of octane must be burned to produce a kilocalorie of heat.

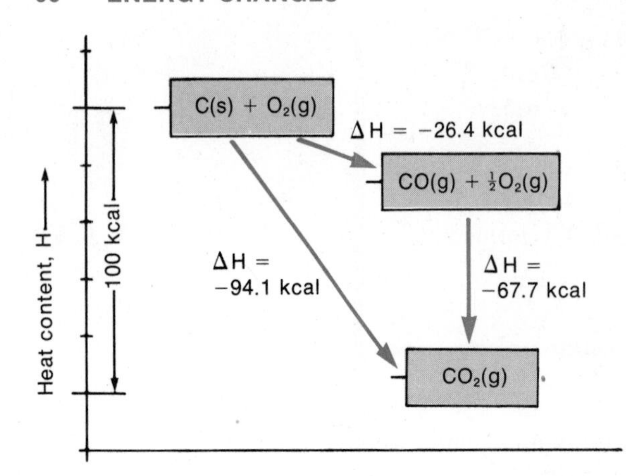

FIGURE 4.7

Carbon dioxide can be formed from carbon and oxygen in one step or can be formed in two steps, by first forming CO and then burning the CO to CO_2. In both cases the change in heat content is the same for the conversion of C(s) and O_2(g) to CO_2(g). ΔH for the one-step process is equal to the sum of the ΔH values for each part of the two-step process.

Hess' Law

In many ways, thermochemical equations resemble ordinary algebraic equations. In particular, two equations can be added to give a third, valid equation. This rule, called Hess' Law, is often stated as follows:

If an equation (c) is the sum of two other equations, (a) and (b), then:

$$\Delta\mathbf{H}\ \textbf{for (c)} = \Delta\mathbf{H}\ \textbf{for (a)} + \Delta\mathbf{H}\ \textbf{for (b)} \tag{4.13}$$

This law is similar to the following: the change in altitude in going from Minneapolis to Los Angeles is the same whether you go by way of Denver or Dallas.

To illustrate this rule, consider the equations:

$$C(s) + \tfrac{1}{2}\,O_2(g) \rightarrow CO(g); \qquad \Delta H = -26.4 \text{ kcal}$$

$$CO(g) + \tfrac{1}{2}\,O_2(g) \rightarrow CO_2(g); \qquad \Delta H = -67.7 \text{ kcal}$$

If we add these two equations, cancelling the CO, we obtain the equation for the combustion of carbon to CO_2:

	ΔH
$C(s) + \tfrac{1}{2}O_2(g) \rightarrow CO(g)$	−26.4 kcal
$CO(g) + \tfrac{1}{2}O_2(g) \rightarrow CO_2(g)$	−67.7 kcal
$C(s) + O_2(g) \rightarrow CO_2(g)$	−94.1 kcal

We conclude that the heat content change for the overall reaction (−94.1 kcal) must be the sum of the ΔH's for the other two reactions. This reasoning is shown graphically in Figure 4.7.

Example 4.7 applies this rule in a somewhat more complex case.

EXAMPLE 4.7

Using data in Table 4.2, calculate ΔH for the reaction:

$$NO(g) + \tfrac{1}{2}\,O_2(g) \rightarrow NO_2(g)$$

SOLUTION

Examining Table 4.2, we do not find this reaction listed directly. We do, however, find:

(a) $\frac{1}{2} N_2(g) + \frac{1}{2} O_2(g) \rightarrow NO(g)$; $\Delta H = +21.6$ kcal

(b) $\frac{1}{2} N_2(g) + O_2(g) \rightarrow NO_2(g)$; $\Delta H = +8.1$ kcal

The problem is to combine these two equations in such a way as to arrive at the one we want. Noting that NO(g) appears as a product in Equation (a) and as a reactant in the equation we want, it seems reasonable to reverse Equation (a). This requires that we change the sign of ΔH.

Sometimes you can solve a big problem if you can see it's really just two small problems laid end to end.

$$NO(g) \rightarrow \frac{1}{2} N_2(g) + \frac{1}{2} O_2(g); \quad \Delta H = -21.6 \text{ kcal}$$

If we now add this equation to (b), we obtain the desired equation.

$$NO(g) \rightarrow \frac{1}{2} \cancel{N_2(g)} + \frac{1}{2} O_2(g); \quad \Delta H = -21.6 \text{ kcal}$$

$$\frac{1}{2} \cancel{N_2(g)} + O_2(g) \rightarrow NO_2(g); \quad \Delta H = +8.1 \text{ kcal}$$

$$NO(g) + \frac{1}{2} O_2(g) \rightarrow NO_2(g); \quad \Delta H = -13.5 \text{ kcal}$$

We will find this rule (Hess' Law) to be very useful in later chapters. Mainly, we will use it to break down complex reactions into two or more simpler steps. Knowing the value of ΔH for the individual steps, we can obtain ΔH for the overall reaction. Usually this analysis is very simple, as in the CO-CO_2 case discussed on p. 96. Sometimes, we have to do a bit more juggling, as in Example 4.7.

Calorimetry The value of ΔH for a reaction can be measured using a device called a *calorimeter*. A simple calorimeter is shown in Fig. 4.8, p. 98. It consists of a styrofoam cup with cover and a thermometer. The cup is partially filled with water. The reaction, taking place inside the cup, transfers heat to the water. If the reaction evolves heat (ΔH negative), the temperature of the water rises. If, on the other hand, the reaction is endothermic (ΔH positive), heat is absorbed from the water. In this case, the temperature of the water drops.

Knowing the amount of water in the cup and the temperature change, we can calculate ΔH. Suppose, for example, that the temperature rises 10.0°C when there is 425 g of water in the cup. We know that 1.00 cal is required to raise the temperature of one gram of water one degree Celsius. (We say that the *specific heat* of water is 1.00 cal/g°C). Hence, the amount of heat transferred to the water is:

$$1.00 \frac{\text{cal}}{\text{g}\cdot{}^\circ\text{C}} \times 10.0^\circ\text{C} \times 425 \text{ g} = 4250 \text{ cal} = 4.25 \text{ kcal}$$

This means that the reaction must have evolved 4.25 kcal. In other words:

$$\Delta H = -4.25 \text{ kcal}$$

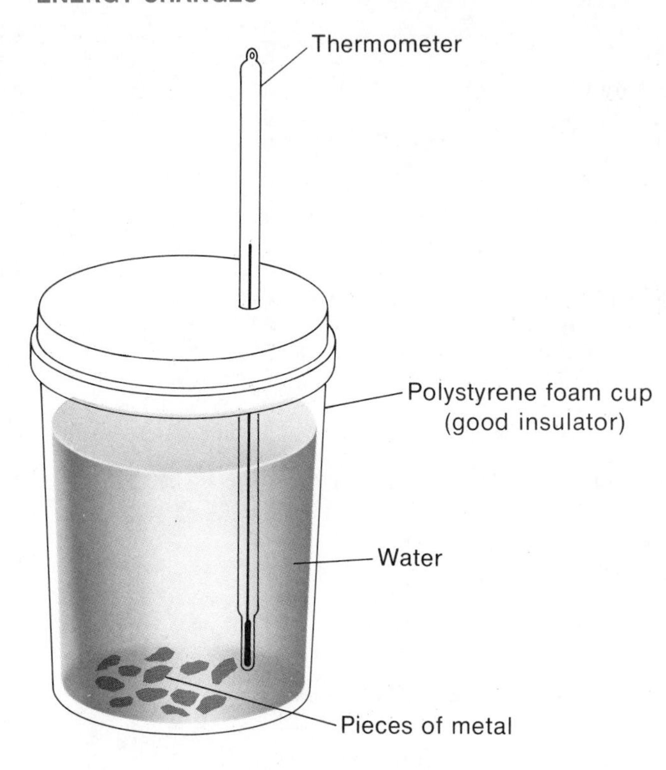

FIGURE 4.8

A coffee cup calorimeter can be used to measure heat flow for changes that can be studied in water. There is very little heat flow through the insulating wall of the cup, so any heat flow must be between the water and the reacting mixture. If the reaction is exothermic, heat will flow from the reaction mixture into the water, and the water temperature will go up. From the change in temperature and the mass of the water you can find the amount of heat that went into the water. This will equal the loss in heat content in the reaction mixture.

The calorimeter shown in Figure 4.8 is suitable for simple reactions in water solution. For reactions involving hot gases, such as combustion, a sealed container is necessary. Commonly, a strong-walled metal container is used. The reaction takes place within the container, which is immersed in water. Most of the heat is transferred to the water. The rest of it goes to increase the temperature of the metal container. The calculation of ΔH is a bit more involved, but the principle is the same.

4.6 SUMMARY

In this chapter, we have considered the heat flow in changes of state and chemical reactions. We have discussed such quantities as *heat of fusion, heat of vaporization, heat of formation,* and *heat of combustion.* The heat flow in any process can be related to the difference in heat content, ΔH, between products and reactants. Endothermic processes are ones for which ΔH is positive. Exothermic processes result in a negative value of ΔH.

Thermochemical equations are ones in which the value of ΔH is specified, usually in kilocalories. In working with these equations, remember that the value of ΔH:

—depends upon the physical states of reactants and products.
—is directly proportional to the amounts of reactants or products.

You should also know that:

—when the equation is reversed, the sign of ΔH changes.
—when two equations are added, their ΔH values are also added.

NEW TERMS

calorie
endothermic
exothermic
heat content (H)
heat content diagram
heat flow
heat of combustion
heat of formation
heat of fusion
heat of vaporization
Hess' Law
joule
kilocalorie
kilowatt hour
kinetic energy
Law of Conservation of Energy
potential energy
thermochemical equation

QUESTIONS

1. Give examples of processes in which a chemical reaction is used to produce

   a. heat b. mechanical energy c. electrical energy

2. Which of the following processes are accompanied by an increase in potential energy? a decrease? no change?

   a. An airplane lands at JFK airport.
   b. $Cl_2(g) \rightarrow 2\ Cl(g)$
   c. A boat crosses a lake.
   d. $2\ O(g) \rightarrow O_2(g)$

3. State the Law of Conservation of Energy.

4. For each of the following pairs of energy units, state which is the larger.

   a. calorie or kilocalorie b. calorie or joule
   c. kilowatt hour or kilocalorie

5. Which of the following reactions would you expect to be endothermic? exothermic?

   a. $CH_4(g) + 2\ O_2(g) \rightarrow CO_2(g) + 2\ H_2O(l)$
   b. $CO_2(g) \rightarrow C(s) + O_2(g)$
   c. $CaCO_3(s) \rightarrow CaO(s) + CO_2(g)$
   d. $2\ Na(s) + Cl_2(g) \rightarrow 2\ NaCl(s)$

6. Give the sign of ΔH for each of the reactions in Question 5.

7. The reaction: $2\ CO(g) + O_2(g) \rightarrow 2\ CO_2(g)$ is exothermic.

   a. What is the sign of ΔH for this reaction?
   b. What is the sign of ΔH for the reverse reaction?
   c. Is the heat content of the product (2 mol CO_2) greater or less than that of the reactants (2 mol CO + 1 mol O_2)?

8. For the reaction: $2\ Mg(s) + O_2(g) \rightarrow 2\ MgO(s)$; $\Delta H = -288$ kcal

   a. Is this reaction exothermic or endothermic?
   b. What is the value of ΔH for the reverse reaction?
   c. How does the heat content of 2 mol MgO compare to that of 2 mol Mg + 1 mol O_2?

9. The reaction involved in photosynthesis is:

$$6\ CO_2(g) + 6\ H_2O(l) \rightarrow C_6H_{12}O_6(s) + 6\ O_2(g)$$

In this reaction, green plants use the sun's energy to produce sugars.

a. Would you expect this reaction to be endothermic or exothermic?
b. What is the sign of ΔH for this reaction?
c. On a heat content diagram, would the products lie above or below the reactants?

10. Consider the following heat content diagram:

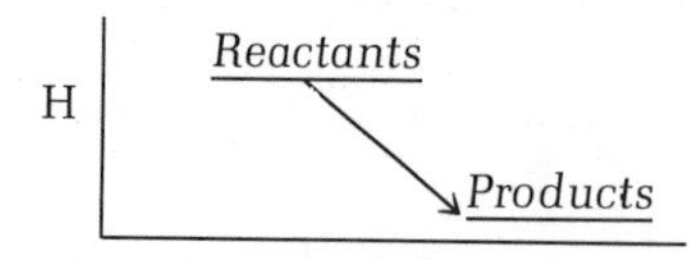

a. Is this reaction endothermic or exothermic?
b. What is the sign of ΔH?

11. Write thermochemical equations for the following changes in state.

a. The vaporization of benzene (ΔH = 7.36 kcal/mol)
b. The freezing of water (ΔH = −1.44 kcal/mol)
c. The condensation of water vapor (ΔH = −9.72 kcal)

12. Which of the following is ordinarily a positive quantity? a negative quantity?

a. heat of vaporization b. heat of fusion c. heat of formation
d. heat of combustion

13. Translate the following statements into thermochemical equations.

a. The heat of formation of Ag_2O is −7.3 kcal/mol.
b. The heat of combustion of carbon is −94.1 kcal/mol.
c. The heat of combustion of propane, C_3H_8, is −531 kcal/mol.

14. Explain, in your own words, why ΔH for forward and reverse reactions have the same magnitude but opposite signs.

15. State Hess' Law.

16. Using Table 4.2, obtain ΔH for the decomposition of one mole of each of the following compounds to the elements.

a. $H_2O(g)$ b. $H_2O(l)$ c. $CO(g)$

PROBLEMS

1. *Conversion of Energy Units (Example 4.1)* Using conversion factors given on p. 82, change each of the following to kilocalories.

a. 219 cal b. 6.283 kJ c. 4.19 kwh

2. *Conversion of Energy Units (Example 4.1)* Suppose electrical energy costs you 3.5¢ per kilowatt hour. How much are you paying for:

a. a kilocalorie of energy? b. 58.2 kcal?

3. *Heat Content Diagrams (Example 4.3)* Draw a heat content diagram

for the vaporization of mercury (Hg). The heat of vaporization of mercury is 14.2 kcal/mol.

4. *Heat Content Diagrams (Example 4.3)* Draw a heat content diagram for the combustion of acetylene, C_2H_2. Use the data in Table 4.3.

5. *Relation between ΔH and Amount of Reactant or Product (Examples 4.5, 4.6)* Using the thermochemical equation: $CuO(s) + H_2(g) \rightarrow Cu(s) + H_2O(l)$; $\Delta H = -31.2$ kcal, calculate ΔH when:

   a. 2 mol of CuO reacts b. 1.00 g of CuO reacts
   c. 1.00 g of Cu is formed

6. *Relation between ΔH and Amount of Reactant or Product (Examples 4.5, 4.6)* The heat of vaporization of $Br_2(l)$ is 7.16 kcal/mol. Calculate ΔH for the vaporization of:

   a. 0.210 mol of $Br_2(l)$ b. 15.2 g of $Br_2(l)$

7. *Relation between ΔH and Amount of Reactant or Product (Examples 4.5, 4.6)* Using the thermochemical equation: $CaO(s) + SO_3(g) \rightarrow CaSO_4(s)$; $\Delta H = -96.0$ kcal, calculate the mass in grams of $CaSO_4$ which must be formed to evolve 1.00 kcal of heat.

8. *Hess' Law (Example 4.7)* Given:

$$H_2O(s) \rightarrow H_2O(l); \quad \Delta H = +1.44 \text{ kcal}$$

$$H_2O(l) \rightarrow H_2O(g); \quad \Delta H = +9.72 \text{ kcal}$$

   Calculate ΔH for: $H_2O(s) \rightarrow H_2O(g)$

9. *Hess' Law (Example 4.7)* Using Table 4.2 and Hess' Law, calculate ΔH for:

$$SO_2(g) + \tfrac{1}{2} O_2(g) \rightarrow SO_3(g)$$

* * * * *

10. Using the data in Table 4.2, construct heat content diagrams for the reactions:

    a. $Ca(s) + \tfrac{1}{2} O_2(g) \rightarrow CaO(s)$
    b. $2\ HF(g) \rightarrow H_2(g) + F_2(g)$

11. For the processes:

$$A \rightarrow B; \quad \Delta H = +2.0 \text{ kcal}$$

$$B \rightarrow C; \quad \Delta H = +3.0 \text{ kcal}$$

    Calculate ΔH for:

    a. A → C b. B → A c. C → B d. C → A

12. Using data in Table 4.2, calculate ΔH for:

    a. $H_2S(g) \rightarrow H_2(g) + S(s)$
    b. the reaction of 1.00 g of calcium with oxygen to form CaO.
    c. the formation of 1.00 g of CO from the elements.

13. Glucose, $C_6H_{12}O_6$, is often eaten as a source of quick energy. For its reaction in the body, ΔH is −673 kcal/mol. How much energy is produced by a candy bar which contains 40.0 g of glucose?

14. Uranium reacts with fluorine as follows:

$$U(s) + 3\ F_2(g) \rightarrow UF_6(g); \qquad \Delta H = -505\ \text{kcal}$$

Calculate the heat released when 1.00 kg of uranium reacts with excess fluorine.

15. The two forms of carbon, graphite (gr) and diamond (dia) have slightly different heats of combustion:

$$C(dia) + O_2(g) \rightarrow CO_2(g); \qquad \Delta H = -94.50\ \text{kcal}$$

$$C(gr) + O_2(g) \rightarrow CO_2(g); \qquad \Delta H = -94.05\ \text{kcal}$$

From this information, calculate ΔH for the conversion of graphite to diamond.

16. The thermite reaction:

$$Fe_2O_3(s) + 2\ Al(s) \rightarrow Al_2O_3(s) + 2\ Fe(s)$$

is so exothermic that it often produces molten iron. Determine ΔH for this reaction, given:

$$2\ Al(s) + \tfrac{3}{2}\ O_2(g) \rightarrow Al_2O_3(s); \qquad \Delta H = -400\ \text{kcal}$$

$$2\ Fe(s) + \tfrac{3}{2}\ O_2(g) \rightarrow Fe_2O_3(s); \qquad \Delta H = -200\ \text{kcal}$$

17. Hydrazine, N_2H_4, is used as a rocket fuel. For the reaction:

$$N_2H_4(g) + O_2(g) \rightarrow N_2(g) + 2\ H_2O(l); \qquad \Delta H = -150\ \text{kcal}$$

How many kilograms of N_2H_4 would be required to produce 1.0×10^6 kcal of heat?

*18. In a certain reaction carried out in a calorimeter, 1.26 kg of water is heated from 20.0°C to 25.8°C. What is ΔH for this reaction?

*19. The specific heat of a substance is the amount of heat that must be absorbed to raise the temperature of one gram one degree Celsius. It is found that 0.214 kcal is required to raise the temperature of 40.0 g of iron from 10.0 to 60.0°C. What is the specific heat of iron, in calories per gram per degree Celsius?

*20. Using Table 4.3, compare the heating values of acetylene, C_2H_2, and ethane, C_2H_6.

*21. Suppose electrical energy costs 3.5¢ per kilowatt hour and one mole of methane, CH_4, costs 0.52¢. Using data in Table 4.3, calculate the cost per kilocalorie of the heat produced by burning methane; the cost per kilowatt hour. On this basis, would it be cheaper to heat a house electrically as opposed to using natural gas, which is mostly methane?

Jacques Charles, who discovered the relation between the volume of a gas and its temperature, was really more interested in balloons. In 1783, he ascended to a height of 3.2 km in this hot-air balloon. *Bettman Archive.*

5 THE PHYSICAL BEHAVIOR OF GASES

The states of matter, solid, liquid, and gas, were introduced briefly in Chapter 1. In this chapter we begin a more detailed study of the states of matter by looking at the gaseous state. We will delay treating the solid and liquid states (Chapter 12) until chemical bonding has been covered. Each state is strikingly different and each state plays an important part in our lives. While most matter on earth is solid, the earth also holds large quantities of liquid water and is surrounded by a gaseous atmosphere.

The gaseous state was the first whose properties could be explained by simple laws. Later in this chapter we will look at these laws. They relate the volume to the pressure, temperature, and amount of gas. Let us start by considering how these quantities are measured in the laboratory.

5.1 MEASUREMENTS ON GASES

The *mass* in grams of a gas in a closed container can be determined by weighing on an analytical balance. Since a gas always fills any container in which it is placed, the *volume* of a gas is that of its container. When we say that "the volume of a sample of $O_2(g)$ is 250 cm^3" we mean that the vessel containing it has a volume of 250 cm^3. We will ordinarily express gas volumes either in cubic centimeters or in liters.

$$1\ \ell = 1000\ cm^3 \tag{5.1}$$

Temperature

The temperature of a gas can be measured using an ordinary mercury-in-glass thermometer. The laws governing the behavior of gases are expressed most simply in terms of the *Kelvin* or *absolute* scale. On this scale, zero is taken to be the lowest possible temperature, that is, "absolute zero". Both theory and experiment confirm that this temperature is −273°C.

$$0\ K = -273°C$$

The size of the Kelvin degree is the same as that of the Celsius degree. The relationship between the two scales is:

$$T = t + 273 \tag{5.2}$$

where T is the temperature in K and t the corresponding temperature in °C.

Not surprisingly, many good hockey players come from International Falls.

EXAMPLE 5.1

A meteorologist measures the air temperature on a cold winter morning in International Falls, Minnesota. Using a mercury-in-glass thermometer, she reads a value of −38°C. Express this in K.

SOLUTION

Taking t to be −38, we substitute into Equation 5.2 and find that:

$$T\ (K) = -38 + 273 = 235$$

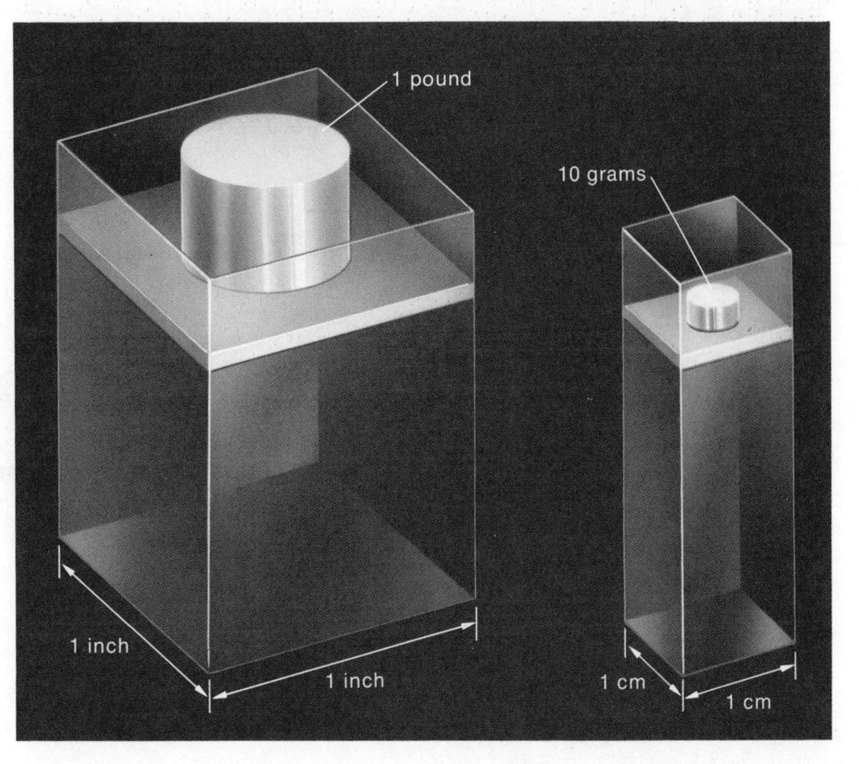

FIGURE 5.1

Pressure is equal to force divided by area. A mass of one pound resting on one square inch of area will exert a pressure downward due to gravity of one pound per square inch. Ten grams on a surface of 1 cm^2 will exert a pressure of about one kilopascal on that surface.

Pressure

Pressure is defined as force per unit area. In the English system, pressure is ordinarily expressed in pounds per square inch (lb/in^2). A mass of one pound resting on a surface one square inch in area exerts a pressure of 1 lb/in^2 (Figure 5.1). A metric pressure unit which is becoming popular nowadays is the *kilopascal* (kPa). A kilopascal is approximately equal to the pressure exerted by a mass of 10 g resting on a surface one square centimeter in area.

We live at the bottom of a sea of air. Air has mass and hence exerts a pressure on the earth's surface and upon our bodies. This pressure is ordinarily greatest at sea level. At higher altitudes, the pressure of the atmosphere is lower because of the smaller mass of air overhead. At Colorado Springs (altitude 1905 m or 6250 ft), normal pressure is only about 80% of that at sea level. Air pressure also varies with weather conditions. It is highest on bright sunny days and drops as a storm approaches.

Ordinarily we are not aware of air pressure, since it is the same inside and outside our bodies. However, if the pressure changes suddenly, we become painfully aware of it. When an airplane takes off, your ears may "pop". Temporarily, the pressure on the inside of your ear drum is greater than that outside. A mountain climber who ascends too quickly is subject to "altitude sickness" because of the pressure difference. Changes in air pressure due to weather patterns have a more subtle effect. Fish bite most eagerly when the pressure is rising (after a thunderstorm, for example). The worst day to fish is just before a storm when the air pressure is falling.

In a tornado, the pressure over a roof may drop suddenly. If it does, the roof flies off.

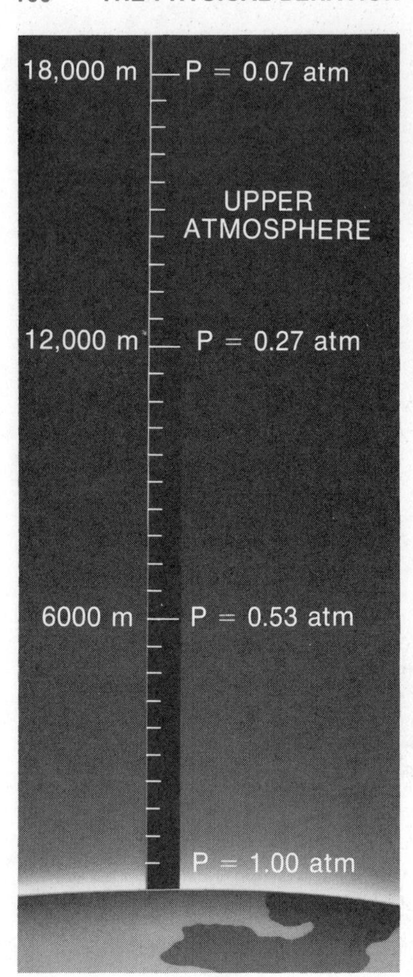

FIGURE 5.2

The pressure at or above the surface of the earth is produced by the mass of the air column that is above that region. Above each square inch of the earth's surface there is a column of air that weighs just about 14.7 lbs.

Torricelli, an Italian physicist, developed in 1643 the first instrument to measure atmospheric pressure accurately. A simple version of this device, called a *barometer,* is shown in Figure 5.3. You can construct your own barometer by taking a glass tube 80 cm long, sealed at one end, and filling it with mercury. Put a stopper in the open end. Carefully invert the tube in a pool of mercury, making sure no air gets into the tube. When the stopper is removed, some of the mercury will flow out of the tube. This continues until the mercury level is a certain height above that of the mercury in the pool. Torricelli realized that this height is a measure of the atmospheric pressure. The pressure of the column of mercury in the tube (P_{Hg}) is exactly balanced by the pressure of the air column above the mercury in the pool (P_{atm}):

$$P_{Hg} = P_{atm}$$

Because of the way atmospheric pressure is measured, it is often expressed in terms of the height of a mercury column. Thus we might say that on a certain day the atmospheric pressure is "752 mm Hg". By this we mean that the pressure of the atmosphere is equal to that

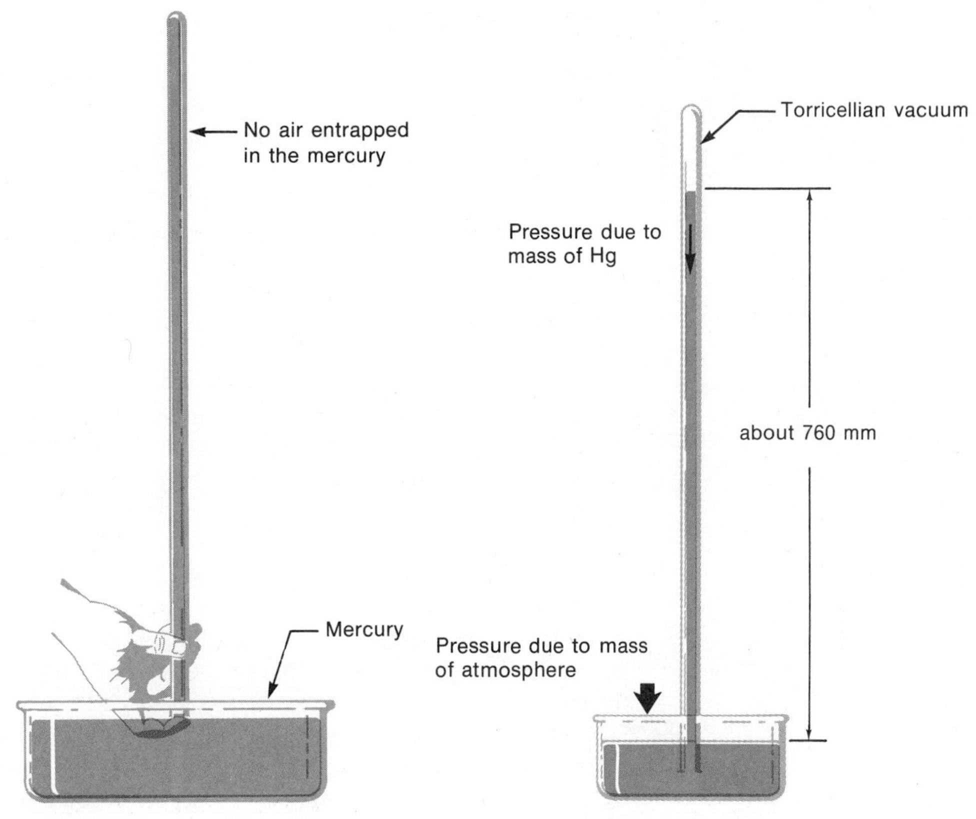

FIGURE 5.3

Making a barometer. The mercury is held up in the tube by the pressure of the atmosphere. The height of the column is a direct measure of the air pressure, since at the lower mercury surface the pressure both inside and outside the tube must have the same value or else the mercury would flow. That pressure is the pressure of the air.

exerted by a column of mercury 752 mm high. Another unit commonly used to express pressure is the *standard atmosphere*, usually referred to simply as an "atmosphere". An atmosphere is defined as the pressure exerted by a column of mercury 760 mm high. In other words:

The force on your body due to air is about equal to that exerted by a 1000 kg mass against the floor on which it rests.

$$1 \text{ atm} = 760 \text{ mm Hg} \tag{5.3}$$

In relation to the pressure units mentioned earlier:

$$1 \text{ atm} = 14.7 \text{ lb/in}^2 = 101.3 \text{ kPa} \tag{5.4}$$

EXAMPLE 5.2

If the air pressure in Colorado Springs is 612 mm Hg, express this in:

a. atmospheres b. kilopascals

SOLUTION

a. $P\ (\text{atm}) = 612\ \text{mm Hg} \times \dfrac{1\ \text{atm}}{760\ \text{mm Hg}} = 0.805\ \text{atm}$

b. $P\ (\text{kPa}) = 0.805\ \text{atm} \times \dfrac{101.3\ \text{kPa}}{1\ \text{atm}} = 81.5\ \text{kPa}$

Partial Pressure

Gases in a container act independently. Each one does its own thing.

Frequently the gases that we work with in the laboratory are mixtures. They contain two or more different substances. The pressure that we measure in this case is the total pressure of the gas mixture. This pressure in turn is equal to the sum of the *partial pressures* of the individual gases. For a mixture of two gases, 1 and 2:

$$P_{tot} = P_1 + P_2 \tag{5.5}$$

Here, P_1 and P_2 are partial pressures. They are defined as *the pressures the gases would exert if they occupied the entire volume of the container by themselves.*

Equation 5.5 was first suggested by John Dalton; it is referred to as Dalton's Law. It is particularly useful in dealing with gases which are collected over water. Color plate 2A, center of book shows a common experiment by which hydrogen gas is obtained in the laboratory. The hydrogen bubbles into the collection bottle, displacing water from it. In the process, the hydrogen becomes saturated with water vapor. The gas in the bottle is a mixture of hydrogen and water vapor. Applying Equation 5.5:

$$P_{tot} = P_{H_2} + P_{H_2O}$$

To calculate the partial pressure of the hydrogen, we subtract the pressure of the water vapor from the total pressure. That is:

$$P_{H_2} = P_{tot} - P_{H_2O}$$

If, for example, $P_{tot} = 754$ mm Hg and $P_{H_2O} = 24$ mm Hg:

$$P_{H_2} = 754\ \text{mm Hg} - 24\ \text{mm Hg} = 730\ \text{mm Hg}$$

In other words, if the hydrogen were "dry" it would exert a pressure of 730 mm Hg.

In the general case, for any gas collected over water:

$$P_{gas} = P_{tot} - P_{H_2O} \tag{5.6}$$

Experimentally, we find that the pressure of water vapor under these conditions is a constant at a given temperature. Hence, we can use the data in Table 5.1 to make the correction called for in Equation 5.6.

TABLE 5.1 VAPOR PRESSURE OF WATER*

t(°C)	15	16	17	18	19	20	21	22	23	24	25
P_{H_2O}(mm Hg)	13	14	15	15	16	18	19	20	21	22	24

*a more extensive table is given on p. 291.

EXAMPLE 5.3

A sample of oxygen gas is collected by bubbling through water at 16°C. The total pressure of the wet gas is equal to the barometric pressure, which is 758 mm Hg. What is the partial pressure of the oxygen?

SOLUTION

At 16°C, P_{H_2O} is 14 mm Hg

$$P_{O_2} = P_{tot} - P_{H_2O} = 758 \text{ mm Hg} - 14 \text{ mm Hg} = 744 \text{ mm Hg}$$

5.2 A MODEL OF THE GAS STATE: THE KINETIC MOLECULAR THEORY

The properties of gases can be explained in terms of the behavior of molecules. In this section we will consider a model of the gas state based on molecular motion. To develop this model, we start by considering an important feature of the gas state. This is the large distance between molecules.

Distance between Molecules

There is a great deal of evidence to indicate that molecules in a gas are much farther apart than in a liquid or solid. In particular, we find that:

1. *Gases, at ordinary temperatures and pressures, have much lower densities* than liquids or solids. A flask filled with air weighs much less than the same flask filled with water or table salt. To quote some numbers, consider the element oxygen. In the solid, where the O_2 molecules are touching one another, the density is about 1.43 g/cm^3. In the gas, on the other hand, the density at 0°C and 1 atm is only 1/1000 as great, 0.00143 g/cm^3. This suggests that the molecules in the gas must be widely separated. Only about 1/1000 of the total volume in gaseous oxygen is occupied by the molecules themselves. In other words, 99.9% of the volume in the gas is empty space.

2. *Gases are much more compressible than liquids or solids.* Doubling the pressure on a gas from 1 to 2 atm reduces its volume by

one half. The same increase in pressure decreases the volume of liquid water by only 0.004%. In a gas, where most of the space is unoccupied to begin with, an increase in pressure produces a drastic decrease in volume. Liquids, where the molecules are in contact with each other, are almost incompressible. You become painfully aware of this if you do a "belly flop" off a diving board into the water of a swimming pool.

3. *Gases diffuse into each other more rapidly than do liquids or solids.* Mixing occurs quickly with gases because the widely separated molecules can move past one another readily. In the liquid state, where the molecules are close together, mixing is slowed down by frequent collisions (Color plate 2B, center of book). There is very little free space for molecules to move through. Diffusion in solids is extremely slow because the molecules are fixed in position as well as being close together.

These observations suggest that, at room temperature and atmospheric pressure, molecules in a gas are perhaps 10 molecular diameters apart. The distance between molecules can be reduced by increasing the pressure on the gas or by cooling it. Conversely, if the pressure is decreased or the temperature raised, the molecules move farther apart.

Molecular Motion

The fact that gases diffuse readily implies that their molecules are in rapid motion. The nature of this motion can be shown simply (Figure 5.4). Here, a microscope is used to follow the movement of finely divided smoke particles suspended in a gas such as air. The smoke particles bounce around in an irregular pattern. This type of movement is known as Brownian motion after its discoverer, Robert Brown, a Scottish botanist. Brownian motion can be explained most simply by taking a gas to be a system of molecules in motion. The molecules frequently collide with the larger smoke particles and set them in motion. The zig-zag path of the smoke particles indicates that the molecules themselves are moving randomly in all directions.

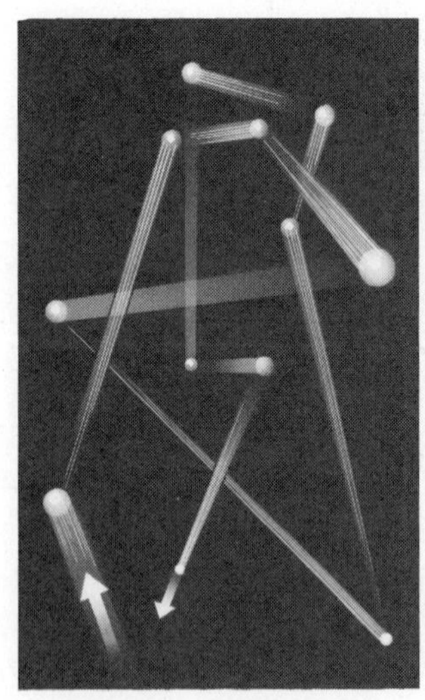

FIGURE 5.4

In still air a smoke particle will move in a random manner like that in the sketch. This is called Brownian motion and is caused by collisions of the smoke particle with air molecules.

As pointed out in Chapter 4, a gas molecule in motion has translational kinetic energy. This type of energy is related to mass, m, and speed, v, by the equation:

$$\text{translational kinetic energy} = mv^2/2 \qquad (5.7)$$

Gas molecules, by ordinary standards, are moving at enormous speeds. For example, at 0°C oxygen molecules on the average are moving at a speed of 461 m/s, about 1000 miles per hour. However, since gas molecules have very small masses (m O_2 molecule = 5.3×10^{-23} g), their translational kinetic energies are very small.

As the molecules of a gas move about, they collide with each other and with the walls of their container. In this sense, they resemble billiard balls moving around on a table. However, gas molecules in motion differ from billiard balls in one important respect. Molecular collisions in a

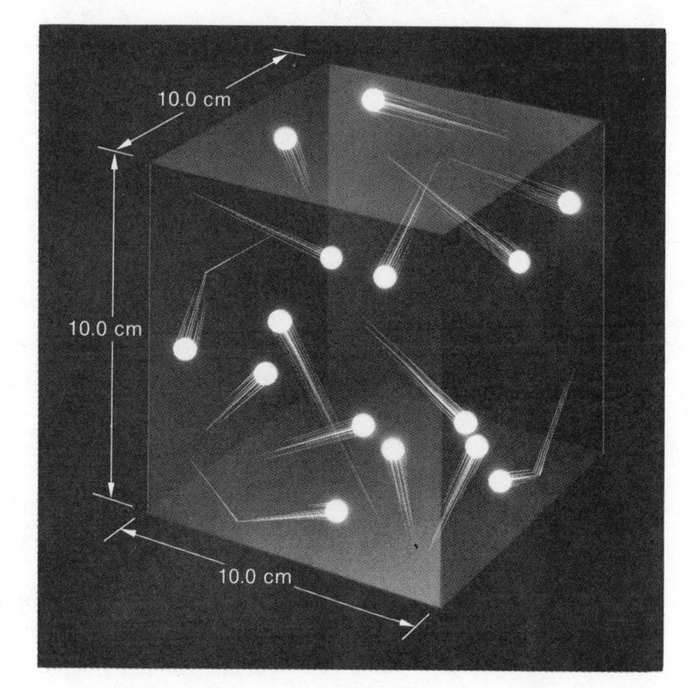

FIGURE 5.5

The pressure exerted by a gas on its container is the same in all directions and is caused by collisions of the gas molecules with the container walls. The force per collision is very small but there are a great many collisions per second on any given area.

gas are perfectly *elastic*. That is, they occur without any overall loss in translational kinetic energy. As the result of collisions, some molecules may speed up while others slow down. However, the total translational kinetic energy after many collisions is the same as it was before.

Since energy of motion is conserved in collisions, molecules, unlike billiard balls, do not eventually come to rest. This is shown by the fact that the pressure of a gas does not change with time. From a molecular standpoint, gas pressure results from collisions of molecules with the walls of a container (Figure 5.5). If such collisions led to a net loss in translational kinetic energy, we would expect pressure to decrease as the molecules slow down.

In air, one molecule undergoes about 10 billion collisions each second.

Our model for molecular motion in gases suggests that, at any given instant, different molecules are moving at different speeds. Like electric bumper cars at an amusement park, some molecules are slowed down by collisions while others are speeded up. In a sample of a gas, a few molecules are moving very slowly, others very rapidly, and still others at intermediate speeds. As we mentioned, the average speed of an O_2 molecule at 0°C is 461 m/s. Most of the molecules (nearly 90% of them) have speeds in the range 200–800 m/s. A very few molecules (about 0.5% of the total) have speeds greater than 1000 m/s.

When the temperature of a gas is increased, it absorbs heat. This raises the translational kinetic energy of the gas molecules. In other words, the molecules move faster at higher temperatures. The effect on average speed is rather small. Consider, for example, $O_2(g)$. When the temperature rises from 0 to 25°C, the average speed goes from 461 to 482 m/s, an increase of less than 5%. (However, an increase in temperature does strongly affect the fraction of molecules having very high energies. We will have more to say on this topic in Chapter 17.)

The Kinetic Molecular Theory of Gases

The model of the gas state just discussed was developed in the last half of the 19th century. It is the basis of the kinetic molecular theory and explains the behavior of gases in terms of molecular motions and distances. It has been one of our most successful models in helping to understand the behavior of matter.

The basic postulates of the kinetic molecular theory are summarized below. All of these were discussed earlier in this section.

As with atomic theory, the postulates of the kinetic theory are quite simple.

1. The molecules in a gas are relatively far apart. At ordinary temperatures and pressures, a gas is mostly empty space.

2. Gas molecules are in continuous, rapid, linear motion. They have a translational kinetic energy, $mv^2/2$, directly related to their speed, v.

3. Collisions between molecules and with the walls of their container are perfectly elastic. The pressure of a gas results from collisions with the walls.

4. In a gas sample, molecules are moving at a variety of different speeds. An increase in temperature increases the average speed.

Average Speed of Gas Molecules

We have described the kinetic theory of gases in a general way, using no mathematics. For the purposes of this course, that is sufficient. However, it is possible to obtain some very useful equations from the theory. One of these is an expression for the average speed of a gas molecule:

$$u = 158\left(\frac{T}{MM}\right)^{1/2} \qquad (5.8)$$

Here: u is the average speed in meters per second
T is the temperature in K
MM is the molecular mass

This equation can be used to calculate the average speed of a gas molecule at a given temperature. Consider, for example, an O_2 molecule, where MM = 32.0.

$$u\ O_2(0°C) = 158\left(\frac{273}{32.0}\right)^{1/2} = 158(2.92) = 461 \text{ m/s}$$

$$u\ O_2(25°C) = 158\left(\frac{298}{32.0}\right)^{1/2} = 158(3.05) = 482 \text{ m/s}$$

There is one particularly important feature of Equation 5.8. It tells us that the average speed is *inversely* related to molecular mass. At a given temperature, light molecules move faster than heavy ones. This is shown by the data in Table 5.2. Notice that the average speed of an H_2 molecule is about four times that of an O_2 molecule. Conversely, molecules of CO_2 are moving somewhat more slowly than O_2 molecules.

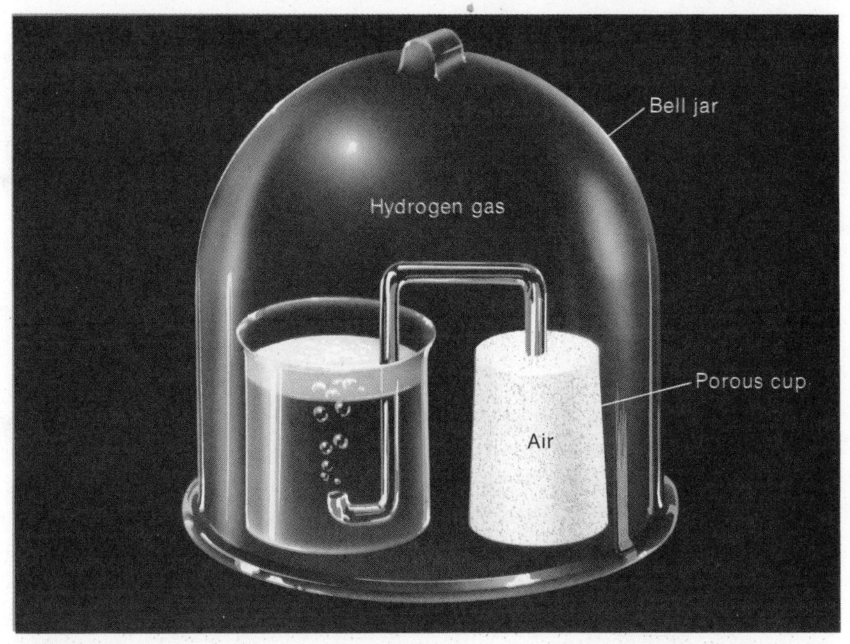

FIGURE 5.6

Diffusion of gases. If a bell jar full of hydrogen is brought down over a porous cup full of air, hydrogen will diffuse into the cup faster than the oxygen and nitrogen in the air can diffuse out of the cup. This causes an increase in pressure in the cup sufficient to produce bubbles in the water in the beaker.

TABLE 5.2 AVERAGE SPEEDS OF GAS MOLECULES AT 0°C

	MM	$(T/MM)^{1/2}$	u (m/s)
H_2	2.02	11.6	1830
He	4.00	8.26	1300
CH_4	16.0	4.13	653
N_2	28.0	3.12	493
O_2	32.0	2.92	461
CO_2	44.0	2.49	394

The fact that lighter molecules move faster than heavy ones is easily demonstrated in the laboratory (Figure 5.6). A bell jar containing H_2 is brought down over a porous cup originally full of air (N_2 and O_2). The hydrogen diffuses into the cup faster than the air moves out. This raises the pressure in the cup. A stream of gas leaves the cup and produces bubbles in the water.

5.3 RELATION BETWEEN VOLUME, PRESSURE, AND TEMPERATURE

We have seen how certain properties of gases can be explained by the kinetic molecular theory. In this section we will look at the properties of gases from a different point of view. We will consider how they can be expressed by simple equations which can be tested in the laboratory. In particular, we will develop equations to relate the volume occupied by a gas to its pressure and temperature.

The relationship between the volume of a sample of gas and its pressure was first studied by an English scientist, Robert Boyle, late in the 17th century. About 100 years later, two French scientists, Jacques Charles and Joseph Gay-Lussac, studied the effect of temperature upon the volume of a gas.

Things happened more slowly in those days.

Volume *versus* Pressure

Experience tells us that the volume of a sample of gas at constant temperature is *inversely* related to its pressure. If we increase the pressure on the air in a bicycle pump by pushing down on the piston, the volume of the air decreases. Again, suppose we open the valve on a cylinder of compressed gas, thereby decreasing the pressure. The gas rushes out into the room where it occupies a larger volume. In both cases, the volume changes in a direction opposite to that of the change in pressure.

The inverse relation between gas volume and pressure is simply explained by the kinetic molecular theory (Figure 5.7). When the volume available to gas molecules is reduced, they strike the walls more frequently. Since gas pressure is caused by wall collisions, it must increase when the volume is reduced.

To develop the relationship between volume and pressure, we change the pressure on a gas sample and see how the volume changes. In general, we find that the volume decreases as the pressure increases. Typical data are plotted in Figure 5.8.

The graph in Figure 5.8 shows that:

The volume occupied by a gas sample at constant temperature is inversely proportional to the pressure exerted upon it.

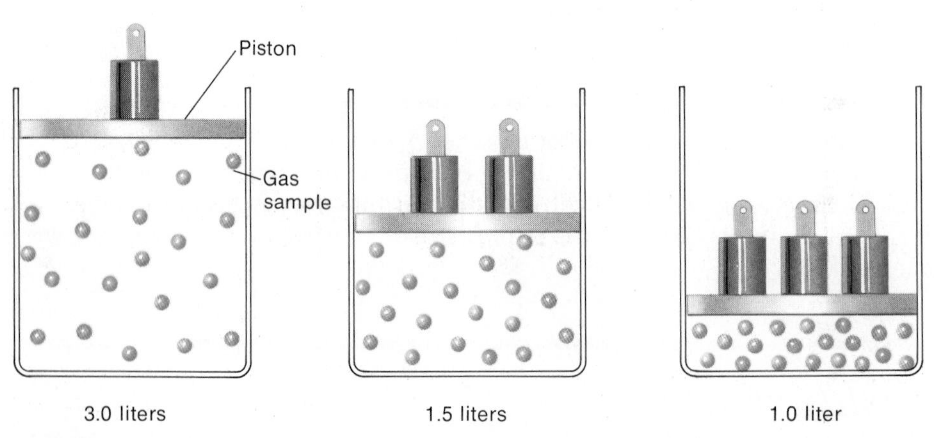

FIGURE 5.7

If the volume of a gas is decreased, its pressure goes up. If the volume is cut in half, there will be twice as many molecules in a given volume as there were before. This produces twice as many collisions on any surface of the container and so increases the pressure by a factor of two.

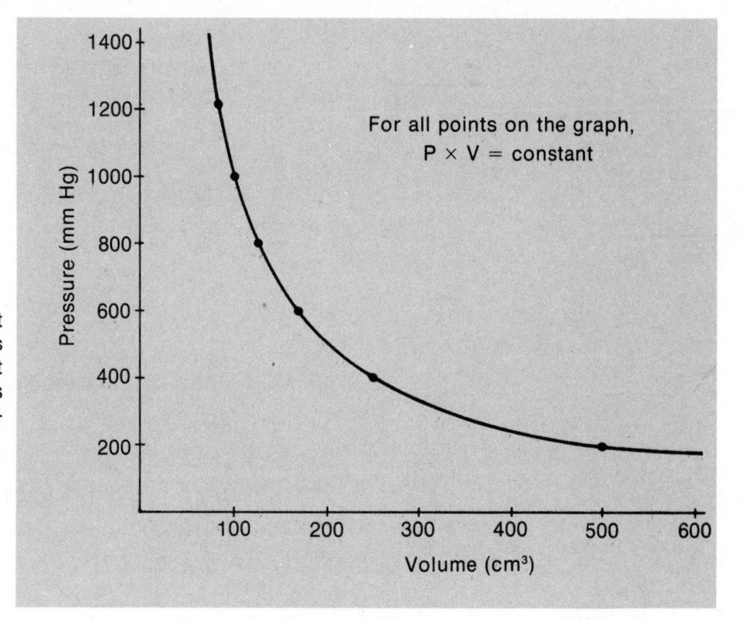

FIGURE 5.8

Boyle's Law. When a gas is compressed at constant temperature, its pressure increases as the volume decreases in such a way that the product of pressure and volume remains constant. You can verify this Law by applying it to the points on the graph.

Mathematically, we can express this relation in any of several ways. We can say that:

$$V = \frac{k_1}{P} \quad \text{(T and amount of gas are constant)}$$

where k_1 is a constant, independent of the values of P and V. We can then multiply both sides of this equation by P and write:

$$PV = k_1 \quad \text{(T and amount of gas are constant)} \qquad (5.9)$$

Equation 5.9 is referred to as Boyle's Law. For use in calculations, we express the Law in a somewhat different form. To relate pressure and volume at two different points, 1 and 2, we write:

Boyle found his law by experimenting with air. In 1660, that was the only gas known.

$$P_1V_1 = k_1 \qquad \text{and} \qquad P_2V_2 = k_1$$

Since P_1V_1 and P_2V_2 are both equal to k_1, they must be equal to each other. That is:

$$P_2V_2 = P_1V_1 \quad \text{(T and amount of gas are constant)} \qquad (5.10)$$

where the subscripts 1 and 2 refer to two different sets of conditions.

EXAMPLE 5.4

A weather balloon is filled with helium to a volume of 250 ℓ at sea level where the pressure is 760 mm Hg. Assuming no temperature change, what will be the volume when the balloon rises to a height of 25 km, where the pressure is only 10 mm Hg?

SOLUTION

Using the subscript 1 to represent initial conditions and 2 for final conditions, we have:

$$P_1 = 760 \text{ mm Hg}; \; V_1 = 250 \; \ell; \; P_2 = 10 \text{ mm Hg}; \; V_2 = ?$$

Clearly we need to solve Equation 5.10 for V_2. To do this, we divide both sides by P_2 and obtain:

$$V_2 = V_1 \times \frac{P_1}{P_2} = 250 \; \ell \times \frac{760 \text{ mm Hg}}{10 \text{ mm Hg}} = 1.9 \times 10^4 \; \ell$$

Two comments are in order about Example 5.4.

1. It doesn't matter what units are used for pressure in a calculation of this type, provided both units are the same. P_2 and P_1 could both be expressed in millimeters of mercury, in atmospheres, and so on. We could not, however, express P_2 and P_1 in different units (P_1 in atmospheres, P_2 in millimeters of mercury). In that case, the units would not cancel.

2. Since the final pressure is less than the initial pressure, we would expect a volume increase. This is indeed the case; $1.9 \times 10^4 \; \ell > 250 \; \ell$. In solving problems like this, it is always a good idea to check the magnitude of the answer. That way you can be sure you haven't made a gross error in arithmetic.

Volume *versus* Temperature

A simple balloon experiment shows that the volume of a gas sample maintained at constant pressure is directly related to temperature. If the air in the balloon is heated, it expands. In other words, an increase in temperature produces an increase in volume. Conversely, if the balloon is cooled (perhaps by putting it in the refrigerator) it contracts. That is, a decrease in temperature results in a decrease in volume.

The effect of temperature upon the volume of a gas can be explained in terms of the kinetic molecular theory. As we saw in Section 5.2, an increase in temperature raises the speed of gas molecules. Hence, they strike the walls of the container more often and with greater force. If we held the volume constant, the pressure of the gas would increase. To maintain a constant pressure at the higher temperature, we must move the walls of the container farther apart. That is, the volume available to the gas molecules must increase.

The relationship between temperature and volume can be obtained from data such as those given in Table 5.3. Here we measure the volume of a gas sample, maintained at constant pressure, as the temperature is increased by 50° intervals. In the second column, temperatures are listed in °C. These values are converted, in the third column, to the absolute scale (K) by adding 273.

TABLE 5.3 DEPENDENCE OF GAS VOLUME, V, ON ABSOLUTE TEMPERATURE, T

V (cm^3)	t (°C)	T (K)
100	0	273
118	50	323
137	100	373
155	150	423
173	200	473

In Figure 5.9, the data listed in Table 5.3 are plotted. The graph is a straight line which, if continued, would pass through the origin. That is, V = 0 when T = 0. The equation of the straight line is:

$$V = k_2T \qquad \text{(P and amount of gas are constant)}$$

Absolute temperature was really defined so as to make this equation valid.

where k_2 is the slope of the straight line. In general, we can say that:

The volume of a gas sample maintained at constant pressure is directly proportional to its absolute temperature (K).

If we divide both sides of the equation written above by T, we obtain:

$$\frac{V}{T} = k_2 \qquad (5.11)$$

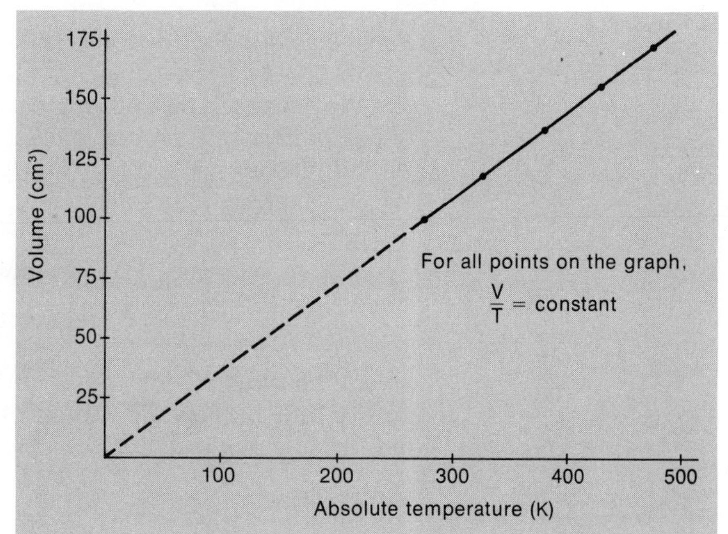

FIGURE 5.9

Charles' Law. When a gas is heated at constant pressure, its volume increases in such a way that the ratio of the volume to the absolute temperature remains constant. If the line joining the points obtained experimentally is extended to low temperatures, it will intersect the V = 0 point at 0 K.

Equation 5.11 is referred to as Charles' Law. To obtain a relation between volume and temperature at two different points:

$$\frac{V_1}{T_1} = k_2 \qquad \text{and} \qquad \frac{V_2}{T_2} = k_2$$

or: $$\frac{V_1}{T_1} = \frac{V_2}{T_2} \qquad \text{(P and amount of gas are constant)} \qquad (5.12)$$

EXAMPLE 5.5

Consider the 250 ℓ weather balloon referred to in Example 5.4. What would its volume become if the temperature at the launch site dropped from 27°C to −13°C?

SOLUTION

$$V_1 = 250\ \ell;\ T_1 = 27 + 273 = 300\ \text{K}$$

$$V_2 = \ ?\ ;\ T_2 = -13 + 273 = 260\ \text{K}$$

To solve Equation 5.12 for V_2, we multiply both sides by T_2 and obtain:

$$V_2 = V_1 \times \frac{T_2}{T_1} = 250\ \ell \times \frac{260\ \text{K}}{300\ \text{K}} = 217\ \ell$$

If you work Example 5.5 using °C, the volume will be negative, which is ridiculous.

Notice that in Example 5.5:

1. The final volume (217 ℓ) must be less than the initial volume (250 ℓ) because the temperature decreased.

2. In working this problem or others of this type, *temperatures must be expressed in K.*

Volume *versus* Pressure and Temperature

We have seen that, at constant temperature, the volume of a sample of gas is inversely proportional to its pressure. At constant pressure, the volume of a gas is directly proportional to its absolute temperature. We can express these relations as:

$$V \propto \frac{1}{P} \qquad \text{(T and amount of gas are constant)}$$

$$V \propto T \qquad \text{(P and amount of gas are constant)}$$

where the symbol ∝ means "is directly proportional to."

To find out how the volume of a gas varies with both pressure and temperature, we combine these two relations to obtain:

$$V \propto \frac{T}{P}$$

or: $$V = k_3\frac{T}{P} \quad \text{(constant amount of gas)} \qquad (5.13)$$

where k_3, like k_2 and k_1, is a proportionality constant.

To obtain a relationship between V, T, and P under two sets of conditions, let us solve Equation 5.13 for k_3. To do this, we multiply both sides of the equation by P/T:

$$V \times \frac{P}{T} = k_3\frac{T}{P} \times \frac{P}{T} = k_3$$

For an "initial" set of conditions (subscript 1):

$$\frac{V_1 \times P_1}{T_1} = k_3$$

and for a "final" set of conditions (subscript 2):

$$\frac{V_2 \times P_2}{T_2} = k_3$$

Hence, the general relation is:

$$\frac{V_2 \times P_2}{T_2} = \frac{V_1 \times P_1}{T_1} \quad \text{(constant amount of gas)} \qquad (5.14)$$

Since both expressions equal k_3, they are equal to each other.

Notice that Equation 5.14 reduces to Boyle's Law ($P_2V_2 = P_1V_1$) when T is constant ($T_2 = T_1$). Moreover, it reduces to Charles' Law ($V_2/T_2 = V_1/T_1$) when P is constant ($P_2 = P_1$).

Equation 5.14 can be used to solve for a final volume, V_2, knowing the initial volume, V_1, and both final and initial pressures and temperatures. This type of calculation is shown in Example 5.6.

EXAMPLE 5.6

A sample of gas occupies a volume of 26.0 cm³ at a pressure of 1.16 atm and a temperature of 0°C. What will be its volume at a pressure of 740 mm Hg and a temperature of 50°C?

SOLUTION

Before attempting to solve this problem, you must realize that:

—temperatures must be in K.
—the two pressures must be in the same units. We could convert 1.16 atm to mm Hg or 740 mm Hg to atm. Let's do the latter:

$$P_1 = 1.16 \text{ atm}; P_2 = 740 \text{ mm Hg} \times \frac{1 \text{ atm}}{760 \text{ mm Hg}} = 0.974 \text{ atm}$$

$$T_1 = 0 + 273 = 273 \text{ K}; T_2 = 50 + 273 = 323 \text{ K}$$

$$V_1 = 26.0 \text{ cm}^3; V_2 = ?$$

Clearly, we need to solve Equation 5.14 for V_2. To do that, we multiply both sides of the equation by T_2/P_2:

$$V_2 \times \frac{P_2}{T_2} \times \frac{T_2}{P_2} = V_1 \times \frac{P_1}{T_1} \times \frac{T_2}{P_2}$$

or:

$$V_2 = V_1 \times \frac{P_1}{P_2} \times \frac{T_2}{T_1}$$

Substituting numbers:

$$V_2 = 26.0 \text{ cm}^3 \times \frac{1.16 \text{ atm}}{0.974 \text{ atm}} \times \frac{323 \text{ K}}{273 \text{ K}} = 36.6 \text{ cm}^3$$

Notice that both the decrease in pressure (1.16 atm to 0.974 atm) and the increase in temperature (273 K to 323 K) tend to increase the volume. Hence you shouldn't be surprised to find that the final volume, 36.6 cm^3, is considerably larger than the initial volume, 26.0 cm^3.

In general, Equation 5.14 can be used to solve for any one quantity, knowing the values of the other five quantities in the equation. Example 5.7 illustrates such a calculation.

EXAMPLE 5.7

An automobile tire at 15°C contains 106 ℓ of air at a total pressure of 42.7 lb/in^2. After driving for some time, the tire heats up to 52°C and expands slightly to a volume of 108 ℓ. What is the air pressure inside the tire under these conditions?

SOLUTION

As usual, it is helpful to start by writing down the value of each of the quantities in Equation 5.14.

$$V_1 = 106\ \ell;\ V_2 = 108\ \ell$$

$$T_1 = 15 + 273 = 288 \text{ K};\ T_2 = 52 + 273 = 325 \text{ K}$$

$$P_1 = 42.7 \text{ lb/in}^2;\ P_2 = ?$$

This is a general system for setting up gas law problems.

To solve Equation 5.14 for P_2, we multiply both sides by T_2/V_2:

$$V_2 \times \frac{P_2}{T_2} \times \frac{T_2}{V_2} = V_1 \times \frac{P_1}{T_1} \times \frac{T_2}{V_2}$$

$$P_2 = P_1 \times \frac{V_1}{V_2} \times \frac{T_2}{T_1}$$

$$= 42.7 \text{ lb/in}^2 \times \frac{106\ \ell}{108\ \ell} \times \frac{325 \text{ K}}{288 \text{ K}} = 47.3 \text{ lb/in}^2$$

Notice that the final pressure is somewhat greater than it was originally. The effect of the increase in temperature, which tends to raise the pressure, is greater than that of the expansion, which tends to lower the pressure. The result agrees with experience. Tire pressure ordinarily increases after you drive for some time.

Incidentally, you may have worried that both pressures, 42.7 and 47.3 lb/in^2, seem high. A tire gauge measures pressure in excess of atmospheric pressure, 14.7 lb/in². The gauge pressures that you would read here are:

$$42.7 \text{ lb/in}^2 - 14.7 \text{ lb/in}^2 = 28.0 \text{ lb/in}^2$$

$$47.3 \text{ lb/in}^2 - 14.7 \text{ lb/in}^2 = 32.6 \text{ lb/in}^2$$

5.4 MOLAR VOLUMES OF GASES

In Section 5.3, we considered how the volume of a *gas sample of fixed mass* varies with pressure and temperature. As you know, the volume of a gas is also related to its mass. Indeed, at constant pressure and temperature, volume is directly proportional to mass. The volume of two grams of oxygen is twice as great as that of one gram. By the same token, the mass of air in a 10-ℓ box is twice as great as that in a 5-ℓ box at the same temperature and pressure.

Not too surprising, when you think about it.

The relationship between volume and mass or amount of gas could be expressed in many different ways. The most useful approach is to consider the molar volume, that is, the volume occupied by one mole of gas. The molar volume is related to the molar mass and density of the gas by the equation:

$$\text{molar volume } (\ell) = \frac{\text{molar mass (g)}}{\text{density } (\text{g}/\ell)} \qquad (5.15)$$

For O_2 (MM = 32.00) at 0°C and 1 atm, where the density is 1.429 g/ℓ, we have:

$$\text{molar volume } O_2 \text{ at 0°C, 1 atm} = \frac{32.00 \text{ g}}{1.429 \text{ g}/\ell} = 22.39 \ \ell$$

Table 5.4 lists the densities at 0°C and 1 atm and the molar masses of a series of different gases. In each case, we can calculate a molar volume at 0°C and 1 atm, using Equation 5.15.

TABLE 5.4 DENSITIES AND MOLAR VOLUMES OF GASES AT 0°C AND 1 ATM

Gas	Density (g/ℓ)	Molar Mass (g)	Molar Volume (ℓ)
Ar	1.783	39.95	22.41
CH_4	0.7168	16.04	22.38
C_2H_6	1.354	30.07	22.21
CO	1.250	28.01	22.41
CO_2	1.979	44.01	22.24
H_2	0.0898	2.016	22.45
He	0.1785	4.003	22.43
N_2	1.250	28.01	22.41
NO	1.340	30.01	22.40
O_2	1.429	32.00	22.39

Looking at Table 5.4, we see that, within about one percent:

This is remarkable, considering the range of molar masses of common gases.

All gases at 0°C and 1 atm have the same molar volume, 22.4 ℓ.

This temperature and pressure, 0°C and 1 atm, are sometimes referred to as *standard conditions* or STP. Thus we would say that the volume of one mole of any gas at STP is 22.4 ℓ or:

$$V(\text{molar}) = 22.4\ \ell \text{ at STP} \qquad (5.16)$$

Equation 5.16 allows us to relate the volume of any gas at STP to its mass. Example 5.8 illustrates how this is done.

EXAMPLE 5.8

Consider ammonia, NH_3, which has a molecular mass of 17.0. Determine:

a. the volume at STP of 2.56 g of NH_3.
b. the mass of 1.00 ℓ of NH_3 at STP.

SOLUTION

At STP (0°C, 1 atm): 22.4 ℓ NH_3 = 17.0 g NH_3
This relation gives us the conversion factors we need.

a. $2.56 \text{ g } NH_3 \times \frac{22.4\ \ell\ NH_3}{17.0 \text{ g } NH_3} = 3.37\ \ell\ NH_3$

b. $1.00\ \ell\ NH_3 \times \frac{17.0 \text{ g } NH_3}{22.4\ \ell\ NH_3} = 0.759 \text{ g } NH_3$

The relationship we have just discussed can be made more general. **All gases, at the same temperature and pressure, have the same molar volume.** The molar volume is 22.4 ℓ at 0°C and 1 atm. At other temperatures and pressures, it will have different values. For example, at 25°C and 1 atm:

$$V(\text{molar}) = 22.4\ \ell \times \frac{298 \text{ K}}{273 \text{ K}} = 24.5\ \ell$$

The equality of molar volumes of gases was first recognized more than 150 years ago by Avogadro. He expressed it somewhat differently. Avogadro's Law states that *equal volumes of all gases at the same temperature and pressure contain the same number of molecules.* Clearly this requires that one mole (6.02×10^{23} molecules) of different gases must occupy the same volume at a given temperature and pressure.

Relation between Molecular Mass and Density

The fact that all gases have the same molar volume at the same temperature and pressure has several important applications. We will explore only one of these. This is the relationship between the molecular mass of a gas and its density.

Since volume is mass/density, the quantity:

$$\frac{MM}{d}$$

(MM = molecular mass, d = density) must be the same for all gases at a given temperature and pressure. Thus for the two gases, 1 and 2:

$$\frac{(MM)_2}{d_2} = \frac{(MM)_1}{d_1}$$

or, rearranging to a more convenient form:

$$\frac{(MM)_2}{(MM)_1} = \frac{d_2}{d_1} \quad \text{(constant T and P)} \qquad (5.17)$$

Equation 5.17 tells us that **the molecular masses of two gases are in the same ratio as their densities.** This relationship suggests a simple approach to the determination of the molecular mass of a gas (Example 5.9).

EXAMPLE 5.9

The density of O_2 at 0°C and 1 atm is 1.43 g/ℓ. Using Equation 5.17, determine the molecular mass of a gas which has a density of 1.70 g/ℓ at the same temperature and pressure.

SOLUTION

Let us use the subscript 1 for O_2, 2 for the other gas. Solving Equation 5.17 for $(MM)_2$, we have:

$$(MM)_2 = (MM)_1 \times \frac{d_2}{d_1}$$

But, $(MM)_1$ = MM of O_2 = 32.0; d_2 = 1.70 g/ℓ; d_1 = 1.43 g/ℓ

$$(MM)_2 = 32.0 \times \frac{1.70\ g/\ell}{1.43\ g/\ell} = 38.0$$

Ideal and Real Gases

The equations we have developed in this chapter for the volume of a gas apply strictly to an *ideal* gas. An ideal gas can be defined as one for which the volume, V, is:

—directly proportional to the number of moles, n
—directly proportional to the absolute temperature, T
—inversely proportional to the pressure, P

Combining these relations, we can say that an ideal gas is one for which:

$$V = \frac{\text{constant} \times n \times T}{P}$$

This equation is ordinarily written in a different form. Both sides of the equation are multiplied by P to clear of fractions. Moreover, the "constant" in the equation, which is the same for all gases, is given the symbol R. In this way, we arrive at what is known as the Ideal Gas Law:

$$PV = nRT \tag{5.18}$$

The constant R can be evaluated. We know that one mole ($n = 1$) of an ideal gas occupies a volume of 22.4 ℓ ($V = 22.4\ \ell$) at STP ($T = 273$ K, $P = 1$ atm). Substituting these quantities in Equation 5.18:

$$(1\ \text{atm})(22.4\ \ell) = (1\ \text{mol})\ R\ (273\ \text{K})$$

Solving for R:

$$R = \frac{(1\ \text{atm})(22.4\ \ell)}{(1\ \text{mol})(273\ \text{K})} = 0.0821\ \frac{\ell \cdot \text{atm}}{\text{mol} \cdot \text{K}} \tag{5.19}$$

Using this value of R, we can calculate the volume of an ideal gas under any conditions. For example, if we want to know the volume occupied by 0.100 mol of gas at 25°C (298 K) and 2.00 atm:

$$V = \frac{nRT}{P} = \frac{(0.100\ \text{mol})(0.0821\ \ell \cdot \text{atm/mol} \cdot \text{K})(298\ \text{K})}{2.00\ \text{atm}} = 1.22\ \ell$$

In practice no gas is entirely ideal. (The same can be said for teachers, parents, and perhaps even students). Real gases deviate at least slightly from Equation 5.18. Under ordinary conditions, the deviations are quite small. Looking back at Table 5.4, we see that the molar volumes of all the gases listed are close to the ideal value of 22.4 ℓ. Consider, for example, O_2, where the real value is 22.39 ℓ. This is about as ideal as you can get. In a couple of cases (CO_2, C_2H_6), the agreement is not quite as good. The real value (22.24 ℓ, 22.21 ℓ) is about 1% lower than the ideal.

Experimentally, we find that real gases behave less ideally at:

—*high pressures.* At 100 atm and 0°C, the molar volume of O_2 is about 8% less than that calculated from the Ideal Gas Law.

—*low temperatures.* At the boiling point (−183°C, 1 atm) $O_2(g)$ has a molar volume about 3% less than that calculated from the Ideal Gas Law.

These effects are readily explained. At high pressures the molecules in a gas are rather close together. At low temperatures, the molecules are moving relatively slowly. Under these conditions our simple model of gases based on the kinetic theory begins to break down. In particular, we can no longer neglect two factors:

1. *The volume occupied by the molecules themselves.* This is a tiny fraction of the total volume under ordinary conditions. When the molecules are close together (high P), the fraction becomes significant.

2. *Attractive forces between molecules.* Under ordinary conditions, the molecules are so far apart that these forces are negligible. As the temperature is lowered, the molecules move more slowly and so are more influenced by attractive forces. Indeed, if we cool a gas sufficiently, it condenses to a liquid. When that happens, the Ideal Gas Law goes out the window.

5.5 VOLUMES OF GASES INVOLVED IN REACTIONS

In Chapter 3 we considered how the masses of reactants and products in a reaction are related to each other. When we deal with gases, it is more convenient to measure volumes rather than masses. It is possible to develop a simple relationship between reacting volumes of gases. We will now consider the nature of this relation.

Law of Combining Volumes

When a direct electric current is passed through water to which a small amount of sulfuric acid has been added, a chemical reaction takes place. Two different gases are produced. Elementary hydrogen is formed at one electrode, oxygen at the other (Figure 5.10, p. 126). The balanced equation for the reaction is:

$$2\ H_2O(l) \rightarrow 2\ H_2(g) + O_2(g)$$

From the coefficients of the equation, we see that two moles of hydrogen are formed for every mole of oxygen. In other words, the mole ratio of H_2 to O_2 is 2:1.

Looking at Figure 5.10, it appears that the volume of hydrogen formed is about twice that of oxygen. Indeed, if we measure the volumes carefully, we find that they are in a 2:1 ratio. There is a simple explanation for this. Remember that the molar volume is the same for all gases at the same temperature and pressure. At STP, it is 22.4 ℓ. If, by electrolysis, we produce 2 mol of H_2 and 1 mol of O_2 at STP:

This simple experiment supports both the atomic theory and the laws of gas behavior.

$$\text{Volume } H_2 = 2 \times 22.4\ \ell = 44.8\ \ell$$

$$\text{Volume } O_2 = 1 \times 22.4\ \ell = 22.4\ \ell$$

So:

$$\frac{\text{Volume } H_2}{\text{Volume } O_2} = \frac{44.8\ \ell}{22.4\ \ell} = 2{:}1$$

The same argument can be applied at any temperature and pressure. Indeed, however we conduct the electrolysis, the volume ratio of H_2 to O_2 is the same as the mole ratio, 2:1. If enough water is electrolyzed to form 120 cm^3 of H_2, 60 cm^3 of O_2 is produced at the same time. If one liter of O_2 is formed, we get two liters of H_2. Moreover, we can extend this argument to any reaction involving gases:

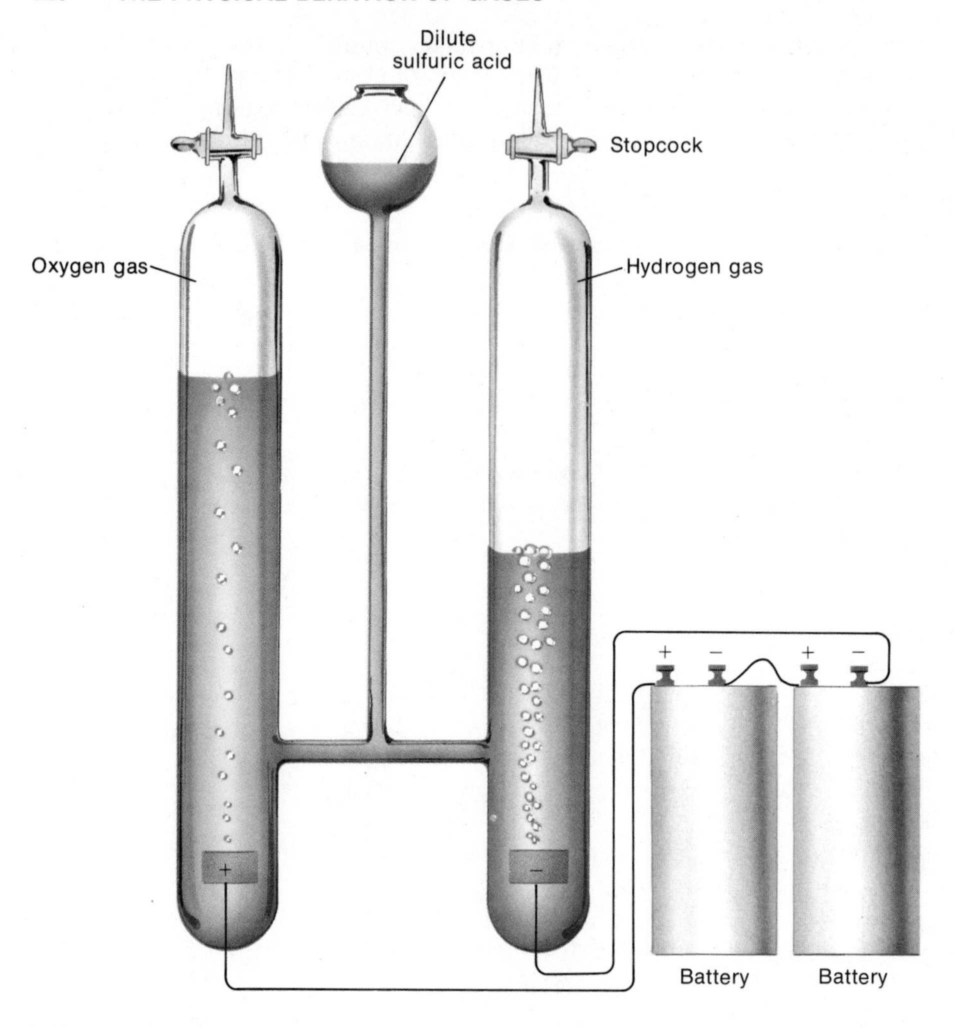

FIGURE 5.10

If an electric current is passed through water, as in the sketch, the water is decomposed to the elements hydrogen and oxygen. The ratio of the volumes of hydrogen and oxygen is found to be 2:1, which is equal to the ratio of number of moles of H_2 and O_2 produced in the decomposition reaction: $2\ H_2O(l) \rightarrow 2\ H_2(g) + 1\ O_2(g)$.

The volumes of gases involved in chemical reactions, measured at the same temperature and pressure, are in the same ratio as their coefficients in the balanced equation.

John Dalton didn't believe this law. His arguments against it held chemistry back for 50 years.

This general principle, known as the Law of Combining Volumes, was discovered by Gay-Lussac in 1808. Its usefulness is illustrated in Example 5.10.

EXAMPLE 5.10

Ammonia, NH_3, is made industrially by the reaction of nitrogen with hydrogen:

$$N_2(g) + 3\ H_2(g) \rightarrow 2\ NH_3(g)$$

In order to produce 24.0 ℓ of NH_3, what volume of N_2 must be used at the same temperature and pressure? what volume of H_2 is required?

SOLUTION

Since the volume ratio is given by the coefficients of the balanced equation, we have:

$$\text{volume } N_2 = 24.0\ \ell\ NH_3 \times \frac{1 \text{ volume } N_2}{2 \text{ volumes } NH_3} = 12.0\ \ell\ N_2$$

$$\text{volume } H_2 = 24.0\ \ell\ NH_3 \times \frac{3 \text{ volumes } H_2}{2 \text{ volumes } NH_3} = 36.0\ \ell\ H_2$$

in other words: $12.0\ \ell\ N_2(g) + 36.0\ \ell\ H_2(g) \rightarrow 24.0\ \ell\ NH_3(g)$
(all volumes measured at the same T and P)

Volume-Mass Relations

The Law of Combining Volumes allows us to relate the volumes of different gases involved in a reaction. Using the concept of molar volume, we can go one step further. That is, we can relate the volume of gas formed in a reaction to the amounts (moles or grams) of reactants. To show how this is done, consider the reaction that occurs when calcium carbonate (limestone) is heated. The products are calcium oxide and gaseous carbon dioxide. The balanced equation is:

$$CaCO_3(s) \rightarrow CaO(s) + CO_2(g)$$

From the equation, we see that:

$$1 \text{ mol } CaCO_3 \rightarrow 1 \text{ mol } CO_2(g)$$

But, since the molar volume of CO_2, or any other gas, is 22.4 ℓ at 0°C and 1 atm, we could say that:

$$1 \text{ mol } CaCO_3 \rightarrow 22.4\ \ell\ CO_2(g) \text{ (at STP)}$$

This relation allows us to determine the volume of $CO_2(g)$ produced from a given amount of $CaCO_3$ (Example 5.11). The same kind of reasoning can be applied to any reaction involving gases. We will use it frequently in Chapter 6.

EXAMPLE 5.11

What volume of $CO_2(g)$, measured at STP, is produced by the decomposition of:

a. 1.35 mol of $CaCO_3$? b. 63.2 g of $CaCO_3$?

SOLUTION

a. Following the unit analysis approach, we note that:

$$1 \text{ mol } CaCO_3 \simeq 22.4 \ \ell \ CO_2(g)$$

To "convert" 1.35 mol of $CaCO_3$ to a volume of CO_2:

$$\text{volume } CO_2 = 1.35 \text{ mol } CaCO_3 \times \frac{22.4 \ \ell \ CO_2}{1 \text{ mol } CaCO_3} = 30.2 \ \ell \ CO_2 \text{ (at STP)}$$

b. Here, one additional step is involved. We must first convert grams of $CaCO_3$ to moles. Noting that the formula mass of $CaCO_3$ is:

$$40.1 + 12.0 + 3(16.0) = 100.1$$

we see that 1 mol $CaCO_3$ weighs 100.1 g. So:

$$100.1 \text{ g } CaCO_3 \simeq 22.4 \ \ell \ CO_2(g)$$

Thus:

$$\text{volume } CO_2 = 63.2 \text{ g } CaCO_3 \times \frac{22.4 \ \ell \ CO_2}{100.1 \text{ g } CaCO_3} = 14.1 \ \ell \ CO_2 \text{ (STP)}$$

5.6 SUMMARY

From a molecular standpoint, the gaseous state is the best understood state of matter. The physical behavior of gases is explained quite simply by the kinetic molecular model. This tells us that the molecules are far apart and in constant motion. The relationships between the volume of a gas and pressure, temperature and amount can be described by simple laws, applicable to all gases. The gas laws can be demonstrated by experiment and are consistent with the kinetic molecular theory.

The quantitative principles introduced in this chapter can be summarized as follows.

If you can solve problems using these equations, you understand this chapter.

1. The volume of a gas sample is inversely proportional to pressure and directly proportional to absolute temperature. For two sets of conditions, 1 and 2:

$$\frac{V_2P_2}{T_2} = \frac{V_1P_1}{T_1}$$

2. At a given temperature and pressure, all gases have the same molar volume. For two gases 1 and 2:

$$V_1 \text{ (molar)} = V_2 \text{ (molar)}$$

At 0°C and 1 atm (STP), the molar volume of a gas is 22.4 ℓ.

3. In a chemical reaction, the volumes of gases, at a given T and P, involved are in the same ratio as their coefficients in the balanced equation.

NEW TERMS

atmosphere (atm)
average speed (molecule)
Avogadro's Law
barometer
Boyle's Law
Charles' Law
Dalton's Law
diffusion
elastic collision
Kelvin temperature (K)
kilopascal (kPa)
kinetic theory
Law of Combining Volumes
millimeter of mercury (mm Hg)
molar volume
partial pressure
pound per square inch (lb/in^2)
pressure
STP

QUESTIONS

1. Explain why
   a. the temperature of a gas in K is greater than in °C.
   b. the pressure of a gas in mm Hg is greater than in atmospheres.

2. Suppose that in making a barometer some air was trapped above the mercury. How would this affect the barometer readings?

3. Explain what is meant by the partial pressure of a gas in a mixture.

4. State, in words and then in the form of an equation:
   a. Dalton's Law b. Boyle's Law c. Charles' Law

5. State in your own words the basic postulates of the kinetic theory.

6. What experimental evidence is there to indicate that:
   a. gas molecules are relatively far apart?
   b. gas molecules are in rapid motion?
   c. collisions between gas molecules are elastic?

7. Explain, in terms of a molecular model, why:
   a. gases are less dense than liquids.
   b. gases are more compressible than liquids.
   c. gases diffuse more rapidly than liquids.
   d. gases exert pressure on the walls of their container.

8. Consider the equation: $P_2V_2 = P_1V_1$. Solve this equation for:
   a. P_2 b. V_2 c. P_1 d. V_1

9. Consider the equation: $V_2P_2/T_2 = V_1P_1/T_1$. Solve this equation for:
   a. V_2 b. P_2 c. T_2 d. V_1 e. P_1 f. T_1

10. Consider the following equations. In which is y directly proportional to x? inversely proportional to x? neither?
    a. $y = 6.2x$
    b. $y = 12/x$
    c. $y = 6.2x + 2.0$
    d. $y/x = 8.1$
    e. $yx = 16$
    f. $y = 2x^2$

11. A container is filled with hydrogen gas at 25°C to give a pressure of 1.50 atm. What will happen to the pressure if:

    a. more hydrogen is added?
    b. the volume of the container is increased?
    c. the temperature increases?

12. Consider Example 5.6. In calculating V_2, could you have expressed:

    a. both pressures in mm Hg?
    b. both temperatures in °C?

13. Explain, in terms of kinetic molecular theory, why:

    a. the pressure of a gas increases when its volume is reduced (constant T).
    b. the pressure of a gas increases when temperature increases (constant V).
    c. the volume of a gas increases when temperature is increased (constant P).

14. How does the molar volume of N_2 at 0°C and 1 atm compare to:

    a. the molar volume of $N_2O(g)$ at 0°C and 1 atm?
    b. the molar volume of N_2 at 25°C and 1 atm?
    c. the molar volume of N_2 at 0°C and 2 atm?

15. Arrange the following in order of increasing volume at STP:

    a. 1 mol $H_2(g)$ b. 1 mol $O_2(g)$ c. 1 g $H_2(g)$ d. 1 mol $H_2O(l)$

16. Arrange the following gases in order of increasing density at the same temperature and pressure:

    a. N_2 b. O_2 c. Ar d. H_2O

17. Consider the following reaction: $C_3H_8(g) + 5\ O_2(g) \rightarrow 3\ CO_2(g) + 4\ H_2O(l)$ Which of the following statements are true?

    a. At 25°C and 1 atm, 5 ℓ of O_2 is required to react with 1 ℓ of C_3H_8.
    b. At 0°C and 1 atm, 3 ℓ of CO_2 is produced from 1 mol of C_3H_8.
    c. At the same temperature and pressure, 4 ℓ of H_2O is produced from 1 ℓ of C_3H_8.

PROBLEMS

1. *Temperature Conversions (Example 5.1)* Express in K:

    a. the normal boiling point of benzene, 80°C.
    b. normal body temperature, 98.6°F (Recall that °F = 1.8[°C] + 32).

2. *Pressure Conversions (Example 5.2)* On a certain day, the barometric pressure is 742 mm Hg. Express this pressure in:

    a. atmospheres b. pounds per square inch c. kilopascals

3. *Partial Pressures (Example 5.3)* Using Table 5.1, calculate:

    a. the partial pressure of O_2 in a gas mixture saturated with water vapor at 21°C. The total pressure is 739 mm Hg.
    b. the total pressure of wet hydrogen collected at 18°C. The partial pressure of H_2 is 720 mm Hg.

4. *Relation between V and P (Example 5.4)* What volume will be occupied by the nitrogen in a 50.0 ℓ cylinder at a pressure of 120 atm if it expands into a space where the pressure is 730 mm Hg?

5. *Relation between V and T (Example 5.5)* The natural gas in a storage tank has a volume of 2.10×10^5 m^3 at 20°C and atmospheric pressure. If the temperature drops to −20°C at constant pressure, what does the volume of the gas become?

6. *Relation between P, V, and T (Example 5.6)* A balloon has a volume of 3.82×10^3 m^3 on the surface of the earth where the pressure is 740 mm Hg and the temperature is 22°C. What will be its volume when it rises to a height where the pressure is 219 mm Hg and the temperature is −15°C?

7. *Relation between P, V, and T (Example 5.7)* Work Example 5.7, but take the final volume to be 110 ℓ and the final temperature to be 45°C.

8. *Molar Volume (Example 5.8)* Calculate the volume at STP of:

   a. 0.500 mol O_2 b. 1.00 g O_2 c. 1.00 g N_2

9. *Density vs Molecular Mass (Example 5.9)* What is the molecular mass of a gas which has a density 2.72 times that of N_2?

10. *Law of Combining Volumes (Example 5.10)* If all volumes are measured at the same temperature and pressure, what volume of hydrogen, H_2, will react with 11.0 ℓ of nitrogen, N_2, in the production of ammonia, NH_3? What volume of ammonia will be produced?

11. *Volume-Mass Relations in Reactions (Example 5.11)* Consider the reaction:

$$NH_4Cl(s) \rightarrow NH_3(g) + HCl(g)$$

Calculate the volume of NH_3 produced at STP from

   a. 1.42 mol NH_4Cl b. 16.0 g NH_4Cl

* * * * *

12. Convert:

   a. 0.912 mol $O_2(g)$ to grams
   b. 0.912 mol $O_2(g)$ to volume in liters at STP
   c. 102 kPa to atmospheres
   d. 102 kPa to millimeters of mercury

13. A sample of nitrogen gas is collected over water at 20°C. The wet gas occupies a volume of 244 cm^3 at a total pressure of 759 mm Hg.

   a. What is the partial pressure of the nitrogen?
   b. What volume would the dry nitrogen occupy at 20°C and 760 mm Hg?

14. A gas at a pressure of 1.00 atm is compressed to ⅛ of its original volume. If the temperature remains constant, what is the final pressure?

15. A flask full of air at 22°C and 740 mm Hg is stoppered. It is then heated to 100°C. What is the pressure of the air inside the flask at 100°C?

16. A thermometer is made using hydrogen gas at a pressure of one atmosphere. It has a volume of 162 cm^3 at 25°C. When placed in boiling benzene its volume is 192 cm^3. What is the boiling point of benzene in °C?

17. The gauge pressure in an automobile tire was 28.2 lb/in^2 in the morning when the air temperature was 15°C. After driving for some time, the temperature of the air in the tire rose to 45°C. Assuming the volume

remained constant, what did the gauge pressure become? (The gauge pressure is the pressure in excess of atmospheric pressure, which may be taken to be 14.7 lb/in^2).

18. A 1.60 g sample of a certain gas occupies a volume of 1.41 ℓ at 25°C and 1 atm.

    a. What is the volume of the sample at STP?
    b. What is the molecular mass of the gas?

19. Calculate the density of $SO_2(g)$ at

    a. STP   b. 25°C, 1 atm (first, determine the molar volume)

20. Calculate the density of uranium hexafluoride, UF_6, at STP.

21. Given that the density of NO at 25°C and 1.00 atm is 1.23 g/ℓ, calculate the density of NO_2 at the same temperature and pressure.

22. Referring to Problem 11, calculate the mass in grams of NH_4Cl required to form:

    a. 1.00 ℓ of NH_3 at STP   b. 1.00 ℓ of HCl at STP

23. A sample of an unknown gas filling a 0.500 ℓ flask at 0°C and 1.00 atm weighed 0.670 g.

    a. Determine the mass of 22.4 ℓ of this gas at STP.
    b. Which of the following gases could it be: N_2, O_2, NO, or CO?

24. When slaked lime is heated above 100°C, the following reaction occurs:

$$Ca(OH)_2(s) \rightarrow CaO(s) + H_2O(g)$$

What volume of water vapor is formed, at 110°C and 1.00 atm, from 5.00 g of $Ca(OH)_2$?

*25. In a reaction carried out above 100°C, one liter of a gas containing only carbon and hydrogen was burned in oxygen. The reaction produced two liters of carbon dioxide and two liters of water vapor (all volumes measured at the same T and P). Determine the formula of the unknown gas and write an equation for its reaction with O_2 to give CO_2 and H_2O.

*26. A concentration of CO of 0.40% by volume in air will lead to death in a short time. What mass of CO will produce this effect in a closed laboratory whose volume is 2.0×10^5 ℓ? (Assume STP.)

*27. Draw graphs of

    a. pressure vs temperature in K (constant V and amount)
    b. volume vs temperature in °C (constant P and amount)
    c. $P \times V$ vs T (constant amount)

*28. The density of a certain gas at 50°C and 720 mm Hg pressure is 1.619 g/ℓ. Determine its molecular mass.

Air separation plant of Airco Industrial Gases in Arroyo, West Virginia. *Courtesy of Airco Industrial Gases.*

6

THE CHEMICAL BEHAVIOR OF GASES: H_2, O_2, N_2, and the Noble Gases

We devoted Chapter 5 to a discussion of the physical behavior of gases. The relationships we developed, the gas laws, are valid for all gases. In contrast, different gases vary greatly from one another in their chemical behavior. Each gas has a unique set of chemical properties which distinguishes it from all other gases.

As pointed out earlier, relatively few elements exist as gases at room temperature and atmospheric pressure. To be exact, there are 11 such elements. Of these, six are monatomic (He, Ne, Ar, Kr, Xe, and Rn). Five form stable diatomic molecules (H_2, O_2, N_2, F_2, Cl_2). In this chapter, we will consider the chemical properties of all of these gaseous elements except fluorine and chlorine. Those two elements will be covered in Chapter 8.

In our discussion of the chemical behavior of gases, we will make frequent use of principles developed in previous chapters. These include the gas laws (Chapter 5), ΔH in reactions (Chapter 4), and mass relations (Chapter 3). This chapter should serve to review and tie together much of the material you have learned up to this point.

6.1 HYDROGEN: H_2

Occurrence

H_2 is the lightest of all known substances.

Elemental hydrogen is not found in nature. The H_2 molecule is so light (MM = 2.016) and moving so fast (1830 m/s at 0°C) that it escapes from the earth's atmosphere. The hydrogen that you make in the laboratory today may be somewhere in outer space tomorrow.

By far the most abundant compound of hydrogen is water, H_2O. There is about 1.3×10^{24} g of water above, below, but mostly on the surface of the earth. Of this, 11.2% by mass is hydrogen. Smaller amounts of hydrogen are found combined with carbon in the compounds called hydrocarbons. Examples include CH_4 (methane), C_8H_{18} (octane), and $C_{10}H_8$ (naphthalene). Our fossil fuels natural gas, petroleum, and coal, are made up mostly of hydrocarbons.

Plant and animal tissue contain compounds of hydrogen with carbon and oxygen. Cellulose, which has the simplest formula $C_6H_{10}O_5$, is present in the cells of all plants. Carbohydrates (starches and sugars) have a similar composition. Fats are quite different chemically from carbohydrates, as we will see in later chapters. However, they too contain the three elements hydrogen, carbon, and oxygen. All in all, hydrogen forms more compounds than any other element.

EXAMPLE 6.1

What is the mass percentage of hydrogen in:

a. octane (molecular formula C_8H_{18})

b. cellulose (simplest formula $C_6H_{10}O_5$)

SOLUTION

We follow the procedure discussed in Chapter 3 (recall Example 3.3, p. 58).

a. The molecular mass of octane (AM C = 12.01, H = 1.008) is:

$$8(12.01) + 18(1.008) = 96.08 + 18.14 = 114.22$$

So, 114.22 g of octane (1 mol of C_8H_{18}) contains 18.14 g of hydrogen.

$$\% \text{ H} = \frac{18.14 \text{ g}}{114.22 \text{ g}} \times 100 = 15.88$$

b. The formula mass of $C_6H_{10}O_5$ is:

$$FM = 6(12.01) + 10(1.008) + 5(16.00)$$
$$= 72.06 + 10.08 + 80.00 = 162.14$$

So, 162.14 g of cellulose contains 10.08 g of hydrogen.

$$\% \text{ H} = \frac{10.08 \text{ g}}{162.14 \text{ g}} \times 100 = 6.217$$

Preparation

In preparing hydrogen or any other substance, we distinguish between two types of procedure. One of these is a *laboratory* preparation, which can be carried out by students in the chemical laboratory. It should involve rather simple apparatus and pose no major safety hazards. The laboratory preparation should go relatively quickly and give a good yield of pure product. Within reason, cost is not a major factor of concern.

The other type is an *industrial* or commercial preparation. Here, the process is designed to give large amounts of product. The equipment may be, and often is, large and complex. Cost is all-important. The starting materials must be relatively cheap. Energy costs must be held to a minimum. If the product is too expensive, customers are unlikely to buy it.

Rarely is a preparation done the same way in the lab as in industry.

Laboratory Preparation

There are two common methods by which H_2 is made in the chemistry laboratory.

1. *Electrolysis of water.* This was mentioned in Chapter 5 (recall Fig. 5.10, p. 126). A direct electric current is passed through the water to which a little acid has been added to raise the conductivity. The equation for the reaction is:

$$2\ H_2O(l) \rightarrow 2\ H_2(g) + O_2(g) \qquad (6.1)$$

Oxygen gas is a "byproduct". Two moles (or two volumes) of H_2 are formed for every one mole (or one volume) of O_2.

EXAMPLE 6.2

Using Equation 6.1, calculate, for the electrolysis of 1.00 g of H_2O:

a. the number of moles of H_2 produced.

b. the volumes of H_2 and O_2 formed at STP.

SOLUTION

a. Here, we follow the procedure discussed in Chapter 3 (see Example 3.9, p. 71). Equation 6.1 tells us that:

$$2 \text{ mol } H_2O \simeq 2 \text{ mol } H_2$$

or:

$$1 \text{ mol } H_2O \simeq 1 \text{ mol } H_2$$

Since the molecular mass of H_2O is 18.0, 1 mol H_2O weighs 18.0 g. Hence the relation between grams of H_2O and moles of H_2 is:

$$18.0 \text{ g } H_2O \simeq 1 \text{ mol } H_2$$

To find the number of moles of H_2 produced from 1.00 g of H_2O:

$$\text{moles } H_2 = 1.00 \text{ g } H_2O \times \frac{1 \text{ mol } H_2}{18.0 \text{ g } H_2O} = 0.0556 \text{ mol } H_2$$

b. To calculate the volume of H_2, we need only recall from Chapter 5 that the molar volume of a gas at STP is 22.4 ℓ.

$$1 \text{ mol } H_2 = 22.4 \ \ell \ H_2 \text{ (STP)}$$

$$\text{Volume } H_2 = 0.0556 \text{ mol } H_2 \times \frac{22.4 \ \ell \ H_2}{1 \text{ mol } H_2} = 1.25 \ \ell \ H_2 \text{ (at STP)}$$

According to the Law of Combining Volumes (Chapter 5):

$$2 \text{ volumes } H_2 \simeq 1 \text{ volume } O_2$$

$$\text{Volume } O_2 = 1.25 \ \ell \ H_2 \times \frac{1 \ \ell \ O_2}{2 \ \ell \ H_2} = 0.625 \ \ell \ O_2 \text{ (at STP)}$$

2. *Reaction of metals with acids.* The metal most commonly used is zinc. Hydrochloric acid, a water solution of HCl, is usually the other reagent. The acid is allowed to drop on pieces of mossy zinc (Recall Figure 3.1, p. 63). Hydrogen gas is evolved rapidly. It may be collected by bubbling through water.

The reaction involves the H^+ ion (proton) common to water solutions of all acids. The other reactant is a zinc atom. The equation is best written as:

$$Zn(s) + 2 \ H^+(aq) \rightarrow Zn^{2+}(aq) + H_2(g) \qquad (6.2)$$

The symbol (aq) is used here to emphasize that the H^+ ions reacting and the Zn^{2+} ions formed are both in water solution. The solution also contains Cl^- ions to make it electrically neutral. Since these ions take no part in the reaction, they do not appear in the equation.

Industrial Preparation

From an economic point of view, neither of the reactions cited is suitable for producing hydrogen on a large scale. Too much electrical energy is used in Reaction 6.1. The zinc used in Reaction 6.2 is rather costly. Most of the hydrogen produced in the United States today is made by the reaction of steam above 100°C with hydrocarbons such as methane, CH_4:

$$CH_4(g) + H_2O(g) \rightarrow CO(g) + 3\ H_2(g);\ \Delta H = +49.3\ \text{kcal} \quad (6.3)$$

Hydrogen can also be made from hot coke and steam: $C(s) + H_2O(g) \rightarrow CO(g) + H_2(g)$

Some hydrogen is also formed as a byproduct in processes used to obtain gasoline from petroleum (Chapter 14).

Reactions

With Non-Metals

Contrary to popular belief, elementary hydrogen is rather unreactive. At room temperature, it reacts readily with only one other element, fluorine:

$$H_2(g) + F_2(g) \rightarrow 2\ HF(g); \qquad \Delta H = -128.4\ \text{kcal} \quad (6.4)$$

Mixtures of hydrogen with chlorine or oxygen can be stored indefinitely with no evidence of reaction. With chlorine, reaction can be started by exposure to ultraviolet light, such as that from burning magnesium (Figure 6.1). The reaction is similar to that with fluorine, although less heat is evolved:

$$H_2(g) + Cl_2(g) \rightarrow 2\ HCl(g); \qquad \Delta H = -44.2\ \text{kcal} \quad (6.5)$$

A mixture of hydrogen and oxygen can be ignited by a spark or flame. An exothermic reaction takes place, sometimes with explosive violence.

$$2\ H_2(g) + O_2(g) \rightarrow 2\ H_2O(l); \qquad \Delta H = -136.6\ \text{kcal} \quad (6.6)$$

The same reaction can occur in air containing as little as 4% hydrogen. For this reason, SAFETY precautions are essential when you work with

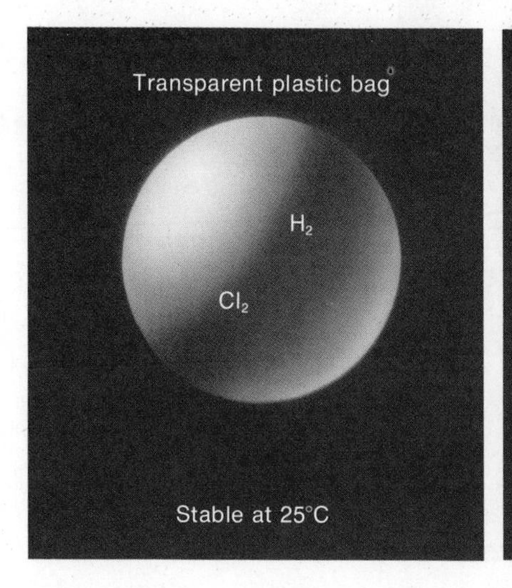

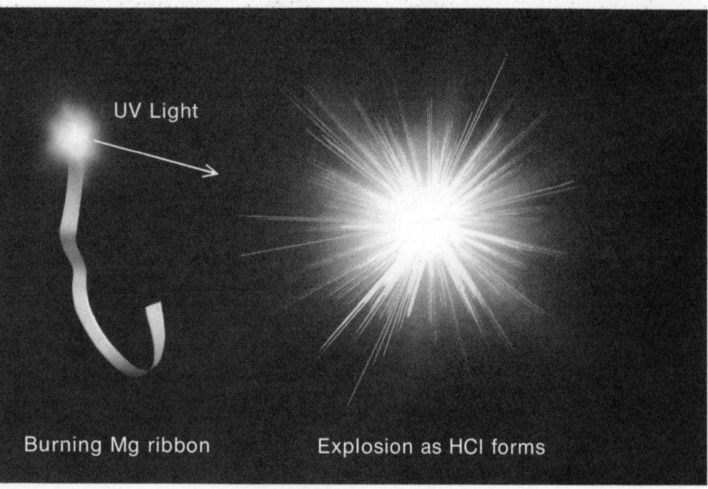

FIGURE 6.1

If a mixture of H_2 and Cl_2 is put in a plastic bag, it will not undergo reaction. However, if exposed to ultraviolet light from a strip of burning Mg ribbon, the reaction mixture will explode.

HYDROGEN in the laboratory. This is particularly true if burners are being used.

The products of Reactions 6.4–6.6, HF, HCl, and H_2O, are all molecular substances. In their molecules, hydrogen atoms are joined by chemical bonds to nonmetal atoms. Indeed, all the compounds of hydrogen with nonmetals are of this type.

With Metals

When hydrogen gas, at temperatures of 150 to 300°C, is passed over certain very reactive metals, it reacts to form ionic compounds. The reaction with sodium is typical:

$$2\ Na(s) + H_2(g) \rightarrow 2\ NaH(s) \qquad (6.7)$$

Chemically, NaH is much more reactive than NaCl.

The product of this reaction, sodium hydride, contains positive Na^+ ions and negative H^- ions, called *hydride ions*. In appearance and physical properties, sodium hydride resembles sodium chloride (Na^+, Cl^- ions).

Several other ionic hydrides are known. They all contain the H^- ion. The positive ion is derived from a metal atom by the loss of electrons.

Uses

At one time, hydrogen, which has a density only about 1/15 that of air, was used to fill balloons and dirigibles. Because of its flammability, hydrogen has been replaced for this purpose by helium. Although helium is heavier and more expensive, it is a great deal safer to work with.

FIGURE 6.2

The *Hindenburg*, shown burning in the air over New Jersey in 1937, was filled with hydrogen gas. Thirty-six people died in this disaster. UPI photo.

Most of the hydrogen produced today is used to make compounds of the element, notably ammonia, NH_3 (p. 148). Liquid fats (oils) are converted to solids by treatment with hydrogen gas under pressure. Smaller amounts of H_2 are used to extract metals in very pure form from their ores. For example, tungsten, used as a filament in electric light bulbs, is prepared by heating the oxide of the element, WO_3, with hydrogen:

$$WO_3(s) + 3\ H_2(g) \rightarrow W(s) + 3\ H_2O(l) \qquad (6.8)$$

The Hydrogen Economy It has been suggested that within the next 10–20 years, hydrogen may replace natural gas as a gaseous fuel. In what has been called the "hydrogen economy" of the future, hydrogen gas would be produced by the electrolysis of water:

$$H_2O(l) \rightarrow H_2(g) + \tfrac{1}{2}\ O_2(g)$$

The electrical energy required to make this reaction go could be supplied by a coal-fired or nuclear power plant. Possibly it could be obtained by the conversion of solar energy. The hydrogen produced could be pumped through natural gas pipelines to homes or industries. There it would be burned in air to produce heat.

$$H_2(g) + \tfrac{1}{2}\ O_2(g) \rightarrow H_2O(l); \qquad \Delta H = -68.3\ kcal$$

The overall process, represented by these two equations, does not, of course, "create" energy. It does, however, offer an efficient way to convert electrical energy into heat without consuming natural gas or petroleum.

As a synthetic fuel, hydrogen has one important advantage. Its combustion product, water, is non-polluting. On the other hand, the heating value per unit volume of hydrogen is rather small. Compare, for example, the thermochemical equations for the combustion of H_2 and CH_4:

$$H_2(g) + \tfrac{1}{2}\ O_2(g) \rightarrow H_2O(l); \qquad \Delta H = -68.3\ kcal$$

$$CH_4(g) + 2\ O_2(g) \rightarrow CO_2(g) + 2\ H_2O(l); \qquad \Delta H = -212.8\ kcal$$

We see that, per mole, more than three times as much heat is given off with methane:

$$\frac{212.8}{68.3} = 3.12$$

The same ratio holds on a volume basis, since a mole of H_2 or CH_4 occupies the same volume at a given temperature and pressure. This means that for a given volume of gas passing through a pipeline, only about ⅓ as much heat would be given off with H_2 as with CH_4.

Hydrogen has been suggested as a future fuel for automobiles as well. The neighborhood "gas station" would supply you with a high pressure tank of hydrogen instead of gasoline. There are some problems here too. Among these is the fact that hydrogen gas has a very low density (0.090 g/ℓ at STP). Even at a pressure of 100 atm, the density of the gas would be only about 1% that of ordinary gasoline. An 80 ℓ tank (20 gal) full of H_2 at this pressure would supply only enough energy to drive a car about 40 km (25 miles).

6.2 OXYGEN: O_2

Occurrence

On a mole basis, elementary oxygen accounts for about 21% of the earth's atmosphere (Table 6.1). Since the O_2 molecule (MM = 32.00) is somewhat heavier than N_2 (MM = 28.01), the mass percentage of oxygen in the air is a little higher than its mole percentage.

Most of the world's oxygen is found as chemical compounds on or beneath the earth's surface. The most abundant of these is water, which contains 88.8% by mass of oxygen. All of our common rocks contain compounds of oxygen with such elements as silicon and aluminum. Many of the ores of our most important metals, including iron, are oxides.

Preparation

Industrial Preparation

The people who make liquid air can't complain about the cost of the raw material.

Oxygen is prepared commercially by the *fractional distillation* of liquid air. The air is first treated to remove carbon dioxide and water vapor. It is then compressed to about 200 atm and cooled to room temperature. The compressed air flows down a long pipe with a small hole at the end. The air expands as it leaves the tube and cools as it does so. The low pressure air flows back outside the pipe, cooling the high pressure gas inside. As this process is repeated the temperature drops until, finally, liquid air at −200°C drips from the tube (Figure 6.3).

When liquid air is allowed to warm up slowly, the gas that comes off first is mostly nitrogen, which has a lower boiling point than oxygen (−196°C vs −183°C). The liquid remaining behind in the distilling flask is mostly oxygen. By repeated fractionation, nearly pure oxygen can be obtained.

Laboratory Preparation

Joseph Priestley, an English chemist, is given credit for the first laboratory preparation of oxygen (and the first use of solar energy to carry out a reaction). By focussing sunlight through a magnifying glass on a sample of the oxide of mercury, HgO, he reached a temperature of

TABLE 6.1 MAJOR COMPONENTS OF CLEAN, DRY AIR AT SEA LEVEL*

Component	MM	Mole %	Mass %
N_2	28.01	78.08	75.51
O_2	32.00	20.95	23.15
Ar	39.95	0.934	1.29
CO_2	44.01	0.031	0.047

*The mole % of water in the air varies from 0.05 to 3.

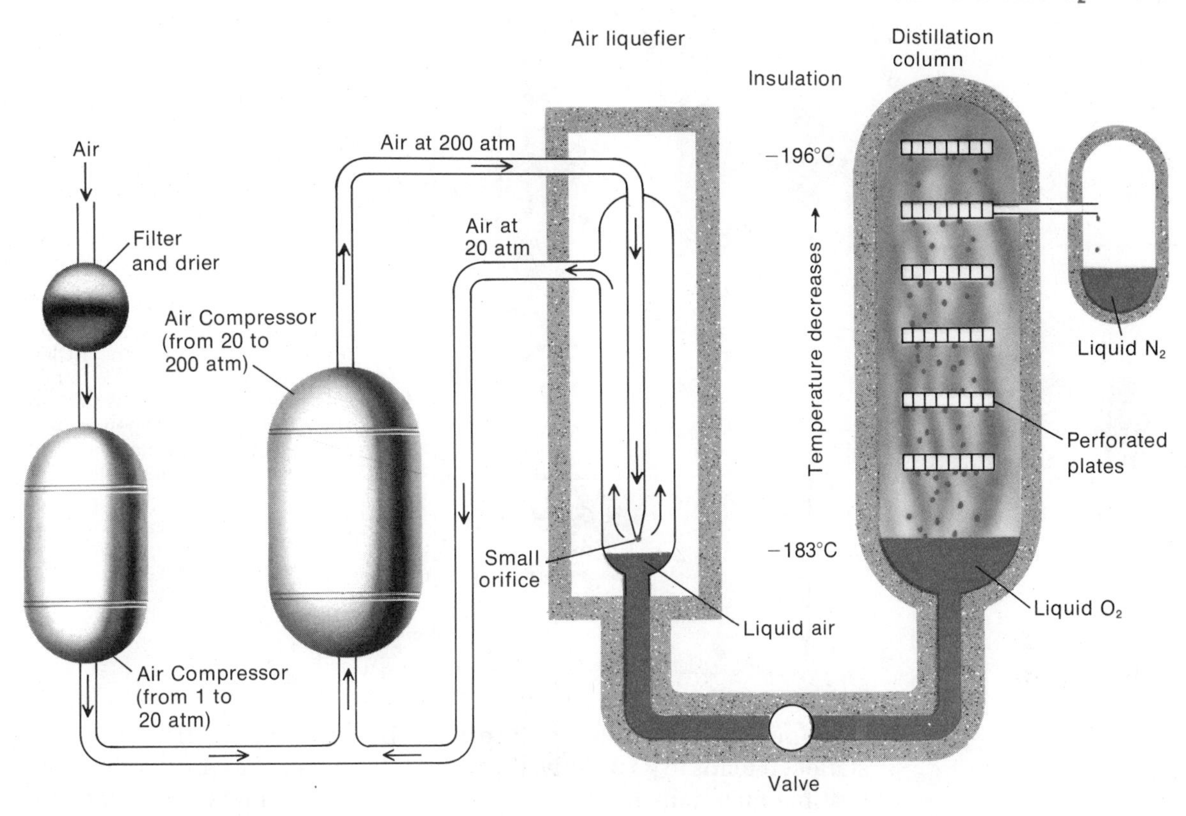

FIGURE 6.3

Liquid air is made in enormous quantities by the process outlined in the drawing. The method is based on the fact that air tends to cool when expanded rapidly. Air thus cooled is passed by incoming gas, cooling it further, until finally liquid air forms. The liquid air is then distilled on a fractionating column, where N_2, which boils at a lower temperature than O_2, can be separated from the liquid. Both N_2 and O_2 have large industrial applications.

about 600°C. This was no mean feat in 1774. At this temperature, the oxide decomposes to the elements by an endothermic reaction.

$$2\ HgO(s) \rightarrow 2\ Hg(l) + O_2(g); \qquad \Delta H = +43.4\ kcal \qquad (6.9)$$

Priestley's experiment should not be repeated in a chemistry laboratory today. The mercury vapor produced is extremely toxic. There are simpler and safer ways of preparing small quantities of oxygen from its compounds. These include:

1. *Electrolysis of water.* This was discussed under the preparation of hydrogen (p. 135).

2. *Heating of $KClO_3$.* This is perhaps the most common method. Potassium chlorate, $KClO_3$, is heated in a test tube to about 200°C (Fig. 6.4, p. 142). A small amount of manganese dioxide, MnO_2, is mixed with the $KClO_3$. It acts as a **catalyst,** speeding up the reaction without itself being consumed. The MnO_2 can be recovered after the reaction is over. For this reason, we do not include it in the equation.

$$2\ KClO_3(s) \rightarrow 2\ KCl(s) + 3\ O_2(g); \qquad \Delta H = -21.4\ kcal \quad (6.10)$$

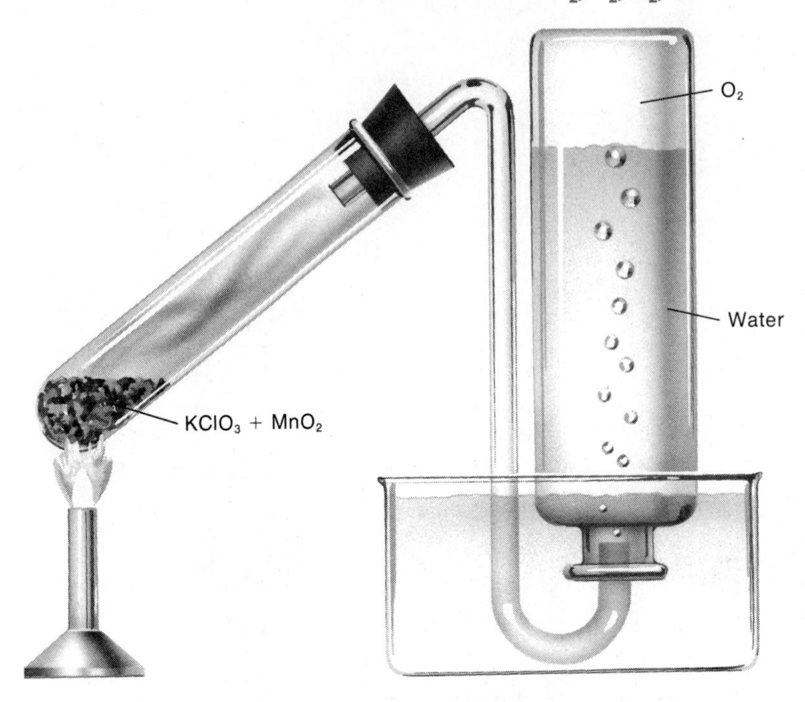

FIGURE 6.4

One method for the laboratory preparation of oxygen. $KClO_3$ when mixed with a little MnO_2 will evolve O_2 when heated to about 200°C. The oxygen is usually collected over water and used in subsequent experiments.

You will notice that this reaction is exothermic ($\Delta H < 0$). Once started, it tends to go rapidly. If you use this method, be sure that no combustible material comes in contact with the molten mixture in the test tube.

3. *Decomposition of H_2O_2.* A simple way to prepare $O_2(g)$ in the laboratory is to add a trace of MnO_2 to a 3% water solution of hydrogen peroxide, H_2O_2. The manganese dioxide catalyzes the decomposition of hydrogen peroxide to water and oxygen:

$$2\ H_2O_2(aq) \rightarrow 2\ H_2O + O_2(g) \tag{6.11}$$

EXAMPLE 6.3

What volume of O_2, at 0°C and 1 atm, can be obtained from 100 g of a 3.0% by mass solution of H_2O_2?

SOLUTION

The water solution must contain:

$$100\ g \times 0.030 = 3.0\ g\ of\ H_2O_2$$

Hence, we need to find the volume of O_2 produced from 3.0 g of H_2O_2. From the coefficients of Equation 6.11:

$$2\ mol\ H_2O_2 \simeq 1\ mol\ O_2$$

To obtain the relation we need, we note that the molecular mass of H_2O_2 is 34.0. So, 1 mol of H_2O_2 weighs 34.0 g. Furthermore, 1 mol O_2 at STP occupies 22.4 ℓ. Thus:

$$2(34.0\ g)\ H_2O_2 \simeq 22.4\ \ell\ O_2$$

or:

$$68.0\ g\ H_2O_2 \simeq 22.4\ \ell\ O_2$$

$$\text{volume } O_2 = 3.0\ g\ H_2O_2 \times \frac{22.4\ \ell\ O_2}{68.0\ g\ H_2O_2} = 1.0\ \ell \text{ of } O_2 \text{ at STP}$$

Reactions

Oxygen is one of the most reactive of all elements. Indeed, under the proper conditions, it reacts with nearly all other elements and with a great many compounds. We will consider only a few of the more important reactions of oxygen.

With Metals

When oxygen reacts with a metal, the product is usually a solid, ionic oxide. The positive ion (Li^+, Mg^{2+}, Al^{3+}) is derived from a metal atom by the loss of electrons. The negative ion is ordinarily the *oxide ion*, O^{2-}. The reactions are always exothermic.

$$4\ Li(s) + O_2(g) \rightarrow 2\ Li_2O(s); \qquad \Delta H = -284\ kcal \qquad (6.12)$$

$$2\ Mg(s) + O_2(g) \rightarrow 2\ MgO(s); \qquad \Delta H = -288\ kcal \qquad (6.13)$$

$$4\ Al(s) + 3\ O_2(g) \rightarrow 2\ Al_2O_3(s); \qquad \Delta H = -798\ kcal \qquad (6.14)$$

Often, part of the energy is given off as light. Magnesium powder is used in flash bulbs and flares where it burns to give off a brilliant white light.

EXAMPLE 6.4

Using Equation 6.14, calculate the amount of heat evolved when 1.00 g of aluminum burns in oxygen.

SOLUTION

You will recall that this type of calculation was carried out in Chapter 4 (Example 4.5, p. 94). From Equation 6.14

$$4\ mol\ Al \simeq -798\ kcal$$

Since the atomic mass of Al is 27.0, one mole of Al weighs 27.0 g

$$4(27.0\ g)\ Al \simeq -798\ kcal$$

or:

$$108\ g\ Al \simeq -798\ kcal$$

So:

$$\Delta H = 1.00\ g\ Al \times \frac{-798\ kcal}{108\ g\ Al} = -7.39\ kcal$$

We conclude that 7.39 kcal of heat is evolved when a gram of aluminum burns.

This problem just involves the conversion of a mass to its equivalent in heat for a given reaction.

Metal oxides, like all ionic compounds, are solids at room temperature. Their melting points are very high, often above 1000°C. One of the highest melting is magnesium oxide (mp MgO = 2800°C). This compound is used in making fire brick and as the lining of high temperature furnaces.

Aluminum oxide is an extremely hard substance with a melting point of 2050°C. In the form of corundum, it is used as an abrasive for grinding and polishing. Emery dust, used for the same purpose, is a mixture of Al_2O_3 and Fe_2O_3. Several precious stones consist of aluminum oxide with small amounts of impurities that give them their color. Included among these are rubies and sapphires (Color plate 2C, center of book). These can now be made synthetically by melting Al_2O_3 with the appropriate impurity. They can be distinguished from the natural gems by the presence of tiny air bubbles, visible only under a microscope.

With Non-Metals

A few of the more familiar compounds of oxygen with the nonmetals are listed in Table 6.2. Most of these are gases at room temperature and atmospheric pressure. All are molecular; nonmetal atoms are bonded to oxygen. As you can see from the Table, many nonmetals form more than one oxide. Nitrogen forms six (N_2O_5, N_2O_4, NO_2, N_2O_3, NO, N_2O), of which only two are listed.

The reaction of oxygen with nonmetals follows a typical pattern. The product formed at high temperatures is the lower oxide, containing the smallest amount of oxygen. Cooling in the presence of excess oxygen or air leads to further reaction. The final product is the higher oxide, containing more oxygen. Consider, for example, the reaction of oxygen with nitrogen. At the high temperatures reached in a car engine, some of the nitrogen and oxygen from the air are converted to nitric oxide, NO:

$$N_2(g) + O_2(g) \rightarrow 2\ NO(g) \tag{6.15}$$

TABLE 6.2 PROPERTIES OF SOME COMMON NONMETAL OXIDES

Formula	Common Name	Melting Point (°C)	Boiling Point (°C)	Physical State 25°C, 1 atm
CO	Carbon monoxide	−207	−192	gas
CO_2	Carbon dioxide	*	*	gas
NO	Nitric oxide	−161	−151	gas
NO_2	Nitrogen dioxide	−9	21	gas
P_4O_6	Phosphorus trioxide	22	173	liquid
P_4O_{10}	Phosphorus pentoxide	*	*	solid
SO_2	Sulfur dioxide	−76	−10	gas
SO_3	Sulfur trioxide	17	45	liquid

*Liquid CO_2 is not stable at 1 atm; the solid (Dry Ice) sublimes (passes directly to vapor) at −79°C. Similarly, P_4O_{10} sublimes at 250°C.

Nitric oxide passing out of the exhaust is slowly converted to nitrogen dioxide, a serious air pollutant.

$$2\ NO(g) + O_2(g) \rightarrow 2\ NO_2(g) \qquad (6.16)$$

Carbon behaves similarly. The lower oxide, CO, is formed at high temperatures with a limited supply of oxygen. These conditions exist in an automobile engine, particularly with a fuel-rich mixture. The exhaust contains significant amounts of carbon monoxide, which is highly toxic. As little as 0.02 mol percent of CO can cause unconsciousness. At the 0.1% level, carbon monoxide is fatal if inhaled for only a few minutes.

Carbon monoxide is responsible for a great many deaths. Most commonly, these come about when car engines are left running in an enclosed space such as a garage. They may also result if the exhaust system is defective or clogged with snow. On a larger scale, CO is responsible for contaminating the air in areas where there is a lot of car traffic. In principle, the carbon monoxide should react with air to form nontoxic carbon dioxide:

$$2\ CO(g) + O_2(g) \rightarrow 2\ CO_2(g) \qquad (6.17)$$

Unfortunately, this reaction occurs very slowly under ordinary conditions. Gas masks used by people exposed to high concentrations of CO (including some traffic police) contain a catalyst which increases the rate of Reaction 6.17.

EXAMPLE 6.5

Using Table 6.2, write a balanced equation for the reaction of

a. phosphorus with a limited supply of oxygen.

b. the oxide formed in (a) with excess oxygen.

SOLUTION

There are two oxides of phosphorus, $P_4O_6(l)$ and $P_4O_{10}(s)$. We would expect the lower oxide, P_4O_6, to be formed first. It should then react with excess oxygen to give the higher oxide.

a. Recall from Table 3.2, p. 66, that elementary phosphorus is a solid with the formula P_4. The balanced equation is:

$$P_4(s) + 3\ O_2(g) \rightarrow P_4O_6(l)$$

b. $P_4O_6(l) + 2\ O_2(s) \rightarrow P_4O_{10}(s)$

Uses

Oxygen in the air is essential to all forms of life. Exothermic reactions with foods such as glucose, $C_6H_{12}O_6$:

$$C_6H_{12}O_6(aq) + 6\ O_2(g) \rightarrow 6\ CO_2(g) + 6\ H_2O(l);\quad \Delta H = -673\ kcal \qquad (6.18)$$

or sucrose (sugar), $C_{12}H_{22}O_{11}$:

$$C_{12}H_{22}O_{11}(aq) + 12\ O_2(g) \rightarrow 12\ CO_2(g) + 11\ H_2O(l);\quad \Delta H = -1350\ kcal \qquad (6.19)$$

Combustion reactions supply the heat we use to warm our homes and operate machines.

supply the energy required to maintain life. Oxygen is also required for the combustion of fuels such as methane (natural gas) or octane (gasoline):

$$CH_4(g) + 2\ O_2(g) \rightarrow CO_2(g) + 2\ H_2O(l);\quad \Delta H = -213\ kcal \qquad (6.20)$$

$$C_8H_{18}(l) + \tfrac{25}{2}\ O_2(g) \rightarrow 8\ CO_2(g) + 9\ H_2O(l);\quad \Delta H = -1308\ kcal \qquad (6.21)$$

The space program required large quantities of liquid oxygen (LOX) to burn rocket fuels to produce the energy required for lift-off. A single Saturn V engine consumed more than two million liters of liquid oxygen. Great care must be taken in storing and handling liquid oxygen. Fires and explosions can easily result if it comes in contact with combustible materials. A tragic fire of this sort took the lives of three U.S. astronauts in 1967.

Pure oxygen or oxygen-enriched air is being used as a substitute for air in many industrial processes. More than 8×10^9 kg of oxygen are used each year in the United States in the manufacture of steel (Chapter 25). Within the past few years, the operators of sewage treatment plants have begun to use pure oxygen instead of air for oxidizing organic matter.

6.3 NITROGEN: N_2

Occurrence

Nitrogen compounds are rather scarce in nature. Combined nitrogen accounts for only about 0.03% of the total mass of the earth's crust. Elementary nitrogen is far more abundant. As indicated in Table 6.1, N_2 accounts for about 78 mol percent of air. All told, the air contains 4×10^{18} kg of elementary nitrogen.

Preparation

The *industrial* preparation of nitrogen starts with liquid air. The N_2 is separated by fractional distillation as described earlier (Fig. 6.3).

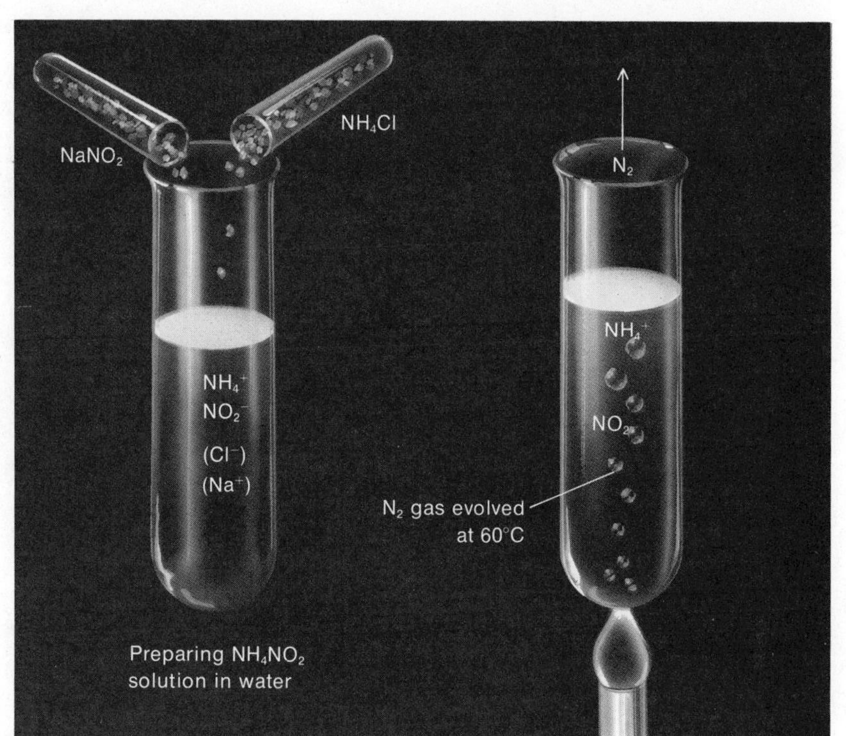

FIGURE 6.5

Preparation of nitrogen gas. Ammonium nitrite, NH_4NO_2, is unstable, so is prepared indirectly by mixing $NaNO_2$ and NH_4Cl in water. The solution, which contains NH_4^+ and NO_2^- ions, gives off N_2 when heated gently.

In the *laboratory*, nitrogen can be made by heating certain compounds containing the *ammonium* ion, NH_4^+. One such compound is ammonium nitrite, NH_4NO_2. The solid is unstable so the reaction is usually carried out in water solution (Figure 6.5). Upon heating the solution to about 60°C, the following reaction takes place:

$$NH_4^+(aq) + NO_2^-(aq) \rightarrow N_2(g) + 2\ H_2O \qquad (6.22)$$

A more spectacular method of preparing nitrogen involves the compound known as ammonium dichromate. This is an ionic compound containing NH_4^+ (ammonium) and $Cr_2O_7^{2-}$ (dichromate) ions. Its formula is written as $(NH_4)_2Cr_2O_7$ to indicate that there are two NH_4^+ ions for every $Cr_2O_7^{2-}$ ion. The reaction that occurs on heating is:

$$(NH_4)_2Cr_2O_7(s) \rightarrow N_2(g) + 4\ H_2O(l) + Cr_2O_3(s) \qquad (6.23)$$

(see Color Plate 2D, center of book).

EXAMPLE 6.6

How many grams of ammonium dichromate, $(NH_4)_2Cr_2O_7$, are required to give one liter of N_2, measured at STP?

SOLUTION

Equation 6.23 tells us that one mole of $(NH_4)_2Cr_2O_7$ produces one mole of N_2. In turn, we know that one mole of N_2 occupies 22.4 ℓ at 0°C and 1 atm. Hence:

$$\text{moles } (NH_4)_2Cr_2O_7 = 1.00\ \ell\ N_2 \times \frac{1\ \text{mol } (NH_4)_2Cr_2O_7}{22.4\ \ell\ N_2}$$

$$= 0.0446\ \text{mol } (NH_4)_2Cr_2O_7$$

To convert to grams, we need the formula mass of $(NH_4)_2Cr_2O_7$. Since one formula unit contains 2 nitrogen, 8 hydrogen, 2 chromium, and 7 oxygen atoms:

$$\text{FM } (NH_4)_2Cr_2O_7 = 2(14.0) + 8(1.0) + 2(52.0) + 7(16.0) = 252$$

one mole of $(NH_4)_2Cr_2O_7$ must weigh 252 g.

$$\text{mass } (NH_4)_2Cr_2O_7 \text{ required} = 0.0446\ \text{mol} \times \frac{252\ \text{g}}{1\ \text{mol}} = 11.2\ \text{g}$$

Reactions

Nitrogen is much less reactive than either hydrogen or oxygen. At room temperature and atmospheric pressure, nitrogen refuses to react with any other element. A few active metals, heated in a stream of N_2 gas, react to form ionic compounds. These compounds are called nitrides. They contain a positive metal ion (Li^+, Mg^{2+}) and the negatively charged *nitride ion,* N^{3-}.

$$6\ Li(s) + N_2(g) \rightarrow 2\ Li_3N(s) \qquad (6.24)$$

$$3\ Mg(s) + N_2(g) \rightarrow Mg_3N_2(s) \qquad (6.25)$$

By far the most important reaction of elementary nitrogen is that with hydrogen. The product is gaseous ammonia, NH_3.

$$N_2(g) + 3\ H_2(g) \rightarrow 2\ NH_3(g) \qquad (6.26)$$

Industrially, this reaction is carried out at high temperatures (400–500°C) and pressures (200–600 atm). A solid catalyst is used to speed up the reaction. We will have more to say about this reaction in Chapter 18.

Uses

Many of the uses of elementary nitrogen reflect its lack of reactivity under ordinary conditions. Instant coffee and potato chips are often packaged under nitrogen rather than air. This prevents reaction with oxygen, which would lead to loss of flavor. Liquid N_2 (bp = −196°C) is often used as a coolant. It is much safer to work with than liquid air.

FIGURE 6.6

This farmer is applying liquid ammonia under pressure to the soil. More commonly, the ammonia is converted to compounds such as urea or ammonium phosphate, which are used as solid fertilizers. USDA—Soil Conservation Service.

Nitrogen gas is being used more and more as a pressurizing gas in aerosol containers. It has certain advantages over other gases formerly used. In particular, it is nontoxic and does not pollute the atmosphere.

Most of the nitrogen produced industrially is converted to ammonia by Reaction 6.26. Some 15 billion kilograms of NH_3 are made annually in the United States. Some of it is now used directly as a fertilizer, as a liquid under pressure (Figure 6.6). All of the solid fertilizers used on lawns and gardens contain nitrogen compounds made from ammonia. One such compound is urea, $CO(NH_2)_2$. It is often referred to as "organic nitrogen". Urea is made from carbon dioxide and ammonia by heating under pressure (100 atm at 180°C).

$$CO_2(g) + 2\ NH_3(g) \rightarrow CO(NH_2)_2(s) + H_2O(l) \qquad (6.27)$$

This is a rather rare example of a reaction where two gases combine to give a solid.

EXAMPLE 6.7

A certain fertilizer, made with urea, contains 11% by mass of nitrogen. How many grams of urea are there in 100 g of fertilizer?

SOLUTION

Clearly the mass of nitrogen is:

$$0.11 \times 100\ g = 11\ g\ N$$

But: MM $CO(NH_2)_2 = 12.0 + 16.0 + 2(14.0) + 4(1.0) = 60.0$
So, one mole of urea weighs 60.0 g and contains 28.0 g of nitrogen. That is:

$$60.0\ g\ urea \simeq 28.0\ g\ N$$

To find the number of grams of urea corresponding to 11 g of nitrogen:

$$11 \text{ g N} \times \frac{60.0 \text{ g urea}}{28.0 \text{ g N}} = 24 \text{ g urea}$$

We conclude that there is 24 g of urea in 100 g of fertilizer. In other words, the fertilizer contains about 24 mass % of urea.

Nitrogen Fixation

Compounds of nitrogen are essential to life. Proteins, which make up the muscles, skin, hair, and vital body organs, are complex nitrogen compounds. On the average, they contain about 17% by mass of this vital element. Ultimately, this nitrogen has to come from the atmosphere. There are several processes by which relatively unreactive N_2 molecules can be converted into useful compounds. They are referred to as methods of *nitrogen fixation.*

Some nitrogen is fixed by a natural process taking place in the atmosphere. Lightning discharges during thunderstorms produce nitric oxide, NO, from N_2 and O_2 (Reaction 6.15). The NO is converted to nitrogen dioxide, NO_2, by Reaction 6.16. Nitrogen dioxide is water soluble, so it is washed into the soil. There it is gradually converted into more useful compounds of nitrogen. Perhaps as much as 5 kg of combined nitrogen per acre is formed in this way each year.

Another natural process of nitrogen fixation takes place in the roots of certain plants. Attached to the roots of clover and alfalfa are nodules (Figure 6.7) which contain colonies of bacteria. These bacteria have the unusual ability to fix atmospheric nitrogen. How they do this is unknown, except that ammonia, NH_3, appears to be the first product. By planting a crop of clover or alfalfa, about 50 kg per acre of combined nitrogen is put into the soil. This is done at the cost of taking the land out of production for one growing season.

Natural processes such as these cannot supply the nitrogen requirements of the world's population. For centuries, fertilizers have been used to restore nitrogen to the soil. Until 1900, virtually all of this was animal waste, which contains, among other things, urea. During the 20th century, "chemical" fertilizers have taken over. All of them are based on ammonia, produced by Reaction 6.26. The use of such fertilizers is essential to our way of life. Without them, mass starvation would be inevitable.

In the past few years, fertilizer prices have risen sharply. This reflects the increased cost of the hydrogen used to produce ammonia. The result has been to drive up the cost of food at the supermarket. In underdeveloped countries such as India, the effect has been much more serious. Clearly, we need to find a cheaper source of combined nitrogen. At least two approaches are possible. One is to find a more economical way to make $H_2(g)$. Another is to develop a new process for fixing nitrogen which does not involve hydrogen as a reactant.

FIGURE 6.7

Bacteria in the nodules of these black-eyed pea roots are able to convert N_2 in the atmosphere into useful nitrogen compounds. USDA—Soil Conservation Service.

6.4 THE NOBLE GASES (He, Ne, Ar, Kr, Xe, Rn)

Occurrence

In 1892, Lord Rayleigh removed the O_2, CO_2 and H_2O from a sample of air. Assuming that the remaining gas was pure nitrogen, he measured its density. To his surprise, he found a value (1.256 g/ℓ at STP) slightly greater than that of N_2 prepared chemically (1.250 g/ℓ at STP). Two years later, in 1894, Lord Rayleigh and Sir William Ramsey found an explanation for this difference. They discovered in air a previously unknown gas, *argon*, whose atoms are heavier than N_2 molecules (AM Ar = 39.9; MM N_2 = 28.0).

Sometimes small discrepancies lead to important discoveries.

Argon is one member of a group of extremely unreactive gaseous elements. These are known as the *inert* or **noble** gases. All told, there are six such gases. Argon is the most abundant; its mole percentage in air is nearly 1%. Four others, helium, neon, krypton and xenon are found in trace amounts in the atmosphere (mole % He = 0.000524, Ne = 0.00182, Kr = 0.000114, Xe = 0.0000087). The last member of the group, radon, cannot be detected in the air. Helium and radon are found among the products of radioactive decomposition reactions (Chapter 2). Helium is an impurity in the natural gas from some wells. All industrial helium is obtained from this source.

Properties

The physical properties of the noble gases are listed in Table 6.3.

TABLE 6.3 PHYSICAL PROPERTIES OF THE NOBLE GASES

Element	Atomic Mass	Melting Point (°C)	Boiling Point (°C)	Density (g/ℓ) 0°C, 1 atm
He	4.003	−272	−269	0.1785
Ne	20.183	−249	−246	0.900
Ar	39.948	−189	−186	1.783
Kr	83.80	−157	−152	3.733
Xe	131.30	−112	−107	5.887
Rn	222*	−71	−62	9.7

*atomic mass of most stable isotope

EXAMPLE 6.8

Referring back to the discussion in Section 5.4, Chapter 5, calculate the density of neon gas, in grams per liter, at 0°C and 1 atm and compare with the value in Table 6.3.

SOLUTION

One mole of neon, which is monatomic, weighs 20.183 g. Neon, like all gases, has a molar volume of 22.4 ℓ at 0°C and 1 atm. Its density should be:

$$\text{density} = \frac{\text{mass}}{\text{volume}} = \frac{20.183\ \text{g}}{22.4\ \ell} = 0.901\ \text{g}/\ell$$

As a group, the noble gases are unique in two important respects.

1. Unlike other gaseous elements, they do not form diatomic molecules. In the noble gases, the basic building block is an individual atom.

2. They are, by all odds, the most unreactive of all elements. Until quite recently, they were believed to be completely inert. Since 1962, several different compounds of xenon have been isolated. These include a chloride, $XeCl_2$, three different fluorides (XeF_2, XeF_4, XeF_6) and two oxides, XeO_3 and XeO_4. A few compounds of krypton and radon with fluorine have also been made. So far, no one has been able to prepare a stable compound of helium, neon, or argon.

For many years, these elements were called the inert gases.

The fact that noble gas atoms do not form diatomic molecules suggests that they are extremely unreactive. This is confirmed by the reluctance of these elements to take part in chemical reactions. We will examine the reason for the stability of noble gas atoms when we discuss their electronic structures in Chapter 9.

Uses

Helium, because of its low density and chemical inertness, is used in "blimps" and weather balloons (Figure 6.8). Liquid helium is used

FIGURE 6.8

Huge helium-filled stratospheric balloons such as this one are launched at Palestine, Texas. Courtesy of National Center for Atmospheric Research, Boulder, CO.

as a cooling agent to reach very low temperatures (bp = −269°C = 4 K). "Neon signs" contain gaseous neon at a low pressure. When an electric current is passed through the neon, an orange-red glow is given off. By mixing other gases with neon, it is possible to get different colors.

He is the lowest boiling of all substances.

Argon is used to give an inert atmosphere in the arc welding of aluminum. It is also used in light bulbs, where it retards evaporation of the tungsten filament. Nitrogen was once used for this purpose, but argon is more effective. Radon, which is radioactive, is sometimes used in cancer therapy.

6.5 SUMMARY

In this chapter, we have looked at the chemical behavior of several gaseous elements. These are:

1. Oxygen (O_2), which reacts with almost all elements. It forms molecular compounds (NO, CO_2, SO_3, and so on) with nonmetals. With metals it forms ionic oxides (Li_2O, MgO, Al_2O_3) containing the O^{2-} ion.

2. Hydrogen (H_2), which reacts with relatively few elements. With nonmetals it forms molecular compounds such as H_2O and HF. With a few very reactive metals, it forms ionic hydrides (NaH, CaH_2) containing the H^- ion.

3. Nitrogen (N_2), an unreactive element. The most important reaction of nitrogen is that with hydrogen to form ammonia, NH_3. Ammonia is the starting material for a host of nitrogen compounds used in fertilizers.

4. The noble gases (He, Ne, Ar, Kr, Xe, and Rn), which are extremely unreactive. They exist as individual atoms rather than diatomic molecules.

NEW TERMS

ammonia (NH_3)
ammonium ion (NH_4^+)
catalyst
fractional distillation
hydride ion (H^-)
industrial preparation
laboratory preparation
mole percentage
nitride ion (N^{3-})
noble gas
oxide ion (O^{2-})

QUESTIONS

1. Of the following gases: H_2, O_2, N_2, He, Ar
   a. Which three are found in the atmosphere?
   b. Which three form compounds?
   c. Which two are monatomic?

2. What factors must be considered in devising a laboratory preparation for a gas like H_2? an industrial method of preparation? Why are Reactions 6.1 and 6.2 unsuitable from an industrial point of view? Why is Reaction 6.3 unsatisfactory as a laboratory preparation of H_2?

3. Write balanced equations for:

   a. two different laboratory preparations of H_2.
   b. three different laboratory preparations of O_2.
   c. a laboratory preparation of N_2.

4. Hydrogen gas can be prepared in the laboratory by reacting aluminum with hydrochloric acid. The reactants are Al(s) and H^+ ions in aqueous solution. The other product, besides H_2, is $Al^{3+}(aq)$. Write a balanced equation for this reaction.

5. Write a balanced equation, similar to 6.5, for the reaction of hydrogen with bromine. The product, HBr, is a gas.

6. Write balanced equations for the reaction of H_2 with

   a. $Cl_2(g)$ b. $O_2(g)$ c. $N_2(g)$

7. Write a balanced equation, similar to 6.8, for the reaction of hydrogen with the oxide of tin, $SnO_2(s)$. The products are tin metal and liquid water.

8. List at least three different uses for elementary hydrogen.

9. Consider Table 6.1. Explain why

   a. the mass % of N_2 is less than its mole %.
   b. the mass % of CO_2 is greater than its mole %.

10. Give the formulas of at least three compounds of oxygen that are found in nature.

11. In the fractional distillation of liquid air, why does N_2 boil off before O_2?

12. What are the safety hazards involved in preparing O_2 from HgO? from $KClO_3$?

13. Give balanced equations for the preparation of O_2 from

    a. $KClO_3$ b. H_2O c. H_2O_2

14. Write balanced equations for the reactions by which the following compounds are formed from oxygen.

    a. Li_2O b. Al_2O_3 c. CO d. NO

15. Write balanced equations for the reaction of O_2 with

    a. CO b. NO c. SO_2 (see Table 6.2)

16. Write a balanced equation for the reaction of each of the following compounds with oxygen. The products are $CO_2(g)$ and $H_2O(l)$.

    a. $C_6H_{12}O_6(s)$ b. $CH_4(g)$ c. $C_5H_{12}(l)$

17. Why is liquid oxygen more dangerous to work with than liquid nitrogen? Would liquid argon be more or less dangerous than liquid oxygen?

18. Write the formulas of

    a. the nitride ion b. the ammonium ion c. the ammonia molecule

19. Give the formulas of ionic compounds containing the NH_4^+ ion and

a. Cl^- b. $Cr_2O_7^{2-}$ c. NO_3^- d. SO_4^{2-}

20. Write balanced equations for the reaction of nitrogen with

a. Li b. Mg c. Ca (product is Ca_3N_2)

21. Write balanced equations for the reactions of nitrogen with

a. O_2 b. H_2 c. Cl_2 (product is gaseous NCl_3)

22. Describe two uses for helium and two for argon.

23. Which one of the noble gases

a. has the highest density at 0°C and 1 atm?
b. is most abundant in the atmosphere?
c. is used in dirigibles?
d. is most reactive?

24. Of the elements: H_2, O_2, N_2, Ar, which one(s) is/are

a. found in the atmosphere?
b. obtained from the electrolysis of water?
c. reactive toward lithium?
d. diatomic?

PROBLEMS

1. *Percentage Composition (Chapter 3 and Example 6.1)* Give the mass percentages of the elements in

a. cellulose, $C_6H_{10}O_5$ b. ammonium nitrate, NH_4NO_3
c. urea, $CO(NH_2)_2$

2. *Mass Relations in Equations (Chapter 3 and Example 6.2)* Consider the equation:

$$2\ KClO_3(s) \rightarrow 2\ KCl(s) + 3\ O_2(g)$$

Using 10.0 g of $KClO_3$, calculate:

a. the number of moles of O_2 produced.
b. the number of grams of O_2 produced.

3. *Mass Relations in Equations (Chapter 3 and Example 6.2)* For the reaction:

$$6\ Li(s) + N_2(g) \rightarrow 2\ Li_3N(s)$$

determine:

a. the number of grams of lithium required to form 1.00 g of Li_3N.
b. the number of moles of N_2 required to form 1.00 g of Li_3N.

4. *Thermochemical Equations (Chapter 4 and Example 6.4)* Given the equation:

$$4\ Li(s) + O_2(g) \rightarrow 2\ Li_2O(s); \qquad \Delta H = -284\ kcal$$

calculate the amount of heat evolved when 1.00 g of lithium burns in oxygen.

5. *Volume-Mass Relations in Equations (Chapter 5 and Examples 6.3, 6.6)* Given:

$$Zn(s) + 2\ H^+(aq) \rightarrow Zn^{2+}(aq) + H_2(g)$$

Using 20.0 g of Zn:

a. How many moles of H_2 are produced?
b. What volume of H_2 is formed at STP?

6. *Volume-Mass Relations in Equations (Chapter 5 and Examples 6.3, 6.6)* Given:

$$(NH_4)_2Cr_2O_7(s) \rightarrow N_2(g) + 4\ H_2O(l) + Cr_2O_3(s)$$

a. What volume of N_2 at STP would be formed from 1.00 g of $(NH_4)_2Cr_2O_7$?
b. What volume of N_2 at 25°C and 1 atm would be formed from 1.00 g of $(NH_4)_2Cr_2O_7$?

7. *Densities of Gases (Chapter 5 and Example 6.8)* Taking the molar volume at STP to be 22.4 ℓ, calculate the density (g/ℓ) of helium and compare to the value given in Table 6.3.

* * * * *

8. Given the equation:

$$C_8H_{18}(l) + 12.5\ O_2(g) \rightarrow 8\ CO_2(g) + 9\ H_2O(l); \qquad \Delta H = -1308\ kcal$$

determine the number of grams of octane which must be burned to produce 1.00 kcal of heat.

9. Referring to the reaction in Problem 2, calculate the mass in grams of $KClO_3$ required to form one liter of O_2 at STP.

10. Consider the reaction: $CH_4(g) + H_2O(g) \rightarrow CO(g) + 3\ H_2(g)$
Suppose we start with 20.0 ℓ of CH_4 at 120°C and 1 atm:

a. What volume of $H_2O(g)$, at the same temperature and pressure, would be required to react with the CH_4?
b. What is the total volume of gases produced, at 120°C and 1 atm?

11. For the reaction in Problem 10, $\Delta H = +49.3$ kcal.

a. How much heat is absorbed per mole of H_2 formed?
b. How much H_2 is formed when 1.00 kcal of heat is absorbed?

12. Consider the reaction: $WO_3(s) + 3\ H_2(g) \rightarrow W(s) + 3\ H_2O(g)$. What volume of $H_2(g)$ at STP is required to react with a sample of WO_3 weighing 1.25 g? What mass of tungsten (W) is produced?

13. Using Equations 6.9–6.11, calculate the number of moles and the volume of O_2 at STP produced by the decomposition of 1.00×10^2 g of

a. HgO b. $KClO_3$ c. H_2O_2

14. Consider Equations 6.12–6.14. Of the three metals, Li, Mg, and Al, which requires the largest volume of O_2 to react with it

a. per mole of metal? b. per gram of metal?

15. How many grams of nitrogen are required to form one gram of

a. NH_3 b. HNO_3 c. NH_4NO_2

16. A certain fertilizer, made with ammonium nitrate, NH_4NO_3, contains 11% by mass of nitrogen. How many grams of NH_4NO_3 are there in 100 g of this fertilizer?

17. Consider the reaction:

$$2\ CO(g) + O_2(g) \rightarrow 2\ CO_2(g); \qquad \Delta H = -135\ kcal$$

When 16.4 g of CO reacts:

a. How much heat is evolved?
b. What mass in grams of CO_2 is formed?
c. What volume of O_2 at STP is required?

*18. The label on a fertilizer bag says that it contains 9.4% by mass of "available nitrogen". If all the nitrogen is in the form of ammonium phosphate, $(NH_4)_3PO_4$, how many grams of this compound are there in a 50 kg bag of the fertilizer?

*19. Using the data in Table 6.1, calculate:

a. the mass of argon in one kilogram of air.
b. the mole % of N_2 in air from which all the O_2 (but none of the other gases) has been removed.
c. the molecular mass of air.

*20. Calculate the density of NO(g) and NO_2(g) at

a. 0°C and 1 atm b. 25°C and 740 mm Hg

*21. It was mentioned in this chapter that nitrogen is now being used as a pressurizing gas in aerosol cans. Why not use:

a. H_2 b. O_2 c. He d. freon, CF_2Cl_2

Таблица II.

Вторая попытка Менделѣева найти естественную систему химическихъ элементовъ. Перепечатана безъ измѣненій изъ „Журнала Русскаго Химическаго Общества“, т. III, стр. 31 (1871 г.).

	Группа I.	Группа II.	Группа III.	Группа IV.	Группа V.	Группа VI.	Группа VII.	Группа VIII, переходъ къ группѣ I.
	H=1							
Типическіе элементы.	Li=7	Be=9,4	B=11	C=12	N=14	O=16	F=19	
1-й періодъ. Рядъ 1-й.	Na=23	Mg=24	Al=27,3	Si=28	P=31	S=32	Cl=35,5	
— 2-й.	K=39	Ca=40	?=44	Ti=50?	V=51	Cr=52	Mn=55	Fe=56, Co=59 Ni=59,
2-й періодъ. — 3-й.	(Cu=63)	Zn=65	?=68	?=72	As=75	Se=78	Br=80	
— 4-й.	Rb=85	Sr=87	Yt?=88?	Zr=90	Nb=94	Mo=96	– =100	Ru=104, Rh=104 Pd=104,
3-й періодъ. — 5-й.	(Ag=108)	Cd=112	In=113	Sn=118	Sb=122	Te=128?	J=127	
— 6-й.	Cs=133	Ba=137	—=137	Ce=138?	—	—	—	— — — —
4-й періодъ. — 7-й.	—	—	—	—	—	—	—	
— 8-й.	—		—	—	Ta=182	W=184	—	Os=199?, Ir=198? Pt=197,
5-й періодъ. — 9-й.	(Au=197)	Hg=200	Tl=204	Pb=207	Bi=208	—	—	
—10-й.		—	—	Th=232	—	Ur=240	—	
Высшая соляная окись	R_2O	R_2O_2 или RO	R_2O_3	R_2O_4 или RO_2	R_2O_3	R_2O_6 или RO_3	R_2O_7	R_2O_8 или RO_4
Высшее водородное соединеніе . . .			(RH_3)	RH_4	RH_3	RH_2	RH	

The Periodic Table of Mendeleev.

7

THE PERIODIC TABLE

In Chapter 6 we considered the chemical properties of a few gaseous elements: hydrogen, oxygen, nitrogen, and the noble gases. In Chapter 8, we will study the chemical behavior of several other elements, about 20 in all. To do this in a meaningful way, we need some method of organizing and correlating the properties of elements. Otherwise, a description of their chemical reactivity becomes little more than a list of facts, difficult to learn and soon forgotten.

The approach we will follow here was first suggested in the mid-19th century. Lothar Meyer (1830–1895) in Germany published a series of papers starting in 1868 in which he pointed out the periodic nature of the properties of elements. Dmitri Mendeleev (1834–1907) in Russia developed an early version of the Periodic Table of the elements. Mendeleev's table first appeared in 1869. The work of these men and others led to the modern Periodic Table shown on the inside back cover of this book. This table is by all odds the most valuable predictive device in all of chemistry.

7.1 THE PERIODIC LAW

The Periodic Table is based upon a simple principle known as the Periodic Law. This states that:

Many of the properties of elements vary in a periodic way with their atomic number. That is, as atomic number increases, the property goes through successive cycles. It first increases to a maximum, then falls to a minimum. This behavior is repeated again and again as we go to higher atomic numbers. To illustrate what this means, let us look at some of the physical and chemical properties of the lighter elements.

Physical Properties *versus* Atomic Number

Table 7.1 p. 160 lists physical properties of the elements of atomic number 1–20. In the column at the far right are listed the **ionization energies** of the elements. These numbers tell us how much energy, in kilocalories, must be absorbed to remove a single electron from each atom in one mole of the element. That is, the ionization energy is the energy change, in kcal/mol, for the process

$$M(g) \rightarrow M^+(g) + e^-$$

where the symbol M refers to one mole of the element involved.

In Figure 7.1, p. 160 ionization energy is plotted *versus* atomic number, using the data in Table 7.1. The graph shows that ionization energy is a periodic function of atomic number. That is, ionization energy first increases, then decreases, increases again, and so on. The curve repeats itself at regular intervals as we go to higher atomic numbers.

The graph in Figure 7.1 resembles that shown in Figure 7.2, where we plot temperature *versus* time of day. During each day, the temperature rises to a maximum around noon and then falls to a minimum near dawn. This pattern is repeated periodically day after day. We would say that temperature is a periodic function of time of day. In the same way, ionization energy is a periodic function of atomic number. True, the progression in Figure 7.1 is not as smooth as that in Figure 7.2. However, the essential feature is the same. The curve can be divided into successive periods, all of which follow the same pattern.

It appears from Figure 7.1 that the "periods" (intervals at which the plot repeats itself) are eight elements long. The high points occur at elements of atomic number 2 (helium), 10 (neon), and 18 (argon). Likewise, successive low points are separated by eight elements. They fall at atomic number 3 (lithium), 11 (sodium), and 19 (potassium). Periods of the same length are found when any of the other properties in Table 7.1 are plotted against atomic number. However, the curves are not as smooth as the one shown.

TABLE 7.1 SOME PHYSICAL PROPERTIES OF THE LIGHTER ELEMENTS

Element	Symbol	Atomic Number	d(g/cm³) of Solid	Melting Point (°C)	Boiling Point (°C)	Ionization Energy (kcal/mol)
Hydrogen	H	1	0.088*	−259	−253	315
Helium	He	2	0.21*	−272	−269	568
Lithium	Li	3	0.53	186	1326	126
Beryllium	Be	4	1.85	1283	2970	216
Boron	B	5	2.34	2300	2550	193
Carbon	C	6	2.25	3570	4200	261
Nitrogen	N	7	1.0*	−210	−196	337
Oxygen	O	8	1.4*	−218	−183	316
Fluorine	F	9	1.5*	−220	−188	403
Neon	Ne	10	1.4*	−249	−246	499
Sodium	Na	11	0.97	98	889	120
Magnesium	Mg	12	1.74	650	1120	178
Aluminum	Al	13	2.70	660	2327	140
Silicon	Si	14	2.33	1414	2355	189
Phosphorus	P	15	1.83	44	280	244
Sulfur	S	16	2.07	119	444	240
Chlorine	Cl	17	2.4*	−101	−34	302
Argon	Ar	18	1.62*	−189	−186	365
Potassium	K	19	0.86	64	774	102
Calcium	Ca	20	1.55	845	1420	142

*Values given at the melting point; others are at 25°C

Note the tremendous variation in properties such as melting and boiling points.

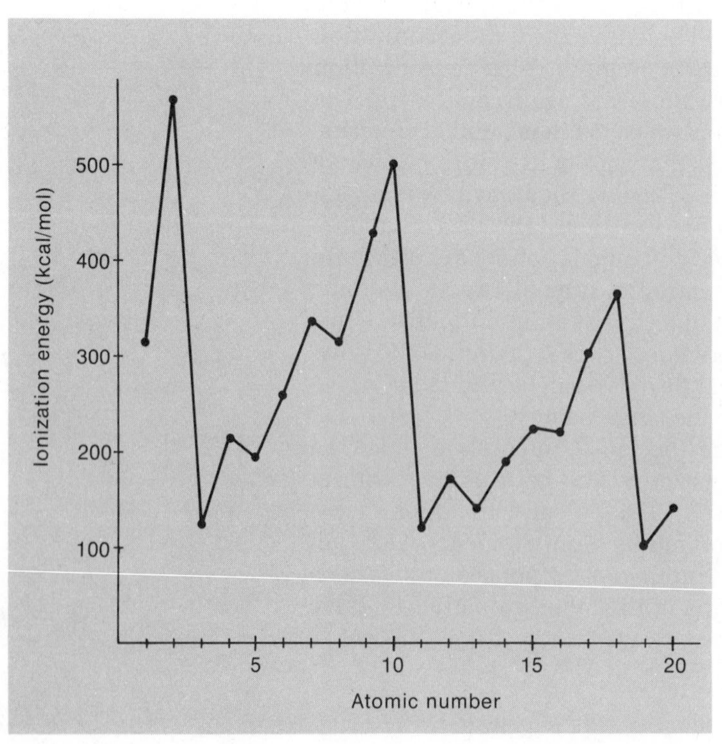

FIGURE 7.1

The ionization energies of the first twenty elements. Note the similarities in the pattern of the values for elements 3 through 10 to that for the elements 11 through 18.

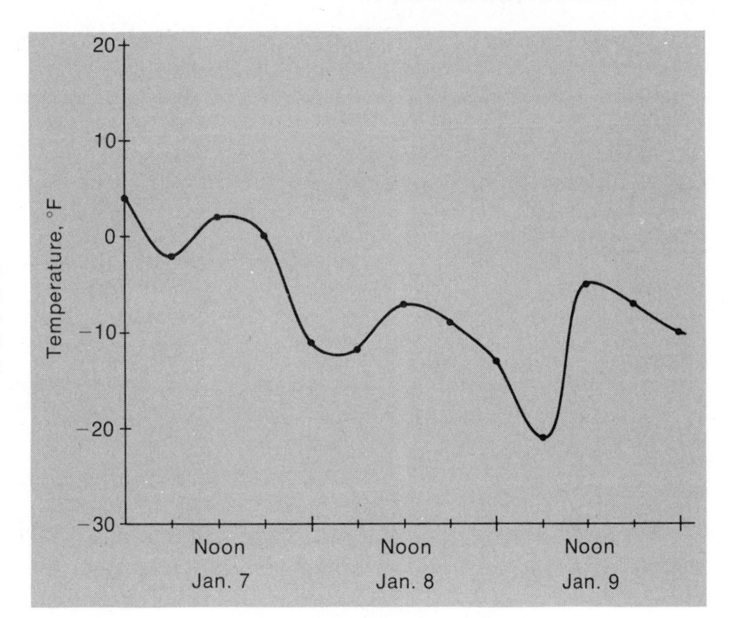

FIGURE 7.2

Variation of temperature with time of day in Minneapolis, Minn. in January, 1979. Here the temperature changes in cycles that are roughly 24 hours long. As with ionization energies, the pattern shows deviations from one cycle to another.

Chemical Properties *versus* Atomic Number

The chemical properties of elements, like their physical properties, vary periodically with atomic number. To illustrate this point, let us focus on a particular property. Consider the simplest formulas of the chlorides of the first 20 elements. We see from Figure 7.3 p. 162 that the number of atoms of chlorine in the formula varies regularly from 1 to 4 and back again.

The data in Figure 7.3 can be presented in a somewhat more useful way (Table 7.2). Here, we have arranged the elements in horizontal rows eight elements long. Notice that when we do this, *elements whose chlorides have similar formulas fall directly beneath one another.* Thus elements of atomic number 3, 11, and 19 form chlorides of simplest formula MCl (LiCl, NaCl, KCl). Those of atomic number 4, 12, and 20 form chlorides of simplest formula MCl_2, and so on.

To illustrate the usefulness of Table 7.2, suppose for a moment

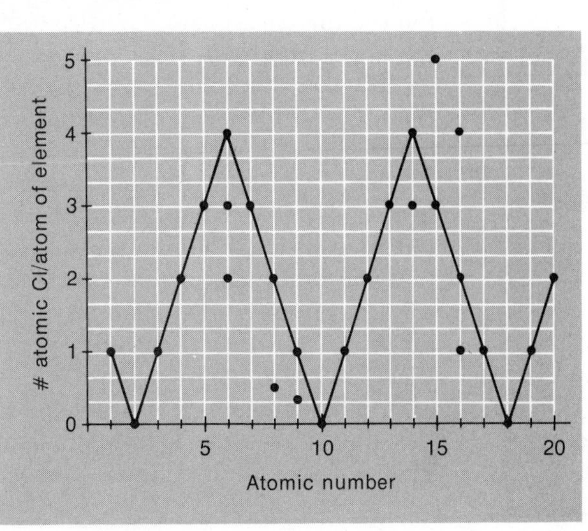

FIGURE 7.3

Formulas of the chlorides of the first 20 elements. For these elements there are chloride compounds that allow one to draw the very regular pattern that is shown. There are, however, other chloride compounds for some of these elements that do not fall on the line. These are indicated with dots.

TABLE 7.2 FORMULAS OF CHLORIDES OF SOME OF THE LIGHTER ELEMENTS (atomic numbers in boldface)

						1	**2**
						HCl	–
3	**4**	**5**	**6**	**7**	**8**	**9**	**10**
LiCl	$BeCl_2$	BCl_3	CCl_4	NCl_3	OCl_2	FCl	–
11	**12**	**13**	**14**	**15**	**16**	**17**	**18**
NaCl	$MgCl_2$	$AlCl_3$	$SiCl_4$	PCl_3	SCl_2	Cl_2	–
19	**20**						
KCl	$CaCl_2$						

(Several other chlorides are known, including PCl_5, ClO_2, C_2Cl_6, C_2Cl_4, Si_2Cl_6, S_2Cl_2, SCl_4, ClF_3)

The regularities in these formulas very strongly imply that the elements fall into a pattern like that in this table.

that the formula of magnesium chloride were unknown. Noting that the formulas of beryllium chloride and calcium chloride are $BeCl_2$ and $CaCl_2$, we could predict with some confidence that the formula of magnesium chloride should be $MgCl_2$. Indeed Table 7.2 is a primitive form of the modern Periodic Table, which will be discussed in Section 7.2.

Mendeleev's Periodic Table

Mendeleev is usually given credit for discovering the arrangement of elements illustrated in Table 7.2. However, the idea was not original with him. It was suggested, in a vague sort of way, by several earlier chemists. They recognized the periodic similarities in the properties of elements and tried to arrange the elements so as to show these similarities. One person who did this was an Englishman, J. A. R. Newlands. His table, presented in 1864, met only ridicule. Mendeleev was more fortunate than Newlands. He was shrewd enough to leave gaps in the table where elements seemed to be missing. In this way, he obtained a regular structure in which elements with similar properties fell directly beneath one another.

Mendeleev's table, shown at the beginning of this chapter, is quite different from the version we are familiar with today. The most obvious difference is in the length of the horizontal rows. Mendeleev's table was condensed. For example, Cu, Ag, and Au were squeezed into the same vertical column as Li, Na, and K. You may also notice that the noble gases are missing from his table. Mendeleev can hardly be blamed for that. These elements were unknown in 1869. Finally, Mendeleev arranged the elements in order of increasing atomic mass. The concept of atomic number (number of protons in the nucleus) was not clearly established until 1914. Fortunately, the order is nearly the same whether atomic mass or atomic number is used. There are a few exceptions. Can you find them, using a modern Periodic Table?

To convince other chemists of the value of his Periodic Table, Mendeleev went out on a very long limb. He dared to predict the properties of three as yet unknown elements (Sc, Ga, Ge), reasoning from the known properties of their neighbors. As it happened, each of these elements was isolated shortly thereafter. They were shown to have properties almost identical with Mendeleev's predictions (Table 7.3). The spectacular agreement between prediction and theory established the validity of Mendeleev's ideas.

TABLE 7.3 PROPERTIES OF GERMANIUM

Property	Predicted, Mendeleev	Observed
Atomic mass	72	72.6
Density (g/cm^3)	5.5	5.36
Formula of oxide	XO_2	GeO_2
Density of oxide (g/cm^3)	4.7	4.70
Formula of chloride	XCl_4	$GeCl_4$
Density of chloride (g/cm^3)	1.9	1.88

The predictions in this table must be among the most remarkable in all of science.

7.2 DESCRIPTION OF THE MODERN PERIODIC TABLE

In Figure 7.4, we have reproduced the modern version of the Periodic Table (atomic masses are rounded off to save space). Quite simply, **the Periodic Table is an arrangement of elements in order of increasing atomic number in horizontal rows. The rows are of such length that elements with similar physical and chemical properties fall directly beneath one another.** Notice for example, that the noble gases, whose properties were discussed in Chapter 6, all fall in the same vertical column or *group* at the far right of the Table. At the far left is another

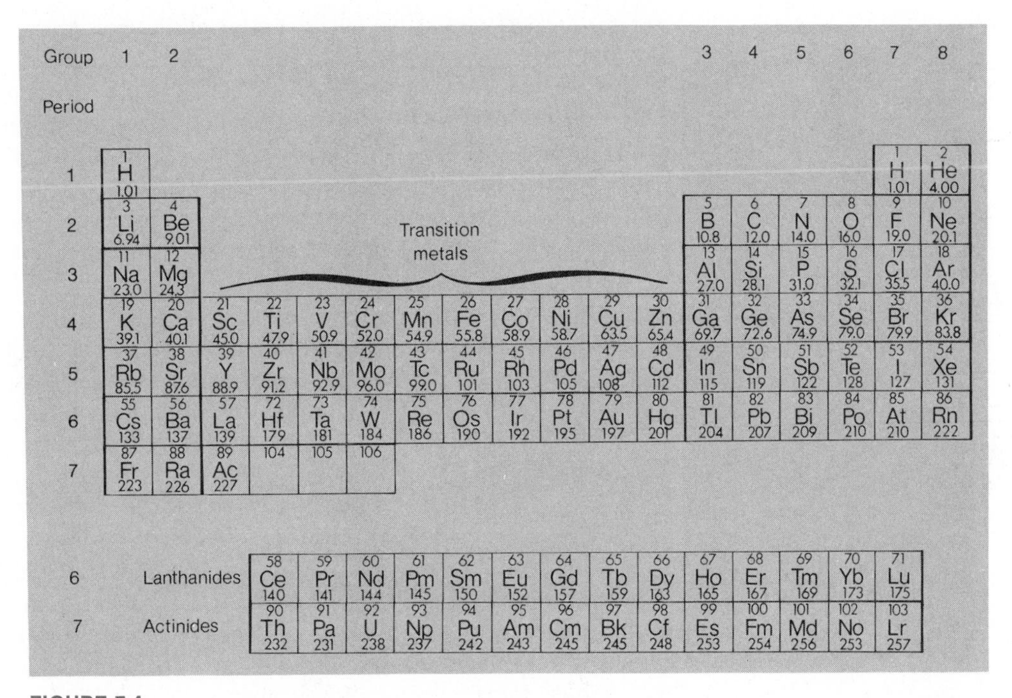

FIGURE 7.4

A modern version of the Periodic Table.

group of elements which resemble each other very closely. These very reactive elements include three metals mentioned earlier (Li, Na, K) and three heavier metals (Rb, Cs, Fr) with similar properties.

Horizontal Periods

You will notice from Figure 7.4 that successive horizontal rows in the Table, called **periods,** differ in length.

—the *first period* is a very short one, containing only two elements, hydrogen and helium. Hydrogen is sometimes placed above lithium, since many of its compounds have formulas similar to those of the Group 1 metals. Compare, for example, HCl with LiCl, NaCl, KCl, and so on. Others prefer to place hydrogen in the block above fluorine in Group 7. Like the other elements in this group, hydrogen forms a negative ion with a −1 charge (H^-; F^-, Cl^-, Br^-, I^-). In truth, the properties of hydrogen are unique. It is difficult to assign it clearly to Group 1, Group 7, or indeed to any specific group of elements. The other element in this period, helium, is clearly the first member of the noble gases, Group 8.

—the *second* and *third periods* each contain 8 elements. They start with a Group 1 metal and end with a noble gas in Group 8.

$$\text{2nd period: } {}_{3}\text{Li} \rightarrow {}_{10}\text{Ne}$$

$$\text{3rd period: } {}_{11}\text{Na} \rightarrow {}_{18}\text{Ar}$$

—the *fourth* and *fifth periods* consist of 18 elements each. The first two members of each period fall in Groups 1 and 2. They are followed by a series of 10 elements known collectively as *transition metals,* which do not appear in previous periods. The transition metals in the fourth period start with scandium (at. no. = 21) and end with zinc (at. no. = 30). In the fifth period, the series extends from yttrium (at. no. = 39) to cadmium (at. no. = 48). Both periods end with a series of six elements which fill out Groups 3 through 8 (${}_{31}\text{Ga} \rightarrow {}_{36}\text{Kr}$ and ${}_{49}\text{In} \rightarrow {}_{54}\text{Xe}$).

—the *sixth period* is the longest complete period, containing 32 elements in all. Of these, 18 elements fall in groups established in previous periods (Groups 1 and 2; 10 transition metals; Groups 3, 4, 5, 6, 7, and 8). The 14 "extra" elements, of atomic number 58–71, are sandwiched between lanthanum (at. no. = 57) and hafnium (at. no. = 72). These 14 elements are referred to as *inner transition metals.* To save space, they are listed in a separate row below the main body of the Table.

—the *seventh period* is incomplete. Like the sixth period, it should contain 32 elements. It starts with a Group 1 metal (Fr, at. no. = 87) and should presumably end with a noble gas of atomic number 118. So far, only 20 of these elements are known, through atomic number 106. Fourteen of these form another inner transition series (${}_{90}\text{Th} \rightarrow {}_{103}\text{Lw}$), listed separately at the bottom of the Table.

TABLE 7.4 HORIZONTAL STRUCTURE OF THE PERIODIC TABLE

Period	First Element	Last Element	Number of Elements
1	$_{1}H$	$_{2}He$	2
2	$_{3}Li$	$_{10}Ne$	8
3	$_{11}Na$	$_{18}Ar$	8
4	$_{19}K$	$_{36}Kr$	18
5	$_{37}Rb$	$_{54}Xe$	18
6	$_{55}Cs$	$_{86}Rn$	32
7	$_{87}Fr$	–	–

There is a symmetry in the periodic table, but it is not the simplest possible symmetry.

Main-Group Elements

The vertical columns of elements in the Periodic Table are referred to as **groups** or **families.** Eight of these, the **main groups,** are located toward the left and right edges of the Table. Groups 1 and 2 are at the left, Groups 3 through 8 at the right.

At least four of the families of elements in the main groups are given special names:

Group 1—alkali metals

Group 2—alkaline earth metals

Group 7—halogens

Group 8—noble gases

All the nonmetals are found in the main group elements.

We considered the properties of one of these families, the noble gases, in Chapter 6. Later, in Chapter 8, we will look at the chemistry of some of the other main groups (1, 2, 6, and 7).

Transition Metals

Located near the center of the Periodic Table, in the 4th, 5th, and 6th periods, are three series of elements known as transition metals. For historical reasons, dating back to Mendeleev, the transition metals are sometimes referred to as "B subgroups". Thus the group headed by scandium (Sc, Y, La) is listed in some versions of the Periodic Table as "Group 3B". We will not use this numbering system, because it can be confusing. Instead, if we wish to refer to a particular group of transition metals, we will give it the name of the first member. Thus we would refer to the *scandium subgroup* (Sc, Y, La) or the *copper subgroup* (copper, silver, gold).

The transition elements include some of our most useful and familiar metals. Those in the 4th period ($_{21}Sc$ to $_{30}Zn$) are particularly common. Iron, the most abundant and the most important of the transition metals, accounts for about 4.7% of the mass of the earth. Curiously, the next most abundant transition metal is titanium (0.58%). It occurs as a shiny black mineral, $FeTiO_3$, found in granite. Titanium metal is stronger and less dense than iron. Unfortunately, it is extremely difficult to extract from its ore. As a result, it is quite expensive and has few uses.

The heavier transition metals in the 5th and 6th periods are much less abundant. However, some of them have special properties which make them very useful. All of us are familiar with silver, gold, and mercury, even though together they make up less than 0.0001% of the mass of the earth. The properties of these and other familiar transition metals are given in Table 7.5.

The transition metals show many pronounced differences from metals in the main groups of the Periodic Table. In particular, the transition metals:

1. *are generally less reactive.* The metals in the copper subgroup (Cu, Ag, Au) do not react with water and react only slowly if at all with O_2. In contrast, all of the metals in Group 1 (and most of those in Group 2) react, often violently, with both oxygen and water.

2. *are higher melting.* Of the 30 transition metals, only four (zinc, silver, cadmium, and mercury) melt below 1000°C. One transition metal,

TABLE 7.5 PROPERTIES OF THE MORE FAMILIAR TRANSITION METALS

	Chromium	Manganese	Iron	Nickel	Copper	Zinc
Symbol	Cr	Mn	Fe	Ni	Cu	Zn
Atomic Number	24	25	26	28	29	30
Density (g/cm^3)	7.19	7.43	7.87	9.91	8.94	7.13
Melting Point (°C)	1857	1244	1535	1453	1083	420
Elect. Conductivity (Pb = 1)	1.5	4.4	2.1	3.1	13	3.6
Abundance (% of earth's crust, oceans, atm)	0.020	0.10	4.7	0.008	0.007	0.013
Annual Production (10^9 kg)	2	8	400	0.3	9	5
Price per kg, 1977	$5.0	$0.8	$0.25	$4.4	$1.4	$0.7

	Silver	Cadmium	Tungsten	Platinum	Gold	Mercury
Symbol	Ag	Cd	W	Pt	Au	Hg
Atomic Number	47	48	74	78	79	80
Density (g/cm^3)	10.49	8.65	19.3	21.5	19.3	13.6
Melting Point (°C)	961	321	3410	1769	1063	−39
Elect. Conductivity (Pb = 1)	13	3.5	3.8	2.0	8.8	0.22
Abundance (% of earth's crust, oceans, atm)	1×10^{-5}	2×10^{-5}	0.0034	5×10^{-7}	5×10^{-7}	3×10^{-5}
Price per kg, 1977	$140	$27	$12	$5500	$5100	$8.7

TABLE 7.6 COMPOSITION AND PROPERTIES OF SOME COMMON ALLOYS

	Composition (Mass %)*	Properties
Alnico	Fe(51), Co(14), Ni(14), Al(8), Cu(3)	Magnetic (electromagnets)
Brass	Cu(70), Zn(30)	Easily machined
Bronze	Cu(88), Sn(8), Zn(4)	Hard, brittle
Coinage metal	Cu(70), Ni(30)	Cheaper than Ag (5¢, 10¢, 25¢ coins)
Gold, 18 carat	Au(75), Cu(25)	Harder than pure gold (24 carat)
Monel metal	Ni(72), Cu(25), Fe(3)	Resists corrosion
Nichrome	Ni(80), Cr(20)	High electrical resistance
Pewter	Sn(85), Pb(15)	Inexpensive, malleable
Silver, sterling	Ag(92.5), Cu(7.5)	Harder than pure silver
Solder, soft	Pb(62), Sn(38)	Low melting
Steel	Fe(99), C(1)	Stronger than pure Fe
Steel, stainless	Fe(76), Cr(15), Ni(8), C(1)	Resists corrosion
Wood's metal	Bi(50), Pb(25), Cd(12.5), Sn(12.5)	Melts at 70°C

*Several of the compositions are approximate.

Most metal products are actually alloys.

tungsten, has a melting point well above 3000°C. Of the eleven metals in Groups 1 and 2, ten melt below 1000°C, three below 100°C.

3. *form many more colored compounds.* All of the transition metals in the first series except scandium and zinc form a variety of brilliantly colored compounds. Some of these are shown in Color Plate 3A, center of book. Virtually all of the compounds of the metals in Groups 1 and 2 are colorless.

4. *form alloys with other metals.* Most, although not all, of our common alloys contain one or more transition metals. Alloys are ordinarily made by heating the metals together until melting occurs. The solid that forms on cooling has properties quite different from those of the pure metals. Often alloys are prepared to give a stronger product. In other cases, they may be lower melting or more resistant to corrosion. The properties of some familiar alloys are given in Table 7.6.

Inner Transition Metals

The 14 elements (at. no. 58–71) which follow lanthanum in the Periodic Table are referred to as **lanthanides** or *rare earths.* Some of these elements are not all that rare. There is more cerium (at. no. = 58) in the earth's crust than boron, arsenic, bromine, silver, tin, iodine, gold, mercury, lead, and uranium *combined.* Another lanthanide, neodymium (at. no. = 60) is more common than cobalt. One of the lanthanides, promethium (at. no. = 61) is radioactive. Its "discovery" was reported several times prior to 1950. We now know that it is the only lanthanide not found in nature.

The lanthanides occur together in monazite sands, found in North Carolina. The elements resemble each other very closely in their chemical properties. For many years, their salts were separated from one another by a tedious process of fractional crystallization. Within the past 20 years, more effective methods of separation have been developed. Oxides and chlorides of several lanthanides are now available in high purity at a reasonable cost.

The lanthanides and their compounds have a variety of uses. The "flints" used in cigarette lighters are actually an alloy of iron with cerium and other lanthanides. A brilliant red phosphor used in TV receivers contains europium oxide, Eu_2O_3. The oxides of lanthanum, praseodymium, and neodymium (La_2O_3, Pr_2O_3, Nd_2O_3) are used in tinted sunglasses. They absorb harmful ultraviolet radiation and reduce the intensity of the sunlight.

A second series of inner transition metals is found in the 7th period. These 14 elements (at. no. 90 → 103) follow actinium and are called **actinides.** All of the actinides are radioactive. Only two, thorium (at. no. = 90) and uranium (at. no. = 92) are found in appreciable amounts in nature. Interest in the actinide elements centers upon uranium. The isotope $^{235}_{92}U$ is capable of splitting into smaller fragments, giving off large amounts of energy in the process. The "fission" of $^{235}_{92}U$ is used to generate electrical energy in nuclear power plants. We will discuss this process in Chapter 27.

The actinides beyond uranium are "man-made" elements. They are prepared in the laboratory by nuclear reactions. By bombarding heavy nuclei with high-energy particles, it is possible to bring about changes in the composition of the nucleus. Frequently, this results in an increase in the number of protons and hence in atomic number. Most of these elements have been made over the past 40 years by a group at the University of California at Berkeley. This is reflected in the names of three of the actinides: berkelium (at. no. = 97), californium (at. no. = 98), and lawrencium (at. no. = 103). E. O. Lawrence was for many years the director of the laboratory at Berkeley. Several other actinides have names derived from those of famous chemists or physicists. These include:

curium (at. no. = 96) from Madame Curie, who did pioneer work in radioactivity
einsteinium (at. no. = 99) from Albert Einstein
fermium (at. no. = 100) from Enrico Fermi, a theoretical physicist
mendeleevium (at. no. = 101) from Dmitri Mendeleev

7.3 PREDICTIONS BASED ON THE PERIODIC TABLE

It is possible to use the Periodic Table as a predictive device, in much the way that Mendeleev did. The properties of elements vary smoothly as we go down a given group or across a given period. Working with the known properties of neighboring elements, it is possible to estimate the properties of an "unknown" element or one of its compounds. The accuracy of such predictions is difficult to estimate in advance. In general, the more data we have to work with, the more confident we can be of the prediction.

The principle to use is that properties vary smoothly if one goes down a group or across a period.

We will consider two simple types of predictions. These involve:

1. The physical properties of an element, given those of its neighbors.

2. The formulas of compounds of an element, knowing those of similar compounds of other elements in the same group.

Physical Properties

Here, we can use the properties of neighboring elements:

—in the same period, to the left and right of the element in question.
—in the same group, above and below the element in question.

Example 7.1 illustrates the reasoning involved.

EXAMPLE 7.1

Consider the element cobalt, Co (at. no. = 27). It lies between Fe (at. no. = 26) and Ni (at. no. = 28) in the first transition series. Given the information below, predict the corresponding properties of cobalt.

	Fe	*Co*	*Ni*
Density (g/cm^3)	7.87	–	9.91
Melting Point (°C)	1535	–	1453
Boiling Point (°C)	2750	–	2732
Ionization Energy (kcal/mol)	182	–	176
Heat of Vaporization (kcal/mol)	84.6	–	91.0

SOLUTION

With the limited amount of information given, about all we can do is to take an average of the properties of iron and nickel. We would predict a density of

$$\frac{7.87\ g/cm^3 + 9.91\ g/cm^3}{2} = 8.89\ g/cm^3$$

The other physical properties are handled in the same way. The predictions are given below, along with the observed properties.

	Predicted	*Observed*
Density (g/cm^3)	8.89	8.92
Melting Point (°C)	1494	1495
Boiling Point (°C)	2741	2870
Ionization Energy (kcal/mol)	179	181
Heat of Vaporization (kcal/mol)	87.8	88.4

Except for the boiling point, the agreement is excellent.

We should emphasize two points in connection with Example 7.1:

a. The agreement between prediction and experiment is unusually good. Many times, predictions of this sort give very approximate values.

b. If enough data are available, there are better ways of predicting physical properties than simple averaging. For example, we might construct graphs similar to that in Figure 7.1, using data for a large number of elements. From such a graph, we could get a better estimate of the properties of the element we are interested in.

Formulas of Compounds

Recall for a moment Table 7.2, p. 161. As we pointed out at the time, chlorides of elements in the same group usually have the same formula. The same is generally true of other compounds such as oxides and sulfides. Example 7.2 illustrates the sort of predictions that can be made based on this principle.

Prediction of chemical formulas from formulas of related compounds usually works quite well.

EXAMPLE 7.2

Consider the zinc subgroup in the Periodic Table (Zn, Cd, Hg). The formulas of certain zinc compounds are:

$ZnCl_2$	zinc chloride
ZnO	zinc oxide
ZnS	zinc sulfide
$Zn(NO_3)_2$	zinc nitrate

Predict the formulas of the corresponding compounds of cadmium and mercury.

SOLUTION

Assuming the formulas are similar, we have:

$CdCl_2$	$HgCl_2$
CdO	HgO
CdS	HgS
$Cd(NO_3)_2$	$Hg(NO_3)_2$

All of these compounds are known. However, mercury also forms another series of compounds which we would not have predicted, based on those of zinc. Their simplest formulas are: $HgCl$, Hg_2O, Hg_2S, $HgNO_3$. All of these, except for HgCl (calomel), are somewhat unstable.

As Example 7.2 indicates, the Periodic Table offers a valuable but not completely reliable method of predicting formulas. Sometimes, as here, one element in a group behaves somewhat differently from the others. A classic example is nitrogen in Group 5. All of the other elements in this group (P, As, Sb, Bi) form two different oxides of simplest formula M_2O_3 and M_2O_5. These oxides (N_2O_3, N_2O_5) are formed by nitrogen as well. However, it also forms four other, better known, oxides with the molecular formulas: N_2O_4, NO_2, NO, N_2O.

7.4 METALS AND NONMETALS IN THE PERIODIC TABLE

Of the 106 elements now known, about 81 can be classified as metals. All metals possess to varying degrees the following properties.

1. *High electrical conductivity.* Metals as a group have electrical conductivities much greater than those of nonmetals. Electrical wiring is usually made of copper, one of the best metallic conductors. (Silver is even better, but too expensive for general use.) Lead is one of the poorest metallic conductors. Yet it has a conductivity nearly 100 times as great as carbon (graphite), the best nonmetallic conductor. Conduction in metals involves the flow of electrons. Clearly, they move readily from one atom to another in a metal.

2. *High thermal conductivity.* Of all solids, metals are by far the best conductors of heat. Frying pans and kettles are made of such metals as copper, iron, and aluminum. Their insulating handles are made of nonmetallic materials such as wood, glass, or plastic.

If someone gave you a powdered substance, how could you find out for sure if it was a metal?

3. *Luster.* Polished metal surfaces are excellent reflectors of light. This accounts for their shiny appearance. Most metals reflect light of all wavelengths and hence have a silvery-white color. Gold and copper absorb light of certain wavelengths and hence have a distinctive color (yellow, red). Plate 3B, center of book, shows these colors.

4. *Ductility and malleability.* Most metals are ductile (capable of being drawn out into wire). They are also malleable (capable of being hammered into thin sheets). Nonmetallic solids tend to shatter if drawn out or hammered.

5. *Ability to give up electrons readily.* All metals, if heated to high enough temperatures, emit electrons. This process is referred to as *thermionic* emission. It is used in the "electron gun" of TV picture tubes, where electrons are emitted by a hot metal filament. A few metals, particularly those in Group 1, give off electrons at room temperature when exposed to light. This *photoelectric effect* is used in an electric-eye photocell to convert light into electrical energy (Fig. 7.5, p. 172).

When a metal takes part in a chemical reaction, it ordinarily loses electrons and forms positively charged ions. Nonmetals tend to gain electrons to form negatively charged ions. A typical metal-nonmetal reaction is that between magnesium and sulfur. In this reaction, there is a transfer of electrons from metal to nonmetal atoms.

$$Mg(s) \rightarrow Mg^{2+} + 2e^-$$

$$S(s) + 2e^- \rightarrow S^{2-}$$

$$Mg(s) + S(s) \rightarrow MgS(s)$$

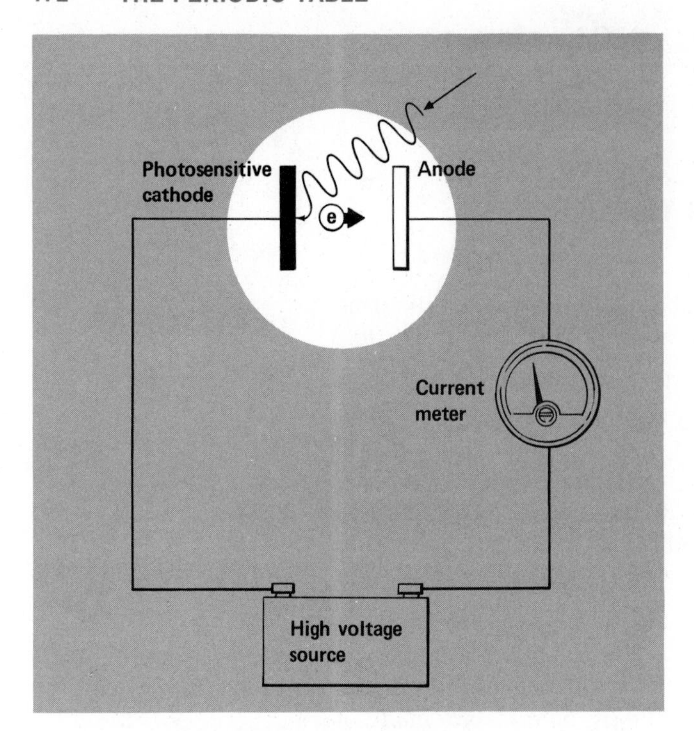

FIGURE 7.5

Schematic diagram showing the photoelectric effect. Light, shown by the wavy line, strikes the surface of an active metal such as cesium. Electrons are emitted from the surface and travel to the anode, causing an electric current that can be detected by the meter. A device of this sort can be used to activate door openers, toll gate controls, and burglar alarms. In such devices a steady light beam is typically broken, which decreases the electric current and so turns on the operating unit.

The most common nonmetals are C, H, O, N, S, and Cl.

Clustered toward the upper right corner of the Periodic Table are 17 elements normally classified as nonmetals. Of these, 11 are gases at room temperature and atmospheric pressure. These include the six noble gases and the elements H_2, F_2, Cl_2, O_2, and N_2. One nonmetal is a liquid (Br_2, bp = 59°C). The rest are solids, nearly all of which are low-melting (mp I_2 = 114°C, S = 119°C, Se = 217°C, P_4 = 44°C, but C = 3570°C).

Located along a diagonal line in the Periodic Table (Figure 7.7) are eight elements which have properties intermediate between those of metals and nonmetals. They include boron in group 3, silicon and germanium in group 4, arsenic and antimony in group 5, tellurium and polonium in group 6, and astatine in group 7. These elements are known collectively as **metalloids.** All of them show metallic luster, yet none form positive ions in reactions. Typically, the metalloids are semiconductors of electricity. Silicon and germanium are used commercially in such semiconductor devices as the transistor (Figure 7.6).

From this discussion, and from Figure 7.7, it should be clear that **metallic character increases as we move down in the Periodic Table and decreases as we move across from left to right.** To illustrate the horizontal trend, consider the elements in the third period, sodium through argon. The first three elements in the period, (Na, Mg, Al) are all typical metals with high electrical conductivities. They are followed by a metalloid—silicon—a semiconductor. The last four elements at the right of the period (P_4, S, Cl_2, Ar) are nonmetals and poor conductors. The vertical trend is illustrated by Group 4. Starting at the top, we have a nonmetal, carbon. This is followed by two metalloids, silicon and germanium. The last two members of the group, tin and lead, are distinctly metallic.

FIGURE 7.6

A silicon semi-conductor chip used in computers. Its tiny size is shown by comparison with the contact lenses surrounding the chip. Texas Instruments.

EXAMPLE 7.3

Which of the following elements would you expect to be the most metallic? the most nonmetallic?

Be (at. no. = 4), B (at. no. = 5), Mg (at. no. = 12), Al (at. no. = 13)

SOLUTION

These four elements form a block in the Periodic Table:

$$_{4}\mathrm{Be} \qquad _{5}\mathrm{B}$$

$$_{12}\mathrm{Mg} \qquad _{13}\mathrm{Al}$$

Metallic character increases as we move down in the Table and decreases as we move across from left to right. Hence, the most metallic element will be magnesium, Mg, at the lower left. The most nonmetallic will be boron, B, at the upper right.

1	2	3	4	5	6	7	8
						H	He
		B	C	N	O	F	Ne
Na	Mg	Al	Si	P	S	Cl	Ar
			Ge	As	Se	Br	Kr
			Sn	Sb	Te	I	Xe
			Pb		Po	At	Rn

FIGURE 7.7

Metallic character and the Periodic Table. As one moves across the Table in any given period, the metallic character of the elements decreases. In the third period, sodium is the most metallic element, argon the least metallic. Among the elements in a given group, the top one is least metallic. Metallic character gradually increases as one goes down any group. Carbon is the least metallic element in Group 4; lead is the most metallic.

7.5 TRENDS IN THE PROPERTIES OF ATOMS

The variation of metallic character with position in the Periodic Table affects other properties. We will now consider how three related properties change as we move down or across the Table. These are *ionization energy, electronegativity,* and *atomic radius.*

Ionization Energy

As we mentioned earlier, the ionization energy of an element is the energy change, ordinarily expressed in kilocalories per mole, for the process:

$$M(g) \rightarrow M^+(g) + e^-$$

As you might expect, ionization energies are always positive quantities. Energy must be absorbed to tear an electron away from the nucleus of an atom.

If you learn the trends of properties in the Periodic Table you can make a lot of useful predictions.

The ionization energy of an element is a measure of the ease or difficulty with which it gives up electrons. Metals, which lose electrons readily, have relatively low ionization energies. Nonmetals, which are reluctant to give up electrons, have high ionization energies. Looking at Figure 7.8, we see the relation between metallic character and ionization energy. As we move from left to right across a period, the elements become less metallic and the ionization energy increases. In the third period, we go from 120 kcal/mol for sodium to 365 kcal/mol for argon. As we move down a given group, metallic character increases and the ionization energy becomes smaller. In Group 4, the ionization energy drops from 261 kcal/mol for carbon to 173 kcal/mol for lead.

1A	2A	3A	4A	5A	6A	7A	8A
						H 315	He 568
Li 126	Be 216	B 193	C 261	N 337	O 316	F 403	Ne 499
Na 120	Mg 178	Al 140	Si 189	P 244	S 240	Cl 302	Ar 365
K 102	Ca 142	Ga 140	Ge 184	As 228	Se 226	Br 274	Kr 324
Rb 98	Sr 133	In 135	Sn 171	Sb 201	Te 210	I 243	Xe 281
Cs 90	Ba 122	Tl 142	Pb 173	Bi 170	Po 196	At	Rn 249

FIGURE 7.8

Ionization energies of the main group elements. The lowest ionization energies are associated with the most metallic elements. Ionization energy decreases as one goes down any column and increases going across any period from left to right. Cesium has the lowest ionization energy of any element; helium has the highest.

EXAMPLE 7.4

Using only the Periodic Table, arrange the three elements Li (at. no. = 3), Be (at. no. = 4), and Na (at. no. = 11) in order of decreasing ionization energy. Check your predictions against Figure 7.8.

SOLUTION

Locating these elements in the Table, we see that they are arranged in the pattern:

Li Be

Na

Sodium, at the lower left of this block, should lose electrons most readily and have the lowest ionization energy. Lithium, directly above sodium, should have a somewhat higher ionization energy. Beryllium, to the right of lithium, should have a still higher ionization energy. The predicted order is:

$$\text{Be} > \text{Li} > \text{Na}$$

From Figure 7.8, we see that this order is correct. The values are 120 kcal/mol for Na, 126 kcal/mol for Li, and 216 kcal/mol for Be.

Electronegativity

We have seen that the tendency of an element to lose electrons is described by its ionization energy. There are several different ways to describe the reverse process, the tendency of an element to gain electrons. The most useful, and the only one we will consider, is electronegativity. In general, the more electronegative an element is, the greater its attraction for electrons.

Electronegativity is more qualitative than ionization energy.

Unlike ionization energy, electronegativity cannot be measured directly. Neither can it be expressed in units such as kcal/mol. The numbers listed in Fig. 7.9, p. 176 are based on a relative scale first proposed by Linus Pauling. They range from 4.0 for the most electronegative element, fluorine, to 0.7 for the element having the least attraction for electrons, cesium.

As you can see from Figure 7.9, **electronegativity generally increases moving across the Periodic Table from left to right. It decreases moving down the Table.** You will recall that the same trends hold for "nonmetallic character". The nonmetals are clustered together at the upper right of the Periodic Table. Indeed, we might say that high electronegativities are characteristic of nonmetals. In the same way, low ionization energies are typical of metals.

Metals have low ionization energies (and low electronegativities).
Nonmetals have high electronegativities (and high ionization energies).

H 2.1						
Li 1.0	Be 1.5	B 2.0	C 2.5	N 3.0	O 3.5	F 4.0
Na 0.9	Mg 1.2	Al 1.5	Si 1.8	P 2.1	S 2.5	Cl 3.0
K 0.8	Ca 1.0	Ga 1.6	Ge 1.8	As 2.0	Se 2.4	Br 2.8
Rb 0.8	Sr 1.0	In 1.7	Sn 1.8	Sb 1.9	Te 2.1	I 2.5
Cs 0.7	Ba 0.9	Tl 1.8	Pb 1.8	Bi 1.9	Po 2.0	At 2.2

FIGURE 7.9

Electronegativities of the main group elements. As metallic character goes down, electronegativity goes up. The most non-metallic elements have the highest electronegativities. Fluorine is the element with the highest electronegativity; cesium has the lowest.

EXAMPLE 7.5

Using only the Periodic Table, arrange the three elements N (at. no. = 7), O (at. no. = 8), and P (at. no. = 15) in order of decreasing electronegativity.

SOLUTION

The location of the three elements in the Table is:

N O

P

Since oxygen is at the upper right of the block, it should have the highest electronegativity. Nitrogen, to the left of oxygen, should be less electronegative. Phosphorus, at the lower left, should have the lowest electronegativity of the three elements. The predicted order is:

$$O > N > P$$

From Figure 7.9, we see that the electronegativities of oxygen, nitrogen, and phosphorus are 3.5, 3.0, and 2.1 in that order. Of the three elements, oxygen is the most strongly nonmetallic, phosphorus the least.

Sizes of Atoms

One of the most important properties of an atom is its atomic radius. The radius of an atom is taken to be one half of the distance between centers of touching atoms in the element (Figure 7.10). It may be ex-

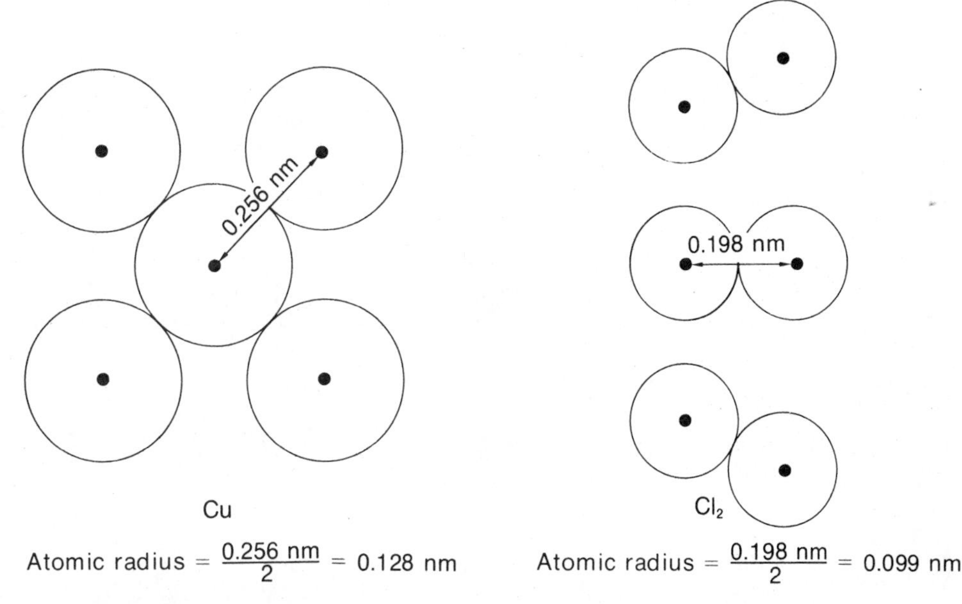

FIGURE 7.10

Atomic radii are defined by assuming that the atoms in the element which are closest to one another are actually touching. In the copper metal crystal the closest atoms are 0.256 nm apart. Since this distance equals two atomic radii, the radius of the copper atom is reported as 0.128 nm.

pressed in *nanometers* (1 nm = 10^{-9} m). In general, the radii of atoms vary from about 0.05 nm (atomic radius H = 0.037 nm) to nearly 0.3 nm (atomic radius Cs = 0.262 nm).

Remember cesium, Cs:
most metallic
smallest ionization energy
smallest electronegativity
largest radius.

In Figure 7.11, p. 178, we show the sizes of atoms of the main-group elements. You will notice that atomic radius:

—*increases as we move down a given group.* Among the Group 1 elements, the atomic radius increases steadily from Li (0.152 nm) to Cs (0.262 nm).

—*decreases as we move across a given period.* In the third period we start with a relatively large atom, Na (0.186 nm). The atoms become progressively smaller as we move across (Mg: 0.160 nm, Al: 0.143 nm, . . ., Ar: 0.094 nm).

You will recall from Section 7.4 that metallic character follows these same trends. Like atomic radius, it increases as we move down the Table and decreases as we move across. We conclude that these two properties are directly related to each other. As atomic radius increases, the elements become more metallic. The smallest atoms are those of the most nonmetallic elements.

The direct relationship between atomic radius and metallic character is perhaps not too surprising. We might expect large atoms, where the electrons are far away from the nucleus, to lose electrons readily. Elements containing such atoms should have low ionization energies and behave as metals. In small atoms, the nucleus is relatively close to the outer edge of the electron cloud. Such atoms should attract electrons strongly, have high electronegativities, and behave as nonmetals.

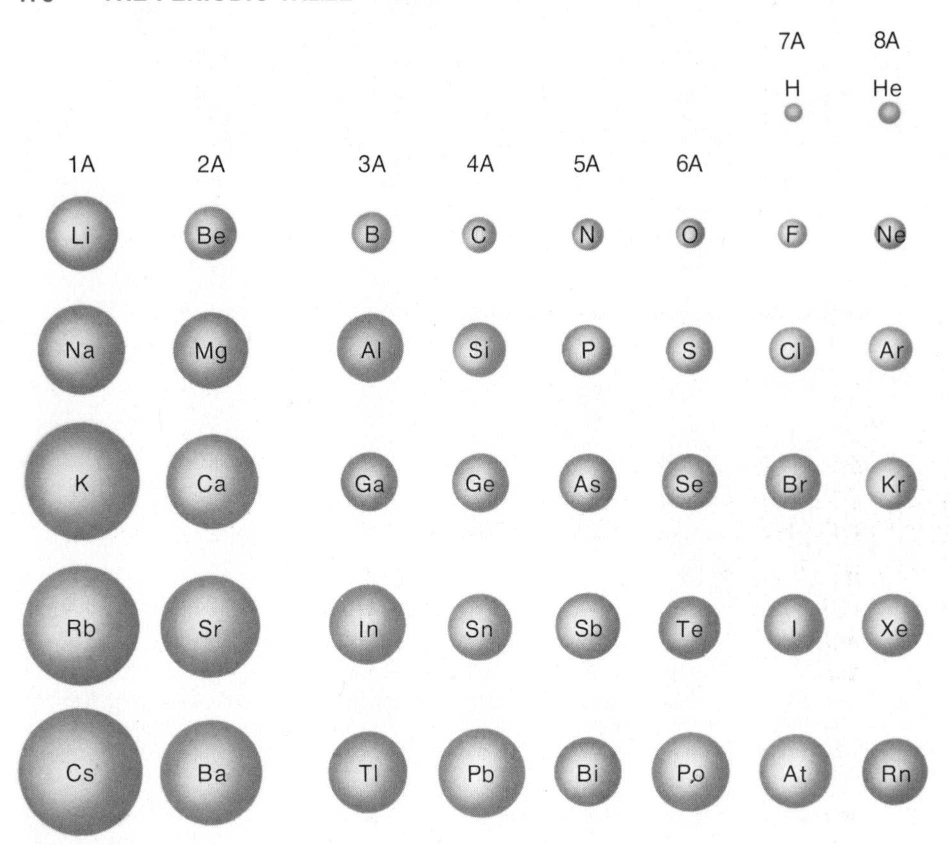

FIGURE 7.11

Atomic radii of the main group elements. The radii increase going down any group and in general decrease going across any row. The largest atom is that of cesium, and the smallest is hydrogen.

EXAMPLE 7.6

Using only the Periodic Table, predict which of the following elements should have the largest atomic radius and which the smallest.

As, Se, Sb, Te

SOLUTION

Locating these elements in the Table, we find that they form a square.

As	Se
Sb	Te

Since Sb (antimony) is at the lower left, it should be the largest of the four atoms. Se (selenium), at the upper right, should be the smallest. The actual radii are:

Sb = 0.141 nm, Te = 0.137 nm, As = 0.121 nm, Se = 0.117 nm

Note the relation to metallic character. Of the four elements, antimony is the most metallic and selenium the least metallic.

7.6 SUMMARY

In the introduction to this chapter, it was stated that the Periodic Table is "the most valuable predictive device in all of chemistry". We have seen how the Table can be used to predict:

1. The physical properties of an element, knowing those of others near it in the Periodic Table (Example 7.1).
2. The formulas of compounds of an element, knowing those of other elements in the same group (Example 7.2).
3. Whether a given element is a metal, nonmetal, or metalloid (Figure 7.7).
4. The relative ionization energies of different elements (Example 7.4).
5. The relative electronegativities of different elements (Example 7.5).
6. The relative sizes of atoms of different elements (Example 7.6).

The Periodic Table can also be used to correlate the chemistry of elements in the same group. We will discuss this topic in some detail in Chapter 8. Later, in Chapter 9, we will consider the theoretical basis for the Periodic Table. This relates to the electronic structure of atoms of the elements.

NEW TERMS

actinide
atomic radius
ductility
electronegativity
group (Periodic Table)
inner transition element
ionization energy
lanthanide
luster
main-group element
malleability
metallic character
metalloid
nonmetal
period (Periodic Table)
Periodic Law
Periodic Table
photoelectric effect
subgroup (Periodic Table)
thermionic emission
transition element

QUESTIONS

1. Explain what is meant by a periodic function. Which of the following graphs represent periodic functions (y *versus* x)?

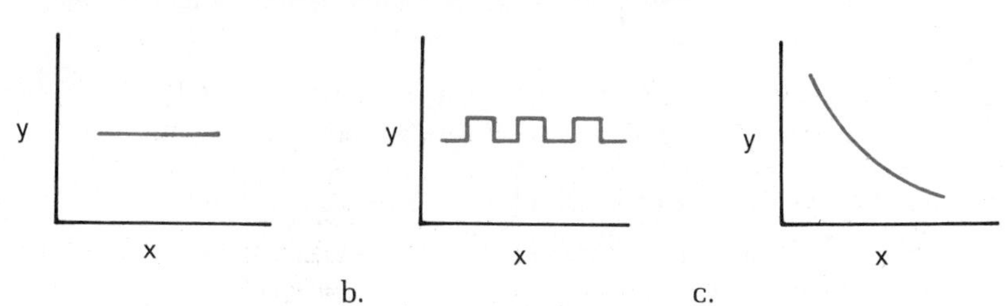

2. Which of the following are periodic functions of atomic number?

   a. ionization energy b. atomic mass c. electronegativity

3. Explain what is meant by ionization energy. Why is it a positive quantity?

4. How many elements are there in

   a. the 1st period? b. the 2nd and 3rd periods?
   c. the 4th and 5th periods? d. the 6th period?

5. How many elements are there in each transition series? What are the atomic numbers of the elements in the first transition series?

6. How many lanthanide elements are there? How many actinides?

7. What would be the atomic number of the element that completes the 7th period? What group would it be in?

8. Which of the following elements would be transition metals? Which would be main-group elements?

   a. at. no. = 105 b. at. no. = 110 c. at. no. = 115

9. Without referring to the text, give the names and symbols of as many transition metals as you can. How many of these metals have you seen?

10. Of the 106 known elements, about 76% are metals. Would you expect this percentage to increase or decrease as more elements are discovered, beyond atomic number 106? (Consider the region of the Periodic Table where these elements will fall.)

11. Classify each of the following statements as true or false.

    a. The physical properties of Ti (at. no. = 22) are expected to be intermediate between those of Sc (at. no. = 21) and V (at. no. = 23).
    b. The formula of the chloride of titanium is expected to be the same as those of scandium and vanadium.
    c. The formula of the oxide of titanium is expected to be the same as those of Zr (at. no. = 40) and Hf (at. no. = 72).

12. Which of the following statements are true? Which are false?

    a. There are fewer metallic elements than nonmetallic elements.
    b. Metals tend to have low electronegativities.
    c. Metalloids lie along a diagonal line in the Periodic Table which runs from the upper right to the lower left.
    d. Nonmetals are located toward the upper right corner of the Periodic Table.

13. On your worksheet, complete the following sentences.

    a. Ionization energy (increases, decreases) as metallic character increases.
    b. Ionization energy (increases, decreases) as electronegativity increases.
    c. Two elements, X and Y, are in the same period. X has an ionization energy of 400 kcal/mol; Y, 200 kcal/mol. Is X to the left or right of Y?

14. List four general ways in which transition metals differ from metals in the main groups.

15. List four general ways in which metals differ from nonmetals (physical properties).

16. As one moves down in the Periodic Table, which of the following properties increase? Which decrease?

    a. metallic character b. ionization energy c. electronegativity
    d. atomic radius

17. As one moves across from left to right in the Periodic Table, which of the following properties increase? Which decrease?

    a. metallic character b. ionization energy c. electronegativity
    d. atomic radius

PROBLEMS

1. *Prediction of Physical Properties (Example 7.1)* Given the following properties of Ca (at. no. = 20) and Ba (at. no. = 56), predict the corresponding properties of Sr (at. no. = 38).

	Ca	*Sr*	*Ba*
Density (g/cm^3)	1.55	?	3.51
Melting Point (°C)	845	?	725
Boiling Point (°C)	1420	?	1640
Ionization Energy (kcal/mol)	142	?	122

The observed values are 2.60 g/cm^3, 770°C, 1380°C, 133 kcal/mol

2. *Prediction of Formulas (Example 7.2)* The formulas of certain compounds of calcium are: $CaCl_2$, $CaCO_3$, $Ca_3(PO_4)_2$. Predict the formulas of the corresponding compounds of radium (at. no. = 88).

3. *Metallic Character (Example 7.3)* Arrange the three elements Ga (at. no. = 31), Ge (at. no. = 32), and In (at. no. = 49) in order of increasing metallic character.

4. *Ionization Energy (Example 7.4)* Using only the Periodic Table, arrange the three elements in Problem 3 in order of decreasing ionization energy. Check your predictions against Figure 7.8.

5. *Electronegativity (Example 7.5)* Using only the Periodic Table, arrange the three elements in Problem 3 in order of decreasing electronegativity. Check your predictions against Figure 7.9. (Surprise!)

6. *Atomic Radius (Example 7.6)* Using only the Periodic Table, arrange the three elements in Problem 3 in order of decreasing atomic radius. Check your predictions against Figure 7.11.

* * * * *

7. Using the data in Table 7.5, plot density *versus* atomic number for the elements of atomic number 74–80. Use your graph to predict the densities of

    a. Os (at. no. = 76) b. Tl (at. no. = 81)

8. The formulas of two different oxides of carbon were given in Chapter 6. What are these formulas? Based on these, predict the formulas of two:

    a. sulfides of carbon b. oxides of lead (at. no. = 82)

    (One of these compounds does not exist!)

9. Refer back to Table 6.3, Chapter 6. Plot the densities at STP of the noble gases against atomic number. As best you can, draw a straight line through the points. Which elements, if any, appear to lie above the line? Which below?

10. Using only the Periodic Table, arrange the elements Mg (at. no. = 12), S (at. no. = 16), and K (at. no. = 19) in order of increasing:

    a. ionization energy   b. electronegativity   c. atomic radius

11. Which two of the following elements would you expect to be closest to each other in metallic character? Ge, Sn, Pb, Sb

12. Given the formulas K_2CrO_4 and $K_2Cr_2O_7$, predict the formulas of

    a. two compounds containing the elements sodium (Na), chromium, and oxygen.
    b. two compounds containing the elements sodium, tungsten (W), and oxygen.

13. The specific heat of a solid element is given by the following approximate relation:

$$\text{specific heat (cal/g}\cdot{}^{\circ}\text{C)} = 6/\text{atomic mass}$$

Sketch a plot of specific heat *versus* atomic mass. Is specific heat a periodic function of atomic mass? of atomic number?

14. Consider the first 103 elements in the Periodic Table. How many of these are

    a. inner transition elements?   b. transition elements
    c. main-group elements?

*15. Using Table 7.1, plot melting point *versus* atomic number. Draw a smooth curve through the points. Which elements appear to be out of line (too high or too low)?

*16. Suppose that in another universe there is a completely different set of elements from the ones we know (26). The inhabitants, logically enough, assign symbols running A, B, C, . . . in order of increasing atomic number, using all the letters of the alphabet. They find that the following elements closely resemble one another in their properties:

A, C, G, K, Q, and Y

B, F, J, P, and X

Construct a Periodic Table, similar to ours, using this information.

*17. Look up, in a reference book, the sources of the names of the following elements.

    a. Np (at. no. = 93)   b. Po (at. no. = 84)   c. Fr (at. no. = 87)

The Taj Mahal is made of marble, a form of calcium carbonate. The Indian Embassy.

8

CHEMICAL BEHAVIOR OF THE MAIN-GROUP ELEMENTS

In Chapter 7 we discussed the structure of the Periodic Table. Perhaps the most important use of the Table is to tie together the descriptive chemistry of the elements. In particular, the Periodic Table helps us to relate chemical properties within a given vertical group. One such group or family, the noble gases, was considered in Chapter 6. Here we will discuss the chemistry of four other families:

1. Group 1 *(alkali metals)*. We will emphasize the chemistry of the most familiar and commercially important member of this group, sodium.

2. Group 2 *(alkaline earth metals)*. Our emphasis will be on the two most abundant elements in this group, magnesium and calcium.

3. Group 6. Of these elements, the chemistry of oxygen was described in Chapter 6. Here we will deal mainly with the second member of the group, sulfur.

4. Group 7 *(halogens)*.

8.1 GROUP 1: Li, Na, K, Rb, Cs

Occurrence

The elements in Group 1 are much too reactive to occur as the free metals. Instead, they are found in ionic compounds. The alkali metal is present in these compounds as a +1 ion (Li^+, Na^+, K^+, Rb^+, Cs^+). Sodium and potassium are by far the most abundant members of the group. Together, they account for about 5% of the total mass of the earth. Compounds of the lightest element, lithium, and the two heavier members, rubidium and cesium, are relatively rare. All the isotopes of francium (at. no. = 87) are radioactive. The element does not occur in nature and very little is known about its chemistry.

The principal ore of sodium is sodium chloride, NaCl. Huge underground deposits of sodium chloride are common in the United States. Presumably, they were formed by the evaporation of ancient seas. A bed of "rock salt", impure NaCl, 100 m thick and 300,000 km^2 in area underlies parts of Kansas, Texas, and Oklahoma. At the current rate of consumption, this bed could supply the world's need for sodium chloride for 100,000 years. Potassium salts, mainly KCl and K_2SO_4, are recovered from brines and salt lake beds in the deserts of California and New Mexico.

It doesn't look like we will run out of NaCl.

FIGURE 8.1

Salt deposits in a dried-up lake bed in the Mojave desert of California.

Properties of the Metals

Table 8.1 lists some of the properties of the alkali metals. You will note two of the trends mentioned in Chapter 7. As we move down the Periodic Table in Group 1:

—the atoms get larger (at. rad. Li = 0.152 nm; Cs = 0.262 nm).
—ionization energy decreases (ion. energy Li = 126 kcal/mol; Cs = 90 kcal/mol).

All of the Group 1 elements are soft, shiny metals, easily cut with a knife. They are among the lowest melting of all metals. All except the first member, lithium, melt below 100°C. Samples of cesium are ordinarily liquid at room temperature. Impurities lower its melting point below that of the pure metal, 29°C. Only two other metallic elements, gallium (mp = 30°C) and mercury (mp = −39°C) are commonly found as liquids.

Sodium and potassium are by far the most common Group 1 metals.

Compounds of the alkali metals impart brilliant colors to a Bunsen flame. These colors are listed in Table 8.1. The yellow color you see when glass tubing is heated is due to sodium compounds in the glass. This color is so intense that traces of sodium tend to mask the colors produced by other elements. Compounds of lithium and potassium are used in fireworks to produce brilliant displays of red or violet light.

TABLE 8.1 PROPERTIES OF THE ALKALI METALS

	Lithium	Sodium	Potassium	Rubidium	Cesium
Atomic Number	3	11	19	37	55
Atomic Mass	6.941	22.990	39.098	85.468	132.905
Melting Point (°C)	186	98	64	39	29
Boiling Point (°C)	1326	889	774	688	690
Density (g/cm^3)	0.534	0.971	0.862	1.53	1.87
Ionization Energy (kcal/mol)	126	120	102	98	90
Atomic Radius (nm)	0.152	0.186	0.231	0.244	0.262
Abundance (% of earth's crust, oceans, atm.)	0.0065	2.6	2.4	0.031	0.0007
Price per kilogram of metal, 1977*	$118	$17	$55	$2800	$3200
Price per kilogram of chloride, 1977*	$15	$1.7	$3.8	$390	$216
Flame Color	bright red	yellow	violet	purple	blue

*Lowest price from chemical supplier. Li, Na, LiCl, NaCl, KCl are available in larger quantities at lower prices from industrial suppliers.

Preparation of the Metals

All the alkali metals are made by electrolysis.

All of the alkali metals can be prepared by electrolysis of their molten chlorides. When a direct electric current is passed through liquid NaCl (mp = 800°C), a pool of liquid sodium forms at one electrode. Chlorine gas is formed at the other electrode. The equation for the overall reaction is:

$$2\ NaCl(l) \rightarrow 2\ Na(l) + Cl_2(g) \qquad (8.1)$$

A considerable amount of electrical energy must be supplied to bring about this reaction. About five kilowatt hours (4400 kcal) is required to form a kilogram of sodium metal.

A clean alkali metal surface in air will be quickly coated with an oxide film.

Reactions of the Metals

FIGURE 8.2

Sodium under mineral oil. Sodium will react with air or water, and so must be protected from both. It is ordinarily stored in mineral oil, a high molecular mass hydrocarbon with which it does not react. Even in the oil, however, traces of oxygen slowly cause the surface of the metal to become crusty.

Sodium is ordinarily stored under mineral oil (Figure 8.2). This prevents it from reacting with oxygen or water vapor in the air. In a limited supply of air, sodium reacts with oxygen to form an ionic oxide (Na^+, O^{2-} ions):

$$4\ Na(s) + O_2(g) \rightarrow 2\ Na_2O(s) \qquad (8.2)$$

With water the products are hydrogen gas, H_2, and Na^+ and OH^- ions in water solution.

$$2\ Na(s) + 2\ H_2O(l) \rightarrow 2\ Na^+(aq) + 2\ OH^-(aq) + H_2(g) \qquad (8.3)$$

Evaporation of this solution gives solid sodium hydroxide, NaOH. Reaction 8.3 is strongly exothermic ($\Delta H = -88$ kcal) and occurs very rapidly. Typically, the sodium melts, catches fire, and ignites the hydrogen. For this reason, large samples (more than about 1 g) of sodium or other alkali metals should not be added to water.

The other alkali metals react in a similar way with oxygen and water. Recall that metallic character increases as we move down a group. Hence, the heavier metals in Group 1 are even more reactive than sodium. Potassium reacts violently with ice at temperatures as low as −100°C (Figure 8.3).

EXAMPLE 8.1

Write a balanced equation for the reaction of potassium with ice.

SOLUTION

Since both potassium and sodium are in Group 1, these metals react in much the same way with water (Equation 8.3). The only

FIGURE 8.3

Potassium reacts violently with water, even with ice. Here we show a small 5 mm cube of potassium burning on an ice cube that it was dropped on just a moment before.

differences are in the symbols for the elements (K *vs.* Na) and in the physical state of water (s *vs.* l). The equation is:

$$2\ K(s) + 2\ H_2O(s) \rightarrow 2\ K^+(aq) + 2\ OH^-(aq) + H_2(g)$$

Uses of the Metals and their Compounds

Sodium is the only alkali metal produced in large quantities. About 2×10^7 kg are made each year in the United States. Most of it is consumed in making sodium compounds. Smaller amounts are used in "sodium vapor" highway lamps. The yellow light given off by the sodium falls in the region to which the human eye is most sensitive. So, it provides good visibility over a wide area of roadway. Because of its low melting point and high thermal conductivity, sodium is also used as a heat transfer fluid in nuclear reactors.

Many sodium compounds are found in familiar household products. These include the following.

All of these sodium compounds are important industrial chemicals.

1. *Sodium chloride, NaCl (Na^+, Cl^- ions).* Table salt is nearly pure sodium chloride. It is essential to the diet of humans and animals. However, too much salt in foods can cause various disorders, including high blood pressure. Rock salt is an impure form of sodium chloride. It is used to melt ice and snow on driveways and highways. Here also, too much can be harmful, killing shrubs, lawns, and trees along highways, as well as making cars rust out faster.

2. *Sodium hydroxide, NaOH (Na^+, OH^- ions).* Sodium hydroxide, called lye, is effective in dissolving fats and grease. It is the major component of most drain cleaners and is used in some products used to

clean ovens. Solutions of sodium hydroxide are extremely caustic, burning the skin. Whenever you use lye, you should wear gloves to protect your hands. Industrially, NaOH is used to make soap (Chapter 15), paper, and textiles.

3. *Sodium carbonate, Na_2CO_3 (Na^+, CO_3^{2-} ions).* This compound is often called washing soda. It is found in many commercial products such as water softeners and in some detergents. Although less caustic than NaOH, it can have an irritating effect upon the skin. A major industrial use of Na_2CO_3 is in making glass.

4. *Sodium hydrogen carbonate, $NaHCO_3$ (Na^+, HCO_3^- ions).* Known as baking soda, $NaHCO_3$ is present in most baking powders. Another component of the powder supplies H^+ ions. These react with sodium hydrogen carbonate to form carbon dioxide.

$$NaHCO_3(s) + H^+(aq) \rightarrow Na^+(aq) + CO_2(g) + H_2O \qquad (8.4)$$

The gaseous CO_2 causes bread or pastries to "rise".

EXAMPLE 8.2

What volume of $CO_2(g)$ at STP is produced from a baking powder containing 1.00 g of $NaHCO_3$?

SOLUTION

From Equation 8.4, we see that:

$$1 \text{ mol } NaHCO_3 \simeq 1 \text{ mol } CO_2$$

But, one mole of CO_2 at STP has a volume of 22.4 ℓ. One mole of $NaHCO_3$ weighs 84.0 g:

$$\text{FM } NaHCO_3 = 23.0 + 1.0 + 12.0 + 3(16.0) = 84.0$$

So:

$$84.0 \text{ g } NaHCO_3 \simeq 22.4\ \ell\ CO_2 \text{ (at STP)}$$

This relation gives us the conversion factor we need to obtain the volume of CO_2 produced from 1.00 g of $NaHCO_3$.

$$1.00 \text{ g } NaHCO_3 \times \frac{22.4\ \ell\ CO_2}{84.0 \text{ g } NaHCO_3} = 0.267\ \ell\ CO_2 \text{ (at STP)}$$

That's just about enough CO_2 to make a nice cake.

The only potassium compound used in large quantities is potassium nitrate, KNO_3. It is a major component of many chemical fertilizers. There it furnishes combined nitrogen and the K^+ ions needed for plant growth. Compounds of the other alkali metals (Li, Rb, Cs) are too expensive for large-scale use. Lithium salts are used to some extent in medicine. It has been found that lithium carbonate, Li_2CO_3, is effective in treating certain mental disorders.

Peroxides and Superoxides

The alkali metals form three different types of compounds with oxygen:

1. "Normal" *oxides*, which have the general formula M_2O (Li_2O, Na_2O, K_2O, Rb_2O, Cs_2O). These compounds contain the oxide ion, O^{2-}. Each oxide ion is balanced by two alkali metal ions, M^+.

2. *Peroxides*, which have the general formula M_2O_2 (example: Na_2O_2). In these compounds, the negative ion is the peroxide ion, O_2^{2-}. Each peroxide ion is balanced by two +1 ions, M^+.

3. *Superoxides*, general formula MO_2 (example: KO_2). Here the negative ion is the superoxide ion, O_2^-. Each superoxide ion, with its −1 charge, is balanced by a +1 ion, M^+.

The type of compound formed by a given metal depends upon the conditions under which it reacts with oxygen. An excess of oxygen favors the formation of the peroxide or superoxide. Sodium peroxide is formed by heating sodium in a current of air at 300°C:

$$2\ Na(s) + O_2(g) \rightarrow Na_2O_2(s) \qquad (8.5)$$

In the laboratory, sodium peroxide is a convenient source of hydrogen peroxide, H_2O_2. This compound is formed by adding Na_2O_2 to water:

$$Na_2O_2(s) + 2\ H_2O(l) \rightarrow 2\ Na^+(aq) + 2\ OH^-(aq) + H_2O_2(aq) \qquad (8.6)$$

The most important of the superoxides is that of potassium, KO_2. This compound reacts readily with water vapor in the air:

$$4\ KO_2(s) + 2\ H_2O(g) \rightarrow 3\ O_2(g) + 4\ KOH(s) \qquad (8.7)$$

This is one of the few ways of forming oxygen spontaneously at room temperature. For this reason, KO_2 is used in self-contained breathing apparatus for firemen and miners. The KOH formed by the reaction with water absorbs the CO_2 in the exhaled air:

$$KOH(s) + CO_2(g) \rightarrow KHCO_3(s) \qquad (8.8)$$

Hence, a person using a mask charged with KO_2 can rebreathe the same air over and over again. This makes it possible to work in an area where there are poisonous vapors. Heavy tanks of compressed air or oxygen are not needed.

8.2 GROUP 2: Be, Mg, Ca, Sr, Ba

Occurrence

The alkaline earth metals, like the alkali metals, are too reactive to occur free in nature. They are found as +2 ions in ionic compounds such as carbonates (M^{2+}, CO_3^{2-}) or sulfates (M^{2+}, SO_4^{2-}). Seawater con-

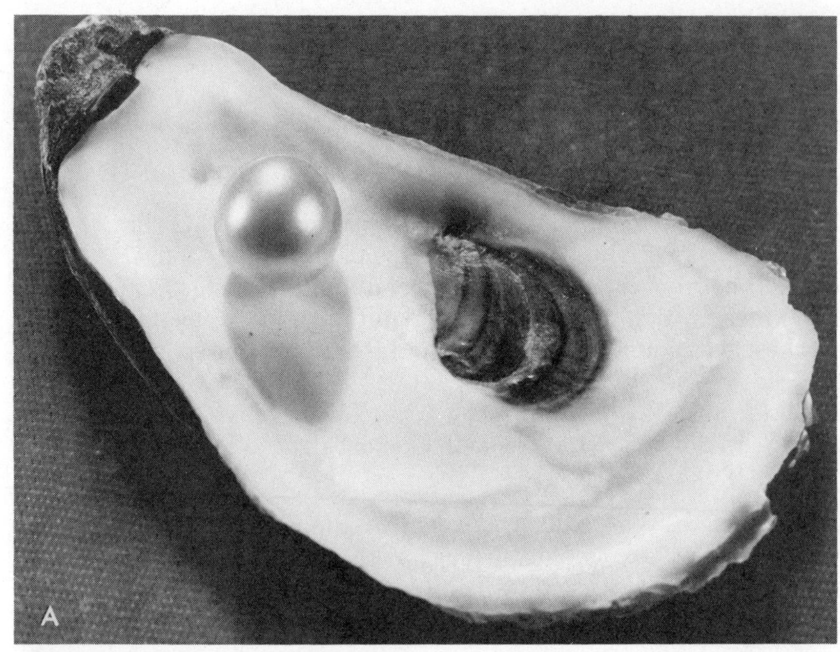

FIGURE 8.4

Two forms of calcium carbonate. Both the pearl formed in the oyster and the limestone cliffs on the English Channel at Dover are nearly pure $CaCO_3$. *A,* H. Armstrong Roberts; *B,* British Tourist Authority.

tains a high concentration of Mg^{2+}. Indeed, Mg^{2+} is the second most abundant positive ion, after Na^+, in seawater. It is also found in salt brines along with Na^+, K^+, and Cl^- ions.

The most abundant alkaline earth compound in the earth's crust is calcium carbonate, $CaCO_3$. It occurs in many different forms. Two common rocks, limestone and marble, are principally calcium carbonate. Their colors are due to impurities; pure $CaCO_3$ is white. In coastal areas, seashells are a major source of calcium carbonate. Coral reefs, found in the Pacific Ocean, and chalk cliffs, such as those found along the English Channel, were formed from the skeletons of tiny shellfish. Oysters form pearls by covering a foreign particle, such as a tiny grain of sand, with successive layers of calcium carbonate.

Radium can also cause cancer, and did in the days before the danger was recognized.

The heavier members of Group 2 are much less abundant than magnesium and calcium. Strontium and barium occur as both the sulfate ($SrSO_4$, $BaSO_4$) and carbonate ($SrCO_3$, $BaCO_3$). Radium compounds are extremely rare. The element was first isolated by Marie Curie in 1898. She obtained a fraction of a gram of radium from more than a ton of pitchblende ore. Like all of its neighbors in the 7th period of the Periodic Table, radium is radioactive. Compounds of the element have been used for many years in the treatment of cancer. The radiation given off by the radium destroys malignant cells.

The lightest of the alkaline earths, beryllium is a relatively rare element. Its principal ore is beryl, which has a formidable simplest formula: $Be_3Al_2Si_6O_{18}$. The precious stones emerald and aquamarine are nearly pure beryl. Traces of impurities give these gems a green or light blue color. Soluble beryllium compounds such as $BeCl_2$ or $BeSO_4$ are extremely toxic. The Be^{2+} ion attacks protein molecules in the body, often with fatal results.

Properties of the Metals

The properties of the metals in Group 2 are summarized in Table 8.2, p. 192. Notice again the inverse relationship between ionization energy and atomic radius. Ionization energy decreases steadily as atomic radius increases. Atomic radius, and metallic character, increase as we move down the Group, from Be to Ba.

Comparing the entries in Table 8.2 to those in Table 8.1, we see that, as a group, the alkaline earth metals have

—*higher ionization energies than the alkali metals.* Compare, for example, Mg with Na (178 kcal/mol *vs.* 120 kcal/mol). As pointed out in Chapter 7, ionization energy increases as we move across the Periodic Table.

—*smaller atomic radii than the alkali metals.* The atomic radius of magnesium (0.160 nm) is considerably smaller than that of sodium (0.186 nm). Similarly, the beryllium atom is smaller than the lithium atom. Indeed, the atomic radius of Be, 0.111 nm, is the smallest of all the metallic elements. Beryllium is the least metallic of all the elements in Groups 1 and 2.

Very hard water might contain as much as 0.5 g Ca^{2+} per liter.

—*higher densities than the alkali metals.* The density of magnesium (1.74 g/cm³) is nearly twice that of sodium (0.971 g/cm³). This is explained in part by the greater mass of the Mg atom (AM Mg = 24.31, AM Na = 22.99). A more important factor is the smaller atomic radius of Mg (0.160 nm *vs.* 0.186 nm for Na).

—*higher melting points than the alkali metals.* The lowest melting Group 2 metal, magnesium, melts at 650°C. In contrast, the highest melting point among the Group 1 metals, that of lithium, is only 186°C.

Compounds of the metals in Group 2 tend to be less soluble in water than those of the Group 1 metals. As a result, Mg^{2+} and Ca^{2+} ions in water frequently form precipitates such as $Mg(OH)_2$, $CaCO_3$, and $CaSO_4$. "Hard water" contains relatively large amounts of Mg^{2+} or Ca^{2+}. Unless treated ("softened") to remove these ions, hard water can cause problems. Grayish stains on laundered clothes are one symptom. More seriously, deposits of magnesium or calcium compounds can clog hot water pipes, teakettles, and boiler tubes (Figure 8.5).

FIGURE 8.5

The pipe elbow on the left is clogged with calcium salts which form when hard water is heated or evaporates. Ecodyne, The Lindsay Division.

TABLE 8.2 PROPERTIES OF THE ALKALINE EARTH METALS

	Beryllium	Magnesium	Calcium	Strontium	Barium
Atomic Number	4	12	20	38	56
Atomic Mass	9.012	24.31	40.08	87.62	137.34
Melting Point (°C)	1283	650	845	770	725
Boiling Point (°C)	2970	1120	1420	1380	1640
Density (g/cm^3)	1.85	1.74	1.55	2.60	3.51
Ionization Energy (kcal/mol)	216	178	142	133	122
Atomic Radius (nm)	0.111	0.160	0.197	0.215	0.217
Abundance (% of earth's crust, oceans, atm.)	0.0006	1.9	3.4	0.030	0.025
Price per kilogram of metal (1977)*	$630	$12	$35	$74	$108
Price per kilogram of chloride (1977)*	$156	$6.4	$5.2	$18	$6.7
Flame Color	–	–	brick red	crimson	green

*Lowest price from chemical supplier. Mg, Ca, $MgCl_2$, $CaCl_2$, and $BaCl_2$ are available in larger quantities at lower prices from industrial suppliers.

Compounds of calcium, strontium, and barium, like those of the alkali metals, produce bright colors when added to a flame. The characteristic colors are listed in Table 8.2. They are sometimes used to test for the presence of Ca^{2+}, Sr^{2+}, and Ba^{2+} in a solid sample or in water solution. The yellow color of Na^+ tends to mask those of the alkaline earth metals. Hence, sodium compounds must be absent if the test is to be conclusive.

Preparation of the Metals

The metals in Group 2, like those in Group 1, can be prepared by electrolysis of their molten chlorides. The general reaction is:

$$MCl_2(l) \rightarrow M(s) + Cl_2(g) \tag{8.9}$$

The metal (Mg, Ca, Sr, or Ba) is formed at one electrode. Chlorine gas collects around the other electrode. High temperatures (500–1000°C) are required to keep the chloride melted. Needless to say, Reaction 8.9, like 8.1, is not a suitable laboratory preparation. It is, however, used on a large scale industrially to prepare magnesium. We will have more to say about the preparation of magnesium in Chapter 25.

EXAMPLE 8.3

Write an equation for the electrolysis of calcium chloride. How many grams of calcium chloride are required to form 1.00 g of Ca?

SOLUTION

Following Equation 8.9, we have:

$$CaCl_2(l) \rightarrow Ca(s) + Cl_2(g)$$

The equation tells us that:

$$1 \text{ mol } CaCl_2 \simeq 1 \text{ mol Ca}$$

The atomic masses of Ca and Cl are 40.1 and 35.5 in that order. So, one mole of $CaCl_2$ has a mass of 40.1 g + 2(35.5 g) = 111.1 g

$$111.1 \text{ g } CaCl_2 \simeq 40.1 \text{ g Ca}$$

$$\text{mass } CaCl_2 \text{ required} = 1.00 \text{ g Ca} \times \frac{111.1 \text{ g } CaCl_2}{40.1 \text{ g Ca}} = 2.77 \text{ g } CaCl_2$$

Reactions of the Metals

As we move down Group 2, metallic character and hence chemical reactivity increases. Barium and strontium resemble the Group 1 metals in that they react readily with oxygen or air at room temperature. Calcium must be warmed before it will react with oxygen. Magnesium is even less reactive. It must be heated electrically or in a flame before it will burn. The general reaction is:

$$2 \text{ M(s)} + O_2(g) \rightarrow 2 \text{ MO(s)}; \qquad (\text{M} = \text{Be, Mg, Ca, Sr, Ba}) \quad (8.10)$$

The products are ionic, containing the O^{2-} ion and the +2 alkaline earth metal ion (Be^{2+}, Mg^{2+}, Ca^{2+}, Sr^{2+}, Ba^{2+}). They are among the highest melting of all known compounds (2000–3000°C). Reaction 8.10 is strongly exothermic in each case. ΔH is −292 kcal for Be, −288 kcal for Mg, −304 kcal for Ca, −282 kcal for Sr, and −267 kcal for Ba.

The same trend in reactivity appears in the behavior of the Group 2 metals towards water. Calcium, strontium, and barium, like the alkali metals, react with cold water to form hydrogen gas:

$$\text{M(s)} + 2 \; H_2O(l) \rightarrow M^{2+}(aq) + 2 \; OH^-(aq) + H_2(g); \quad (\text{M} = \text{Ca, Sr, Ba}) \quad (8.11)$$

The water solution formed contains the OH^- ion and the +2 ion of the alkaline earth metal (Ca^{2+}, Sr^{2+}, Ba^{2+}). Evaporation gives the metal hy-

In the laboratory the common Group 2 ions are Mg^{2+}, Ca^{2+}, and Ba^{2+}.

droxides, $Ca(OH)_2$, $Sr(OH)_2$, and $Ba(OH)_2$. In the case of calcium, Reaction 8.11 occurs slowly at room temperature. A steady stream of bubbles of hydrogen is formed. With the more active metal, barium, Reaction 8.11 occurs so rapidly that the hydrogen produced often catches fire.

The two lighter members of the group, beryllium and magnesium, do not react with water under ordinary conditions. However, at 100°C magnesium does react with steam:

$$Mg(s) + H_2O(g) \rightarrow MgO(s) + H_2(g); \qquad \Delta H = -86.0 \text{ kcal} \quad (8.12)$$

This reaction produces enough heat to ignite the hydrogen. Firemen, attempting to put out a magnesium fire by spraying water on it, have discovered this reaction, often with tragic results. The best way to put out a magnesium fire is to dump dry sand on it.

Uses of the Metals and Their Compounds

Of the Group 2 metals, only magnesium is produced in large quantities. About a million kilograms are produced annually in the United States. Beryllium is too expensive and too toxic to find any large-scale use. Calcium, strontium, and barium are too reactive. Magnesium is used principally as a structural metal, usually as an alloy with aluminum or other metals. Here, the low density of magnesium is a distinct advantage (d Mg = 1.74 g/cm^3, Al = 2.70 g/cm^3, Fe = 7.87 g/cm^3). Magnesium alloys are widely used in aircraft parts, where low density is critical. Wheels for racing cars ("mag wheels") are also made from a Mg-Al alloy.

The most widely used compound of the alkaline earth metals is calcium carbonate, $CaCO_3$. Marble statues consist of nearly pure $CaCO_3$, as does limestone. The Romans discovered that when calcium carbonate is heated in a fire it decomposes to calcium oxide, CaO, and carbon dioxide. A temperature of at least 800°C is required:

$$CaCO_3(s) \rightarrow CaO(s) + CO_2(g) \quad (8.13)$$

The calcium oxide (quicklime) produced reacts with water to produce calcium hydroxide (slaked lime):

CaO is called quicklime because this reaction goes fast and is so exothermic it can boil the water.

$$CaO(s) + H_2O(l) \rightarrow Ca(OH)_2(s) \quad (8.14)$$

Slaked lime is mixed with water and sand to give a paste which we know as mortar. As the paste dries it becomes hard ("sets") by reacting with carbon dioxide in the air:

$$Ca(OH)_2(s) + CO_2(g) \rightarrow CaCO_3(s) + H_2O(l) \quad (8.15)$$

EXAMPLE 8.4

Reasoning by analogy with the corresponding calcium compounds, write equations for the reactions that occur when:

a. barium carbonate is heated.

b. magnesium hydroxide reacts with CO_2.

c. strontium oxide reacts with water.

SOLUTION

a. $BaCO_3(s) \rightarrow BaO(s) + CO_2(g)$; compare 8.13

b. $Mg(OH)_2(s) + CO_2(g) \rightarrow MgCO_3(s) + H_2O(l)$; compare 8.15

c. $SrO(s) + H_2O(l) \rightarrow Sr(OH)_2(s)$; compare 8.14

8.3 GROUP 6: O, S, Se, Te

Occurrence

Of the elements in Group 6 of the Periodic Table, oxygen is by far the most abundant. Since its chemistry was discussed earlier, in Chapter 6, we will have little more to say on that subject here. The heavier members of the family, selenium, tellurium, and polonium, are extremely rare. All of the isotopes of polonium are radioactive and little is known about its chemistry. The element itself was discovered by Marie Curie, who named it for her native country, Poland.

Sulfur is widespread in nature, both as the free element and in combined form. Many of the less active metals, including copper, silver, mercury, and lead, occur as sulfide ores (Cu_2S, Ag_2S, HgS, PbS). Large deposits of elementary sulfur are found in Texas and Louisiana, several hundred feet below the ground. Water under pressure at a temperature of 170°C is pumped down to melt the sulfur. The liquid is then mixed with compressed air to form a froth which rises to the surface (Fig. 8.6, p. 196). Upon cooling, the sulfur solidifies, filling huge vats which may be half a kilometer long. This method of mining sulfur is referred to as the Frasch process, after the American chemist who devised it.

The sulfur produced by this method is very pure.

Properties of the Elements

Some of the properties of the elements in Group 6 are listed in Table 8.3. Notice that, as expected (Chapter 7), atomic radius *increases* and electronegativity *decreases* as we move from oxygen to tellurium. Of the four elements, oxygen has by far the strongest attraction for electrons. Oxygen, sulfur, and selenium are nonmetals while tellurium is a metalloid, on the boundary between metals and nonmetals.

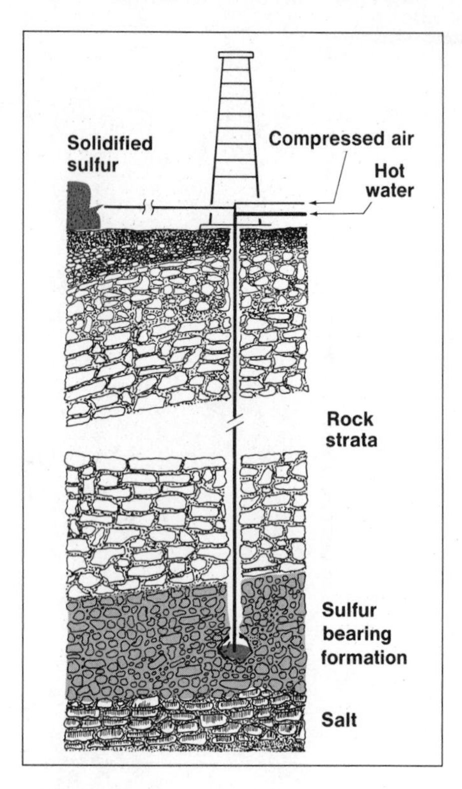

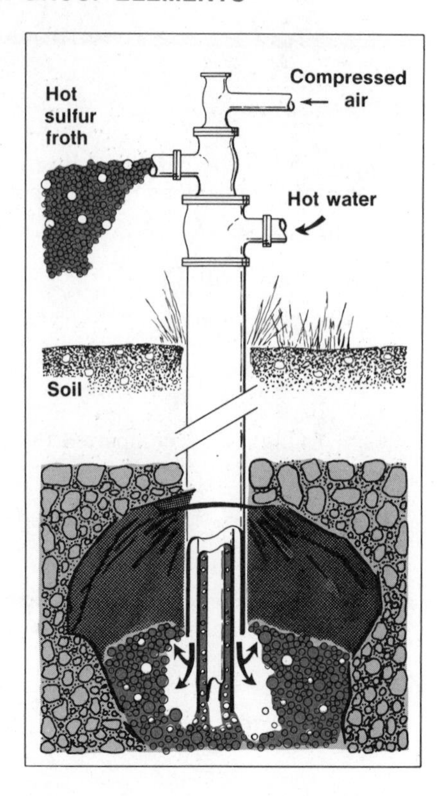

FIGURE 8.6

The Frasch process for sulfur mining. There are three concentric pipes used. Superheated water is pumped down one, air down the second; a froth containing hot molten sulfur comes up the third. The sulfur obtained by this process is very pure.

TABLE 8.3 PROPERTIES OF THE GROUP 6 ELEMENTS

	Oxygen	Sulfur	Selenium	Tellurium
Atomic Number	8	16	34	52
Atomic Mass	15.9994	32.06	78.96	127.60
Melting Point (°C)	−218	119	217	450
Boiling Point (°C)	−183	444	685	990
Density (O_2 at STP) in g/cm^3	0.00143(g)	2.07(s)	4.79(s)	6.24(s)
Electronegativity	3.5	2.5	2.4	2.1
Atomic Radius	0.066	0.104	0.117	0.137
Abundance (% of earth's crust, oceans, atm.)	49.5	0.060	9×10^{-6}	2×10^{-9}
Price per kilogram (1977)	$0.02	$0.20	$46	$26

TABLE 8.4 PROPERTIES OF O_2, O_3

	O_2	O_3
Molecular mass	32.0	48.0
Melting point, °C	−218	−192
Boiling point, °C	−183	−112
Density at STP (g/ℓ)	1.429	2.145

Ozone is poisonous. It has a characteristic odor you may have noticed near electric motors or generators.

Allotropy

All the elements in Group 6 show *allotropy*. That is, they exist in two different forms in the same physical state. In the case of oxygen, two different molecules are known in the gas state. One of these is the familiar diatomic molecule, O_2. A less stable allotrope of oxygen has the molecular formula O_3. It is known as *ozone* and can be formed by passing ordinary oxygen between two electrically charged plates. The formation of ozone is strongly endothermic:

$$3\ O_2(g) \rightarrow 2\ O_3(g); \qquad \Delta H = +68 \text{ kcal} \qquad (8.16)$$

Its properties are quite different from those of ordinary oxygen (Table 8.4).

Two different allotropic forms of solid sulfur are known, *rhombic* and *monoclinic* (Figure 8.7). When liquid sulfur freezes at 119°C, needle-like crystals of the monoclinic allotrope form. Upon further cooling,

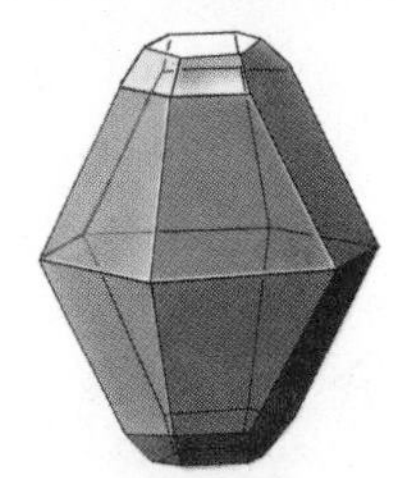

Rhombic sulfur

Monoclinic sulfur

FIGURE 8.7

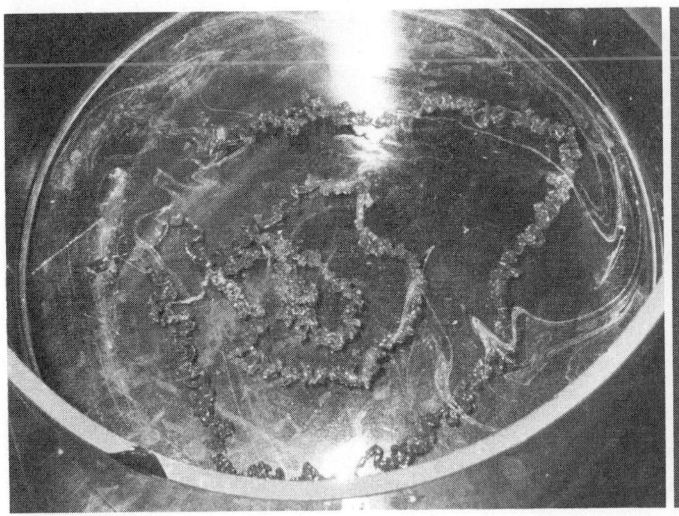

FIGURE 8.8

Plastic sulfur. The plastic sulfur was obtained by heating sulfur until it became so viscous it was reluctant to flow. It was then poured into the water in the large tank, where it cooled to a rubbery material. In the photo on the right the material shows its elastic properties. Within about 24 hours this material reverted to hard crystalline sulfur.

these are slowly converted to rhombic sulfur, which is stable below 96°C.

Liquid sulfur at the melting point is a free-flowing, straw-colored liquid. Upon heating to 160°C, the liquid becomes extremely viscous and turns to a deep reddish-brown color. If cooled quickly by pouring into cold water (Figure 8.8), it forms a sticky, plastic solid called amorphous sulfur (amorphous solids are ones which do not have a distinct crystal form). Upon standing for some time at room temperature, the amorphous sulfur slowly converts to rhombic crystals.

Reactions of Sulfur

Because of its lower electronegativity (2.5 *vs.* 3.5), sulfur is somewhat less reactive toward metals than is oxygen. It does, however, react upon heating with many metals, including magnesium and iron. The product is an ionic solid containing the sulfide ion, S^{2-}, and a positive metal ion (Mg^{2+}, Fe^{2+}).

$$Mg(s) + S(s) \rightarrow MgS(s) \tag{8.17}$$

$$Fe(s) + S(s) \rightarrow FeS(s) \tag{8.18}$$

When sulfur is heated strongly in air, it ignites and burns with a blue flame. The principal product is sulfur dioxide, a gas with a choking odor.

If you happen to have walked near the volcano on Hawaii Island, you know what SO_2 smells like.

$$S(s) + O_2(g) \rightarrow SO_2(g) \tag{8.19}$$

EXAMPLE 8.5

Noting that selenium and sulfur are in the same group of the Periodic Table, write balanced equations for the reaction of selenium with:

a. Fe b. O_2

SOLUTION

a. $Fe(s) + Se(s) \rightarrow FeSe(s)$; compare Eqn. 8.18

b. $Se(s) + O_2(g) \rightarrow SeO_2(s)$; compare Eqn. 8.19

Uses of Sulfur and Selenium

The principal use of sulfur is in the preparation of sulfuric acid, to be described in Chapter 23. Smaller amounts are used in vulcanizing rubber, in insecticides and fertilizers, and in preparing other sulfur

Sulfuric acid is H_2SO_4.

compounds. Traces of selenium compounds are added to glass to offset the green color caused by Fe^{2+} ions. Larger amounts of selenium give a deep, ruby-red color to glass. The red glass used in traffic lights contains selenium compounds.

Elementary selenium, which is near the borderline between metals and nonmetals in the Periodic Table, is a semiconductor. Its conductivity increases by a factor of 1000 upon exposure to light. This effect was used in some of the early photocells used to operate mechanical devices. More recently, selenium compounds containing the Se^{2-} ion have been developed for use in solar cells.

The Ozone Layer

In the upper atmosphere, at an altitude of 20–40 km, there is a relatively high concentration of ozone, O_3. This is formed by the action of ultraviolet light on oxygen molecules (Reaction 8.16). The production of ozone by this reaction plays a vital role so far as life on earth is concerned. Ozone absorbs harmful ultraviolet radiation from the sun. If this radiation were to reach the earth's surface, it would have several adverse effects. Among other things, it would drastically increase the incidence of skin cancers.

In recent years, fears have been expressed that the ozone layer might be at least partially destroyed by human activities. One natural process which consumes ozone is:

$$O_3(g) + O(g) \rightarrow 2\ O_2(g) \qquad (8.20)$$

This reaction involves the collision between an ozone molecule and a free oxygen atom. Fortunately, it occurs rather slowly. Under ordinary conditions, ozone is produced at the same rate that it decomposes. So, the ozone layer is maintained.

It is known that Reaction 8.20 is catalyzed (speeded up) by several foreign molecules and atoms. One such species is the chlorine atom, Cl. If there were appreciable numbers of chlorine atoms in the upper atmosphere, Reaction 8.20 would occur more rapidly. Until 1974, no one worried very much about this because there was no known source of free Cl atoms near the ozone layer. Then it was discovered that compounds such as CF_2Cl_2 can decompose in sunlight to give Cl atoms:

$$CF_2Cl_2(g) \rightarrow CF_2Cl(g) + Cl(g) \qquad (8.21)$$

Compounds of this type, which go under the trade name of Freon, were widely used in aerosol sprays. They are now being phased out for this purpose, even though there is no clear evidence that they produce enough chlorine atoms to be harmful. The judgment is that the ozone layer is too important to take any chances with it.

8.4 GROUP 7: F, Cl, Br, I

Occurrence

The halogens, like the alkali and alkaline earth metals, are too reactive to occur as the free elements. Instead, they are found in ionic compounds as -1 ions (F^-, Cl^-, Br^-, I^-). The principal source of fluorine is the mineral fluorite, CaF_2. Chlorine is present as the Cl^- ion in such

In the laboratory, the common Group 7 elements are Cl, Br, and I. We usually encounter them as negative ions.

solids as sodium chloride, NaCl, and potassium chloride, KCl. It is also the most abundant negative ion in seawater. Even though the concentration of Br^- ions in seawater is quite low (8×10^{-4} mol/ℓ), this is a major source of the element. Iodine is obtained from certain oil-well brines where the concentration of I^- ions is relatively high. Astatine is not found in nature. Like the heaviest members of Groups 1, 2, and 6, all of its isotopes are radioactive.

Properties of the Halogens

Trends in the properties of the halogens are shown in Table 8.5. Notice the steady increase in melting point and boiling point with molecular mass. This trend is generally observed with all molecular substances, for reasons we will consider in Chapter 12. The first two members of the family, fluorine and chlorine, are gases at room temperature and atmospheric pressure. Bromine is a deep-red, volatile liquid with a vapor pressure of about 0.3 atm at 25°C. Iodine is a shiny, black solid. Upon heating, it is readily converted to a violet vapor (see color plate 3C, center of book).

Comparing the entries in Table 8.5 to those in Table 8.3, we see that the halogens as a group are more strongly nonmetallic than the elements in Group 6. Consider, for example, their electronegativities. Chlorine (EN = 3.0) has a stronger attraction for electrons than its neighbor in Group 6, sulfur (EN = 2.5). The same is true of fluorine (EN = 4.0) *versus* oxygen (EN = 3.5). Halogen atoms acquire electrons very readily

TABLE 8.5 PROPERTIES OF THE HALOGENS

	Fluorine	Chlorine	Bromine	Iodine
Atomic Number	9	17	35	53
Molecular Formula	F_2	Cl_2	Br_2	I_2
Molecular Mass	38.0	70.9	159.8	253.8
Melting Point (°C)	−220	−101	−7	114
Boiling Point (°C)	−188	−34	59	184
Density (F_2, Cl_2 at STP) g/cm³	0.00170(g)	0.00321(g)	3.12(l)	4.94(s)
Color	pale yellow	yellow-green	dark red	black (violet)
Electronegativity	4.0	3.0	2.8	2.5
Atomic Radius, nm	0.064	0.099	0.114	0.133
Abundance (% of earth's crust, oceans, atm.)	0.027	0.19	0.00016	0.00003
Price per kilogram, 1977	$45	$0.15	$1.7	$8.8

from metal atoms, forming −1 ions. The first member of the group, fluorine, is by far the most reactive of all nonmetals.

In fact, it is the most reactive element.

All of the free halogens are toxic and corrosive to the skin. Fluorine is so reactive that it must be stored and handled with special equipment. It is seldom if ever used in the teaching laboratory. Chlorine and bromine are usually found in the laboratory as saturated water solutions. These elements produce painful burns (or worse) if they come in contact with the skin or mucous membranes. Iodine is much safer to work with, but even here you should avoid direct contact with the crystals or vapor.

Preparation of the Halogens

The two heavier halogens, Br_2 and I_2, can be prepared by bubbling chlorine gas through a water solution containing Br^- or I^- ions. The more electronegative Cl_2 molecule (EN Cl = 3.0, Br = 2.8, I = 2.5) picks up electrons from Br^- or I^- ions and is itself converted to Cl^- ions.

$$Cl_2(g) + 2\ Br^-(aq) \rightarrow 2\ Cl^-(aq) + Br_2(l) \qquad (8.22)$$

$$Cl_2(g) + 2\ I^-(aq) \rightarrow 2\ Cl^-(aq) + I_2(s) \qquad (8.23)$$

The same reactions are used commercially to extract bromine and iodine from salt brines.

Elementary chlorine, because of its strong attraction for electrons, is more difficult to prepare. Commercially, it is obtained by electrolysis of sodium chloride, either molten (Reaction 8.1) or in water solution. In the laboratory, small amounts of chlorine are prepared by reacting manganese dioxide, MnO_2, with hydrochloric acid (H^+, Cl^- ions).

$$MnO_2(s) + 4\ H^+(aq) + 2\ Cl^-(aq) \rightarrow Mn^{2+}(aq) + Cl_2(g) + 2\ H_2O \qquad (8.24)$$

Br_2 and I_2 can also be made by a reaction like this.

This reaction should be carried out in a well-ventilated hood. Breathing fumes of chlorine gas is unpleasant at best and can be a great deal more serious. Chlorine was the first poison gas used in warfare, in 1917.

EXAMPLE 8.6

If you want to prepare 1.00 ℓ of Cl_2 gas at 0°C and 1 atm by Reaction 8.24, how many grams of MnO_2 will you need?

SOLUTION

According to Equation 8.24:

$$1\ \text{mol}\ MnO_2 \simeq 1\ \text{mol}\ Cl_2$$

To relate grams of MnO_2 to liters of Cl_2, we note that:

$$1\ \text{mol}\ MnO_2 = 54.9\ \text{g} + 2(16.0\ \text{g}) = 86.9\ \text{g}$$

$$1\ \text{mol}\ Cl_2 = 22.4\ \ell\ \text{(STP)}$$

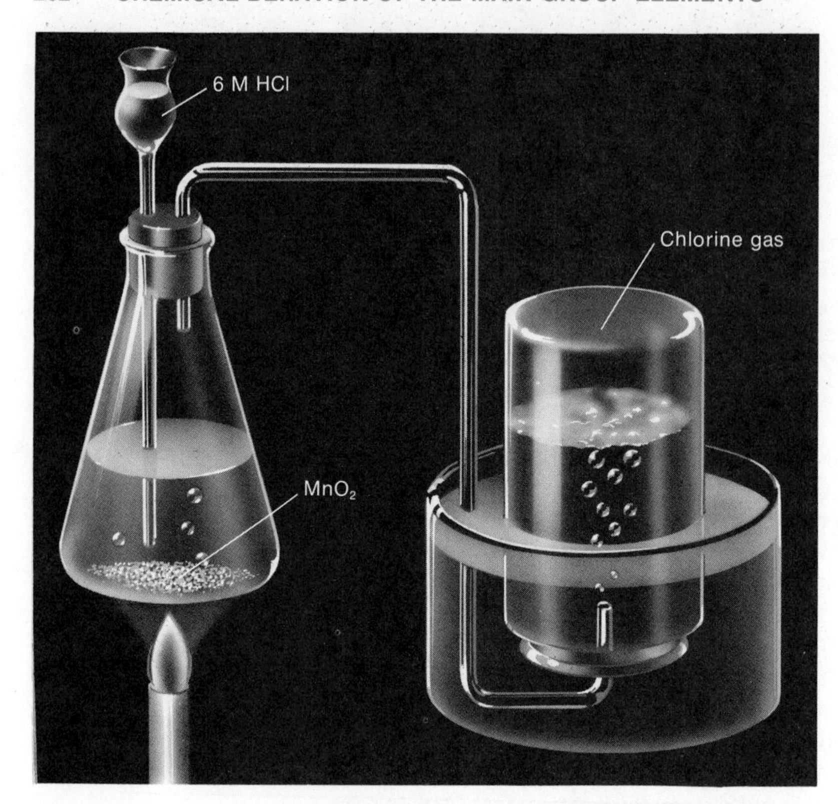

FIGURE 8.9

Preparation of chlorine. If MnO_2 in hydrochloric acid is heated, one obtains Cl_2 gas, which can be collected over water and used for other reactions.

Hence: $$86.9 \text{ g MnO}_2 \simeq 22.4 \ \ell \text{ Cl}_2$$

Thus: $$\text{mass MnO}_2 = 1.00 \ \ell \text{ Cl}_2 \times \frac{86.9 \text{ g MnO}_2}{22.4 \ \ell \text{ Cl}_2} = 3.88 \text{ g MnO}_2$$

Fluorine, the most electronegative of the halogens, cannot be made from F^- ions in water solution. The free element, if it were produced, would decompose water to liberate oxygen gas. Commercially, fluorine is made by the electrolysis of liquid hydrogen fluoride, containing dissolved potassium fluoride.

F_2 is usually made where it is to be used, and is used as fast as it is prepared.

$$2 \text{ HF(l)} \rightarrow \text{H}_2\text{(g)} + \text{F}_2\text{(g)} \quad (8.25)$$

Reactions of the Halogens

In reacting with metals, the halogens pick up electrons and are converted to negatively charged ions (F^-, Cl^-, Br^-, I^-). The reactions with sodium and magnesium are typical:

$$2 \text{ Na(s)} + \text{X}_2 \rightarrow 2 \text{ NaX(s)} \quad (8.26)$$

(X = F, Cl, Br, or I; similar reactions with other Group 1 elements)

$$\text{Mg(s)} + \text{X}_2 \rightarrow \text{MgX}_2\text{(s)} \quad (8.27)$$

(X = F, Cl, Br, I; similar reactions with other Group 2 elements)

In the case of fluorine, the reactions are violent and explosive. With chlorine, they can be spectacular (Figure 8.10) but dangerous. With the less electronegative elements bromine and iodine, reaction occurs more slowly.

The reactions of the halogens with hydrogen were discussed in Chapter 6. The general equation is:

$$H_2(g) + X_2 \rightarrow 2\ HX(g);\ X = F, Cl, Br, I \qquad (8.28)$$

In practice, these reactions are seldom used to prepare the *hydrogen halides,* HF, HCl, HBr, and HI. Reaction 8.28 is too difficult to control with F_2 and Cl_2, too slow with Br_2, and gives a poor yield of product with I_2. Instead, the hydrogen halides are made by adding an acid (H^+ ions) to *halide* (X^-) ions. The reaction is:

$$H^+(aq) + X^-(aq) \rightarrow HX(g) \qquad (8.29)$$

In the case of HCl, sulfuric acid acts as a source of H^+ ions. Sodium chloride, either as a solid or in concentrated water solution, provides the Cl^- ions (Figure 8.11).

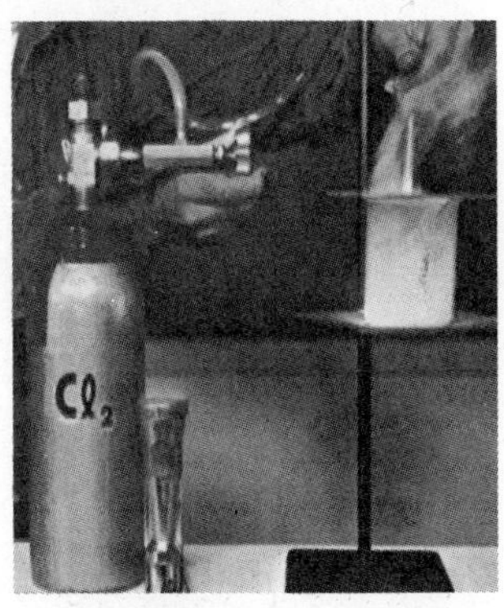

FIGURE 8.10

Sodium metal, in the beaker, ignites upon exposure to chlorine gas. The product of this exothermic reaction is sodium chloride, NaCl. From the CHEM Study Film, *Chemical Families.*

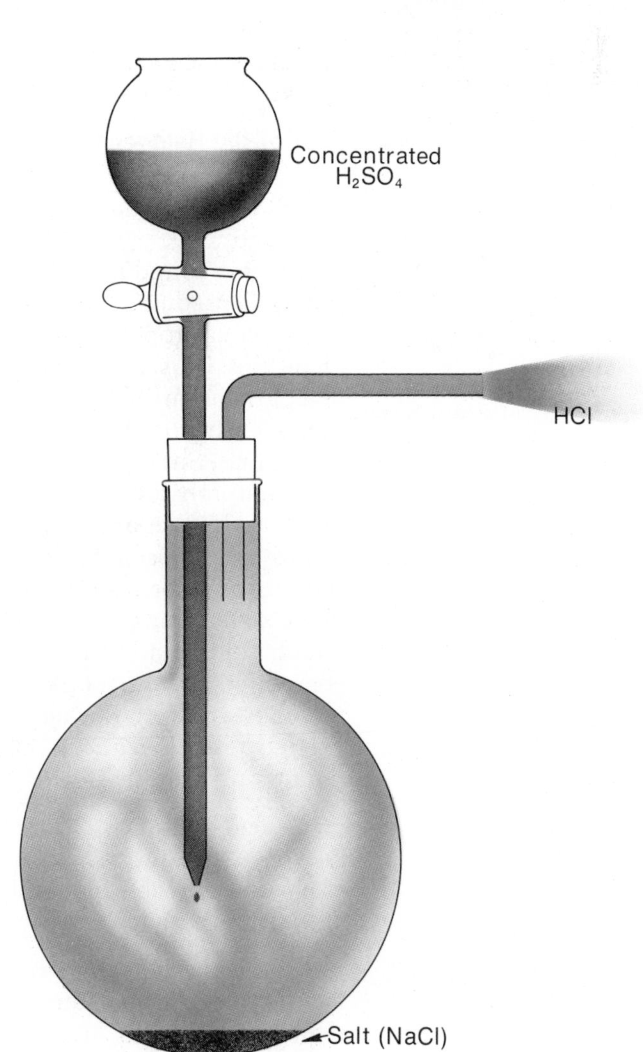

FIGURE 8.11

Preparation of HCl. Hydrogen chloride gas can be made by adding sulfuric acid, H_2SO_4, to solid NaCl. The gas must be used as it is prepared, since it is too soluble in water to be collected by the method used for several other gases (see Figure 8.9).

EXAMPLE 8.7

Using Equations 8.26–8.28 as a reference, write balanced equations for the reactions of iodine with

a. potassium b. calcium c. hydrogen

SOLUTION

a. Recalling that potassium, like sodium, is in Group 1, we use Reaction 8.26 as a model. Remember that iodine is a solid.

$$2\ K(s) + I_2(s) \rightarrow 2\ KI(s)$$

b. Calcium is a Group 2 metal, so will react like magnesium in 8.27:

$$Ca(s) + I_2(s) \rightarrow CaI_2(s)$$

c. $H_2(g) + I_2(s) \rightarrow 2\ HI(g)$

Uses of the Halogens

Fluorine, F_2, has few commercial uses, largely because it is so dangerous to work with. Several organic compounds containing fluorine atoms have found a market. One of these is "Freon", CCl_2F_2. This compound is used as a coolant in refrigerators, and, at least until recently, as a gaseous propellant in aerosol sprays. Sodium fluoride, in very small quantities, is used to "fluoridate" water supplies. At a concentration of one part per million (one gram per 10^6 g of water), F^- ions are effective in preventing tooth decay. Many toothpastes contain a nearly insoluble compound, SnF_2, which has the same function.

Nearly all dentists agree that fluoridation of water is a good idea.

You probably associate the acrid odor of chlorine with swimming pools. In large municipal pools, chlorine gas is added to the water to destroy algae, bacteria, and other organic matter. Much lower concentrations of chlorine are added to drinking water supplied to cities. In home swimming pools, water is "chlorinated" by adding solid tablets which contain a chemical such as sodium hypochlorite, NaOCl. This compound and others like it react with water to produce free chlorine.

Industrially, chlorine is used as a bleach for wood pulp and cotton. It is also required in the manufacture of many familiar organic compounds. Among these are:

—carbon tetrachloride, CCl_4, and chloroform, $CHCl_3$, used as solvents.

—insecticides, including DDT (dichlorodiphenyltrichloroethane).

—plastics such as polyvinylchloride (PVC).

For many years the principal use of bromine has been in the manufacture of dibromoethane, $C_2H_4Br_2$. This compound is used with lead tetraethyl, $Pb(C_2H_5)_4$, to improve the antiknock quality of gasoline. However, the anti-pollution devices ("catalytic converters") installed in cars since 1975 are fouled by lead compounds. For this reason, the market for dibromoethane and hence for bromine itself is declining.

Large quantities of bromine and iodine are used to make the silver compounds of these halogens, AgBr and AgI. Both of these compounds are light-sensitive and so find use in photography. Silver iodide is also used on a limited scale in cloud seeding. It promotes the formation of ice crystals in clouds, perhaps because its crystal structure is similar to that of ice. The ice crystals serve as nuclei for the formation of rain drops.

Iodine is perhaps most familiar in the form of its alcohol solution, tincture of iodine, used as an antiseptic. Iodide ions, in small amounts, are essential to our diet. They are required to form the growth-regulating hormone, thyroxine, $C_{15}H_{11}O_4NI_4$. This compound is produced by the thyroid gland. If iodide ions are lacking, the gland expands to give the condition called goiter. To prevent this, about 0.01% of sodium iodide, NaI, is added to ordinary table salt to give "iodized salt".

8.5 SUMMARY

In this chapter, we have discussed the chemistry of four different groups of elements. Two of these, Groups 1 and 2, are metals. As you would expect from their position in the Periodic Table:

—the Group 1 elements are more strongly metallic than their neighbors in Group 2. As a result, the Group 1 metals are more reactive toward nonmetals and water.

—the heavier metals in each group are more reactive than the lighter ones.

In all of their compounds, the Group 1 metals are present as +1 ions, the Group 2 metals as +2 ions.

The other two groups considered, 6 and 7, consist of nonmetals. As you would expect from their positions in the Periodic Table:

—the Group 7 elements are more electronegative than those in Group 6 and hence more reactive toward metals and hydrogen.

—the lighter nonmetals in each group are more electronegative than the heavier ones and hence more reactive.

In their compounds with the metals, the Group 6 and 7 elements are present as −2 and −1 ions respectively. The compounds of these elements with oxygen and hydrogen are molecular.

Of the elements we have considered, oxygen is by far the most abundant (49.5%). Fourteen of the other elements (Li, Na, K, Rb, Cs in Group 1; Mg, Ca, Sr, Ba in Group 2; S in Group 6; F, Cl, Br, I in Group 7) are abundant enough to have a well-developed chemistry and some commercial use. Only one of these, sulfur, occurs as the free element. The others must be prepared from their compounds. You should be familiar with the methods used to do this.

There are about 30 different chemical equations in this chapter.

All of them represent reactions that occur in the "real world"; in industry, in the home, or in the teaching laboratory. By this time, you should have developed the ability to write such equations on your own and use them. The several Examples presented in this chapter are designed to help you do this. For additional practice, try working the questions and problems that follow.

NEW TERMS

alkali metals
alkaline earth metals
allotropy
baking soda ($NaHCO_3$)
electrolysis
flame color
Frasch process
halide ion (F^-, Cl^-, Br^-, I^-)
halogen (F_2, Cl_2, Br_2, I_2)
hard water
limestone ($CaCO_3$)
lye ($NaOH$)
monoclinic sulfur
ozone (O_3)
quicklime (CaO)
rhombic sulfur
rock salt
slaked lime ($Ca(OH)_2$)
washing soda (Na_2CO_3)

QUESTIONS

1. What is the charge of the ions formed by the elements in

   a. Group 1 b. Group 2 c. Group 6 d. Group 7?

2. Which of the elements occur in the uncombined state in nature in

   a. Group 1 b. Group 2 c. Group 6 d. Group 7?

3. What is the most abundant element in

   a. Group 1 b. Group 2 c. Group 6 d. Group 7?

4. Give the symbol of an element all of whose isotopes are radioactive in

   a. Group 1 b. Group 2 c. Group 6 d. Group 7

5. Which element has the largest atomic radius in

   a. Group 1 b. Group 2 c. Group 6 d. Group 7?

6. Which element has the largest ionization energy in

   a. Group 1 b. Group 2 c. Group 6 d. Group 7?

7. Which of the following elements are ordinarily obtained from their compounds by electrolysis?

   a. Na b. Mg c. Ca d. S e. Cl f. F

8. Which of the following elements form diatomic molecules which are stable at 25°C and 1 atm?

   a. Na b. Ca c. O d. F e. Cl f. I

9. Explain why the metals in Group 2 have higher densities than those in Group 1.

10. Describe at least one use for:

    a. Na b. NaCl c. NaOH d. Na_2CO_3 e. $NaHCO_3$

11. Give the formula of the principal component of

    a. limestone b. marble c. pearl d. seashells

12. What is meant by hard water? What are some of the problems associated with it?

13. Why are alloys of magnesium used in aircraft parts?

14. Describe the process by which sulfur is obtained from underground deposits.

15. Explain what is meant by allotropy. Illustrate by describing the allotropy shown by oxygen and sulfur.

16. Which two of the halogens are *not* obtained commercially from seawater?

17. Of the four halogens, which one has the highest

    a. electronegativity b. density c. atomic radius
    d. melting point?

18. Would you expect Br_2 to react with I^- ions in water solution? Justify your answer.

19. Give the formulas of

    a. sodium hydroxide b. sodium carbonate
    c. sodium hydrogen carbonate

20. Give the formulas of:

    a. calcium hydroxide b. calcium carbonate c. calcium oxide

21. Give the formulas of:

    a. magnesium sulfide b. sulfur dioxide

22. Give the formulas of:

    a. potassium chloride b. magnesium chloride
    c. hydrogen chloride

PROBLEMS

The first 17 problems in this set deal with writing equations. Problems 1 through 6 involve writing equations which already appear in the text. Problems 7 through 13 require that you write equations similar to those in the text (see Examples 8.1, 8.3, 8.4, 8.5, 8.7). In Problems 14 through 17, you are asked to carry out calculations involving balanced equations (see Examples 8.2, 8.6).

1. Write balanced equations for the reaction of sodium with

    a. O_2 b. H_2O

2. Write a balanced equation for the reaction of magnesium with steam.

3. Write a balanced equation to explain what happens when

   a. calcium carbonate is heated.
   b. quicklime is added to water.
   c. mortar hardens.

4. Write balanced equations for the reaction of sulfur with

   a. Mg b. Fe c. O_2

5. Write balanced equations for the reactions that occur when Cl_2 gas is bubbled through water solutions containing:

   a. Br^- ions b. I^- ions

6. Write balanced equations for

   a. a laboratory preparation of Cl_2.
   b. a commercial preparation of Cl_2.

7. Write balanced equations (compare Eqns. 8.2, 8.3) for the reaction of lithium with

   a. O_2 b. H_2O

8. Write balanced equations (compare Eqns. 8.1, 8.9) for the preparation of

   a. K from KCl b. Sr from $SrCl_2$

9. Write balanced equations (compare Eqns. 8.10, 8.11) for the reaction of barium with

   a. O_2 b. H_2O

10. Write balanced equations (compare Eqns. 8.17, 8.19) for the reaction of tellurium with

    a. Mg b. O_2

11. On your worksheet complete and balance the following equations.

    a. $Mg(s) + I_2(s) \rightarrow ?$ b. $Na(s) + I_2(s) \rightarrow ?$

12. Complete and balance the following equations.

    a. $Rb(s) + H_2O \rightarrow ?$ b. $Ca(s) + H_2O \rightarrow ?$ c. $Sr(s) + O_2(g) \rightarrow ?$

13. Complete and balance the following equations.

    a. $H_2(g) + Br_2(l) \rightarrow ?$ b. $Br_2(l) + I^-(aq) \rightarrow ?$
    c. $MnO_2(s) + Br^-(aq) + H^+(aq) \rightarrow ?$

14. Consider Equation 8.3. When 0.120 g of Na reacts with excess H_2O:

    a. how many moles of H_2 are produced?
    b. what volume of H_2 is produced at STP?

15. Consider Equation 8.13. In order to produce 2.00 ℓ of CO_2 at STP, how many moles of $CaCO_3$ should be used? how many grams of $CaCO_3$?

16. Consider Example 8.6. If you wanted to make 1.00 ℓ of $Cl_2(g)$ at 25°C and 1 atm, how many grams of MnO_2 should you use?

17. Consider Equation 8.27. What mass of iodine, in grams, is required to form 1.00 g of MgI_2?

* * * * *

18. What are the percentages by mass of the elements in the mineral beryl, whose formula is given on page 191?

19. Sodium hydroxide is made from sodium chloride. What mass, in kilograms, of NaCl is required to make the 1.0×10^{10} kg of NaOH produced annually in the United States?

20. Referring to Equation 8.10, how much heat is evolved when 1.00 g of Sr burns in oxygen?

21. Compare the density of ozone given in Table 8.4 to that predicted from the gas laws.

22. Compare the density of chlorine gas given in Table 8.5 to that predicted from the gas laws.

*23. When dibromoethane reacts with lead tetraethyl and air in an automobile engine, the products are $PbBr_2(s)$, $CO_2(g)$, and $H_2O(l)$. Write a balanced equation for the reaction.

*24. A certain mortar mix contains 40% by mass of $Ca(OH)_2$. How many kilograms of limestone, $CaCO_3$, were consumed to produce a 20-kg bag of this mix?

*25. Consider the element sodium, which has an atomic mass of 22.990 and an atomic radius of 0.186 nm.

a. What is the mass in grams of a sodium atom?
b. What is the radius in centimeters of a sodium atom?
c. What is the volume in cubic centimeters of a sodium atom ($V = \frac{4}{3}\pi r^3$)
d. What is the density of a sodium atom in g/cm^3?

An artist's conception of the motion of electrons in atoms. Since we have no way of knowing what these tiny particles are doing, his ideas are as good as any.

9 ELECTRONIC STRUCTURE OF ATOMS

The existence of the Periodic Table indicates that the elements are related in some very fundamental way.

In Chapters 7 and 8, we showed how the Periodic Table is used to correlate the properties of elements. Our approach was a factual one. We described *how* elements behave without attempting to explain *why* they react as they do. Chemists, from the time of Mendeleev, searched for a rational explanation of the Periodic Table. In particular, they sought to learn more about the structure of atoms. They believed that a knowledge of atomic structure would tell them why certain atoms such as F, Cl, Br, and I behave chemically in a very similar way.

Early in this century it became clear that the key to understanding the chemical behavior of atoms lay in their electronic structures. You will recall (Chapter 2) that atoms consist of tiny, positively charged nuclei surrounded by a "cloud" of electrons. The number of electrons in that cloud is exactly equal to the number of protons in the nucleus. This number is called the atomic number of the element. Thus we have:

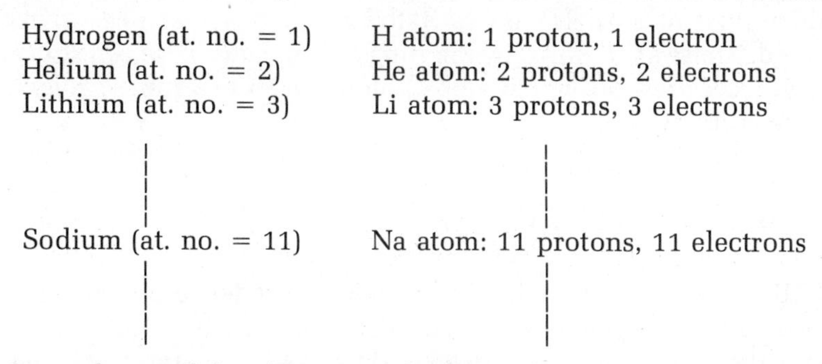

To understand the chemistry of these and other elements, we need to know how the electrons in an atom are distributed about the nucleus. In particular, we need to know something about the relative energies of these electrons. These aspects of the electronic structure of atoms will be considered in this chapter. We start with the simplest atom, that of hydrogen, which contains a single electron.

9.1 THE BOHR MODEL OF THE HYDROGEN ATOM

During the first decade of this century, several different models were proposed for the hydrogen atom. All of these were based on the classical laws of physics. None of them agreed with experiment. Then, in 1913, Niels Bohr, a young Danish physicist working with Rutherford, made a revolutionary suggestion. He maintained that the electron in the hydrogen atom need not obey all the laws of physics that apply to the much larger systems studied in the laboratory. Following that idea, Bohr constructed a simple model of the hydrogen atom. Unlike previous models, Bohr's was in excellent agreement with experiment.

Quantum Theory of the Electron

The basic postulate of the Bohr model was a simple one. He applied to the electron in the hydrogen atom a more general idea proposed a few years earlier by Max Planck.

The hydrogen electron can only exist in certain states (energy levels) associated with definite amounts of energy. When the electron changes its state, it must absorb or emit the exact amount of energy which will bring it from the initial to the final state.

According to this postulate, an electron in a hydrogen atom can have only certain definite amounts of energy. Under normal conditions, in its *ground state*, it has a certain energy which we might call E_1. When it moves to an *excited state* where its energy is E_2, it absorbs an amount of energy exactly equal to the difference between E_2 and E_1.

The ground state is the lowest energy state available to the electron.

$$\Delta E_{electron} = E_2 - E_1 \qquad (9.1)$$

FIGURE 9.1

A man on a ladder changes his energy by fixed amounts as he moves from one rung to another. An atom has energy levels similar to those fixed by the ladder rungs. In an atom the energy levels (rungs) are not equally spaced.

One way to describe this situation is to say that the energy of the electron is quantized. It can change only by certain discrete amounts, "quantum jumps". We can draw an analogy to a person climbing a ladder (Figure 9.1). He is restricted to certain distances above the ground. These are defined by the rungs of the ladder. He can stop at any rung he chooses, but not at a position between two rungs (at least, not for very long!).

Bohr Orbits

Bohr assumed that in the normal hydrogen atom, the electron moved about the nucleus in a circular orbit, much as the earth moves about the sun. So long as the electron stays in this orbit, its energy is constant. However, if the electron becomes excited by absorbing energy, it can jump out to another orbit farther away from the nucleus. In this orbit, it has a higher energy. The fact that electron energies are quantized he explained by assuming that only certain orbits of fixed radius are available to the electron. A few of these orbits are shown in Figure 9.2. The smallest one, closest to the nucleus, is occupied by the electron in its normal or ground state. The other orbits, farther out from the nucleus, represent excited, high-energy states.

Energies of the Hydrogen Electron

Using this model, Bohr developed an equation for the energy levels (states) available to the electron in the hydrogen atom:

$$E_{\text{mole electrons}} = \frac{-313.6\ \text{kcal}}{n^2} \qquad (9.2)$$

In this equation n, often referred to as a *quantum number*, can be any integer:

$$n = 1, 2, 3, 4, \ldots$$

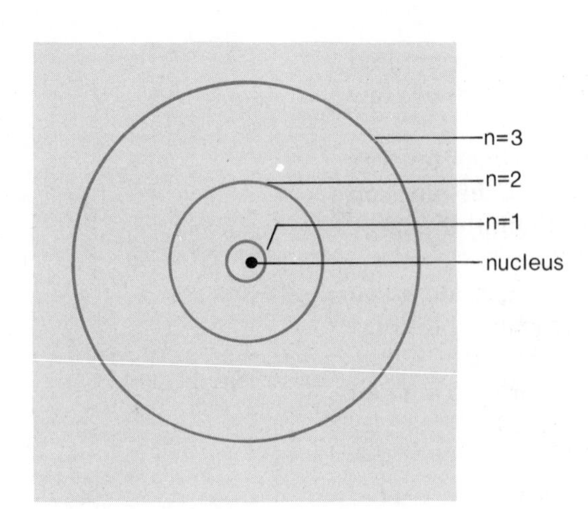

FIGURE 9.2

Bohr orbits in an H atom. By Bohr's theory, which was later discarded, the electron in the H atom moves in a circular orbit around the nucleus. The radii of the allowed orbits increase as n^2, so in the n = 2 orbit the radius is 4 times that in the n = 1 orbit. For high n values the atom is much larger than it is in the ground state.

When n = 1, Equation 9.2 gives the energy of a mole of electrons in the ground state:

$$E\ (n = 1) = -313.6 \text{ kcal}$$

Note that −313.6 kcal is a *lower* energy than −78.4 kcal.

$$E\ (n = 2) > E\ (n = 1)$$

If we set n = 2, we obtain the molar energy of electrons in the first excited state:

$$E\ (n = 2) = \frac{-313.6 \text{ kcal}}{4} = -78.4 \text{ kcal}$$

By substituting n = 3, 4, . . ., we can obtain molar energies of electrons in successively higher excited states, corresponding to orbits farther and farther out from the nucleus.

The values calculated from Equation 9.2 for several energy levels of the hydrogen electron are shown in Figure 9.3. Two features of this figure (and of Equation 9.2) are worth noting.

1. The fact that all the energy levels have negative values (−313.6, −78.4, . . .) is a result of the way in which Bohr chose his zero of energy. He took the energy to be zero when the electron is completely removed from the atom. Since energy must be absorbed to remove the electron, this means that the energy of the electron while it is in the atom must be less than zero.

2. As n increases, the energy levels become closer and closer together. As n becomes very large, approaching infinity, E approaches a limit of zero.

$$E\ (n = \infty) = \frac{-313.6 \text{ kcal}}{(\infty)^2} = 0$$

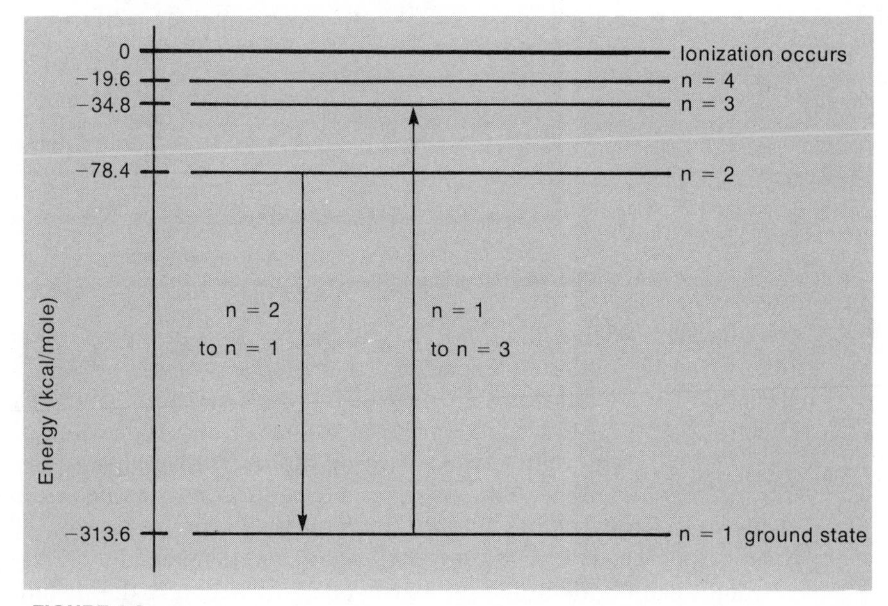

FIGURE 9.3

Energy levels in the H atom. These levels are the same in modern quantum theory. As n gets larger the levels get closer together. As n approaches infinity the energy of the level approaches zero. Atomic spectra arise from transitions from higher levels to lower ones.

Actually, for an atom 1 mm is essentially an infinite distance.

Physically, this corresponds to a situation where the electron is at an infinite distance from the nucleus. In other words, the electron has been completely removed from the atom. Ionization has occurred:

$$H(g) \rightarrow H^+(g) + e^-$$

Using Equation 9.2, it is possible to calculate the energy change when the hydrogen electron moves from one state to another. The way in which this is done is illustrated in Example 9.1.

EXAMPLE 9.1

Calculate ΔE, per mole of electrons, going from

a. n = 2 to n = 1 b. n = 1 to n = 3

SOLUTION

a. As we have seen, E(n = 1) = −313.6 kcal and E(n = 2) = −78.4 kcal.

$$\Delta E = E(n = 1) - E(n = 2)$$
$$= -313.6 \text{ kcal} - (-78.4 \text{ kcal}) = -235.2 \text{ kcal}$$

ΔE is *negative* because energy is *evolved* when an electron moves from an excited state (n = 2) back to the ground state (n = 1).

b. $E(n = 3) = \dfrac{-313.6 \text{ kcal}}{9} = -34.8 \text{ kcal}$

$$\Delta E = E(n = 3) - E(n = 1)$$
$$= -34.8 \text{ kcal} - (-313.6 \text{ kcal}) = +278.8 \text{ kcal}$$

ΔE is *positive*, because energy must be *absorbed* to raise an electron from the ground state (n = 1) to an excited state (n = 3).

The Quantum Theory and Atomic Spectra

The quantum theory of the electron in the hydrogen atom is well supported by experimental evidence. Indeed, we now know that electrons in all atoms are restricted to certain definite energies. The best evidence for this statement comes from a study of *atomic spectra*. These refer to the light given off when atoms of elements are excited by absorbing energy. This energy can come from a flame, as is the case with the elements in Groups 1 and 2. More commonly, the energy is supplied by a spark or electrical discharge, as in a fluorescent light or neon sign.

Light can be identified by its *wavelength*, which is the distance between successive peaks of the light wave. We always find that excited atoms give off light at definite, discrete wavelengths. With sodium, the color is due to light given off at two wavelengths in the yellow region of the spectrum at 589.0 and 589.6 nm. On both sides of these lines are dark regions where no light is being emitted. Other elements behave similarly (see Color Plate 4, center of book).

Quantum theory offers a simple explanation for discrete atomic spectra. The light given off corresponds to the energy evolved when an excited electron drops back to a lower energy level. Since the energy difference, ΔE, can have only certain definite values, the same must be true of the wavelength, λ, of the light emitted. The relationship between these two quantities is:

$$\Delta E = \frac{2.858 \times 10^4 \text{ kcal} \cdot \text{nm}}{\lambda} \tag{9.3}$$

Thus, the line at 589.0 nm in the sodium spectrum corresponds to an energy difference, per mole of electrons, of

$$\Delta E = \frac{28580 \text{ kcal} \cdot \text{nm}}{589.0 \text{ nm}} = 48.52 \text{ kcal}$$

There must then be two electronic energy levels in the sodium atom which differ in energy by exactly 48.52 kcal/mol.

Using this approach, it is possible to check the Bohr model of the hydrogen atom against experiment. The wavelengths of the lines in the atomic spectrum of hydrogen are known very precisely (Figure 9.4). Each of these lines must correspond to an electron moving from a high to a low energy level. Consider, for example, a transition from $n = 2$ to $n = 1$. We saw in Example 9.1 that, for this case, 235.2 kcal of energy is evolved per mole of electrons. Substituting in Equation 9.3:

$$235.2 \text{ kcal} = \frac{2.858 \times 10^4 \text{ kcal} \cdot \text{nm}}{\lambda}$$

Solving,

$$\lambda = \frac{28580 \text{ kcal} \cdot \text{nm}}{235.2 \text{ kcal}} = 121.5 \text{ nm}$$

Looking at the wavelength just calculated, we see that it corresponds exactly to one of the lines in the hydrogen spectrum shown in Figure 9.4. The other lines can be explained similarly. (See what happens, for example, if you try a transition from $n = 3$ to $n = 1$). The excellent agreement between theory and experiment gave impressive support for the Bohr model of the hydrogen atom. More precisely, it supported the validity of Equation 9.2.

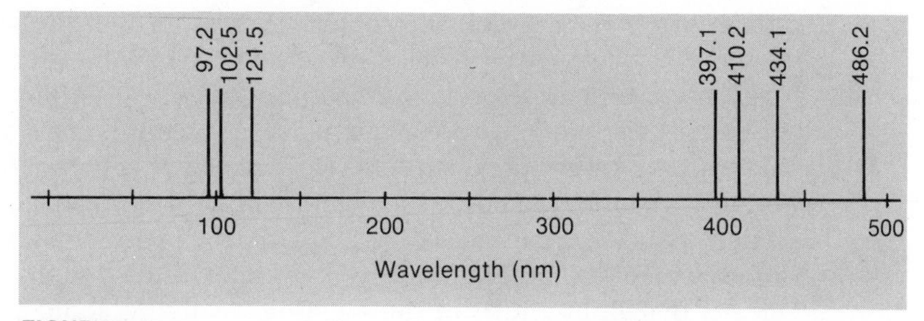

FIGURE 9.4

Some lines in the atomic spectrum of hydrogen. The lines appear in bunches in different parts of the light spectrum. The lines in the right-hand bunch are in the visible region; they are part of the Balmer series. The lines on the left are in the ultraviolet region and belong to the Lyman series. Both series have many more lines than we have shown, and there are several other series that lie in the infrared region. All of the wavelengths of all of the lines in the hydrogen spectrum can be predicted by using Equations 9.2 and 9.3, which were first obtained by Bohr.

9.2 THE QUANTUM MECHANICAL ATOM

At the time it was introduced, the Bohr model of the hydrogen atom appeared to be a success. The extension of Bohr's model to more complex atoms containing more than one electron seemed obvious. Presumably, as electrons were added, they would fill the levels available to an excited electron in the hydrogen atom. Consider, for example, the sodium atom, with 11 electrons. We might expect these electrons to be arranged in concentric "shells" about the nucleus. Some of them would be in the Bohr orbit with $n = 1$, others in the orbit with $n = 2$, and so on. Scientists of 60 years ago believed they were on the verge of unraveling the electronic structure of all atoms.

Unfortunately, the extension of the Bohr model to many-electron atoms didn't work. At best, predictions were only in qualitative agreement with experiment. Gradually it became clear that a simple shell structure for atoms, based on Equation 9.2 is a vast oversimplification.

Actually, the failure of the Bohr model for many-electron atoms was, in a sense, a blessing in disguise. It led to the development of a whole new science known as quantum mechanics. The quantum mechanical model of the atom is a highly mathematical one. We will not attempt to discuss the complex equations upon which it is based. Instead, we will concentrate upon certain qualitative features of the model.

A central postulate of quantum mechanics is that small particles confined to very small regions of space behave quite differently from large, visible objects. The motion of electrons in atoms is basically different from that of billiard balls on a pool table or stock cars on a racetrack. Perhaps the most important difference, and certainly the most surprising, is that *we cannot specify precisely where an electron is, relative to the nucleus, at a particular instant. The best we can do is to quote the odds, or the "probability" that we can find the electron in a particular region of space.* For example, we might say that there is a 50-50 chance of finding the hydrogen electron somewhere in a region within 0.05 nm of the nucleus.

This situation bears at least some resemblance to ones we meet in everyday life. Suppose a high-school teacher wants to locate one of his students on a Saturday afternoon. Unless he has unusual psychic powers, he can't say with absolute certainty where the student is. However, based on past experience, he might be able to estimate the chance of finding the student in various places. Table 9.1 gives a probability description for a typical student. How does yours compare?

If we cannot pin down the position of an electron at a given instant, it follows that we can't describe its path over a period of time. (If I don't know where you are, I certainly don't know how you got there.) This means that Bohr's idea of the electron revolving about the hydrogen nucleus in a circular orbit of fixed radius has to be abandoned. We have to look for other ways to describe the general location of the electron in the hydrogen atom. A graphical model that attempts to do this is shown in Figure 9.5.

At the left of Figure 9.5, we show the probability of finding the hydrogen electron at a given distance from the nucleus in its ground state ($n = 1$ in Eqn. 9.2). In the bright region there is a relatively high

TABLE 9.1 LOCATION OF TYPICAL STUDENT ON SATURDAY AFTERNOON

Location	Probability
Girl-friend's house	25%
Boy-friend's house	10%
Somebody's Hamburger Palace	15%
Home in bed	20%
Cruising	15%
Athletic field	15%
Working in chem lab	0%
	100%

probability of finding the electron. Regions far out from the nucleus, where we are unlikely to find the electron, are dark.

There are several features of the "probability map" at the left of Figure 9.5 which are of interest. One is particularly important. As we move out from the nucleus of the H atom a given distance in any direction, the brightness is the same. This means, for example, that the electron is as likely to be found 0.05 nm "north" of the nucleus as it is 0.05 nm "south" of the nucleus (or east, or west, . . .). In other words, the density of the electron cloud is symmetrical about the nucleus. We

In general, atoms tend to be spherical.

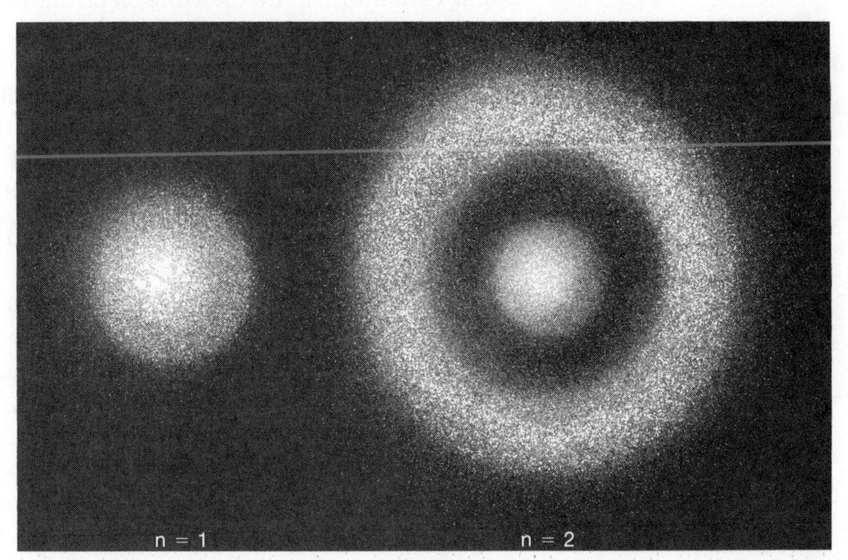

FIGURE 9.5

Electron charge clouds in the H atom. In the spherical n = 1 cloud on the left, the probability of finding the electron is largest at the nucleus and drops off gradually as the distance from the nucleus increases. In the n = 2 state the cloud exists as two concentric spheres. The probability of finding the electron is high near the nucleus and also high in the outer region where the cloud is bright. Note the change in size of the atom as the quantum number changes from 1 to 2.

might say that in three dimensions, the electron cloud is "spherically symmetrical".

At the right of Figure 9.5 is a probability map for the hydrogen electron in its first excited state (n = 2 in Eqn. 9.2). Here, as in the ground state, the electron cloud is spherically symmetrical. However, there is a much higher probability of finding the electron farther out from the nucleus. Notice that a region quite far from the nucleus which is dark in the figure at the left is quite bright in the diagram at the right. Qualitatively, this agrees with the Bohr model. When the hydrogen electron moves to a higher energy state, it is more likely to be found at greater distances from the nucleus.

Actually most of our information about electrons relates to their energies, and there the theory works quite well.

The fact that we cannot specify exactly the position of electrons in atoms makes our discussion of electron distribution in atoms a bit fuzzy. However, the quantum mechanical model of the atom is not quite as nebulous as it may seem. The model does lead to equations which describe the energy states available to electrons in atoms. From a practical standpoint, it is much more important to be able to obtain the energy of an electron than it is to locate it exactly. Electron energies have a direct bearing on experiments that we can carry out in the laboratory. These include the determination of atomic spectra, ionization energies, and, ultimately, energy changes in chemical reactions.

9.3 ELECTRON CONFIGURATIONS OF ATOMS; PRINCIPAL LEVELS AND SUBLEVELS

In this section, we will extend the quantum mechanical model to atoms containing more than one electron, such as He, Li, and so on. The model considers that electrons in atoms are distributed among a series of **principal energy levels** of successively higher energies. These are quite similar to the levels proposed by Bohr for the hydrogen atom. However each principal level (except the first) is divided into **sublevels,** which differ at least slightly from one another in energy.

In discussing electron energy levels, both principal levels and sublevels, we will be concerned with two factors:

—their relative energies
—their capacities for holding electrons.

If these two factors are known, we can predict the ground-state **electron configurations** of atoms. That is, we can predict how the electrons in a normal, unexcited atom are distributed among the various levels and sublevels.

Principal Energy Levels

You will recall that the Bohr model of the hydrogen atom established a series of energy levels of increasing energy. The first level (n = 1) was the one normally occupied by the electron in its ground state. Higher levels such as the 2nd level (n = 2), the 3rd level (n = 3), . . . were available to an excited electron. The quantum mechanical model retains these levels, now described as principal energy levels. For many-electron atoms, several of these levels may be occupied by electrons in the ground state.

Each principal level has a limited capacity for electrons. The maximum number of electrons that can fit into a given principal level is **$2n^2$**, where n is the number of the level. Thus the total number of electrons that can occupy various principal energy levels is:

number of level (n)	1	2	3	4	...
capacity of level ($2n^2$)	2	8	18	32	...

There is a lot of symmetry in the system by which electrons are arranged in atoms.

We see, for example, that the 1st principal energy level can hold only two electrons. Atoms with more than two electrons such as lithium (at. no. = 3) must put some of these electrons into higher levels.

In many-electron atoms, the energies of the various principal levels are quite different from those predicted by Equation 9.2. However, they remain in the same order relative to one another. An electron in the $n = 2$ level has a higher energy than one in the $n = 1$ level. An electron in the 3rd level has a higher energy than one in the 2nd level, and so on.

We can relate the energies of electrons in various principal levels to their distances from the nucleus. As we move to higher principal levels, electrons are more likely to be found further out from the nucleus. On the average, an electron in the 2nd level is further from the nucleus than one in the 1st level (recall Figure 9.5). An electron in the $n = 3$ level will be still more distant from the nucleus, and so on.

Energy Sublevels

The quantum mechanical model of the atom tells us that not all the electrons in a given principal energy level need have exactly the same energy. Within a principal level, there are one or more sublevels which differ from one another in energy. These sublevels are designated by the letters s, p, d, and f*. There are two important principles that apply here.

1. The *relative energies* of these sublevels increase in the order

$$s < p < d < f$$

Within a given principal level, the s sublevel is always the lowest in energy. The p sublevel is always higher in energy than the s, the d sublevel higher than the p, and the f higher than the d.

2. Each sublevel, regardless of the principal level in which it is found, has a fixed *capacity* for electrons.

an s sublevel can hold a maximum of 2 electrons
a p sublevel can hold a maximum of 6 electrons
a d sublevel can hold a maximum of 10 electrons
an f sublevel can hold a maximum of 14 electrons

*This rather peculiar set of letters is a carry-over from the study of atomic spectra. Spectroscopists distinguish between "sharp", "principal", "diffuse", and "fundamental" spectral lines.

With these factors in mind, let us consider the sublevel structure of successive principal energy levels.

—the first principal level (n = 1), with a total capacity of 2 electrons, must contain only an s sublevel. This sublevel, capable of holding 2 electrons, is designated as **1s.**

n = 1 1s $(2\ e^-)$ total capacity = $2\ e^-$

—the second principal level (n = 2) has a total capacity of 8 electrons. It contains an s sublevel (2 electrons) and a p sublevel (6 electrons). These sublevels are designated **2s** and **2p** respectively. The 2s sublevel is lower in energy than the 2p.

2p $(6\ e^-)$

n = 2 total capacity = $8\ e^-$

2s $(2\ e^-)$

The system is sort of like a hotel with different floors (levels) and different size rooms on each floor (sublevels).

—the third principal level (n = 3) has a total capacity of 18 electrons. It is divided into three sublevels. The lowest of these is the **3s** (2 electrons). At a slightly higher energy, we have the **3p** sublevel (6 electrons). Finally, there is a **3d** sublevel (10 electrons) at a still higher energy.

3d $(10\ e^-)$

n = 3 3p $(6\ e^-)$ total capacity = $18\ e^-$

3s $(2\ e^-)$

—the fourth principal level (n = 4) has a total capacity of 32 electrons. It is divided into four sublevels: **4s** (2 electrons), **4p** (6 electrons), **4d** (10 electrons), and **4f** (14 electrons). These sublevels increase in energy in the order:

$$4s < 4p < 4d < 4f$$

4f $(14\ e^-)$

4d $(10\ e^-)$

n = 4 4p $(6\ e^-)$ total capacity = $32\ e^-$

4s $(2\ e^-)$

Table 9.2 summarizes the discussion we have just gone through. The numbers in parentheses indicate the capacity for electrons of each sublevel. The total capacity of each principal level, indicated at the right of the table, is the sum of the sublevel capacities. From the table you will notice that *the total number of different sublevels within a given principal level is given by the "quantum number" n of that level.* Thus when n = 1, we have one sublevel (1s). With n = 2, there are two different

TABLE 9.2 SUBLEVEL STRUCTURE OF THE PRINCIPAL ENERGY LEVELS

Principal Level	Sublevels	Total Capacity
1	1s(2)	2
2	2s(2) 2p(6)	8
3	3s(2) 3p(6) 3d(10)	18
4	4s(2) 4p(6) 4d(10) 4f(14)	32

sublevels (2s, 2p). For n = 3, there are three sublevels (3s, 3p, 3d), and for n = 4 there are four (4s, 4p, 4d, 4f).

Higher principal levels, beyond those listed in Table 9.2 also have s, p, d, and f sublevels. Thus, we can have a 5s, 5p, 5d, or 5f sublevel. In principle, there should be a fifth sublevel in this principal level, perhaps designated as 5g and capable of holding 18 electrons. In practice, this sublevel turns out to have an energy so high that it is not populated by electrons in any atom presently known.

EXAMPLE 9.2

a. How many electrons can go into a 5p sublevel? a 5f sublevel?

b. Which sublevel is higher in energy, 5p or 5f?

SOLUTION

a. 6 electrons in 5p; 14 electrons in 5f.

b. 5f; an f sublevel is always higher in energy that a p sublevel in the same principal level.

Electron Configurations

A common way to describe the electronic structure of an atom is to give the number of electrons in each energy sublevel. This is done by writing the **electron configuration.** Here, the number of electrons in each sublevel is indicated by a superscript immediately after the symbol of the sublevel. The electron configuration of an atom in which there are two electrons in the 1s sublevel and one in the 2s would be:

$$1s^2\ 2s^1$$

In another case, for an atom in which there are two 1s electrons, two 2s electrons, and five 2p electrons, we would write:

$$1s^2\ 2s^2\ 2p^5$$

We can derive the ground state electron configuration of an atom of an element by a quite simple process. **Electrons are added to sublevels of increasingly higher energy, filling each sublevel to capacity before moving on to the next.** This process continues until all the electrons in the atom are accounted for.

In order to carry out this process, we must know the relative energies of various sublevels. At least for the first 18 electrons, there are no surprises.

The electron configuration of the H atom is 1s.

1. *The 1s sublevel fills first.* Since this sublevel has a capacity of only two electrons, the 1s sublevel fills with He (at. no. = 2). The electron configuration of helium is

$$_{2}He \qquad 1s^2$$

There are no other sublevels within the n = 1 level, so we move to n = 2.

2. Within the 2nd principal energy level, *the 2s sublevel fills first.* This accommodates two more electrons, taking us to Be (at. no. = 4).

$$_{4}Be \qquad 1s^2\ 2s^2$$

Now, *the 2p sublevel fills.* Recall that a p sublevel has a capacity of 6 electrons. So, the 2p sublevel is full with Ne (at. no. = 10).

$$_{10}Ne \qquad 1s^2\ 2s^2\ 2p^6$$

There are no other sublevels within the n = 2 level, so we move on to n = 3.

3. Within the 3rd principal level, *the 3s sublevel fills first.* This requires two more electrons, bringing us to Mg (at. no. = 12).

$$_{12}Mg \qquad 1s^2\ 2s^2\ 2p^6\ 3s^2$$

Then *the 3p sublevel fills,* with a capacity of 6 electrons. This takes us to argon (at. no. = 18).

$$_{18}Ar \qquad 1s^2\ 2s^2\ 2p^6\ 3s^2\ 3p^6$$

In summary, the sublevels filled at this point, in order of increasing energy, are:

$$1s < 2s < 2p < 3s < 3p$$

With this filling order in mind, we can readily deduce the electron configuration of any element through argon (18 electrons). Example 9.3 shows how this is done.

EXAMPLE 9.3

Give the electron configurations of atoms of:

a. Li (at. no. = 3) b. F (at. no. = 9) c. S (at. no. = 16)

SOLUTION

Note that in each case the total number of electrons in the neutral atom is equal to the atomic number (the number of protons in the nucleus).

a. The first two electrons fill the 1s sublevel to capacity. The third electron must enter the 2s sublevel:

$$_{3}Li: 1s^2\ 2s^1$$

b. Two electrons are required to fill the 1s sublevel. Two more fill the 2s. That leaves five electrons, all of which can enter the 2p sublevel (capacity = 6 electrons).

$$_{9}F: 1s^2\ 2s^2\ 2p^5$$

c. Of the 16 electrons to be accommodated, we put two in the 1s sublevel, two in the 2s, and six in the 2p. This accounts for 10 electrons. The next two electrons fill the 3s sublevel. The remaining four electrons go into the 3p sublevel.

$$_{16}S: 1s^2\ 2s^2\ 2p^6\ 3s^2\ 3p^4$$

Writing electron configurations is easy once you know the rules. Learn the rules.

As we have pointed out, Ar (at. no. = 18) fills the 3p sublevel. The next element, K (at. no. = 19) has to put its extra electron into the next highest sublevel beyond the 3p. Here we meet with a surprise. All the evidence that we have indicates that the 19th electron in potassium enters the 4s sublevel rather than the 3d. The electron configuration is:

$$_{19}K \qquad 1s^2\ 2s^2\ 2p^6\ 3s^2\ 3p^6\ 4s^1$$

It appears that, so far as the atoms in the 4th period of the Periodic Table are concerned, there is an "overlap" between the 3rd and 4th principal energy levels. That is, the *lowest* sublevel for n = 4, 4s, fills before the *highest* sublevel for n = 3, 3d. Only when the 4s sublevel is filled, with calcium, do electrons start to enter the 3d sublevel.

$$_{20}Ca \qquad 1s^2\ 2s^2\ 2p^6\ 3s^2\ 3p^6\ 4s^2$$

$$_{21}Sc \qquad 1s^2\ 2s^2\ 2p^6\ 3s^2\ 3p^6\ 4s^2\ 3d^1$$

$$_{22}Ti \qquad 1s^2\ 2s^2\ 2p^6\ 3s^2\ 3p^6\ 4s^2\ 3d^2$$

| | | | | | |

$$_{30}Zn \qquad 1s^2\ 2s^2\ 2p^6\ 3s^2\ 3p^6\ 4s^2\ 3d^{10}$$

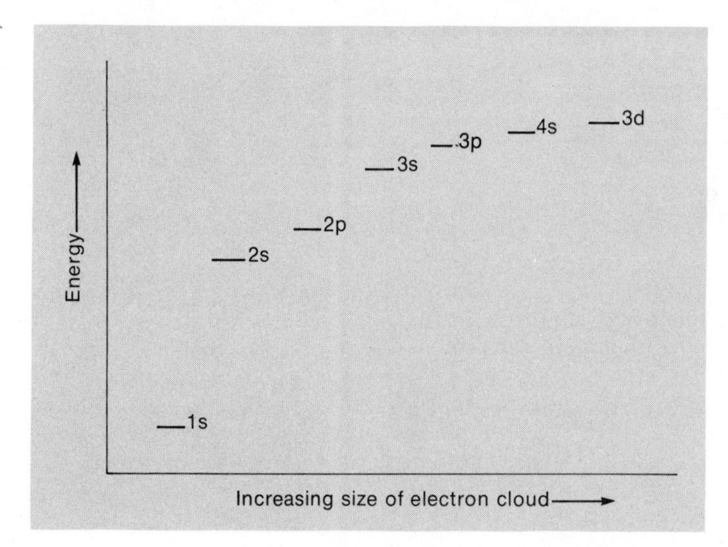

FIGURE 9.6

An energy level diagram for many electron atoms. The energies of the sublevels for all atoms follow the pattern shown. Electrons fill the sublevels in order of increasing energy.

EXAMPLE 9.4

Give the electron configuration of an atom of nickel (at. no. = 28).

SOLUTION

The first 18 electrons fill the levels 1s through 3p. The next 2 electrons enter the 4s sublevel. The last 8 electrons go into the 3d sublevel.

$$1s^2\ 2s^2\ 2p^6\ 3s^2\ 3p^6\ 4s^2\ 3d^8$$

In Table 9.3 we indicate the electron configurations of the first 30 elements in the Periodic Table. To save space, we have used the symbols [He], [Ne], and [Ar] as an abbreviation:

for [He], read $1s^2$

for [Ne], read $1s^2\ 2s^2\ 2p^6$

for [Ar], read $1s^2\ 2s^2\ 2p^6\ 3s^2\ 3p^6$

You will note that with two minor exceptions, chromium and copper, the electron configurations listed in the table follow the order we have described.

TABLE 9.3 ELECTRON CONFIGURATIONS OF THE FIRST 30 ELEMENTS

$_{1}H$	$1s^1$	$_{16}S$	[Ne] $3s^2\ 3p^4$
$_{2}He$	$1s^2$	$_{17}Cl$	[Ne] $3s^2\ 3p^5$
$_{3}Li$	[He] $2s^1$	$_{18}Ar$	[Ne] $3s^2\ 3p^6$
$_{4}Be$	[He] $2s^2$	$_{19}K$	[Ar] $4s^1$
$_{5}B$	[He] $2s^2\ 2p^1$	$_{20}Ca$	[Ar] $4s^2$
$_{6}C$	[He] $2s^2\ 2p^2$	$_{21}Sc$	[Ar] $4s^2\ 3d^1$
$_{7}N$	[He] $2s^2\ 2p^3$	$_{22}Ti$	[Ar] $4s^2\ 3d^2$
$_{8}O$	[He] $2s^2\ 2p^4$	$_{23}V$	[Ar] $4s^2\ 3d^3$
$_{9}F$	[He] $2s^2\ 2p^5$	* $_{24}Cr$	[Ar] $4s^1\ 3d^5$
$_{10}Ne$	[He] $2s^2\ 2p^6$	$_{25}Mn$	[Ar] $4s^2\ 3d^5$
$_{11}Na$	[Ne] $3s^1$	$_{26}Fe$	[Ar] $4s^2\ 3d^6$
$_{12}Mg$	[Ne] $3s^2$	$_{27}Co$	[Ar] $4s^2\ 3d^7$
$_{13}Al$	[Ne] $3s^2\ 3p^1$	$_{28}Ni$	[Ar] $4s^2\ 3d^8$
$_{14}Si$	[Ne] $3s^2\ 3p^2$	* $_{29}Cu$	[Ar] $4s^1\ 3d^{10}$
$_{15}P$	[Ne] $3s^2\ 3p^3$	$_{30}Zn$	[Ar] $4s^2\ 3d^{10}$

*The somewhat unusual configurations of Cr and Cu are believed to reflect the special stability of half-filled sublevels (s^1, d^5) or completely filled sublevels (d^{10}).

9.4 ELECTRON CLOUD GEOMETRIES; ORBITALS

In Section 9.2, we discussed the nature of the charge cloud associated with the single electron in the hydrogen atom. This was done by drawing a probability diagram (Figure 9.5). Here, the brightness indicated the chance of finding the electron in a particular region. In principle, we could draw similar diagrams for electron clouds in atoms containing more than one electron. However, this is seldom done. As the number of electrons increases, such diagrams become very complex and difficult to interpret. Instead, we commonly represent electron clouds by simple, 3-dimensional figures. These enclose the volume in which there is a 90% chance of finding a particular electron. They are referred to as **orbitals.** The shape of an orbital differs depending upon the sublevel, s, p, d, or f, in which the electron is located.

An orbital encloses the region where the electron charge density is high.

s Orbitals

In Figure 9.7a, p. 226, we show the orbital associated with the 1s electron in hydrogen. Compare it with the probability diagram at the left of Figure 9.5. Clearly all we have done is to enclose the bright region by a sphere. We can say that there is a 90% probability of finding the 1s electron in hydrogen within this spherical orbital.

You will recall that in helium (at. no. = 2), there are two 1s electrons. The shape of the cloud associated with the second electron is

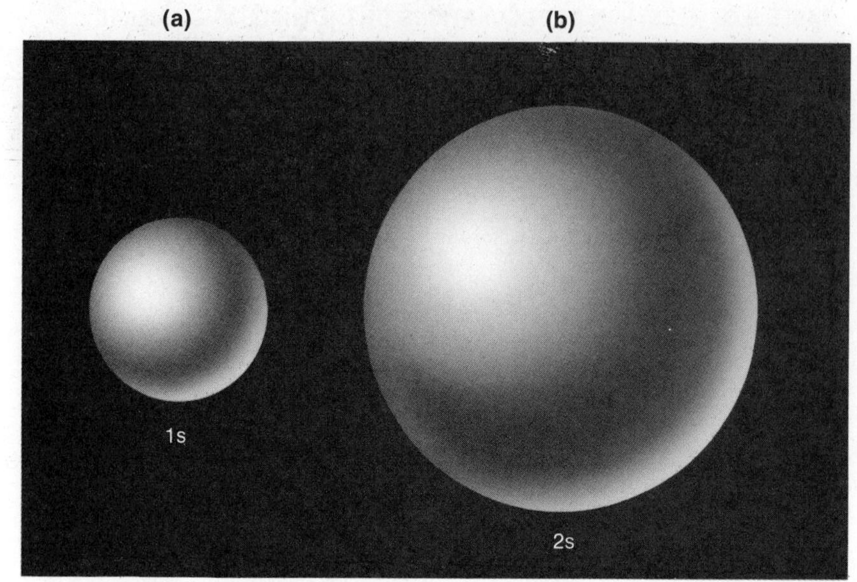

FIGURE 9.7

Representations of the 1s and 2s orbitals. The spheres are drawn to include 90% of the charge clouds in Figure 9.5, so the odds on finding the electron in the region shown are 9 out of 10 when the atom is in the 1s or the 2s state.

The geometry of the orbitals comes from theory.

identical with that of the first. Putting it another way, the 1s orbital shown in Figure 9.7a is capable of holding two electrons. In helium and in heavier atoms, both of the 1s electrons spend most of their time within this orbital.

Orbitals associated with s sublevels are always spherical. This remains true regardless of the principal level involved. As the number of the principal level increases, so does the radius of the sphere representing the s orbital. This reflects the fact that as n increases, the electron is more likely to be found at a greater distance from the nucleus. Thus the 2s orbital is represented by a sphere with a larger radius than the 1s orbital (Figure 9.7b). This orbital contains one electron with lithium (at. no. = 3) and fills with two electrons in the beryllium atom (at. no. = 4).

The radius of the sphere associated with the 3s orbital would be larger than that of the 2s, and so on. Each of these spherical orbitals (1s, 2s, 3s, . . .) has its center at the nucleus. Moreover, each s orbital is capable of holding two electrons.

p Orbitals

The orbitals associated with p sublevels have a more complex shape than those of s sublevels. As shown in Figure 9.8, a p electron spends

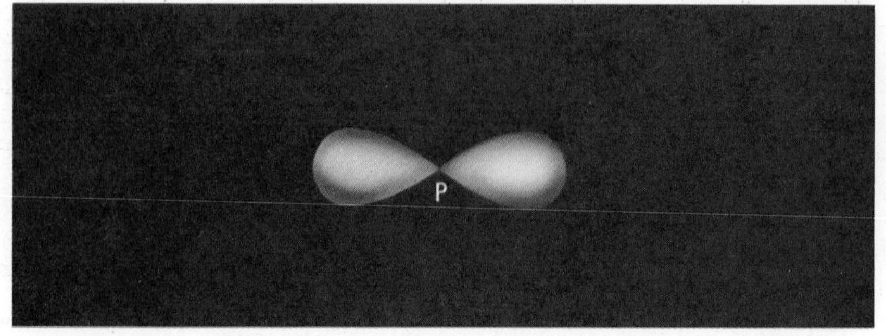

FIGURE 9.8

Representation of a p orbital. Electrons in p orbitals have charge clouds like that shown. The electron is most likely to be found in the dumbbell region (Figures 9.7 and 9.8 are *not* drawn to the same scale).

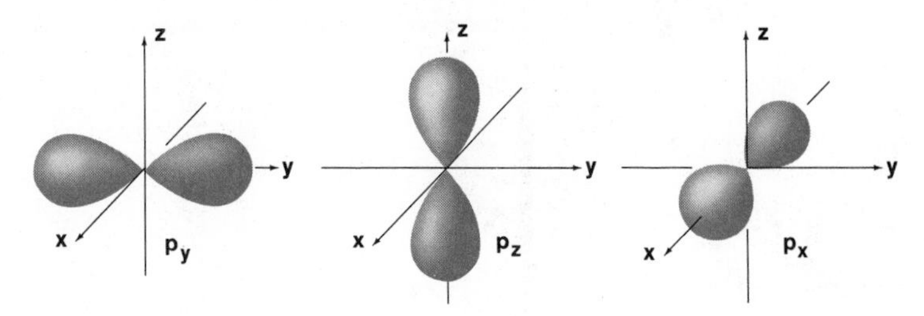

FIGURE 9.9

Three p orbitals available to an electron. Electrons in different p orbitals are relatively far apart. Their charge clouds are concentrated along perpendicular directions.

most of its time within a dumbbell-shaped orbital. The "dumbbell" has its center at the nucleus. It consists of two lobes, directed at a 180° angle to one another.

Within any given p sublevel, there are three p orbitals, oriented at right angles to one another. For example, there are three different 2p orbitals. As shown in Figure 9.9, one of these is centered along the x axis, another along the y axis, and the third along the z axis. Each of these orbitals, like the s orbitals, is capable of holding two electrons. Consider the neon atom, which has 6 electrons in the 2p sublevel (electron configuration = $1s^2\ 2s^2\ 2p^6$). Two of the 2p electrons are in the p_x orbital, the one directed along the x axis. Two more are in the p_y orbital and two more in the p_z orbital.

Orbital Structure of Sublevels

The picture just presented can be extended to describe the geometries of the clouds associated with electrons in d or f sublevels. We will not attempt to show the geometries of d or f orbitals; they are rather complex. However, the same principle applies as before. **Each orbital is capable of holding two electrons.** Since a d sublevel has a capacity of 10 electrons, there must be:

In a full orbital the two electrons are said to be paired.

$$10/2 = 5$$

different d orbitals within a given d sublevel. Again, there must be seven different f orbitals to account for the total capacity of an f sublevel, 14 electrons:

$$7 \text{ orbitals} \times 2\ \frac{\text{electrons}}{\text{orbital}} = 14 \text{ electrons}$$

TABLE 9.4 ORBITAL STRUCTURE OF SUBLEVELS

Sublevel	Number of Orbitals	Capacity of Each Orbital	Total Capacity Sublevel
s	1	2	2
p	3	2	6
d	5	2	10
f	7	2	14

How many electrons can fit in a 2p orbital?

EXAMPLE 9.5

What is the total number of orbitals in the principal level for which n = 3?

SOLUTION

The 3s sublevel consists of one orbital. There are three 3p orbitals and five 3d orbitals:

$$1 + 3 + 5 = 9$$

Thus we conclude that there is a total of 9 orbitals within the 3rd principal energy level. Each orbital can hold two electrons, so the total capacity of the level is 18 e^-.

In Table 9.5, we summarize the electron capacities of the various principal levels, sublevels and orbitals through n = 4. Remember that:

—each orbital has a capacity of 2 e^-.

—each s sublevel has a capacity of 2 e^-, each p sublevel a capacity of 6 e^-, each d sublevel a capacity of 10 e^-, each f sublevel a capacity of 14 e^-.

—each principal level of quantum number n has a capacity of $2n^2$ electrons and contains n sublevels.

TABLE 9.5 ELECTRON CAPACITIES

Principal Level	Sublevel	Number of Orbitals	Number Electrons per Orbital	Number Electrons in Sublevel	Number Electrons Principal Level
1	1s	1	2	2	2
2	2s	1	2	2	
	2p	3	2	6	8
3	3s	1	2	2	
	3p	3	2	6	
	3d	5	2	10	18
4	4s	1	2	2	
	4p	3	2	6	
	4d	5	2	10	
	4f	7	2	14	32

Order of Filling Orbitals Within a Sublevel With the p, d, and f sublevels, where more than one orbital is available, a question arises as to the order in which these orbitals are filled. Consider, for example, the carbon atom. It has two electrons in the 2p sublevel (electron configuration: $1s^2\ 2s^2\ 2p^2$). There are two possible distributions for these two electrons among the three orbitals in the 2p sublevel:

—both of them might go into the same orbital, perhaps the p_x orbital.
—they might go into different orbitals, perhaps one in the p_x, the other in the p_y orbital.

It turns out, as you might guess, that the two electrons go into different orbitals. Remember that electrons carry a negative charge and hence repel each other. It seems reasonable that they would spread out over as large a volume as possible rather than crowd into the same orbital.

The principle just illustrated applies in all cases. In p, d, and f sublevels, electrons tend to spread out among the several orbitals (3, 5, or 7) so as to half-fill as many orbitals as possible. Consider the elements boron (at. no. = 5) through neon (at. no. = 10). If we assume that electrons go into the p_x, p_y, and p_z orbitals in that order, we have the situation shown in Table 9.6.

TABLE 9.6 POPULATION OF 2p ORBITALS

Element	Electron Configuration	No. e^- in p_x	No. e^- in p_y	No. e^- in p_z
$_5B$	$1s^2\ 2s^2\ 2p^1$	1	–	–
$_6C$	$1s^2\ 2s^2\ 2p^2$	1	1	–
$_7N$	$1s^2\ 2s^2\ 2p^3$	1	1	1
$_8O$	$1s^2\ 2s^2\ 2p^4$	2	1	1
$_9F$	$1s^2\ 2s^2\ 2p^5$	2	2	1
$_{10}Ne$	$1s^2\ 2s^2\ 2p^6$	2	2	2

(The order in which the p orbitals are filled is arbitrary. We could just as well put the single 2p electron in boron in the p_y or p_z orbital.)

9.5 ELECTRON CONFIGURATIONS AND THE PERIODIC TABLE

Having considered the distribution of electrons among energy levels, we can now understand the basis of the Periodic Table. We start by asking the question, "How do the electron configurations of elements in a given group of the Table compare to one another?" To answer this question, we turn back to Table 9.3 and write down the configurations of the elements in Groups 1, 2, 7 and 8:

$_3Li$ [He] $2s^1$ $_4Be$ [He] $2s^2$ $_9F$ [He] $2s^2\ 2p^5$ $_{10}Ne$ [He] $2s^2\ 2p^6$

$_{11}Na$ [Ne] $3s^1$ $_{12}Mg$ [Ne] $3s^2$ $_{17}Cl$ [Ne] $3s^2\ 3p^5$ $_{18}Ar$ [Ne] $3s^2\ 3p^6$

$_{19}K$ [Ar] $4s^1$ $_{20}Ca$ [Ar] $4s^2$

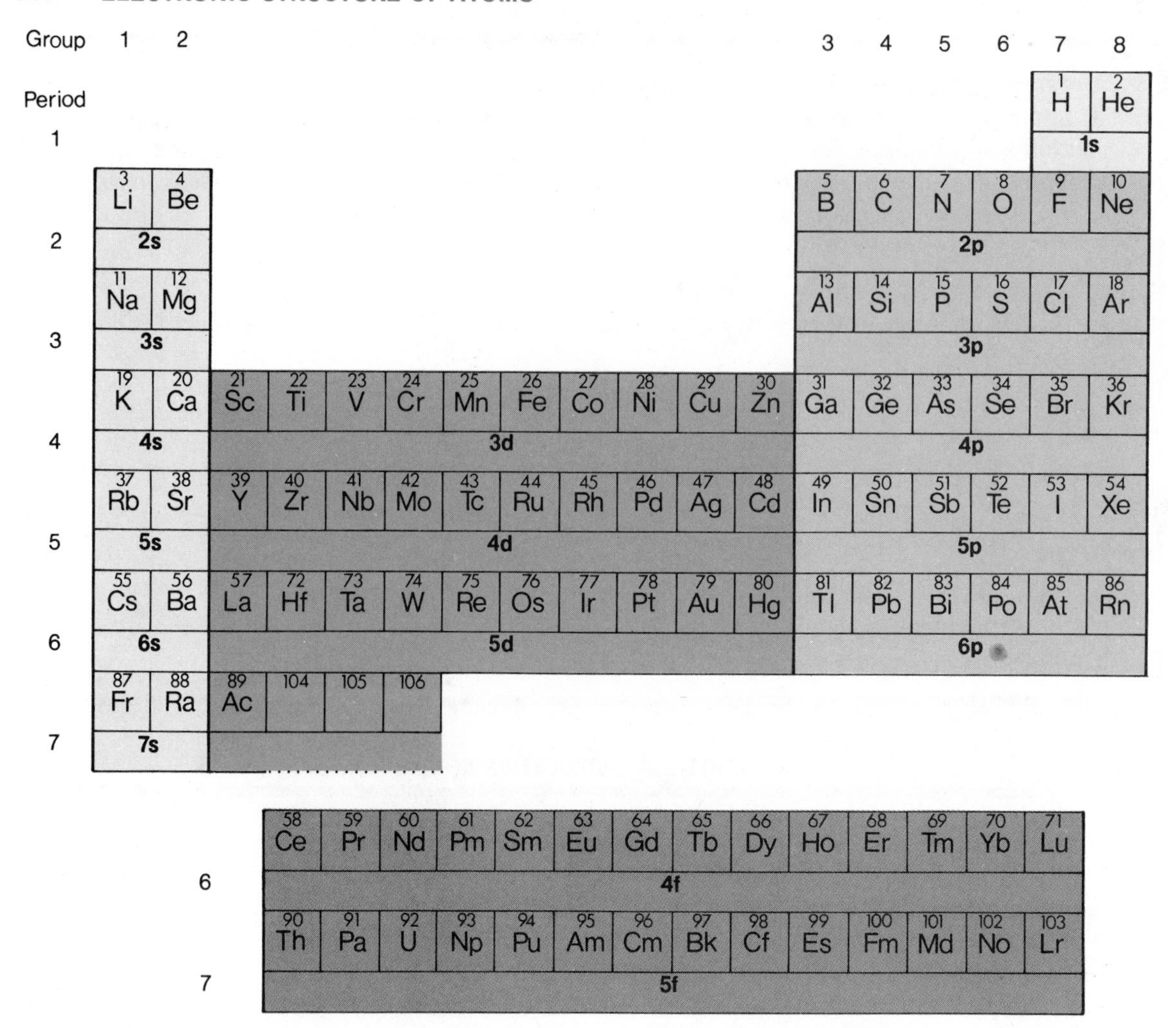

FIGURE 9.10

Diagram showing the sublevels that are being filled in different parts of the Periodic Table. Using the diagram one can find the electron configuration of an atom very easily, since all sublevels up to the one being filled are full. In Mn, for example, the highest sublevel, 3d, has five electrons; the configuration must be $1s^22s^22p^63s^23p^64s^23d^5$.

Amazingly, electron configurations allow us to explain both the existence and the structure of the Periodic Table in a remarkably simple way.

Looking at these structures, it should be clear that **elements in the same group of the Periodic Table have the same outer electron configuration.** We see, for example, that the Group 1 metals have a single s electron in the outermost principal energy level. All of the Group 2 metals have two outer s electrons, and so on.

As we will see in Chapter 10, it is the outer electrons which are involved in chemical reactions. This explains why elements in the same group of the Periodic Table have similar chemical properties. In the case of the alkali metals, the outer s electron is lost in reactions with nonmetals. This explains why each of these elements forms +1 ions (Li^+, Na^+, K^+, . . .).

In Figure 9.10, we indicate the regions of the Periodic Table where various sublevels are filling. Notice that:

1. The elements in Groups 1 and 2 are filling an s sublevel. Thus,

Li and Be in the 2nd period fill the 2s sublevel. Na and Mg in the 3rd period fill the 3s sublevel, and so on.

2. The elements in Groups 3 through 8 (6 elements in each period) fill p sublevels, which have a capacity of 6 electrons. In the second period, the 2p sublevel starts to fill with B (at. no. = 5) and is completed with Ne (at. no. = 10). In the third period, the elements Al (at. no. = 13) through Ar (at. no. = 18) fill the 3p sublevel.

3. The 10 transition metals fill d sublevels, which have a capacity of 10 electrons. The elements Sc (at. no. = 21) through Zn (at. no. = 30) fill the 3d sublevel. In the 5th period, the elements Y (at. no. = 39) through Cd (at. no. = 48), fill the 4d sublevel. The 10 transition metals in the 6th period fill the 5d sublevel.

4. The 14 inner transition metals fill f sublevels, which have a capacity of 14 electrons. The lanthanides fill the 4f sublevel. In the next period, the actinides fill the 5f sublevel.

We could, if we had to, use Figure 9.10 to deduce the complete electron configuration of any element. More commonly, we use the Periodic Table itself for a less ambitious purpose. *We use it to predict the outer electron configuration of the main-group elements.* To do this, we note that the distribution of electrons in the outermost principal energy level for each of these elements is:

Group	1	2	3	4	5	6	7	8
outer config.	ns^1	ns^2	$ns^2\ np^1$	$ns^2\ np^2$	$ns^2\ np^3$	$ns^2\ np^4$	$ns^2\ np^5$	$ns^2\ np^6$

where n is the number of the period in which the element is located. That is, n = 2 for the second period, n = 3 for the third period, and so on.

EXAMPLE 9.6

Using the Periodic Table, give the outer electronic configurations of

a. Sr (at. no. = 38) b. At (at. no. = 85)

SOLUTION

a. We locate strontium, Sr, in the 5th period, in Group 2. Its outer electron configuration must be $5s^2$.

b. Astatine, At, is in the 6th period, in Group 7. Its outer electron configuration is $6s^2\ 6p^5$.

One final point concerning the electronic structures of the main-group elements is worth noting. The total number of electrons in the outermost principal energy level, often referred to as **valence electrons,** is given by the group number. To illustrate this rule, consider the elements Sr in Group 2 and astatine in Group 7 (Example 9.6). Since the outer electron configuration of Sr is $5s^2$, there are 2 electrons in the outermost (5th) level. In astatine with the configuration $6s^2\ 6p^5$, there are 2 + 5 = 7 valence electrons. In general:

Group	1	2	3	4	5	6	7	8
no. valence e^-	1	2	3	4	5	6	7	8

We will find this general rule to be of value when we discuss the bonding in compounds of the main-group elements in Chapters 10 and 11.

9.6 SUMMARY

In this chapter, we have considered the energy levels available to electrons in atoms. In the case of the hydrogen atom, it is possible to obtain values for the energy levels, using the Bohr equation:

$$E_{\text{mole electrons}} = \frac{-313.6\ \text{kcal}}{n^2}$$

where n is a whole number (1, 2, 3, . . .).

The Bohr model, which worked so well for the hydrogen atom, could not be applied quantitatively to any other atom. A major problem is that we cannot specify exactly the position of an electron in an atom. The best we can do is to give the probability of finding the electron in a particular region around the nucleus. This is most often done by drawing an orbital (Figures 9.7, 9.8) which encloses the volume in which a particular electron is most likely to be found. Each orbital is capable of holding two electrons. Where more than one orbital of equal energy is available, electrons tend to split up, half-filling as many orbitals as possible.

The energies of electrons in atoms are described by citing the principal level and sublevel in which they are located. The principal level is designated by the value of n, the integer in the Bohr equation. The lowest principal level is the one for which n = 1. The principal levels n = 2, n = 3, . . . have successively higher energies. The total number of electrons that can fit into a given principal level is $2n^2$ (2, 8, 18, 32, . . .).

Each principal level contains a total of n sublevels. When n = 1, there is one sublevel. When n = 2, there are two sublevels, and so on. The sublevels are designated, in order of increasing energy, by the letters s, p, d, and f. An s sublevel has a capacity of 2 electrons, a p sublevel a capacity of 6 electrons, a d sublevel a capacity of 10, and an f sublevel a capacity of 14.

Electron configurations such as $1s^2\ 2s^2\ 2p^6$ tell us directly the number of electrons in each sublevel in an atom. They can be deduced by remembering the order in which the various sublevels fill (1s, 2s, 2p, 3s, 3p, 4s, 3d, . . .). Perhaps more simply, they can be obtained from the

Periodic Table, provided you understand how the Table is constructed (Figure 9.10). Remember that:

—all elements in the same group have the same outer electron configuration.

—for the main-group elements, the number of valence electrons is given by the group number.

While Mendeleev and others did not realize it at the time, electron configurations of atoms are the basis of the Periodic Table. Atoms with the same outer electron configuration have similar properties.

NEW TERMS

Bohr equation
d sublevel
d orbital
electron cloud
electron configuration
energy level
excited state
f sublevel
f orbital
ground state
orbital
outer electron configuration
p sublevel
p orbital
principal energy level
quantum mechanics
quantum number
quantum theory
s sublevel
s orbital
sublevel
valence electron

QUESTIONS

1. What is the sign of ΔE when an electron moves from a lower to a higher energy level, such as $n = 1$ to $n = 2$? What is the sign of ΔE when the electron moves in the reverse direction?

2. In the Bohr model, what is n when the electron is in the ground state? when the electron is completely removed from the atom?

3. How does:
   a. the distribution of energy levels in the hydrogen atom differ from the spacing between rungs in an ordinary ladder?
   b. the motion of an electron in a hydrogen atom differ from that of the earth revolving about the sun?

4. An example of quantization is the gearing in stick-shift automobiles. Can you think of other examples of quantization in familiar processes?

5. Explain, in your own words, what is meant by the probability map shown in Figure 9.5.

6. A baseball game is being played in a circular stadium with a sell-out crowd. Draw a probability diagram similar to that shown in Figure 9.5, to express the probability of finding a spectator at a given distance from the pitcher's mound, taken as the center. Use heavy shading to indicate where there is a heavy density of spectators. How would the diagram change if:
   a. there were a riot on the field in which spectators took part.
   b. the stadium were only half-full of spectators.

7. Give the total capacity for electrons of the principal level for which

   a. n = 1 b. n = 2 c. n = 3 d. n = 4 e. n = 5

8. Which has the higher energy, an electron in the

   a. 1st or 2nd principal energy level?
   b. 1s or 2s sublevel?
   c. 2s or 2p sublevel?

9. What is the highest sublevel in the principal level for which n is

   a. 1 b. 2 c. 3 d. 4

10. What is the capacity for electrons of each of the following sublevels?

    a. s b. p c. d d. f

11. What is the total capacity for electrons of:

    a. an orbital? b. an s sublevel? c. a p sublevel?
    d. a d sublevel? e. the 1st principal level?
    f. the 3rd principal level?

12. Which of the following sublevels do not exist?

    a. 2p b. 1p c. 2d d. 3d e. 3f

13. Arrange the following sublevels in order of increasing energy

    a. 1s b. 3p c. 2s d. 2p e. 3s

14. In the 3rd principal energy level, what is the total electron capacity? How many different sublevels are in this level? List them, in order of increasing energy. One of these is actually higher in energy than the lowest sublevel in the 4th principal level. Which one?

15. Draw a 1s orbital; a 2s orbital; a 2p orbital. How do you interpret the shapes of these orbitals?

16. How many orbitals are there in the following sublevels?

    a. s b. p c. d d. f

17. What is the total number of orbitals in the principal level for which n is:

    a. 1 b. 2 c. 3 d. 4

18. What is the total number of orbitals filled in:

    a. Ar, at. no. = 18 b. Zn, at. no. = 30

    Example: Ne, at. no. = 10 5 orbitals

19. Give the outer electron configuration of the elements in the following groups of the Periodic Table.

    a. 3 b. 5 c. 7 d. 8

    Example: Group 2 ns^2

20. Which group in the Periodic Table has the outer electron configuration:

    a. ns^2 b. $ns^2\ np^2$ c. $ns^2\ np^5$

21. Explain in terms of electronic structure why:

    a. There are 2 elements in the first period of the Periodic Table.
    b. There are 8 elements in the second period.
    c. There are 18 elements in the 4th period.
    d. There are 32 elements in the 6th period.

22. Give the total number of valence electrons of an element in Group

    a. 1 b. 3 c. 4 d. 6 e. 7

23. In what group of the Periodic Table do all the elements have:

    a. 2 valence electrons b. 5 valence electrons c. 6 valence electrons

PROBLEMS

1. *Bohr Equation (Example 9.1)* Using Equation 9.2, calculate E in kcal/mol for n =

   a. 1 b. 2 c. 3 d. 4 e. 5

2. *Bohr Equation (Example 9.1)* Calculate ΔE, in kcal/mol, for the transition:

   a. n = 1 to n = 3 b. n = 3 to n = 2 c. n = 2 to n = 5

3. *Electron Configurations (Example 9.3)* Without referring to Table 9.3, give the configuration of:

   a. B (at. no. = 5) b. Ne (at. no. = 10) c. Al (at. no. = 13)
   d. Cl (at. no. = 17)

4. *Electron Configurations (Example 9.4)* Without referring to Table 9.3, give the configuration of:

   a. Ca (at. no. = 20) b. Sc (at. no. = 21) c. Fe (at. no. = 26)
   d. Zn (at. no. = 30)

5. *Outer Electron Configurations (Example 9.6)* Give the outer electron configuration of:

   a. As (at. no. = 33) b. Rb (at. no. = 37) c. Sn (at. no. = 50)
   d. I (at. no. = 53)

6. *Outer Electron Configurations (Example 9.6)* Identify the atoms which have the following outer electron configurations:

   a. $3s^2\ 3p^3$ b. $4s^1$ c. $6s^2\ 6p^5$ d. $5s^2\ 5p^4$

* * * * *

7. Calculate, using the Bohr Equation, the ionization energy of the hydrogen atom. That is, calculate ΔE in kcal/mol when an electron moves from the ground state to n = ∞.

8. Give the atomic number of the element in which there is:

   a. a single 2s electron.
   b. two 3p electrons.
   c. an equal number of 4s and 3d electrons.

9. Give the complete electron configurations of atoms which have the following "condensed configurations".

   a. [Ne] $3s^2\ 3p^1$ b. [Ar] $4s^2$ c. [Ar] $4s^1\ 3d^5$

10. Write the electron configuration of each element in the third period, from Na to Ar.

11. In a certain atom, the first and second principal energy levels are completely filled, as is the 3s sublevel. There are three electrons in the 3p sublevel. How many electrons are there in the atom?

12. Give the electron configuration of the atom in Problem 11. What is the symbol of the element?

13. For each of the atoms in Problems 5 and 6, give the total number of electrons in the outermost principal energy level.

14. Which of the following electron configurations corresponds to an atom in its ground state? Which could be formed if one or more electrons were excited to higher sublevels? Which are impossible?

   a. $1s^2\ 2s^2\ 2p^4$ b. $1s^2\ 2s^2\ 2p^3\ 2d^1$ c. $1s^2\ 2s^2\ 2p^3\ 3s^1$
   d. $1s^2\ 2s^1\ 2p^2$ e. $1s^2\ 2s^2\ 2p^1$ f. $1s^2\ 2s^3$

*15. Using Equations 9.2 and 9.3, calculate the wavelength of the line produced when an electron moves from n = 3 to n = 2.

*16. One of the lines in the hydrogen spectrum has a wavelength of 102.5 nm. Calculate ΔE, using Equation 9.3. Using Figure 9.3, identify the levels between which the electron is moving.

*17. Using Figure 9.10, give the complete electron configuration of

   a. Tc (at. no. = 43) b. I (at. no. = 53) c. La (at. no. = 57)

*18. Give the number of electrons in each of the 3p orbitals in each element from Al (at. no. = 13) through Ar (at. no. = 18). Example: P 1, 1, 1

*19. Give the total number of half-filled orbitals in atoms which have the following electron configurations:

   a. [Ar] $4s^2\ 3d^3$ b. [Ar] $4s^1\ 3d^5$ c. [Ar] $4s^2\ 3d^7$

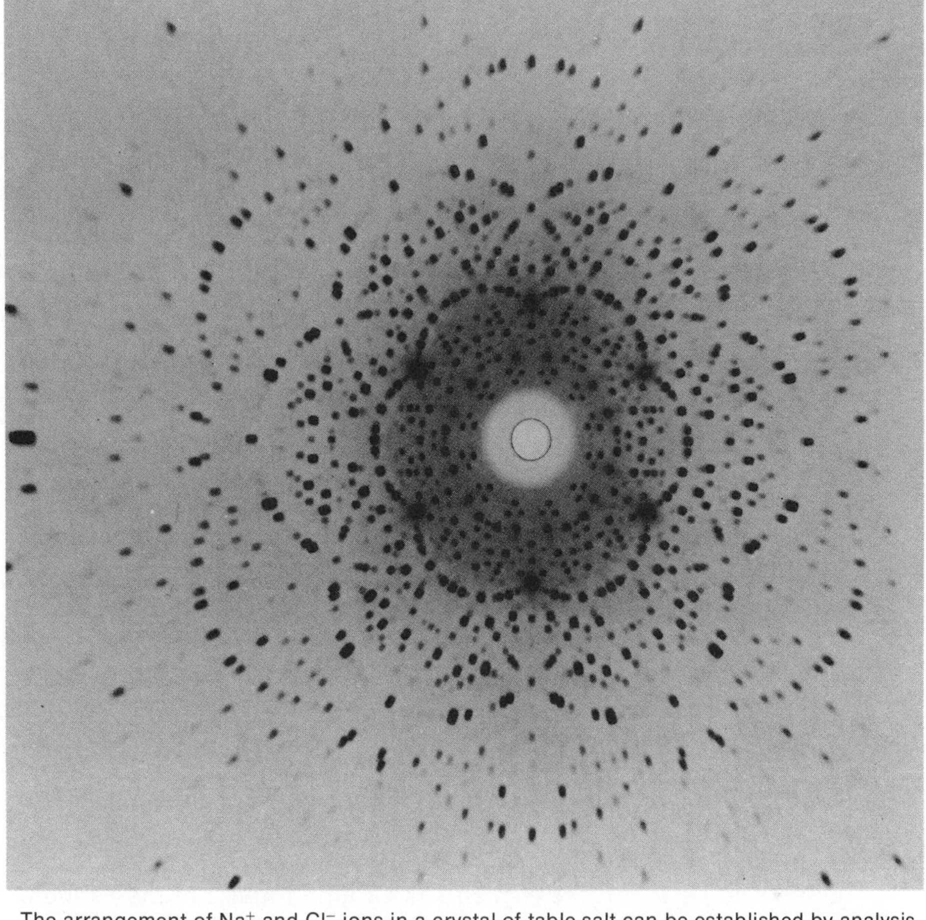

The arrangement of Na^+ and Cl^- ions in a crystal of table salt can be established by analysis of the X-ray diffraction pattern of NaCl. *Courtesy of Eastman Kodak.*

10 IONIC BONDING

Chemistry would not be very interesting if all matter were made up of isolated atoms. Indeed, it would hardly exist at all. There are only about 100 elementary substances. In contrast, millions of different compounds are known and more are being discovered every day. Clearly atoms prefer to combine with one another rather than remaining by themselves. They do so by forming *chemical bonds.* These are the strong forces that bind atoms to one another in virtually all pure substances.

The formation of bonds between atoms involves changes in their electronic structures. In particular, **it is ordinarily the valence electrons, those in the outermost principal energy level, which interact to form chemical bonds.** In this chapter, we will consider one such interaction. It involves a transfer of electrons from one atom to another.

Inner electrons in atoms are too stable to participate in chemical reactions.

This leads to the formation of charged ions. We can represent this process as:

$$M\cdot + N \rightarrow M^+ + N\cdot^-$$

where M and N represent two different atoms and the dot stands for an electron that is transferred from M to N. The forces between oppositely charged ions in the compound MN are called ionic bonds.

10.1 FORMATION AND PROPERTIES OF IONIC COMPOUNDS

Elements that Form Ionic Compounds

Ion formation is most likely to occur when one of the atoms involved gives up electrons readily and the other atom has a strong attraction for electrons. You will recall from Chapter 7 that ionization energy is a measure of the ease with which an atom gives up electrons. Metals with low ionization energies lose electrons readily. By the same token, electronegativity is a measure of the tendency of an atom to acquire electrons. Nonmetals of high electronegativity gain electrons readily. On this basis, we expect an ionic compound to be formed when *a metal of low ionization energy reacts with a nonmetal of high electronegativity.*

Every ionic compound contains a metallic and a nonmetallic element.

In practice, most metal atoms lose electrons in chemical reactions to form ions with a positive charge. Among the metals that do this are the following.

1. All the elements in Groups 1 and 2 of the Periodic Table. The Group 1 metals form +1 ions (Na^+, K^+). The Group 2 metals form +2 ions (Mg^{2+}, Ca^{2+}).
2. Aluminum in Group 3, which forms a +3 ion, Al^{3+}.
3. Certain of the heavier elements in Groups 4 and 5. These include tin (Sn^{2+}), lead (Pb^{2+}), and bismuth (Bi^{3+}).
4. Most of the transition and inner transition elements (lanthanides and actinides). Transition metal ions may have charges of +1 (Ag^+), +2 (Zn^{2+}), or +3 (Cr^{3+}). The inner transition metals ordinarily form +3 ions.

Nonmetals which gain electrons in reactions to form ions with a negative charge include the following.

1. The halogens (Group 7). They form −1 ions (F^-, Cl^-).
2. The elements in Group 6, which form −2 ions (O^{2-}, S^{2-}).
3. Nitrogen in Group 5, which forms a −3 ion, N^{3-}.

Energy Changes in the Formation of Ionic Compounds

To illustrate how ionic compounds are formed, consider the reaction between sodium (Group 1) and chlorine (Group 7). The product of the reaction is the solid compound NaCl, which consists of Na^+ and Cl^- ions. The overall reaction can be represented by the equation:

$$Na(s) + \frac{1}{2} Cl_2(g) \rightarrow NaCl(s); \qquad \Delta H = -98 \text{ kcal} \qquad (10.1)$$

This reaction is strongly exothermic ($\Delta H = -98$ kcal). Typically, large amounts of energy are given off when ionic compounds are formed from the elements.

To understand where this energy comes from, we might imagine that the reaction takes place in three steps.

(a) *The elements are converted to gaseous atoms.*

$$Na(s) + \frac{1}{2} Cl_2(g) \rightarrow Na(g) + Cl(g); \qquad \Delta H = +55 \text{ kcal} \qquad (10.1a)$$

This step is endothermic ($\Delta H = +55$ kcal). Energy has to be absorbed to break the bonds holding the atoms together in the elementary substances.

(b) *An electron is transferred from a sodium to a chlorine atom.*

$$Na(g) + Cl(g) \rightarrow Na^+(g) + Cl^-(g); \qquad \Delta H = +31 \text{ kcal} \qquad (10.1b)$$

This step is also endothermic ($\Delta H = +31$ kcal).

(c) *The ions combine to form an ionic solid.*

$$Na^+(g) + Cl^-(g) \rightarrow NaCl(s); \qquad \Delta H = -184 \text{ kcal} \qquad (10.1c)$$

This is the exothermic step ($\Delta H = -184$ kcal). A large amount of energy is given off when the ions of opposite charge come together. This reflects the strength of the ionic bonds between Na^+ and Cl^- ions.

184 kcal is the energy needed to break a mole of ionic bonds in NaCl.

By Hess' Law (Chapter 4), ΔH for Reaction 10.1 is the sum of the ΔH's for the three steps we have written,

$$\Delta H = +55 \text{ kcal} + 31 \text{ kcal} - 184 \text{ kcal} = -98 \text{ kcal}$$

We see that the exothermic nature of the reaction is due to the third step. The value of ΔH in step (c) is known as the *lattice energy*. It is a large

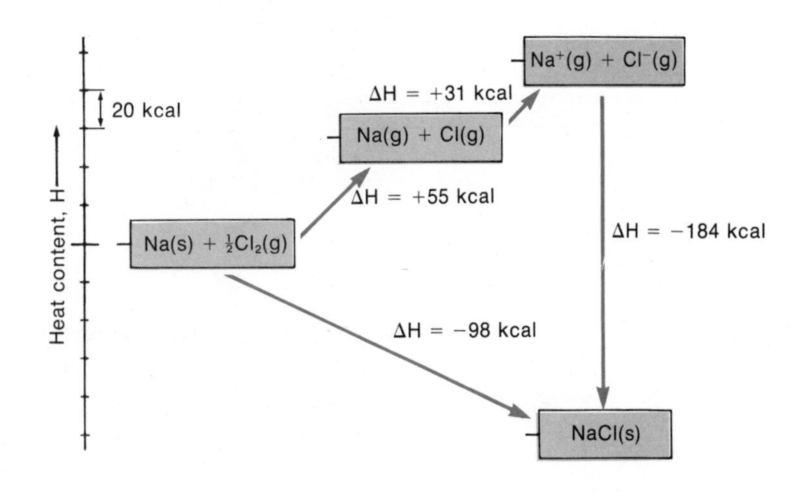

FIGURE 10.1

Enthalpy diagram for the formation of NaCl. If NaCl is formed from the elements, the reaction gives off 98 kcal per mole of NaCl. We could also form NaCl by the alternate path shown, in which we convert the elements to free atoms ($\Delta H = +55$ kcal), convert the atoms to Na^+ and Cl^- ions ($\Delta H = +31$ kcal), and then combine the ions to form solid NaCl ($\Delta H = -184$ kcal). The high attraction between the oppositely charged Na^+ and Cl^- ions gives rise to the large negative value of ΔH in the last step and is the source of the ionic bonds between the ions in the solid.

The ionic bond is due to the attraction between the + and − ions.

enough negative number to make the overall reaction exothermic. This is generally true for other reactions involving the formation of ionic compounds from the elements.

Properties of Ionic Compounds

We referred briefly to certain of the properties of ionic compounds in Chapter 2. Here we emphasize those properties which are directly related to their structure and to the nature of the ionic bonds that hold them together.

1. *Ionic compounds form crystal lattices in which positive ions are surrounded by negative ions and vice versa.* In the sodium chloride crystal, each Na^+ ion is surrounded by Cl^- ions and vice versa. This is a very stable arrangement, since it puts ions of opposite charge very close together. In a crystal of sodium chloride there is no such thing as an NaCl "molecule". No pair of ions can be singled out as a distinct structural unit.

A substance which is liquid at 25°C is not ionic.

2. *Ionic compounds are solids at room temperature.* Typically, they have melting points ranging from several hundred degrees Celsius to over 2000°C. The high melting point reflects the strong electrical forces that hold the crystal together. To melt sodium chloride all the ionic bonds must be broken. This can happen only at high temperatures (800°C for NaCl). Only then do the oppositely charged ions have sufficient kinetic energy to break away from one another.

3. *Ionic compounds are good conductors of electricity,* either in the molten state or in water solution. In either case, the ions are free to move and thereby conduct an electric current (Figure 10.2). In the solid state, the ions cannot flow past one another and hence do not conduct.

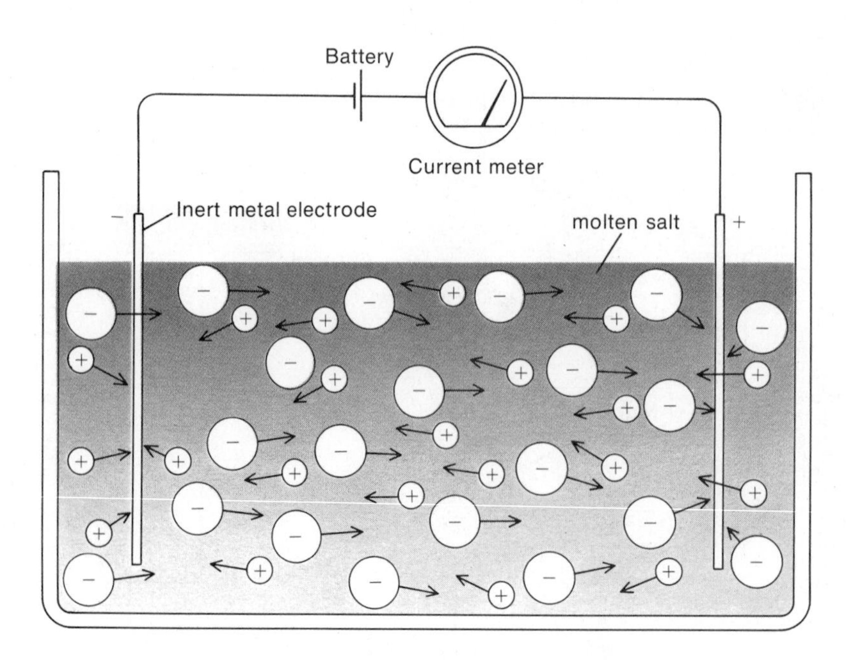

FIGURE 10.2

In a molten salt the electric current is carried by the ions. The positive ions are attracted to the negative electrode, the negative ions to the positive electrode. Motion of the ions through the molten salt effectively moves electric charge from one electrode to the other and makes the salt a conductor of electricity.

10.2 SIZES OF IONS

Figure 10.3 compares the radii of the ions formed by the elements in Groups 1 and 7 to those of the corresponding atoms. You will note that:

—*positive ions are smaller than the atoms from which they are derived.*

—*negative ions are larger than the atoms from which they are derived.*

These effects are typical of all ions. For example, a Mg^{2+} ion (radius = 0.065 nm) is smaller than a magnesium atom (radius = 0.160 nm). In contrast, an O^{2-} ion (radius = 0.140 nm) is larger than an oxygen atom (radius = 0.066 nm).

It is hardly surprising that the loss of electrons by an atom should lead to a decrease in radius. The excess nuclear charge (+1, +2, +3) should draw electrons in closer to the nucleus. This should make the positive ion smaller than the atom. By the same argument, size should increase when electrons are gained. A negative charge (−1 or −2) should expand the electron cloud of the entire atom. As a result, a negative ion should be somewhat larger than the atom from which it is formed.

These effects lead to an interesting result. Although metal atoms as a rule are larger than nonmetal atoms, the reverse is true for ions. That is, *positive ions are usually smaller than negative ions.* In sodium chloride, for example, most of the space in the crystal lattice is taken up

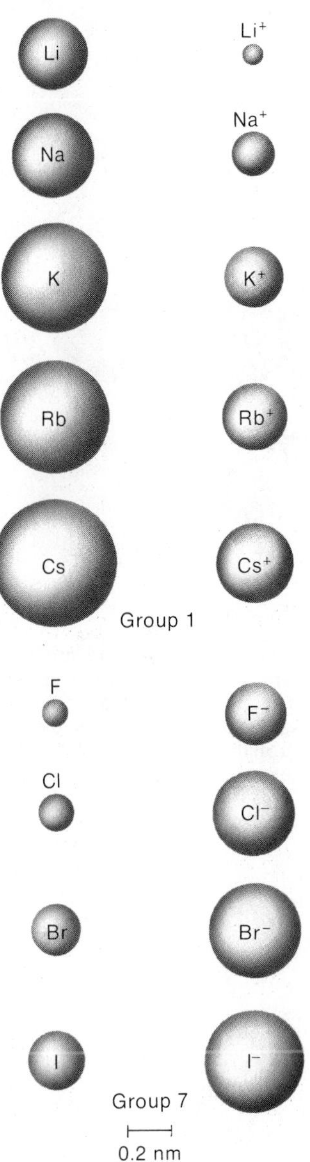

FIGURE 10.3

Sizes of the atoms and ions in Groups 1 and 7. Positive ions are always smaller than the atoms from which they are derived. Negative ions are larger than the atoms from which they are formed.

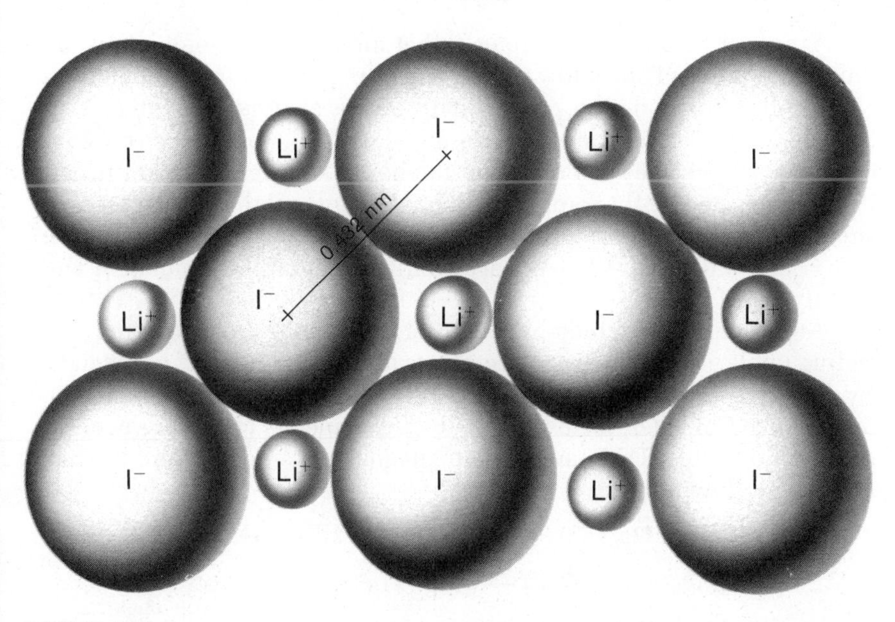

FIGURE 10.4

Face of the LiI crystal. In LiI the I^- ions are much larger than the Li^+ ions. In the crystal the I^- ions are touching, and the Li^+ ions are relatively free to rattle about. The distance between ions in this crystal is determined by the I^- ions and so can be used to determine the radius of the I^- ion. Once that radius is known, other ionic radii can be found. The radius of the I^- ion in the above figure is 0.216 nm.

by Cl^- ions. They are considerably larger than Na^+ ions. Lithium iodide, LiI, is an extreme case of this sort. The I^- ions (radius = 0.216 nm) are so much bigger than Li^+ ions (radius = 0.060 nm) that they actually touch one another in the crystal. The Li^+ ions rattle around in the cage formed by neighboring I^- ions (Figure 10.4).

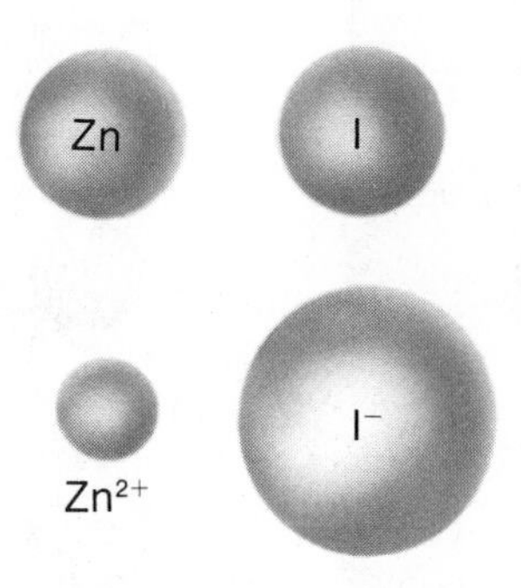

FIGURE 10.5

The zinc and iodine atoms are just about the same size. Since the size of a particle goes down if it loses electrons and goes up if it gains electrons, it is very reasonable that the Zn^{2+} ion is much smaller than the I^- ion.

EXAMPLE 10.1

The zinc atom is almost exactly the same size as the iodine atom (atomic radius = 0.133 nm). How would you expect the Zn^{2+} ion to compare in radius to the I^- ion?

SOLUTION

The Zn^{2+} ion should be smaller than the Zn atom. The I^- ion should be larger than the I atom. So, we would expect Zn^{2+} to be considerably smaller than I^-. Indeed it is.

$$\text{ionic radius } Zn^{2+} = 0.074 \text{ nm}$$
$$\text{ionic radius } I^- = 0.216 \text{ nm}$$

Even though the atoms are the same size, the I^- ion has a radius nearly three times that of the Zn^{2+} ion.

10.3 IONS WITH NOBLE-GAS STRUCTURES

In the old days, the nobility didn't mingle with the masses.

You will recall from Chapter 6 that atoms of the noble gases are unusually stable. They show no tendency to combine with each other. Neither do they combine readily with atoms of other elements. This implies that noble-gas atoms have an unusually stable electron configuration which they are reluctant to change. Going one step further, it seems reasonable that atoms of other elements might tend to react in such a way as to acquire a noble-gas configuration. This is indeed the case. Such a configuration is very common among both positive and negative ions.

Atoms of metals which have 1, 2, or 3 electrons more than the preceding noble gas tend to lose these electrons in reacting with nonmetals. By doing so, they form positive ions (+1, +2, +3) with noble-gas configurations. Consider, for example, the three metals following neon in the Periodic Table. These are sodium, magnesium, and aluminum. They form the following ions:

$$_{11}Na(1s^22s^22p^63s^1) \rightarrow {}_{11}Na^+(1s^22s^22p^6) + e^-$$

$$_{12}Mg(1s^22s^22p^63s^2) \rightarrow {}_{12}Mg^{2+}(1s^22s^22p^6) + 2\ e^-$$

$$_{13}Al(1s^22s^22p^63s^23p^1) \rightarrow {}_{13}Al^{3+}(1s^22s^22p^6) + 3\ e^-$$

Each of the ions formed, Na^+, Mg^{2+} and Al^{3+}, has the electron configuration of neon:

$$_{10}Ne(1s^22s^22p^6)$$

At the opposite side of the Periodic Table, nonmetals with 1 or 2 electrons less than the following noble gas tend to acquire these electrons when they react with metals. By doing so, they form negative ions with noble-gas configurations. Thus we have:

$$_{8}O(1s^2 2s^2 2p^4) + 2\ e^- \rightarrow O^{2-}(1s^2 2s^2 2p^6)$$

$$_{9}F(1s^2 2s^2 2p^5) + e^- \rightarrow F^-(1s^2 2s^2 2p^6)$$

Both the O^{2-} and the F^- ion, like Na^+, Mg^{2+}, and Al^{3+}, have the neon configuration.

Table 10.1 lists several simple ions which have noble-gas structures. Several features of this table are of interest. Most important, you will note that:

The reason many ions have noble-gas structures is that, in terms of energy, such structures are favored over others.

—all the Group 1 metals form +1 ions
—all the Group 2 metals form +2 ions
—all the nonmetals in Group 6 form −2 ions
—all the nonmetals in Group 7 form −1 ions (including hydrogen, H^-)

These general rules were first presented in Chapter 8. They can now be given a simple explanation. **The charges correspond to the number of electrons which atoms of these elements must lose or gain to acquire noble-gas structures.**

Three other points concerning Table 10.1 are worth noticing.

—Al in Group 3 forms a +3 ion with a noble-gas structure. The heavier elements in this group are too far removed from the preceding noble gas to acquire its structure by losing electrons. In principle, the first member of the group, boron, could form a +3 ion with a noble-gas structure. In practice, boron shows no tendency to do this. Its ionization energy (193 kcal/mol) is too high, close to that of most nonmetals.

—the elements in the scandium subgroup form +3 ions with noble-gas structures.

—nitrogen in Group 5 forms a −3 ion (Ne structure) when it reacts with a few very active metals such as lithium and magnesium. None of the heavier elements in Group 5 show any tendency to do this. They have electronegativities close to 2, considerably lower than nitrogen (3.0).

TABLE 10.1 MONATOMIC IONS WITH NOBLE-GAS STRUCTURES

Period	Group							
	1	2		3	4	5	6	7
1								H^-
2	Li^+	Be^{2+}		–	–	N^{3-}	O^{2-}	F^-
3	Na^+	Mg^{2+}		Al^{3+}	–	–	S^{2-}	Cl^-
4	K^+	Ca^{2+}	Sc^{3+}	–	–	–	Se^{2-}	Br^-
5	Rb^+	Sr^{2+}	Y^{3+}	–	–	–	Te^{2-}	I^-
6	Cs^+	Ba^{2+}	La^{3+}					

EXAMPLE 10.2

Consider the elements polonium (at. no. = 84), astatine (at. no. = 85), francium (at. no. = 87), and radium (at. no. = 88). Atoms of these elements form ions which have a noble-gas structure. What are their charges?

SOLUTION

Po is in Group 6; forms a −2 ion, Po^{2-}
At is in Group 7; forms a −1 ion, At^{-}
Fr is in Group 1; forms a +1 ion, Fr^{+}
Ra is in Group 2; forms a +2 ion, Ra^{2+}

10.4 OTHER TYPES OF IONS

So far, all of the ions we have considered have been *monatomic* (derived from a single atom). Moreover, they all have electron configurations identical with that of a noble gas. Such ions are very common in chemical compounds. However there are many relatively common ions, such as OH^{-} and Zn^{2+}, which do not fit this description. We will now consider the charges and other properties of ions which either do not have a noble-gas structure or are composed of more than one atom.

Monatomic Positive Ions

The metals to the right of scandium in the Periodic Table cannot acquire a noble-gas structure by losing electrons. They have too many electrons in their outer energy levels to do this. However, metals such as silver, zinc, and chromium react with nonmetals such as oxygen or chlorine to form ionic compounds. The metals in these compounds are present as ions with charges of +1, +2, or +3 (Ag^{+}, Zn^{2+}, Cr^{3+}).

There is no way to predict in advance the charges of positive ions in this category. Some of the more common ions are listed in Table 10.2. They include positive ions of the transition metals and the heavy metals in Groups 4 and 5. You will meet these ions frequently throughout this course, so you should learn their charges.

Transition metal ions usually have charges of either +2 or +3.

TABLE 10.2 POSITIVE IONS OF THE TRANSITION AND POST-TRANSITION METALS

+1	+2	+3
Ag^{+}	Fe^{2+}, Ni^{2+}	Cr^{3+}
	Cu^{2+}, Zn^{2+}	Fe^{3+}
	Sn^{2+}, Pb^{2+}	Bi^{3+}

TABLE 10.3 FORMULAS OF COMPOUNDS OF IRON

	Chlorides (Cl^-)	Hydroxides (OH^-)	Oxides (O^{2-})	Sulfates (SO_4^{2-})
Fe^{2+}	$FeCl_2$	$Fe(OH)_2$	FeO	$FeSO_4$
Fe^{3+}	$FeCl_3$	$Fe(OH)_3$	Fe_2O_3	$Fe_2(SO_4)_3$

You will note from Table 10.2 that iron forms two different positive ions, Fe^{2+} and Fe^{3+}. Which ion is formed depends upon the nature of the reaction. When iron reacts with H^+ ions in water solution, it forms the Fe^{2+} ion:

$$Fe(s) + 2\ H^+(aq) \rightarrow Fe^{2+}(aq) + H_2(g) \qquad (10.2)$$

On the other hand, the direct reaction of iron with chlorine gas gives a solid, $FeCl_3$, in which there are Fe^{3+} ions.

$$2\ Fe(s) + 3\ Cl_2(g) \rightarrow 2\ FeCl_3(s) \qquad (10.3)$$

In general, iron forms two different types of ionic compounds, depending upon whether it is present as a +2 or +3 ion (Table 10.3). The colors of some of these compounds are shown in Color Plate 5A, center of book.

EXAMPLE 10.3

What are the mass percentages of iron in its two sulfates?

SOLUTION

The atomic masses of Fe, S, and O are 55.85, 32.06, and 16.00 in that order.

$$\text{FM } FeSO_4 = 55.85 + 32.06 + 4(16.00) = 151.91$$

One mole of $FeSO_4$ weighs 151.91 g and contains 55.85 g of Fe. So:

$$\%\ Fe\ in\ FeSO_4 = \frac{55.85\ g}{151.91\ g} \times 100 = 36.77$$

$$\text{FM } Fe_2(SO_4)_3 = 2(55.85) + 3(32.06) + 12(16.00)$$
$$= 111.70 + 96.18 + 192.00 = 399.88$$

One mole of $Fe_2(SO_4)_3$ weighs 399.88 g and contains 111.70 g of Fe. So:

$$\%\ Fe\ in\ Fe_2(SO_4)_3 = \frac{111.70\ g}{399.88\ g} \times 100 = 27.93$$

In general, the percentage of iron in the compound containing Fe^{2+} ions is always higher than that in the Fe^{3+} compound.

TABLE 10.4 SOME COMMON POLYATOMIC IONS

+1		−1		−2		−3	
NH_4^+	ammonium	OH^-	hydroxide	CO_3^{2-}	carbonate	PO_4^{3-}	phosphate
		NO_3^-	nitrate	SO_4^{2-}	sulfate		
		ClO_3^-	chlorate	CrO_4^{2-}	chromate		
		ClO_4^-	perchlorate	$Cr_2O_7^{2-}$	dichromate		
		MnO_4^-	permanganate				
		HCO_3^-	hydrogen carbonate				

Although iron is perhaps the best known metal that forms more than one positive ion, it is by no means the only one. Other examples include:

chromium	$\mathbf{Cr^{3+}}$	Cr^{2+}
copper	$\mathbf{Cu^{2+}}$	Cu^+
mercury	$\mathbf{Hg^{2+}}$	Hg^+

(The more stable, or more common, positive ion is shown in bold type.)

Polyatomic Ions

Many familiar ionic compounds contain *polyatomic ions,* charged species containing more than one atom. An example is the ammonium ion, NH_4^+. This is the positive ion found in such compounds as ammonium chloride, NH_4Cl, and ammonium sulfide, $(NH_4)_2S$. A common example of a polyatomic negative ion is the hydroxide ion, OH^-. The hydroxide ion is found in such common compounds as sodium hydroxide (lye), NaOH, and calcium hydroxide (lime), $Ca(OH)_2$.

The structures of some common polyatomic ions are shown in Figure 10.6. Within the ion, the atoms are held together by covalent bonds, to be discussed in Chapter 11. Structurally, polyatomic ions are very similar to molecules. Indeed, we can think of polyatomic ions as "charged molecules".

Table 10.4 gives the names and formulas of several important polyatomic ions. Many of these were mentioned in earlier chapters. A few may be new to you. You should become familiar with their names and charges. You will meet them frequently, both in classroom discussions and in the laboratory.

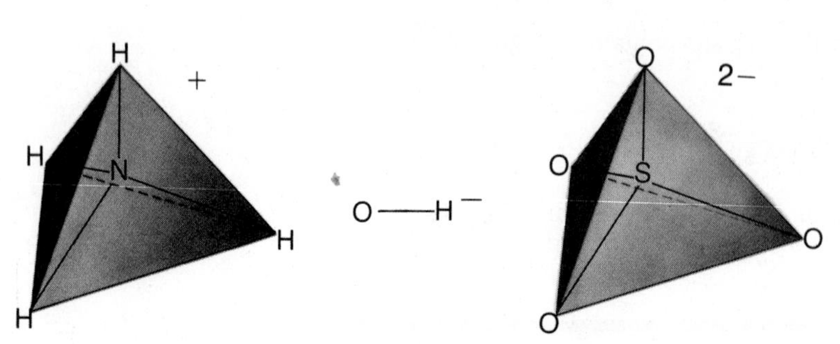

FIGURE 10.6

Structures of some common polyatomic ions. In these ions the bonds between the atoms are strong. In NH_4^+ and SO_4^{2-}, a central atom (N, S) is bonded symmetrically to four other atoms (H, O).

10.5 FORMULAS OF IONIC COMPOUNDS

Principle of Electrical Neutrality

The simplest formula of an ionic compound is readily obtained if the charges of the ions are known. All you need do is to apply the principle that the compound itself must be electrically neutral. That is, the total positive charge must be equal to the total negative charge. To show how this works, consider the compound aluminum fluoride. This compound consists of Al^{3+} and F^- ions (Table 10.1). Clearly, we must have three F^- ions to balance one Al^{3+} ion. The simplest formula must then be AlF_3. In another case, suppose we wish to determine the formula of ammonium sulfate. From Table 10.4, we see that the two ions involved are NH_4^+ and SO_4^{2-}. For the compound to be electrically neutral, we need two NH_4^+ ions for every SO_4^{2-} ion. To indicate this we write the formula of ammonium sulfate as $(NH_4)_2SO_4$. The subscript 2 outside the parenthesis shows that there are two NH_4^+ ions per formula unit.

Further examples of this principle are shown in Example 10.4. Notice that in order to apply the rule of electrical neutrality, you must know the charges of the ions involved. In particular, you must know:

—the charges of the ions with noble-gas structures formed by elements in various groups of the Periodic Table (Table 10.1).

—the charges of the ions formed by the transition and post-transition metals (Table 10.2).

—the charges and names of the polyatomic ions listed in Table 10.4.

EXAMPLE 10.4

Give the simplest formula of:

a. magnesium oxide

b. two different bromides of iron

c. scandium hydroxide

d. aluminum sulfate

SOLUTION

a. Since magnesium is in Group 2, it should form a +2 ion, Mg^{2+}. Oxygen, in Group 6, forms a −2 ion, O^{2-}. For electrical neutrality, these ions must combine in a 1:1 ratio. Hence, the simplest formula of magnesium oxide must be MgO.

b. From Table 10.2, we recall that iron can form either Fe^{2+} or Fe^{3+} ions. One compound will have the formula $FeBr_2$ (two Br^- ions per Fe^{2+}). The other will be $FeBr_3$ (three Br^- ions per Fe^{3+}).

c. Scandium forms the Sc^{3+} ion (Table 10.1). The hydroxide ion is OH^- (Table 10.4). We write the formula $Sc(OH)_3$ to indicate that there are three OH^- ions for every Sc^{3+} ion.

d. The ions present are Al^{3+} (Table 10.1) and SO_4^{2-} (Table10.4). To achieve electrical neutrality, we need two Al^{3+} ions (total charge = +6) to balance three SO_4^{2-} ions (total charge = −6). The simplest formula is $Al_2(SO_4)_3$.

Just a little practice on problems like these helps a lot.

Equations for Formation of Ionic Compounds

If you know the formulas of ionic compounds, you can readily write balanced equations for their formation by the reaction of metals with nonmetals. This is illustrated in Example 10.5. Note that, in addition to knowing the formulas of the compounds formed, you must know:

—**the physical states of elements at room temperature.**

—**the molecular formulas of certain nonmetals** (H_2, O_2, F_2, Cl_2, Br_2, I_2).

These properties of elements were described in earlier chapters (Chapters 3, 6, and 7). You should know them by now. Finally, you must remember that **all ionic compounds are solids at room temperature.**

EXAMPLE 10.5

Write balanced chemical equations to represent the formation of the following ionic compounds from the elements.

a. MgO b. aluminum chloride c. silver sulfide

SOLUTION

a. Unbalanced equation: $Mg(s) + O_2(g) \rightarrow MgO(s)$
Balanced equation: $2\ Mg(s) + O_2(g) \rightarrow 2\ MgO(s)$

Note that the equation: $Mg(s) + O(g) \rightarrow MgO(s)$ would not be correct because elementary oxygen under ordinary conditions exists as diatomic molecules, not as isolated atoms.

b. Here we work "backwards". We first deduce the formula of aluminum chloride. The ions involved are Al^{3+} and Cl^-. The compound must then have the formula $AlCl_3$.

unbalanced equation: $Al(s) + Cl_2(g) \rightarrow AlCl_3(s)$
balanced equation: $2\ Al(s) + 3\ Cl_2(g) \rightarrow 2\ AlCl_3(s)$

Note that although we needed to know the charges of the ions in order to predict the formula of the product, these charges do not appear in the final equation.

c. The ions are Ag^+ (Table 10.2) and S^{2-} (Table 10.1). The formula of silver sulfide must be Ag_2S. Both elements are solid and are shown as monatomic. The equation is:

$$2\ Ag(s) + S(s) \rightarrow Ag_2S(s)$$

Formulas of Hydrates

Frequently, when an ionic solid crystallizes from water solution, water molecules are retained in the crystal lattice. The compounds formed are called **hydrates.** The composition of a hydrate is shown by a formula which indicates the amount of water present. We write

$$BaCl_2 \cdot 2H_2O(s)$$

to indicate that in hydrated barium chloride there are two moles of H_2O for every mole of $BaCl_2$. Similarly, in the common hydrate of copper sulfate:

In a hydrate like this, water molecules are actually present in the crystal.

$$CuSO_4 \cdot 5H_2O(s)$$

there are five moles of H_2O per mole of $CuSO_4$.

EXAMPLE 10.6

What is the percentage by mass of H_2O in $CuSO_4 \cdot 5H_2O$?

SOLUTION

The atomic masses of Cu, S, and O are 63.55, 32.06, and 16.00 in that order. The molecular mass of H_2O is 18.02.

$$\text{FM } CuSO_4 \cdot 5H_2O = 63.55 + 32.06 + 64.00 + 5(18.02)$$

$$= 159.61 + 90.10 = 249.71$$

Thus, one mole of $CuSO_4 \cdot 5H_2O$ weighs 249.71 g and contains 90.10 g of H_2O.

$$\%\ H_2O = \frac{90.10\ g}{249.71\ g} \times 100 = 36.08$$

Properties of Hydrates

Hydrates ordinarily have a quite different appearance than the corresponding anhydrous substances. Commonly, they form as large, often transparent crystals. This is the case with the hydrate of sodium carbonate, $Na_2CO_3 \cdot 10H_2O$. Anhydrous sodium carbonate, Na_2CO_3, is a white powder. Hydrates of many transition metal compounds are brightly colored (Color Plate 5B, center of book). For example, crystals of $CuSO_4 \cdot 5H_2O$ are a deep blue. In contrast, pure $CuSO_4$ is white.

When hydrates are heated, they decompose, giving off water vapor. The solid remaining is most often the anhydrous ionic compound. Such is the case with barium chloride:

$$BaCl_2 \cdot 2H_2O(s) \rightarrow BaCl_2(s) + 2\ H_2O(g) \qquad (10.4)$$

Sometimes, water is lost in successive steps as the hydrate is heated to higher temperatures. This happens with copper sulfate.

$$CuSO_4 \cdot 5H_2O(s) \rightarrow CuSO_4 \cdot 3H_2O(s) + 2\ H_2O(g) \qquad \text{(at about 50°C)} \qquad (10.5)$$

$$CuSO_4 \cdot 3H_2O(s) \rightarrow CuSO_4 \cdot H_2O(s) + 2\ H_2O(g) \qquad \text{(at about 100°C)} \qquad (10.6)$$

$$CuSO_4 \cdot H_2O(s) \rightarrow CuSO_4(s) + H_2O(g) \qquad \text{(at about 200°C)} \qquad (10.7)$$

Loss of water by a hydrate is referred to as *efflorescence*. This process can occur upon heating, as above. Water can also be lost if a hydrate is exposed to dry air at room temperature. Some of the gadgets used to indicate relative humidity in the home operate this way. They contain an ionic solid which changes color when it gains or loses water of hydration. One such compound is cobalt chloride, $CoCl_2$. In moist air, the hydrate, $CoCl_2 \cdot 6H_2O$ forms. This compound has a pink color. If the air is quite dry, the stable hydrate is $CoCl_2 \cdot 4H_2O$, which is blue.

$$CoCl_2 \cdot 6H_2O(s) \rightarrow CoCl_2 \cdot 4H_2O(s) + 2\ H_2O(g) \qquad (10.8)$$

pink; high humidity — *blue; low humidity*

Efflorescence is always an endothermic process. Heat must be supplied to drive water molecules out of the crystal lattice. Often, the heat effect is appreciable. With $Na_2SO_4 \cdot 10H_2O$ (commonly called Glauber's salt), 19 kcal must be absorbed per mole to drive off all the water as liquid:

$$Na_2SO_4 \cdot 10H_2O(s) \rightarrow Na_2SO_4(s) + 10\ H_2O(l); \qquad \Delta H = +19 \text{ kcal} \qquad (10.9)$$

Reaction 10.9 has been suggested as a way of absorbing energy from the sun in a solar heating system. When the sun is shining, crystals of $Na_2SO_4 \cdot 10H_2O$ are converted to the anhydrous salt. At night or on a cloudy day, water is added to Na_2SO_4 to reverse the reaction. In this way, 19 kcal of heat is given off per mole of sodium sulfate.

10.6 NAMING IONIC COMPOUNDS

So far in this chapter, we have referred by name to a great many ionic compounds. These names are arrived at by a simple process which you may already have guessed. Ionic compounds such as NaCl or KNO_3 are named by giving first the name of the positive ion and then the name of the negative ion. Thus we have:

Compound	*Positive Ion*	*Negative Ion*	*Name of Compound*
NaCl	Na^+ (sodium)	Cl^- (chloride)	sodium chloride
KNO_3	K^+ (potassium)	NO_3^- (nitrate)	potassium nitrate

With this basic idea in mind, let us see how the names of the ions themselves are obtained.

Positive Ions

Monatomic positive ions which have noble-gas structures take the names of the metals from which they are derived:

Na^+ sodium $\qquad$ Mg^{2+} magnesium $\qquad$ Al^{3+} aluminum

When a metal forms more than one ion, it is necessary to distinguish between them. To do this, we indicate the charge of the ion by a Roman numeral in parentheses* immediately following the name of the metal:

If speaking, we would say "iron two" or "iron three".

Fe^{2+} iron(II) Fe^{3+} iron(III)

In practice, we often do this with all ions derived from transition and post-transition metals (Table 10.2). Thus we may write:

Cu^{2+} copper(II) Ni^{2+} nickel(II)

As Table 10.4 indicates, we need consider only one positive polyatomic ion:

NH_4^+ ammonium

Negative Ions

Monatomic negative ions are named simply. All we do is to add the suffix *-ide* to the stem of the name of the nonmetal. Examples include:

O^{2-}	oxide	F^-	fluoride	H^-	hydride
S^{2-}	sulfide	Cl^-	chloride		

Polyatomic negative ions are given special names as indicated in Table 10.4. There are a couple of general rules used:

1. Where a nonmetal combines with oxygen to form two different polyatomic ions, the suffixes *-ate* and *-ite* are used to distinguish between them. The "ate" ion is the one containing the greater number of oxygen atoms.

SO_4^{2-}	sulfate	NO_3^-	nitrate
SO_3^{2-}	sulfite	NO_2^-	nitrite

With an element such as chlorine, which forms more than two polyatomic ions with oxygen, the prefixes *per-* and *hypo-* are also used.

ClO_4^-	perchlorate
ClO_3^-	chlorate
ClO_2^-	chlorite
ClO^-	hypochlorite

*An older system uses the suffixes *-ous* and *-ic* for the ions of lower and higher charge. These are added to the Latin stem of the name of the metal. Thus, for the two ions derived from iron (Latin: ferrum), we would write:

Fe^{2+} ferrous Fe^{3+} ferric

Similarly, $FeCl_2$ would be called ferrous chloride; $FeCl_3$ would be called ferric chloride. The newer system presented here is simpler and will be used throughout this text. However, you may encounter the older system, particularly in reading the labels on reagent bottles.

2. Polyatomic negative ions that contain hydrogen as well as nonmetal and oxygen atoms are named as illustrated below:

HCO_3^- hydrogen carbonate	; compare CO_3^{2-}	carbonate
HSO_4^- hydrogen sulfate	; compare SO_4^{2-}	sulfate
HPO_4^{2-} hydrogen phosphate	; compare PO_4^{3-}	phosphate
$H_2PO_4^-$ dihydrogen phosphate		

Ionic Compounds

The naming system is simple, *but* you have to know the names of the ions.

As we have pointed out, an ionic compound is named in the same order in which the ions appear in the formula. The name of the positive ion is given first followed by that of the negative ion.

NaBr	sodium bromide
$Mg(NO_3)_2$	magnesium nitrate
NH_4ClO_3	ammonium chlorate
FeI_2	iron(II) iodide

EXAMPLE 10.7

Name the following compounds:

a. $BaCl_2$ b. $Al(NO_3)_3$ c. $Cr(NO_3)_3$ d. $CuSO_4$

SOLUTION

In each case, we first decide what ions are present.

a. Ba^{2+}, Cl^- ions; barium chloride

b. Al^{3+}, NO_3^- ions; aluminum nitrate

c. Cr^{3+}, NO_3^- ions; chromium(III) nitrate. The Roman numeral is used since chromium is a transition metal, capable of forming more than one positive ion.

d. Cu^{2+}, SO_4^{2-} ions; copper(II) sulfate. Here, in common usage, the Roman numeral II is often dropped.

EXAMPLE 10.8

Give the formulas of ionic compounds which have the following names:

a. potassium sulfate b. sodium hydrogen carbonate
c. cobalt(II) chloride

SOLUTION

Here in each case we start by deducing, from the name, the charges of the ions present. Then we use the principal of electrical neutrality to obtain the formula (Recall Example 10.4).

a. K^+, SO_4^{2-} ions; K_2SO_4

b. Na^+, HCO_3^- ions; $NaHCO_3$ (commonly called "sodium bicarbonate")

c. Co^{2+}, Cl^- ions; $CoCl_2$

10.7 SUMMARY

When a metal of low ionization energy reacts with a nonmetal of high electronegativity, the product is a solid ionic compound. The oppositely charged ions are held together in the crystal by ionic bonds. The positive ion, derived from the metal, often has a noble-gas structure. If it does, its charge is readily determined from the group number (+1 in Group 1, +2 in Group 2, +3 in Group 3). Metals to the right of scandium in the 4th, 5th, and 6th periods form positive ions whose charges cannot be deduced from their positions in the Periodic Table. Monatomic negative ions always have noble-gas structures. Their charges are predictable (−2 in Group 6, −1 in Group 7).

Polyatomic ions (Table 10.4) can be regarded as "charged molecules". Commonly, they have charges of −1 or −2. Charges of +1 or −3 are much less common. Like monatomic ions, they act as structural units in the crystal lattice. For example, the crystal structure of NaOH is very similar to that of NaCl. The only difference is that the OH^- ion has replaced the Cl^- ion.

Much of this chapter was devoted to the formulas and names of ionic compounds. From now on, you will be expected to be able to name an ionic compound, given its formula. Conversely, you should be able to write the formula of an ionic compound, given its name.

NEW TERMS

carbonate ion, CO_3^{2-}
chlorate ion, ClO_3^-
chromate ion, CrO_4^{2-}
dichromate ion, $Cr_2O_7^{2-}$
hydrate
hydrogen carbonate ion, HCO_3^-
ionic bond
ionic radius
lattice energy
monatomic ion
nitrate ion, NO_3^-
noble gas structure
perchlorate ion, ClO_4^-
permanganate ion, MnO_4^-
polyatomic ion
sulfate ion, SO_4^{2-}

QUESTIONS

1. Which of the following elements form positive ions when they react?

   a. Li in Group 1 b. S in Group 6 c. Ni, a transition metal
   d. C in Group 4

2. Explain, in terms of metallic character, why aluminum forms a +3 ion much more readily than boron in the same group.

3. Explain why nitrogen forms a −3 ion, N^{3-}, while bismuth, in the same group, forms a +3 ion, Bi^{3+}.

4. Why, in an ionic crystal, are + ions surrounded by − ions rather than ions of the same charge?

5. Why do ionic compounds usually have high melting points?

6. How do you explain the fact that solid ionic compounds do not conduct an electrical current, yet they become conductors when melted?

7. In general, which is larger:
   a. a positive ion or the corresponding atom?
   b. a negative ion or the corresponding atom?
   c. a metal atom or a nonmetal atom in the same period of the Periodic Table?
   d. a metal ion or a nonmetal ion in the same period?

8. Give the electron configuration of:
   a. Na^{+} b. Ca^{2+} c. Al^{3+} d. Cl^{-} e. O^{2-}

9. What charges would you expect for positive ions of the following metals?
   a. K b. Sc c. Sr d. Li e. Al

10. What charges would you expect for negative ions of the following nonmetals?
    a. S b. F c. N d. I e. O

11. Which of the following ions have noble-gas configurations?
    a. Cs^{+} b. Be^{2+} c. In^{3+} d. Mn^{2+} e. Ca^{+}

12. List as many ions as you can that have the electron configuration of argon.

13. Give the charges of ions formed by the following transition metals:
    a. Ag b. Zn c. Cu d. Fe

14. Name the following ions:
    a. NH_4^{+} b. CO_3^{2-} c. NO_3^{-} d. ClO_3^{-}

15. Give the charges of the following polyatomic ions:
    a. sulfate b. phosphate c. hydrogen carbonate d. chromate

16. Give the formulas of 7 different positive ions with a charge of +1 (include transition metal and polyatomic ions).

17. Give the formulas of 11 different negative ions with a charge of −1.

18. Give the simplest formula of a compound containing:
    a. A^{+} and X^{-} ions c. C^{3+} and X^{-} ions e. B^{2+} and Y^{2-} ions
    b. B^{2+} and X^{-} ions d. A^{+} and Y^{2-} ions f. C^{3+} and Y^{2-} ions

PROBLEMS

1. *Sizes of Ions (Example 10.1)* The radii of the Co and S atoms are 0.125 nm and 0.104 nm in that order. Which of the following values are most reasonable for the radii of the Co^{2+} and S^{2-} ions in that order?

   a. 0.125 nm and 0.104 nm
   b. 0.125 nm and 0.184 nm
   c. 0.072 nm and 0.104 nm
   d. 0.072 nm and 0.184 nm
   e. 0.184 nm and 0.072 nm

2. *Percentage Composition (Example 10.3)* What are the percentages of chromium in chromium(II) sulfate, $CrSO_4$, and chromium(III) sulfate, $Cr_2(SO_4)_3$?

3. *Percentage Composition (Example 10.6)* What is the percentage of water in $Na_2CO_3 \cdot 10H_2O$?

4. *Formulas of Ionic Compounds (Example 10.4)* Give the simplest formulas of the ionic compounds formed by:

   a. calcium and iodine   d. strontium and oxygen
   b. potassium and sulfur   e. rubidium and chlorine
   c. aluminum and oxygen   f. aluminum and fluorine

5. *Formulas of Ionic Compounds (Example 10.4)* Give the simplest formulas of the ionic compounds formed by nickel with:

   a. chlorine   b. oxygen   c. sulfur   d. bromine

6. *Formulas of Ionic Compounds (Example 10.4)* Give the simplest formulas of the chlorides of:

   a. Ag   b. Cu   c. Fe   d. Zn

7. *Balanced Equations (Example 10.5)* Write balanced equations for the reactions of elements which yield the ionic compounds listed in Problem 4.

8. *Balanced Equations (Example 10.5)* Write balanced equations for the reaction of nickel with the elements listed in Problem 5.

9. *Names of Ionic Compounds (Example 10.7)* Name the compounds formed by the elements listed in Problem 4.

10. *Names of Ionic Compounds (Example 10.7)* Name the following compounds:

    a. $FeCl_2$ b. $FeCl_3$ c. $CoBr_2$ d. CoF_3 e. $CuNO_3$

11. *Names of Ionic Compounds (Example 10.8)* Give the formulas of ionic compounds which have the following names:

    a. potassium hydrogen carbonate   b. magnesium sulfate
    c. cobalt(III) nitrate

* * * * *

12. Arrange the following species in order of increasing size:

    a. Fe atom   b. Fe^{2+} ion   c. Fe^{3+} ion

13. Give the simplest formulas of:

    a. sodium sulfide   b. iron(II) nitrate   c. zinc sulfate
    d. barium carbonate

14. Write balanced equations for the reaction of potassium with:

    a. fluorine   b. chlorine   c. iodine   d. oxygen   e. sulfur

15. Write balanced equations for the reaction of oxygen with:

    a. lithium   b. zinc   c. scandium   d. bismuth

16. On your worksheet, complete and balance the following equations:

    a. $Rb(s) + S(s) \rightarrow ?$ b. $Sr(s) + O_2(g) \rightarrow ?$
    c. $Cu(s) + Cl_2(g) \rightarrow ?$ d. $Ag(s) + O_2(g) \rightarrow ?$
    e. $Zn(s) + Br_2(l) \rightarrow ?$

17. Name the following ionic compounds.

    a. K_3PO_4 b. $Fe(NO_3)_3$ c. $(NH_4)_2CO_3$ d. $CrCl_3$ e. Al_2S_3

18. Give the percentages by mass of the elements in:

    a. sodium nitrate b. silver oxide c. ammonium phosphate

19. Give the percentage by mass of the metal in:

    a. $Al_2(SO_4)_3$ b. $BaCl_2 \cdot 2H_2O$

20. Calculate the percentage by mass of chlorine in:

    a. $CaCl_2$ b. $CaCl_2 \cdot 6H_2O$ c. $Ca(ClO_3)_2$

21. A sample of $CuSO_4 \cdot 5H_2O$ is heated until all the water is driven off. Starting with 1.00 g of the hydrate, what mass of $CuSO_4$ remains?

22. Scandium reacts with oxygen to form an ionic compound, scandium oxide.

    a. Write a balanced equation for the reaction.
    b. How many grams of oxygen are required to react with 1.00 g of scandium?
    c. What volume of $O_2(g)$ at STP is required to react with 1.00 g of Sc?

23. Given:

$$Li(s) + \tfrac{1}{2} F_2(g) \rightarrow Li(g) + F(g); \quad \Delta H = +55 \text{ kcal}$$

$$Li(g) + F(g) \rightarrow Li^+(g) + F^-(g); \quad \Delta H = +43 \text{ kcal}$$

$$Li^+(g) + F^-(g) \rightarrow LiF(s); \quad \Delta H = -244 \text{ kcal}$$

Calculate ΔH for: $Li(s) + \tfrac{1}{2} F_2(g) \rightarrow LiF(s)$

*24. Given:

$$Ag(s) + \tfrac{1}{2} Cl_2(g) \rightarrow AgCl(s); \quad \Delta H = -30 \text{ kcal}$$

$$Ag(s) + \tfrac{1}{2} Cl_2(g) \rightarrow Ag(g) + Cl(g); \quad \Delta H = +90 \text{ kcal}$$

$$Ag(g) + Cl(g) \rightarrow Ag^+(g) + Cl^-(g); \quad \Delta H = +87 \text{ kcal}$$

Calculate the lattice energy of AgCl.

*25. A certain ionic compound contains 60.7% O, 17.7% N, 15.2% C, and 6.37% H. Find the simplest formula of the compound and identify the ions present.

*26. When a certain hydrate of $CoCl_2$ is heated to drive off water, the solid remaining weighs 0.643 times as much as the starting material. Assuming that the residue is pure $CoCl_2$, what is the formula of the hydrate?

Molecules have a particular geometry which in some cases is similar to that of well-recognized structures. The properties of pyramids and molecules are influenced by their shape. *Russell A. Thompson/Taurus Photos.*

11 COVALENT BONDING

In Chapter 10, we considered ionic bonding, found in compounds of a metal with a nonmetal. There are many substances in which ionic bonds are not present. Ionic bonding cannot occur in elements, where all the atoms are identical. For example, a hydrogen atom would hardly transfer an electron to another hydrogen atom with exactly the same electronegativity. Moreover, the properties of many compounds indicate that they do not consist of ions. The fact that water is a liquid at room temperature suggests that it is not ionic. This is confirmed by another property of water. Unlike all ionic compounds, water is a nonconductor of electricity in the pure, liquid state.

The type of bond found in hydrogen, water (and many other substances) is called a *covalent* or electron-pair bond. It is formed when two atoms share a pair of electrons. These electrons, called *valence electrons*, come from the outermost principal energy level. In general, we can show the formation of a covalent bond as:

In a covalent bond, two electrons are shared by the bonded atoms.

$$M\cdot + N\cdot \rightarrow M:N$$

where M and N represent two atoms. The dots represent valence electrons. The two dots are written between the two atoms in the product to indicate that they are shared. Together, they comprise the covalent bond holding the atoms together. Usually, the covalent bond is shown as a straight line rather than a pair of dots. Thus we write:

$$M\cdot + N\cdot \rightarrow M—N$$

The understanding is that the dash between the two atoms represents a shared electron pair.

Since the formation of a covalent bond involves sharing electrons, it occurs when the two atoms involved have similar electronegativities. In practice, **it takes place when both atoms are derived from nonmetals.** All the nonmetallic elements form covalent bonds with themselves and with other nonmetals. Thus we find covalent bonding in the elements:

—in Group 7 of the Periodic Table (including hydrogen)
—in Group 6
—nitrogen and phosphorus in Group 5
—carbon and silicon in Group 4

In addition, the compounds formed by these elements with each other are covalently bonded.

All molecules, whether of elements or compounds, are held together by covalent bonds between the atoms. Throughout this chapter, we will emphasize covalent bond formation in molecules. Indeed, this chapter could well have been titled, "Molecular Structure". You should not conclude, however, that covalent bonding is found only in molecular substances. The atoms that make up polyatomic ions (Chapter 10) are held to one another by covalent bonds. The same is true in macromolecular substances, to be discussed in Chapter 12.

11.1 FORMATION AND PROPERTIES OF MOLECULAR SUBSTANCES

When isolated atoms combine to form a molecule, the process is always exothermic. Consider, for example, what happens when two hydrogen atoms approach each other. A total of 104 kcal of energy is evolved for every mole of H_2 formed:

$$H(g) + H(g) \rightarrow H_2(g); \qquad \Delta H = -104 \text{ kcal} \qquad (11.1)$$

This suggests that the covalent bond holding the H_2 molecule together is very strong. Indeed it is; as a general rule, covalent bonds are about as strong as ionic bonds.

Orbital Overlap

It is by no means obvious why sharing a pair of electrons between two atoms should give a stable molecule. One model of the covalent bond considers the effect of bond formation on the electron cloud surrounding the nucleus of an atom. You will recall from Chapter 9 that in the H atom there is a single electron in a 1s orbital. When two hydrogen atoms approach each other closely, their 1s orbitals **overlap** (Figure 11.1). The two electrons, now attracted by the two nuclei, spend more time between the nuclei than they do at the far ends of the molecule. Under these conditions, the attractive forces between particles of unlike charge (electron-proton) predominate. They are stronger than the repulsive forces between particles of like charge (electron-electron and proton-proton). Consequently, the molecule is stable, to the extent of 104 kcal/mol.

This model "explains" covalent bonding in many familiar molecular substances. It predicts that, *in order to form a covalent bond, both atoms involved must have a half-filled orbital.* Only if this is true can two orbitals, one from each atom, overlap to form a stable, electron-pair bond. As we have seen, this occurs when two hydrogen atoms, each with a half-filled 1s orbital, approach each other.

Covalent bonds result from the overlap of half-filled orbitals.

$$H\ (1s^1) + H\ (1s^1) \rightarrow H_2$$

The product is the H_2 molecule, held together by an electron-pair bond. In contrast, consider the helium atom which has the configuration $1s^2$. Since the 1s orbital is filled with two electrons, overlap cannot occur.

$$He\ (1s^2) + He\ (1s^2) \rightarrow \text{no reaction}$$

No bond forms between He atoms. The He_2 molecule is unknown.

Single, Double, and Triple Bonds

A bond which consists of a single pair of electrons is called a *single bond.* It is possible for two atoms to share more than one pair of electrons. If two pairs of electrons are shared, we say there is a *double bond* between the atoms. If three pairs of electrons are shared, we have a *triple bond.* Double and triple bonds, like single bonds, are shown by dashes which stand for electron pairs. Thus we have:

Single Bond	*Double Bond*	*Triple Bond*
A—A	A═A	A≡A

Double or triple bonds are formed by only a few kinds of atoms. C, O, and N are the common ones.

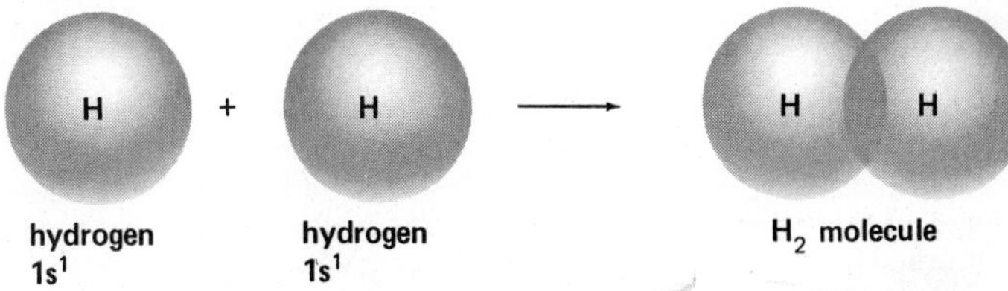

FIGURE 11.1

When two H atoms combine to form an H_2 molecule, the 1s orbitals on the two H atoms overlap, forming a new orbital which contains the two H atom electrons. The total energy of the H_2 molecule is 104 kcal/mole less than the energy of the separated atoms, so the molecule does not tend to break up into atoms.

H + H → H H

hydrogen $1s^1$ | hydrogen $1s^1$ | H_2 molecule

To illustrate the distinction between single, double, and triple bonds, consider the three hydrocarbons containing two carbon atoms. These have different molecular formulas: C_2H_6, C_2H_4, C_2H_2. They also have very different chemical and physical properties. In each of these molecules, the hydrogen atoms are joined to carbon by single bonds. In ethane, C_2H_6, there is a single bond between the carbon atoms. In ethylene, C_2H_4, the two carbon atoms are joined by a double bond. Finally, in acetylene, C_2H_2, there is a triple bond between the two carbon atoms. The structures of these molecules are:

```
    H  H          H         H
    |  |           \       /
 H—C—C—H            C═C           H—C≡C—H
    |  |           /       \
    H  H          H         H
   ethane          ethylene         acetylene
```

Properties of Molecular Substances

Molecules tend to interact only weakly with other molecules of the same kind.

Molecular substances differ in many ways from the ionic compounds discussed in Chapter 10. Two differences are particularly important.

1. *Molecular substances as a group have much lower melting and boiling points than ionic compounds.* Typically, molecular substances are gases, liquids, or low-melting solids at room temperature and atmospheric pressure. This property reflects the fact that it is not necessary to break chemical bonds to melt or boil a molecular substance. All we have to do is to separate molecules from one another. This re-

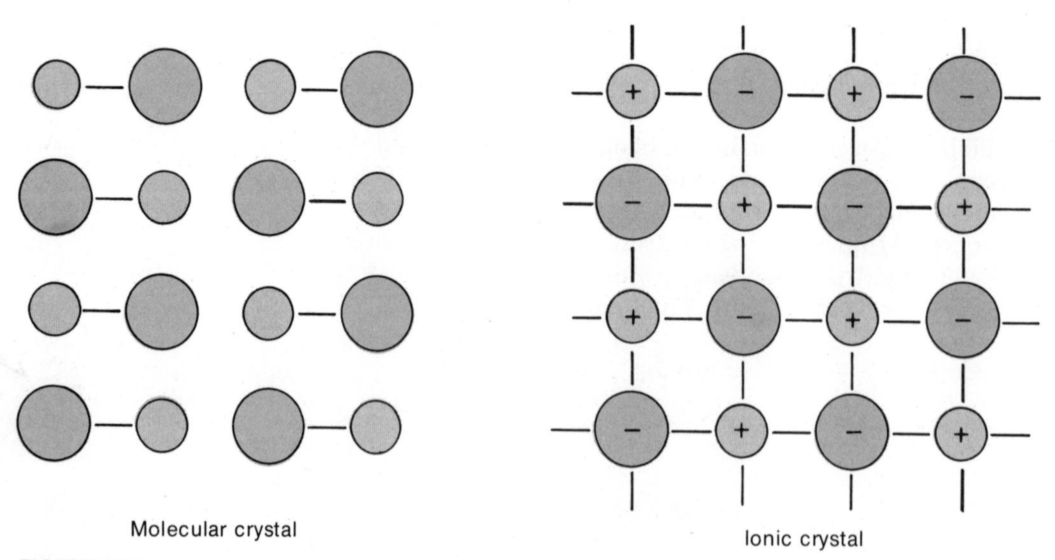

FIGURE 11.2

In a molecular crystal there are no strong bonds between molecules, so the crystal is relatively easily melted, or even vaporized. In an ionic crystal the ions are all attached together by ionic bonds. The crystal is not readily melted and is very hard to vaporize.

quires relatively little energy. In ionic compounds, on the other hand, chemical bonds between oppositely charged ions must be broken (Figure 11.2).

2. *Molecular substances, in the pure state, do not conduct electricity,* since they consist of uncharged molecules. For example, pure water, containing only neutral H_2O molecules, is a nonconductor.

11.2 LEWIS STRUCTURES; THE OCTET RULE

You will recall from Chapter 10 that ions formed from nonmetal atoms have noble-gas structures. For example, a hydrogen atom, by acquiring an electron, forms the H^- ion.

$$H\ (1s^1) + e^- \rightarrow H^-\ (1s^2)$$

The hydride ion has the electron configuration of helium, $1s^2$. In 1916, the American chemist G. N. Lewis pointed out that nonmetal atoms can acquire noble-gas structures in a quite different way. They do this by sharing electrons with other atoms. Consider, again, the hydrogen atom. By forming an electron-pair bond with another atom, it gains a share in a second electron. This electron enters the 1s orbital. In effect, the hydrogen atom in molecules such as H_2 or HF has the $1s^2$ configuration of helium. Thus in H_2 we have:

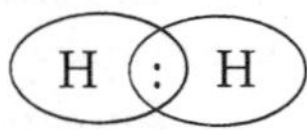

In a covalent bond, both atoms have both electrons.

Each H atom is surrounded by two electrons.

The idea that nonmetal atoms reach a noble-gas structure by covalent bond formation is a very useful one. In a sense, it helps to explain the stability of the covalent bond in many simple molecules. It can also be helpful in predicting the formulas of molecular substances. Finally, it enables us to correlate and explain the geometry and polarity of molecules (Sections 11.3, 11.4).

To show how atoms acquire noble-gas structures by forming covalent bonds, Lewis developed a symbolism which we still use today. This involves drawing what are known as Lewis structures. We will consider first how these structures are drawn for individual atoms. Then we will look at the problem of drawing Lewis structures for molecules containing atoms joined by covalent bonds.

Lewis Structures of Atoms

The Lewis structure of an atom is a diagram which shows the number of valence electrons possessed by the atom. Recall (Chapter 9) that the valence electrons are those in the outermost principal energy level. In the Lewis structure of an atom, valence electrons are shown as dots surrounding the symbol of the atom. The number of dots is equal to the number of valence electrons. Consider, for example, the hydrogen atom. It has the electron configuration $1s^1$. Its Lewis structure is simply:

$$H\cdot$$

TABLE 11.1 LEWIS STRUCTURES OF THE ATOMS OF THE 2ND PERIOD ELEMENTS

Element	Group	Electron Configuration	Number of Valence Electrons	Lewis Structure
Li	1	$1s^2\ 2s^1$	1	Li·
Be	2	$1s^2\ 2s^2$	2	·Be·
B	3	$1s^2\ 2s^2\ 2p^1$	3	·Ḃ·
C	4	$1s^2\ 2s^2\ 2p^2$	4	·Ċ̣·
N	5	$1s^2\ 2s^2\ 2p^3$	5	·Ṇ̈·
O	6	$1s^2\ 2s^2\ 2p^4$	6	·Ö̤·
F	7	$1s^2\ 2s^2\ 2p^5$	7	:F̤̈·
Ne	8	$1s^2\ 2s^2\ 2p^6$	8	:N̤̈e:

Table 11.1 gives the Lewis structures of the elements in the second period of the Periodic Table. Note that, in general:

1. Only the valence electrons appear in the Lewis structure. For elements in this period, there are two electrons in the level n = 1. These inner 1s electrons are not shown.

2. In the Lewis structure, no distinction is made between s and p electrons. Putting it another way, the number of valence electrons is the sum of the outer s and p electrons.

3. The electrons are put in four positions around the symbol of the atom (above, below, left, right). They are not paired until necessary. This first occurs with the N atom (five electrons).

One other principle should be obvious from Table 11.1. For these and all other main-group elements beyond the first period:

The number of valence electrons is equal to the Group number in the Periodic Table (In the 1st period, hydrogen has 1 valence electron; helium has 2).

EXAMPLE 11.1

Write Lewis structures for atoms of

a. I b. P c. Ar

SOLUTION

a. I is in Group 7. Its Lewis structure is similar to that of fluorine.

:Ï̤·

b. P, in Group 5, has a Lewis structure with 5 valence electrons.

·P̣̈·

c. Ar, like Ne, is in Group 8. It has 8 valence electrons.

:Ä̤r:

Lewis Structures of Simple Molecules

The Lewis structures of atoms can be combined to give Lewis structures for molecules. To show how this is done, consider the HF molecule. We start with a hydrogen atom (1 valence electron) and a fluorine atom (7 valence electrons). They combine to give a molecule, HF, with 8 valence electrons.

$$H\cdot + :\ddot{\underset{..}{F}}\cdot \rightarrow H—\ddot{\underset{..}{F}}:$$

Notice that in the Lewis structure of the HF molecule:

1. The covalent bond between the atoms is shown as a dash, representing a pair of electrons. This electron pair is formed by combining the single electron of hydrogen with one of the valence electrons of fluorine.
2. The remaining 6 valence electrons (3 pairs) of fluorine are shown as dots around the F atom. These are referred to as *unshared electrons*.

The formation of molecules such as H_2O, NH_3, and CH_4 from the corresponding atoms can be shown in a similar way.

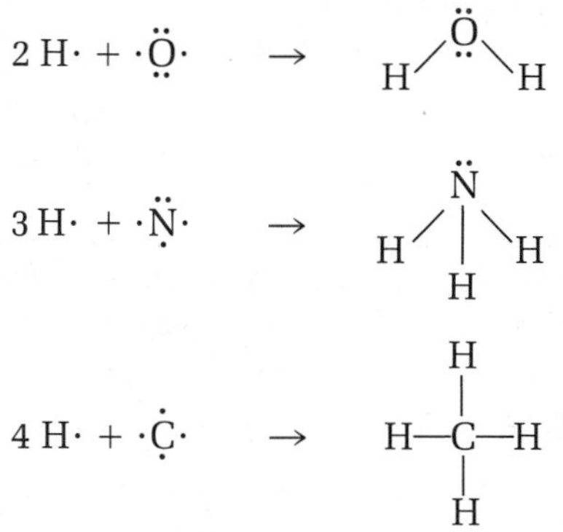

$$2\,H\cdot + \cdot\ddot{\underset{..}{O}}\cdot \rightarrow H—\ddot{\underset{..}{O}}—H \text{ (bent)}$$

$$3\,H\cdot + \cdot\ddot{\underset{\cdot}{N}}\cdot \rightarrow H—\ddot{N}(—H)—H$$

$$4\,H\cdot + \cdot\dot{\underset{\cdot}{C}}\cdot \rightarrow H—C(H)(H)—H$$

Looking at the Lewis structures we have drawn, you will note that each atom in the molecule has a noble-gas structure. In HF, H_2O, NH_3, and CH_4, the hydrogen atom is surrounded by 2 valence electrons. These are the ones forming the covalent bond that joins each H atom to a nonmetal atom. You will recall that the noble gas helium has 2 valence electrons. In each of these molecules the other nonmetal atom (F, O, N, C) is surrounded by 8 valence electrons. That is:

F in HF: 2 in the H—F bond, 6 unshared
O in H_2O: 4 in the two H—O bonds, 4 unshared
N in NH_3: 6 in the three H—N bonds, 2 unshared
C in CH_4: 8 in the four H—C bonds

This number of valence electrons, 8 (four pairs), is that found in neon and indeed in all the noble gases except helium.

The eight electrons will always exist as four pairs.

The general rule that atoms, through covalent bond formation, tend to acquire noble-gas structures is often referred to as the **octet rule.** In most stable molecules, all the atoms except hydrogen are surrounded by eight valence electrons, an "octet". We will find this rule very helpful in writing reasonable electronic structures for many different molecules.

Writing Lewis Structures for Molecules

To write a reasonable Lewis structure for a molecule, you follow a simple process. You start with the valence electrons supplied by the atoms involved. In the molecule, these electrons are either shared to form bonds or left unshared. In distributing the valence electrons, your objective is to give each atom a noble-gas structure. This means that:

1. H atoms should be surrounded by 2 valence electrons.
2. All other nonmetal atoms should be surrounded by 8 valence electrons.

For very simple molecules, you may be able to do this by inspection. For HF, H_2O, NH_3, and CH_4, you could probably have arrived, without too much trouble, at the structures we have shown. However, it is safer to follow a logical, step-by-step procedure which can be applied to all molecules. Such a procedure becomes essential when many atoms are involved. It consists of the following steps.

1. **Count the number of valence electrons available.** To do this, simply add up those supplied by each atom. Remember that:

—a H atom has 1 valence electron
—a Group 4 atom (C, Si, . . .) has 4 valence electrons
—a Group 5 atom (N, P, . . .) has 5 valence electrons
—a Group 6 atom (O, S, . . .) has 6 valence electrons
—a Group 7 atom (F, Cl, . . .) has 7 valence electrons

2. **Draw a "skeleton structure" for the molecule, joining atoms by single bonds.** The skeleton structure for a molecule X_2 is:

X—X

For the molecule XY, the skeleton structure is:

X—Y

For more complex molecules, more than one skeleton structure is possible. Thus for the molecule XY_2 we might write:

X—Y—Y or Y—X—Y

Here, experimental evidence must be used to decide which arrangement of atoms is correct. In this course, you will either be given the correct structure or general rules which will help you find it.

3. **From the total number of valence electrons calculated in (1), subtract two for each single bond in the skeleton structure.** This tells you how many valence electrons you have left to distribute.

4. **Distribute the remaining valence electrons, calculated in (3), as unshared pairs around the various atoms.** Try to do this in such a way that you get 8 electrons around each atom (except hydrogen, which should have 2).

To illustrate this procedure, let's see how it works for the SCl_2 (sulfur dichloride) molecule.

Step 1 Since sulfur is in Group 6 and chlorine in Group 7:

$$\text{number valence electrons} = 6 + 2(7) = 20$$

Step 2 Two different skeletons could be written:

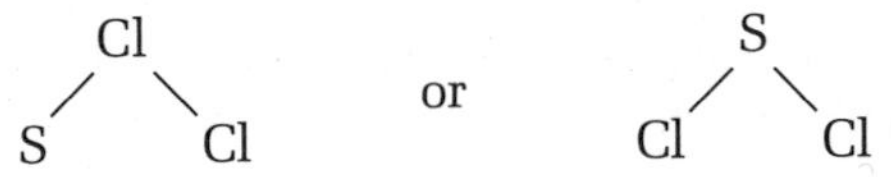

```
     Cl                     S
    /  \         or        / \
   S    Cl               Cl   Cl
```

In molecules such as this, the like atoms (two Cl atoms in this case) *are usually bonded to a central atom* (sulfur) *rather than to each other.* The correct skeleton is:

Nature likes symmetry when she can get it.

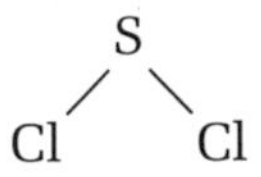

```
     S
    / \
  Cl   Cl
```

Step 3 Four electrons were consumed by the two single bonds.

$$\text{number of valence electrons left} = 20 - 4 = 16$$

Step 4 To fill out the skeleton structure of SCl_2, we need to put 6 more electrons around each chlorine atom. They have only 2 at the moment and need 8. Similarly, we can give the sulfur atom an octet by putting 4 more electrons around it (it has 4 already). So we need:

$$6 + 6 + 4 = 16$$

valence electrons. Fortunately, as calculated in Step 3, this is precisely the number we have left. The correct Lewis structure of SCl_2 is:

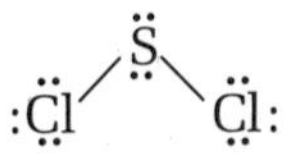

```
        ..
        S
      / .. \
   :Cl:    :Cl:
```

To make sure that you understand how to draw Lewis structures, you may wish to try a couple more (Example 11.2). It would be a good idea to try these on your own (and *then* peek at the answer).

The eight electrons around the S atom come from the four electrons in the two bonds and the four unshared electrons.

EXAMPLE 11.2

Draw Lewis structures for:

a. ClF b. CCl_4

SOLUTION

a. *Step 1* Chlorine and fluorine are in Group 7:

no. of valence e^- in ClF = 7 + 7 = 14

Step 2 There is only one possible skeleton structure for ClF:

Cl—F

Step 3 In the skeleton written for ClF, 2 valence electrons were used (1 bond).

When counting electrons around an atom, add two electrons for each bond to that atom.

no. of valence e^- left in ClF = 14 − 2 = 12

Step 4 Looking at the skeleton drawn for ClF, we see that each atom needs 6 more electrons to complete its octet. Fortunately, we have just enough left, 12, to do the job. The Lewis structure of ClF is:

```
 ..  ..
:Cl—F:
 ..  ..
```

b. *Step 1* Carbon is in Group 4, chlorine in Group 7:

no. of valence e^- in CCl_4 = 4 + 4(7) = 32

Step 2 In CCl_4, the four Cl atoms are bonded to a central C atom:

```
     Cl
     |
 Cl—C—Cl
     |
     Cl
```

Step 3 There are four bonds in the CCl_4 skeleton, using 8 valence electrons.

no. of valence e^- left in CCl_4 = 32 − 8 = 24

Step 4 In CCl_4, the C atom already has its octet. However each of the Cl atoms needs 6 more electrons. Again, we have just enough electrons left, 24. We put 6 of them around each Cl atom.

```
       ..
      :Cl:
       |
 ..    |    ..
:Cl—C—Cl:
 ..    |    ..
       |
      :Cl:
       ..
```

Lewis Structures Involving Multiple Bonds

Sometimes when you reach the last step in writing a Lewis structure, you find that you have too few electrons to go around. That is, the number of valence electrons left is less than the number needed to give each atom an octet. When this happens, it is a signal that you must economize on valence electrons. To do this, you use one or more pairs of unshared electrons to form an additional bond. This way, these electrons do "double duty". They are counted in the octet of each of the bonded atoms. The rules here are simple:

1. Forming a double bond "saves" two electrons.
2. Forming a triple bond "saves" four electrons.

To show how this process works, consider the SO_2 molecule. Since both sulfur and oxygen are in Group 6:

$$\text{no. of valence e}^- = 6 + 2(6) = 18$$

Both oxygen atoms are bonded to a central sulfur atom:

```
   S
  / \
 O   O
```

Since this skeleton uses two pairs of valence electrons:

$$\text{no. of valence e}^- \text{ left} = 18 - 4 = 14$$

This simply isn't enough valence electrons to give each atom a noble-gas structure. The best we can do is to use these 14 electrons to give a structure such as

```
     ..
     S
  .. / \ ..
 :O     O:
  ..     ..
```

This leaves only 6 valence electrons around the sulfur atom, an unhappy situation. To solve the electron shortage, we move one of the unshared pairs of electrons sitting on the oxygen atom. We put it between the sulfur and oxygen atoms, creating a double bond.

It would be nice if other shortages could be solved as easily.

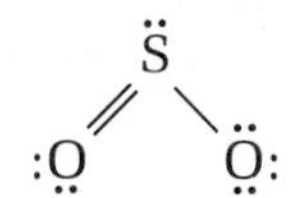

Now each atom has 8 valence electrons around it. (Count them!)

EXAMPLE 11.3

Write a Lewis structure for the N_2 molecule.

SOLUTION

Since nitrogen is in Group 5,

$$\text{no. of valence e}^- = 2 \times 5 = 10$$

There is only one possible skeleton:

N—N

Since the bond between the nitrogen atoms uses two electrons,

$$\text{no. of valence e}^- \text{ left} = 10 - 2 = 8$$

There is no way we can satisfy the octet rule if we "waste" all these electrons as unshared pairs. At best, we might arrive at

:N̈—N̈:

Clearly, we are four electrons short. Each nitrogen atom is surrounded by only 6 valence electrons rather than 8. We have to move in two pairs of unshared electrons, forming a triple bond.

For each N atom, $3 \times 2 + 2 = 8$ electrons.

:N≡N:

Exceptions to the Octet Rule

Although the octet rule is very useful, there are some molecules where it doesn't work. A rather obvious case is the NO molecule. Counting the number of valence electrons, we find:

$$5 + 6 = 11$$

We cannot possibly use an odd number of valence electrons, 11, to obtain a structure which has an even number, 8, around each atom. The best we can do is to write a Lewis structure such as:

·N̈=Ö:

which clearly does not follow the octet rule.

Many of the molecules formed by beryllium, in Group 2, and boron, in Group 3, "violate" the octet rule. Frequently, Be is surrounded by two pairs of electrons rather than four. This is the case, for example, in the BeF_2 molecule:

:F̤̈—Be—F̤̈:

In BF_3, there are only three pairs of electrons around the boron atom.

```
:F̤̈\   /F̤̈:
     B
     |
    :F̤:
```

At the opposite extreme, in a few molecules there are more than four electron pairs around the central atom. In PCl_5, the phosphorus atom in the center of the molecule is bonded to all five Cl atoms. This puts five electron pairs around the phosphorus. In SF_6, the sulfur atom forms a total of six bonds, one to each of the fluorine atoms. This means that sulfur is surrounded by 6 pairs, or 12 electrons.

These exceptions to the octet rule show that our simple model of covalent bonding does not tell the whole story. We will leave to later courses a more rigorous model capable of explaining these exceptions. Meanwhile, it is well to point out an important fact. Before you can recognize that a molecule does not obey the octet rule, you have to know what the rule is. The octet rule does enable us to arrive at reasonable electronic structures for most simple molecules.

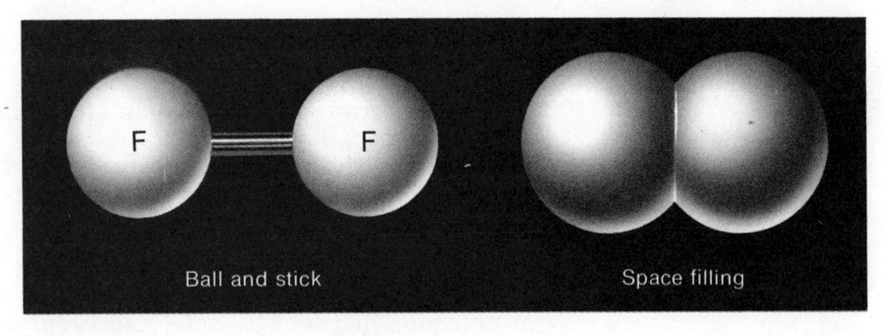

FIGURE 11.3

The F_2 molecule consists simply of two F atoms attached by a covalent bond. In the ball and stick model, the F atoms are represented by balls and the bond by a stick. In the space-filling model the electron clouds around the F nuclei are shown and define the contours of the molecule as it might appear if we could magnify it sufficiently.

11.3 MOLECULAR GEOMETRY

The physical and chemical properties of molecular substances depend to a considerable extent upon the geometric pattern in which the atoms are arranged. Just as the bouncing properties of a basketball differ from those of a football, so the shape of molecules can affect such properties as boiling point and chemical reactivity.

Diatomic molecules such as H_2, F_2, and HF have a very simple shape. If we think of the atoms as spheres, their centers must lie on a straight line. This is clear from Figure 11.3, which shows two different models of the F_2 molecule. In the "ball and stick" model at the left, the two balls represent the fluorine atoms. The stick stands for the covalent bond joining them. The "space-filling" model at the right actually gives a more accurate picture of the molecule. It shows the two atoms as overlapping spheres. The bond itself is not shown.

When a molecule contains more than two atoms, more than one geometry is possible. Consider a molecule AX_2 where two atoms of X are bonded to a central atom, A. If these bonds are directed at an angle of 180° to each other, then all three atoms will be in a straight line. We would describe such a molecule as being *linear*.

X—A—X
180°

If, on the other hand, the angle between the two bonds is something less than 180°, the three atoms will not be in a straight line. Instead, the molecule will be *bent*.

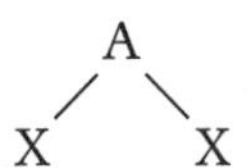

As we go on to more complex molecules, the number of possible geometries increases. Clearly, we need to develop a general principle which will allow us to predict molecular geometries.

Electron Pair Repulsion

Since electrons are negatively charged, we find that the electron pairs around an atom in a molecule repel each other (like charges repel).

This fact leads to a simple principle which allows us to predict the geometry of a great many molecules.

The electron pairs around an atom in a molecule are directed so as to be as far apart as possible.

The "electron pairs" referred to here may be single bonds or unshared electron pairs. (We will not worry in this chapter about the geometry of molecules containing multiple bonds.)

This principle is used in a very simple way to predict the geometry of molecules that obey the octet rule. Here, we are dealing with four pairs of electrons. All we need do is to decide how these four pairs should be oriented to get them as far from each other as possible.

Tetrahedral Geometry

It turns out that to get four electron pairs as far apart as possible, they should be directed toward the corners of a regular tetrahedron. Figure 11.4 shows the geometry involved. Notice that a tetrahedron is a three-dimensional figure with four sides and four corners. In a regular tetrahedron, each side is an equilateral triangle. Furthermore:

—the center of the tetrahedron is on a straight line from one corner to the center of the equilateral triangle formed by the other three corners.

—the so-called "tetrahedral angle" (BAC or BAD or . . .) is 109.5°.

With this figure in mind, let us consider the geometry of some simple molecules in which the central atom is surrounded by four electron pairs.

CH_4

The geometry of the CH_4 molecule is quite simple. Remember that there are four pairs of electrons around the carbon atom. These electron pairs will be directed toward the corners of a tetrahedron. The carbon atom is at the center, with a hydrogen atom at each of the four corners. The molecule itself has the shape of a tetrahedron. We describe it as *tetrahedral*. The four C—H bonds are directed at an angle of 109.5° to each other. That is, each H—C—H bond angle is the tetrahedral angle referred to above. Figure 11.5 shows "ball and stick" and space-filling models of the CH_4 molecule.

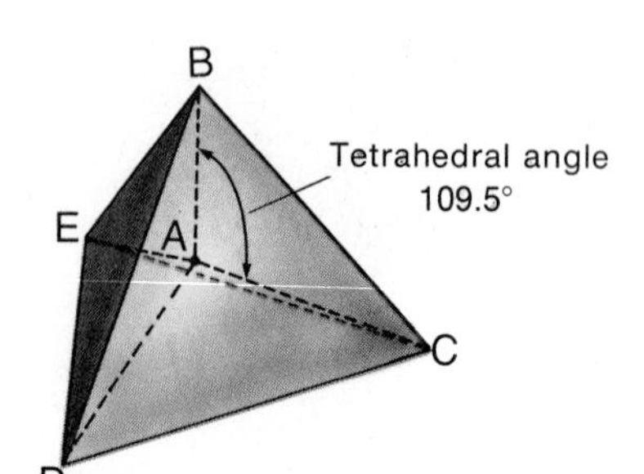

FIGURE 11.4

A tetrahedron is a solid figure in which all four faces are regular triangles. The four corners are equivalent. The angle formed by drawing lines from any two corners to the center of the figure is called the tetrahedral angle and is equal to 109.5°.

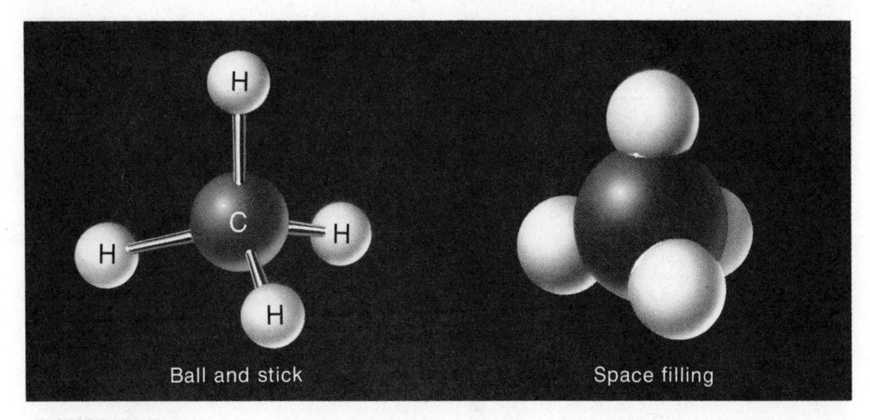

FIGURE 11.5

The CH_4 molecule. In methane, the carbon atom is at the center of a tetrahedron and the H atoms are at the corners. The molecule is tetrahedral and has a great deal of symmetry. Its geometry follows from the fact that the four pairs of electrons around the C atom are directed toward the corners of a tetrahedron.

NH_3

Consider now the structure of the NH_3 molecule. Here again, the central atom (nitrogen) is surrounded by four pairs of electrons. As in CH_4, these electron pairs are directed toward the corners of a tetrahedron. The nitrogen atom is at the center. Three of the four corners are occupied by hydrogen atoms. There is an unshared pair of electrons at the fourth corner. This structure is shown in Figure 11.6a. The unshared pair is at the top of the molecule.

Ordinarily, in describing the geometry of NH_3 or other molecules, we do not include the unshared electron pair. We consider only the positions of the atoms. Note from Figure 11.6a that the nitrogen atom is sitting above the center of an equilateral triangle formed by the three hydrogen atoms. Taken together, the four atoms form a pyramid. The nitrogen atom is at the top; the three H atoms form the base of the pyramid. We describe the NH_3 molecule as *pyramidal* (Figure 11.7, p. 272).

Experimentally, we find that the H—N—H bond angle in NH_3 is slightly less than the tetrahedral angle. It is about 107° rather than 109.5°. The unshared pair of electrons on the N atom occupies a slightly larger volume than the electron pairs used for the N—H bonds. This "crowds" the H atoms a bit, reducing the bond angle.

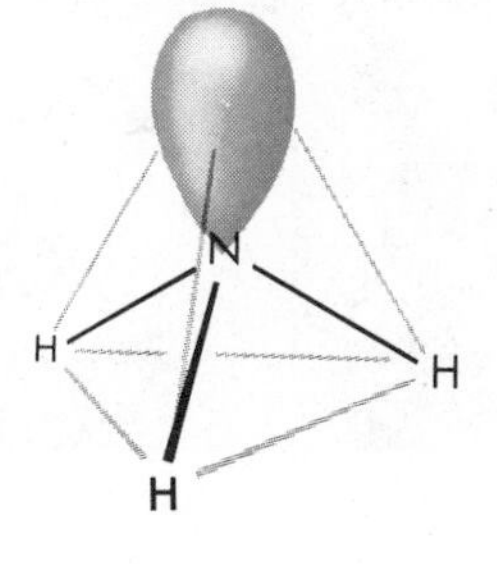

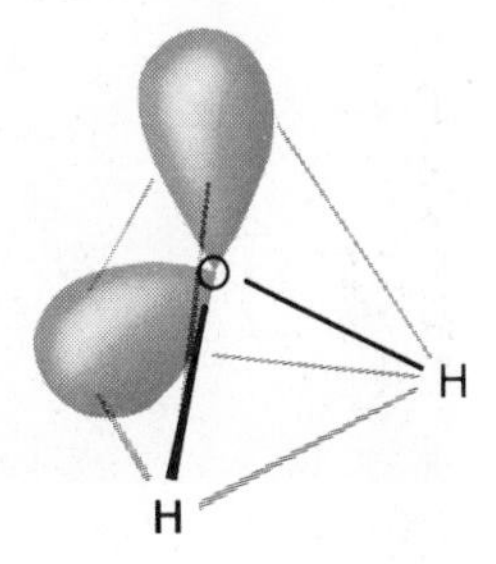

FIGURE 11.6

a. In the NH_3 molecule the four electron pairs around the N atom are directed toward the corners of a tetrahedron. Three of these pairs are involved in N—H bonds. The fourth pair is unshared. b. In H_2O, two of the electron pairs around the O atom form O—H bonds, and two pairs are unshared.

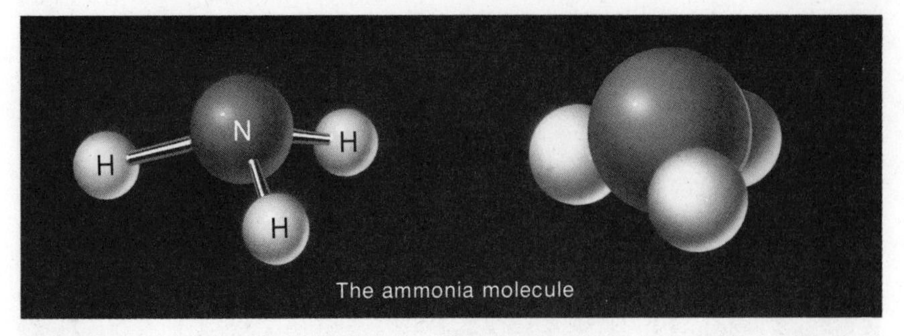

FIGURE 11.7

In describing the geometry of a molecule, we indicate the positions of the atoms and *not* of the electron pairs. In the ammonia molecule, the N atom lies above an equilateral triangle formed by the three H atoms, forming a pyramid. The molecule is a trigonal pyramid, or pyramidal in shape.

H_2O

Going one step further, consider the water molecule. Here again, the central atom (oxygen) is surrounded by four pairs of electrons. These are directed toward the corners of a tetrahedron. This structure is shown in Figure 11.6b. Note that the oxygen atom is at the center. Two of the corners are occupied by H atoms. There are unshared electron pairs at the other two corners.

To describe the geometry of the H_2O molecule, we consider only the atoms. Note from Figure 11.6b that the molecule is *bent*. That is, the three atoms are not in a straight line (Figure 11.8). We might expect the H—O—H bond angle to be 109.5°, the tetrahedral angle. Actually, it is a bit less than that, about 105°. As in NH_3, the unshared pairs crowd the H atoms a bit.

Other Molecules

The idea that the four electron pairs surrounding an atom are directed toward the corners of a regular tetrahedron is a very useful one. As we have seen, it allows us to predict the geometry of many simple molecules. It is most readily applied to molecules where a central atom is bonded to 2, 3, or 4 other atoms. Consider, for example, the methyl

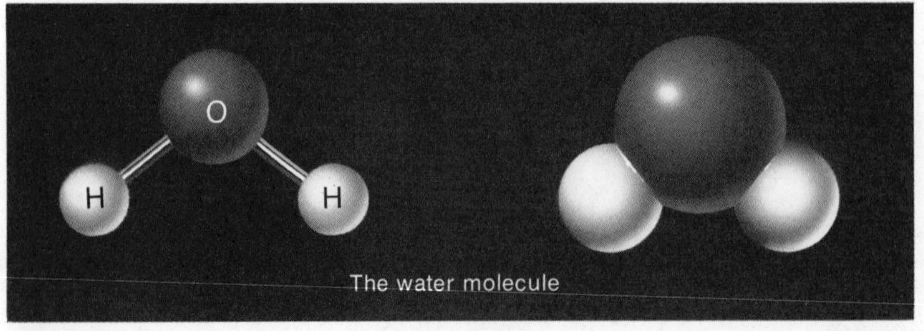

FIGURE 11.8

In the water molecule the H—O—H bond angle is about 105°. The H_2O molecule is said to be bent.

chloride molecule, CH_3Cl. The bonding could be shown in two dimensions as:

```
      Cl
      |
   H—C—H
      |
      H
```

We would expect this molecule to have a geometry similar to that of CH_4. The carbon atom is in the center of a tetrahedron. There are hydrogen atoms at three corners and a chlorine atom at the fourth corner.

Example 11.4 gives a further illustration of the prediction of molecular geometry.

EXAMPLE 11.4

Consider the SCl_2 molecule, whose Lewis structure was shown on p. 265. Describe its geometry.

SOLUTION

The SCl_2 molecule resembles the H_2O molecule in that the central atom, S, is bonded to two other atoms. Moreover, there are two unshared pairs around sulfur in SCl_2, just as there are around oxygen in H_2O. The four pairs of electrons around the sulfur atom should be directed towards the corners of a regular tetrahedron. We predict a bent molecule with a bond angle of about 109°.

The electrons around S are arranged tetrahedrally. The SCl_2 molecule is bent, not tetrahedral, since there are only two Cl atoms bonded to S.

11.4 POLARITY OF MOLECULES

An important property of a molecule is its polarity. Certain molecules are *polar*. That is, there is a separation of charge within the molecule. Electrons are shifted away from one end of the molecule, toward the other end. This produces a positive pole where the electron density is low. At the opposite end of the molecule, where the electron density is relatively high, there is a negative pole. Other molecules are *nonpolar*. In such a molecule, there is no separation of charge. The electrons are evenly distributed and there are no positive and negative poles.

The presence or absence of polarity affects many of the physical properties of molecular substances. In particular, it affects their behavior in an electrical field. Polar molecules tend to line up in a direction opposite to that of the field (Figure 11.9, p. 274). In contrast, nonpolar molecules are not affected by the field. They are arranged in a perfectly random way. As we will see in Chapter 12, this difference in behavior affects such properties as melting point and boiling point.

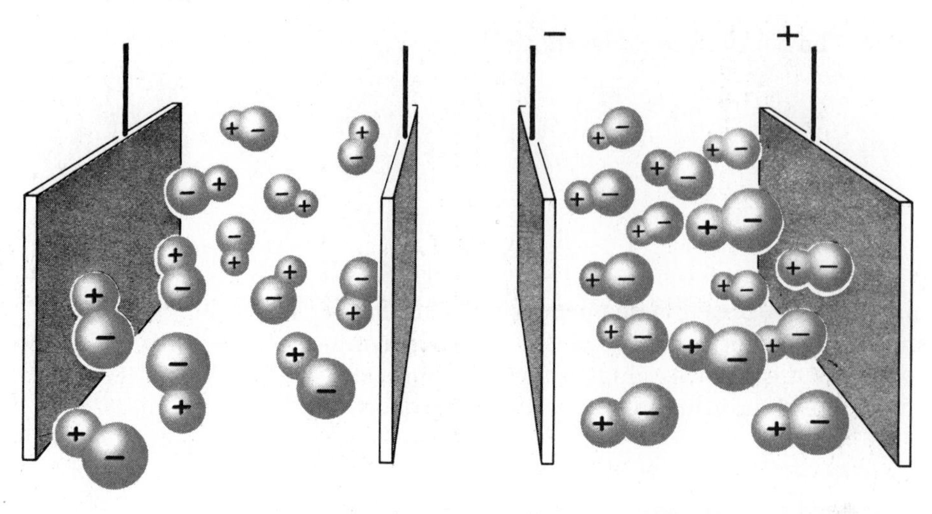

FIGURE 11.9

If polar molecules are subjected to an electric field, they tend to line up in a direction opposite to that of the field. This minimizes the electrostatic energy of the molecules. Nonpolar molecules are not oriented by an electric field.

Diatomic Molecules

It is easy to determine whether a diatomic molecule is polar or nonpolar. If the two atoms joined by the covalent bond are identical, as in H_2, the molecule is nonpolar. If the atoms joined are different, as in HF, the molecule is polar.

To understand the basis of these rules, consider the H_2 molecule, structural formula:

$$H{-}H$$

Note the symmetry of H_2. There is no way one atom could take on more charge than the other.

Since the two hydrogen atoms are identical, they both have an equal share of the bonding electrons. There is no tendency for the electrons to shift in one direction or the other. Hence, there are no + and − poles, and the H_2 molecule is nonpolar.

The situation is quite different in HF, which has the skeleton structure:

$$H{-}F$$

Fluorine is more electronegative than hydrogen (E.N. F = 4.0 *vs.* 2.1 for H). Hence the F atom will tend to pull the bonding electrons closer to it, away from the H atom (Figure 11.10). This creates polarity in the molecule. The more electronegative fluorine atom acts as a negative pole. The electron density around it is relatively high. The H atom acts as a positive pole. The electron density around it is relatively low.

FIGURE 11.10

In the HF molecule the F atom (shown at the right) is more electronegative than H. It will tend to attract electrons and so take on a negative charge. This leaves a positive charge on the H atom and makes the HF molecule polar. Any diatomic molecule containing different kinds of atoms will be polar.

EXAMPLE 11.5

Classify the following molecules as polar or nonpolar.

a. Cl_2 b. ClF

SOLUTION

a. nonpolar; the two atoms are identical.

b. polar; the two atoms are different. From Fig. 7.9, p. 176, we see that the electronegativities of Cl and F are 3.0 and 4.0 in that order. Hence, the bonding electrons are shifted toward the fluorine atom. In ClF, the Cl atom acts as a positive pole, the F atom as a negative pole.

Other Simple Molecules

When there are more than two atoms in a molecule, we must consider its geometry to decide whether it is polar. Consider, for example, the molecule AX_2. Suppose that the central atom, A, is more electronegative than X. Both A—X bonds will be polar; the bonding electrons will be shifted toward A. If the molecule is bent:

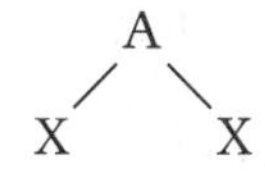

polar

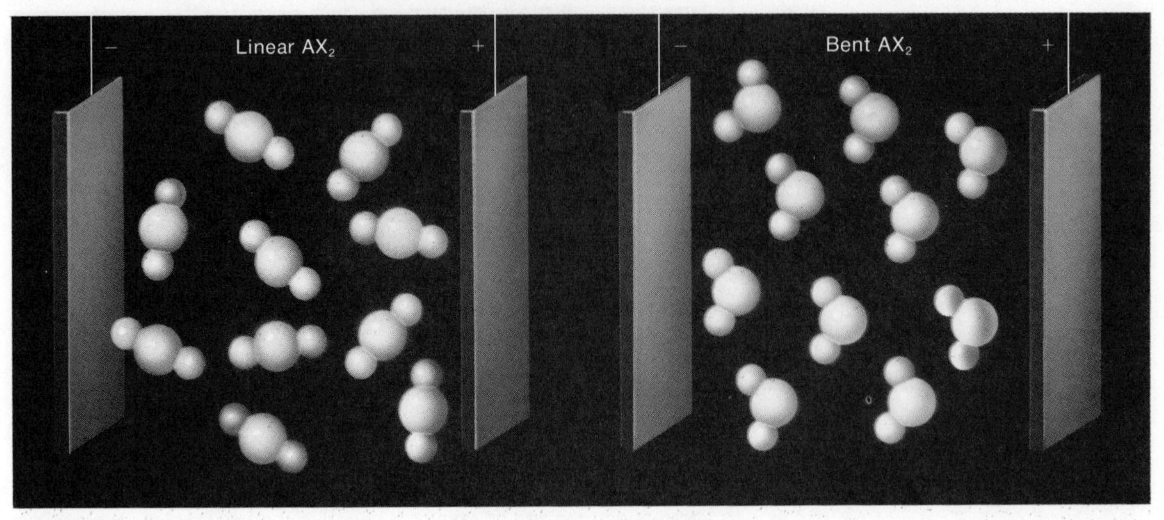

FIGURE 11.11

Orientation of AX_2 molecules in an electric field. If the AX_2 molecule is bent, it will be polar and tend to be oriented by an electric field. In the drawing, A atoms are taken to be more electronegative than X, so A will be attracted to the + electrode. If the AX_2 molecule is linear, the two X atoms will experience the same shift of charge. There will be no positive or negative end of the molecule, so the molecule will not be oriented in an electric field.

it will be polar. There will be a negative pole located at atom A, where the electron density is relatively high. The positive pole will be midway between the two X atoms. Molecules of AX_2 will line up in an electrical field as shown in Figure 11.11.

Suppose, however, that the molecule AX_2 is linear.

nonpolar

X—A—X

In this case, even though the bonds are polar, the molecule itself is nonpolar. The two polar bonds are in opposite directions, at a 180° angle. In effect, they cancel each other. There is no way that this molecule can line up in an electrical field (Figure 11.11).

Both types of triatomic molecules are known. The most familiar example of a bent molecule containing three atoms is H_2O.

```
   O
  / \
 H   H
```

The molecule is polar. The negative pole is located at the oxygen atom (E.N. = 3.5). The positive pole is midway between the two hydrogen atoms (E.N. = 2.1). An example of a triatomic molecule of the type AX_2 which is nonpolar is BeF_2. This molecule is linear:

F—Be—F

The two polar Be—F bonds cancel each other.

To make our discussion more general, we can say that *completely symmetrical molecules are nonpolar, even if they contain polar bonds.* The methane molecule shown in Figure 11.5 is nonpolar. The four polar C—H bonds cancel each other. If a chlorine atom is substituted for hydrogen, the symmetry is destroyed. Methyl chloride, CH_3Cl, is polar.

EXAMPLE 11.6

Is the CCl_4 molecule polar? The SCl_2 molecule?

SOLUTION

Carbon tetrachloride, CCl_4, resembles methane in that it is a symmetrical molecule. The four polar C—Cl bonds cancel one another and the molecule is nonpolar. In SCl_2, the situation is quite different. You will recall from Example 11.4 that the molecule is bent:

```
   S
  / \
Cl   Cl
```

Since chlorine is more electronegative than sulfur (3.0 *vs.* 2.5), the two S—Cl bonds are polar. The molecule itself is polar. There is a positive pole at the sulfur atom and a negative pole midway between the two chlorine atoms.

Would $CHCl_3$ be polar? Ans.: Yes.

11.5 NAMES OF MOLECULAR COMPOUNDS

The system used to name molecular compounds depends to some extent upon the number of elements present. We will consider first how to name molecular compounds containing only two nonmetals (binary compounds). Then we will discuss the system used to name one particular type of ternary compound (three elements): oxygen-containing inorganic acids. These two types of molecular compounds are the ones that we deal with most frequently in general chemistry.

Binary Compounds of the Nonmetals

When a pair of nonmetals form only one compound, it is named quite simply. The name of the element whose symbol appears first in the formula (the least electronegative element) is written first. The second part of the name is formed by adding the suffix *-ide* to the stem of the name of the second nonmetal. Thus we have:

HF	hydrogen fluoride
HCl	hydrogen chloride
H_2S	hydrogen sulfide

These names usually refer to the compounds in the gas phase.

One special note: a *water solution* of hydrogen chloride is called *hydrochloric acid.* A similar system is used with the other hydrogen halides:

water solution of hydrogen fluoride (HF) : hydrofluoric acid
water solution of hydrogen chloride (HCl): hydrochloric acid
water solution of hydrogen bromide (HBr): hydrobromic acid
water solution of hydrogen iodide (HI) : hydriodic acid

If more than one compound is formed by a pair of nonmetals, as is often the case, the Greek prefixes:

di = 2
tri = 3
tetra = 4
penta = 5
hexa = 6

are used to indicate the number of atoms of each element present. To illustrate, consider the several oxides of nitrogen:

N_2O_5	dinitrogen pentoxide
N_2O_4	dinitrogen tetroxide
NO_2	nitrogen dioxide
N_2O_3	dinitrogen trioxide
NO	nitrogen oxide
N_2O	dinitrogen oxide

(Note that when the prefixes "tetra" and "penta" precede a vowel, the final "a" is dropped to make the word easier to say. Thus we write "tetroxide" instead of tetraoxide and "pentoxide" instead of pentaoxide.)

EXAMPLE 11.7

Following the rules just described, name:

a. SO_2 and SO_3 b. SF_4 and SF_6

SOLUTION

a. sulfur dioxide; sulfur trioxide

b. sulfur tetrafluoride; sulfur hexafluoride

Many of the best known binary compounds of the nonmetals have common names which are widely used. Indeed, in some cases the common name is the only one used. No one, for example, ever refers to H_2O as "dihydrogen oxide". Other common names include:

H_2O_2	hydrogen peroxide
N_2H_4	hydrazine
NH_3	ammonia
PH_3	phosphine
AsH_3	arsine
NO	nitric oxide
N_2O	nitrous oxide

In inorganic chemistry, we have relatively few common names.

Oxygen-Containing Acids

Several of our most important acids contain three elements. That is, they contain hydrogen, oxygen, and a nonmetal such as carbon, sulfur, and so on.

H_2CO_3	carbonic acid	$HClO_4$	perchloric acid
H_3BO_3	boric acid	$HClO_3$	chloric acid
H_2SO_4	sulfuric acid	$HClO_2$	chlorous acid
H_2SO_3	sulfurous acid	HClO	hypochlorous acid
HNO_3	nitric acid		
HNO_2	nitrous acid		

Notice that:

1. When a nonmetal forms only one oxyacid, the name is formed by adding the suffix *-ic* to the name of the nonmetal (carbonic acid, boric acid).

2. When a nonmetal forms two different oxyacids, the suffix *-ic* is used for the acid containing the larger number of oxygen atoms. The suffix *-ous* is applied to the acid having fewer oxygen atoms. Thus we have nitric acid (HNO_3) and nitrous acid (HNO_2); sulfuric acid (H_2SO_4) and sulfurous acid (H_2SO_3).

3. When more than two oxyacids are formed, the prefixes *per-* and *hypo-* are used. These designate the acids containing the greatest and least numbers of oxygen atoms respectively. (See the four oxyacids of chlorine.)

EXAMPLE 11.8

Name:

a. H_3PO_4 and H_3PO_3 b. HNO

SOLUTION

a. phosphoric acid and phosphorous acid

b. Since HNO contains fewer oxygen atoms than HNO_2 (nitrous acid), it should be named hyponitrous acid.

You may recall from Chapter 10 that the system used for naming polyatomic ions containing oxygen was very similar to the one just described. Thus we have:

$HClO_4$	perchloric acid	ClO_4^-	perchlorate
$HClO_3$	chloric acid	ClO_3^-	chlorate
$HClO_2$	chlorous acid	ClO_2^-	chlorite
$HClO$	hypochlorous acid	ClO^-	hypochlorite

Clearly, the name of the oxyacid is closely related to that of the ion derived from it by loss of an H^+ ion. Ions whose names end in *-ate* are derived from acids whose names end in *-ic*. Ions whose names end in *-ite* are derived from acids whose names end in *-ous*.

11.6 SUMMARY

When two nonmetals combine with one another, they form covalent bonds. A single covalent bond consists of an electron pair shared between two atoms. It is also possible to have double or triple covalent bonds. Here, two or three electron pairs in that order are shared between the atoms.

In this chapter, we have focused upon covalent bonding in molecular substances. We have considered:

—the general properties of molecular substances (Section 11.1)

—the procedure used to write the Lewis structure of a molecule that obeys the octet rule (Section 11.2)

—the geometry and polarity of molecules (Sections 11.3, 11.4)

—the nomenclature of molecular substances (Section 11.5)

NEW TERMS

bent molecule
covalent bond
double bond
electron pair
electron pair repulsion
Lewis structure
linear molecule
nonpolar molecule
octet rule
orbital overlap
polar molecule
pyramidal molecule
single bond
skeleton structure
tetrahedral molecule
tetrahedron
triple bond
unshared electron

QUESTIONS

1. What did Charlie Brown mean when he said, "Happiness is a shared pair of electrons"?

2. Explain what is meant by the following types of bonds.

   a. covalent b. single c. marital d. double e. triple f. ionic

3. Explain, in terms of orbital overlap, why H_2 is a stable molecule but He_2 is not.

4. The electron configurations of Li and Be are $1s^2 2s^1$ and $1s^2 2s^2$ in that order. Which molecule would you expect to be the more stable, Li_2 or Be_2?

5. Carbon tetrachloride, CCl_4, boils below 100°C; sodium chloride, NaCl, boils above 1000°C. Explain this difference in terms of the bonding involved.

6. Why is molten NaCl a much better electrical conductor than liquid CCl_4?

7. Explain what is meant by a skeleton structure. How does the Lewis structure of H_2O differ from its skeleton structure? Why are the two the same in the case of CH_4?

8. Give the total number of single, double, and triple bonds in:
a. CH_4 b. C_2H_6 c. C_2H_4 d. C_2H_2

9. How is the Lewis structure of an atom related to its position in the Periodic Table?

10. What is meant by the octet rule? Hydrogen in its reactions always acquires a noble-gas structure. Yet it does not obey the octet rule. Explain.

11. Which of the following Lewis structures represent molecules in which the octet rule is followed?
a. $:\ddot{X}{=}\ddot{X}:$ b. $:\ddot{X}{-}\ddot{Y}:$ c. $:\ddot{X}{=}\ddot{Y}$ d. $:X{\equiv}Y:$

12. In a certain molecule there are 4 fewer electrons than the number required for a completely single-bonded structure. Which of the following could account for this deficiency of electrons?
a. 1 double bond b. 1 triple bond c. 2 double bonds
d. 2 triple bonds

13. A molecule AX_2 can have either a linear or a bent structure. Explain what this statement means, using diagrams. Which of these structures leads to a polar molecule? a nonpolar molecule?

14. State the electron pair repulsion principle. How does it apply to the CH_4 molecule?

15. Describe the geometries of the following molecules.
a. CH_4 b. NH_3 c. H_2O

16. How many faces does a regular tetrahedron have? What kind of triangle is each side? How many corners are there in a tetrahedron? What is the tetrahedral angle?

17. Explain what is meant by a polar molecule; a nonpolar molecule. Which one lines up in an electrical field?

18. Explain how it is possible for a molecule to be nonpolar even though the bonds within it are polar.

19. Consider a molecule AX_3. Which of the following geometries would give a polar molecule? a nonpolar molecule?

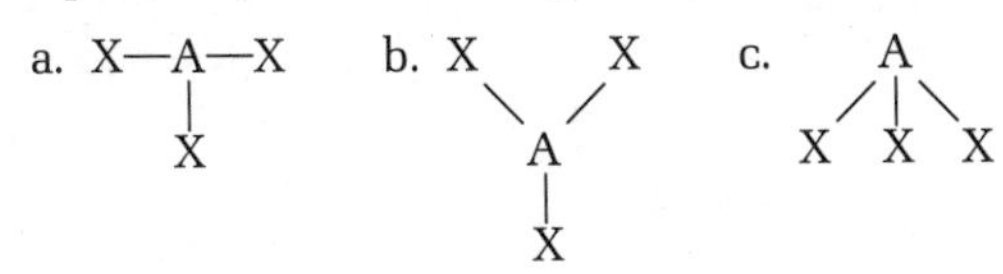

20. Explain why the H_2O molecule is polar while the BeF_2 molecule is nonpolar.

21. What are the special names given to water solutions of:

    a. HCl b. HBr c. HF d. HI

22. Which of the following represent systematic names for binary molecular compounds?

    a. carbon dioxide b. water c. nitrous oxide
    d. dinitrogen oxide

23. Give the names of the following acids.

    a. HNO_3 b. H_2SO_4 c. H_3BO_3

24. Phosphorus forms three oxygen acids. Their names are phosphoric acid, phosphorous acid, and hypophosphorous acid. Which molecule contains the smallest number of oxygen atoms? the largest?

PROBLEMS

1. *Lewis Structures of Atoms (Example 11.1)* Consider an atom X. What is its Lewis structure if it is in Group

   a. 3 b. 7 c. 5 d. 6

2. *Lewis Structures of Molecules (Example 11.2)* Give the Lewis structures of:

   a. HCl b. PH_3 c. H_2S d. SiH_4

3. *Lewis Structures of Molecules (Example 11.3)* Write Lewis structures for each of the following molecules. In each case one or more multiple bonds is involved.

   a. SO_2 b. SO_3 c. S_2

4. *Molecular Geometry (Example 11.4)* Describe the geometry of the molecules whose Lewis structures you wrote in Problem 2.

5. *Polarity of Molecules (Example 11.5)* Which of the following diatomic molecules are polar? nonpolar?

   a. N_2 b. HI c. ICl d. H_2

6. *Polarity of Molecules (Example 11.6)* Which of the molecules considered in Problem 4 are polar? nonpolar?

7. *Naming of Molecular Compounds (Example 11.7)* Name the following compounds in a systematic way.

   a. H_2O b. PCl_5 c. N_2O_3 d. S_2Cl_2

8. *Naming of Molecular Compounds (Example 11.8)* On a separate sheet of paper, name the following species:

H_3PO_4 phosphoric acid	PO_4^{3-} ion	Na_3PO_4
H_3PO_3	PO_3^{3-} ion	Na_3PO_3
H_3PO_2	PO_2^{3-} ion	

* * * * *

9. Write Lewis structures for each of the following atoms:

   a. Si b. As c. Kr d. Br

10. Give the Lewis structures of:

    a. PCl_3 b. F_2O c. ICl

11. Describe the geometry of the molecules whose Lewis structures you wrote in Problem 10.

12. Which of the molecules considered in Problem 11 are polar? nonpolar?

13. Convert the following skeleton structures into Lewis structures. Make sure you use the correct number of valence electrons.

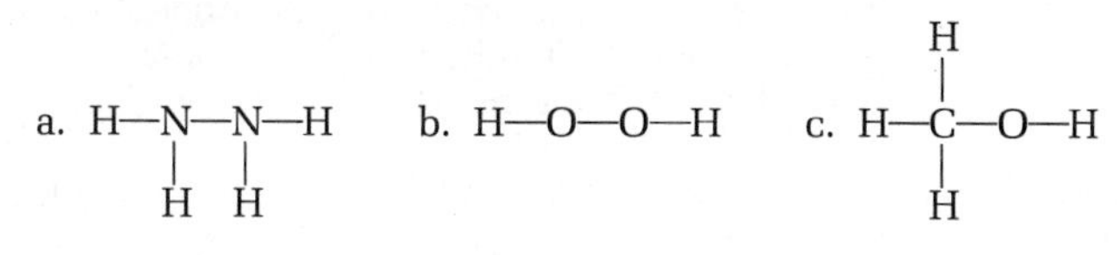

14. Write Lewis structures for each of the following molecules. In each case, one or more multiple bonds is involved.

    a. CO_2 b. HCN c. N_2F_2

15. Give the Lewis structures of:

    a. C_2H_6 b. C_2H_6O c. C_2H_4O

16. Consider the CH_3OH molecule shown in Problem 13. Sketch its geometry, assuming all the bond angles are 109°.

17. Which of the following molecules are polar? nonpolar?

    a. CH_4 b. CH_3Cl c. CH_2Cl_2 d. $CHCl_3$ e. CCl_4

18. Consider the following polyatomic ions, which have the Lewis structures shown. Describe the geometry of each ion.

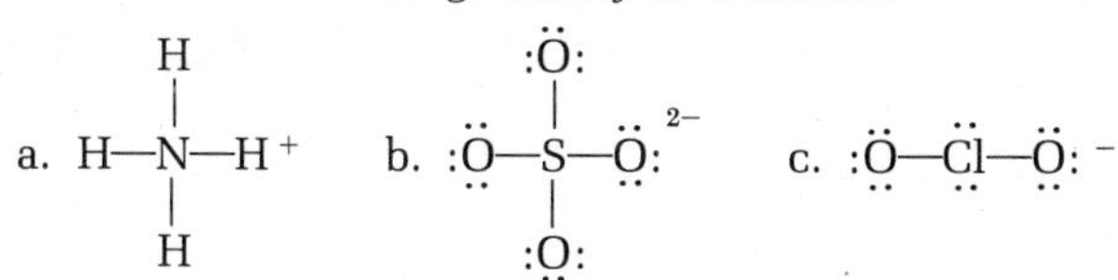

19. Draw Lewis structures and predict the geometry and polarity for:

    a. Cl_2O b. NH_2Cl c. $SiCl_3Br$

20. The skeleton structure of HNO_3 is:

H—O—N—O
|
O

Count the number of valence electrons and draw the Lewis structure.

21. Write the formulas of:

    a. sulfur dichloride b. tetraphosphorus hexoxide
    c. dinitrogen pentoxide

22. Name the following compounds:

    a. H_2SeO_4 b. H_2SeO_3 c. As_2O_5 d. H_2O_2

*23. The two molecules NO_2 and ClO_2 do not obey the octet rule. Suggest reasonable Lewis structures for these molecules.

*24. Lewis structures can be written for polyatomic ions in much the same way as for molecules. The only difference is that you have to take into account the charge of the ion in counting valence electrons. For negative ions, the charge is added to the valence electrons supplied by the neutral atoms. Write the Lewis structures of:

    a. OH^- b. ClO^- c. CO_3^{2-}

*25. A certain hydrocarbon contains 10.1% by mass of hydrogen and 89.9% carbon. Its molecular mass is about 40. Determine the molecular formula and suggest a reasonable Lewis structure.

*26. White phosphorus has the molecular formula P_4. Suggest a possible Lewis structure for the molecule.

Photo by Peter Southwick/Stock Boston.

12

LIQUIDS AND SOLIDS

In Chapter 5, we discussed the physical properties of gaseous substances. You will recall that we were able to write simple equations (Boyle's Law, Charles' Law, . . .) to describe the behavior of all gases. These laws are readily explained in terms of the kinetic molecular model. This model considers that the molecules in a gas are in rapid, chaotic motion and are relatively far removed from one another.

In this chapter, our goals are more modest. We cannot express the physical properties of liquids and solids by equations which apply to all substances. The basic difficulty is that in the "condensed states" of matter, the particles (atoms, molecules or ions) are in close contact. The best we can do here is to develop a molecular model which explains the general properties of liquids and solids.

12.1 MOLECULAR MODEL OF LIQUIDS AND SOLIDS

Liquids and solids differ from gases in two important ways. In the first place, the particles (atoms, molecules, or ions) are much closer together. Moreover, attractive forces between these particles are much stronger. Let us see how these factors affect the properties of liquids and solids.

Distance Between Particles

A major difference between the gas state and the condensed states is the distance between particles. At 25°C and 1 atm, the average distance between molecules in a gas is about 10 molecular diameters. In contrast, particles in a liquid or solid are virtually touching each other. This means that whereas a gas is mostly empty space, there is very little "free volume" in a liquid or solid.

This basic difference between the states of matter shows up when we compare molar volumes. You will recall from Chapter 5 that all gases have the same molar volume at the same temperature and pressure. At 0°C and 1 atm, the molar volume of any gas is about 22.4 ℓ (22.4×10^3 cm^3). In contrast, the molar volumes of liquids and solids are much smaller. Consider, for example, liquid water. A mole of water (H_2O) weighs 18.0 g. At 0°C and 1 atm, liquid water has a density of 1.00 g/cm^3. Hence:

$$\text{molar volume } H_2O(l) = \frac{\text{molar mass}}{\text{density}} = \frac{18.0 \text{ g}}{1.00 \text{ g/cm}^3} = 18.0 \text{ cm}^3$$

This is less than 1/1000 of the molar volume of a gas at the same temperature and pressure.

Attractive Forces Between Particles

In the gas state, where the molecules are far apart, attractive forces between molecules are of minor importance. We often assume that gas molecules behave as independent particles with no interactions between them. This explains why gases can expand to occupy the entire volume of their container. In the condensed states of matter, the situation is quite different. Since the particles are close together, attractive forces between them are quite strong.

Attractive forces have their greatest effect in the solid state. There, they cause the particles to line up in a regular pattern. Particles in the solid are held so firmly that they cannot move past one another. Hence, solids typically have a fixed volume and a fixed shape, independent of the nature of their container.

In the liquid state, attractive forces between particles are nearly as strong as in the solid. Although the particles can move past one another, they never get very far apart. As a result, liquids have a fixed volume but are fluid enough to take on the shape of their container (Figure 12.1).

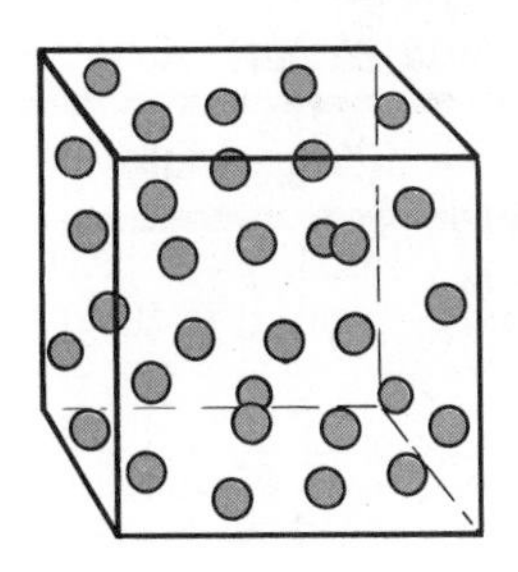
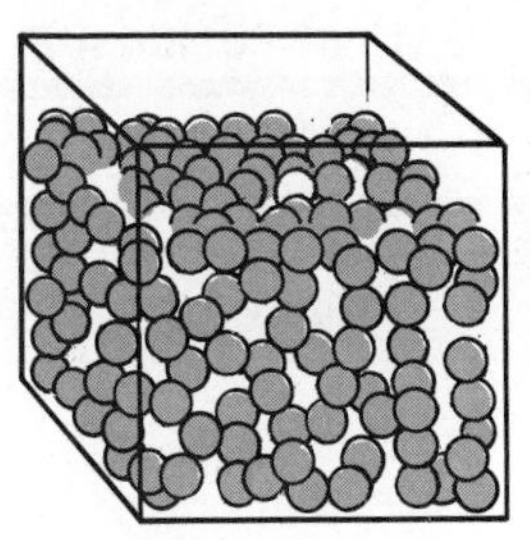
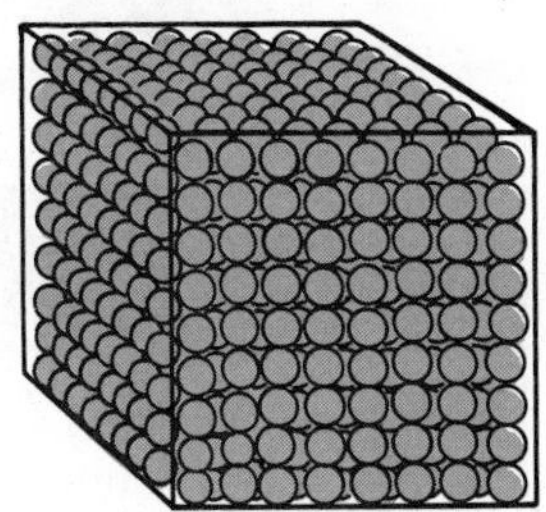

FIGURE 12.1

The three states of matter. The densities of liquids and solids are nearly equal and much larger than those of gases at one atm. In solids the particles are usually packed in definite geometric arrangements characteristic of crystals.

Some particles in gaseous state (not drawn to scale); particles far apart, completely fill container.

Some particles in a liquid state; not in fixed positions (liquid flows), definite volume

Particles in a solid; fixed positions, definite volume.

Energy has to be absorbed to overcome the attractive forces between particles in the solid or liquid state. The melting of a solid is an endothermic process. Ice melts only if heat is supplied to it, perhaps by the sun on a warm spring day. Heat must also be absorbed to vaporize a liquid. You feel cool when you step out of a shower because the heat required to evaporate the water is absorbed from your skin.

In Chapter 4, we discussed the heat effects involved in changes of physical state. In particular, we talked about the *heat of fusion* of a solid, ΔH_{fus}:

$$\text{solid} \rightarrow \text{liquid}; \qquad \Delta H = \Delta H_{fus}$$

and the *heat of vaporization* of a liquid, ΔH_{vap}:

$$\text{liquid} \rightarrow \text{gas}; \qquad \Delta H = \Delta H_{vap}$$

It always takes energy to melt a solid or to vaporize a liquid.

$\Delta H > 0$

Heats of fusion or vaporization are usually given either in kcal/mol or cal/g.

$$\Delta H \text{ (in kcal/mol)} = \frac{\text{molar mass (in g)}}{1000} \times \Delta H \text{ (in cal/g)} \qquad (12.1)$$

Table 12.1, p. 288, lists values of ΔH_{fus} and ΔH_{vap} for several different substances. Notice that:

1. Heats of fusion and vaporization differ widely from one substance to another. This shows that the attractive forces between particles depend upon the nature of the particle. The stronger the attractive forces, the higher the heat of fusion or vaporization. On a calorie per gram basis, water has very large values of ΔH_{fus} and ΔH_{vap}. Indeed the heat of vaporization of water, in cal/g, is larger than that of any other substance.

2. For a given substance, the heat of fusion is only a small fraction of the heat of vaporization. This suggests that the forces between particles are nearly as strong in the liquid as they are in the solid. Only when the molecules are widely separated in the gas state do these forces become of minor importance.

TABLE 12.1 HEATS OF FUSION AND VAPORIZATION

		Solid		Liquid	
SUBSTANCE	MOLAR MASS	ΔH_{fus} (cal/g)	ΔH_{fus} (kcal/mol)	ΔH_{vap} (cal/g)	ΔH_{vap} (kcal/mol)
Mercury	201 g	2.8	0.56	71	14.2
Bromine	160 g	15.8	2.52	45	7.16
Naphthalene	128 g	36.0	4.61	76	9.7
Benzene	78.1 g	30.1	2.35	94	7.36
Water	18.0 g	80.0	1.44	540	9.72

12.2 PROPERTIES OF LIQUIDS

Liquids have certain properties which distinguish them from gases or solids. In this section, we will look at a few such properties, relating them to the model of the liquid state. We will be particularly interested in the properties of liquid water. Water is both the most common and the most unusual of liquids.

Vapor Pressure

In a liquid, as in a gas, the molecules are moving at a variety of speeds. In a beaker of water at 25°C, some of the molecules are moving very slowly. Most are moving at speeds close to the average. A few are moving very rapidly. They have enough kinetic energy to break away from the surface and escape into the air above the beaker. As time passes, more and more molecules leave the liquid. The level of water in the beaker drops, slowly but steadily. This process, called *evaporation,* continues until all the liquid is gone.

The situation is quite different if water is placed in a closed container (Figure 12.2). As before, some of the faster moving molecules escape from the liquid to enter the space above it. This time, however, they cannot get out into the surrounding air. Some of them bounce around, strike the surface of the liquid, and condense in it. At first, this process of condensation (vapor → liquid) is much slower than the reverse process of vaporization (liquid → vapor). However, as more water molecules enter the vapor, the rate of condensation increases. Eventually, it becomes equal to the rate of vaporization. When this happens, there is a **dynamic equilibrium** between liquid and vapor, indicated by the double arrow:

$$\text{liquid} \rightleftharpoons \text{vapor}$$

$$\text{rate of vaporization} = \text{rate of condensation}$$

From this time on, the number of molecules condensing exactly balances the number vaporizing. The concentration of water molecules

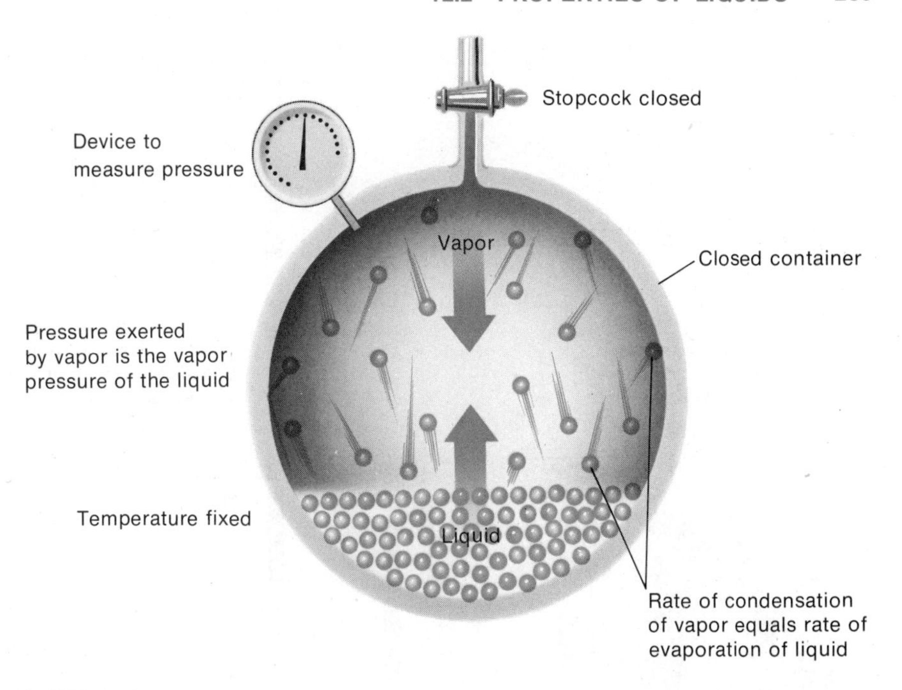

FIGURE 12.2

Vapor pressure. At equilibrium at a given temperature the vapor pressure of a liquid is constant and does not depend on either the amount of liquid or the volume of the container.

in the vapor remains constant as time passes. This condition is characteristic of all equilibrium processes. To an outside observer, the system appears to have settled down. There is no net change in either direction.

In the container, water exists both as a liquid and as a gas. At equilibrium, the amount of each phase remains constant.

The pressure exerted by a vapor in equilibrium with a liquid is called the **equilibrium vapor pressure** (or simply the "vapor pressure") of the liquid. This quantity can be measured in an apparatus such as that shown in Figure 12.3, p. 290. When water is squirted into the flask from an eye-dropper, the mercury rises in the right arm of the manometer and falls in the left. Eventually, when equilibrium is reached, the levels stop moving. If this experiment is carried out with water at 25°C, the difference in the mercury levels is about 24 mm. Thus we would say that at 25°C the equilibrium vapor pressure of water is 24 mm Hg.

An important point concerning the equilibrium between liquid and vapor can be related to the experiment shown in Figure 12.3. The pressure read on the manometer is the same regardless of

—the amount of water added
—the volume of the flask

These relations hold provided there is enough liquid to establish equilibrium with the vapor. In other words, if there is liquid present at the end of the experiment, the pressure difference will be 24 mm Hg. In general, we can say that the *equilibrium vapor pressure of a liquid is independent of the amount of liquid present or the volume occupied by the vapor.*

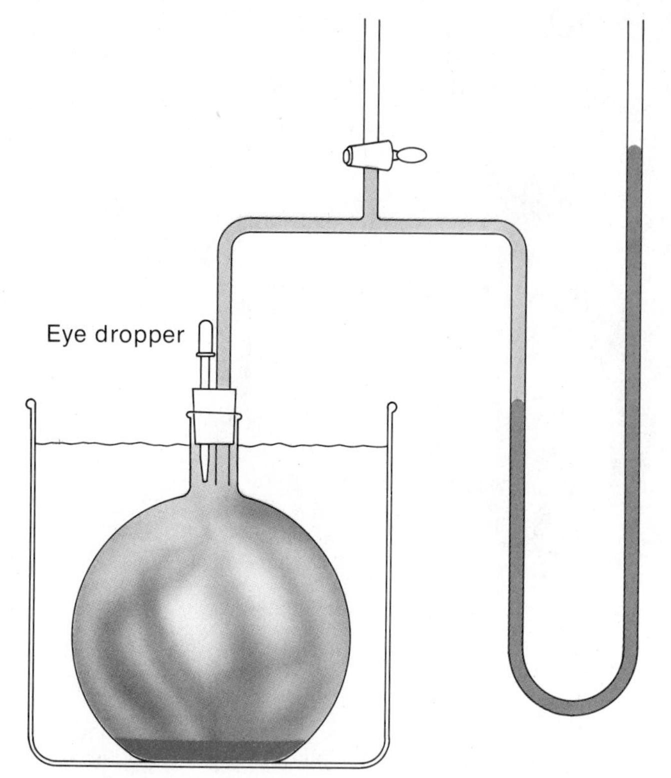

FIGURE 12.3

Measurement of vapor pressure. If a little liquid is added to a closed container full of air, the pressure in the container rises. If the mercury levels in the manometer were equal before the liquid was added, the difference between the levels at equilibrium will equal the vapor pressure of the liquid.

We find that the equilibrium vapor pressure of a liquid depends upon two factors.

1. **The nature of the liquid.** Many liquids have vapor pressures higher than that of water. The vapor pressure of benzene at 25°C is 92 mm Hg, nearly four times that of water. Diethyl ether has a vapor pressure at 25°C of more than half an atmosphere (537 mm Hg).

The vapors of organic liquids, unlike that of water, usually have distinct odors. Sometimes these are pleasant, as in perfumes or shaving lotion. Other times they are decidedly unpleasant. Such is the case with the organic compounds (mercaptans) responsible for the odor of skunk. Liquids such as these need not and usually do not have high vapor pressures. The nose is extremely sensitive to certain odors, detecting them even in trace amounts.

2. **The temperature.** We always find that vapor pressure increases as the temperature rises. The stoppers in bottles of volatile liquids such as diethyl ether pop out if the laboratory gets too warm. The odor of perfume becomes more pronounced on a hot day. At higher temperatures, more molecules have kinetic energies large enough to allow them to overcome the attractive forces and escape from the liquid. Since there are more molecules in a given volume of vapor, they exert a higher pressure.

The vapor pressure of water nearly doubles each time the temperature is raised by 10°C. Thus it increases from 24 mm Hg at 25°C to 42 mm Hg at 35°C to 72 mm Hg at 45°C, and so on. In Figure 12.4 we show a graph of the vapor pressure of water *versus* temperature. The data for this graph are taken from Table 12.2.

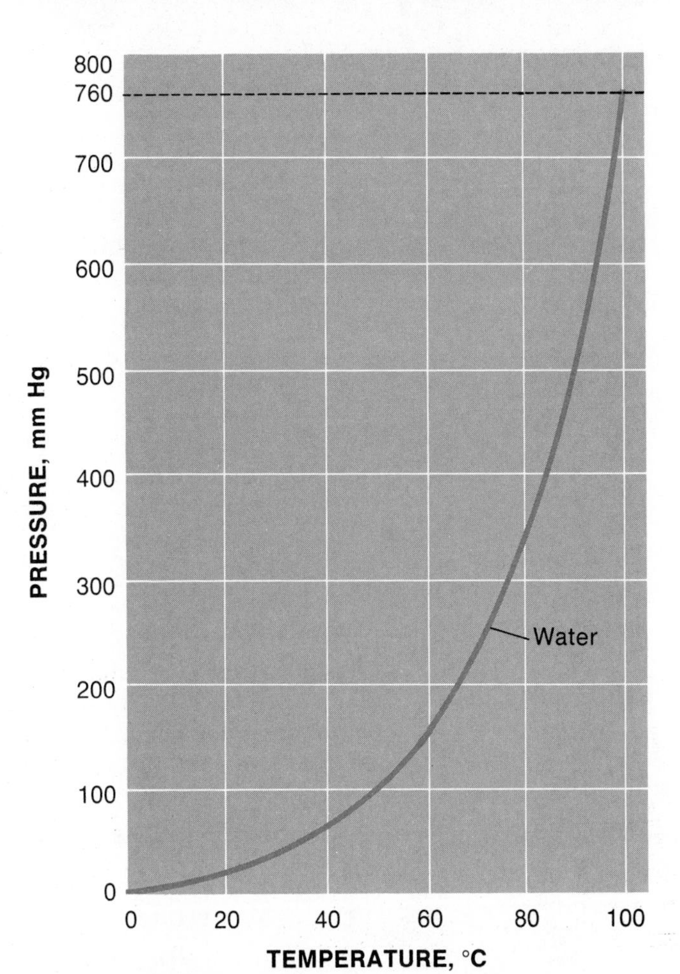

FIGURE 12.4

Effect of temperature on the vapor pressure of water. The vapor pressure of a liquid increases ever more rapidly as the temperature increases. At temperatures above 100°C the vapor pressure of water rises rapidly, reaching 15 atm at 200°C and 85 atm at 300°C.

TABLE 12.2 VAPOR PRESSURE OF WATER (MM HG) AT VARIOUS TEMPERATURES

T (°C)	vp	T (°C)	vp	T (°C)	vp
0	4.6	26	25.2	92	567
5	6.5	27	26.7	94	611
10	9.2	28	28.4	96	658
12	10.5	29	30.0	98	707
14	12.0	30	31.8	100	760
16	13.6	35	42.2	102	816
17	14.5	40	55.3	104	875
18	15.5	45	71.9	106	938
19	16.5	50	92.5	108	1004
20	17.5	55	118	110	1075
21	18.6	60	149	112	1149
22	19.8	65	188	114	1227
23	21.1	70	234	116	1310
24	22.4	80	355	118	1397
25	23.8	90	526	120	1490

Relative Humidity

The vapor pressure of water is involved in several quantities of interest so far as our weather and climate are concerned. One of these, which you hear quoted on TV weather reports, is the *relative humidity*, RH. This is a measure of the concentration of water vapor in the air, defined as:

$$\% \text{ RH} = \frac{\text{pressure of water vapor in air}}{\text{equil. v.p. of water at same T}} \times 100 \qquad (12.2)$$

Suppose that on a day when the temperature is 25°C, the pressure of water vapor in the air is 20.0 mm Hg. From Table 12.2, we see that the equilibrium vapor pressure of water is 23.8 mm Hg. Consequently:

$$\% \text{ relative humidity} = \frac{20.0}{23.8} \times 100 = 84.0$$

We might say that under these conditions the air is 84% saturated with water vapor. It contains 84% of the water that would be present if the air were in equilibrium with liquid water and hence saturated with water vapor. The maximum value of relative humidity is 100%. At that point, air contains all the water it can hold at a particular temperature. When it is raining, the relative humidity is 100%.

Can the relative humidity be 100%, and no rain happen?

EXAMPLE 12.1

On a clear autumn day, the temperature is 22°C and the pressure of water vapor in the air is 8.8 mm Hg. Calculate the relative humidity.

SOLUTION

From Table 12.2, the equilibrium vapor pressure of water at 22°C is 19.8 mm Hg. Hence:

$$\% \text{ relative humidity} = \frac{8.8}{19.8} \times 100 = 44$$

Boiling Point

If we heat water in an open beaker, nothing spectacular happens until we reach a temperature of about 100°C. At that point, vapor bubbles suddenly form in the body of the liquid, rise to the surface, and break. When this happens, we say that the water is boiling (Figure 12.5).

For a vapor bubble to form in a liquid, the pressure within the bubble must be at least equal to that outside it. Otherwise, any bubble formed would collapse before it reached the surface. In other words, the condition for boiling is:

$$P_1 = P_2$$

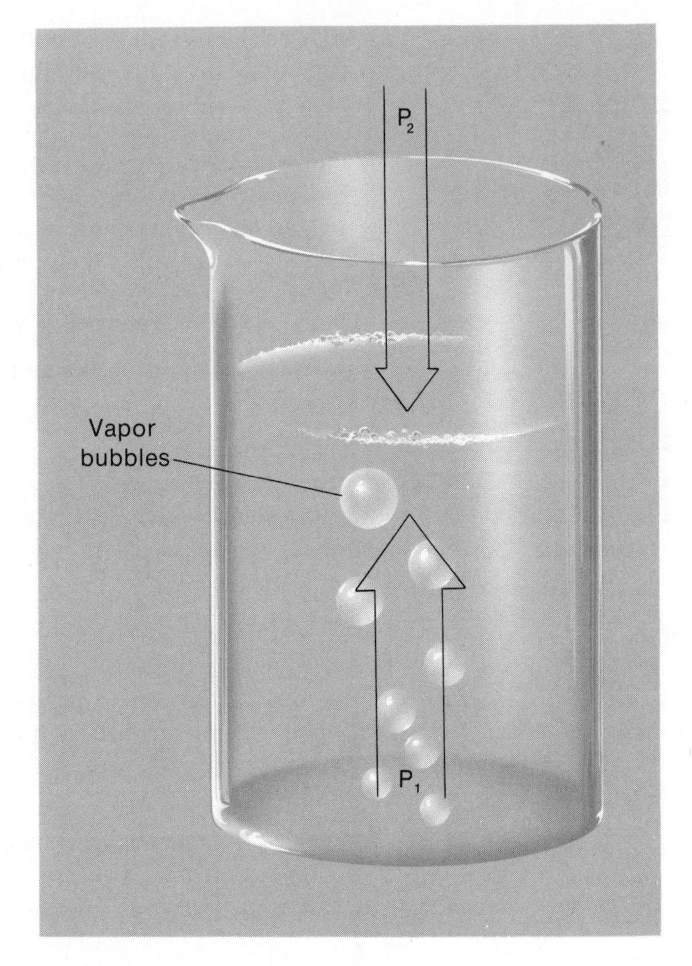

FIGURE 12.5

A liquid will boil when its vapor pressure becomes slightly greater than the pressure applied to it. At that point bubbles can form and remain stable until they reach the surface of the liquid.

where P_1 is the pressure inside the bubble and P_2 the pressure outside. But, since the bubble contains only vapor, P_1 must be the vapor pressure of the liquid at the boiling temperature. The pressure outside the bubble, P_2, is the pressure of the air or other gas above the liquid. We see then that **a liquid boils when its vapor pressure becomes equal to the pressure over its surface.**

This condition means that the temperature at which a liquid boils (its *boiling point*) will depend upon the external pressure. When we state the boiling point of a liquid, we must specify the pressure. The *normal boiling point* is defined as the temperature at which a liquid boils when the external pressure is one atmosphere (760 mm Hg). Since the vapor pressure of water reaches 760 mm Hg at 100°C, the normal boiling point of water is 100°C.

The effect of pressure on boiling point can be demonstrated with the apparatus shown in Figure 12.6, p. 294. By creating a vacuum, we reduce the pressure over the water in the thick-walled flask. Quite quickly, we reach the equilibrium vapor pressure at room temperature, about 24 mm Hg. At this point, the water starts to boil vigorously. A similar effect is observed at high altitudes, where the pressure is commonly well below 760 mm Hg. At an elevation of 3000 m, the boiling point of water

In the Rocky Mountains, a three-minute egg would be raw.

FIGURE 12.6

In the photograph water is boiling at room temperature. The flask was attached to a vacuum pump which lowered the pressure below the vapor pressure of the water. No heat was needed. Actually, the temperature of the water fell as the boiling proceeded. Can you suggest why?

may be as low as 90°C. Foods cook more slowly at this temperature. The directions on many food products specify longer cooking times at high altitudes.

By increasing the pressure over a liquid, it is possible to reach temperatures above the normal boiling point. You (or, more likely, your mother) make use of this effect with a pressure cooker. In this sealed container, the pressure builds up, perhaps to as much as 2 atm (1520 mm Hg). This raises the boiling point of water to slightly above 120°C. At this temperature, foods cook more rapidly. Potatoes become edible in perhaps ¼ the time required at 100°C.

EXAMPLE 12.2

Using Table 12.2, estimate the boiling point of water, to the nearest degree Celsius, at a pressure of:

a. 650 mm Hg b. 1.50 atm

SOLUTION

a. From the Table, we see that water has a vapor pressure of 658 mm Hg at 96°C. Hence water should boil a bit below 96°C when the pressure above it is 650 mm Hg.

b. Here we first convert 1.50 atm to mm Hg:

$$P\ (\text{mm Hg}) = 1.50\ \text{atm} \times \frac{760\ \text{mm Hg}}{1\ \text{atm}} = 1140\ \text{mm Hg}$$

From Table 12.2 we see that the vapor pressure of water is 1149 mm Hg at 112°C. Hence the boiling point should be slightly below 112°C.

12.3 TYPES OF FORCES BETWEEN MOLECULES

Virtually all of the substances that are liquids at temperatures of 25°C or below are built of molecules. When we talk about "attractive forces" in liquids, we are ordinarily dealing with forces between molecules. As pointed out in Chapter 11, these forces are rather weak, at least in comparison with ionic or covalent bonds. However, they are strong enough to hold molecules close to one another in the liquid state. Indeed, such forces exist between molecules in all states of matter. In this section, we will discuss two different types of intermolecular forces, Van der Waals forces and hydrogen bonds.

When we boil water, we overcome forces between the molecules, but we don't break any O—H bonds within the molecule.

Van der Waals Forces

The general type of attractive force that exists between all molecules is called a Van der Waals force. With polar molecules, we can readily see how such an attractive force arises (Figure 12.7). The polar molecules tend to line up so that the positive pole of one molecule is next to the

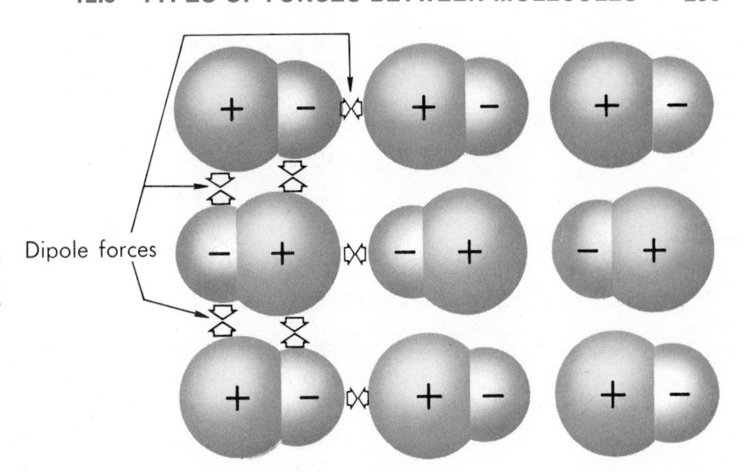

FIGURE 12.7

In a polar solid, the positive and negative ends of the molecules attract one another. This attraction leads to a type of Van der Waals force called a "dipole force."

negative pole of its neighbor. This results in an electrical attraction between adjacent molecules. In nonpolar molecules, the origin of Van der Waals forces is less obvious. In general, they result from the charges created by movement of electrons within molecules. The charges on two neighboring molecules interact to give a net attractive force.

The strength of Van der Waals forces ordinarily increases with molecular mass. Since these forces must be overcome to boil a molecular liquid, we might expect boiling point to follow this same trend. This is indeed true. Among molecular liquids with similar structures, boiling point increases with molecular mass. This effect is illustrated in Table 12.3.

TABLE 12.3 EFFECT OF MOLECULAR MASS ON NORMAL BOILING POINT

Substance	MM	BP (°C)	Substance	MM	BP (°C)	Substance	MM	BP (°C)
CH_4	16	−162	CH_4	16	−162	F_2	38	−188
SiH_4	32	−112	CH_3Cl	50	−24	Cl_2	71	−34
GeH_4	77	−90	CH_2Cl_2	85	40	Br_2	160	59
SnH_4	123	−52	$CHCl_3$	119	61	I_2	254	184
			CCl_4	154	77			

EXAMPLE 12.3

Consider the three elements whose properties were described in Chapter 6: H_2, O_2, and N_2. Arrange these molecular substances in order of increasing normal boiling point.

SOLUTION

The molecular masses are as follows:

MM H_2 = 2.0 MM O_2 = 32.0 MM N_2 = 28.0

We predict the order of increasing boiling point to be:

$$H_2 < N_2 < O_2$$

The observed values are $-253°C$ for H_2, $-196°C$ for N_2, and $-183°C$ for O_2.

Hydrogen Bonds

The rule illustrated in Table 12.3 and Example 12.3 is a valuable one for predicting relative boiling points of molecular substances. Like most rules, however, this one doesn't always work. The most striking exceptions occur with polar molecules in which hydrogen atoms are bonded to nitrogen, oxygen, or fluorine. Consider the graphs of boiling point *versus* molecular mass shown in Figure 12.8. Clearly, the three compounds NH_3, H_2O and HF have abnormally high boiling points.

If the molecules in a compound form hydrogen bonds, it will boil at a much higher temperature than it would if its molecules didn't form them.

Ammonia, water, and hydrogen fluoride have one factor in common. They consist of polar molecules in which hydrogen atoms are bonded to a small, highly electronegative atom (N, O, or F). At close range, there is a strong electrical attraction between the positive pole (H atom) of one molecule and the negative pole (N, O, or F atom) of another. This type of attractive force is referred to as a hydrogen bond. It is perhaps easiest to visualize in HF (Figure 12.9), but is equally important in H_2O and NH_3. Hydrogen bonds hold molecules of these substances together in the liquid. These bonds must be broken to boil the substance. This explains the high boiling points of hydrogen fluoride, water, and ammonia.

Hydrogen bonding is by no means limited to the three compounds we have mentioned, NH_3, H_2O, and HF. Indeed it is very common among organic compounds. In general, hydrogen bonding can occur between molecules of any compound in which hydrogen is bonded to nitrogen, oxygen, or fluorine.

EXAMPLE 12.4

Which of the following molecules would show hydrogen bonding?

```
     H                                   H
     |                                   |
a. H—C—H     b. H—N—N—H      c. H—C—F
     |              |   |                |
     H              H   H                H
```

SOLUTION

Only (b), hydrazine, can show hydrogen bonding. Here, hydrogen is bonded to nitrogen. In the other molecules, hydrogen is bonded to carbon.

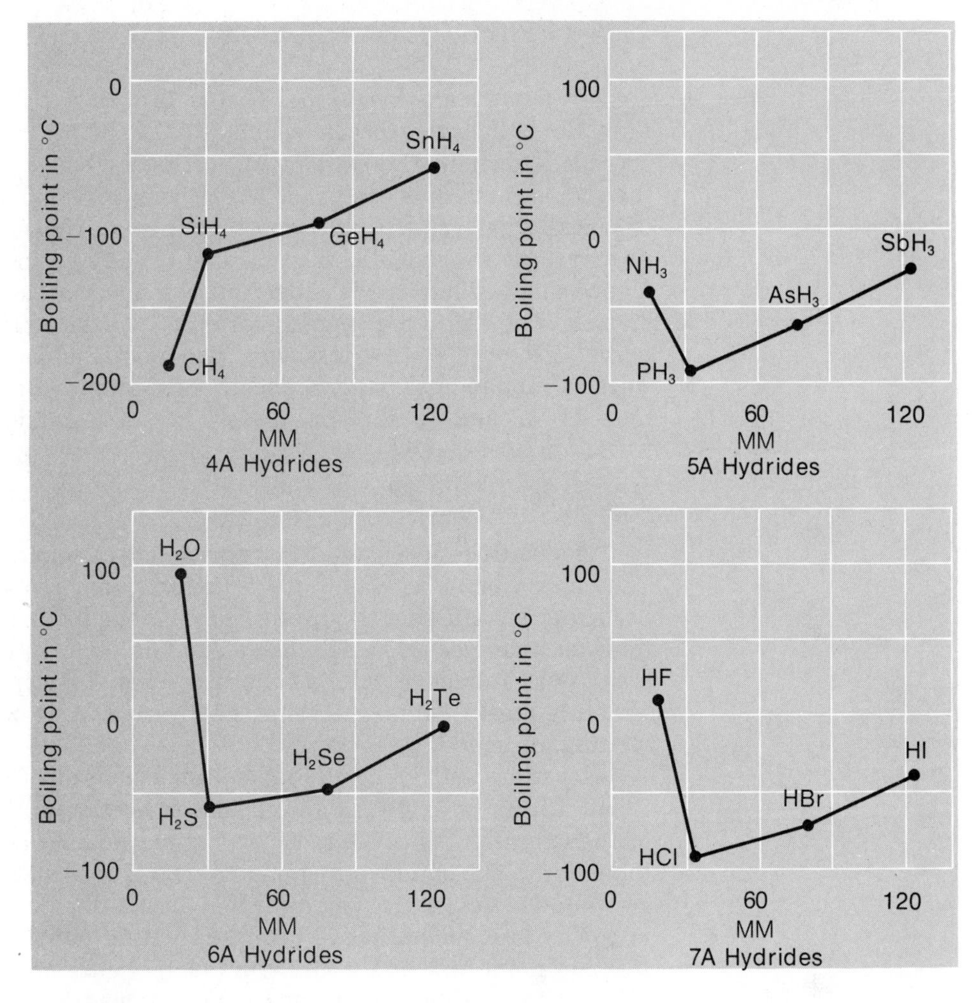

FIGURE 12.8

Boiling points of some nonmetal hydrides. Hydrogen bonding is responsible for the high boiling points of NH_3, H_2O and HF. CH_4 does not have hydrogen bonding because carbon atoms are not very electronegative.

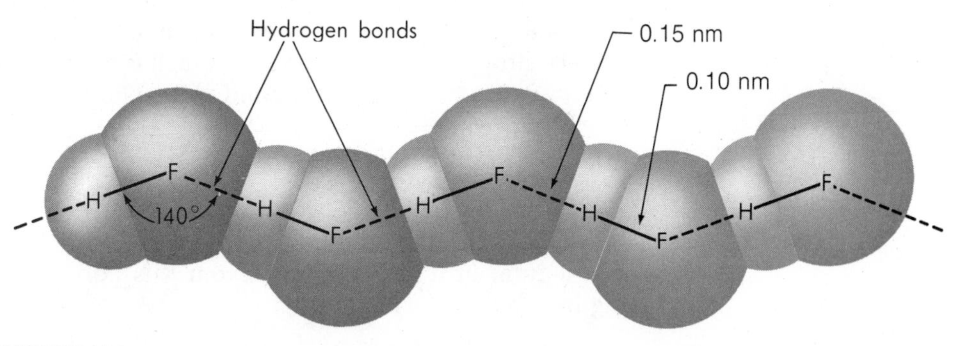

FIGURE 12.9

In HF the H atom on one molecule is strongly attracted to the F atom on another. The molecules tend to link up in chains, making for a high boiling point in liquid HF. Note that the bond distance in the hydrogen bond is considerably longer than in the molecule.

12.4 PROPERTIES OF SOLIDS

Most solids are crystalline. That is, they have a distinct geometric form. Sometimes this form is visible to the eye. This is the case with the crystals shown in Color Plate 5C, center of book. More frequently, the geometric form is obscured. The crystals may be too tiny to be seen or several small crystals may grow together. Looking at a sample of zinc dust or a steel bar, it is hard to distinguish any geometric pattern. Yet if you examine these solids under a microscope, you can see tiny distinct crystals. The definite geometric form of a crystal reflects the way the particles themselves are arranged. The atoms, molecules, or ions are lined up in a regular pattern. Recall, for example, the regular arrangement of Na^+ and Cl^- ions in the NaCl crystal (Chapter 10).

Melting Points of Solids

The particles in a crystal are restricted to certain definite positions. They may vibrate back and forth. However, they remain centered at a particular spot. If heat is supplied to raise the temperature, the vibrations become greater. Suddenly, at a certain temperature, the particles break away and move past one another. When this happens, the crystal structure breaks down; the solid is converted to a liquid. We say that melting, or fusion, has occurred.

Experimentally, we find that melting occurs at a fixed temperature. When a sample of solid naphthalene is heated, nothing spectacular happens until a temperature of 80°C is reached. At that point, we see liquid form. The temperature remains at 80°C until all the solid has been converted to liquid. We would say that the melting point of naphthalene is 80°C. More generally, we take the melting point of a solid, or the freezing point of a liquid, to be the temperature at which liquid and solid are in equilibrium. The melting point of a solid changes very little with pressure, much less so than the boiling point of a liquid. Unless otherwise specified, the pressure is understood to be one atmosphere.

The melting point of a substance, like its boiling point, is directly related to the strength of the forces between particles. Low melting points are typical of substances where the attractive forces are very weak, such as hydrogen (mp = −259°C). At the opposite extreme, we associate high melting points with substances where the forces between particles are very strong. Examples include silicon dioxide (mp = 1700°C), magnesium oxide (mp = 2800°C), and elementary carbon (mp $>$ 3500°C).

12.5 TYPES OF SOLIDS

Perhaps the most useful way to classify solids is to specify the type of structural unit they contain. From this point of view, we can distinguish four types of solids:

1. Ionic (NaCl, $KClO_3$, . . .)
2. Molecular (I_2, H_2O, . . .)
3. Metallic (Na, Mg, . . .)
4. Covalent network (C, SiO_2, . . .)

Ionic Solids

As pointed out in Chapter 10, ionic solids have high melting points, typically above 600°C. This reflects the strong electrical forces between positive and negative ions in the crystal lattice. Chemical (ionic) bonds must be broken to separate the ions from each other to form a liquid.

Certain ionic compounds decompose when heated rather than melting. Among these are hydrates, such as $BaCl_2 \cdot 2H_2O$. As mentioned in Chapter 10, water of hydration is lost, often upon gentle heating. A more drastic type of decomposition takes place when compounds containing polyatomic ions are heated. An example is potassium chlorate, $KClO_3$. At temperatures above 200°C, the following reaction occurs:

Most hydrates lose water near 100°C.

$$2\ KClO_3(s) \rightarrow 2\ KCl(s) + 3\ O_2(g) \qquad (12.3)$$

The chlorate ion, ClO_3^-, decomposes to the simple chloride ion, Cl^-. The oxygen atoms in the polyatomic ion are given off as $O_2(g)$.

Among the most common polyatomic ions are the OH^- (hydroxide) and CO_3^{2-} (carbonate) ions. Upon heating, these decompose to the simple O^{2-} ion. Thus at 600°C, calcium *hydroxide* decomposes to calcium *oxide:*

$$Ca(OH)_2(s) \rightarrow CaO(s) + H_2O(g) \qquad (12.4)$$

At about 800°C, calcium *carbonate* decomposes to calcium *oxide:*

$$CaCO_3(s) \rightarrow CaO(s) + CO_2(g) \qquad (12.5)$$

These reactions are typical of most hydroxides and carbonates (Example 12.5), although the temperature required for decomposition varies from one compound to another.

EXAMPLE 12.5

a. What are the formulas of aluminum hydroxide and aluminum carbonate?

b. Write balanced equations for the reactions that you expect to occur when these compounds are heated.

SOLUTION

a. As pointed out in Chapter 10, the aluminum ion has a charge of +3, Al^{3+}. The formulas must then be $Al(OH)_3$ and $Al_2(CO_3)_3$.

b. When a hydroxide is heated it decomposes to an oxide; the other product is water vapor. The unbalanced equation is:

$$Al(OH)_3(s) \rightarrow Al_2O_3(s) + H_2O(g)$$

Balancing the equation, as in Chapter 3, gives:

$$2\ Al(OH)_3(s) \rightarrow Al_2O_3(s) + 3\ H_2O(g)$$

In the case of aluminum carbonate, the products should be aluminum oxide and carbon dioxide. The balanced equation is:

$$Al_2(CO_3)_3(s) \rightarrow Al_2O_3(s) + 3\ CO_2(g)$$

Molecular Solids

Those molecular substances in which the intermolecular forces are strongest are solids at ordinary temperatures. These include substances of high molecular mass such as iodine, I_2, naphthalene, $C_{10}H_8$, and others in which hydrogen bonds are present such as ice (H_2O). Molecular solids as a group have rather low melting points, almost always below 300°C.

Many molecular solids are volatile. That is, they have an appreciable vapor pressure at ordinary temperatures. Sometimes we can see or smell the vapor formed. In a capped bottle of iodine, there is a purple vapor in the space above the shiny black crystals. Naphthalene has a strong odor which we associate with mothballs.

The process by which a solid vaporizes directly without passing through the liquid state is called **sublimation.** It can occur with any solid which has an appreciable vapor pressure at temperatures below the melting point. Examples include naphthalene (vp = 7.4 mm Hg at 80°C) and iodine (vp = 90.1 mm Hg at 114°C). Both of these molecular substances are often purified by sublimation (see Color Plate 6A, center of book). Carbon dioxide ("dry ice") is another molecular solid that sublimes readily. When dry ice warms up, it "disappears" by vaporizing directly, leaving no residue.

Ordinary ice has a high enough vapor pressure (4.6 mm Hg at 0°C) to undergo sublimation. On a cold winter day when the temperature is below freezing, ice and snow "disappear" by sublimation. The same process is responsible for the drying of clothes hung out on the line in the winter. Commercially, freeze-dried foods are prepared by reducing the pressure at temperatures well below 0°C. Under these conditions, ice crystals sublime, leaving a dry food which still retains its flavor. The food is later reconstituted by adding the proper amount of water.

Some foods freeze-dry better than others.

Water is unusual in that the solid, ice, is less dense than the liquid. This is explained by the peculiar crystal structure of ice, shown in Figure 12.10. The H_2O molecules are arranged in an open, hexagonal pattern. They are held to one another by hydrogen bonds. The large amount of "empty space" in this structure explains the low density of ice (0.917 g/cm^3 at 0°C). When ice melts, this open structure collapses to give a more closely packed pattern. The density of liquid water at 0°C, 0.99987 g/cm^3, is considerably higher than that of ice. The density of liquid water increases slightly as the temperature rises above 0°C. This suggests that some of the ice structure remained upon melting and is breaking down as the temperature rises. At 4°C, liquid water reaches its maximum density, 1.00000 g/cm^3. Above this temperature, the normal

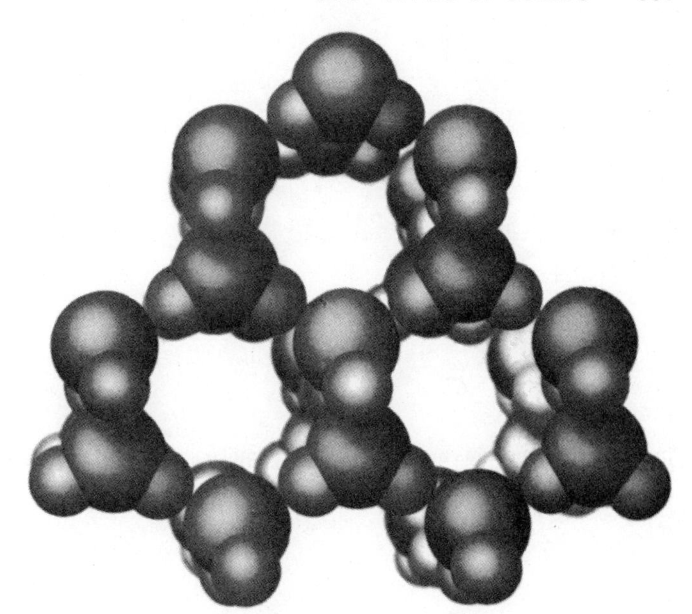

FIGURE 12.10

Crystal structure of ice. The structure of ice is relatively "open", so that on melting the volume actually decreases. Most solids increase in volume on melting.

expansion on heating overrides the collapse of the open structure. The density of water decreases above 4°C.

The fact that water has its greatest density at 4°C has profound effects upon our environment. In winter, at the bottom of lakes and rivers, there is a layer of water at this temperature, in which fish and other marine life can find food and oxygen. Ice collects at the surface. If ice were more dense than water, it would sink to the bottom. This would make it easier for lakes or rivers to freeze solidly, destroying marine life.

Ice on the pond acts as a heat insulator for the water below.

Metallic Solids

Of the 81 elements which can be clearly classified as metals, all but one (mercury) are solids at room temperature. The properties of metals, discussed in Chapter 7, depend upon the mobility of electrons. Thus metals:

—are excellent conductors of electricity.
—have low ionization energies.
—form positive ions when they react with nonmetals.

A model of metallic bonding that explains these properties is shown in Figure 12.11, p. 302. In this "electron sea" model, metal ions occupy lattice sites in the solid. They are surrounded by electrons. However, the electrons are spread out over the entire crystal rather than being tied down to a given ion. Hence the electrons are relatively free to move in an electrical field. This explains the high conductivity of metals. Moreover, only a small amount of energy is required to remove these electrons completely. This explains the low ionization energy of metals and their ability to form positive ions in reactions.

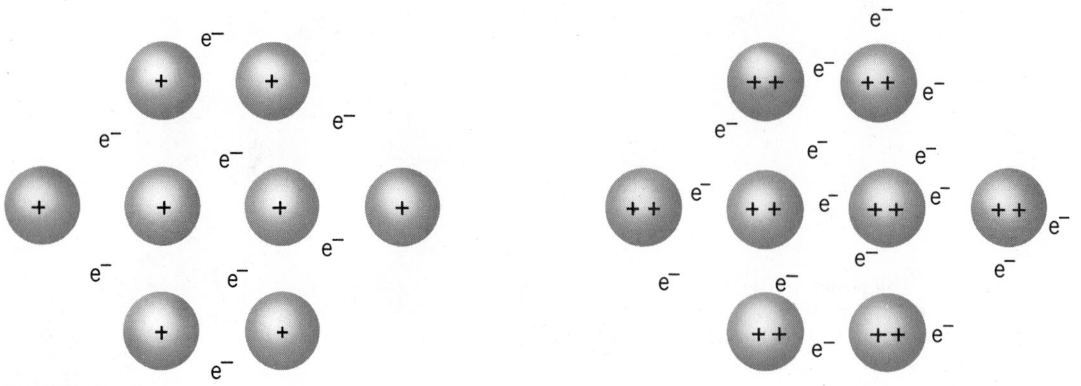

FIGURE 12.11

In a metal the electrons are relatively free to move about. This explains the high electrical conductivity of metals. The metal ions remain fixed in a definite crystal pattern. The drawing on the left might represent metallic sodium, that on the right magnesium.

Covalent Network Solids

We often think of covalent bonding as leading to the formation of small, discrete molecules. However, as mentioned in Chapter 11, this is not always the case. One possible result of covalent bonding is the formation of a network of bonds extending through an entire crystal. Solids which have such a structure, shown in Figure 12.12, can be either elements or compounds. Ordinarily, the atoms making up the network are those of nonmetals.

Substances which have structures of this type are invariably solids at room temperature. They are often referred to as *covalent network* solids, because of the network of covalent bonds. The word *macromolecular* is also used to describe such solids. In a sense, the entire crystal consists of one huge molecule (macro = large). Typically, covalent network solids have very high melting points, above 1500°C. The strong covalent bonds holding atoms together must be broken for melting to occur. In this respect, they resemble ionic solids. In both types of solids, the particles are held together by strong chemical bonds. These bonds are covalent in one case, ionic in the other.

Covalent networks usually involve one or more of the nonmetals in Group 4 of the Periodic Table. With elementary carbon, two different structures are known. These correspond to the two allotropic forms of carbon, diamond and graphite. As you can see from Figure 12.13 the two structures are quite different. In graphite, the atoms are covalently bonded in two dimensions. There are only weak Van der Waals forces between the layers. Diamond, in contrast, has a three-dimen-

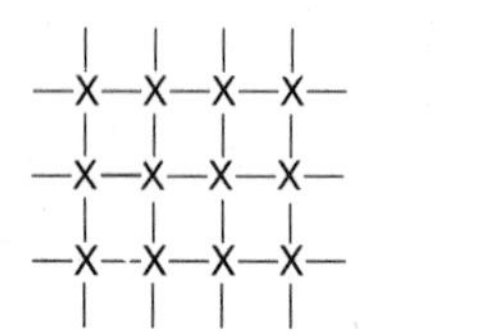

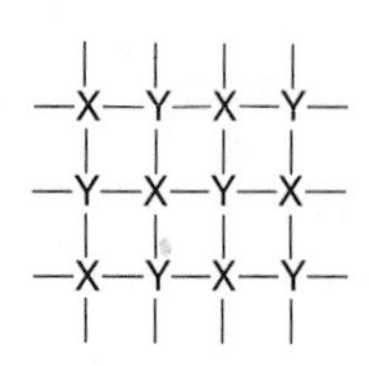

FIGURE 12.12

Schematic drawings of two macromolecular crystals. In a covalent network solid all of the atoms are linked together by covalent bonds, so the crystal is really one large molecule. In a molecular solid, the atoms are linked together into (small) molecules, but the forces between molecules are weak.

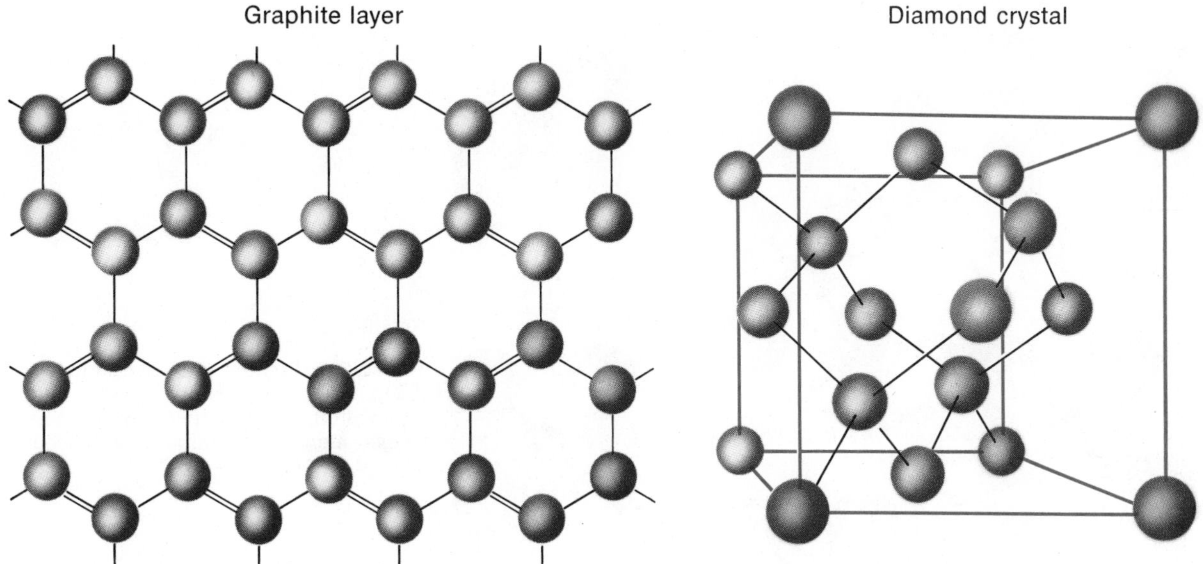

FIGURE 12.13

The structures of graphite and diamond. In graphite the atoms are linked together in hexagons, forming a layer containing many atoms. The forces linking the layers together are weak, but the layers themselves are macromolecular. Sliding of layers is responsible for the softness of graphite. In diamond each carbon atom is at the center of a tetrahedron, on each corner of which is another carbon atom. The structure of diamond is very strong and makes the crystal the hardest of all known substances.

sional network of bonds. Each carbon atom is at the center of a tetrahedron, covalently bonded to four other carbon atoms.

The differences and similarities between diamond and graphite can be explained in terms of their structures. Since both are macromolecular, they are very high melting (>3500°C). In both diamond and graphite, covalent bonds must be broken before the crystal can melt. However, it is relatively easy to separate layers from one another in graphite, since there are no bonds between atoms in adjacent layers. This accounts for the use of graphite in "lead" pencils. As you write, a few billion layers of carbon atoms rub off on the paper. In contrast, diamond, with its three-dimensional structure, is the hardest of all known substances. High-speed cutting tools are often diamond-tipped. Diamond dust is a very effective (but expensive) abrasive.

Silicon, both as the free element and in many of its compounds, forms covalent network solids. One of these, silicon carbide (SiC) has the same structure as diamond. Commonly known as carborundum, it is used as an abrasive, in grinding wheels as a powder, or as a coating on certain types of "sand paper". Its melting point, like that of all macromolecular solids, is extremely high, 2700°C.

Graphite contains a two-dimensional network.

One of the most common silicon compounds is the oxide, SiO_2. Sand is largely silicon dioxide, as are the minerals flint and obsidian (see Color Plate 6B, center of book). The most familiar form of pure crystalline SiO_2 is quartz. A crystal of quartz is essentially one huge molecule, with countless billions of atoms. The arrangement of silicon and oxygen atoms is shown in Figure 12.14. Notice that each silicon atom is

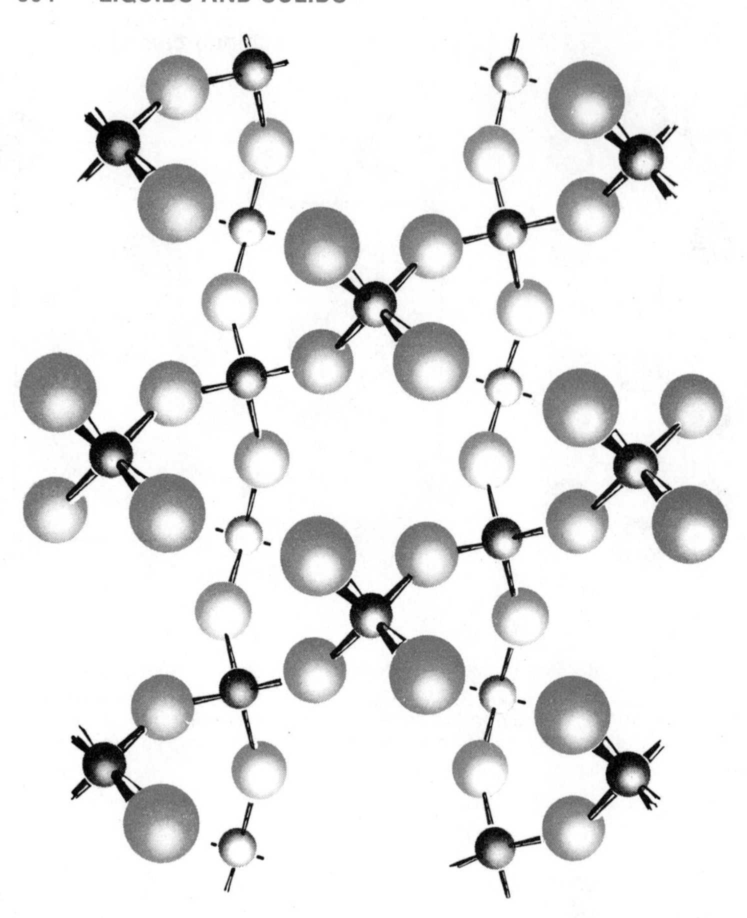

FIGURE 12.14

The crystal structure of quartz. In quartz (SiO_2), every silicon is at the center of a tetrahedron, on the corner of which is an oxygen atom. The solid is hard, but not as hard as diamond.

at the center of a tetrahedron, surrounded by four oxygen atoms. Each tetrahedron is bonded into the network through the oxygen atoms. Overall, there are two oxygen atoms for every silicon. This explains the simplest formula, SiO_2.

Silicon dioxide is one of the components of ordinary glass. A special type of glass, known as silica glass, is made from pure quartz. Unlike ordinary glass, quartz is transparent to ultraviolet light. Hence, silica glass is used in optical instruments that operate in the ultraviolet region. Several years ago, another interesting property of quartz was discovered. Quartz crystals of the proper thickness can be made to vibrate at a very precise frequency, using an electric current. Such crystals are now used in CB radio transmitters and in very precise timing instruments ("quartz watches").

12.6 SUMMARY

Liquids and solids differ from gases in that the particles are much closer together. In liquids, these particles can flow past each other. In solids, the particles are restricted to a particular position about which they can vibrate.

TABLE 12.4 STRUCTURES AND PROPERTIES OF DIFFERENT TYPES OF SOLIDS

	Ionic	Molecular	Metallic	Covalent Network
Structural Unit	+, − ions	molecules	+ ions, e^-	atoms
Forces between Units	ionic bonds	VdW forces, H bonds	metallic bond	covalent bond
Strength of Forces	strong	weak	intermediate	strong
Melting Point	high	low	variable	very high
Conductivity,				
solid	no	no	yes	no
liquid	yes	no	yes	no
Examples	NaCl KOH	I_2 H_2O	Na Mg	C SiO_2

Most substances which are liquids at room temperature or below consist of molecules. The molecules are attracted to each other by Van der Waals forces, which increase with the mass of the molecule. In a few liquids, notably water, there is an additional attractive force, the hydrogen bond. The strength of the attractive forces in a liquid are reflected by its:

—*vapor pressure* (pressure exerted by a vapor in equilibrium with its liquid).
—*normal boiling point* (boiling point at one atmosphere pressure).
—*heat of vaporization* (amount of heat required to vaporize a unit amount of liquid).

The properties of crystalline solids are dependent upon the type of particles making up the lattice and the strength of the forces between these particles. The four major types of solid structures are listed with their properties in Table 12.4.

EXAMPLE 12.6

A certain solid has a melting point of 90°C. The solid and its melt are nonconductors of electricity. Identify the type of solid present, using Table 12.4.

SOLUTION

The low melting point suggests that it is probably molecular, though it could be a low-melting metal. Since it does not conduct electricity, it cannot be a metal. We conclude that it is a molecular solid.

NEW TERMS

covalent network
dynamic equilibrium
electron sea
equilibrium
evaporation
hydrogen bond
macromolecular
relative humidity
sublimation
vaporization
vapor pressure
Van der Waals force

QUESTIONS

1. Compare the arrangement and mobility of students during a lecture period, a laboratory period, and at dismissal to the three states of matter.

2. In which of the states of matter, gas, liquid, or solid, are:

   a. the particles farthest apart? closest together?
   b. attractive forces strongest? weakest?
   c. particles able to move past one another?

3. Why must heat be absorbed to vaporize a liquid? to melt a solid?

4. Explain what is meant by:

   a. evaporation b. dynamic equilibrium
   c. equilibrium vapor pressure

5. A sample of liquid water weighing 10.0 g is placed in a stoppered 125 cm^3 flask at 25°C. It establishes a vapor pressure of 24 mm Hg. Does the pressure increase, decrease, or remain the same if:

   a. more water is added to the flask?
   b. the temperature drops to 10°C?
   c. the volume of the flask is increased to 250 cm^3?

6. What is meant by the statement that the relative humidity is 50%?

7. On a certain day, the temperature outside is 0°C and the relative humidity is 80%. When this air enters a house at 20°C, the relative humidity drops sharply. Why?

8. Distinguish between boiling and evaporation.

9. The normal boiling point of benzene is 80°C. Will the boiling point be greater than 80°C, less than 80°C, or equal to 80°C when the pressure is:

   a. 720 mm Hg b. 760 mm Hg c. 800 mm Hg

10. Why are pressure cookers particularly useful at high altitudes?

11. What experimental evidence leads us to believe that Van der Waals forces increase with molecular mass?

12. What is a hydrogen bond? What three elements form hydrogen bonds when bonded to hydrogen?

13. Which of the following ionic solids would you expect to decompose before melting?

    a. NaCl b. $KClO_3$ c. KF d. $MgCO_3$

14. Explain why ionic solids have higher melting points than molecular solids.

15. Cite two examples of processes, natural or commercial, that involve the sublimation of ice.

16. When a frozen water pipe thaws, it often leaks. Explain what happens in terms of the molecular structures of ice and liquid water.

17. What is meant by the electron sea model of metals? Draw a diagram, similar to Figure 12.11, for the structure of potassium metal.

18. Explain, in terms of the electron sea model, why metals have a

    a. high electrical conductivity b. low ionization energy

19. How does a covalent network solid differ in particle structure from

    a. a molecular solid? b. an ionic solid?

20. Why do covalent network solids have such high melting points?

21. Explain, in terms of structure, why diamond is so much harder than graphite.

22. Consider the four types of solids discussed in Section 12.5. Which type, in general, would you expect to:

    a. have the highest melting point?
    b. conduct electricity?

23. You are given a white solid. Describe some tests you might carry out to decide whether the solid is ionic, molecular, metallic, or a covalent network.

PROBLEMS

1. *Relative Humidity (Example 12.1)* The pressure of water vapor in the air is 6.5 mm Hg at 12°C. Calculate the relative humidity.

2. *Effect of Pressure on Boiling Point (Example 12.2)* Using Table 12.2, estimate the boiling point of water at a pressure of 300 mm Hg.

3. *Effect of Pressure on Boiling Point (Example 12.2)* Answer Problem 2, using Figure 12.4 instead of Table 12.2.

4. *Van der Waals Forces and Boiling Point (Example 12.3)* Arrange the following substances in order of increasing boiling point.

    a. C_4H_{10} b. C_3H_8 c. C_2H_6 d. CH_4

5. *Van der Waals Forces and Boiling Point (Example 12.3)* Which two of the following substances would you expect to be closest to each other in boiling point?

    a. H_2 b. N_2 c. CO d. CO_2

6. *Hydrogen Bonds (Example 12.4)* Which of the following molecules would you expect to form hydrogen bonds?

    a. H—O—O—H (hydrogen peroxide)
    b. H—N(—H)—C(=O)—N(—H)—H (urea)
    c. H—C(=O)—H (formaldehyde)
    d. H_3C—C(=O)—O—H (acetic acid)

7. *Decomposition of Solids Containing Polyatomic Ions (Example 12.5)* Write balanced equations for the reaction that occurs when each of the following compounds is heated.

   a. $Ni(OH)_2$ b. AgOH c. $NiCO_3$ d. Ag_2CO_3

8. *Types of Solids (Example 12.6)* A certain element is a solid which melts at 1000°C to give a conducting liquid. Would you expect it to be ionic, molecular, metallic, or covalent network?

9. *Types of Solids (Example 12.6)* A certain element melts above 2000°C to give a liquid which does not conduct electricity. Classify it as to type of solid.

* * * * *

10. For Cl_2, $\Delta H_{fus} = 23.0$ cal/g and $\Delta H_{vap} = 67.4$ cal/g. Calculate the heats of fusion and vaporization of Cl_2 in kcal/mol.

11. Using the data in Table 12.1, calculate ΔH for the vaporization of:

    a. 1.50 mol of Hg(l) b. 1.29 g of liquid benzene

12. Using Table 12.2, estimate the:

    a. boiling point of water when the pressure above it is 100 mm Hg.
    b. pressure above a sample of water boiling at 105°C.

13. On a certain day, the relative humidity is 42% and the temperature is 22°C. Using Table 12.2, calculate the pressure of water vapor in the air.

14. The dew point is defined as the temperature to which air must be cooled to make the relative humidity 100%. What is the dew point under the conditions described in Problem 13?

15. For each of the following pairs of substances, choose the one that you would expect to have the higher boiling point. Explain your reasoning.

    a. N_2 or N_2O b. N_2O or N_2O_4 c. CH_4 or H_2O

16. Which of the following would you expect to form hydrogen bonds?

    a. H—S—H b. H—O—Cl

17. Write balanced equations, analogous to Equation 12.3, for the decomposition of

    a. $NaClO_3$ b. $Ca(ClO_3)_2$

18. Classify each of the following solids as ionic, molecular, metallic, or covalent network.

    a. Cr b. I_2 c. C d. $BaCl_2$ e. SiO_2 f. Al g. H_2O
    h. $Al(NO_3)_3$

19. A certain solid is a nonconductor. However when it melts, at 600°C, it conducts an electric current. The solid is most likely:

    a. Cr b. LiCl c. $C_{10}H_8$ d. SiC

20. A certain solid when heated gives off water vapor. The solid could be:

    a. $BaCl_2 \cdot 2H_2O$ b. $MgCO_3$ c. $Mg(OH)_2$ d. SiO_2 e. CI_4

*21. For each of the following pairs of substances, select the one that you would expect to have the higher melting point. Explain your reasoning.

a. I_2 or MgO b. CO_2 or SiO_2 c. H_2O or Fe d. NaF or MgO

*22. Using Hess' Law and Table 12.1, estimate the heat of sublimation of mercury in kcal/mol.

*23. How many molecules are there in 1.00 cm^3 of liquid water at 0°C? in 1.00 cm^3 of a gas at 0°C and 1 atm?

Bubbles of CO_2 come out of a solution of soda water when the pressure drops. *Fredrik D. Bodin/Stock Boston.*

13 WATER SOLUTIONS

So far in our study of chemistry, we have been concerned for the most part with the structure, properties, and reactions of pure substances. In this chapter, we turn our attention to solutions. As pointed out earlier, a solution may be defined as a homogeneous mixture of two or more substances.

To understand what is meant by a "homogeneous mixture", consider what happens when sugar is added to water. The sugar dissolves as molecules less than 1 nm (10^{-9} m) in diameter. As the sugar molecules dissolve, they become surrounded by water molecules. Provided the mixture is thoroughly stirred, the sugar molecules are spaced evenly throughout it. Suppose now that we withdraw small samples from vari-

TABLE 13.1 TYPES OF SOLUTIONS

Physical States of Components	Examples
Gas–Liquid	Soda water (CO_2, H_2O)
Liquid–Liquid	Alcoholic beverages, antifreeze, gasoline
Solid–Liquid	Seawater, coffee, maple syrup
Gas–Solid	Hydrogen in platinum
Solid–Solid	Brass (Cu, Zn), wedding ring (Au, Cu)
Gas–Gas	Air, natural gas

ous parts of the mixture. We find that the concentration of sugar is the same in all samples. The mixture is homogeneous. We are dealing with a true solution.

The components of a solution can be gases, liquids, or solids. All possible combinations are known (Table 13.1). In this chapter, we will concentrate upon a particular type of solution. This is one in which a gas, liquid, or solid, referred to as the *solute*, is dissolved in liquid water, which we call the *solvent*. We emphasize water solutions because water is the solvent in nearly all the solutions you work with in the laboratory. Moreover, much of the rest of this text will deal with reactions taking place in water solution.

13.1 CONCENTRATION

When we deal with solutions, we must know not only the total amount but also the relative amounts of the different components. It would be useless to talk about "one liter of sugar solution" without saying how much of it is sugar and how much is water. To do this, we specify the *concentration*. The concentration of a solution tells us how much solute there is relative to solvent.

A variety of different units are used to express concentration. Here, we will consider only two such units. One is the **mass percent** of solute. It tells us what fraction of the total mass of solution is solute. The other unit we will work with is **molarity.** The molarity of a solution tells us how many moles of solute there are in a liter of solution.

Mass Percent

The mass percent of a component, X, in a solution is:

$$\text{mass \% of X} = \frac{\text{mass of X}}{\text{total mass solution}} \times 100 \qquad (13.1)$$

In a water solution in which the mass percent of NaCl is 2.0, there are:

2.0 g NaCl + 98.0 g water in 100.0 g of solution

or 20 g NaCl + 980 g water in 1000 g of solution

or 0.20 g NaCl + 9.80 g water in 10.00 g of solution

and so on.

In another solution, made by adding 10 g of sugar to 90 g of water:

$$\text{mass \% sugar} = \frac{10\text{ g}}{10\text{ g} + 90\text{ g}} \times 100 = 0.10(100) = 10$$

$$\text{mass \% water} = \frac{90\text{ g}}{10\text{ g} + 90\text{ g}} \times 100 = 0.90(100) = 90$$

It should be clear from these examples that the sum of the mass percents of all the components of a solution must be 100. That is, if we have several components, A, B, C, . . ., then:

$$\text{mass \% A} + \text{mass \% B} + \text{mass \% C} + \ldots = 100 \qquad (13.2)$$

The use of Equations 13.1 and 13.2 in practical calculations is further shown in Examples 13.1 and 13.2.

In this problem, we look at the various quantities we can calculate using Equation 13.1.

EXAMPLE 13.1

The water solution of hydrogen peroxide, H_2O_2, sold in drugstores as an antiseptic, contains 3.0% by mass of H_2O_2.

a. What is the mass percent of water in this solution?

b. How many grams of H_2O_2 are there in a one-liter bottle of this solution, which has a density of 1.01 g/cm³?

c. How many grams of solution are required to furnish 1.0 g of H_2O_2?

SOLUTION

a. Using Eqn. 13.2: mass % H_2O_2 + mass % water = 100.0

Solving: $$\text{mass \% water} = 100.0 - \text{mass \% } H_2O_2$$
$$= 100.0 - 3.0 = 97.0$$

b. From the information given, we can calculate the total mass of the solution. Since one liter is 1000 cm³:

$$\text{total mass solution} = 1000\text{ cm}^3 \times 1.01\text{ g/cm}^3 = 1010\text{ g}$$

The mass of H_2O_2 can then be found from its mass percent:

$$\text{mass } H_2O_2 = \text{total mass} \times \text{mass \% } H_2O_2$$

$$= 1010\text{ g} \times \frac{3.0}{100} = 30\text{ g}$$

c. Here, we know both the mass % of H_2O_2 (3.0) and the mass of H_2O_2 (1.0 g). We need to find the total mass of solution. Substituting in Eqn. 13.1:

$$3.0 = \frac{1.0 \text{ g}}{\text{total mass solution}} \times 100$$

Solving: $3.0 \times \text{total mass solution} = 1.0 \text{ g} \times 100$

$$\text{total mass solution} = 1.0 \text{ g} \times \frac{100}{3.0} = 33 \text{ g}$$

EXAMPLE 13.2

For a certain laboratory experiment, you are asked to prepare 480 g of a water solution in which the mass percent of potassium chromate, K_2CrO_4, is 6.2.

a. How many grams of K_2CrO_4 and water should you use to prepare this solution?

b. How many grams of K_2CrO_4 will there be in 150 g of this solution?

SOLUTION

a. Using Equation 13.1:

$$\text{mass } K_2CrO_4 = \frac{\text{mass \% } K_2CrO_4}{100} \times \text{total mass solution}$$

$$= \frac{6.2}{100}(480 \text{ g}) = 0.062\,(480 \text{ g}) = 30 \text{ g}$$

$$\text{mass water} = \text{total mass solution} - \text{mass } K_2CrO_4$$

$$= 480 \text{ g} - 30 \text{ g} = 450 \text{ g}$$

b. $\text{mass } K_2CrO_4 = \frac{\text{mass \% } K_2CrO_4}{100} \times \text{total mass solution}$

$$= \frac{6.2}{100} \times 150 \text{ g} = 9.3 \text{ g}$$

Molarity

Concentrations of reagents in the chemistry laboratory are most often given in terms of molarity. Molarity (abbreviation M) is defined as:

$$\text{Molarity (M)} = \frac{\text{number of moles of solute}}{\text{volume of solution in liters}} \qquad (13.3)$$

The bottle is labelled 1.00 M NaCl.

Thus a 1.00 M solution of NaCl contains:

1.00 mol (58.5 g) of NaCl in one liter (1.00 ℓ) of solution
or 10.0 mol (585 g) of NaCl in ten liters (10.0 ℓ) of solution
or 0.100 mol (5.85 g) of NaCl in one tenth of a liter (0.1000 ℓ) of solution and so on.

Equation 13.3 can be used to calculate the molarity of a solution, knowing the volume and the number of moles (or the number of grams) of solute (Example 13.3).

EXAMPLE 13.3

Calculate the molarity of a solution prepared by dissolving:

a. 1.20 mol of NaOH in enough water to give 0.650 ℓ of solution.

b. 50.0 g of NaOH to give 1.60 ℓ of solution.

SOLUTION

a. Substituting directly in Eqn. 13.3:

$$\text{Molarity NaOH} = \frac{1.20\ \text{mol}}{0.650\ \ell} = 1.85\ \text{M}$$

b. One mole of NaOH weighs: 23.0 g + 16.0 g + 1.0 g = 40.0 g
In 50.0 g of NaOH:

$$\text{moles NaOH} = 50.0\ \text{g} \times \frac{1\ \text{mol}}{40.0\ \text{g}} = 1.25\ \text{mol}$$

Substituting in Eqn. 13.3:

$$\text{Molarity NaOH} = \frac{1.25\ \text{mol}}{1.60\ \ell} = 0.781\ \text{M}$$

There are many uses for Equation 13.3. We have used it to find the molarity of a solute. It can also be used to calculate the number of moles of solute in a given volume of a solution of known molarity. To do this, we solve Equation 13.3 for the number of moles of solute:

$$\text{no. moles solute} = \text{Molarity} \times (\text{volume soln. in liters}) \quad (13.4)$$

We can also obtain the volume of a solution that contains a certain number of moles of a solute of known molarity. Here the relation is:

$$\text{volume of solution in liters} = \frac{\text{number of moles of solute}}{\text{Molarity}} \quad (13.5)$$

The use of these equations is shown in Example 13.4.

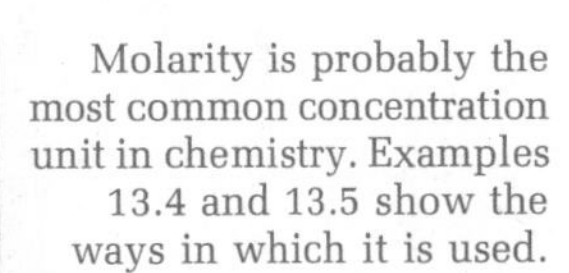

Molarity is probably the most common concentration unit in chemistry. Examples 13.4 and 13.5 show the ways in which it is used.

EXAMPLE 13.4

You are given a large amount of 6.0 M NaOH solution.

a. If you measure out 220 cm^3 of this solution, how many moles of NaOH will it contain? how many grams of NaOH?

b. To obtain 0.24 mol of NaOH, what volume of solution should you take?

SOLUTION

a. Applying Equation 13.4:

$$\text{moles NaOH} = \text{Molarity NaOH} \times (\text{volume of solution in liters})$$

Noting that 1000 cm^3 is one liter, we have:

$$\text{moles NaOH} = 6.0\ \frac{\text{mol}}{\ell} \times \frac{220}{1000}\ \ell = 1.3\ \text{mol}$$

Since one mole of NaOH weighs 40.0 g:

$$\text{grams NaOH} = 1.3\ \text{mol} \times \frac{40.0\ \text{g}}{1\ \text{mol}} = 52\ \text{g}$$

b. Applying Equation 13.5:

$$\text{volume solution in liters} = \frac{\text{no. moles NaOH}}{\text{Molarity NaOH}}$$

$$= \frac{0.24\ \text{mol}}{6.0\ \text{mol}/\ell} = 0.040\ \ell$$

Thus we need 0.040 ℓ or 40 cm^3 of solution.

It is relatively easy to make up a solution to a desired molarity. You start by weighing out the calculated amount of solute. You then dissolve this in enough solvent to give the desired volume of solution. If high accuracy is not required, a graduated cylinder can be used to add the solvent. However, if you want to obtain a particular molarity within 0.1%, the approach shown in Figure 13.1, p. 316 is used. The weighed solute is transferred to a *volumetric flask*. This flask is designed to contain a certain specified volume when filled to a narrow mark on the thin stem. Solvent is added with constant swirling until all of the solute has gone into solution. Then, more solvent is added carefully, with further swirling or shaking, until the liquid level comes exactly up to the mark.

EXAMPLE 13.5

You are asked to prepare 500 cm^3 of a 2.00 molar solution of sodium nitrate, $NaNO_3$, in water. Describe in some detail how you would do this.

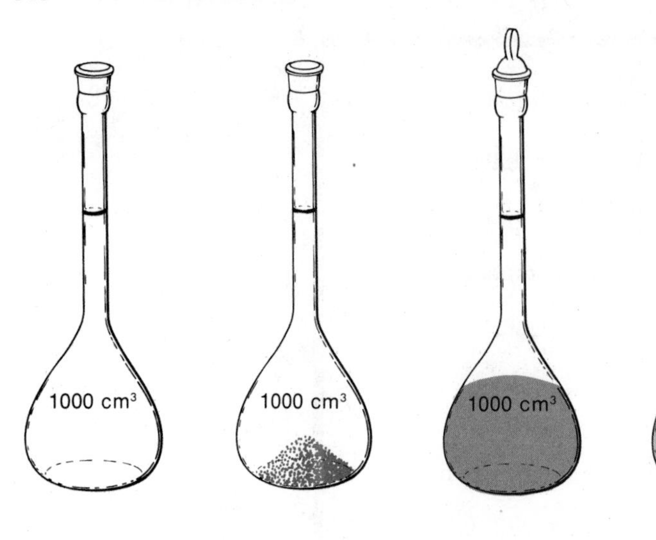

1.
Take a volumetric flask.

2.
Add carefully the weighed amount of solid.

3.
Add some water, shake, and dissolve solid.

4.
Fill flask to one-liter mark and shake until homogeneous solution is obtained.

FIGURE 13.1

Preparation of a solution of known concentration. A measured mass of solute is dissolved in enough solvent to give a measured volume of solution.

SOLUTION

You need to calculate the mass in grams of $NaNO_3$ required. The number of moles of $NaNO_3$ is readily obtained from Equation 13.4:

$$\text{moles } NaNO_3 = \text{molarity } NaNO_3 \times \text{volume of solution in liters}$$

$$= 2.00\,\frac{\text{mol}}{\ell} \times 0.500\ \ell = 1.00\ \text{mol}$$

To obtain the number of grams of $NaNO_3$, we note that:

$$1\ \text{mol } NaNO_3 \text{ weighs: } 23.0\ \text{g} + 14.0\ \text{g} + 48.0\ \text{g} = 85.0\ \text{g}$$

Why is it necessary to mix a solution like this very thoroughly?

You should then weigh out 85.0 g of sodium nitrate. Transfer this to a 500 cm³ volumetric flask and add enough water, with frequent stirring, to bring the solution level up to the mark.

13.2 TYPES OF SOLUTES

There are many ways in which we could classify solutes in water solution. We could distinguish between gaseous, liquid, and solid solutes, as we did in Table 13.1. For most purposes, however, it is more useful to classify solutes in terms of their particle structure rather than their physical state.

Water is sometimes referred to as the "universal solvent". In practice, its solvent ability is really quite limited. Of the four types of substances discussed in Section 12.5 of Chapter 12, no metals or macromolecular species are soluble in water. All of the solutes that dissolve in water are either ionic or molecular.

If we had a truly universal solvent, it would be hard to find a container to store it in.

Ionic Solutes

Many, though by no means all, ionic solids are "soluble" in water. That is, they dissolve to give relatively concentrated solutions, let us say 0.1 M or greater. Lithium chlorate, $LiClO_3$, is an extreme case. Its solubility in water at room temperature is 35 mol/ℓ. Such familiar ionic compounds as sodium chloride, sodium hydroxide, and potassium nitrate dissolve readily in water. The solubilities of these compounds at 25°C are 6.2 M, 20 M, and 4.0 M respectively. On the other hand, some ionic compounds have very low water solubilities. Calcium carbonate, $CaCO_3$, and aluminum oxide, Al_2O_3, are essentially insoluble in water.

To understand why ionic solutes tend to dissolve in water, let us consider what happens when sodium chloride is added to water (Figure 13.2). Since water molecules are polar, they will tend to line up in a particular direction around charged ions. Near the Na^+ ion, the H_2O molecule is oriented so that the electronegative oxygen atom is as close as possible to the + ion. The orientation is quite different near the Cl^- ion. Here, the hydrogen atoms, which carry a slight positive charge, are directed toward the − ion.

The situation shown at the left of Figure 13.2 leads to an electrical

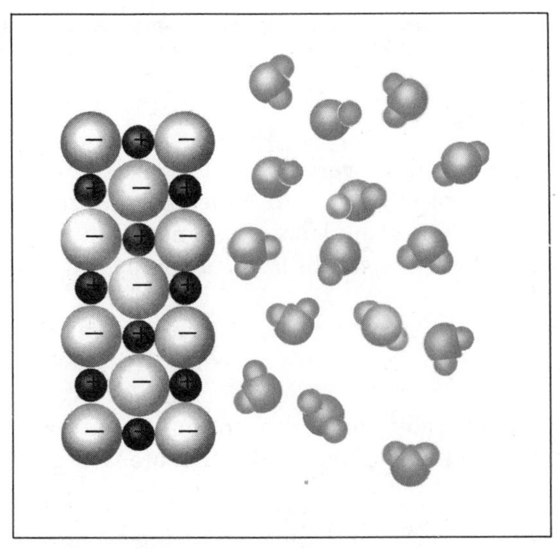

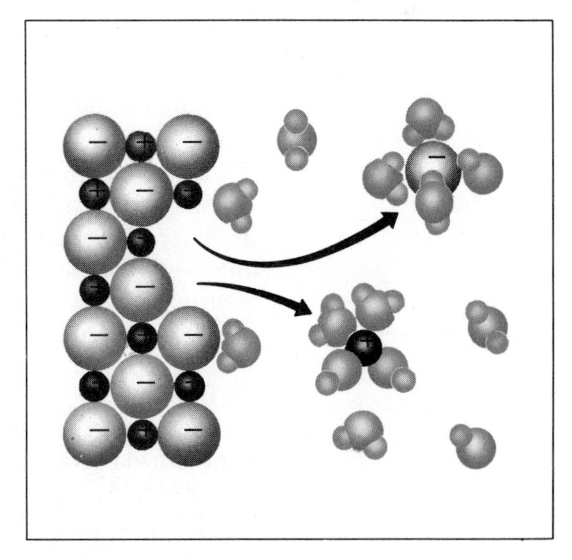

FIGURE 13.2

Dissolving an ionic solute in water. The attraction between the oxygen atoms in the water molecules and the Na^+ ions in the crystal is sufficient to cause the Na^+ ions to dissolve. At the same time Cl^- ions, attracted to the H atoms in the water molecules, also dissolve.

attractive force between H_2O molecules and ions at the surface of the crystal. This force is about as strong as that between ions of opposite charge. As a result, ions are able to move out of the crystal into solution. In water solution, the ions are *hydrated*, as shown at the right of Figure 13.2.

The process we have just described can be represented by a chemical equation. The reactant is the solid ionic compound. The products are ions in water solution. For sodium chloride dissolving in water:

$$NaCl(s) \rightarrow Na^+(aq) + Cl^-(aq)$$

All ionic compounds that dissolve in water break down into ions when they dissolve.

Note that the charges of the ions are shown when they are separated from one another in solution, but not in the solid state. The symbol (aq) is used to emphasize that the ions are hydrated or "aquated" by water molecules.

EXAMPLE 13.6

Write equations to represent the processes by which KNO_3 and $MgCl_2$ dissolve in water.

SOLUTION

Like all ionic compounds, KNO_3 and $MgCl_2$ are solids. The ions present are K^+ and NO_3^- in KNO_3; Mg^{2+} and Cl^- in $MgCl_2$. The balanced equations are:

$$KNO_3(s) \rightarrow K^+(aq) + NO_3^-(aq)$$

$$MgCl_2(s) \rightarrow Mg^{2+}(aq) + 2\ Cl^-(aq)$$

Molecular Solutes

Most molecular substances have very limited water solubilities. Gasoline, which is a mixture of hydrocarbon molecules, is insoluble in water. Elements which consist of molecules ordinarily dissolve to only a small extent. For example, the solubility of N_2 at 25°C and 1 atm is only 6.4×10^{-4} mol/ℓ.

Which compound is more soluble in water?

```
    H           H
    |           |
H—C—H  or  H—C—O—H
    |           |
    H           H
```

It is easy to see why molecular substances should be generally insoluble in water. In order to dissolve, a solute molecule must break into the hydrogen-bonded structure of water. Unless there is a strong attractive force between solute and water molecules, solubility will be low. In practice, there are only two kinds of molecules that are very soluble in water.

1. *Molecules which form hydrogen bonds with water.* This group includes species which contain an N-H or O-H bond, such as NH_3 (28 mol/ℓ at 25°C and 1 atm) or H_2O_2, which is soluble in water in all proportions.

2. *Molecules which form ions on addition to water.* Most of the compounds in this category are molecular acids. Upon dissolving in water, they ionize, forming H^+ ions and negative ions. Examples include three familiar laboratory acids, HCl, HNO_3, and H_2SO_4. In the pure state, these compounds are molecular. In water solution, they exist as ions. The equations for the processes by which they "dissolve" in water are:

$$HCl(g) \rightarrow H^+(aq) + Cl^-(aq)$$

$$HNO_3(l) \rightarrow H^+(aq) + NO_3^-(aq)$$

$$H_2SO_4(l) \rightarrow H^+(aq) + HSO_4^-(aq)$$

In a very real sense, these compounds react with water, in that the H^+ ion is not free, but strongly hydrated.

We will have more to say about these processes when we discuss acids and bases in Chapter 19.

13.3 SOLUTION EQUILIBRIUM

Suppose we add a large amount of NaCl, perhaps 500 g, to a liter of water at room temperature (Figure 13.3). As we stir the mixture, the sodium chloride starts to go into solution. The amount of undissolved solid decreases, at first rapidly then more slowly. After a few minutes, the amount of solid in the flask remains constant. No matter how much we stir the mixture, no more NaCl goes into solution.

We can analyze this situation in terms of two competing processes. When we first add the sodium chloride, the movement of ions is in one direction. They move from the crystal lattice into the solution:

$$NaCl(s) \rightarrow Na^+(aq) + Cl^-(aq)$$

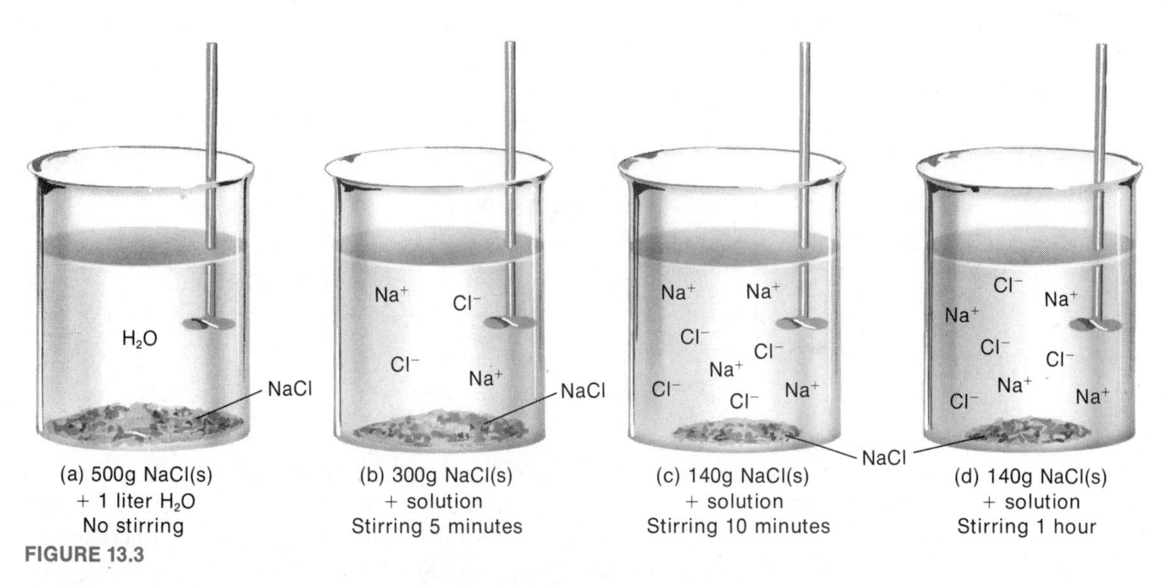

FIGURE 13.3

Formation of a saturated solution of NaCl. When NaCl is added to water, most of the solid goes to the bottom. On stirring, the solid slowly dissolves. After a period of time, the amount of solid in solution becomes constant, and the solution is said to be saturated.

However, as the concentration of ions in solution builds up, the reverse process takes place. Ions return to the lattice from the solution:

$$Na^+(aq) + Cl^-(aq) \rightarrow NaCl(s)$$

Eventually, a state of *dynamic equilibrium* is reached. The rates of the two processes become equal. We represent this condition by the equation:

$$NaCl(s) \rightleftharpoons Na^+(aq) + Cl^-(aq)$$

where the double arrow is used to indicate that both processes, solution and crystallization, are taking place at the same rate.

The equilibrium between NaCl and water is similar to that described in Chapter 12 for water and its vapor. You will recall that, at a given temperature, water establishes an equilibrium pressure of vapor above it. This pressure is independent of the amount of water present, provided there is some left at equilibrium. It is also independent of the volume of the vapor. In the same way, at a given temperature, NaCl establishes an equilibrium concentration in the solution around it. This concentration is independent of the amount of sodium chloride present, provided there is some solid left at equilibrium. It is also independent of the volume of the solution.

The process we have described with sodium chloride is typical of all solids dissolving in water. Potassium nitrate or iodine would behave in much the same way. The concentration of solute in the end would be quite different. However, in each case, an equilibrium is established. Once the rates of solution and crystallization become equal, no more solid dissolves. We describe these equilibria by the equations:

$$KNO_3(s) \rightleftharpoons K^+(aq) + NO_3^-(aq)$$

$$I_2(s) \rightleftharpoons I_2(aq)$$

Here, as always, the "position of the equilibrium", that is the concentration of solute at equilibrium, is independent of:

Can you see what is meant here? Try saying it in your own words.

—the amount of solute present
—the volume of solution

A gas such as carbon dioxide dissolves in water in much the same way as a solid. Once equilibrium is reached, the concentration of gas remains constant.

$$CO_2(g) \rightleftharpoons CO_2(aq)$$

With a gas, equilibrium can be established quickly by bubbling it through water. The solution process can be followed by watching the bubbles. At first, as the gas goes into solution, the bubbles decrease in size or disappear completely. Later, after equilibrium is established, they move through the water without changing size.

When a liquid such as bromine is shaken with water, two layers

50g $Br_2(l)$
+ 1 liter H_2O
Before stirring

14g $Br_2(l)$
+ saturated solution
After stirring

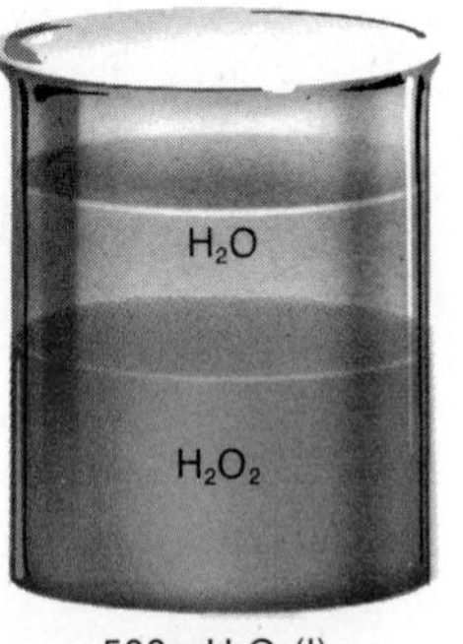

500g $H_2O_2(l)$
+ 500g H_2O
Before stirring

Solution
after stirring

FIGURE 13.4

The solubility of Br_2 in water is limited. One can dissolve up to 36 g of Br_2 in a liter of water at 25°C, at which point the solution is saturated with Br_2. If one adds H_2O_2 to water, no separation of the liquids ever occurs, no matter how much H_2O_2 is added. The solubility of H_2O_2 in water is infinite.

form (Figure 13.4). The upper layer contains water, saturated with a small amount of bromine.

$$Br_2(l) \rightleftharpoons Br_2(aq)$$

Certain liquids behave quite differently from bromine. If we add hydrogen peroxide to water, only one layer is obtained. This is true regardless of how much H_2O_2 we add. All of it dissolves. We never reach equilibrium between solute and solvent. We say that hydrogen peroxide is infinitely soluble in water. Some organic liquids, as we will see in Chapter 15, behave this way. Others, like bromine, have a limited solubility in water. They reach equilibrium, forming a saturated solution. After that point, no more liquid dissolves.

Saturated, Unsaturated, and Supersaturated Solutions

A solution that is saturated has had all it can take.

A solution in equilibrium with undissolved solute is said to be *saturated*. The solutions of NaCl and Br_2 shown in Figures 13.3 and 13.4 fit this description. With NaCl at 25°C, the concentration of the saturated solution is 6.2 mol/ℓ. That is, one liter of the saturated solution contains 6.2 mol of dissolved NaCl. When we refer to the solubility of a substance, we mean its concentration in the saturated solution. The solubility of NaCl in water at 25°C is 6.2 mol/ℓ.

An unsaturated solution can take some more.

Solutions which contain less solute than is present at saturation are said to be *unsaturated*. Any water solution of NaCl at 25°C which contains less than 6.2 mol/ℓ fits this description. In an unsaturated solution, equilibrium has not been reached. If more solute is added, it dissolves to approach a saturated solution.

A supersaturated solution will give some back if it gets the chance.

Under certain conditions, it is possible to have a *supersaturated* solution. Such a solution contains more solute than is present at equilibrium. A supersaturated solution of NaCl in water at 25°C would contain

FIGURE 13.5

Unsaturated, saturated, and supersaturated solutions. An unsaturated solution can dissolve more solute. A saturated solution is in equilibrium with solute and cannot dissolve any more. A supersaturated solution contains more solute than it can hold at equilibrium. It is unstable; if a tiny crystal of solute is added, enough solute will crystallize to bring its concentration to the equilibrium value.

more than 6.2 mol/ℓ of NaCl. A supersaturated solution, like an unsaturated solution, is not at equilibrium. It is not stable in the presence of solute. If a crystal of NaCl is added to a supersaturated solution, the excess solute quickly crystallizes. The solution that remains is saturated.

Supersaturated solutions are most readily formed with solids. An interesting example occurs in making fudge. Directions call for adding 2 cups (about 300 g) of sugar to 1 cup (about 200 g) of milk. At the boiling point, the sugar dissolves to form a viscous solution. Upon cooling, the sugar does not come out of solution, even though its solubility is exceeded. In other words, a supersaturated solution forms. Eventually, fudge crystallizes, restoring equilibrium.

Effect of Temperature

You will recall from Chapter 12 that the equilibrium vapor pressure of a liquid varies with temperature. In a similar way, the equilibrium concentration of a solute changes with temperature. With solids, solubility almost always increases as temperature rises (Figure 13.6). This makes it possible to purify a solid by recrystallization (Chapter 1). When a hot, saturated solution is cooled, solid crystallizes. This happens because less solid can stay in solution at the lower temperature.

With gases, we find that solubility in water decreases as temperature rises (Table 13.2). Air is less soluble in hot water. This explains why gas bubbles form when a beaker of water is heated. The same effect makes carbonated beverages taste "flat" when they warm to room temperature. Here, it is carbon dioxide that comes out of solution as the temperature rises.

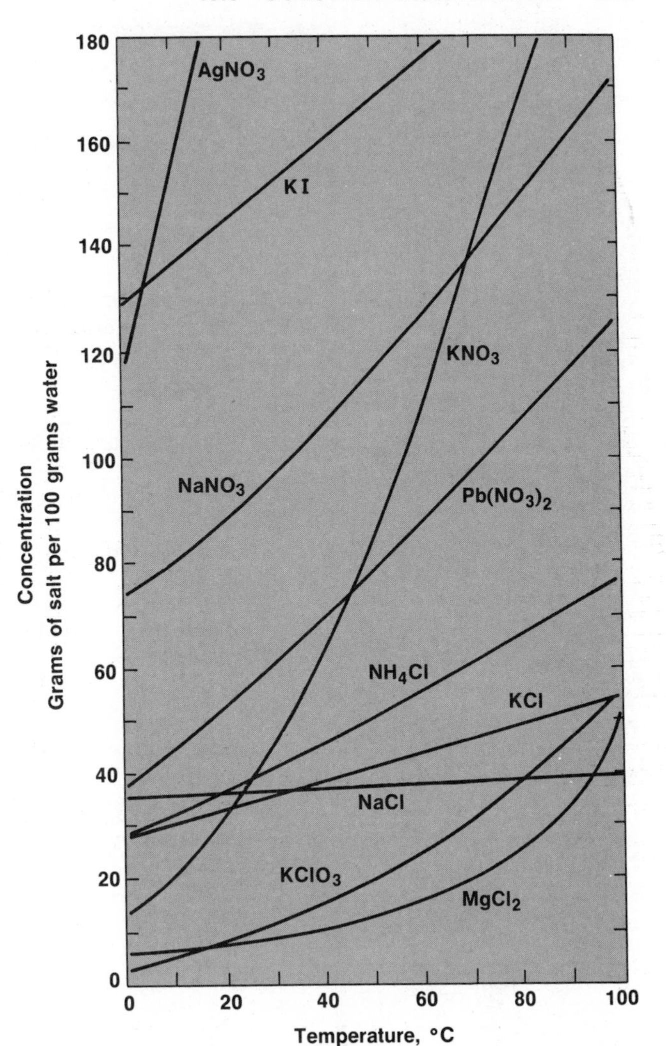

FIGURE 13.6

The water solubilities of some common substances at different temperatures. In general the solubility increases as the temperature goes up. Cooling a hot saturated solution of a salt like KNO_3 will cause solid KNO_3 to crystallize because its solubility is much lower at low temperatures than at high.

TABLE 13.2 SOLUBILITY OF GASES IN WATER AT 1 ATM AND VARIOUS TEMPERATURES

Gas	Solubility (mol/ℓ)		
	0°C	25°C	50°C
N_2	0.00105	0.00064	0.00049
O_2	0.00218	0.00126	0.00093
Ar	0.00236	0.00139	0.00101
CO_2	0.0765	0.0338	0.0195
Cl_2	0.206	0.0888	0.0536

EXAMPLE 13.7

Consider the solubility curve for KNO_3 in Figure 13.6.

a. What are the solubilities of KNO_3 at 85°C and at 0°C?

b. How much water is required to dissolve a 50 g sample of KNO_3 at 85°C?

c. If the solution formed in (b) is cooled to 0°C, how much KNO_3 stays in solution? How much crystallizes out?

SOLUTION

a. From the curve it appears that:

$$\text{solubility at } 85°\text{C} = 180 \text{ g } KNO_3/100 \text{ g water}$$

$$\text{solubility at } 0°\text{C} = 15 \text{ g } KNO_3/100 \text{ g water}$$

b. We can treat the solubility as a conversion factor to find out how much water is required to dissolve 50 g of KNO_3. That is,

$$\text{at } 85°\text{C, } 180 \text{ g } KNO_3 \simeq 100 \text{ g water}$$

$$\text{grams water required} = 50 \text{ g } KNO_3 \times \frac{100 \text{ g water}}{180 \text{ g } KNO_3} = 28 \text{ g water}$$

c. We are asked in effect to calculate the number of grams of KNO_3 dissolved by 28 g of water. At 0°C, the conversion factor expression is:

$$15 \text{ g } KNO_3 \simeq 100 \text{ g water}$$

$$\text{grams } KNO_3 \text{ in solution} = 28 \text{ g water} \times \frac{15 \text{ g } KNO_3}{100 \text{ g water}} = 4.2 \text{ g } KNO_3$$

The rest of the KNO_3 crystallizes:

$$\text{grams } KNO_3 \text{ out of solution} = 50 \text{ g} - 4.2 \text{ g} = 46 \text{ g}$$

Notice that most of the potassium nitrate (over 90%) is recovered. This happens because KNO_3 is so much more soluble at the higher temperature.

The effect of temperature upon gas solubility leads to a problem known as "thermal pollution". A power plant that discharges hot water into a lake or river can be harmful to marine life. As the temperature rises, the concentration of dissolved oxygen decreases. If the effect is large enough, fish can suffocate from lack of oxygen. Brook trout are particularly sensitive to temperature. They cannot live at water temperatures much above 25°C. Rainbow and particularly brown trout are less affected by this and other types of water pollution.

13.4 PHYSICAL PROPERTIES OF WATER SOLUTIONS

We might expect the properties of water solutions to depend both upon the nature of the solute and its concentration. This is true for many physical properties. One of these is color. Solutions of sodium chloride in water are colorless. On the other hand, water solutions of potassium permanganate, $KMnO_4$, are colored purple. The color intensifies as the concentration of $KMnO_4$ increases (see Color Plate 6C, center of book).

In this section we will look at several of the physical properties of water solutions. One of these, electrical conductivity, depends upon the type of solute as well as concentration. In that sense, it resembles color. The other three properties considered here are vapor pressure, boiling point, and freezing point. These are often referred to as **colligative** properties. Unlike color and conductivity, they are virtually independent of the type of solute. In essence, they depend only upon the concentration of solute particles.

Electrical Conductivity

Chemically pure water does not conduct electricity. However, when certain solutes such as NaCl are added to water, the solution conducts. This can readily be demonstrated with the simple device shown in Figure 13.7. When sodium chloride is added to the water, the bulb

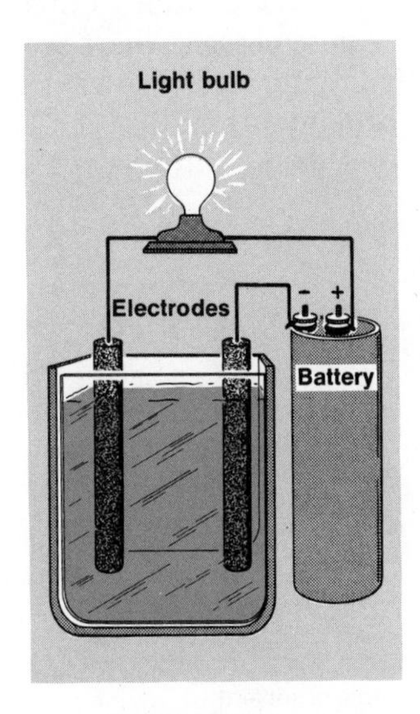

a. Solution of table salt
(an electrolytic solution)

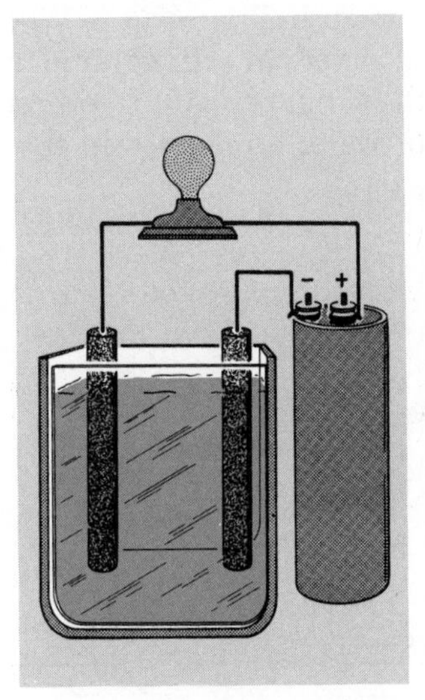

b. Solution of table sugar
(a nonelectrolytic solution)

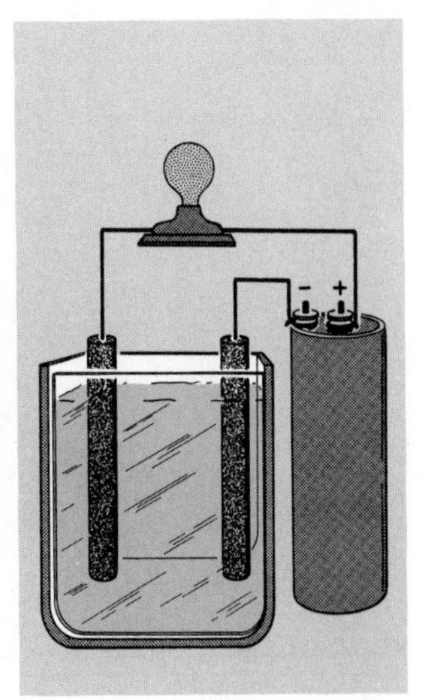

c. Pure water
(a nonelectrolyte)

FIGURE 13.7

An apparatus for testing electrical conductivity. For an electric current to flow, the solution must contain ions, which serve as carriers of electric charge. Pure water contains very few ions, so is not a conductor. If NaCl is dissolved in water, it forms ions. These conduct the current and the light bulb glows. Dissolving sugar in water produces only sugar molecules in the solution, so it does not become a conductor.

lights up. In contrast, a sugar solution does not conduct electricity. The bulb does not light when sugar is added to the water. This indicates that no current is passing through the solution.

Solutes which dissolve as molecules do not conduct. This is the case with sugar, ethyl alcohol, and many other organic compounds. The neutral molecules show no tendency to move in an electrical field. Solutes of this type are referred to as **non-electrolytes.**

Electrolytes form ions when they dissolve in water.

Solutes which exist as ions in water give solutions that conduct electricity. In an electrical field, the positive ions move through the water in one direction. The negative ions move in the opposite direction. This flow of ions amounts to an electrical current. Solutes which behave in this way are called **electrolytes.** They include:

—all ionic solids (NaCl, KNO_3, $CuSO_4$, . . .)
—the few molecular solutes that form ions upon addition to water (HCl, HNO_3, H_2SO_4, . . .).

Vapor Pressure Lowering

The addition of a solute to water always lowers its vapor pressure. We find, for example, that the pressure of water vapor over a solution of sugar or sulfuric acid at 25°C is less than that of pure water (23.8 mm Hg). In concentrated solution, the effect can be quite large. Over 6M H_2SO_4, the pressure of water vapor is about 11.4 mm Hg, less than half that of pure water.

From a molecular standpoint, it is easy to see why the addition of a solute to water should lower its vapor pressure. Consider the situation shown in Figure 13.8. With a solute present, there are fewer molecules of water per unit volume in the liquid. Hence, fewer water molecules in the vapor phase are required to maintain equilibrium. The concentration of water molecules in the vapor drops and so does the pressure they exert.

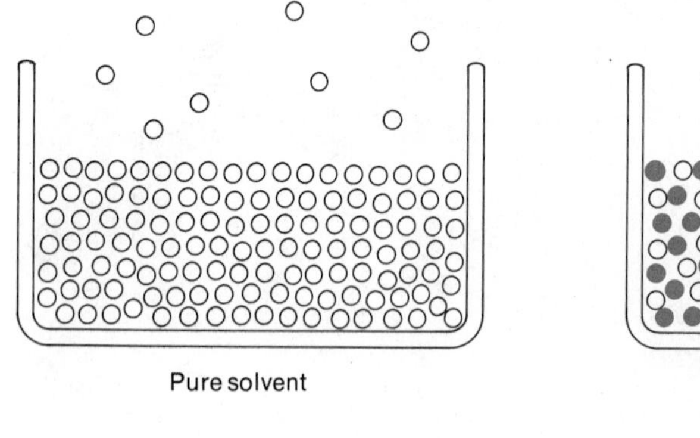

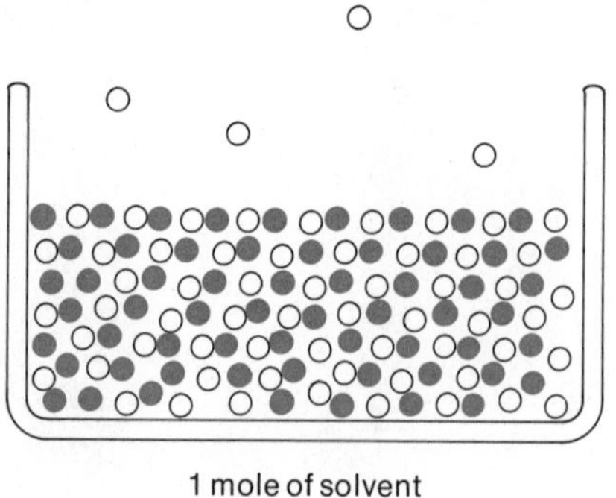

FIGURE 13.8

The effect of a solute on the vapor pressure of its solvent. In the presence of solute the concentration of solvent goes down, both in the bulk liquid and on the surface. This decreases the number of water molecules per unit area on the surface, so fewer molecules have sufficient energy to escape. The vapor pressure of the solvent is always lowered by addition of a solute.

TABLE 13.3 VAPOR PRESSURE LOWERING IN WATER SOLUTION AT 25°C

Concentration (mol/ℓ)	Glucose (mm Hg)	Sucrose (mm Hg)
0.10	0.04	0.04
0.20	0.09	0.09
0.30	0.13	0.14
0.40	0.18	0.19
0.50	0.23	0.24

Vapor pressure lowering is a colligative property. That is, it depends primarily upon the concentration of solute particles and not upon the type of particle. This is shown by the data in Table 13.3, which compares the vapor pressure lowering of two sugars. One of these is glucose, $C_6H_{12}O_6$; the other is sucrose (cane sugar), $C_{12}H_{22}O_{11}$. Notice that:

If we mixed a mole of H_2O with a mole of H_2O_2, the vapor pressure of water would be about one-half the value for pure water.

1. The vapor pressure lowering is directly proportional to concentration. For example, with both solutes the vapor pressure lowering in 0.40 M solution is about twice that in 0.20 M solution.
2. At a given concentration, the vapor pressure lowering is almost exactly the same for the two solutes.

Deliquescence Very soluble solids behave in an interesting way if exposed to moist air. Calcium chloride is one such solid. Suppose we take a few crystals of $CaCl_2$ and put them on a watch glass. Very quickly we notice that the crystals become wet (Figure 13.9). If the air is very moist, all of the solid dissolves to form a water solution. This process, by which a solid picks up water from the air, is called *deliquescence*.

Deliquescence is readily explained in terms of vapor pressure lowering. A thin layer of water from the air condenses on the solid. In this way, a saturated solution forms

FIGURE 13.9

Deliquescence. In the photograph on the left are some dry $CaCl_2$ crystals shortly after being poured on to the watch glass. After about an hour in moist air (right photo), some of the crystals have dissolved in the water absorbed from the air (note drops of liquid). The rest of the solid has caked and the surface has become coated with wet crystals.

at the surface. If the vapor pressure of water in this solution is less than the pressure of water vapor in the air, condensation continues. Eventually, all of the solid dissolves. This happens with $CaCl_2$ and $MgCl_2$, both of which are very soluble in water. The process stops when the solution becomes so dilute that its vapor pressure is equal to that of the water in the air.

In our daily lives, deliquescence is most often observed in the kitchen. The kitchen salt shaker clogs in humid weather because salt contains small amounts of $MgCl_2$. This makes the salt get sticky and refuse to pour. In another case, the stickiness is helpful. Calcium chloride, $CaCl_2$, is spread on a dusty dirt road in summer. This compound picks up water from the air and helps keep the dust down.

Boiling Point Elevation, Freezing Point Lowering

As we would expect, the pressure of water vapor over a solution at 100°C is less than that over pure water. In a 1 M sucrose solution at 100°C, the vapor pressure lowering is 14 mm Hg. So, the pressure of water vapor over the solution is:

$$760 \text{ mm Hg} - 14 \text{ mm Hg} = 746 \text{ mm Hg}$$

This means that the solution will not boil at 100°C when the atmospheric pressure is 760 mm Hg. Instead, the solution must be heated to about 100.5°C to make it boil. Not until this temperature is reached does the vapor pressure of water in the solution become 760 mm Hg.

The effect just described is general for all water solutions of *non-volatile* solutes*. The boiling points of their solutions are higher than that of pure water (Figure 13.10). The extent to which the boiling point is raised depends upon the concentration of solute. For very concentrated solutions, the effect can be quite large. A 50% solution of ethylene glycol, commonly used as an antifreeze, has a boiling point of 108.4°C at 760 mm Hg.

The freezing points of water solutions are always *lower* than that of pure water (0°C). A 1 M solution of a nonelectrolyte like sugar freezes at about −2°C. We can understand this behavior by looking at Figure 13.10. For a water solution to freeze, it must be in equilibrium with ice. This requires that the vapor pressure of water in the solution be equal to that of ice. As you can see from the Figure, the vapor pressures are not equal at 0°C. That of the solution is less than that of ice. They become equal only when the temperature drops below 0°C.

That's what they use in St. Paul, Minnesota.

We take advantage of this effect when we add antifreeze to a car radiator to protect it from freezing. A 50% solution of ethylene glycol does not freeze until the temperature drops to −34°C (−29°F). Another application has to do with keeping roads free of ice in winter. Solids such as NaCl or $CaCl_2$ are mixed with sand and spread on the road. They

*Volatile solutes such as methyl alcohol exert a vapor pressure of their own over a water solution. This contributes to the total pressure, which may then exceed 760 mm Hg at 100°C. Solutions of methyl alcohol in water boil at temperatures below the normal boiling point of water. This explains why antifreeze containing methyl alcohol has a tendency to boil over in an automobile radiator on a hot summer day.

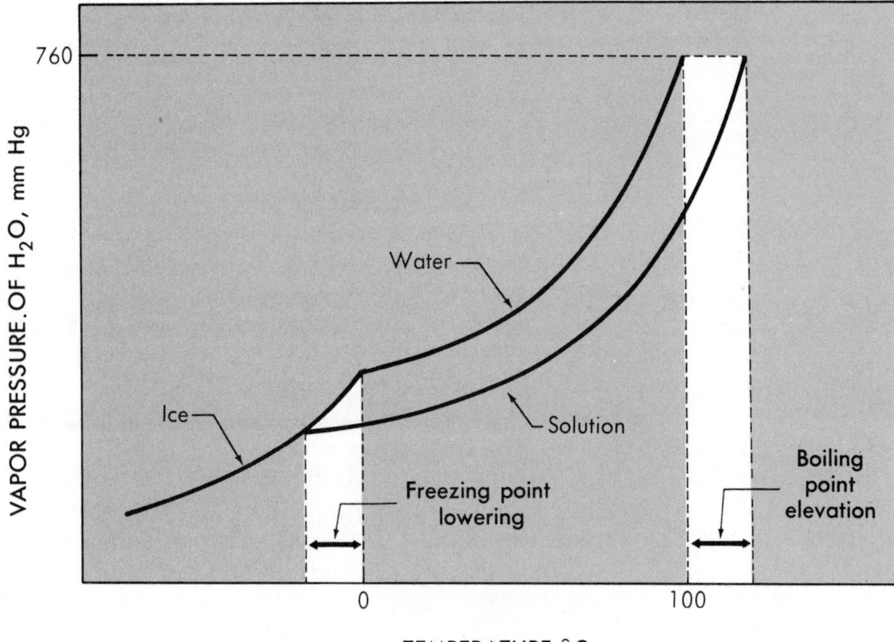

FIGURE 13.10

The vapor pressure of a solution of a nonvolatile solute is lower than that of the solvent. To boil such a solution, its vapor pressure must equal one atmosphere; this requires a higher temperature than is needed to boil the pure solvent. When a solution freezes, pure solid solvent crystallizes. The freezing point occurs at the temperature where the vapor pressures of the solid and the solution are equal. That temperature is lower than the freezing point of the pure solvent.

form water solutions which do not freeze at temperatures as low as −20°C.

Boiling point elevation and freezing point lowering, like vapor pressure lowering, are colligative properties. At least in dilute solution, they are:

1. *independent of the type of nonelectrolyte.* One molar solutions of glucose, sucrose, urea, and so forth boil at about 100.5°C and freeze at about −1.9°C. In other words, the boiling point elevation for a 1 M solution of a nonelectrolyte is about 0.5°C. The freezing point lowering is about 1.9°C.

2. *directly proportional to concentration.* In 0.1 M solutions of nonelectrolytes, the boiling point elevation and freezing point lowering are about 1/10 of that in 1 M solutions. That is, the boiling point is about 100.05°C and the freezing point about −0.19°C.

EXAMPLE 13.8

Consider two water solutions, one of glucose (MM = 180), the other of sucrose (MM = 342). Both contain 100 g of solute in one liter of solution. Which has the higher boiling point? the lower freezing point?

SOLUTION

We start by calculating the molarities of the two solutions:

$$\text{glucose: } M = \frac{(100/180)\text{ mol}}{1.00\ \ell} = 0.556\text{ mol}/\ell$$

$$\text{sucrose: } M = \frac{(100/342)\text{ mol}}{1.00\ \ell} = 0.292\text{ mol}/\ell$$

Since the concentration of glucose, in moles per liter, is higher, that solute will have a greater effect on the boiling point and freezing point. The glucose solution will have a higher boiling point and a lower freezing point.

At the same molarity, electrolytes have a greater effect on boiling point and freezing point than do nonelectrolytes. We can readily see why this should be the case. Compare the equations for the dissolving of glucose, $C_6H_{12}O_6$, sodium chloride, and calcium chloride.

$$C_6H_{12}O_6(s) \rightarrow C_6H_{12}O_6(aq)$$

$$NaCl(s) \rightarrow Na^+(aq) + Cl^-(aq)$$

$$CaCl_2(s) \rightarrow Ca^{2+}(aq) + 2\ Cl^-(aq)$$

We see that when one mole of each of these solutes dissolves:

—glucose forms 1 mol of particles ($C_6H_{12}O_6$ molecules)
—NaCl forms 2 mol of particles (1 mol of Na^+ ions + 1 mol of Cl^- ions)
—$CaCl_2$ forms 3 mol of particles (1 mol of Ca^{2+} ions + 2 mol of Cl^- ions)

When working with colligative properties, we count particles of solute, be they molecules or ions.

Remember that colligative properties such as freezing point lowering depend upon the total concentration of solute particles. An ion is as effective as a molecule in lowering the freezing point. On this basis, we expect 1 mol of NaCl to have twice the effect of 1 mol of glucose. Similarly, 1 mol of $CaCl_2$ should have three times the effect of 1 mol of glucose. Experimentally, we find that this is roughly the case.

fp of 1 M glucose solution = −1.9°C

fp of 1 M NaCl solution = −3.4°C

fp of 1 M $CaCl_2$ solution = −5.8°C

EXAMPLE 13.9

Arrange the following solutions in order of decreasing freezing point.

a. 0.1 M urea (a nonelectrolyte) b. 0.1 M KNO_3

c. 0.1 M $Cu(NO_3)_2$

SOLUTION

Urea dissolves as molecules. The electrolytes KNO_3 and $Cu(NO_3)_2$ dissolve as ions:

$$urea(s) \rightarrow urea(aq) \quad ; 1 \text{ mol of molecules}$$

$$KNO_3(s) \rightarrow K^+(aq) + NO_3^-(aq) \quad ; 2 \text{ mol ions}$$

$$Cu(NO_3)_2(s) \rightarrow Cu^{2+}(aq) + 2\,NO_3^-(aq); 3 \text{ mol ions}$$

Potassium nitrate should have a larger effect than urea. The effect of copper(II) nitrate should be even greater. The order of decreasing freezing point is:

$$urea > KNO_3 > Cu(NO_3)_2$$

Molar Masses of Nonelectrolytes

As we have pointed out, colligative properties are directly related to the concentration of solute. It is possible to write simple equations relating vapor pressure lowering, boiling point elevation, or freezing point lowering to concentration. One such equation, valid for nonelectrolytes dissolved in water is:

$$\Delta T_f = 1.86°C \times m \qquad (13.6)$$

Here, ΔT_f is the freezing point lowering in °C. The symbol m stands for a concentration unit known as molality. This is defined as the number of moles of solute per kilogram of solvent. That is:

$$m = \frac{\text{no. moles solute}}{\text{no. kg solvent}} \qquad (13.7)$$

Equation 13.6 can be used for several purposes. Perhaps most obviously, it allows us to calculate the freezing points of solutions of nonelectrolytes of known molality. For example, it predicts that 1 m solutions of glucose, sucrose, urea, and so on should freeze at −1.86°C.

From the standpoint of chemistry, the most important application of Equation 13.6 is in determining the molar masses of nonelectrolytes. To show how this is done, suppose you are given an "unknown" solid. Your objective is to determine the molar mass of the solid by freezing point lowering. You carefully weigh out a known mass of the solid, let us say 1.00 g. This is dissolved in a weighed amount of water, 20.0 g (0.0200 kg). You now measure the freezing point of the solution, finding it to be −0.93°C. From Equation 13.6, the molality of the solution must be 0.50. That is:

$$0.93°C = 1.86°C \times m$$

$$m = \frac{0.93°C}{1.86°C} = 0.50$$

Now, from Equation 13.7:

$$\text{no. moles solute} = \text{m} \times (\text{no. kg solvent})$$
$$= (0.50) \times (0.0200) = 0.010$$

Hence, the 1.00 g of solute that you weighed out must represent 0.010 mol

$$1.00 \text{ g} = 0.010 \text{ mol}$$

The mass of one mole must then be: $1.00 \text{ mol} \times \frac{1.00 \text{ g}}{0.010 \text{ mol}} = 100 \text{ g}$

In other words, the molar mass of the "unknown solid" must be 100 g. Its molecular mass is 100.

This approach can be used for a variety of nonelectrolytes. The method is not limited to water solutions. Organic solvents can be used as well. An equation similar to 13.6 still applies. However, the number 1.86°C has to be replaced by a number characteristic of the organic solvent. Usually, this number is larger than the freezing point constant for water; for benzene, for example, it is 5.10°C. This means that, at a given concentration, the freezing point lowering will be greater.

13.5 SUMMARY

Solutions are homogeneous mixtures. They are found in all physical states: solid, liquid, and gas. Liquid solutions in which water is the solvent are most common in the laboratory, in the home, and in the world around us. For that reason, we have concentrated upon water solutions. However, many of the ideas presented can be applied to other types of solutions.

Water is a good solvent for many ionic compounds because of the strong attraction between polar water molecules and charged ions. Not very many molecular substances dissolve readily in water. Those that are water-soluble either form ions (HCl) or hydrogen bonds (H_2O_2). Solubility in water is affected by temperature. Solids become more soluble and gases less soluble as the temperature rises.

The concentration of solute in a solution can be expressed in various ways. We have considered two different concentration units, mass percent and molarity (moles per liter of solution). The physical properties of solutions always vary with concentration. In many cases (color, electrical conductivity), they depend upon the nature of the solute as well. Colligative properties, on the other hand, depend primarily upon the concentration of solute particles, not upon their nature.

We have discussed three different colligative properties: vapor pressure, boiling point and freezing point. We ordinarily find that the addition of a solute to water lowers the vapor pressure, raises the boiling point, and lowers the freezing point. The changes in all three cases are directly proportional to the concentration of solute particles.

NEW TERMS

boiling point elevation
colligative property
concentrated solution
concentration
dilute solution
electrolyte
freezing point lowering
hydrated ion
molarity
nonelectrolyte
saturated solution
solute
solvent
supersaturated solution
unsaturated solution
vapor pressure lowering
volumetric flask

QUESTIONS

1. Classify the following solutions as to the state of the solute and solvent.

   a. household ammonia b. brass c. tincture of iodine

   Example: Salt water is a solid dissolved in a liquid.

2. Explain what is meant by the mass percent of solute in solution; by molarity.

3. The mass percent of urea in a water solution is 42.6. What is the mass percent of water?

4. Describe how you would prepare 500 cm^3 of a 1.00 M solution of KCl, using a volumetric flask.

5. Explain why ionic solutes tend to be soluble in water.

6. Explain why HCl is more soluble in water than is Cl_2.

7. Why is H_2O_2 more soluble in water than Br_2?

8. Write equations for the processes by which the following solutes dissolve in water.

   a. $NaNO_3(s)$ b. $KCl(s)$ c. $Br_2(l)$ d. $O_2(g)$

9. Describe a simple test that you could use in the laboratory to tell whether a solution of a solid in water is unsaturated, saturated, or supersaturated.

10. A saturated solution of Br_2 in water at 25°C contains 0.22 mol Br_2 per liter. Which of the following bromine solutions are unsaturated? supersaturated?

    a. 0.10 M b. 0.20 M c. 0.25 M

11. A saturated solution of KNO_3 in water at 20°C is prepared by adding 50 g of KNO_3 to 100 g of water. The solution is stirred until no more solid dissolves. What effect would each of the following changes have on the concentration of potassium nitrate in this solution?

    a. adding another 10 g of KNO_3.
    b. using less water to prepare the solution.
    c. raising the temperature to 50°C.
    d. lowering the temperature to 10°C.

12. Referring to Figure 13.6, which compound:

    a. has the largest solubility at 20°C?
    b. has the smallest solubility at 80°C?
    c. shows the least change in solubility between 20 and 80°C?

13. Using Figure 13.6, state the number of grams of KNO_3 that dissolves in 100 g of water at:

    a. 20°C b. 40°C c. 60°C d. 80°C

14. Explain why air bubbles form when water is heated.

15. The density of a 1.0 M solution of NaCl in water at 20°C is 1.037 g/cm³. At the same temperature, the density of a 1.0 M NH_3 solution is 0.989 g/cm³. Is density a colligative property?

16. Which of the following properties of water solutions are colligative?

    a. color b. vapor pressure c. freezing point
    d. electrical conductivity

17. Describe two tests, other than taste, which would distinguish a 1.0 M salt (NaCl) solution from a 1.0 M sugar ($C_{12}H_{22}O_{11}$) solution.

18. Which of the following solutes are electrolytes? nonelectrolytes?

    a. NaCl b. Br_2 c. KNO_3 d. HCl e. $C_6H_{12}O_6$

19. Explain, in terms of vapor pressure, why the boiling temperature of a solution is ordinarily higher than that of pure water. Why is the freezing temperature lower?

20. How would you expect a 0.2 M solution of sucrose to compare to a 0.1 M solution in:

    a. color b. electrical conductivity c. vapor pressure
    d. boiling point e. freezing point

21. Explain why rock salt is sometimes used on icy streets and sidewalks.

22. How many moles of solute particles are formed when one mole of each of the following solutes dissolves in water?

    a. sucrose b. $CaCl_2$ c. NaCl d. KNO_3 e. $Al(NO_3)_3$

PROBLEMS

1. *Mass Percent (Examples 13.1, 13.2)* A sodium chloride solution weighing 32.4 g was evaporated to dryness; 2.78 g of NaCl was recovered. What was the percent by mass of NaCl in the solution? the percent of water?

2. *Mass Percent (Examples 13.1, 13.2)* For a 2.0% solution of HCl in water, calculate:
    a. the number of grams of HCl in 25 g of solution.
    b. the number of grams of water in 75.0 g of solution.
    c. the number of grams of solution that contains 5.6 g of HCl.

3. *Molarity (Example 13.3)* A student weighed out 25 g of NaOH and dissolved it in enough water to give 250 cm³ of solution. What is the molarity?

4. *Molarity (Example 13.4)* How many moles of K_2CrO_4 are there in 0.10 ℓ of a 0.50 M K_2CrO_4 solution? how many grams?

5. *Molarity (Example 13.4)* You are given a large volume of 0.200 M $Mg(NO_3)_2$ solution. What volume should you measure out to obtain:

   a. 1.00 mol of $Mg(NO_3)_2$? b. 1.00 g of $Mg(NO_3)_2$?

6. *Molarity (Example 13.5)* Write directions, using grams, for preparing the following solutions.

   a. 1.0 ℓ of 1.0 M $AgNO_3$ b. 0.50 ℓ of 3.0 M NH_4Br

7. *Effect of Temperature upon Solubility (Example 13.7)* Referring to Figure 13.6, determine:

   a. the solubilities of KCl in water at 20°C and 80°C.
   b. the number of grams of water required to dissolve 20 g of KCl at 80°C.
   c. the number of grams of KCl that stays in solution when the solution in (b) is cooled to 20°C.

8. *Colligative Properties (Example 13.8)* Solutions A and B both contain 20.0 g of two different nonelectrolytes in one liter of solution. The solute in A has a molecular mass of 80. The solute in B has a molecular mass of 120. Which solution has the higher:

   a. vapor pressure b. boiling point c. freezing point

9. *Colligative Properties (Example 13.9)* Which of the following solutions, all 0.10 M, would you expect to have the lowest freezing point?

   a. KBr b. K_2SO_4 c. urea d. $AgNO_3$

* * * * *

10. You are given 5.02 g of $CaCl_2$ and asked to prepare from it a 3.00% solution in water.

    a. How many grams of solution can you make?
    b. How many grams of water will it take to make the solution?

11. 2.00 g of $NaNO_3$ and 5.60 g of KI are dissolved in 122 g of water. Determine the mass percents of all three components of the solution.

12. 15.0 g of sucrose (MM = 342) is dissolved in enough water to form 250 cm^3 of solution.

    a. What is the molarity of sugar?
    b. What would the molarity be if the solution were evaporated to a volume of 125 cm^3?

13. How many grams of solute are there in:

    a. 220 g of a 6.0% solution of $CaCl_2$?
    b. 220 cm^3 of a 2.50 M solution of $CaCl_2$?
    c. 220 cm^3 of a 2.50 M solution of $Ca(NO_3)_2$?

14. What volume of 0.100 M K_2CrO_4 solution should you take to obtain:

    a. 1.60 mol of K_2CrO_4? b. 1.60 g of K_2CrO_4?

15. A saturated solution is prepared by shaking excess potassium nitrate with 200 g of water at 60°C. Using Figure 13.6, determine

    a. how much KNO_3 dissolves at 60°C.
    b. how much KNO_3 comes out of solution upon cooling to 20°C.

16. Which of the following solutions of ethylene glycol (MM = 62) has the lowest freezing point? the highest freezing point?

    a. 1.0 M b. 60 g/ℓ c. 70 g/ℓ

17. If a 0.10 M solution of NaCl freezes at −0.36°C, what would you expect the freezing points of the following solutions to be?

    a. 0.20 M NaCl b. 0.10 M KNO_3 c. 0.10 M $CaCl_2$

18. Arrange the following solutions in order of decreasing freezing point.

    a. 1.0 M $Mg(NO_3)_2$ b. 1.0 M glucose c. 1.0 M sucrose
    d. 1.0 M $KClO_3$

19. Arrange the following solutions in order of decreasing boiling point.

    a. 0.10 M $Al(NO_3)_3$ b. 0.20 M NaCl c. 0.10 M $CaBr_2$

*20. Using Equation 13.6, calculate the freezing point of a solution containing 15 g of ethylene glycol (MM = 62) in 60.0 g of water.

*21. Using Equation 13.6, calculate the number of grams of methanol, CH_3OH, which must be added to 100 g of water to give a solution that freezes at −5.0°C.

*22. It is found that a solution containing 2.00 g of a certain nonelectrolyte dissolved in 30.0 g of water freezes at −0.61°C. Estimate the molar mass of the nonelectrolyte.

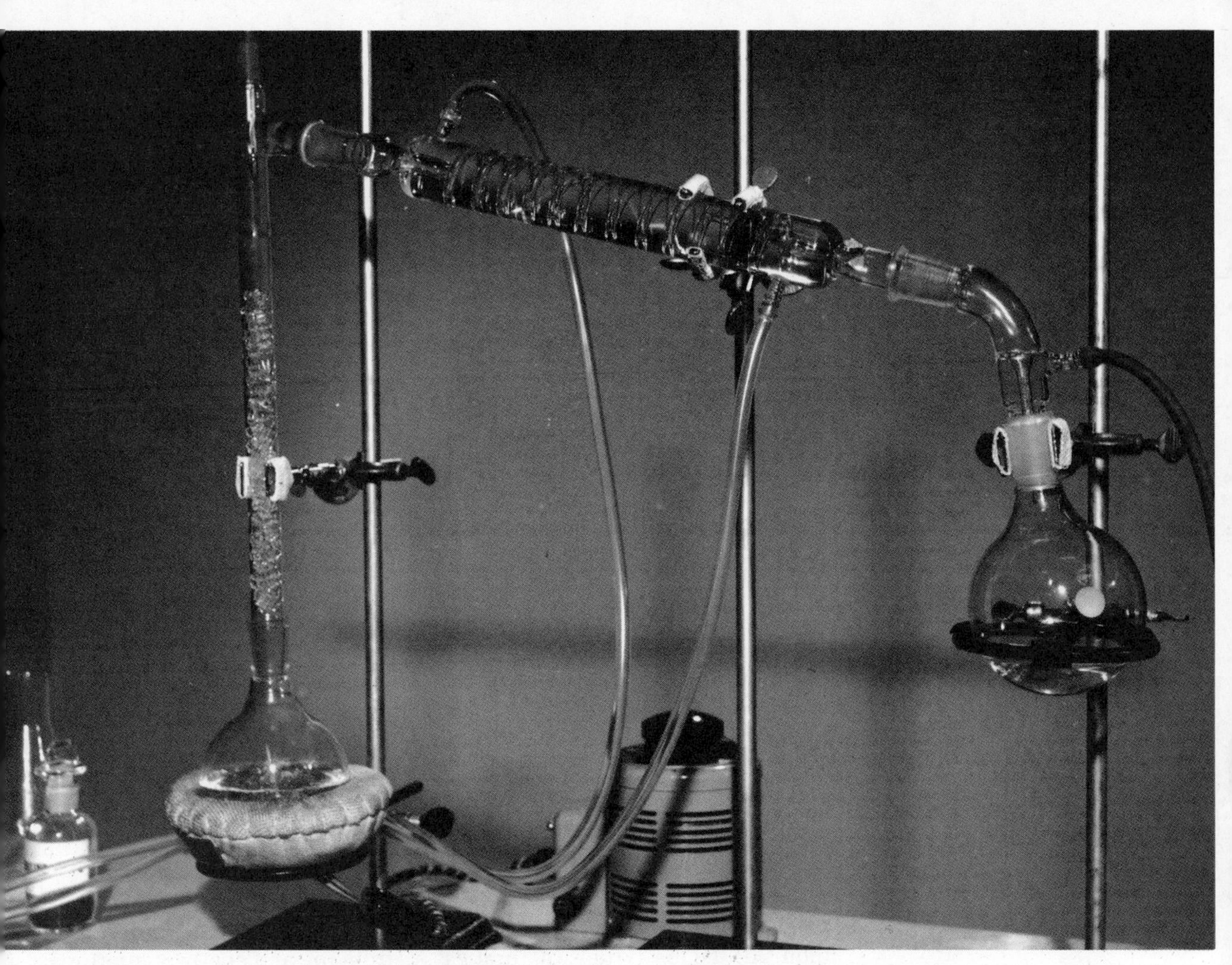

An apparatus used to distill a flammable organic liquid. The flask is heated electrically, avoiding an open flame. From the CHEM study film, Synthesis of an Organic Compound.

14

ORGANIC CHEMISTRY: Hydrocarbons

In this chapter, we begin a study of organic chemistry. This branch of chemistry deals with the properties and reactions of carbon compounds. Until about 150 years ago, it was believed that organic compounds could only be produced by living organisms. A "vital force", present only in living cells, was thought to be required. In 1828 a German chemist, Friedrich Wohler, disproved this theory. He made urea, an organic compound found in urine, from reactants of mineral origin.

Since the time of Wohler, a great many organic compounds have been made in the laboratory. Others have been isolated from natural materials. No one has kept a precise count, but there are more than three million known organic compounds. They represent more than 90% of all known substances.

There are two main reasons for the huge number of carbon compounds.

1. Carbon atoms can bond to one another to form long chains. Organic molecules may contain from one to several thousand carbon atoms.

2. Organic compounds commonly show *isomerism*. For a given molecular formula, there are often several different compounds with different properties. These are called **isomers.** For example, five different compounds all have the molecular formula C_6H_{14}.

Organic chemistry is vital to all aspects of our life. About 95% of the energy we use comes from the combustion of carbon compounds. The clothes we wear, whether made of natural or synthetic fibers, are organic in nature. The same is true of virtually all the foods we eat and the medicines we take. If it were not for carbon compounds, the earth would be as desolate and devoid of life as the moon.

Life is based on the reactions of organic compounds.

As you can imagine, organic chemistry is a huge subject. We have only time, in three chapters, to look at a few types of organic compounds. We start in this chapter by discussing **hydrocarbons.** These substances contain only the two elements hydrogen and carbon.

14.1 HYDROCARBONS; AN OVERVIEW

From a structural point of view, hydrocarbons comprise the simplest class of organic compounds. They are also among the most common. All the fossil fuels (natural gas, petroleum, and coal) are mixtures of hydrocarbons. A whole host of organic compounds are made from hydrocarbons. These include such familiar compounds as ethyl alcohol, used in beverages, and ethylene glycol, used as an antifreeze. Many common plastics such as polyethylene are made from hydrocarbons. It is no exaggeration to say that our entire economy is based on hydrocarbons. The increase in price of natural gas, petroleum, and hydrocarbons derived from them, has been a major cause of inflation in the past several years.

Table 14.1 lists the physical properties of several common hydrocarbons. The structures and reactions of many of these compounds will be discussed later in this chapter. Here and elsewhere, hydrocarbons are grouped into three series depending upon the types of bonds present.

1. *Saturated hydrocarbons,* in which all the bonds between carbon atoms are single bonds. Their molecules contain the maximum number of hydrogen atoms possible. Hence the word "saturated".

2. *Unsaturated hydrocarbons,* in which there are one or more multiple bonds between carbon atoms ($C{=}C$ or $C{\equiv}C$) in the molecule.

3. *Aromatic hydrocarbons,* which have molecular structures (Section 14.4) built around that of benzene. Most of these compounds have strong, pleasant odors. This gives rise to the name "aromatic".

All hydrocarbons, and indeed all organic compounds, are molecular.

TABLE 14.1 PHYSICAL PROPERTIES OF HYDROCARBONS

	Molecular Formula	Molecular Mass	mp (°C)	bp (°C)	Phys. State (25°C, 1 atm)	Vapor Pressure (mm Hg at 25°C)
SATURATED						
Methane	CH_4	16	−183	−162	gas	
Ethane	C_2H_6	30	−172	−89	gas	
Propane	C_3H_8	44	−187	−42	gas	
Butane	C_4H_{10}	58	−135	0	gas	
Pentane	C_5H_{12}	72	−130	36	liquid	507
Hexane	C_6H_{14}	86	−94	69	liquid	152
Heptane	C_7H_{16}	100	−90	98	liquid	46
Octane	C_8H_{18}	114	−57	126	liquid	14
UNSATURATED						
Acetylene	C_2H_2	26	subl	−84	gas	
Ethylene	C_2H_4	28	−169	−104	gas	
Propylene	C_3H_6	42	−185	−48	gas	
AROMATIC						
Benzene	C_6H_6	78	5	80	liquid	92
Toluene	C_7H_8	92	−95	111	liquid	29
Naphthalene	$C_{10}H_8$	128	80	218	solid	0.08
Anthracene	$C_{14}H_{10}$	178	217	340	solid	0.0002

Looking at Table 14.1, we see the behavior expected of molecular substances.

1. Most of the common hydrocarbons are gases or liquids at ordinary temperatures and pressures. Their molecules are held to one another only by weak Van der Waals forces. Hence, the molecules are readily separated from each other to form a liquid or gas.

2. Boiling point increases steadily with molecular mass. The lightest hydrocarbons, with 1 to 4 carbons per molecule, are gases at room temperature and atmospheric pressure. Heavier hydrocarbons are volatile liquids. Hydrocarbons of very high molecular mass are soft, waxy solids.

Given the data in Table 14.1, predict the boiling point of nonane, C_9H_{20}.

The high vapor pressure of such hydrocarbons as pentane, hexane, and benzene poses a safety hazard. These liquids catch fire if exposed to a flame or spark at room temperature. They should never be heated with a burner. A steam bath or heating mantle should be used. The same is true of all volatile organic compounds, most of which are flammable.

Hydrocarbons, like most molecular substances, are virtually insoluble in water. The most soluble, benzene, dissolves to the extent of only 0.01 mol/ℓ. Most hydrocarbons have much lower solubilities. The best solvents for hydrocarbons are other hydrocarbons. Benzene is a cheap and effective solvent for organic compounds. Recently it has been shown that benzene vapor can cause cancer. Hence it is being replaced for many purposes by toluene.

14.2 SATURATED HYDROCARBONS

The saturated hydrocarbons are often referred to as *alkanes*. The first three members of the series as shown in Table 14.1 are:

CH_4	methane
C_2H_6	ethane
C_3H_8	propane

The bonding in these compounds is indicated by drawing their **structural formulas,** shown below.

```
    H           H  H           H  H  H
    |           |  |           |  |  |
  H—C—H       H—C——C—H       H—C——C——C—H
    |           |  |           |  |  |
    H           H  H           H  H  H

 methane       ethane          propane
```

Each of the carbon atoms in these molecules forms a total of four bonds. Each hydrogen atom is joined by a single bond to a carbon atom. **This bonding pattern holds for all hydrocarbon molecules.** In these and all other saturated hydrocarbons, all the carbon-carbon bonds are single.

Structural formulas of organic molecules are often written in somewhat condensed form. To save space, we write:

Structural formulas like these are more informative than molecular formulas, and so are often used if they are known.

ethane $CH_3—CH_3$, or CH_3CH_3

propane $CH_3—CH_2—CH_3$, or $CH_3CH_2CH_3$

In these formulas it is understood that the hydrogen atoms are bonded to the carbon atom that precedes them in the formula.

Since structural formulas are two-dimensional, they seldom give a true picture of molecular geometry. This is true of methane, ethane, and propane, all of which are three-dimensional. Their geometries are shown in Figure 14.1. In each case, the four bonds around each carbon atom are directed toward the corners of a tetrahedron. This means that the bond angles are all 109°. In propane and the higher members of the series, the carbon chain has a "zig-zag" pattern:

```
   C     C
  / \   / \
     \ /
      C
```

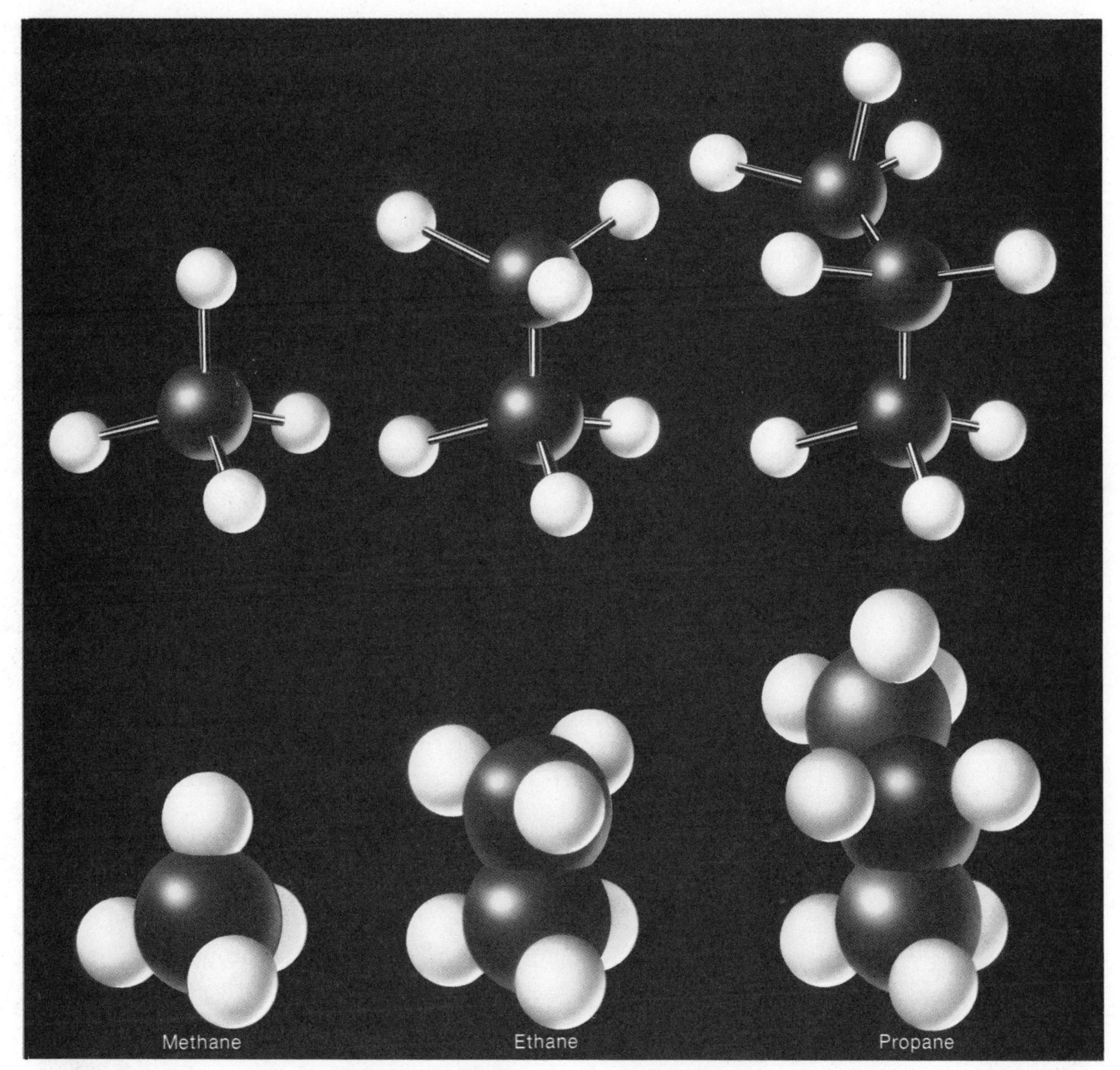

FIGURE 14.1

Models showing the molecular structures of methane, ethane and propane. The bond angles in all of these compounds are equal to 109°, the tetrahedral angle.

It is possible to write a general molecular formula which applies to all saturated hydrocarbons. To obtain this formula, look at the structural formulas of methane, ethane, and propane. Notice that each carbon is bonded to at least two hydrogens, the ones written above and below it. In addition, there are two "extra" hydrogen atoms, one at the far left and one at the far right of the molecule. Hence the general formula is:

$$C_nH_{2n+2}$$

Here, n is the number of carbon atoms in the molecule. For CH_4, n is 1. For C_2H_6, n is 2, and so on. If you check this formula against the entries in Table 14.1, you will find that it works for all the saturated hydrocarbons. For octane, C_8H_{18}:

$$n = 8; 2n + 2 = 2(8) + 2 = 18$$

EXAMPLE 14.1

What is the molecular formula of:

a. decane (10 carbon atoms)?

b. the alkane containing 34 hydrogen atoms?

SOLUTION

a. $n = 10;\ 2n + 2 = 2(10) + 2 = 22;\ C_{10}H_{22}$

b. $2n + 2 = 34;\ 2n = 32;\ n = 16;\ C_{16}H_{34}$

Isomerism in Alkanes

The first member of the alkane series to show isomerism is C_4H_{10}. There are two different hydrocarbons which have this molecular formula. One of these, called butane, is the one listed in Table 14.1. It has a normal boiling point of 0°C. The other, called 2-methylpropane, boils at −10°C. In general, the two isomers have distinctly different chemical and physical properties. They are two different compounds.

Seems like there is always something to make things complicated.

The existence of two isomers of C_4H_{10} is readily explained. Two different structural formulas can be written:

$$\begin{array}{ccccccccc} & & H & & H & & H & & H & & \\ & & | & & | & & | & & | & & \\ H & - & C & - & C & - & C & - & C & - & H \\ & & | & & | & & | & & | & & \\ & & H & & H & & H & & H & & \end{array}$$

butane

$$\begin{array}{ccccccc} & & H & & H & & H & & \\ & & | & & | & & | & & \\ H & - & C & - & C & - & C & - & H \\ & & | & & | & & | & & \\ & & H & H- & C & -H & H & & \\ & & & & | & & & & \\ & & & & H & & & & \end{array}$$

2-methylpropane

or, in condensed form:

$$CH_3{-}CH_2{-}CH_2{-}CH_3$$

butane

$$\begin{array}{ccccc} & & H & & \\ & & | & & \\ CH_3 & - & C & - & CH_3 \\ & & | & & \\ & & CH_3 & & \end{array}$$

2-methylpropane

Note that in butane there is a continuous chain of four carbon atoms. In 2-methylpropane, the longest chain has only three carbon atoms.

The two isomers of C_4H_{10} are referred to as **structural isomers.** They have different structural formulas. Putting it another way, they differ in the way the atoms are bonded to one another. The two molecules also have quite different shapes (Figure 14.2).

As the number of carbon atoms increases, the number of structural isomers goes up rapidly. Table 14.2 gives the calculated number of isomers for various alkanes. For the heavier alkanes, very few of the possible isomers have been isolated. However, the numbers listed give you some idea as to why there are so many organic compounds.

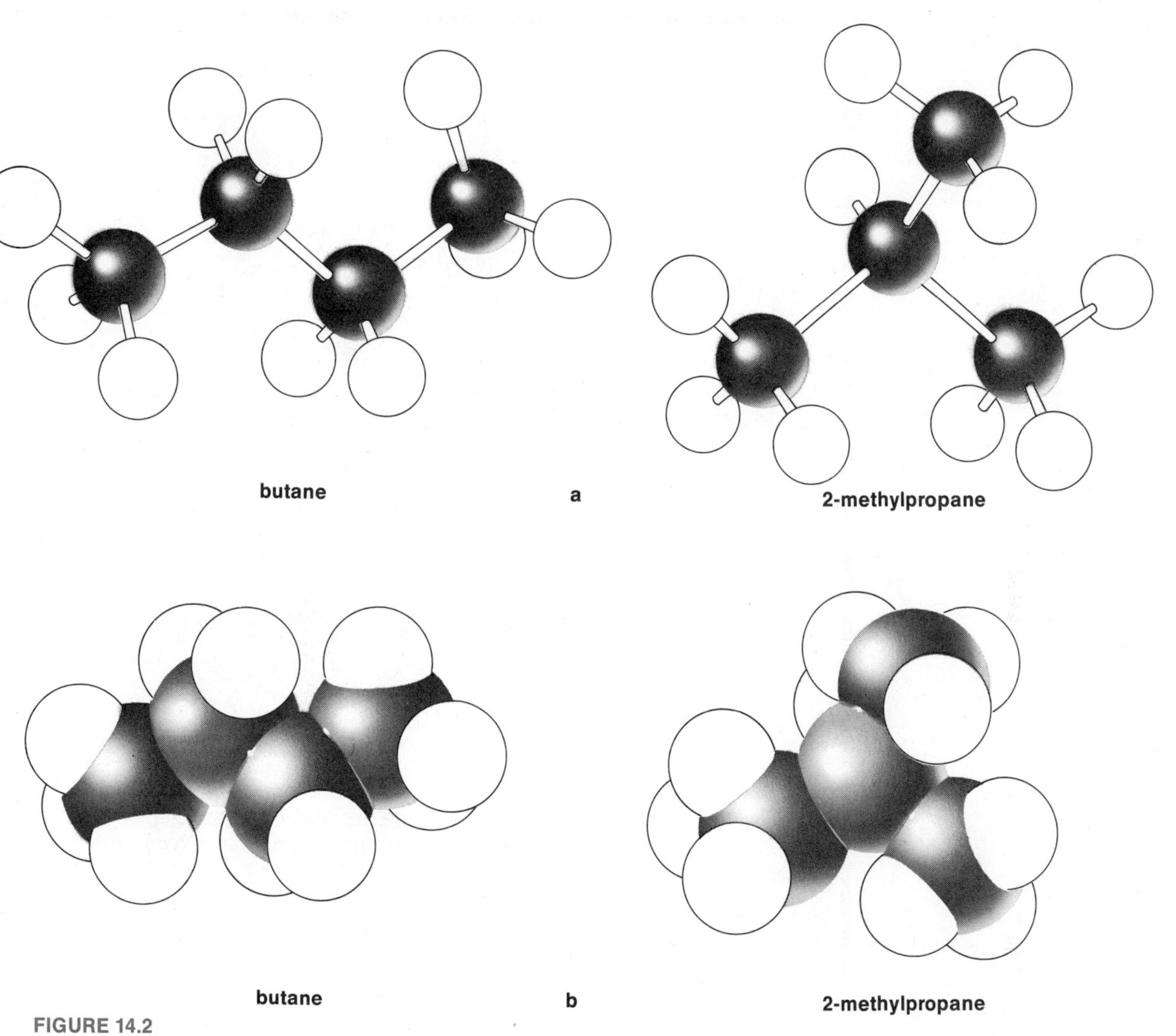

FIGURE 14.2

Models showing the structures of butane and 2-methylpropane. These are the two isomers of C_4H_{10}.

TABLE 14.2 STRUCTURAL ISOMERS OF ALKANES

Molecular Formula	Number of Isomers	Molecular Formula	Number of Isomers
C_4H_{10}	2	C_9H_{20}	35
C_5H_{12}	3	$C_{10}H_{22}$	75
C_6H_{14}	5	$C_{20}H_{42}$	336,319
C_7H_{16}	9	$C_{30}H_{62}$	4,111,846,763
C_8H_{18}	18		

EXAMPLE 14.2

Draw condensed structural formulas for the three isomers of C_5H_{12}.

SOLUTION

It is best to proceed systematically. We start with the isomer in which there is a 5-carbon chain:

$$CH_3{-}CH_2{-}CH_2{-}CH_2{-}CH_3 \qquad (1)$$

Next, we consider a chain which has only 4 carbons. There is one and only one such isomer:

$$\begin{array}{c} \quad H \\ \quad | \\ CH_3{-}C{-}CH_2{-}CH_3 \\ \quad | \\ \quad CH_3 \end{array} \qquad (2)$$

In the third isomer, the longest chain contains only 3 carbon atoms:

$$\begin{array}{c} CH_3 \\ | \\ CH_3{-}C{-}CH_3 \\ | \\ CH_3 \end{array} \qquad (3)$$

In working a problem of this type, the most common error is to draw too many isomers. On paper, it might seem that (A) and (B) represent "new" isomers:

Isomers differ in structure. They can't be converted to one another by turning the molecule or by rotation around a carbon-carbon single bond.

$$\begin{array}{c} CH_3{-}CH_2{-}CH_2{-}CH_2 \\ \qquad\qquad\qquad | \\ \qquad\qquad\qquad CH_3 \\ \text{(A)} \end{array} \qquad \begin{array}{c} \qquad\qquad H \\ \qquad\qquad | \\ CH_3{-}CH_2{-}C{-}CH_3 \\ \qquad\qquad | \\ \qquad\qquad CH_3 \\ \text{(B)} \end{array}$$

However, if you think about it for a moment, you will realize that (A) is identical with (1). Both contain a 5-carbon chain. It really doesn't matter whether we put the end $-CH_3$ group at the "east" or the "south" of the chain. In three dimensions, the two positions are equivalent. By the same token, (B) is identical with (2). In both, the longest continuous chain has four carbon atoms. If we flip (B) 180° through the plane of the paper, it becomes (2).

Naming Alkanes

As organic chemistry developed, it became apparent that some systematic way of naming compounds was needed. About 50 years ago, the International Union of Pure and Applied Chemistry (IUPAC), devised a system which could be used for all organic compounds. With some changes, this system is in use today. We will not attempt to describe all of its features. For the most part, we will refer to organic compounds by their common names because these are still generally used. Thus, we still call the compound C_2H_2 acetylene rather than by its IUPAC name, ethyne. However, we will illustrate how the IUPAC system works by applying it to the alkanes.

Common names are much more common in organic chemistry than they are in inorganic chemistry.

For "straight-chain" alkanes such as:

$CH_3—CH_2—CH_2—CH_3$ butane

$CH_3—CH_2—CH_2—CH_2—CH_3$ pentane

the IUPAC name consists of a single word. These names (up to 8 carbon atoms) are listed in Table 14.1, p. 339. Notice that, beyond four carbons, the name is formed by adding -ane to a Greek prefix which gives the number of carbon atoms. Thus we have *hex*ane (6 carbon atoms), *hep*tane (7 carbon atoms), and so on.

$CH_3—CH_2—CH_2—CH_2—CH_2—CH_3$ hexane

$CH_3—CH_2—CH_2—CH_2—CH_2—CH_2—CH_3$ heptane

With alkanes containing a branched chain, such as

```
        H
        |
  CH3—C—CH3
        |
       CH3
```

2-methylpropane

the name is a bit more complex. It consists of three parts. These are, in reverse order:

1. A **"family name"** or **suffix** which gives the number of carbon atoms in the longest chain. For a 3-carbon chain, the family name is *propane*. This, of course, is the name of the straight chain alkane with 3 carbon atoms. Similarly, a 4-carbon chain is a *butane*. A 5-carbon chain is a *pentane*, and so on.

2. A **"surname"** or **prefix** which denotes the *alkyl group* which forms the branch on the chain. An alkyl group is the residue left when a hydrogen atom is removed from a hydrocarbon molecule. The names of a few alkyl groups are given in Table 14.3.

3. A **number**, which shows the carbon atom in the chain at which branching occurs. In 2-methylpropane, referred to above, the methyl branch is located at the second carbon atom from the end of the chain.

```
C1—C2—C3
    |
```

TABLE 14.3 NAMES OF ALKYL GROUPS

$CH_3—$	methyl
$CH_3—CH_2—$	ethyl
$CH_3—CH_2—CH_2—$	propyl

Following this system, let us consider the names of the isomers of C_5H_{12}.

$$CH_3—CH_2—CH_2—CH_2—CH_3 \qquad \text{pentane}$$

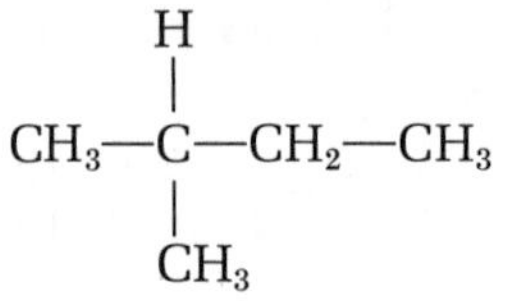

$$\begin{array}{c} \quad H \\ \quad | \\ CH_3—C—CH_2—CH_3 \\ \quad | \\ \quad CH_3 \end{array} \qquad \text{2-methylbutane}$$

$$\begin{array}{c} CH_3 \\ | \\ CH_3—C—CH_3 \\ | \\ CH_3 \end{array} \qquad \text{2,2-dimethylpropane}$$

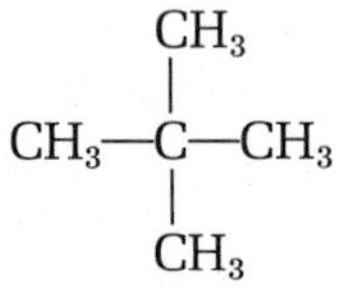

Notice that:

a. If there is the same alkyl group at two branches, the prefix "di" is used (2,2-*di*methylpropane). If there were three methyl branches, we would write *tri*methyl, and so on.

b. The number in the name is made as small as possible. Thus we write 2-methylbutane, numbering the chain from the left

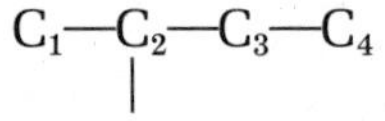

$$\begin{array}{c} C_1—C_2—C_3—C_4 \\ | \end{array}$$

rather than from the right.

EXAMPLE 14.3

Assign IUPAC names to the following:

$$\text{a. } \begin{array}{c} CH_3 \\ | \\ CH_3—C—CH_2—CH_3 \\ | \\ CH_3 \end{array} \qquad \text{b. } \begin{array}{c} H \\ | \\ CH_3—CH_2—C—CH_2—CH_3 \\ | \\ CH_2 \\ | \\ CH_3 \end{array}$$

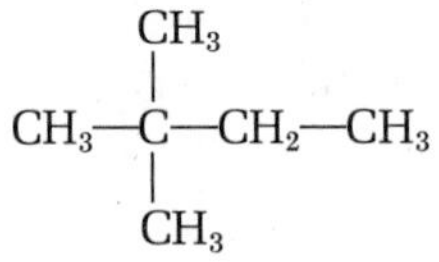

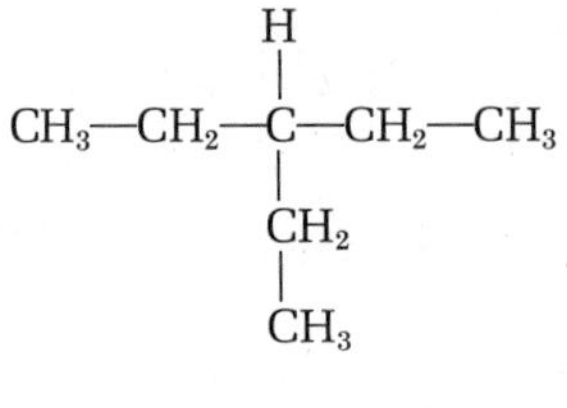

SOLUTION

a. The longest chain contains 4 carbon atoms (-butane). There are two CH_3 (methyl) groups branching off at the second carbon atom from the end of the chain (2). The correct name is:

2,2-dimethylbutane

b. The longest chain, however you count it, contains 5 carbon atoms. There is a $CH_3—CH_2$ branch at the number 3 carbon, whichever end of the chain you start from. The IUPAC name is:

3-ethylpentane

There is one bonus in using IUPAC names. Naming compounds properly can help you identify isomers and avoid writing extra ones. The principle here is a simple one. *Each isomer must have a different name.* If you come up with the same name for two structures, they are identical. Consider, for example, the two structures:

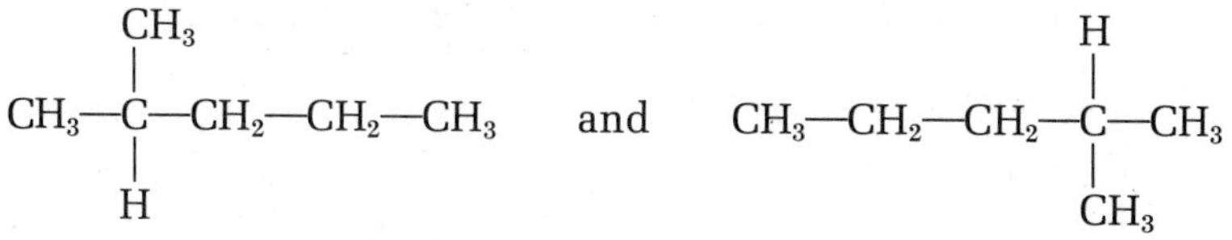

```
    CH3                                          H
    |                                            |
CH3—C—CH2—CH2—CH3      and      CH3—CH2—CH2—C—CH3
    |                                            |
    H                                           CH3
```

If you follow the rules, you will give these two structures the same name, 2-methylpentane. They must then represent the same compound. On the other hand, the structure:

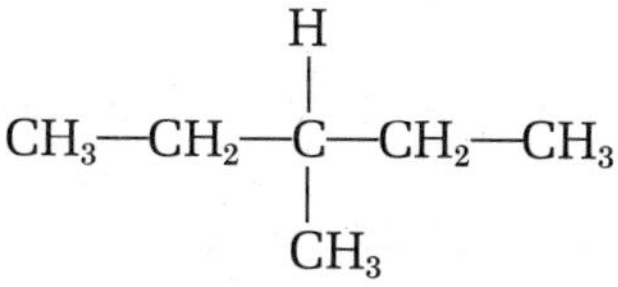

```
            H
            |
CH3—CH2—C—CH2—CH3
            |
           CH3
```

is named properly as 3-methylpentane. It represents a different compound from the two structures above. It, along with one of the above structures, is an isomer of C_6H_{14}.

Which isomer of C_8H_{18} is described in Table 14–1?

14.3 UNSATURATED HYDROCARBONS; ALKENES AND ALKYNES

As pointed out earlier, an unsaturated hydrocarbon has one or more multiple bonds between carbon atoms. There are many different kinds of unsaturated hydrocarbons. Here, we will discuss two of the simplest and most common types.

1. **Alkenes,** in which there is one double bond (C═C) in the molecule.
2. **Alkynes,** where the molecule has one triple bond (C≡C).

Alkenes

By introducing a double bond into a hydrocarbon molecule, we eliminate two hydrogen atoms.

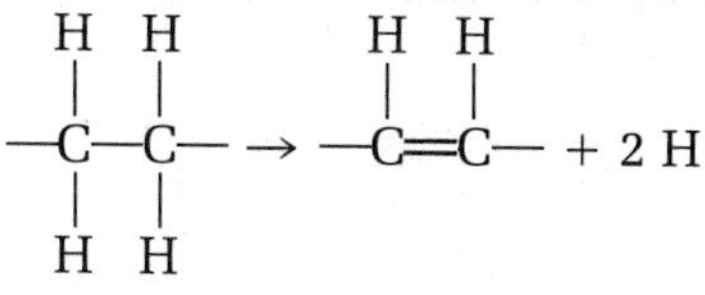

```
 H  H          H  H
 |  |          |  |
—C—C—   →    —C═C—  + 2 H
 |  |
 H  H
```

This means that an alkene molecule should contain two fewer hydrogens than the corresponding alkane. Recall that the general formula of an alkane is C_nH_{2n+2}. The general formula of an alkene must then be:

Is this a molecular formula or a structural formula?

$$C_nH_{2n}$$

In other words, in all alkenes, there are twice as many hydrogen as carbon atoms.

The first member of the alkene series has the molecular formula C_2H_4. This compound is commonly called ethylene (IUPAC name ethene). It is almost unique among organic compounds in one respect. By writing the structural formula in a certain way:

```
H       H
 \     /
  C=C
 /     \
H       H
```

we accurately represent the geometry of the molecule. The molecule is 2-dimensional, as this structure implies. That is, all six atoms are located in a plane. Moreover, all the bond angles are 120°:

Following this pattern, we write the structural formula of propylene, C_3H_6, as:

```
H       CH3
 \     /
  C=C
 /     \
H       H
```

We might say that this molecule is derived from ethylene by replacing one of the hydrogen atoms with a $—CH_3$ group (Figure 14.3).

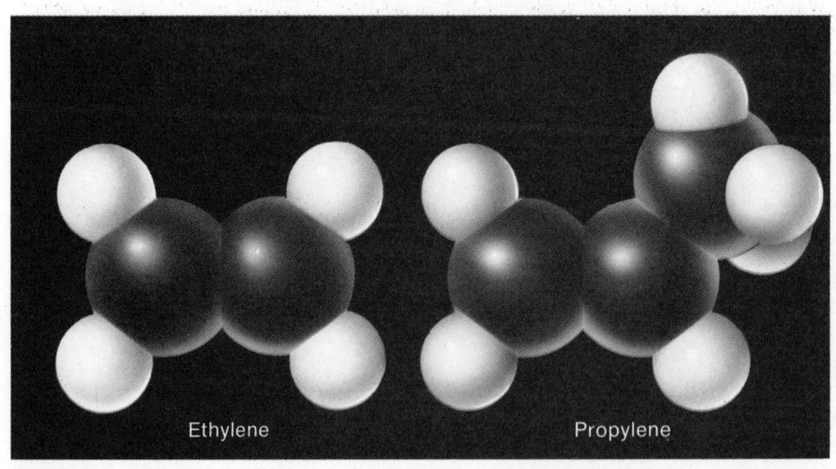

FIGURE 14.3

Space-filling models for ethylene and propylene. The ethylene molecule is planar, since the six nuclei lie in a plane. The carbon-carbon double bond contains four electrons. Propylene contains three carbon atoms, two of which are attached by a double bond.

In the alkenes, as with the alkanes, isomerism first occurs with four carbon atoms (Example 14.4).

EXAMPLE 14.4

Draw condensed structural formulas for the structural isomers of the 4-carbon alkene, C_4H_8.

SOLUTION

We start by drawing a partial structure which includes the double bond.

$$\begin{matrix} \diagdown & & \diagup \\ & C{=}C & \\ \diagup & & \diagdown \end{matrix}$$

With C_4H_8, we have two more carbon atoms to account for. We might attach an ethyl group ($-CH_2-CH_3$) at one position and hydrogen atoms at the other three:

$$\begin{matrix} H & & CH_2{-}CH_3 \\ & C{=}C & \\ H & & H \end{matrix} \qquad \text{isomer (1)}$$

An alternative would be to add two $-CH_3$ groups (and two H atoms). We could do this in either of two ways. In one isomer, we put the two methyl groups on the same carbon atom.

$$\begin{matrix} H & & CH_3 \\ & C{=}C & \\ H & & CH_3 \end{matrix} \qquad \text{isomer (2)}$$

A third isomer* is obtained by putting one methyl group on each carbon atom.

$$\begin{matrix} CH_3 & & CH_3 \\ & C{=}C & \\ H & & H \end{matrix} \qquad \text{isomer (3)}$$

*Actually, there is still another alkene with the formula C_4H_8. It has the structure:

$$\begin{matrix} CH_3 & & H \\ & C{=}C & \\ H & & CH_3 \end{matrix}$$

This compound and (3) are referred to as geometric isomers. They differ only in the distance between the CH_3 groups (or H atoms). We will not consider this type of isomerism further.

Alkynes

The formation of an alkyne from the corresponding alkene involves the loss of two hydrogen atoms.

$$\begin{array}{c}\text{H}\ \ \text{H}\\ |\ \ \ \ |\\ -\text{C}{=}\text{C}-\end{array} \rightarrow -\text{C}{\equiv}\text{C}- + 2\text{ H}$$

Recall that the general formula of an alkene is C_nH_{2n}. The general formula of an alkyne must then be:

$$C_nH_{2n-2}$$

EXAMPLE 14.5

Identify the following as an alkane, alkene, or alkyne.

a. $C_{10}H_{20}$ b. C_9H_{20} c. $C_{10}H_{18}$

SOLUTION

a. alkene ($n = 10, 20 = 2n$)

b. alkane ($n = 9, 20 = 2n + 2$)

c. alkyne ($n = 10, 18 = 2n - 2$)

The simplest alkyne, and the only one we will consider, has the molecular formula C_2H_2. It is commonly called acetylene (IUPAC name: ethyne). The molecule is linear:

$$\text{H}-\text{C}{\equiv}\text{C}-\text{H}$$

That is, the four atoms are arranged in a straight line.

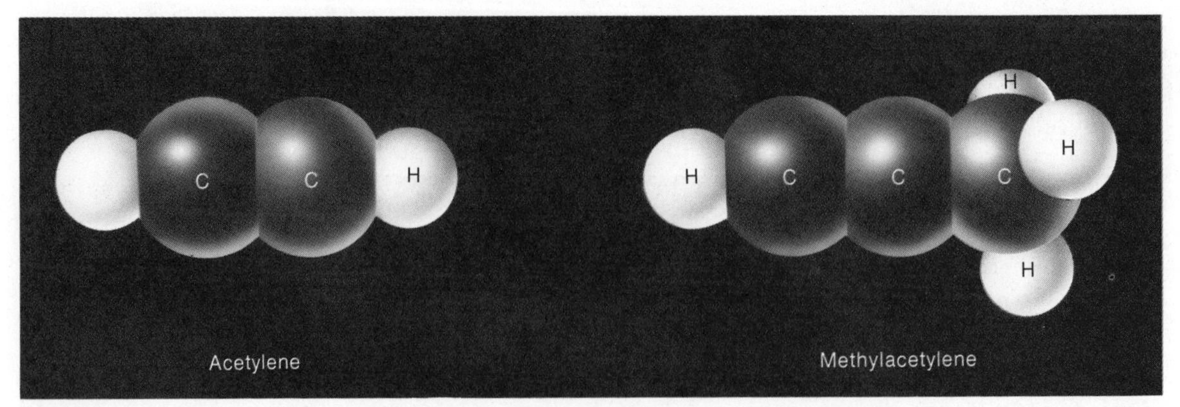

FIGURE 14.4

In acetylene and its simplest derivative, methylacetylene, two carbon atoms are linked by a triple bond. Both molecules contain four atoms on a straight line. Both are very reactive.

Acetylene is one of the most important hydrocarbons industrially. About 5×10^8 kg of C_2H_2 is produced each year in the United States. Half of this is converted to other organic compounds. The rest is used in oxyacetylene torches. These are widely used in welding, where they produce temperatures as high as 2800°C.

FIGURE 14.5

A high-temperature flame for welding is achieved by burning acetylene gas with pure oxygen. *Photo by Robert Shapiro.*

Acetylene cannot be liquefied safely. The liquid decomposes explosively to the elements.

$$C_2H_2(l) \rightarrow 2\ C(s) + H_2(g);\ \Delta H = -54\ \text{kcal} \qquad (14.1)$$

Acetylene for welding is shipped in steel tanks under rather low pressures, about 12 atm. The tanks contain asbestos sheets wet with acetone which is saturated with acetylene gas. When the valve on the tank is opened, acetylene gas (and a little acetone) comes out of solution. One thing a welder does not need is to have his or her tank of acetylene detonate. This system avoids such experiences, which were all too common at one time.

In recent years the principal source of acetylene in the United States has been methane, obtained from natural gas. Methane is heated to 1500°C in a limited amount of air. Under these conditions, part of the CH_4 is converted to acetylene.

$$6\ CH_4(g) + O_2(g) \rightarrow 2\ C_2H_2(g) + 2\ CO(g) + 10\ H_2(g) \qquad (14.2)$$

The shortage of natural gas makes an older synthesis of acetylene look more attractive. If coke, a coal derivative which is mostly carbon, is heated with limestone, the following reaction occurs:

$$CaCO_3(s) + 4\ C(s) \rightarrow CaC_2(s) + 3\ CO(g)$$

The solid product, calcium carbide, is then treated with water to give acetylene.

$$CaC_2(s) + H_2O(l) \rightarrow C_2H_2(g) + CaO(s) \qquad (14.3)$$

This process uses starting materials (coal, limestone, and water) which are abundant and relatively cheap. The garlic-like odor of the gas produced is due to an impurity, phosphine, PH_3. Acetylene itself is colorless and odorless.

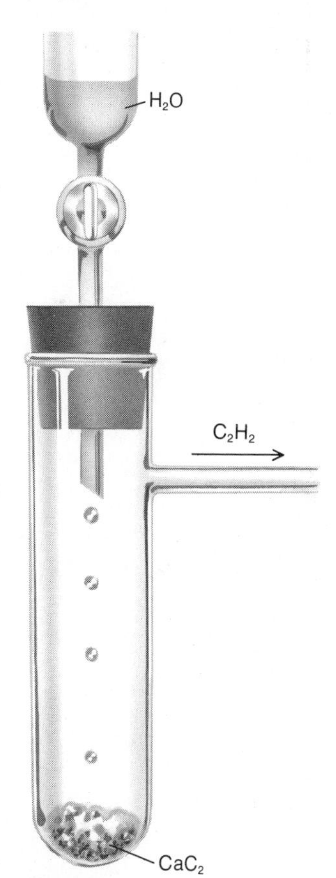

FIGURE 14.6

Preparation of acetylene from calcium carbide. Addition of water to CaC_2 produces C_2H_2 spontaneously. The gas easily forms explosive mixtures with air.

Addition Reactions

Hydrocarbons containing a double or triple bond undergo a reaction called addition. That is, a reactant molecule adds across the multiple bond. With an alkene, the general reaction is:

$$\begin{matrix} \diagdown & & \diagup \\ & C{=}C & \\ \diagup & & \diagdown \end{matrix} \quad + \quad Y{-}Z \quad \rightarrow \quad \begin{matrix} | & | \\ -C- & C- \\ | & | \\ Y & Z \end{matrix}$$

Here, Y and Z represent two atoms or groups of atoms composing a molecule. The double bond is broken in the reaction. Only single bonds are present in the product.

With alkynes, addition can occur in two steps:

Unsaturated hydrocarbons are much more reactive chemically than saturated ones.

$$—C{\equiv}C— \quad + \quad Y—Z \quad \rightarrow \quad \begin{matrix} \diagdown & & \diagup \\ & C{=}C & \\ \diagup & & \diagdown \\ Y & & Z \end{matrix}$$

$$\begin{matrix} \diagdown & & \diagup \\ & C{=}C & \\ \diagup & & \diagdown \\ Y & & Z \end{matrix} \quad + \quad Y—Z \quad \rightarrow \quad \begin{matrix} & Y & Z & \\ & | & | & \\ — & C & — C & — \\ & | & | & \\ & Y & Z & \end{matrix}$$

To illustrate addition reactions, we use ethylene, C_2H_4, as an example. We consider four different molecules that can add to ethylene.

1. *H_2 (Y = H; Z = H)* If the two gases, H_2 and C_2H_4, are mixed in a container, nothing happens. However, in the presence of finely divided nickel, a reaction takes place. The product is the saturated hydrocarbon ethane.

$$\begin{matrix} H & & H \\ & C{=}C & \\ H & & H \end{matrix} \quad + \quad H—H \quad \rightarrow \quad \begin{matrix} & H & H & \\ & | & | & \\ H— & C & — C & —H \\ & | & | & \\ & H & H & \end{matrix}$$

$$C_2H_4(g) \quad + \quad H_2(g) \quad \xrightarrow{Ni} \quad C_2H_6(g) \qquad (14.4)$$

The nickel in this reaction acts as a catalyst. That is, it speeds up the reaction without being consumed by it.

2. *Cl_2 (Y = Cl; Z = Cl)* Chlorine adds readily to ethylene to give a product containing two chlorine atoms (dichloroethane, a common dry-cleaning solvent).

$$\begin{matrix} H & & H \\ & C{=}C & \\ H & & H \end{matrix} \quad + \quad Cl—Cl \quad \rightarrow \quad \begin{matrix} & H & H & \\ & | & | & \\ H— & C & — C & —H \\ & | & | & \\ & Cl & Cl & \end{matrix}$$

$$C_2H_4(g) \quad + \quad Cl_2(g) \quad \rightarrow \quad C_2H_4Cl_2(l) \qquad (14.5)$$

Bromine reacts similarly. A common laboratory test for unsaturation uses a solution of bromine in CCl_4, to which the sample is added. Upon shaking, the deep red color of the Br_2 fades if a double or triple bond is present.

The brominated hydrocarbon is colorless.

3. *HCl (Y = H, Z = Cl)* Reaction occurs directly between the two gases:

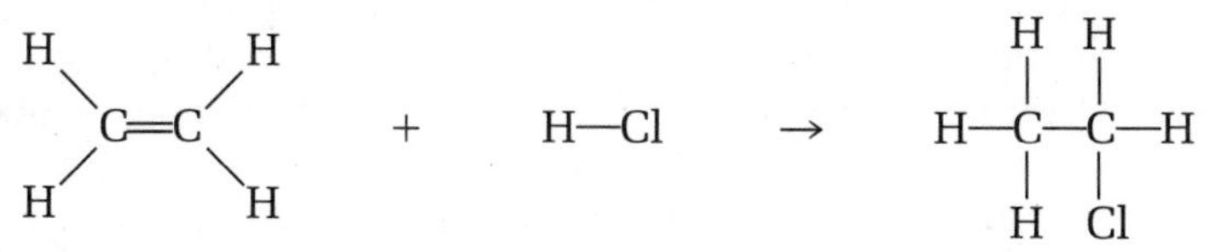

$$\begin{array}{c} H \quad\quad H \\ \diagdown \;\; \diagup \\ C{=}C \\ \diagup \;\; \diagdown \\ H \quad\quad H \end{array} \quad + \quad H{-}Cl \quad \rightarrow \quad \begin{array}{c} H \;\; H \\ | \;\;\; | \\ H{-}C{-}C{-}H \\ | \;\;\; | \\ H \;\; Cl \end{array}$$

$$C_2H_4(g) \quad + \quad HCl(g) \quad \rightarrow \quad C_2H_5Cl(l) \qquad (14.6)$$

In this reaction, a single Cl atom is introduced into the molecule. HBr and HI react similarly.

4. *H_2O (Y = H, Z = OH)* Unsaturated hydrocarbons are insoluble in water, so special conditions are required here. The reaction is ordinarily carried out in dilute sulfuric acid solution. The H^+ ions of the acid act as a catalyst. The product is ethyl alcohol, C_2H_5OH.

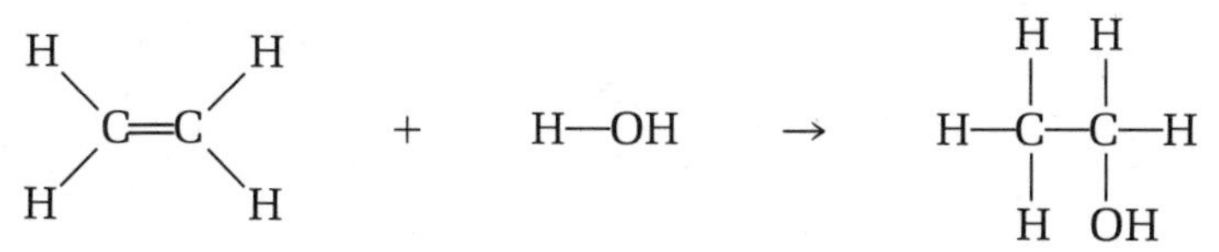

$$\begin{array}{c} H \quad\quad H \\ \diagdown \;\; \diagup \\ C{=}C \\ \diagup \;\; \diagdown \\ H \quad\quad H \end{array} \quad + \quad H{-}OH \quad \rightarrow \quad \begin{array}{c} H \;\; H \\ | \;\;\; | \\ H{-}C{-}C{-}H \\ | \;\;\; | \\ H \;\; OH \end{array}$$

$$C_2H_4(g) \quad + \quad H_2O(l) \quad \xrightarrow{H^+} \quad C_2H_5OH(l) \qquad (14.7)$$

About 10^9 kg of ethyl alcohol are produced each year in the United States by this process.

EXAMPLE 14.6

Starting with acetylene, C_2H_2, how could you prepare:

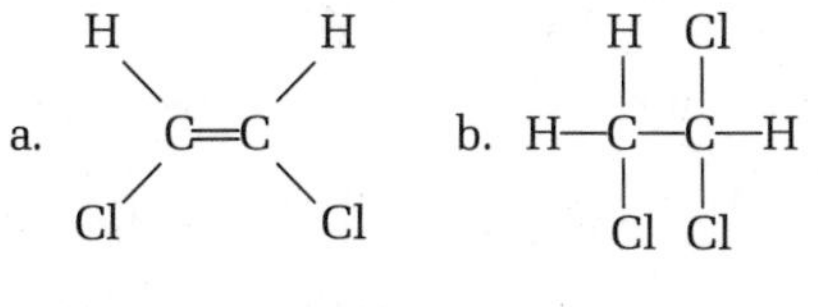

a. $\begin{array}{c} H \quad\quad H \\ \diagdown \;\; \diagup \\ C{=}C \\ \diagup \;\; \diagdown \\ Cl \quad\quad Cl \end{array}$ b. $\begin{array}{c} H \;\; Cl \\ | \;\;\; | \\ H{-}C{-}C{-}H \\ | \;\;\; | \\ Cl \;\; Cl \end{array}$

SOLUTION

a. Add Cl_2, stopping (you hope) after the first step.

b. Isolate (a) and add HCl.

14.4 AROMATIC HYDROCARBONS

We begin our study of aromatic hydrocarbons by looking at the structure of benzene, the parent member of the series.

Benzene

Benzene was first made by Michael Faraday, probably the greatest experimentalist of all time.

The molecular formula of benzene is C_6H_6. Compare this with the 6-carbon alkane, C_6H_{14}. It would seem that there must be a great deal of unsaturation in the benzene molecule. Yet, benzene does not behave chemically like an unsaturated hydrocarbon. Moreover, it appears that all six carbon atoms in the molecule are equivalent. That is, none of the carbon atoms show different properties from any of the others. The same is true of the six hydrogen atoms.

Reflecting on these facts, Friedrich Kekulé in 1865 proposed a ring structure for benzene:

```
        H
        |
        C
      //  \
  H—C      C—H
     |      ||
  H—C      C—H
      \\  /
        C
        |
        H
```

Here, all of the carbon atoms (and all of the hydrogens) are equivalent. Each carbon atom forms a double bond with one other carbon, a single bond with a second carbon atom, and a single bond with a hydrogen atom.

Benzene has the symmetry of a snowflake.

There is other evidence for the Kekulé structure of benzene. The molecule is planar. The six carbon atoms are at the corners of a regular hexagon. All of the bond angles are 120°. To emphasize this, and to save space, the Kekulé structure is usually abbreviated as:

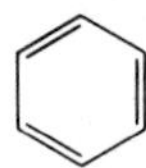

Here, it is understood that there is a carbon atom at each corner of the hexagon. The hydrogen atoms are not shown. There is a hydrogen atom bonded to each ring carbon.

There is one difficulty with the Kekulé structure of benzene. It implies that there are two different kinds of carbon-carbon bonds in the molecule, double and single. Experimental evidence, on the other hand, indicates that all the carbon-carbon bonds are equivalent. For example, all the carbon-carbon distances are the same, 0.139 nm. This is about halfway between the single and double bond distances.

This situation is difficult to account for within the simple theory of bonding we have used. There is no way of showing a bond intermediate between a single and double bond. What is often done is to write two structures with a double arrow between them:

$$\longleftrightarrow$$

The understanding is that the true structure of benzene is midway between the two shown. We might describe benzene as a "hybrid" of these two structures. By way of an analogy, a mule is a hybrid between a horse and a donkey. It, like benzene, is a distinct animal whose properties can be described in terms of two other species.

The situation we have just discussed is not unique to benzene. Many substances have properties which do not agree exactly with a single Lewis structure or structural formula. The general case is referred to as *resonance*. We call the two structures written above for benzene "resonance forms". Note the important distinction between resonance forms and isomers. Isomers are different substances with different physical and chemical properties. There is only one isomer of benzene. It has a structure intermediate between those of the two resonance forms shown.

In quantum mechanics its structure is the average of the two forms.

The structural formula of benzene is sometimes shown in still another way. Here, only the single bonds of the Kekulé structure are retained. The remaining 6 electrons (three electron pairs) are shown as a circle within the benzene ring. In some ways, this structure represents more accurately the electron distribution. The extra electrons are spread over the entire molecule instead of being tied down to a particular atom. However, throughout this text we will use the simple Kekulé formula shown at the left below.

Kekulé formula "Circle formula"

Alkylbenzenes

In the simplest class of aromatic compound, one or more alkyl groups is substituted for a hydrogen atom in the benzene ring. Examples include methylbenzene (toluene) and ethylbenzene:

CH_3

toluene, C_7H_8

$CH_2—CH_3$

ethylbenzene, C_8H_{10}

Toluene is used chiefly as a solvent for other organic compounds. Ethylbenzene is the starting material for the manufacture of styrene, used to make polymers (Chapter 16).

Other Aromatic Hydrocarbons

When two or more alkyl groups are attached to the benzene ring, isomerism is possible. There are three different dimethylbenzenes, commonly called xylenes. They differ in the distance between the two methyl groups.

CH_3 CH_3 | CH_3 CH_3 | CH_3 CH_3

1,2 | 1,3 | 1,4

The compound in which the two —CH_3 groups are attached to adjacent carbons is called 1,2-dimethylbenzene or *ortho* xylene. In the second isomer (1,3-dimethylbenzene or *meta* xylene), the CH_3 groups are separated by one ring carbon. When the methyl groups are as far apart as possible, we have 1,4-dimethylbenzene (*para* xylene).

In another type of aromatic compound, two or more benzene rings are "fused" together. That is, they share one side of a hexagon. The simplest example is naphthalene, sometimes used in mothballs.

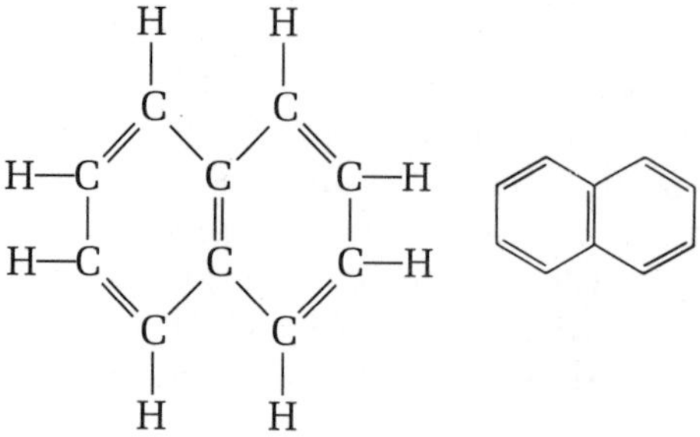

naphthalene, $C_{10}H_8$

Another such compound is anthracene, the starting material for a series of dyes. Here, three benzene rings are fused together.

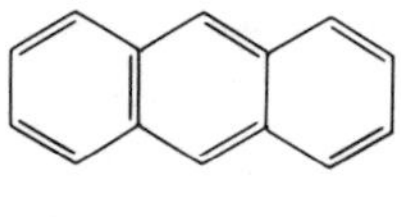

anthracene, $C_{14}H_{10}$

Certain condensed ring compounds are known to be carcinogens (cancer-causing agents). One of these, called benzopyrene, has the structure:

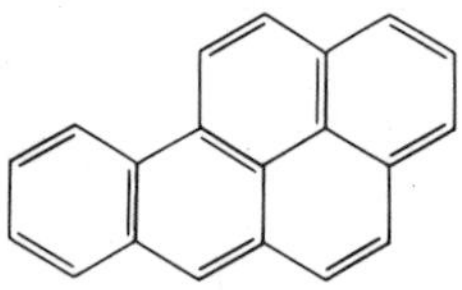

benzopyrene

Benzopyrene has been found in cigarette smoke. This may explain why lung cancer is relatively common among heavy smokers. Benzopyrene has also been isolated, in very small amounts, from the outer layer of charcoal broiled steaks. Apparently, benzopyrene and other compounds like it are formed by the decomposition of fats dripping on hot charcoal.

14.5 SOURCES OF HYDROCARBONS

The fossil fuels coal, natural gas, and petroleum are the only major sources of hydrocarbons. Large deposits of these materials were formed over eons of time by the gradual decomposition of plant and animal life. Natural gas and petroleum are usually trapped underground in rock-covered domes (Figure 14.7). After drilling through the rock, the petroleum and natural gas are brought to the surface through gas pressure or by pumping.

Natural gas and petroleum are often found together.

Coal is obtained from both underground and surface mines. It is found in layers, or seams, which vary in thickness from 1 to 3 m. Automatic machinery is used to obtain the coal from both deep and surface

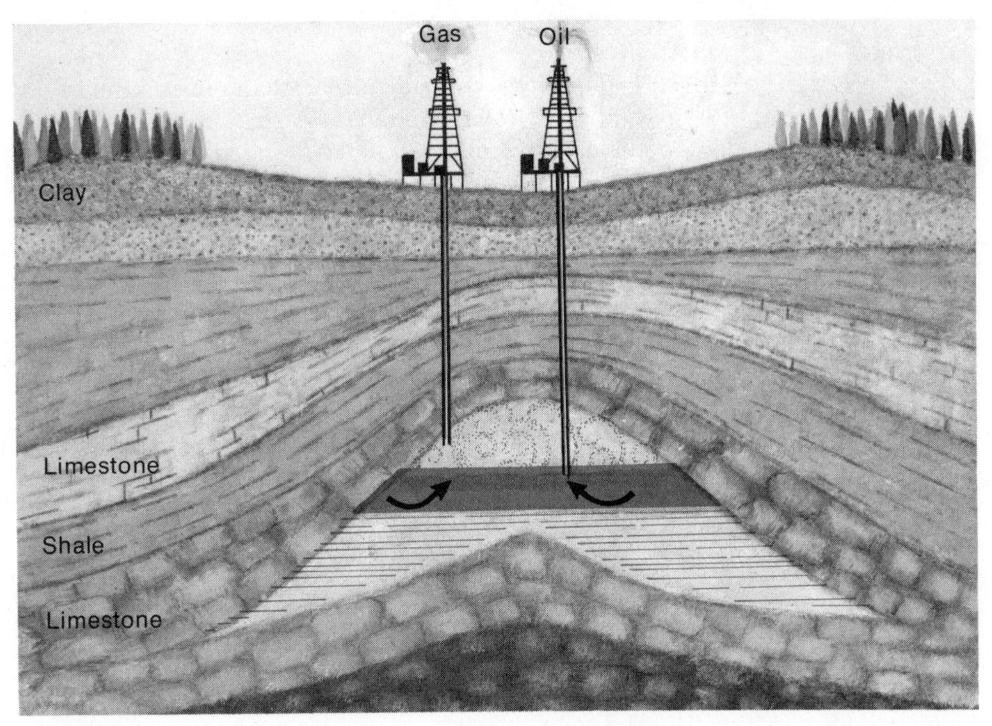

FIGURE 14.7

Oil and natural gas are typically found in oil domes like that shown. Sometimes the gas is under pressure and will force the oil up the well pipe without pumping, but pumping is usually required.

("strip") mines. Coal is found in the United States in many places. There are large deposits in Pennsylvania, West Virginia, Illinois, and the northern great plains.

Natural Gas

Natural gas is an excellent source of the alkanes of low molecular mass. A typical composition of natural gas is given in Table 14.4. Most of the propane and butane are separated by condensing as a liquid mixture. The remaining gas is distributed as commercial natural gas. The propane and butane are sold as liquid fuels under pressure (Bottled Gas). They become gaseous as the pressure drops.

TABLE 14.4 COMPOSITION OF NATURAL GAS

Component	Mole Percent
Methane	82
Ethane	10
Propane	4
Butane	2
Nitrogen, higher hydrocarbons	2

Petroleum

Crude oil is usually a dark-colored, smelly liquid.

Petroleum, or crude oil, is a complex mixture of hydrocarbons. In addition, petroleum contains small amounts of other organic compounds. These compounds contain sulfur, oxygen, or nitrogen atoms in addition to carbon and hydrogen. Petroleum is refined by separating it into portions which are suitable for specific uses. The process used is fractional distillation. The crude oil is heated with a fractionating column (Figure 14.8). The more volatile components are converted to gases and condensed. No attempt is made to separate individual hydrocarbons. Instead, portions which distill within a temperature range are collected along the column. Table 14.5 lists the properties of the distillation fractions from a typical petroleum.

Most of the hydrocarbons in petroleum are alkanes. There are smaller amounts of aromatic hydrocarbons and *cycloalkanes* such as:

cyclopentane C_5H_{10}

cyclohexane C_6H_{12}

Cracking is only one of the processes used to increase the yield of gasoline from petroleum.

The gasoline yield from petroleum can be increased by a process called *cracking*. High molecular mass hydrocarbons in kerosene or fuel oil are broken down into smaller molecules. This is done by heating at about 500°C with a solid catalyst. A typical cracking reaction is:

$$C_{16}H_{34}(l) \rightarrow C_8H_{18}(l) + C_8H_{16}(l) \quad (14.8)$$

TABLE 14.5 DISTILLATION FRACTIONS FROM A TYPICAL PETROLEUM

Fraction	Percent of Total	Composition (C Atoms Per Molecule)	Boiling Range (°C)	Uses
Gas	2	C_1–C_5	<20	gaseous fuel
Petroleum Ether	2	C_5–C_7	30–100	solvents
Gasoline	32	C_6–C_{12}	40–200	motor fuel
Kerosene	18	C_{12}–C_{15}	175–275	diesel, jet fuel
Fuel Oil	20	C_{15}–C_{20}	250–400	heating
Lubricating Oil, Residue	26	>C_{20}	>300	lubricants, paraffin wax, asphalt

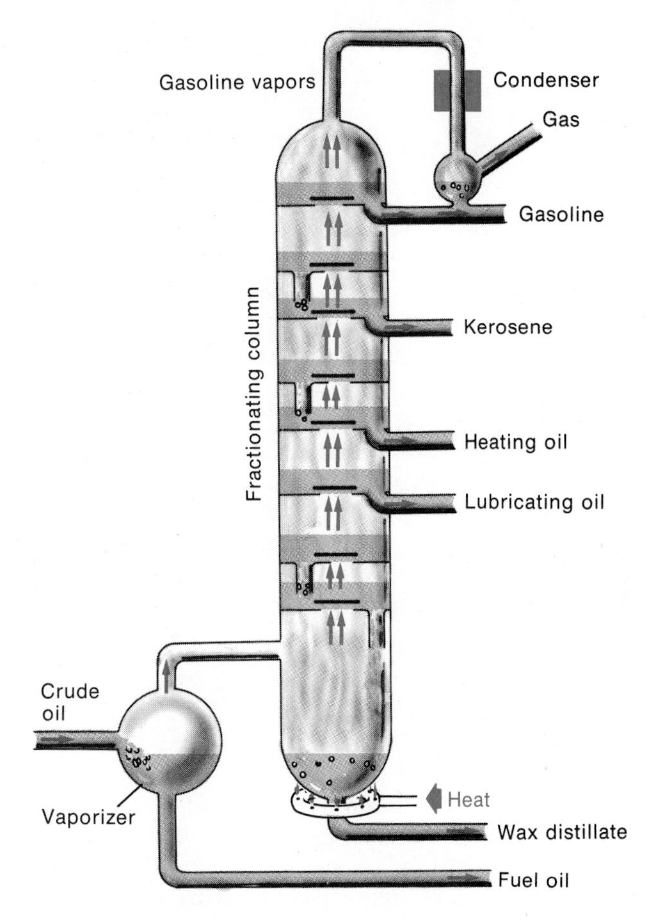

FIGURE 14.8

Fractionation of petroleum. The temperature gradually decreases as one goes up the column. The lighter boiling fractions are removed from the top of the column, the heavier oils near the bottom. Industrial fractionating columns are as tall as a several-story building and are highly automated.

The starting material is an alkane (n = 16; 2n + 2 = 34). The products are an alkane, C_8H_{18}, (n = 8; 2n + 2 = 18) and an alkene, C_8H_{16} (n = 8; 2n = 16).

The cracking process can also be used with gaseous alkanes such as butane:

$$C_4H_{10}(g) \rightarrow C_2H_6(g) + C_2H_4(g) \quad (14.9)$$

This and similar reactions are the principal source of low molecular mass alkenes such as ethylene, C_2H_4. Alkenes are not found in either natural gas or petroleum.

Coal

Coal is a chemically complex material.

Coal consists largely of condensed ring compounds of very high molecular mass. These compounds contain a high proportion of carbon as opposed to hydrogen. There are several different types of coal, differing somewhat in their chemical composition. Lignite (peat) contains about 20% water and has a rather low carbon content, about 50–60 mass %. Soft (bituminous) coal contains from 70–80 mass % of carbon. Hard (anthracite) coal has the highest carbon content, greater than 80%.

When soft coal is heated to about 1000°C in the absence of air, the high molecular mass compounds are largely decomposed. This process is often referred to as the "destructive distillation" of coal. The products, from a kilogram of coal, include:

—*coke* (750 g), a solid material which is mostly carbon with some inorganic ash. The biggest use of coke is in the manufacture of steel. It is also a starting material for making acetylene (Section 14.3) and many other organic chemicals.

—*coal gas* (150 g), which consists mostly of methane (52 mol %) and hydrogen (32 mol %).

—*coal tar* (70 g), a gummy liquid which is an important source of aromatic hydrocarbons such as benzene and naphthalene.

Octane Number of Gasoline

One of the most important properties of gasoline is its octane number. This is a measure of the tendency of the fuel to "knock" (burn unevenly) in a car engine. The octane number of gasoline can be determined using two alkanes as standards:

$$CH_3-\overset{\overset{\large CH_3}{|}}{\underset{\underset{\large CH_3}{|}}{C}}-CH_2-\overset{\overset{\large H}{|}}{\underset{\underset{\large CH_3}{|}}{C}}-CH_3 \qquad\qquad CH_3-CH_2-CH_2-CH_2-CH_2-CH_2-CH_3$$

2, 2, 4-trimethylpentane ("isooctane") heptane

Isooctane, which resists knocking very well, is assigned an octane number of 100. Heptane, which knocks very badly, has an octane number of 0. Gasoline which knocks to the same extent as a mixture of 90% isooctane and 10% heptane would have an octane number of 90.

Until recently, the most common method used to raise octane number was to add tetraethyllead, $Pb(C_2H_5)_4$. With as little as 0.1% of $Pb(C_2H_5)_4$, "high-test" gasoline with an octane number of 100 or greater is produced. (An octane number greater than 100 means that the gasoline has less tendency to knock than isooctane.)

Tetraethyllead cannot be used with most cars built in the United States since 1975. Lead compounds foul the catalytic converter used to reduce air pollution. The octane number of "no-lead" gasoline is raised by:

—increasing the ratio of branched chain to straight chain hydrocarbons. In general, branching improves resistance to knocking. Note, for example, that isooctane (octane number = 100) is a branched hydrocarbon. Heptane (octane number = 0) is a straight chain hydrocarbon.

—adding aromatics such as benzene, which have high octane numbers. Benzene has an octane number of 103; that of toluene is 106.

14.6 SUMMARY

Hydrocarbons are organic compounds containing only the two elements hydrogen and carbon. All hydrocarbons consist of molecules held together by strong covalent bonds. The molecules are held to one another by weak Van der Waals forces. Hydrocarbons are insoluble in water. They burn in excess air to evolve heat and form CO_2 and H_2O.

We have discussed four different types of hydrocarbons.

1. *Alkanes,* whose molecules contain only single bonds. They have the general formula C_nH_{2n+2} (CH_4, C_2H_6, C_3H_8, and so on).
2. *Alkenes,* whose molecules contain one double bond. They have the general formula C_nH_{2n} (C_2H_4, C_3H_6, and so on).
3. *Alkynes,* where there is one triple bond in the molecule. Their general formula is C_nH_{2n-2} (C_2H_2, C_3H_4, and so on).
4. *Aromatics,* which are derivatives of benzene, C_6H_6. In many aromatic hydrocarbons, one or more alkyl groups ($—CH_3$, $—C_2H_5$, and so on) are substituted for an H atom on the benzene ring.

Isomerism, the existence of two or more compounds with the same molecular formula, is common among hydrocarbons. We have considered *structural isomers,* which differ in the way the atoms are bonded to each other. Structural isomers can exist with all types of hydrocarbons. The most reactive hydrocarbons are alkenes and alkynes. They add molecules such as H_2, Cl_2, HCl and H_2O across the double bond.

Low molecular mass alkanes are obtained from natural gas. Petroleum consists mostly of alkanes of higher molecular mass, along with some aromatics. Coal tar is a major source of aromatics. Alkenes and alkynes do not occur naturally. They can be made by cracking alkanes. The only important alkyne, acetylene, can be made by adding calcium carbide to water.

It is possible to name hydrocarbons, and indeed all organic compounds, in a systematic way (IUPAC). We have applied this system to alkanes. For other hydrocarbons, we have relied primarily on common names. We will continue to do this in the next two chapters.

NEW TERMS

addition reaction
alkane
alkene
alkyl group
alkylbenzene
alkyne
aromatic hydrocarbon
coal gas
coal tar
coke
cracking
destructive distillation
hydrocarbon
isomerism
IUPAC name
organic chemistry
resonance
saturated hydrocarbon
structural formula
structural isomers
unsaturated hydrocarbon

QUESTIONS

1. Give two reasons for the large number of different hydrocarbons.
2. What are the three classes of hydrocarbons? How do they differ from each other?

3. Why do the boiling points of straight-chain saturated hydrocarbons increase with molecular mass? Why are they insoluble in water?

4. In all hydrocarbon molecules, how many bonds does a hydrogen atom form to a carbon atom? How many bonds do carbon atoms always form?

5. What is the general formula of an:

   a. alkane b. alkene c. alkyne

6. Explain why C_4H_{10} has isomers but C_3H_8 does not.

7. Give the names of the straight-chain alkanes with 1 to 8 carbon atoms.

8. What is an alkyl group? Give examples of three different alkyl groups.

9. None of the following are correct names for alkanes. Explain what is wrong with each name.

   a. 1-methylpropane b. 3-methylbutane c. 2,2,2-trimethylbutane
   d. 2-ethylbutane

10. How many double and triple bonds are there in a molecule of an:

    a. alkane b. alkene c. alkyne

11. Describe the geometry of the C_2H_4 molecule; the C_2H_2 molecule.

12. Give the structural formulas of the first three alkenes, containing from 2 to 4 carbon atoms.

13. Give the structural formulas of the first three alkynes, containing from 2 to 4 carbon atoms.

14. Write balanced equations for two different methods of preparing acetylene. What is acetylene used for?

15. Draw structural formulas for:

    a. ethylbenzene b. propylbenzene

16. Draw structural formulas for the products formed when the following reagents add to ethylene.

    a. H_2 b. Cl_2 c. Br_2 d. HBr

17. Draw structural formulas for the compounds formed when acetylene, C_2H_2, adds one mole of:

    a. H_2 b. Cl_2 c. Br_2 d. HBr

18. Explain how the following are obtained from underground deposits.

    a. natural gas b. petroleum c. coal

19. Explain what is meant by the fractional distillation of petroleum.

20. Cracking of $C_{12}H_{26}(l)$ gives two hydrocarbons, both of which have six carbon atoms. Write a balanced equation for the reaction.

21. Which of the following compounds would you expect to be found in highest concentration in petroleum?

    a. C_6H_{14} b. C_6H_{12} (an alkene) c. C_6H_6

22. What are the three products of the destructive distillation of coal?

PROBLEMS

1. *General Formulas of Hydrocarbons (Example 14.1)* Give the formula of an alkane with:

   a. 6 carbon atoms b. 12 hydrogen atoms c. 18 carbon atoms

2. *General Formulas of Hydrocarbons (Example 14.5)* Which of the following represent alkanes? alkenes? alkynes?

   a. C_5H_{12} b. C_5H_{10} c. C_5H_8 d. C_5H_6 e. C_6H_{10}

3. *Isomerism (Example 14.2)* Which of the following structural formulas represent the same molecule?

   a. $$\begin{array}{ccccccc} & & H & & & & \\ & & | & & & & \\ CH_3 & - & C & - & CH_2 & - & CH_3 \\ & & | & & & & \\ & & CH_3 & & & & \end{array}$$

   b. $$\begin{array}{ccccccc} & & & & H & & \\ & & & & | & & \\ CH_3 & - & CH_2 & - & C & - & CH_3 \\ & & & & | & & \\ & & & & CH_3 & & \end{array}$$

   c. $$\begin{array}{ccccc} & & CH_3 & & \\ & & | & & \\ CH_3 & - & C & - & CH_3 \\ & & | & & \\ & & CH_3 & & \end{array}$$

   d. $$\begin{array}{ccccccc} & & H & & H & & \\ & & | & & | & & \\ CH_3 & - & C & - & C & - & H \\ & & | & & | & & \\ & & CH_3 & & CH_3 & & \end{array}$$

4. *Isomerism (Example 14.2)* Draw the structural formulas of all the isomers of C_6H_{14}.

5. *Isomerism (Example 14.4)* Which of the following alkynes show structural isomerism?

   a. C_2H_2 b. C_3H_4 c. C_4H_6 d. C_5H_8

6. *Names of Alkanes (Example 14.3)* Give the IUPAC names of all the isomers of hexane shown in Problem 4.

7. *Names of Alkanes (Example 14.3)* Give the structural formulas corresponding to the following IUPAC names.

   a. 2-methylhexane b. 2,2-dimethylpentane
   c. 2-methyl, 3-ethylhexane d. octane e. 4-ethylheptane

8. *Addition Reactions (Example 14.6)* Suggest how the following compounds could be made from propylene, C_3H_6, by addition reactions.

   a. $$\begin{array}{ccccccc} & & H & & H & & \\ & & | & & | & & \\ CH_3 & - & C & - & C & - & H \\ & & | & & | & & \\ & & Cl & & Cl & & \end{array}$$

   b. $CH_3-CH_2-CH_3$

   c. $$\begin{array}{ccccc} & & H & & \\ & & | & & \\ CH_3 & - & C & - & CH_3 \\ & & | & & \\ & & Br & & \end{array}$$

9. *Addition Reactions (Example 14.6)* Starting with acetylene, how could you prepare

   a. C_2H_4 b. C_2H_5Cl

* * * * *

10. Give the structural formula of a 7-carbon

    a. alkane b. alkene c. alkyne d. aromatic hydrocarbon

11. Classify the following as alkanes, alkenes, or alkynes.

    a. C_4H_{10} b. C_4H_{12} c. C_4H_6 d. C_4H_8

12. Draw structural formulas for all the isomers of the alkyne C_4H_6.

13. Give the formula of:

 a. the alkene with 8 carbon atoms
 b. the alkyne with 8 carbon atoms
 c. the alkene with 8 hydrogen atoms
 d. the alkyne with 8 hydrogen atoms

14. By reacting propane, C_3H_8, with Cl_2 it is possible to form compounds of molecular formula $C_3H_5Cl_3$. How many such compounds are there?

15. Alkyl chlorides are named like alkanes except that the prefix "chloro" is used. Name the following alkyl chlorides.

 a. 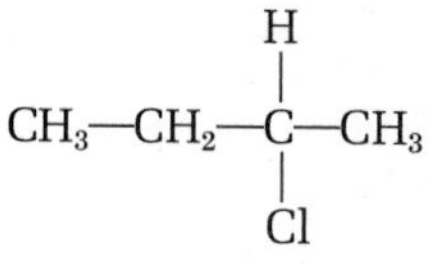 $CH_3—CH_2—CH(H)(Cl)—CH_3$ (central C bears H above and Cl below)

 b. $CH_3—C(H)(Cl)—C(H)(Cl)—CH_3$ (each central C bears H above and Cl below)

 c. $CH_3—CH_2—CH_2—CCl_3$

 d. $H—C(H)(Cl)—CH_2—C(H)(Cl)—H$ (each terminal C bears H above and Cl below)

16. What reagents would you use to prepare the following compounds from propylene?

 a. C_3H_8 b. $C_3H_6Cl_2$ c. C_3H_7Cl d. C_3H_7OH

*17. Draw the structural formulas of all the compounds that can be formed by substituting chlorine atoms for hydrogen in the benzene ring.

*18. Benzene has an octane number of 103, hexane an octane number of 25. What would be the octane number of a fuel containing equal parts of benzene and hexane?

*19. A certain hydrocarbon is found to contain 8.75% by mass of hydrogen. Its molecular mass is about 92. What is its molecular formula? Suggest a likely structural formula.

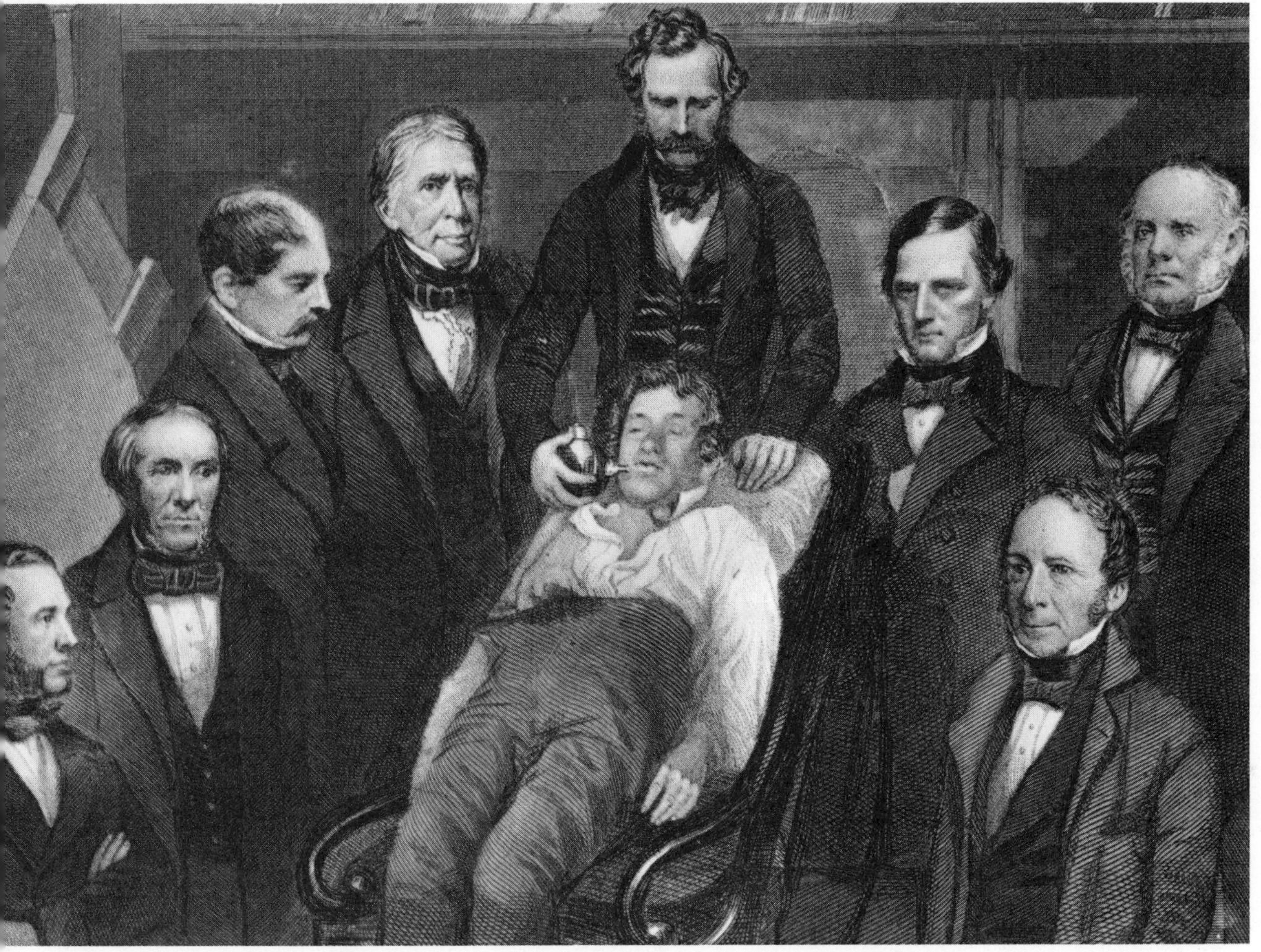

The first use of ether as an anaesthetic in 1846. The doctor standing over the patient is William T. Morton. The other gentlemen took advantage of the occasion to have their pictures taken. Bettman Archive.

15

ORGANIC CHEMISTRY: Oxygen Compounds

Carbon forms strong covalent bonds with several nonmetals besides hydrogen. Many organic compounds contain, in addition to carbon and hydrogen, atoms of:

—a halogen (F, Cl, Br, or I) in Group 7 of the Periodic Table
—oxygen or sulfur in Group 6
—nitrogen or phosphorus in Group 5.

Some organic compounds containing halogens were discussed in Chapter 14. In this chapter, we will consider organic compounds containing oxygen. Next to carbon and hydrogen, oxygen is the element most often found in organic molecules.

Oxygen compounds are classified according to the type of **functional group** present in the molecule. A functional group consists of an atom or group of atoms. It confers certain properties upon an organic compound. Perhaps the simplest functional group is the —OH group. This group of atoms is present in all the organic compounds that we call alcohols. All alcohols have similar structural formulas and resemble one another in their chemical reactions.

In this chapter, we will discuss several different series of organic compounds. Each series has a characteristic functional group. We will write structural formulas for several members of each series. In all cases, we find that an oxygen atom forms a total of two bonds. Sometimes, an oxygen atom forms two single bonds to two different atoms. At least one of these is a carbon atom.

$$\diagdown\!\!-\!\!\!\!\!\!\!\!\!\diagup\,\mathrm{C{-}O{-}}$$

In other compounds, an oxygen atom forms a double bond to a carbon atom.

$$\diagdown\!\!\!\!\diagup\,\mathrm{C{=}O}$$

Actually, all organic molecules obey the octet rule.

The oxygen atom in these molecules obeys the octet rule. There are two unshared pairs of electrons around the oxygen, in addition to the two bonds. These unshared pairs are shown in Lewis structures (Chapter 11). However, they are not shown in the structural formulas used in this chapter.

15.1 ALCOHOLS. FUNCTIONAL GROUP: —OH

The general formula of an alcohol can be written as R—OH. Here, R is most commonly an alkyl group (CH_3—, CH_3—CH_2—, and so on). We might consider alcohols to be derivatives of alkanes in which an H atom has been replaced by an —OH group. From another point of view, an alcohol can be considered as a derivative of water (H—O—H). An H atom of water has been replaced by an alkyl group.

The first two members of this series are well-known compounds. Their common names are methyl alcohol and ethyl alcohol.

The IUPAC names are methanol and ethanol.

```
     H
     |
  H—C—O—H        or    CH3—OH           methyl alcohol
     |
     H

     H  H
     |  |
  H—C—C—O—H      or    CH3—CH2—OH       ethyl alcohol
     |  |
     H  H
```

The geometries of these molecules are shown in Figure 15.1.

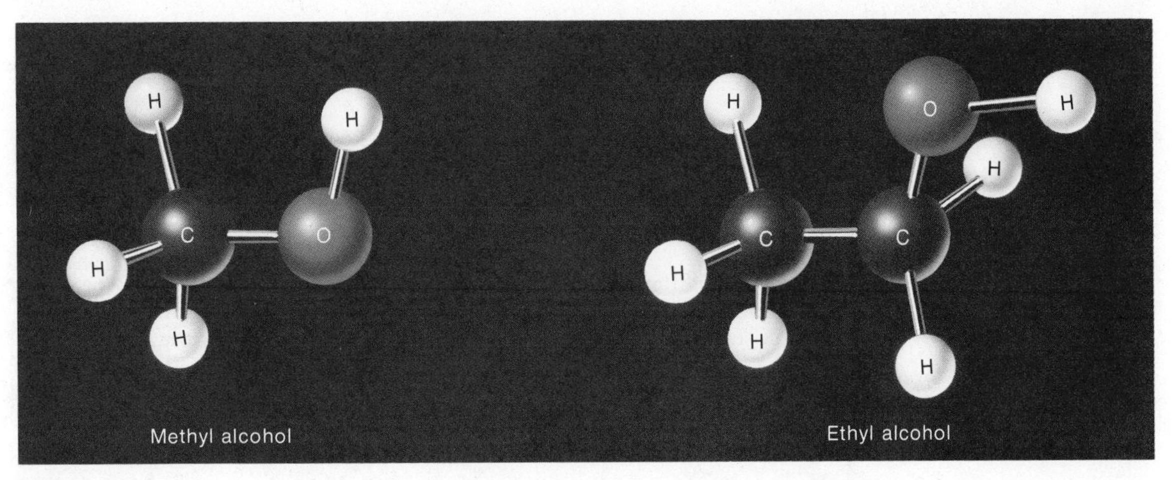

FIGURE 15.1

Structures of methyl and ethyl alcohol. In both of these molecules the bond angles are all 109°, the tetrahedral angle. The geometries follow from the fact that there are four pairs of electrons around each of the C and O atoms and all bonds are single.

There are two isomeric alcohols containing three carbon atoms. In one of these, the —OH group is attached to the carbon atom at the end of the chain. In the other, the —OH group is joined to the carbon atom in the center of the chain.

$$CH_3—CH_2—CH_2—OH \qquad \begin{array}{c} H \\ | \\ CH_3—C—CH_3 \\ | \\ OH \end{array}$$

propyl alcohol   isopropyl alcohol

EXAMPLE 15.1

Draw structural formulas for all the isomeric alcohols containing four carbon atoms, C_4H_9OH.

SOLUTION

You will recall from Chapter 14 that there are two alkanes with four carbon atoms:

$$CH_3—CH_2—CH_2—CH_3 \quad \text{and} \quad \begin{array}{c} H \\ | \\ CH_3—C—CH_3 \\ | \\ CH_3 \end{array}$$

The alcohols are formed by replacing an H atom in one of these hydrocarbons with an —OH group. In the straight-chain hydrocarbon, there are two different types of hydrogen atoms:

—those at the end of the chain (CH_3 groups)
—those in the middle of the chain (CH_2 groups)

Once we add another element to C and H, the number of isomers goes up fast.

So, there are two isomeric alcohols derived from this hydrocarbon:

$$CH_3{-}CH_2{-}CH_2{-}CH_2{-}OH \qquad CH_3{-}\underset{\underset{OH}{|}}{\overset{\overset{H}{|}}{C}}{-}CH_2{-}CH_3$$

(isomer 1) (isomer 2)

A similar situation applies with the branched-chain hydrocarbon. The hydrogen atom bonded to the central carbon differs from those in the CH_3 groups. This gives two more isomeric alcohols:

$$CH_3{-}\underset{\underset{CH_3}{|}}{\overset{\overset{OH}{|}}{C}}{-}CH_3 \qquad CH_3{-}\underset{\underset{CH_3}{|}}{\overset{\overset{H}{|}}{C}}{-}CH_2{-}OH$$

(isomer 3) (isomer 4)

There are four and only four alcohols with the formula C_4H_9OH.

In one type of alcohol, the —OH group is attached to a benzene ring. The simplest example is the compound phenol:

C_6H_5OH (benzene ring bearing OH) phenol

Phenol was one of the first antiseptics. It is effective in killing germs but can cause severe skin burns. Today, most of the phenol produced in the United States (5×10^8 kg/yr) is used to make plastics.

Physical Properties

Table 15.1 lists some physical properties of a series of straight-chain alcohols. In each case, the —OH group is bonded to a carbon atom at the end of the chain. It is interesting to compare the physical properties of these alcohols to those of the alkanes given in Table 14.1, p. 339.

The effect of the —OH group on boiling point is enormous here.

1. The alcohols are much higher boiling than the alkanes. Methyl alcohol, CH_3OH, boils more than 200° above methane, CH_4 (65°C *vs.* −162°C). Indeed, the boiling point of methyl alcohol is about equal to that of C_6H_{14} (69°C). This is true even though the molecular mass of CH_3OH (32) is less than half that of C_6H_{14} (86).

2. The alcohols are much more soluble in water than the alkanes. The first three alcohols are soluble in water in all proportions. In contrast, all of the alkanes are almost insoluble in water.

TABLE 15.1 PHYSICAL PROPERTIES OF ALCOHOLS

Common Name	Formula	State at 25°C	mp (°C)	bp (°C)	Solubility in Water (g/100 g water at 20°C)
Methyl alcohol	CH_3OH	liquid	−98	65	infinite
Ethyl alcohol	CH_3CH_2OH	liquid	−112	78	infinite
Propyl alcohol	$CH_3(CH_2)_2OH$	liquid	−88	82	infinite
Butyl alcohol	$CH_3(CH_2)_3OH$	liquid	−89	118	7.9
Amyl alcohol	$CH_3(CH_2)_4OH$	liquid	−78	138	2.8
Hexyl alcohol	$CH_3(CH_2)_5OH$	liquid	−51	157	0.59

These properties of alcohols can be explained in terms of hydrogen bonding. An alcohol can hydrogen bond to itself (Figure 15.2). This explains the relatively high boiling points of these compounds. It can also hydrogen bond to water. This explains the high water solubility. Notice (Table 15.1) that solubility drops off as the number of carbon atoms increases. In a molecule like hexyl alcohol

$$CH_3—CH_2—CH_2—CH_2—CH_2—CH_2—OH$$

hydrogen bonding does not supply enough energy to allow the long hydrocarbon chain to break into the water structure.

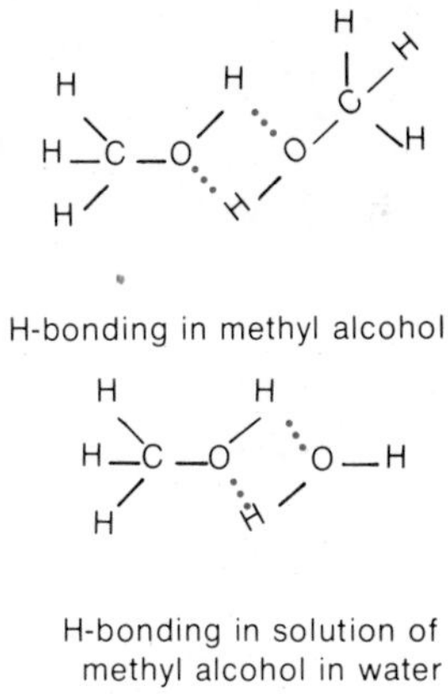

H-bonding in methyl alcohol

H-bonding in solution of methyl alcohol in water

FIGURE 15.2

Hydrogen bonding can occur between alcohol molecules and between alcohol and water molecules. Such bonding gives alcohols relatively high boiling points and increases their solubility in water.

Methyl Alcohol

Methyl alcohol is often referred to as "wood alcohol". At one time it was made by heating hardwoods such as birch or maple to about 300°C in the absence of air. The products include:

—a solid, charcoal, which is mostly carbon.

—a foul-smelling liquid, which contains methyl alcohol, acetic acid (Section 15.5), and many other organic compounds.

—a gas which consists largely of CO and CO_2.

About 250 g of charcoal and 20 g of methyl alcohol are obtained from a kilogram of wood.

This process, known as the destructive distillation of wood, was given up 40 years ago as a source of methyl alcohol. However, there are plans for reviving it now, at least in areas where there are large hardwood forests. It is one of the few organic preparations where the starting material is a renewable resource. Typically, such preparations start with a fossil fuel.

Gasohol contains about 10 percent methyl or ethyl alcohol.

Most methyl alcohol today is made from carbon monoxide and hydrogen. The reaction is carried out at high temperatures (300–400°C) and pressures (200–300 atm) in the presence of a metal oxide catalyst.

$$CO(g) + 2\,H_2(g) \xrightarrow[Cr_2O_3]{ZnO} CH_3OH(l) \qquad (15.1)$$

Natural gas is the starting material for making both CO and H_2.

FIGURE 15.3

Destructive distillation of wood. In this process, wood inside the container is heated, probably by a wood fire, in the absence of air. The distillate is mainly water but contains a significant amount of methyl alcohol. The latter can be recovered by a second distillation. The apparatus is crude, and could be assembled in a lumber yard or out in the woods. Charcoal is the other product.

Methyl alcohol is poisonous. Very small amounts taken internally can cause blindness or death. This property somewhat limits the usefulness of methyl alcohol. However, large quantities of it are used to make other organic compounds, notably formaldehyde (Section 15.3).

Ethyl Alcohol

For industrial uses, ethyl alcohol is made from ethylene as described in Chapter 14:

$$\underset{\text{ethylene}}{C_2H_4(g)} + H_2O(l) \rightarrow \underset{\text{ethyl alcohol}}{C_2H_5OH(l)} \quad (15.2)$$

Ethyl alcohol made by Equation 15.3 is the same substance as is made by Equation 15.2.

The ethyl alcohol used in beverages is made by a different process discovered thousands of years ago. This is the fermentation of grains or other vegetable material. The first step in fermentation involves breaking down organic molecules known as carbohydrates. The product is a simple sugar, glucose, molecular formula $C_6H_{12}O_6$. Glucose is then converted to ethyl alcohol and carbon dioxide.

$$C_6H_{12}O_6(aq) \rightarrow 2\ C_2H_5OH(aq) + 2\ CO_2(g) \quad (15.3)$$

TABLE 15.2 COMMON ALCOHOLIC BEVERAGES

Beverage	Major Source of Carbohydrate	Volume % Alcohol	Proof
Beer	Barley, rice, wheat	3–9	6–18
Wine	Grapes, other fruit	10–12	20–24
Brandy	Wine	40–45	80–90
Whiskey	Barley, rye, corn	40–55	80–110
Rum	Molasses	45	90
Gin	Barley, other grains	35–45	70–90
Vodka	Grains, potatoes, sugar beets	35–45	70–90

Each step in the fermentation process requires a specific organic catalyst known as an *enzyme*. The process stops when the alcohol content reaches 12%. At higher concentrations of alcohol, the enzyme molecules are destroyed.

Table 15.2 lists the carbohydrate source and volume percent of ethyl alcohol in several alcoholic beverages. The last column gives the "proof" of the beverage. This is twice the percentage of ethyl alcohol. In many cases, other ingredients in addition to those listed are present. These may be added to give a desired taste (hops with beer, juniper berries with gin). In other cases, they supply the enzymes for fermentation (yeast with home-made wine).

The last five beverages listed in Table 15.2 are made by distillation so as to increase the concentration of ethyl alcohol. There is an upper limit to the percentage of ethyl alcohol obtained by distillation from water solution. At 78.2°C, a mixture of 95% C_2H_5OH and 5% water boils off. Most of the ethyl alcohol used in the chemistry laboratory has this composition.

Ethyl alcohol is one of the most common reagents in organic chemistry. It is used as a solvent for recrystallization. It is also a starting material for preparing many different organic compounds. Ordinarily, the ethyl alcohol used for these purposes is "denatured". A small amount of a compound such as methyl alcohol is added to make it unfit to drink. This way, the heavy tax on alcoholic beverages is avoided. Undenatured 95% alcohol costs about 30 times as much as the denatured product.

Effects of Alcohol in the Body

When ethyl alcohol is taken internally, it passes directly from the stomach into the bloodstream. If the stomach is empty, this process occurs almost immediately. Food, particularly fatty food, slows it down somewhat. When the alcohol reaches the brain, it produces the effects shown in Table 15.3, p. 372. At very low concentrations, alcohol can act as a stimulant. However, as the concentration increases, the effects are quite different. At 0.15%, the symptoms of intoxication are obvious. This is the level ordinarily used for conviction of drunken drivers. Still higher levels produce unconsciousness and even death.

TABLE 15.3 EFFECT OF ALCOHOL ON THE BRAIN

Amount of Beverage[a]	Concentration of Alcohol in Blood	Effect	Time Required to Drop to 0.02%
1 shot or 1 beer[b]	0.02%	Mild stimulation	
2 shots or 2 beers	0.05%	Relaxation, some loss of skills	2 hr
4 shots or 4 beers	0.10%	Poor coordination, exaggerated behavior	5 hr
6 shots or 6 beers	0.15%	Unsteadiness, loss of coordination	8 hr
10 shots	0.30%	Unconsciousness	
20 shots	0.50%	Coma	

[a]Varies with the mass and habits of the individual. The values quoted are for a moderate drinker weighing 70 kg (154 lb). For people who weigh less or who seldom drink, the amount required may be considerably less.

[b]1 shot = 1 oz = 30 cm^3, of 90-proof whiskey. 1 beer = 12 oz = 360 cm^3.

The body begins to dispose of alcohol as soon as it is taken in. Small amounts are eliminated in the urine and breath (the basis of the "breathalyzer" test). However, most of the alcohol is acted upon by the liver. There it is metabolized, ultimately to CO_2 and H_2O. This is a slow process, as you will note from the times listed in Table 15.3. Moreover, it interferes with other processes going on in the liver. These include the metabolism of glucose and the synthesis of protein. These processes virtually cease while the liver is disposing of large amounts of alcohol. For these reasons among others, drinking large amounts of alcohol quickly is a very dangerous practice.

15.2 ETHERS. FUNCTIONAL GROUP: —O—

In the compounds known as ethers, a single oxygen atom is bonded to two hydrocarbon groups. These groups may be the same or different:

$CH_3—O—CH_3$	$CH_3—O—CH_2—CH_3$
dimethyl ether	methyl ethyl ether

The general formula of an ether is written as

$$R—O—R'$$

with the understanding that R and R′ may be the same or different. We might consider an ether as a derivative of water in which both H atoms have been replaced by hydrocarbon groups. (Recall that in an alcohol only one of the H atoms in water is replaced.)

Ethers, like all organic compounds, can show structural isomerism. There are three different ethers containing four carbon atoms (Example 15.2).

EXAMPLE 15.2

Write structural formulas for all the ethers of molecular formula $C_4H_{10}O$.

SOLUTION

In one isomer, the oxygen atom is bonded to two ethyl groups

$CH_3—CH_2—O—CH_2—CH_3$ diethyl ether

If one of the groups joined to oxygen is a CH_3 group, there are two possibilities.

$CH_3—O—CH_2—CH_2—CH_3$ methyl propyl ether

$$\begin{array}{c} \quad\quad\quad CH_3 \\ \quad\quad\quad | \\ CH_3—O—C—CH_3 \\ \quad\quad\quad | \\ \quad\quad\quad H \end{array}$$ methyl isopropyl ether

The geometry of dimethyl ether is shown in Figure 15.4. Notice that this molecule, like those of other ethers, is "bent". The angle around the oxygen atom is the tetrahedral angle, 109°.

O

109°

This is just what it should be, since O atoms follow the octet rule.

Because of this structure, ether molecules, like water molecules, are polar. The oxygen atom acts as the negative pole in the molecule.

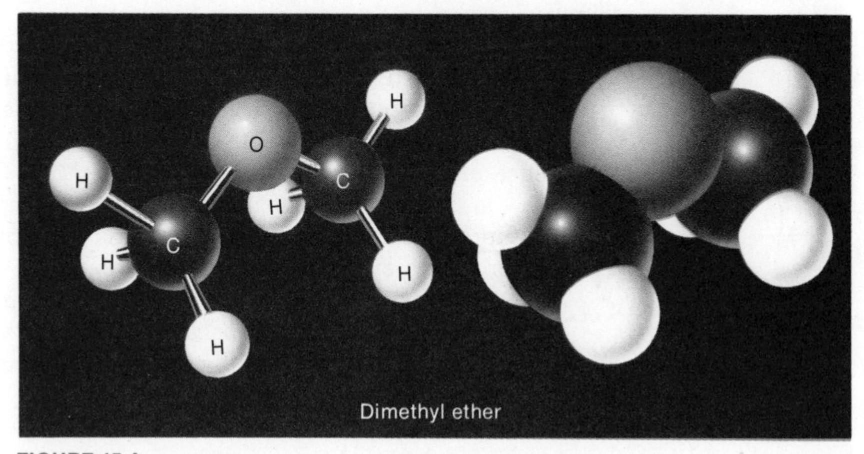

FIGURE 15.4

Models showing the structure of dimethyl ether. The structure resembles that of propane, C_3H_8. Since ethers do not hydrogen bond, dimethyl ether and propane have similar boiling points (−25 and −42°C respectively).

TABLE 15.4 PHYSICAL PROPERTIES OF ETHERS

Common Name	Structure	State (25°C)	bp (°C)	Water Solubility (g/100 g) at 20°C
Dimethyl ether	CH_3—O—CH_3	g	−25	70
Methyl ethyl ether	CH_3—O—CH_2—CH_3	g	8	"soluble"
Methyl propyl ether	CH_3—O—CH_2—CH_2—CH_3	l	40	3
Methyl isopropyl ether	CH_3—O—CH—$(CH_3)_2$	l	33	6
Diethyl ether	CH_3—CH_2—O—CH_2—CH_3	l	35	7.5
Methyl phenyl ether	CH_3—O—C_6H_5 (phenyl ring)	l	154	trace

Unlike alcohols, ether molecules cannot form hydrogen bonds with each other. There is no hydrogen atom bonded to oxygen in an ether. As a result, ethers have lower boiling points than alcohols of similar molecular mass. Notice (Table 15.4) that the first two ethers are gases at 25°C and 1 atm.

Diethyl Ether

The only ether in common use is diethyl ether, often called simply "ether". It is made from ethyl alcohol. Upon heating to 140°C with sulfuric acid, two molecules of alcohol condense with loss of water.

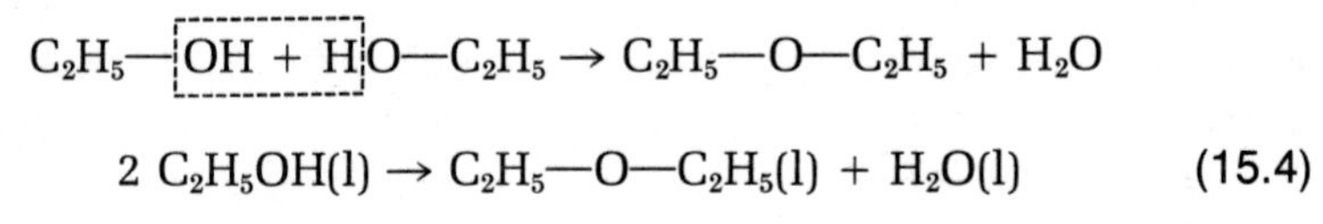

$$C_2H_5—\boxed{OH + H}O—C_2H_5 \rightarrow C_2H_5—O—C_2H_5 + H_2O$$

$$2\ C_2H_5OH(l) \rightarrow C_2H_5—O—C_2H_5(l) + H_2O(l) \quad (15.4)$$

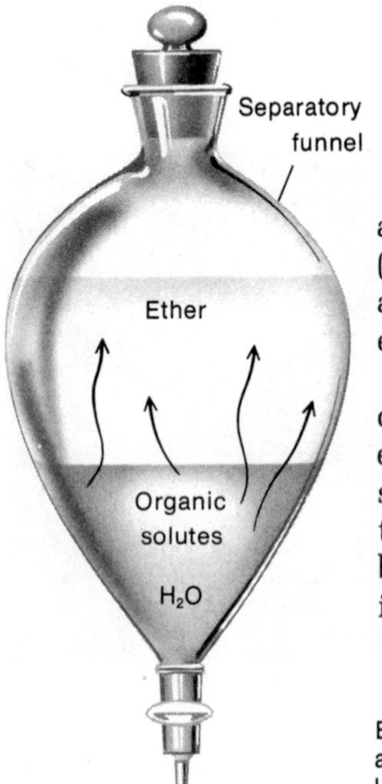

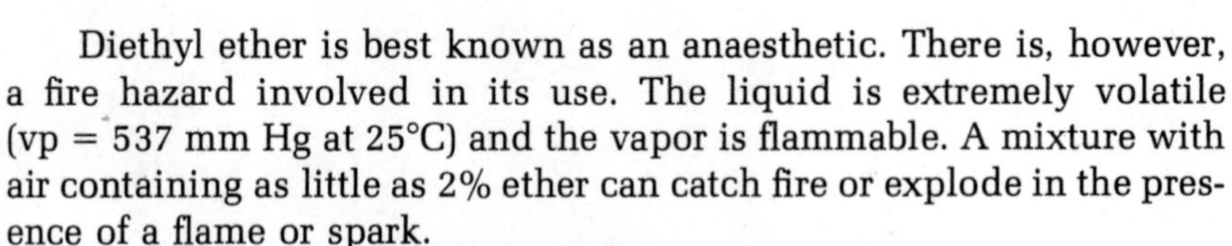

Diethyl ether is best known as an anaesthetic. There is, however, a fire hazard involved in its use. The liquid is extremely volatile (vp = 537 mm Hg at 25°C) and the vapor is flammable. A mixture with air containing as little as 2% ether can catch fire or explode in the presence of a flame or spark.

In the chemistry laboratory, diethyl ether is used to extract organic compounds from water. Most of these compounds are more soluble in ether than they are in water. When the water solution is shaken with a small amount of diethyl ether, most of the organic compound enters the ether layer. The ether, because of its high vapor pressure, can easily be separated from the compound by evaporation. This should be done in a hood, far removed from any flame.

Extraction by a separatory funnel. The solution to be extracted is put in the funnel, along with diethyl ether. Shaking results in a transfer of organic solutes to the ether layer. When the layers separate, the water is drawn off *via* the stopcock.

15.3 ALDEHYDES. FUNCTIONAL GROUP: $-\overset{\displaystyle O}{\overset{\|}{C}}-H$ (abbreviated —CHO)

Aldehydes have the general formula:

$$R-\underset{\underset{\displaystyle O}{\|}}{C}-H$$

where R may be, in the simplest case, a hydrogen atom:

$$H-\underset{\underset{\displaystyle O}{\|}}{C}-H \qquad \text{formaldehyde}$$

In all other cases, R is a hydrocarbon group (CH_3, CH_3-CH_2-, and so on).

$$CH_3-\underset{\underset{\displaystyle O}{\|}}{C}-H \qquad \text{acetaldehyde}$$

$$CH_3-CH_2-\underset{\underset{\displaystyle O}{\|}}{C}-H \qquad \text{propionaldehyde}$$

The $\rangle C{=}O$ group present in aldehydes is referred to as a *carbonyl* group. This group is planar. The carbon atom of the group is in the center of an equilateral triangle. There is an oxygen atom at one corner of this triangle. In all aldehydes, there is an H atom at the second corner of the triangle. In formaldehyde, the third corner is occupied by

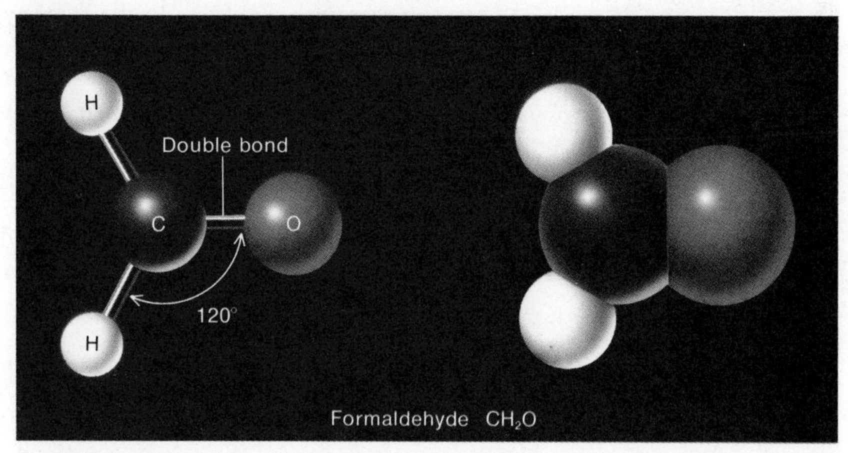

FIGURE 15.5

Models showing the structure of formaldehyde. The C=O bond is a double bond; the geometry is clearest in the ball and stick model. The bond angles are all 120°.

TABLE 15.5 PHYSICAL PROPERTIES OF ALDEHYDES

Common Name	Structure	State (25°C, 1 atm)	bp (°C)	Solubility in Water (g/100 g at 20°C)
Formaldehyde	H—CHO	gas	−21	>55
Acetaldehyde	CH_3—CHO	gas	20	infinite
Propionaldehyde	CH_3—CH_2—CHO	liquid	49	20
Butyraldehyde	CH_3—CH_2—CH_2—CHO	liquid	76	4
Isobutyraldehyde	$(CH_3)_2$—CH—CHO	liquid	63	11
Benzaldehyde	C_6H_5—CHO	liquid	179	0.3

another H atom. In acetaldehyde and all higher aldehydes, there is a carbon atom at this third corner. In all cases, the bond angles around the $\rangle C{=}O$ group are 120°.

Would you expect aldehydes to show hydrogen bonding?

The physical properties of aldehydes follow the trends we have seen with other series. Boiling point increases steadily with molecular mass. Water solubility decreases as molecular mass increases. The first two members of the series (Table 15.5) are gases at 25°C and 1 atm. They are also very soluble in water.

Formaldehyde

You are probably familiar with formaldehyde, at least with its odor. It is sold as a 40% water solution known as "formalin". This solution is used to preserve specimens in the biology laboratory. It slows down the decay of cells. The same property makes formalin useful in embalming. At one time, formaldehyde was used as a food preservative, particularly in milk. However, its use for that purpose was prohibited many years ago because it is poisonous. Today the major use of formaldehyde is in making other organic compounds. Large amounts are used to make plastics and adhesives.

Pure HCHO is chemically very reactive.

Formaldehyde is prepared from methyl alcohol. By one process, methyl alcohol vapor and air are passed over a silver or copper catalyst at 500°C.

$$CH_3OH(g) + \tfrac{1}{2}\,O_2(g) \xrightarrow{Ag} HCHO(g) + H_2O(g) \qquad (15.5)$$

The gaseous mixture formed is dissolved in water. Formalin always contains some unreacted methyl alcohol along with 40% formaldehyde.

15.4 KETONES. FUNCTIONAL GROUP: $-\underset{\underset{\displaystyle O}{\|}}{C}-$

Ketones, like aldehydes, contain the carbonyl group. However, in a ketone, there are no hydrogen atoms attached to the carbon atom of that group. Instead, this carbon is joined to two hydrocarbon groups. The general formula of a ketone is:

$$R-\underset{\underset{\displaystyle O}{\|}}{C}-R'$$

R and R′ may be the same or different. Thus R and R′ might both be methyl groups:

Neither R nor R′ can be H. Why not?

$$CH_3-\underset{\underset{\displaystyle O}{\|}}{C}-CH_3 \qquad \text{dimethyl ketone (acetone)}$$

One might be a methyl group and the other an ethyl group:

$$CH_3-\underset{\underset{\displaystyle O}{\|}}{C}-CH_2-CH_3 \qquad \text{methyl ethyl ketone}$$

EXAMPLE 15.3

Classify each of the following as an aldehyde, ketone, alcohol, or ether.

a. $H-\underset{\underset{\displaystyle O}{\|}}{C}-CH_3$ b. 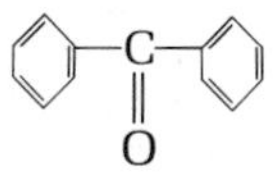 $C_6H_5-\underset{\underset{\displaystyle O}{\|}}{C}-C_6H_5$ c. CH_3-CH_2-OH

d. 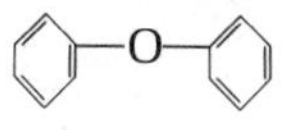 $C_6H_5-O-C_6H_5$ e. $CH_3-\underset{\underset{\displaystyle H-C=O}{|}}{\overset{\overset{\displaystyle H}{|}}{C}}-CH_3$

SOLUTION

Compounds (a) and (e) are aldehydes. They contain the $-\overset{\overset{\displaystyle H}{|}}{C}=O$ group. Compound (b) is a ketone; the C=O group is attached to two hydrocarbon groups. Compound (c) is an alcohol, (d) is an ether.

The geometry of ketone molecules is similar to that of aldehydes. Compare, for example, the geometry of acetone (Figure 15.6) to that of

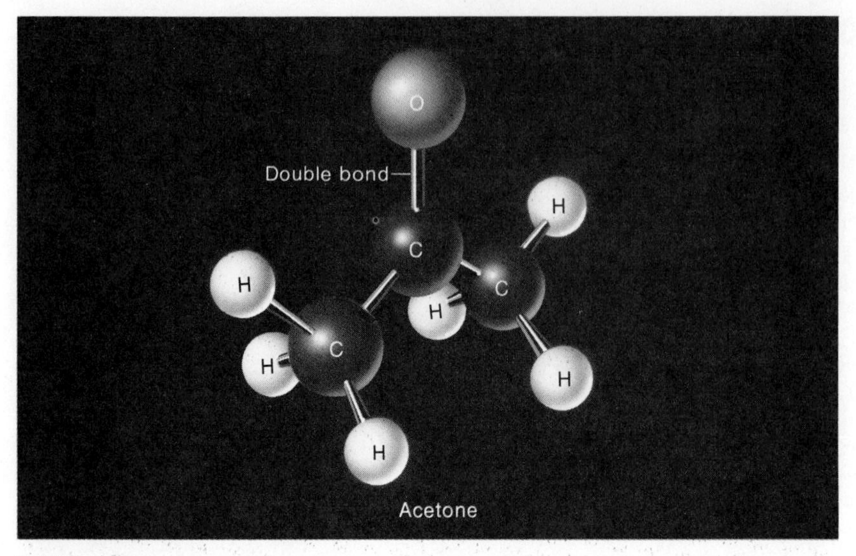

FIGURE 15.6

The geometry of the acetone molecule. The C═O bond is a double bond. The bond angles around the central C atom are all 120°.

formaldehyde or acetaldehyde. In each case, the bond angles around the carbon atom of the carbonyl group are 120°.

Acetone

The ketone of greatest commercial importance is the first one, acetone. It is used as a solvent for waxes, plastics, and lacquers. In the laboratory, acetone is sometimes used to dry glassware. It is both water soluble and volatile (bp = 56°C; vp = 229 mm Hg at 25°C). So, flasks washed with acetone dry quickly. However, acetone vapor is flammable. There must be no flame in the vicinity when you use acetone.

Both aldehydes and ketones are made by controlled oxidation of alcohols. If we just burned the alcohol in air, the oxidation would yield CO_2 and H_2O.

Acetone is made by the reaction of isopropyl alcohol with oxygen. The conditions are similar to those used to prepare formaldehyde from methyl alcohol. A mixture of isopropyl alcohol vapor and air is passed over a metal catalyst at a high temperature.

$$CH_3-\underset{\displaystyle OH}{\overset{\displaystyle H}{\underset{|}{\overset{|}{C}}}}-CH_3(g) + \tfrac{1}{2}\,O_2(g) \xrightarrow{Ag} CH_3-\underset{\displaystyle O}{\underset{\|}{C}}-CH_3(g) + H_2O(g) \qquad (15.6)$$

15.5 CARBOXYLIC ACIDS. FUNCTIONAL GROUP: $-\underset{\displaystyle O}{\underset{\|}{C}}-OH$ (abbreviated —COOH)

The functional group common to all carboxylic acids is called a carboxyl group. It combines the carbonyl group $\left(\diagdown\!\!\!\diagup C{=}O\right)$ found in aldehydes and ketones with the hydroxyl (—OH) group of alcohols. The general formula of a carboxylic acid is:

$$\mathrm{R{-}\underset{\underset{\displaystyle O}{\|}}{C}{-}OH}$$

In the first member of the series, R is a hydrogen atom

$$\mathrm{H{-}\underset{\underset{\displaystyle O}{\|}}{C}{-}OH} \qquad \text{formic acid (R = H)}$$

In all other acids, R is a hydrocarbon group. Examples include:

$$\mathrm{CH_3{-}\underset{\underset{\displaystyle O}{\|}}{C}{-}OH} \qquad \text{acetic acid } (\mathrm{R = CH_3})$$

$$\mathrm{CH_3{-}CH_2{-}\underset{\underset{\displaystyle O}{\|}}{C}{-}OH} \qquad \text{propionic acid } (\mathrm{R = CH_3{-}CH_2{-}})$$

Carboxylic acids in water solution are partially ionized to give H^+ ions. In all cases, it is the hydrogen atom of the —OH group that ionizes:

$$\underset{\text{formic acid}}{\mathrm{H{-}\underset{\underset{\displaystyle O}{\|}}{C}{-}OH(aq)}} \rightarrow \underset{\text{formate ion}}{\mathrm{(H{-}\underset{\underset{\displaystyle O}{\|}}{C}{-}O)^-(aq)}} + \mathrm{H^+(aq)} \qquad (15.7)$$

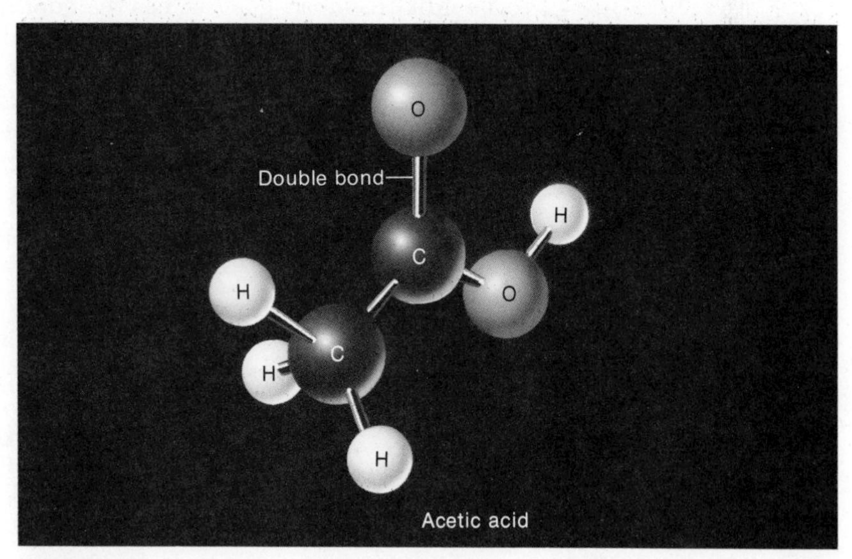

FIGURE 15.7

Model showing geometry of the acetic acid molecule. The C=O bond, which is vertical, is a double bond. The bond angles around the central C atom are all 120°. The C—C—H and C—O—H bond angles are 109°.

TABLE 15.6 PHYSICAL PROPERTIES OF CARBOXYLIC ACIDS

Common Name	Structure	State (25°C, 1 atm)	bp (°C)	Solubility in Water (g/100 g at 20°C)
Formic Acid	$HCOOH$	liquid	100	infinite
Acetic Acid	CH_3COOH	liquid	118	infinite
Propionic Acid	CH_3CH_2COOH	liquid	141	infinite
Butyric Acid	$CH_3(CH_2)_2COOH$	liquid	162	infinite
Valeric Acid	$CH_3(CH_2)_3COOH$	liquid	187	5
Caproic Acid	$CH_3(CH_2)_4COOH$	liquid	205	1
Benzoic Acid	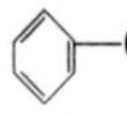 C_6H_5—COOH	solid	249	0.2

$$CH_3—\underset{\underset{O}{\|}}{C}—OH(aq) \rightarrow (CH_3—\underset{\underset{O}{\|}}{C}—O)^-(aq) + H^+(aq) \qquad (15.8)$$

acetic acid   acetate ion

The negative ion formed (formate, acetate), takes the name of the acid (formic, acetic). The suffix *-ic* in the acid is replaced by *-ate* in the ion.

The common names of the carboxylic acids often reflect the source from which they were first prepared. A century ago, formic acid was made by distilling a fluid obtained from ants (L. *formica*, ant). Acetic acid (L. *acer*, sour) is the compound which gives vinegar its sour taste. Butyric acid can be obtained from butter. Rancid butter and stale perspiration owe their odor to this compound. Caproic acid is present in goat's milk (L. *caper*, goat). It has the odor of roquefort cheese.

The physical properties of a few carboxylic acids are given in Table 15.6. Like alcohols, the first members of the series are all liquids and infinitely soluble in water. Carboxylic acids, like alcohols, can form hydrogen bonds readily since they have an —OH group. An acid molecule can hydrogen bond to water or to another acid molecule (Figure 15.8).

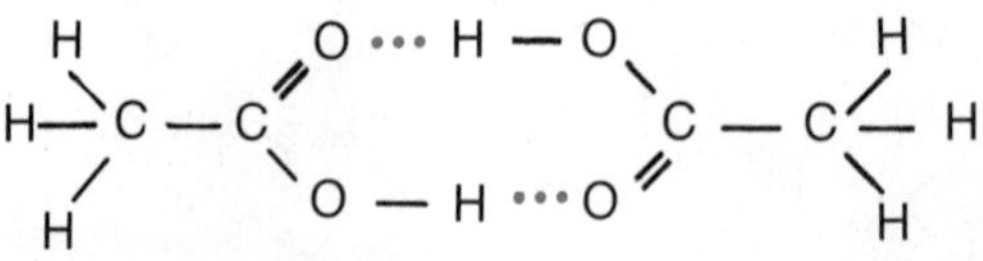

FIGURE 15.8

Acetic acid molecules show strong hydrogen bonding. In the vapor phase the molecules are often dimers $(CH_3COOH)_2$, in which the two units are held together by hydrogen bonds. The acid molecules in water solution are hydrogen-bonded to H_2O molecules.

Acetic Acid

Vinegar contains about 4–5% of acetic acid. The acetic acid is formed from the ethyl alcohol present in the starting material: wine or cider. The ethyl alcohol reacts with oxygen of the air to give acetic acid. The reaction is catalyzed by certain enzymes. These enzymes are contained in "mother of vinegar", a yeastlike material.

Wine makers don't want this reaction to occur. Vinegar makers depend on it.

$$CH_3—CH_2—OH(aq) + O_2(g) \rightarrow CH_3—COOH(aq) + H_2O \qquad (15.9)$$

This same reaction occurs, more slowly, in opened bottles of wine containing less than 12% ethyl alcohol. Higher concentrations of alcohol destroy the enzyme required for Reaction 15.9.

Pure acetic acid is not prepared from vinegar. Its concentration is too low to make this practical. Instead, acetic acid is ordinarily made from acetaldehyde by reaction with oxygen. A small amount of a compound containing Co^{2+} or Mn^{2+} ions is added as a catalyst. Reaction takes place at room temperature.

$$CH_3\text{—}CHO(l) + \tfrac{1}{2}\, O_2(g) \xrightarrow{Co^{2+}} CH_3\text{—}COOH(l) \qquad (15.10)$$

About 10^{10} kg of acetic acid are produced each year in the United States. The pure acid is sold in bottles labelled "glacial" acetic acid. This name reflects the relatively high freezing point of acetic acid, 16.6°C. Acetic acid is used in making cellulose acetate (rayon, photographic film). Smaller amounts are converted to salts such as sodium acetate:

sodium acetate: $NaC_2H_3O_2$ $\qquad Na^+, CH_3\text{—}\underset{\underset{O}{\|}}{C}\text{—}O^-$ ions

15.6 ESTERS. FUNCTIONAL GROUP: $\text{—}\underset{\underset{O}{\|}}{C}\text{—}O\text{—}$ (abbreviated —COO—)

We can consider an ester to be a derivative of a carboxylic acid. The hydrogen atom of the —OH group of the acid is replaced by a hydrocarbon group in the ester. The structures of a few esters of formic and acetic acids are listed in Table 15.7.

TABLE 15.7 ESTERS OF FORMIC AND ACETIC ACIDS

Acid		Ester		
NAME	STRUCTURE	NAME	STRUCTURE	bp (°C)
Formic Acid	$H\text{—}\underset{\underset{O}{\|}}{C}\text{—}OH$	Methyl Formate	$H\text{—}\underset{\underset{O}{\|}}{C}\text{—}O\text{—}CH_3$	32
		Ethyl Formate	$H\text{—}\underset{\underset{O}{\|}}{C}\text{—}O\text{—}CH_2\text{—}CH_3$	54
Acetic Acid	$CH_3\text{—}\underset{\underset{O}{\|}}{C}\text{—}OH$	Methyl Acetate	$CH_3\text{—}\underset{\underset{O}{\|}}{C}\text{—}O\text{—}CH_3$	57
		Ethyl Acetate	$CH_3\text{—}\underset{\underset{O}{\|}}{C}\text{—}O\text{—}CH_2\text{—}CH_3$	77

As the examples in Table 15.7 suggest:

1. The general formula of an ester is:

$$R{-}\underset{\underset{\displaystyle O}{\|}}{C}{-}O{-}R'$$

R′ is always a hydrocarbon group ($-CH_3$, $-C_2H_5$, and so on). R may be a hydrogen atom if the ester is derived from formic acid. Otherwise, it too is a hydrocarbon group which may be the same as R′ or different.

2. The common name of an ester consists of two parts. The first name is that of the hydrocarbon group R′. The last name is derived from that of the corresponding acid, replacing *-ic* by *-ate*.

EXAMPLE 15.4

Classify each of the following compounds as an alcohol, ether, aldehyde, ketone, acid, or ester.

a. $CH_3{-}\underset{\underset{\displaystyle O}{\|}}{C}{-}C_6H_5$ (benzene ring) b. $CH_3{-}\underset{\underset{\displaystyle O}{\|}}{C}{-}O{-}C_6H_5$ (benzene ring) c. $CH_3{-}\underset{\underset{\displaystyle OH}{|}}{C}{=}O$

d. $C_6H_5{-}\underset{\underset{\displaystyle H}{|}}{C}{=}O$ (benzene ring) e. $CH_3{-}O{-}CH_2{-}CH_3$

SOLUTION

In each case, identify the functional group present. On this basis:

a. ketone b. ester c. acid d. aldehyde e. ether

Esters and organic acids are present in many foods. Aldehydes and ketones are not.

Most of the common esters are volatile, pleasant smelling liquids. Many of them are present in fruits or flowers. Artificial fruit essences are made by blending several esters to give the flavor and aroma of the natural product. One recipe for raspberry flavoring combines nine different esters and five other organic compounds. Esters are also used to some extent in perfumes.

Preparation of Esters

Esters are readily prepared in the laboratory. A carboxylic acid, in water solution, is warmed with an alcohol. Hydrochloric acid or sulfuric acid is added to give H^+ ions, which serve as a catalyst. A condensation reaction occurs. Water is split out and the ester forms. The general reaction is:

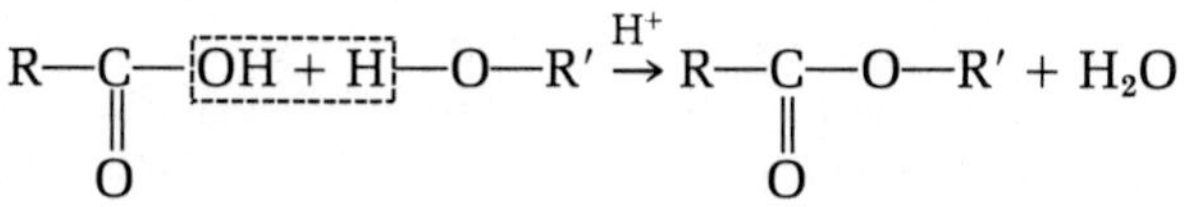

$$\underset{\text{acid}}{R-\underset{\underset{O}{\|}}{C}-\boxed{OH + H}}\underset{\text{alcohol}}{-O-R'} \xrightarrow{H^+} \underset{\text{ester}}{R-\underset{\underset{O}{\|}}{C}-O-R'} + H_2O \qquad (15.11)$$

To prepare methyl acetate, we react acetic acid with methyl alcohol.

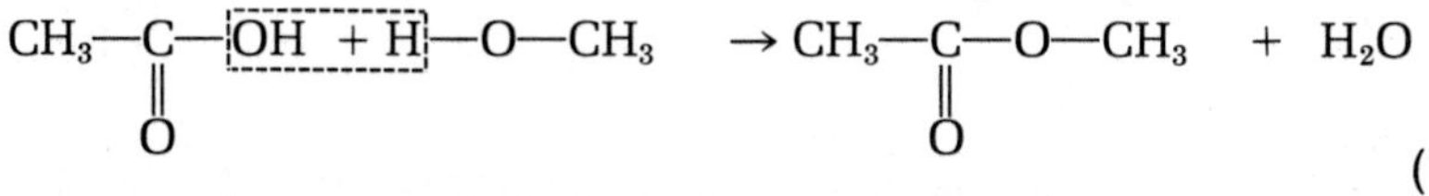

$$\underset{\text{acetic acid}}{CH_3-\underset{\underset{O}{\|}}{C}-\boxed{OH + H}}\underset{\text{methyl alcohol}}{-O-CH_3} \rightarrow \underset{\text{methyl acetate}}{CH_3-\underset{\underset{O}{\|}}{C}-O-CH_3} + H_2O \qquad (15.12)$$

Note how esters get their names.

$$CH_3COOH(aq) + CH_3OH(aq) \rightarrow CH_3-COO-CH_3(aq) + H_2O$$

EXAMPLE 15.5

Give the structure and the name of the ester formed by the reaction of ethyl alcohol with propionic acid.

SOLUTION

The reactants are:

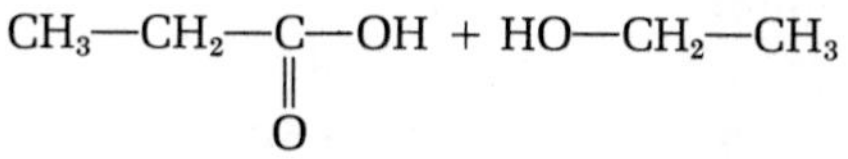

$$\underset{\text{propionic acid}}{CH_3-CH_2-\underset{\underset{O}{\|}}{C}-OH} + \underset{\text{ethyl alcohol}}{HO-CH_2-CH_3}$$

The product is:

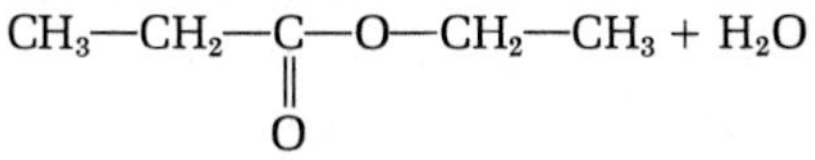

$$\underset{\text{ethyl propionate}}{CH_3-CH_2-\underset{\underset{O}{\|}}{C}-O-CH_2-CH_3} + H_2O$$

15.7 FATS AND SOAPS

Among the most important oxygen-containing organic compounds are fats. These are found in many animal and vegetable products. Fats, along with carbohydrates and proteins are basic components of our diet. Related to fats, and prepared from them, are the substances we call soaps. Both fats and soaps are derived from long-chain carboxylic acids. Fats are esters of these acids with an alcohol containing three —OH groups called glycerol:

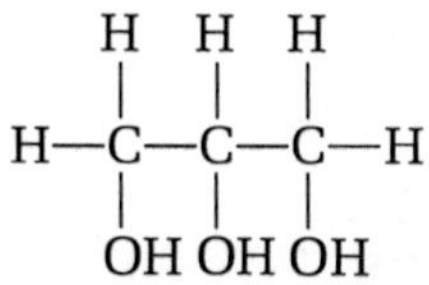

$$H-\underset{\underset{OH}{|}}{\overset{\overset{H}{|}}{C}}-\underset{\underset{OH}{|}}{\overset{\overset{H}{|}}{C}}-\underset{\underset{OH}{|}}{\overset{\overset{H}{|}}{C}}-H$$

Soaps are salts, usually sodium salts, of these same acids.

Fats

All fats, whether of animal or vegetable origin, have similar structures. They are triple esters of glycerol. That is, a given fat molecule contains three carboxylic acid residues. These acids have long carbon chains. Typically, the chain contains 12 to 18 carbon atoms. We might represent the formation of a fat by the equation:

All fats are esters of glycerol.

$$
\begin{array}{lcl}
\begin{array}{l}
R-\overset{\displaystyle O}{\overset{\|}{C}}-\boxed{OH\quad H}O-\overset{\displaystyle H}{\overset{|}{C}}-H \\
\qquad\qquad\qquad\qquad\quad | \\
R'-\overset{\displaystyle O}{\overset{\|}{C}}-\boxed{OH + H}O-C-H \\
\qquad\qquad\qquad\qquad\quad | \\
R''-\overset{\displaystyle O}{\overset{\|}{C}}-\boxed{OH\quad H}O-\underset{\displaystyle H}{\underset{|}{C}}-H
\end{array}
& \rightarrow &
\begin{array}{l}
R-\overset{\displaystyle O}{\overset{\|}{C}}-O-\overset{\displaystyle H}{\overset{|}{C}}-H \\
\qquad\qquad\qquad\quad | \\
R'-\overset{\displaystyle O}{\overset{\|}{C}}-O-C-H + 3\,H_2O \\
\qquad\qquad\qquad\quad | \\
R''-\overset{\displaystyle O}{\overset{\|}{C}}-O-\underset{\displaystyle H}{\underset{|}{C}}-H
\end{array} \\
\text{glycerol} & & \text{fat}
\end{array}
$$

Here, R, R′, and R″ represent hydrocarbon groups. These three groups may be the same. More often, they are different.

Several long-chain acids commonly found in fats are listed in Table 15.8. The first four acids are *saturated*. That is, all the carbon-carbon bonds are single bonds. Under these conditions, the acid has the maximum number of hydrogen atoms. The last three acids listed contain one or more carbon-carbon double bonds. They are said to be *unsaturated*. Notice that as the degree of unsaturation increases, the number of hydrogen atoms in the molecule decreases.

Fat molecules have high molecular masses.

The natural products that we call "fats" are really mixtures of many individual fats. Butter contains at least 100 different types of fat molecules, derived from 14 different acids. Fats containing a high proportion of unsaturated acids, such as oleic acid, are liquids at room temperature. All vegetable fats are of this type. Animal fats contain rather large

TABLE 15.8 SOME LONG-CHAIN CARBOXYLIC ACIDS FOUND IN FATS

Common Name	Number of C Atoms	Number of C=C Bonds	Formula	mp (°C)
Lauric Acid	12	0	$C_{11}H_{23}COOH$	48
Myristic Acid	14	0	$C_{13}H_{27}COOH$	57
Palmitic Acid	16	0	$C_{15}H_{31}COOH$	63
Stearic Acid	18	0	$C_{17}H_{35}COOH$	70
Oleic Acid	18	1	$C_{17}H_{33}COOH$	14
Linoleic Acid	18	2	$C_{17}H_{31}COOH$	−10
Linolenic Acid	18	3	$C_{17}H_{29}COOH$	−5

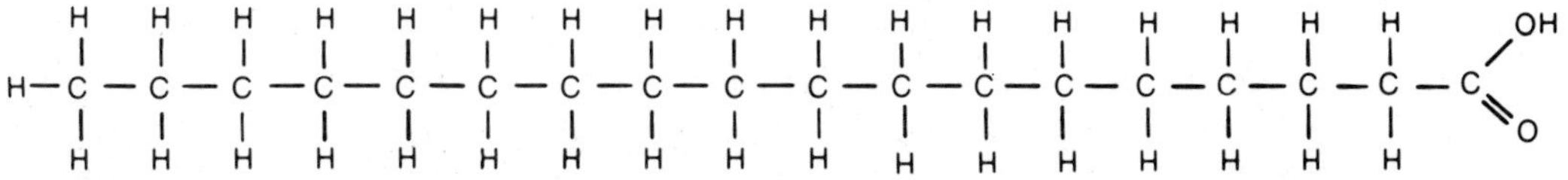

$$
\begin{array}{cccccccccccccccccccccccccccccccccccc}
 & H & & H & & H & & H & & H & & H & & H & & H & & H & & H & & H & & H & & H & & H & & H & & H & & H & & \\
 & | & & | & & | & & | & & | & & | & & | & & | & & | & & | & & | & & | & & | & & | & & | & & | & & | & & \\
H- & C & - & C & - & C & - & C & - & C & - & C & - & C & - & C & - & C & - & C & - & C & - & C & - & C & - & C & - & C & - & C & - & C & - & C\begin{smallmatrix} \nearrow OH \\ \searrow\!\!\searrow O \end{smallmatrix} \\
 & | & & | & & | & & | & & | & & | & & | & & | & & | & & | & & | & & | & & | & & | & & | & & | & & | & & \\
 & H & & H & & H & & H & & H & & H & & H & & H & & H & & H & & H & & H & & H & & H & & H & & H & & H & &
\end{array}
$$

FIGURE 15.9

Structural formula for stearic acid. The carbon chain is unbranched, as are the chains in the other molecules in Table 15.8. The last three acids in Table 15.8 contain one or more double bonds in the chain.

amounts of high molecular mass saturated acids, such as stearic acid. They are solids at room temperature.

Until about 50 years ago, only solid fats of animal origin (lard, butter) were used in cooking. When vegetable fats, such as corn oil or cottonseed oil, were first used, they were converted to solids to gain acceptance. Such products as margarine are made by a process called hydrogenation. The liquid fats are treated with hydrogen gas under pressure with a solid nickel catalyst. Under these conditions, H_2 molecules add across the C=C bonds in the fat molecules. The product is a solid fat.

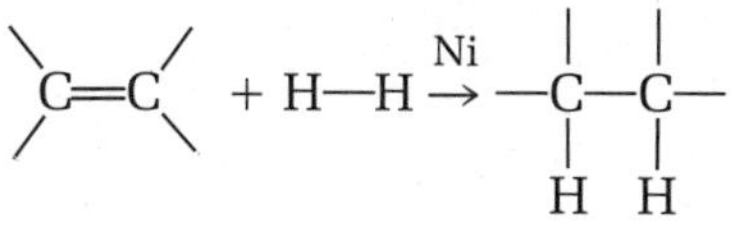

$$>C{=}C< + H{-}H \xrightarrow{Ni} -\underset{H}{\overset{|}{\underset{|}{C}}}-\underset{H}{\overset{|}{\underset{|}{C}}}-$$

Recently, the trend has been toward the use of liquid, unhydrogenated fats in cooking. It appears that saturated fats may raise the amount of cholesterol in the blood to unhealthy levels.

Soaps

Soaps are ionic compounds in which:

—the positive ion is derived from a Group 1 metal. Most commonly, it is Na^+.

—the negative ion is derived (by loss of a proton) from one of the long-chain carboxylic acids listed in Table 15.8. Typical soaps might contain:

sodium stearate: $NaC_{18}H_{35}O_2$ (Na^+, $C_{17}H_{35}-\overset{\overset{\displaystyle O}{\|}}{C}-O^-$ ions)

sodium palmitate: $NaC_{16}H_{31}O_2$ (Na^+, $C_{15}H_{31}-\overset{\overset{\displaystyle O}{\|}}{C}-O^-$ ions)

Soaps are sometimes described as being sodium salts of fatty acids.

Commercial soaps are made from fats by a process known at least since the time of the Romans (Figure 15.10). Beef tallow is most often

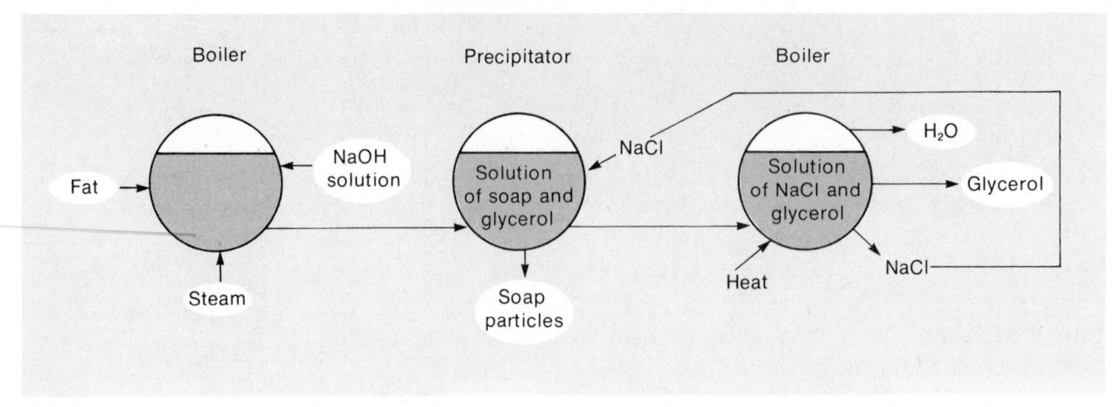

FIGURE 15.10

In making soap, fats are broken down, by heating with NaOH, into their component fatty acids and glycerol. The soap, which is a sodium salt of the acid, is precipitated by addition of NaCl. Glycerol and NaCl can be recovered by evaporation of the resulting solution.

In earlier times, the hydroxide came from wood ashes.

used, although vegetable fats will work as well. The fats are heated with a concentrated solution of sodium hydroxide, NaOH. The general reaction is:

$$
\begin{array}{lcll}
\mathrm{R{-}\overset{\overset{\displaystyle O}{\|}}{C}{-}O{-}\overset{\overset{\displaystyle H}{|}}{C}{-}H} & & \mathrm{R{-}\overset{\overset{\displaystyle O}{\|}}{C}{-}O^-} & \mathrm{HO{-}\overset{\overset{\displaystyle H}{|}}{C}{-}H} \\
\mathrm{R'{-}\overset{\overset{\displaystyle O}{\|}}{C}{-}O{-}\overset{|}{C}{-}H} & + 3\,\mathrm{OH^-} \rightarrow & \mathrm{R'{-}\overset{\overset{\displaystyle O}{\|}}{C}{-}O^-} & + \mathrm{HO{-}\overset{|}{C}{-}H} \\
\mathrm{R''{-}\overset{\overset{\displaystyle O}{\|}}{C}{-}O{-}\underset{\underset{\displaystyle H}{|}}{\overset{|}{C}}{-}H} & & \mathrm{R''{-}\overset{\overset{\displaystyle O}{\|}}{C}{-}O^-} & \mathrm{HO{-}\underset{\underset{\displaystyle H}{|}}{\overset{|}{C}}{-}H} \\
\text{fat} & & \text{negative ions of soap} & \text{glycerol}
\end{array}
$$

Soaps owe their cleansing ability to several factors. For one thing, they lower the surface tension of water. Soapy water can penetrate into cracks and crevices to reach dirt particles. These particles consist largely of water-insoluble, greasy, or oily organic compounds. The soap forms a suspension with these compounds. The long hydrocarbon chain of the soap attaches itself to the organic molecule. The ionic COO^- end of the soap surrounds itself with water molecules (Figure 15.11). The water can then carry away the dirt.

Soap has certain drawbacks as a cleaning agent. It gives problems with "hard" water, which contains Ca^{2+} and Mg^{2+} ions. They form in-

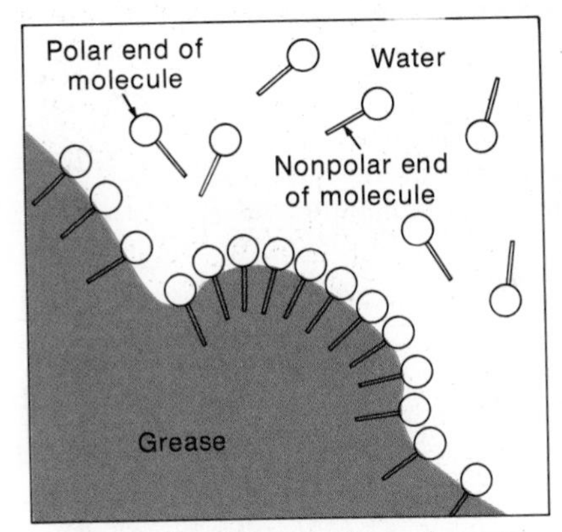

Soap molecules on surface of grease

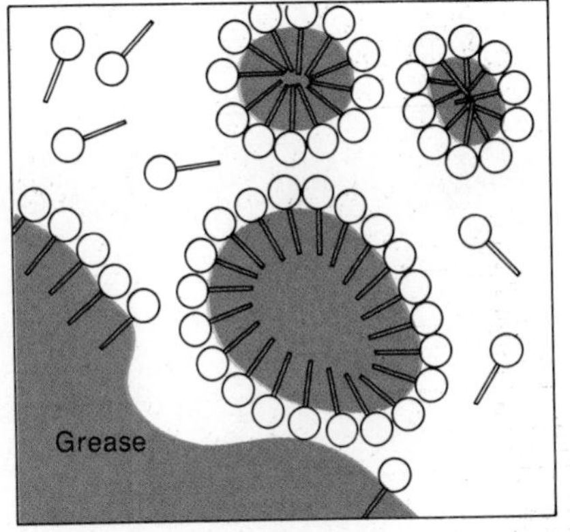

Grease droplets dispersed in wash water

FIGURE 15.11

Diagram showing the cleansing action of soap. The long hydrocarbon chain on the soap ion is a good solvent for organic grease. The polar COO^- end of the ion is soluble in water. In the washing process the grease is dispersed into the water as droplets surrounded by soap ions.

soluble, waxy deposits with the negative ions of the soap. Compounds such as

$Ca(C_{16}H_{31}O_2)_2$	calcium palmitate
$Mg(C_{18}H_{33}O_2)_2$	magnesium oleate

are insoluble in water. They cause bathtub rings and "ring around the collar".

These and other problems with natural soaps led chemists to make "synthetic" soaps, usually called detergents. The structure of a common detergent is shown in Figure 15.12. Notice the similarity to soap. In both, there is a long hydrocarbon chain attached to a water-soluble, ionic group. Detergents do not form precipitates with Ca^{2+} and Mg^{2+} ions. They are also more effective cleaning agents than soap.

15.8 SUMMARY

General formulas for the six classes of oxygen-containing organic compounds are given in Table 15.9, p. 388. Here, R or R′ is most commonly an alkyl group (CH_3—, C_2H_5—, and so on). It can also be an aromatic group. In special cases (formaldehyde, formic acid, and the esters of formic acid), R is a hydrogen atom.

$$\begin{array}{cccccccccccccccccccccccccccc} & & H & & H & & H & & H & & H & & H & & H & & H & & H & & H & & H & & H & & & & O & & \\ & & | & & | & & | & & | & & | & & | & & | & & | & & | & & | & & | & & | & & & & | & & \\ H & - & C & - & C & - & C & - & C & - & C & - & C & - & C & - & C & - & C & - & C & - & C & - & C & - & O & - & S & - & O^-, Na^+ \\ & & | & & | & & | & & | & & | & & | & & | & & | & & | & & | & & | & & | & & & & | & & \\ & & H & & H & & H & & H & & H & & H & & H & & H & & H & & H & & H & & H & & & & O & & \end{array}$$

FIGURE 15.12

Structural formula of a detergent molecule. The difference between a detergent and a soap is in the polar end of the molecule. The introduction of the SO_4 group into the detergent increases its solubility in hard water.

EXAMPLE 15.6

For each of the classes listed in Table 15.9, draw the structural formula of a molecule containing two carbon atoms.

SOLUTION

alcohol: $CH_3{-}CH_2{-}OH$

ether: $CH_3{-}O{-}CH_3$

aldehyde: $CH_3{-}\underset{\underset{O}{\Vert}}{C}{-}H$

acid: $CH_3{-}\underset{\underset{O}{\Vert}}{C}{-}OH$

ester: $H{-}\underset{\underset{O}{\Vert}}{C}{-}O{-}CH_3$

Notice that there is no ketone with two carbon atoms. Why?

All six classes of oxygen compounds have certain properties in common.

1. Their molecules are polar.
2. As molecular mass increases, boiling point increases and water solubility decreases.
3. The higher members show structural isomerism.

TABLE 15.9 CLASSES OF OXYGEN COMPOUNDS

Class	Functional Group	General Formula	First Member
Alcohol	$-OH$	$R{-}OH$	$CH_3{-}OH$
Ether	$-O-$	$R{-}O{-}R'$	$CH_3{-}O{-}CH_3$
Aldehyde	$-\underset{\underset{O}{\Vert}}{C}{-}H$	$R{-}\underset{\underset{O}{\Vert}}{C}{-}H$	$H{-}\underset{\underset{O}{\Vert}}{C}{-}H$
Ketone	$-\underset{\underset{O}{\Vert}}{C}-$	$R{-}\underset{\underset{O}{\Vert}}{C}{-}R'$	$CH_3{-}\underset{\underset{O}{\Vert}}{C}{-}CH_3$
Carboxylic Acid	$\underset{\underset{O}{\Vert}}{C}{-}OH$	$R{-}\underset{\underset{O}{\Vert}}{C}{-}OH$	$H{-}\underset{\underset{O}{\Vert}}{C}{-}OH$
Ester	$-\underset{\underset{O}{\Vert}}{C}{-}O-$	$R{-}\underset{\underset{O}{\Vert}}{C}{-}O{-}R'$	$H{-}\underset{\underset{O}{\Vert}}{C}{-}O{-}CH_3$

Of these six classes:

1. Only alcohols and acids have a hydrogen atom bonded to oxygen. Members of these classes are capable of hydrogen bonding with themselves. The others (ethers, aldehydes, ketones, and esters) are not.

2. Four (aldehydes, ketones, acids, and esters) have the grouping

$$\begin{array}{c} R{-}C{-} \\ \quad\| \\ \quad O \end{array}$$

In aldehydes, the fourth bond is to a H atom. In ketones, it is to an —R group, in acids to an —OH group, and in esters to an —O—R group.

3. Four (alcohols, ethers, aldehydes and ketones) contain a single oxygen atom per molecule. The others (acids, esters) have two oxygens per molecule.

We have considered the properties and methods of preparation of several of the more common oxygen compounds. These include methyl alcohol and ethyl alcohol; diethyl ether, formaldehyde, acetone, and acetic acid. In addition, we have discussed the general method of preparing an ester from the corresponding acid and alcohol.

In the last section of the chapter, we described the structures of fats and soaps. A fat is a triple ester of glycerol with long-chain organic acids. A soap is usually a sodium salt of a long-chain organic acid. In practice, natural fats and commercial soaps are a mixture of many different compounds of the type just described.

NEW TERMS

alcohol	ether
aldehyde	fat
carbonyl group	functional group
carboxyl group	ketone
carboxylic acid	saturated fat
detergent	soap
ester	unsaturated fat

QUESTIONS

1. In all of the compounds considered in this chapter, how many bonds are formed by an atom of hydrogen? carbon? oxygen?

2. What is the functional group present in an alcohol? What is the general formula of an alcohol?

3. Which is more soluble in water, CH_3OH or $CH_3(CH_2)_5OH$? Explain your answer.

4. Explain why methyl alcohol, CH_3OH, has a higher boiling point than dimethyl ether, $CH_3—O—CH_3$.

5. Draw structural formulas for:

   a. methyl alcohol b. ethyl alcohol c. phenol

6. What are the products of the destructive distillation of wood?

7. Describe two different methods of preparing methyl alcohol.

8. Describe two different methods of preparing ethyl alcohol.

9. What is meant by "denatured" ethyl alcohol? What is it used for?

10. What functional group is present in an ether? What is the general formula of an ether?

11. Draw structural formulas for:

    a. dimethyl ether b. diethyl ether

12. Draw the structural formula of acetaldehyde and indicate all the bond angles (109° or 120°).

13. What is the general formula of an aldehyde? a ketone? How do the two classes differ from one another?

14. Draw structural formulas for:

    a. methyl ethyl ketone b. diethyl ketone

15. Draw the structural formulas of all the oxygen compounds discussed in this chapter which contain a single carbon atom per molecule. (There are three of them.)

16. What two classes of compounds discussed in this chapter form hydrogen bonds most readily?

17. Draw structural formulas for:

    a. acetic acid b. acetone

18. Write balanced equations for two different methods of preparing acetic acid.

19. How does the structural formula of an ester differ from that of

    a. an acid b. a ketone c. an aldehyde

20. Draw structural formulas for:

    a. methyl formate b. methyl acetate c. ethyl propionate

21. Fats are esters of a certain alcohol. What alcohol?

22. What is a saturated fat? an unsaturated fat? How is an unsaturated fat converted to a saturated fat?

23. How are soaps prepared from fats?

24. Explain the cleaning action of soap. Why does it give problems with hard water?

PROBLEMS

1. *Isomerism (Example 15.1)* Draw structural formulas for the 3 straight chain alcohols containing 5 carbon atoms.

2. *Isomerism (Example 15.2)* Draw structural formulas for the 2 straight chain ethers containing 5 carbon atoms.

3. *Isomerism (Examples 15.1, 15.2)* Draw structural formulas for the 2 carboxylic acids containing four carbon atoms.

4. *Classes of Compounds (Examples 15.3, 15.4, 15.6)* Identify the class to which each of the following compounds belongs.

   a. $CH_3—CH_2—\overset{\overset{\large O}{\|}}{C}—H$ b. $HO—\overset{\overset{\large O}{\|}}{C}—H$ c. $CH_3—\overset{\overset{\large O}{\|}}{C}—O—CH_3$

   d. $CH_3—O—C_6H_5$ (benzene ring) e. $HO—CH_2—CH_3$

5. *Classes of Compounds (Examples 15.3, 15.4, 15.6)* Draw structural formulas for all the oxygen-containing organic compounds in which R or R′ in the general formula is a $—CH_3$ group. Example: alcohol, $CH_3—OH$.

6. *Classes of Compounds (Examples 15.3, 15.4, 15.6)* Draw structural formulas for a compound containing three carbon atoms which is a(an):

   a. alcohol b. ether c. aldehyde d. ketone e. acid
   f. ester

7. *Preparation of Esters (Example 15.5)* Give the structure of the ester formed by reacting:

   a. formic acid with propyl alcohol
   b. acetic acid with isopropyl alcohol

* * * * *

8. Draw structural formulas for all the aldehydes and ketones containing 4 carbon atoms.

9. Draw structural formulas for all the esters containing 4 carbon atoms.

10. What is the smallest number of carbon atoms that can be present in a molecule of a(an)

    a. alcohol b. ether c. aldehyde d. ketone e. acid
    f. ester

11. Consider the ester ethyl acetate. Draw structural formulas for all of the isomers that can be formed by replacing one hydrogen atom in an ethyl acetate molecule with a chlorine atom.

12. Suppose the oxygen atom in the first member of each class of compound discussed in this chapter were replaced by sulfur. Draw structural formulas for each of these sulfur-containing compounds.

    Example alcohol: $CH_3—SH$

13. Give the structural formulas of all the esters that can be formed by reacting methyl or ethyl alcohol with formic or acetic acid.

14. What alcohol and acid are required to form:

a. $H{-}\underset{\underset{O}{\|}}{C}{-}O{-}CH_2{-}CH_3$ b. $CH_3{-}\underset{\underset{O}{\|}}{C}{-}O{-}CH_2{-}CH_2{-}CH_3$

c. $CH_3{-}\underset{\underset{O}{\|}}{C}{-}O{-}\overset{\overset{H}{|}}{\underset{\underset{CH_3}{|}}{C}}{-}CH_3$

15. Ethylene glycol, $C_2H_6O_2$, is commonly used as an anti-freeze. How many grams of ethylene glycol would be required to produce the same freezing point lowering as 32.0 g of methyl alcohol (using the same amount of water in the two cases)?

16. Calculate the number of grams of glucose required in fermentation (Equation 15.3) to make 1.00 ℓ of wine containing 10% ethyl alcohol by volume. The density of ethyl alcohol is 0.80 g/cm^3.

*17. Draw structural formulas for all compounds with the following molecular formulas. In each case, indicate the kind of compound involved.

a. C_2H_6O b. C_3H_6O c. $C_2H_4O_2$

*18. What is the density at STP of dimethyl ether?

*19. A certain brand of beer contains 6.0% by volume of ethyl alcohol. How many 12-oz bottles of this beer contain the same amount of alcohol as 2.0 oz of 90-proof whiskey?

The Gossamer Condor, which on August 23, 1977 won the Kremer Prize of $86,000 (£50,000) for being the first aircraft to fly under human power alone, was designed and built in California by Dr. Paul B. MacCready, an aeronautic engineer. The wing covering is made from thin sheets of Mylar, a rugged synthetic polymer. *E. I. du Pont de Nemours Company.*

16

ORGANIC CHEMISTRY: Polymers

So far, all the organic molecules we have considered have been small ones. They contain a small number of carbon, hydrogen, and other nonmetal atoms. They range in molecular mass from 16 (methane) to a few hundred (fats). In this chapter we will deal with a different type of organic compound called a *polymer*. Polymer molecules contain large numbers of atoms held together by covalent bonds. Their molecular masses are very high, ranging from a few thousand to several million units.

Many organic polymers occur in nature. They include such familiar substances as starch and cellulose. The proteins in our hair, skin, and body tissue are huge molecules of high molecular mass. These species will be discussed in Sections 16.4 and 16.5. In the first parts of this chapter, we will consider only man-made polymers. These are prepared in the laboratory by chemical reactions. The starting materials are small molecules called *monomers*. Under the proper conditions, these molecules polymerize. That is, they bond together to form huge molecules of high molecular mass.

More industrial chemists work with polymers than with any other material.

The first polymer of this type, bakelite, was put on the market in 1910. Twenty-five years later, in 1935, the discovery of nylon transformed the textile industry. Most of the polymers we use today were developed during and shortly after World War II. The polymer industry is now one of the largest in the United States. About 10^{10} kg of polymeric materials are produced every year.

We are familiar with synthetic polymers in the form of:

—*plastics*, which are used for almost everything from bottles to phonograph records to sandwich bags. (See Color Plate 7A, in the center of the book.)

—*synthetic fibers*, such as dacron and nylon, which compete with natural fibers (wool, silk, and cotton).

—*synthetic rubber*, blended in about equal amounts with natural rubber in automobile tires.

16.1 ADDITION POLYMERS

Many of our most common plastics are addition polymers. They are made by an addition reaction in which small molecules (monomers) add to each other. In all the addition polymers that we will consider in this section, **the monomer molecule contains a carbon-carbon double bond.** Upon polymerization, the double bond breaks. The polymer formed is a long-chain molecule in which all the carbon-carbon bonds are single.

Polyethylene

Perhaps the simplest addition polymer is polyethylene. It is made from the monomer ethylene:

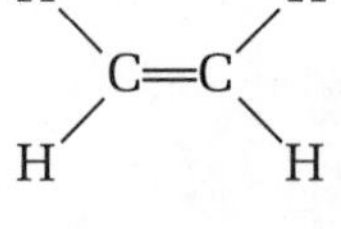

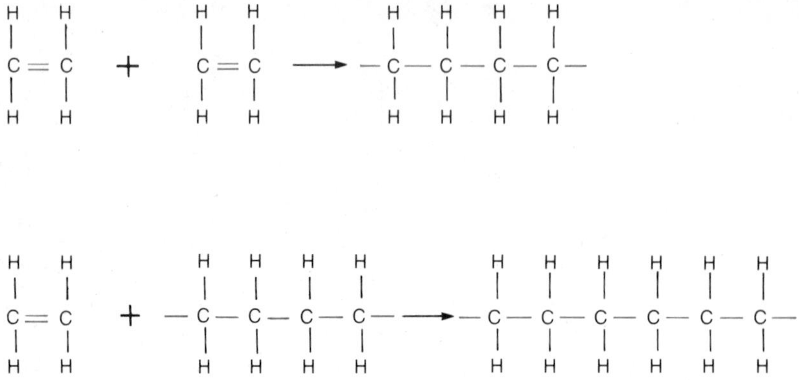

$$\begin{matrix} H & & & & H \\ & \diagdown & & \diagup & \\ & & C{=}C & & \\ & \diagup & & \diagdown & \\ H & & & & H \end{matrix}$$

$$\begin{matrix} H & H \\ | & | \\ C & = & C \\ | & | \\ H & H \end{matrix} + \begin{matrix} H & H \\ | & | \\ C & = & C \\ | & | \\ H & H \end{matrix} \longrightarrow -\begin{matrix} H & H & H & H \\ | & | & | & | \\ C- & C- & C- & C \\ | & | & | & | \\ H & H & H & H \end{matrix}-$$

$$\begin{matrix} H & H \\ | & | \\ C & = & C \\ | & | \\ H & H \end{matrix} + -\begin{matrix} H & H & H & H \\ | & | & | & | \\ C- & C- & C- & C \\ | & | & | & | \\ H & H & H & H \end{matrix}- \longrightarrow -\begin{matrix} H & H & H & H & H & H \\ | & | & | & | & | & | \\ C- & C- & C- & C- & C- & C \\ | & | & | & | & | & | \\ H & H & H & H & H & H \end{matrix}-$$

FIGURE 16.1

When ethylene polymerizes, the double bonds in the monomer units are converted to single bonds linking those units. The final polymer molecule contains no double bonds.

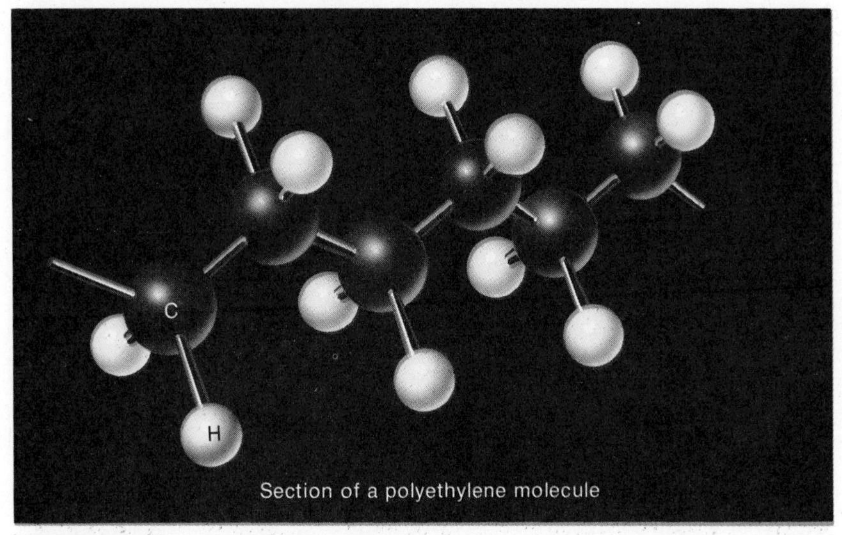

FIGURE 16.2

In polyethylene the carbon chain is zig-zag. The bond angles all have essentially the same size, 109°, the tetrahedral angle.

Upon polymerization, the double bond in the ethylene molecule opens up. The product is a long-chain molecule made up of $—CH_2—$ groups bonded to one another (Figure 16.1). The chemical equation for the reaction might be written:

$$n\ C_2H_4(g) \quad \rightarrow \quad (—CH_2—CH_2)_n(s) \qquad (16.1)$$

Here, n is a very large number, usually 1000 or greater.

The structure shown at the bottom of Figure 16.1 is misleading in one way. The carbon atoms in polyethylene are not in a straight line. Instead, they form a zig-zag chain with a 109° bond angle. This is shown in the ball-and-stick model of Figure 16.2. Only a small portion of the chain is included. A still more realistic picture of a polyethylene molecule is given by the space-filling model of Figure 16.3. Notice that the

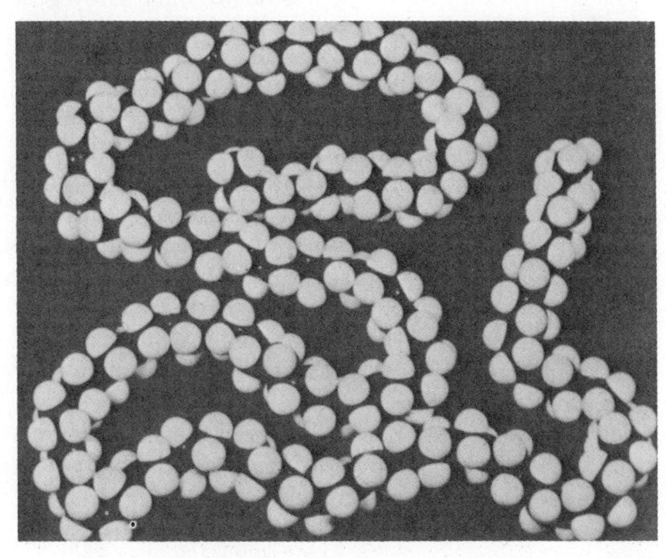

FIGURE 16.3

In polyethylene, the carbon chain is coiled as shown here. This gives a flexible, somewhat elastic polymer.

molecule is not drawn out to its full length in one dimension. Instead, it is folded back upon itself in two or three dimensions.

EXAMPLE 16.1

The molecule of polyethylene shown in Figure 16.3 contains 150 carbon atoms.

a. How many hydrogen atoms are there in this molecule?

b. What is the molecular mass?

c. How many ethylene molecules are required to form this polymeric molecule?

SOLUTION

a. There are two hydrogen atoms per carbon atom. Hence, there must be 300 hydrogen atoms.

b. $MM = 150(12.0) + 300(1.0) = 1800 + 300 = 2100$

c. One C_2H_4 molecule forms two $—CH_2—$ units (Figure 16.1). Hence:

$$\text{number } C_2H_4 \text{ molecules} = 150/2 = 75$$

(Actually, many more C_2H_4 molecules would be involved, probably 1000 or more. The molecules in a polyethylene sample contain many more carbon atoms than the model shown in Figure 16.3.)

Polyvinyl Chloride

Although polymers are safe to use, some monomers, including vinyl chloride, can cause cancer.

The polymer known as polyvinyl chloride (PVC) is made from vinyl chloride:

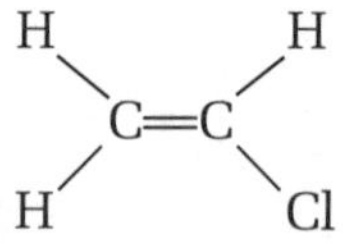

```
H       H
 \     /
  C=C
 /     \
H       Cl
```

Vinyl chloride is a close relative of ethylene. It differs only in the substitution of a chlorine for a hydrogen atom. The equation for the polymerization is:

$$n\ C_2H_3Cl(g) \rightarrow (—CH_2—CHCl)_n(s) \qquad (16.2)$$

A model of the polymerization reaction is shown in Figure 16.4. In polyvinyl chloride, $—CH_2—$ and $—CHCl—$ groups alternate along the chain. This requires that the $—CHCl—$ group of one vinyl chloride molecule add to the $—CH_2—$ group of the next. This structure is typical

```
H   H        H   H            H   H   H   H
|   |        |   |            |   |   |   |
C = C   +    C = C   ——▶   —C— C— C— C—
|   |        |   |            |   |   |   |
H   Cl       H   Cl           H   Cl  H   Cl

H   H         H   H   H   H              H   H   H   H   H   H
|   |         |   |   |   |              |   |   |   |   |   |
C = C   +   —C— C— C— C—   ——▶   —C— C— C— C— C— C—
|   |         |   |   |   |              |   |   |   |   |   |
H   Cl        H   Cl  H   Cl             H   Cl  H   Cl  H   Cl
```

FIGURE 16.4

Formation of polyvinyl chloride. In the polymer molecule, the monomer units are lined up head-to-tail, with a Cl atom on every other carbon atom.

of most addition polymers formed from unsymmetrical molecules such as vinyl chloride. When a monomer of the type

```
H       H
 \     /
  C=C          (R = group other than H)
 /     \
H       R
```

polymerizes, the polymer ordinarily has the structure:

$$—CH_2—CHR—CH_2—CHR—CH_2—CHR—$$

EXAMPLE 16.2

Consider the polymer:

```
  H   H    H   H    H   H
  |   |    |   |    |   |
—C—C—C—C—C—C—
  |   |    |   |    |   |
  H  CH3   H  CH3   H  CH3
```

Assuming this pattern is repeated:

a. Identify the R group.

b. Identify the monomer from which this polymer is formed.

SOLUTION

a. R = CH_3

b. Compare with Figure 16.1. This polymer differs from polyethylene in only one way. A hydrogen atom on every other carbon has been replaced by a CH_3 group. The monomer must be:

```
H       H
 \     /
  C=C          propylene
 /     \
H       CH3
```

Polyvinyl chloride, among addition polymers, is second only to polyethylene in production. Polyvinyl chloride is also flame-resistant. The chlorine atoms in the molecule make it less flammable than polyethylene. On the other hand, when polyvinyl chloride does burn, one of the products is hydrogen chloride, HCl. This can cause serious corrosion if articles made from this polymer are burned in a metal incinerator.

Other Addition Polymers

Each polymer has its own properties and uses for which it is suitable.

Table 16.1 lists some of the more important addition polymers and their uses. Notice that in every case the monomer has a carbon-carbon double bond. Upon polymerization the double bond opens up. In the polymer, all the carbon-carbon bonds are single.

The polymer made from methyl methacrylate goes by the trade name of Plexiglas or Lucite. It is a clear, transparent material, much more resistant to shock than glass. A light source placed behind the plastic makes it luminous. This property makes Lucite useful in outdoor signs and lettering. (See Color Plate 7B, in center of book.) Sheets of Lucite are placed over fluorescent lights to create a luminous ceiling or dome skylight. The shield over automobile taillights and turn signals is also made of this material.

EXAMPLE 16.3

Sketch a portion of the polymer formed from methyl methacrylate.

SOLUTION

Using three monomer units:

```
 H   CH3  H   CH3  H   CH3
 |    |   |    |   |    |
—C  — C — C  — C — C  — C —
 |    |   |    |   |    |
 H   C=O  H   C=O  H   C=O
      |        |        |
      O        O        O
      |        |        |
     CH3      CH3      CH3
```

Perhaps the most unusual of the addition polymers is polytetrafluoroethylene, commonly called Teflon. It is among the most inert of all organic solids. It is not affected by any solvent and will not burn. It is not damaged by exposure to temperatures as low as −250°C or as high as 250°C. These properties make Teflon valves, gaskets, and tubing ideal for working with corrosive chemicals at extreme temperatures. It also has a very low surface friction, which makes it useful as a coating on non-stick fry pans.

TABLE 16.1 SOME COMMON ADDITION POLYMERS

Monomer Formula	Monomer Name	Polymer Name	Uses	Production, kg (U.S., 1975)
$CH_2{=}CH_2$	ethylene	polyethylene	films, coatings, bottles, toys	3×10^9
$CH_2{=}CHCl$ (H and Cl on C)	vinyl chloride	polyvinyl chloride (PVC)	credit cards, phonograph records, bags, floor tile	1.5×10^9
$CH_2{=}CH(CH_3)$ (H and CH_3 on C)	propylene	polypropylene	fibers, films, bottles, beakers	1×10^9
$CH_2{=}CH(C_6H_5)$ (H and phenyl ring on C)	styrene	polystyrene	insulation, packing material, combs	1×10^9
$CH_2{=}CH(CN)$ (H and CN on C)	acrylonitrile	polyacrylonitrile (PAN)	rug fibers (Orlon and Acrilan are copolymers of acrylonitrile with other monomers)	3×10^8
$CH_2{=}CH{-}O{-}C({=}O){-}CH_3$ (H and $O{-}C{=}O{-}CH_3$ on C)	vinyl acetate	polyvinyl acetate	latex paint	2×10^8
$CH_2{=}C(CH_3){-}C({=}O){-}O{-}CH_3$ (CH_3 and $C{=}O{-}O{-}CH_3$ on C)	methyl methacrylate	Plexiglas, Lucite	glass substitutes, costume jewelry	5×10^7
$CF_2{=}CF_2$	tetrafluoroethylene	Teflon	gaskets, bearings, insulation, pan coatings	2×10^7

EXAMPLE 16.4

What are the percentages by mass of carbon and fluorine in Teflon?

SOLUTION

Since Teflon is an addition polymer, its composition must be the same as that of the monomer, C_2F_4. In one mole of C_2F_4:

$$\begin{aligned} \text{number of grams C} &= 2 \times 12.0 \text{ g} = 24.0 \text{ g} \\ \text{number of grams F} &= 4 \times 19.0 \text{ g} = \underline{76.0 \text{ g}} \\ & \qquad\qquad\qquad\quad 100.0 \text{ g} \end{aligned}$$

$$\% \text{ C} = \frac{24.0 \text{ g}}{100.0 \text{ g}} \times 100 = 24.0\%; \quad \% \text{ F} = \frac{76.0 \text{ g}}{100.0 \text{ g}} \times 100 = 76.0\%$$

16.2 NATURAL AND SYNTHETIC RUBBER

About half of the rubber we use today is a natural product obtained from the rubber trees of Southeast Asia. The remainder is made in the laboratory, using a variety of starting materials. Both natural and synthetic rubber are addition polymers. They differ, however, from the addition polymers discussed in Section 16.1 in one important respect. The monomers from which they are derived ordinarily contain **two carbon-carbon double bonds per molecule rather than one.** One of these double bonds is retained in the polymer. The other is converted to a single carbon-carbon bond.

FIGURE 16.5

A tapper makes a cut in a rubber tree on a plantation in Sumatra. The gummy liquid that oozes from the wound in the tree collects in the cup below. *Goodyear Tire and Rubber Co.*

Natural Rubber

Rubber was introduced into Europe by early explorers, who found it growing in South and Central America. It was named by Joseph Priestley, the discoverer of oxygen, who used it to rub out pencil marks. Today, most natural rubber comes from Ceylon and Indonesia. It is grown on plantations of cultivated rubber trees. To obtain the rubber, a cut is made in the trunk of the tree (Figure 16.5). A milky liquid called latex oozes out of the cut. By adding salt and acetic acid, the rubber is precipitated. It makes up about ⅓ of the latex.

The chemical composition of natural rubber was established by Michael Faraday in 1826. He showed that it has the simplest formula C_5H_8. We now know that rubber is derived from the monomer called isoprene:

$$\begin{array}{c} \quad\ CH_3\ H \\ \quad\ \ |\quad\ | \\ CH_2{=}C{-}C{=}CH_2 \end{array} \qquad \text{isoprene}$$

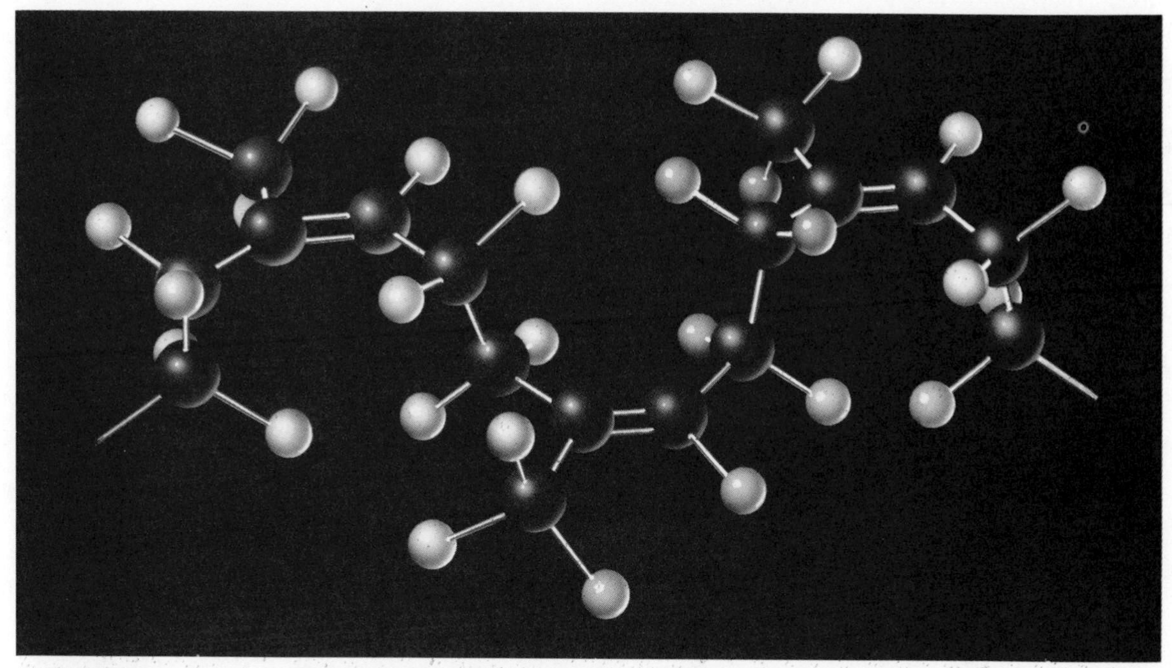

FIGURE 16.6

In the rubber molecule the carbon chain has the structure of a large coil. This gives rubber its elastic properties. Compare the atom arrangement in rubber with that in polyethylene, Figure 16.2.

Unlike the other monomers we have considered, isoprene has *two* carbon-carbon double bonds. Upon polymerization, these bonds open up. A long-chain addition polymer is formed. It contains from 1000 to 30,000 isoprene units. The structure of natural rubber may be shown as:

$$\left| \begin{array}{c} \quad CH_3\ H \\ -CH_2-\overset{|}{C}=\overset{|}{C}-CH_2- \\ \text{isoprene} \\ \text{unit} \end{array} \right| \begin{array}{c} \quad CH_3\ H \\ -CH_2-\overset{|}{C}=\overset{|}{C}-CH_2- \\ \text{isoprene} \\ \text{unit} \end{array} \left| \begin{array}{c} \quad CH_3\ H \\ -CH_2-\overset{|}{C}=\overset{|}{C}-CH_2- \\ \text{isoprene} \\ \text{unit} \end{array} \right|$$

Notice that there are double bonds left in natural rubber. Polymerization reduces the number of such bonds from two to one per isoprene unit.

In the natural state, rubber molecules do not have the linear shape suggested by the structure we have drawn. Instead, they are coiled (Figure 16.6). When a rubber band is expanded, these coils tend to straighten out. The molecules stretch out much like a spring to which a mass is attached. When the tension is released, the molecules snap back to their original position. This effect explains why rubber and rubber-like materials are so elastic.

Synthetic Rubber

Synthetic rubber first became available in 1910. By 1930, it was being used for a few special purposes where it was superior to natural rubber. However, since it was nearly ten times as expensive, synthetic rubber had no large-scale uses. Then came World War II. When the Japanese invaded Southeast Asia, the United States was cut off from its supply of natural rubber. The production of synthetic rubber jumped from nearly zero in 1941 to 7×10^8 kg/year in 1945. Since then it has increased by a factor of four.

Several different monomers can be polymerized to give synthetic rubbers. These include:

1. *Isoprene*, which gives a product nearly identical to natural rubber. Since polyisoprene is more expensive than natural rubber, its uses are limited.

2. *Chloroprene*

$$CH_2{=}\underset{}{\overset{Cl}{\overset{|}{C}}}{-}\overset{H}{\overset{|}{C}}{=}CH_2 \qquad \text{chloroprene}$$

All rubbers are "vulcanized" by heating with sulfur. This cross-links the polymer chains, improving elasticity and resistance to solvents.

The polymer produced from chloroprene is called neoprene. It was one of the first synthetic rubbers, developed during World War II. Neoprene is more resistant to attack by solvents than is natural rubber. This property makes it useful for gasoline fuel lines and hoses. Neoprene is also used in the soles and heels of shoes, since it is highly resistant to abrasion.

The molecular structure of chloroprene is very similar to that of isoprene. Each molecule contains a 4-carbon chain with double bonds at each end. The molecular structures of the polymers formed are also very similar (Example 16.5).

EXAMPLE 16.5

Sketch a portion of the neoprene polymer (include three monomer units).

SOLUTION

$$-CH_2-\overset{Cl}{\overset{|}{C}}{=}\overset{H}{\overset{|}{C}}-CH_2-CH_2-\overset{Cl}{\overset{|}{C}}{=}\overset{H}{\overset{|}{C}}-CH_2-CH_2-\overset{Cl}{\overset{|}{C}}{=}\overset{H}{\overset{|}{C}}-CH_2-$$

Note the similarity to natural rubber. The only difference is the substitution of a Cl atom for a CH_3 group.

3. *Butadiene*

$$CH_2{=}\overset{H}{\overset{|}{C}}{-}\overset{H}{\overset{|}{C}}{=}CH_2 \qquad \text{butadiene}$$

Butadiene can be polymerized by itself to give polybutadiene. This polymer has properties very similar to those of natural rubber. More frequently, butadiene is mixed with another monomer, styrene:

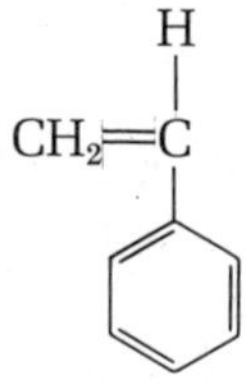

$$CH_2{=}C(H)(C_6H_5) \quad \text{styrene}$$

The product formed on polymerization is referred to as a *copolymer*, since it contains two different monomer units. Typically, the two monomers are mixed in a 1:3 mole ratio (1 mol styrene, 3 mol butadiene). In the polymer, there is one styrene unit for every three butadiene units:

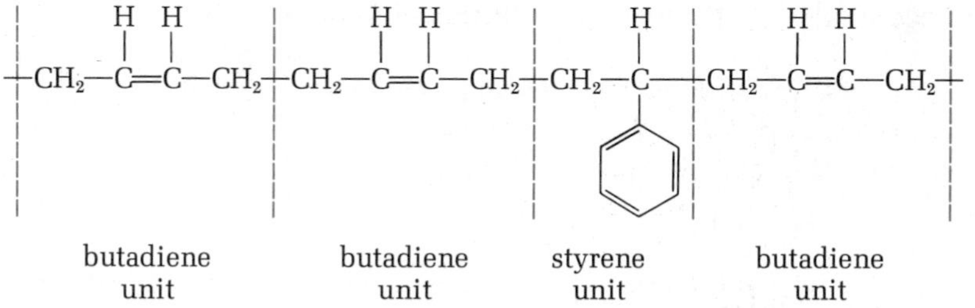

$$\text{—}CH_2\text{—}CH{=}CH\text{—}CH_2\text{—}|\text{—}CH_2\text{—}CH{=}CH\text{—}CH_2\text{—}|\text{—}CH_2\text{—}CH(C_6H_5)\text{—}|\text{—}CH_2\text{—}CH{=}CH\text{—}CH_2\text{—}$$

butadiene unit | butadiene unit | styrene unit | butadiene unit

This copolymer, known as "SBR" (styrene-butadiene rubber) accounts for about 60% of our total production of synthetic rubber. Most of it is mixed with natural rubber to make automobile tires. SBR is more resistant to abrasion and chemical attack by oxygen of the air. However, natural rubber is better for the tread because of superior elastic properties.

A big advantage of SBR rubber is that it is relatively cheap.

16.3 CONDENSATION POLYMERS

A condensation polymer is formed by splitting out small molecules such as H_2O between monomer units. In all of the cases we will consider here:

1. *The two monomers are different compounds.* One monomer might, for example, be a carboxylic acid while the other is an alcohol.
2. *Each monomer has two functional groups in the molecule.* These might be two —COOH groups or two —OH groups.

Polyesters

Suppose that two compounds, one a dicarboxylic acid (2 —COOH groups)

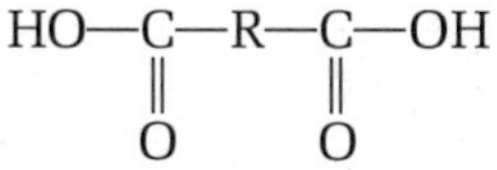

$$HO\text{—}\underset{\underset{O}{\|}}{C}\text{—}R\text{—}\underset{\underset{O}{\|}}{C}\text{—}OH$$

and the other a dialcohol (2 —OH groups)

$$HO{-}R'{-}OH$$

As with addition polymers, there are many possible condensation polymers.

react with each other. (R and R′ represent hydrocarbon groups.) We would expect them to form an ester by splitting out water. This is shown in the equation below, with the ester group in color.

$$HO{-}\underset{\underset{O}{\|}}{C}{-}R{-}\underset{\underset{O}{\|}}{C}{-}\boxed{OH + H}O{-}R'{-}OH \rightarrow HO{-}\underset{\underset{O}{\|}}{C}{-}R{-}\underset{\underset{O}{\|}}{C}{-}O{-}R'{-}OH + H_2O$$

The product of this reaction still has reactive groups at each end of the molecule. At one end, there is a —COOH group which can react with an alcohol molecule. At the other end is an —OH group which can react with another molecule of the carboxylic acid. So, further reaction can occur to form a molecule with four monomer units:

$$HO{-}R'{-}O{-}\underset{\underset{O}{\|}}{C}{-}R{-}\underset{\underset{O}{\|}}{C}{-}O{-}R'{-}O{-}\underset{\underset{O}{\|}}{C}{-}R{-}\underset{\underset{O}{\|}}{C}{-}OH$$

alcohol acid alcohol acid

Again we have a molecule with reactive end groups. Condensation can occur over and over again to yield a polymer containing thousands of monomer units. This polymer is called a **polyester** because the monomers are joined through ester groups. Its structure, with the ester groups shown in color, is:

$${-}\underset{\underset{O}{\|}}{C}{-}R{-}\underset{\underset{O}{\|}}{C}{-}O{-}R'{-}O{-}\underset{\underset{O}{\|}}{C}{-}R{-}\underset{\underset{O}{\|}}{C}{-}O{-}R'{-}O{-}\underset{\underset{O}{\|}}{C}{-}R{-}\underset{\underset{O}{\|}}{C}{-}O{-}R'{-}O$$

Notice that the monomers react in a 1:1 mol ratio. Equal numbers of acid and alcohol molecules are involved. For every ester group formed, a molecule of water is split out (condensed). The general equation may be written:

$$n(HO{-}\underset{\underset{O}{\|}}{C}{-}R{-}\underset{\underset{O}{\|}}{C}{-}OH) + n(HO{-}R'{-}OH) \rightarrow ({-}\underset{\underset{O}{\|}}{C}{-}R{-}\underset{\underset{O}{\|}}{C}{-}O{-}R'{-}O{-})_n + 2n\ H_2O \quad (16.3)$$

where n is a large number, 100 or greater.

One of the most important polyesters is made from the two monomers

$$HO{-}\underset{\underset{O}{\|}}{C}{-}C_6H_4{-}\underset{\underset{O}{\|}}{C}{-}OH \qquad HO{-}CH_2{-}CH_2{-}OH$$

terephthalic acid ethylene glycol

Clearly, here

$$R = {-}C_6H_4{-}; \qquad R' = {-}CH_2{-}CH_2{-}$$

FIGURE 16.7

The polyester formed from terephthalic acid can be made into a thin film (Mylar) for recording or computer tape. As a fiber (Dacron) it is used in wearing apparel and in sails such as this one. *E. I. du Pont de Nemours Company.*

When these two monomers react, they form a long-chain polyester with the structure:

$$-\underset{\underset{O}{\|}}{C}-C_6H_4-\underset{\underset{O}{\|}}{C}-O-CH_2-CH_2-O-\underset{\underset{O}{\|}}{C}-C_6H_4-\underset{\underset{O}{\|}}{C}-O-CH_2-CH_2-O-$$

The polymer can be rolled into very thin sheets, 1/10 the thickness of a human hair. In this form, called Mylar, it is used as a base for magnetic recording tape. More commonly, this polyester is drawn out into fibers. These synthetic fibers, called Dacron, compete with natural fibers such as cotton. Clothes made from Dacron resist stretching and shrinking and hold their press well.

Dacron is usually blended with cotton.

EXAMPLE 16.6

Consider each of the following pairs of compounds. In each case, give the formula of the product formed upon complete reaction. Which pairs can react to give a polyester?

a. CH_3COOH (acetic acid) and CH_3OH (methyl alcohol)

b. HOOC—COOH (oxalic acid) and CH_3OH

c. HOOC—COOH and HO—CH_2—CH_2—OH (ethylene glycol)

What general principle governs whether two monomers can produce a condensation polymer?

SOLUTION

a. The product is a simple ester, methyl acetate:

$$\mathrm{CH_3{-}\underset{\overset{\|}{O}}{C}{-}OH + CH_3{-}OH \rightarrow CH_3{-}\underset{\overset{\|}{O}}{C}{-}O{-}CH_3 + H_2O}$$

Since there are no reactive end groups, reaction stops here.

b. The first product is a simple ester, methyl oxalate:

$$\mathrm{HO{-}\underset{\overset{\|}{O}}{C}{-}\underset{\overset{\|}{O}}{C}{-}OH + CH_3{-}OH \rightarrow HO{-}\underset{\overset{\|}{O}}{C}{-}\underset{\overset{\|}{O}}{C}{-}O{-}CH_3 + H_2O}$$

There is a —COOH group at one end of this molecule. Another molecule of alcohol can react to form dimethyl oxalate:

$$\mathrm{CH_3{-}OH + HO{-}\underset{\overset{\|}{O}}{C}{-}\underset{\overset{\|}{O}}{C}{-}O{-}CH_3 \rightarrow CH_3{-}O{-}\underset{\overset{\|}{O}}{C}{-}\underset{\overset{\|}{O}}{C}{-}O{-}CH_3 + H_2O}$$

There are now no reactive end-groups, so reaction stops.

c. The first reaction produces:

$$\mathrm{HO{-}\underset{\overset{\|}{O}}{C}{-}\underset{\overset{\|}{O}}{C}{-}O{-}CH_2{-}CH_2{-}OH}$$

An alcohol molecule can react with the —COOH group at the left. An acid molecule can react with the —OH group at the right:

$$\mathrm{HO{-}CH_2{-}CH_2{-}O{-}\underset{\overset{\|}{O}}{C}{-}\underset{\overset{\|}{O}}{C}{-}O{-}CH_2{-}CH_2{-}O{-}\underset{\overset{\|}{O}}{C}{-}\underset{\overset{\|}{O}}{C}{-}OH}$$

Again, we have reactive end-groups. Reaction can continue indefinitely to produce a polyester of the type:

$$\mathrm{\left({-}\underset{\overset{\|}{O}}{C}{-}\underset{\overset{\|}{O}}{C}{-}O{-}CH_2{-}CH_2{-}O{-}\right)_n}$$

Polyamides

Another type of condensation polymer is made from a dicarboxylic acid and a monomer having $-NH_2$ groups at both ends. When the two monomers react, a water molecule is condensed out:

$$\mathrm{HO{-}\underset{\overset{\|}{O}}{C}{-}R{-}\underset{\overset{\|}{O}}{C}{-}\boxed{OH + H}{-}\underset{\overset{|}{H}}{N}{-}R'{-}\underset{\overset{|}{H}}{N}{-}H \rightarrow HO{-}\underset{\overset{\|}{O}}{C}{-}R{-}\underset{\overset{\|}{O}}{C}{-}\underset{\overset{|}{H}}{N}{-}R'{-}\underset{\overset{|}{H}}{N}{-}H + H_2O}$$

A compound called an *amide* is formed. It contains the *amide* functional group, shown in color:

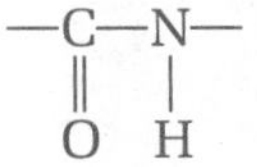

$$-\underset{\overset{\|}{O}}{C}-\underset{\overset{|}{H}}{N}-$$

There are still reactive groups at both ends of the molecule. At one end there is a —COOH group, at the other an $-NH_2$ group. So, reaction can continue. The next product might be a molecule containing four monomer units.

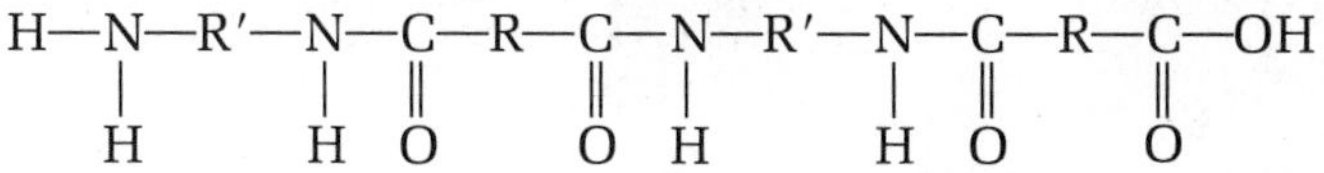

$$H-\underset{\overset{|}{H}}{N}-R'-\underset{\overset{|}{H}}{N}-\underset{\overset{\|}{O}}{C}-R-\underset{\overset{\|}{O}}{C}-\underset{\overset{|}{H}}{N}-R'-\underset{\overset{|}{H}}{N}-\underset{\overset{\|}{O}}{C}-R-\underset{\overset{\|}{O}}{C}-OH$$

The eventual product is a long-chain polymer called a **polyamide.** This contains a large number of amide groups, shown in color.

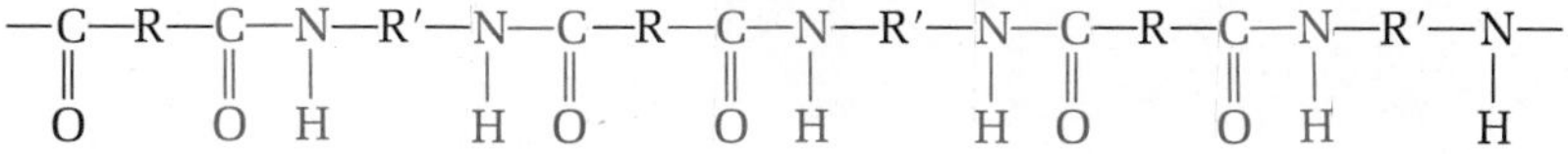

$$-\underset{\overset{\|}{O}}{C}-R-\underset{\overset{\|}{O}}{C}-\underset{\overset{|}{H}}{N}-R'-\underset{\overset{|}{H}}{N}-\underset{\overset{\|}{O}}{C}-R-\underset{\overset{\|}{O}}{C}-\underset{\overset{|}{H}}{N}-R'-\underset{\overset{|}{H}}{N}-\underset{\overset{\|}{O}}{C}-R-\underset{\overset{\|}{O}}{C}-\underset{\overset{|}{H}}{N}-R'-\underset{\overset{|}{H}}{N}-$$

Notice that the two monomers react in a 1:1 mol ratio. For every amide group formed, a water molecule is split out. The general reaction is:

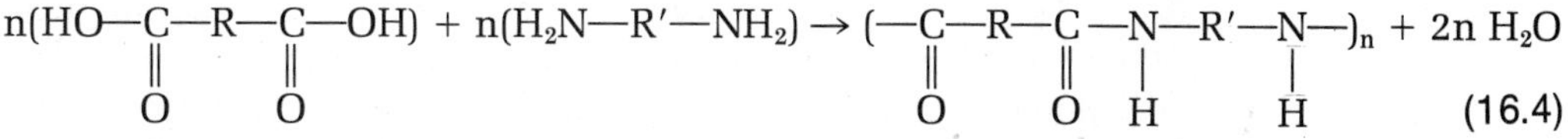

$$n(HO-\underset{\overset{\|}{O}}{C}-R-\underset{\overset{\|}{O}}{C}-OH) + n(H_2N-R'-NH_2) \rightarrow (-\underset{\overset{\|}{O}}{C}-R-\underset{\overset{\|}{O}}{C}-\underset{\overset{|}{H}}{N}-R'-\underset{\overset{|}{H}}{N}-)_n + 2n\ H_2O \qquad (16.4)$$

Here again, n is a large number, 100 or greater.

The first polyamide discovered, nylon, is still the most important. It was prepared by Wallace Carothers of the Du Pont company in 1935. Here, the two monomers are:

Dr. Carothers also discovered neoprene.

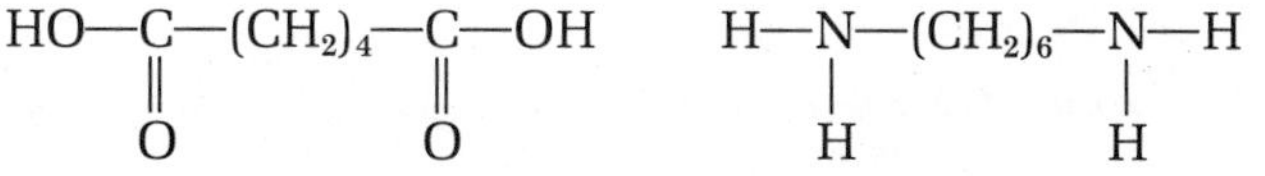

$$HO-\underset{\overset{\|}{O}}{C}-(CH_2)_4-\underset{\overset{\|}{O}}{C}-OH \qquad H-\underset{\overset{|}{H}}{N}-(CH_2)_6-\underset{\overset{|}{H}}{N}-H$$

adipic acid — hexamethylenediamine

In this case, $R = -CH_2-CH_2-CH_2-CH_2-$

and $R' = -CH_2-CH_2-CH_2-CH_2-CH_2-CH_2-$.

The structure of the polyamide formed is:

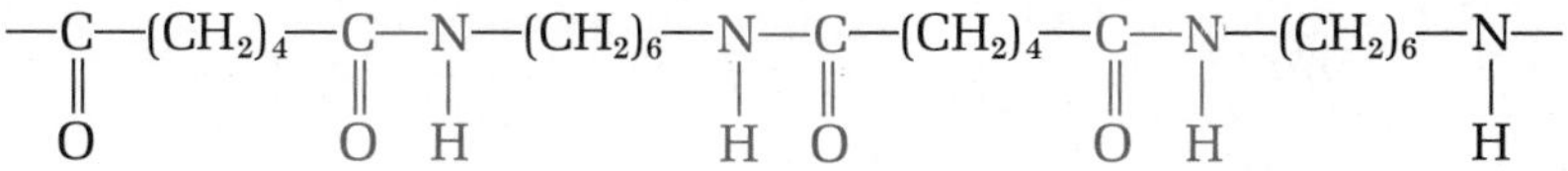

$$-\underset{\overset{\|}{O}}{C}-(CH_2)_4-\underset{\overset{\|}{O}}{C}-\underset{\overset{|}{H}}{N}-(CH_2)_6-\underset{\overset{|}{H}}{N}-\underset{\overset{\|}{O}}{C}-(CH_2)_4-\underset{\overset{\|}{O}}{C}-\underset{\overset{|}{H}}{N}-(CH_2)_6-\underset{\overset{|}{H}}{N}-$$

nylon

As commercially prepared, nylon has a molecular mass of about 10,000. It is insoluble in water and most organic solvents.

The first product made from nylon was women's hosiery. "Nylons"

FIGURE 16.8

Carpeting is one of many products made from nylon, a polyamide polymer of adipic acid and hexamethylenediamine. *E. I. du Pont de Nemours Company.*

The development of nylon resulted in a tremendous increase in industrial chemical research in the U.S.

first went on sale in October of 1939. They were an instant success, largely because nylon is more durable than silk. Commercial production of nylon was suspended during World War II. Nylon hose became a valuable commodity on the black market. When production was resumed after the War, it took seven years to meet the backlog of orders. Not until 1952 did nylon become available for other uses. Now it is one of the most versatile of all polymers, used for a variety of purposes (Figure 16.8). More nylon is used in automobiles than any other polymer.

EXAMPLE 16.7

At least in principle, it would be possible to make a polyamide from the two monomers:

HOOC—COOH	$H_2N—CH_2—NH_2$
oxalic acid	methylenediamine

Sketch the structure of:

a. the molecule formed by condensing one molecule of oxalic acid with one of methylenediamine.

b. the polyamide formed on further polymerization.

SOLUTION

a. A molecule of water is split out between a —COOH group and an $—NH_2$ group. The product is:

$$\begin{array}{l}HO—\underset{\displaystyle \|\atop O}{C}—\underset{\displaystyle \|\atop O}{C}—\underset{\displaystyle |\atop H}{N}—CH_2—\underset{\displaystyle |\atop H}{N}—H\end{array}$$

b. The structure of the polymer is:

$$\begin{array}{ccccccccccc} -\underset{\underset{O}{\|}}{C}- & \underset{\underset{O}{\|}}{C}- & \underset{\underset{H}{|}}{N}- & CH_2- & \underset{\underset{H}{|}}{N}- & \underset{\underset{O}{\|}}{C}- & \underset{\underset{O}{\|}}{C}- & \underset{\underset{H}{|}}{N}- & CH_2- & \underset{\underset{H}{|}}{N}- \end{array}$$

Compare this with the structure of nylon, shown on p. 407.

Processing of Polymers

The polymers we have discussed are made into useful products by a variety of different processes. Synthetic fibers such as nylon are made by extrusion. The molten polymer is compressed and forced (extruded) through tiny holes in a die (Figure 16.9a). Fibers are produced at the rate of about 10 m/s. After cooling, these are drawn out to four times their original length. Plastic tubing (polypropylene, polyvinyl chloride) is made by a similar process (Figure 16.9b). In this case, a rotating screw is used to force the polymer through the die.

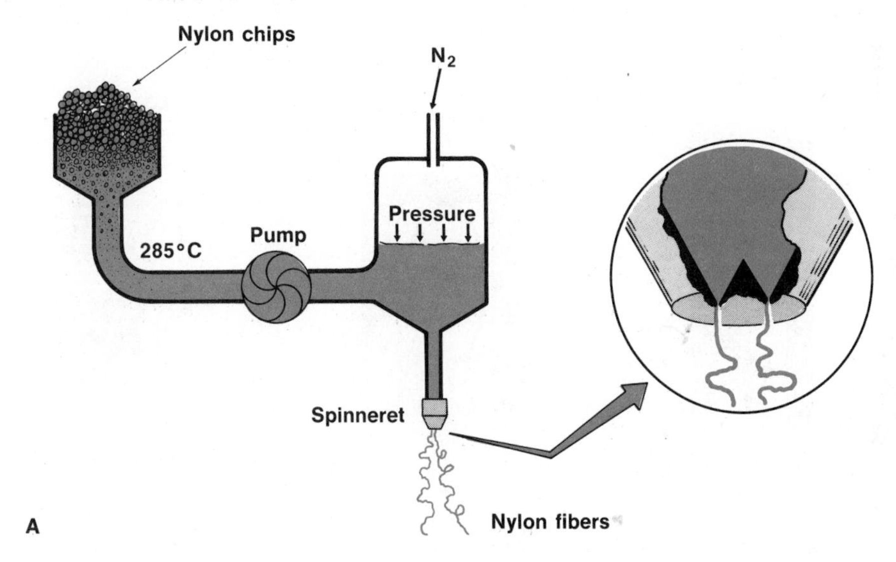

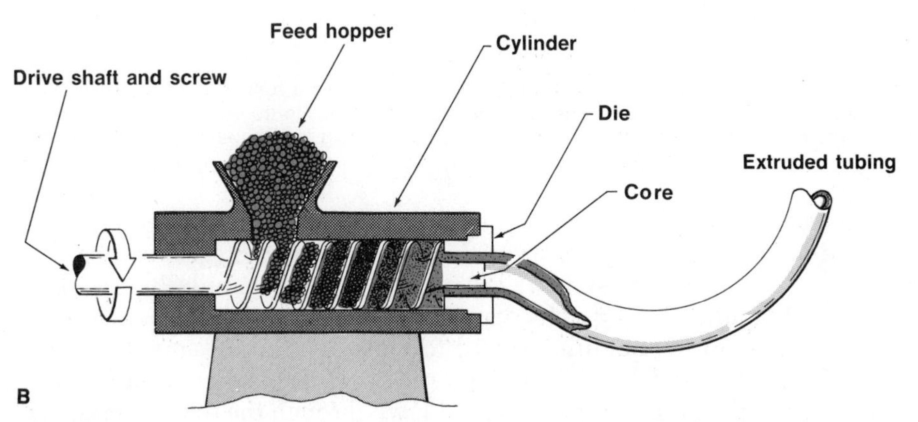

FIGURE 16.9

Fibers of synthetic polymers are made by extrusion of the molten material through fine orifices, called spinnerets (16.9a). Plastic tubing is made by extrusion of the molten polymer through a die with a cross section like that of the tubing (16.9b).

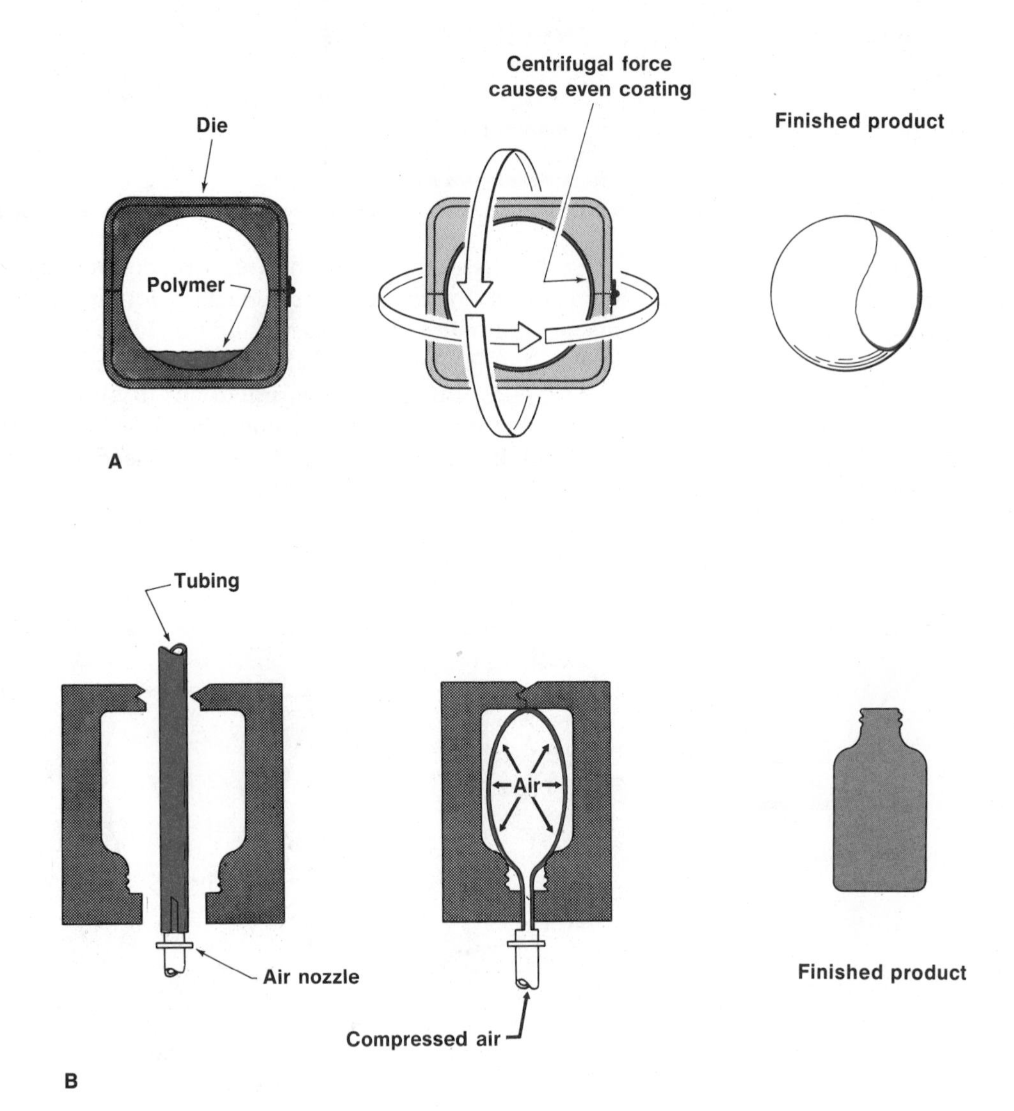

FIGURE 16.10

Plastics can be molded in various ways. In Figure 16.10a is shown a rotational mold that would be used in making tennis balls. Bottles can be made by blow molding as in Figure 16.10b. Large objects and plastic parts are made by forcing the molten polymer into a mold having the shape of the final object.

Plastic objects such as balls or bottles are made by molding. Two processes commonly used are shown in Figure 16.10. Hollow, spherical objects can be formed by adding the proper amount of molten polymer to a two-part mold (Figure 16.10a). The mold is rotated rapidly. The liquid is deposited over the walls of the mold in a layer of uniform thickness. Blow molding is used to make bottles and similar objects. Here (Figure 16.10b), the plastic is inserted into the mold in the form of a tube, which is softened by heating. Compressed air is then blown through the tube, forcing the plastic against the walls of the mold. After cooling, the mold is taken apart, releasing the bottle. Commercial machines using this process can make as many as 400 bottles a minute.

Foamed plastics are made by forming gas bubbles in a molten polymer. The process used to make styrofoam is typical. Styrene is polymerized in a volatile solvent such as pentane (bp = 36°C). The polymer forms as beads which retain 5% or more of the liquid solvent. Upon heating, the solvent vaporizes. The beads expand to one hundred times their original volume and fuse together. Cooling gives a solid product which is full of tiny holes. Styrofoam, used in such objects as cooling chests and coffee cups, has a very low density, about 0.01 g/cm³.

16.4 CARBOHYDRATES

Carbohydrates form a class of organic compounds. Most of them have the general formula:

$$C_x(H_2O)_y$$

where x and y are whole numbers. The name carbohydrate comes from this formula. However, there are no H_2O molecules as such in carbohydrates. Instead, H atoms, O atoms, and OH groups are bonded to carbon atoms in the molecule.

Hundreds of different carbohydrates are known. Some of them are relatively simple molecules of low molecular mass. Others are polymeric with molecular masses of 10^6 or greater. We will consider only five representative carbohydrates. Of these, two (glucose and fructose) might be considered monomers. Another, sucrose, is a dimer of glucose and fructose. The others, starch and cellulose, are polymers derived from glucose.

Glucose

This compound is also known as dextrose or grape sugar. It is found in honey and in many fruits. Glucose has a sweet taste and is very soluble in water (100 g/100 g water at 25°C). Structurally, it is one of the simplest carbohydrates. Its molecular formula is:

$$C_6H_{12}O_6$$

The normal structure of the glucose molecule is shown in Figure 16.11. Notice that the molecule contains a ring in which there are six atoms. Five of these are carbon atoms, one is an oxygen atom. There are five —OH groups in the molecule. Four of these are attached to ring carbons. The other is in the one-carbon sidechain.

Glucose is absorbed very quickly into the bloodstream. In the body, it undergoes a series of reactions whose net effect is the following:

$$C_6H_{12}O_6(aq) + 6\ O_2(g) \rightarrow 6\ CO_2(g) + 6\ H_2O(l);\ \Delta H = -673\ \text{kcal} \quad (16.5)$$

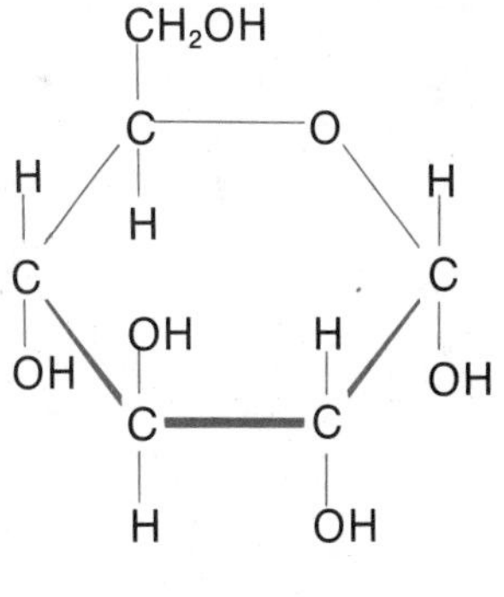

FIGURE 16.11

Glucose. In the glucose molecule there is a six-membered ring containing five carbon atoms and one oxygen atom. The five —OH groups in the molecule can hydrogen-bond with water, making glucose very soluble in aqueous systems. The geometry of the molecule is complex, and is not shown in the drawing.

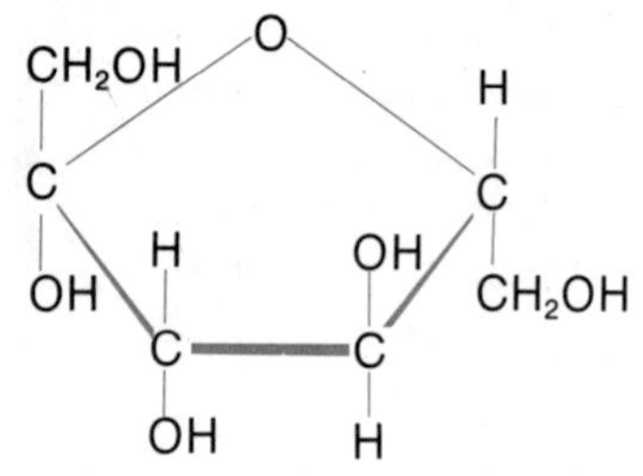

FRUCTOSE

FIGURE 16.12

Fructose. Like glucose, the fructose molecule also exists as a ring, but the structure is not the same as that of glucose, although the two sugars are isomers.

This reaction serves as a quick source of energy. Athletes who need energy in a hurry may eat foods or tablets rich in glucose. Glucose solutions are also used for intravenous feeding of hospital patients.

Fructose

This compound is a structural isomer of glucose. It has the same molecular formula, $C_6H_{12}O_6$. However, as you can see (Figure 16.12), the arrangement of atoms is somewhat different. Like glucose, fructose is found in honey and in many fruits, including pears and plums. It is much sweeter than glucose.

Sucrose

Much of the sucrose used in the United States is obtained from sugar beets.

This is the "sugar" that you buy at the grocery store. It is sold as white, chemically pure crystals. Sucrose is obtained from sugar cane or sugar beets. The sugar present in most pastries, candy, and soft drinks is actually sucrose. On the average, each person in the United States consumes about 50 kg (~100 lb) of sucrose each year.

Sucrose has the molecular formula $C_{12}H_{22}O_{11}$. Its molecular structure is shown in Figure 16.13. Notice that the sucrose molecule contains two rings. One of these is six-membered. The other ring is made up of five atoms. The two rings are joined through an oxygen atom. One of these is six-membered, derived from glucose. The other is a fructose ring, made up of five atoms.

Sucrose, like glucose and fructose, has a sweet taste. It also has a high water solubility (211 g/100 g water at 25°C). Unlike glucose,

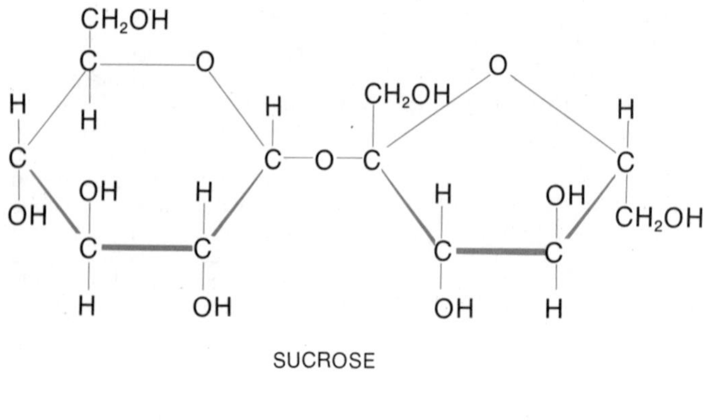

SUCROSE

FIGURE 16.13

Sucrose. Sucrose contains a molecule of glucose linked to a molecule of fructose. In the digestive system, or in acidic solution, the linkage is broken and the two simple sugar molecules are produced.

sucrose does not pass directly into the bloodstream. Instead, it breaks down in the digestive tract into a molecule of glucose and a molecule of fructose. The equation for the reaction may be written:

$$C_{12}H_{22}O_{11}(aq) + H_2O \rightarrow 2\ C_6H_{12}O_6(aq) \qquad (16.6)$$

The $C_6H_{12}O_6$ molecules then react with oxygen to release energy.

Starch

Starch is a polymeric carbohydrate found in many grains and vegetables. Corn, wheat, rice, and potatoes are major sources. "Starchy" foods, such as bread, spaghetti, and pizza, are made from these products. Starch has a very low water solubility and does not taste sweet.

Starch is a condensation polymer of glucose. The structure of a starch molecule is indicated in Figure 16.14. As many as 500 glucose units may be linked together. The simplest formula of starch is $C_6H_{10}O_5$. We might represent its molecular formula as:

Glucose, in its pure or condensed forms, is the most abundant organic compound.

$$(C_6H_{10}O_5)_n$$

where n is very large, of the order of 500. In the digestive tract, starch is broken down into smaller molecules. The final product is glucose. The equation may be written as:

$$(C_6H_{10}O_5)_n(s) + n\ H_2O \rightarrow n\ C_6H_{12}O_6(aq) \qquad (16.7)$$

Cellulose

Another carbohydrate found in small amounts in certain foods is cellulose. Like starch, it is a polymer. Its simplest formula is the same as that of starch, $C_6H_{10}O_5$. However, the glucose units are linked together

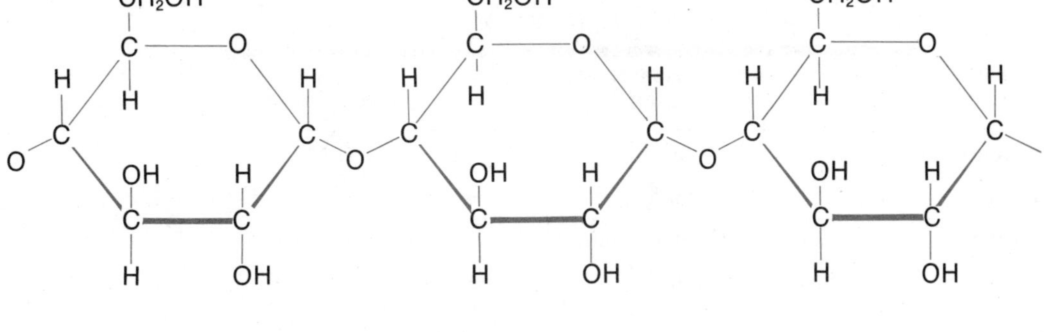

Portion of Starch Molecule

FIGURE 16.14

Starch. The starch molecule is a condensation polymer in which the monomer unit is glucose. Starch is broken down to glucose molecules in the digestive system.

in a somewhat different way. Unlike starch, cellulose does not break down to simpler molecules in our digestive system. Instead it passes through the body undigested (it's what we call roughage). For this reason, cellulose is usually not included as a carbohydrate in giving the composition of a food.

16.5 PROTEINS

Of the major components of foods, proteins have the most complex structures. They are polymers containing many different monomer units. All of the monomers found in proteins belong to the class of organic compounds known as **α-amino acids.** Alpha amino acids have the general structure:

amino—NH_2
α—"on carbon atom next to COOH group"

$$\begin{array}{ccccccc} & & & & H & & \\ & & & & | & & \\ H & — & N & — & C & — & C—OH \\ & & | & & | & & \| \\ & & H & & R & & O \end{array}$$

They differ from each other in the nature of the R group. As you can see from Table 16.2, R may be:

—a hydrogen atom, in the simplest amino acid, glycine.
—a hydrocarbon group (alanine, valine, . . .)
—a more complex group, containing oxygen atoms (serine, . . .), sulfur atoms (cysteine, . . .), or nitrogen atoms (lysine, . . .).

About 20 different amino acids are present in the proteins in foods.

TABLE 16.2 STRUCTURES OF A FEW OF THE SIMPLER AMINO ACIDS

General Structure:

$$\begin{array}{ccccccc} & & & & H & & \\ & & & & | & & \\ H & — & N & — & C & — & C—OH \\ & & | & & | & & \| \\ & & H & & R & & O \end{array}$$

Amino Acid	R =	Amino Acid	R =
Glycine	—H	Methionine	$—CH_2—CH_2—S—CH_3$
Alanine	$—CH_3$	Aspartic acid	$—CH_2—COOH$
Serine	$—CH_2—OH$	Glutamic acid	$—CH_2—CH_2—COOH$
Cysteine	$—CH_2—SH$	Lysine	$—CH_2—CH_2—CH_2—CH_2—NH_2$
Valine	$—\underset{CH_3}{\overset{H}{C}}—CH_3$	Phenylalanine	$—CH_2—C_6H_5$ (benzene ring)

Proteins are formed from amino acids by condensation polymerization. This process is very similar to that involved in the formation of a polyamide (Section 16.3). The —NH_2 group of one molecule combines with the —COOH group of another molecule. A molecule of water is condensed out. For example, a molecule of glycine might condense with one of alanine:

$$
\begin{array}{*{35}{c}}
&&&&H&&&&&&&&&&H&&&&&&&&&&H&&&&&&H \\
&&&&\mid&&&&&&&&&&\mid&&&&&&&&&&\mid&&&&&&\mid \\
H&-&N&-&C&-&C&-&OH&+&H&-&N&-&C&-&C&-&OH&\rightarrow&H&-&N&-&C&-&C&-&N&-&C&-&C&-&OH \\
&&\mid&&\mid&&\|&&&&&&\mid&&\mid&&\|&&&&&&\mid&&\mid&&\|&&\mid&&\mid&&\| \\
&&H&&H&&O&&&&&&H&&CH_3&&O&&&&&&H&&H&&O&&H&&CH_3&&O
\end{array}
$$

glycine   alanine   glycine unit   alanine unit

Since there are reactive groups at both ends of the molecule formed, polymerization can continue. This time, we might add a serine molecule at the right of the chain to give:

$$
\begin{array}{*{21}{c}}
&&&&H&&&&&&H&&&&&&H \\
&&&&\mid&&&&&&\mid&&&&&&\mid \\
H&-&N&-&C&-&C&-&N&-&C&-&C&-&N&-&C&-&C&-&OH \\
&&\mid&&\mid&&\|&&\mid&&\mid&&\|&&\mid&&\mid&&\| \\
&&H&&H&&O&&H&&CH_3&&O&&H&&CH_2&&O \\
&&&&&&&&&&&&&&&&\mid \\
&&&&&&&&&&&&&&&&OH
\end{array}
$$

glycine unit   alanine unit   serine unit

Eventually, we form a long-chain polymer with the general structure:

It is possible that Carothers used the structure of proteins as a guide in his development of nylon.

$$
\begin{array}{*{25}{c}}
&&&H&&&&&&H&&&&&&H&&&&&&H \\
&&&\mid&&&&&&\mid&&&&&&\mid&&&&&&\mid \\
-&N&-&C&-&C&-&N&-&C&-&C&-&N&-&C&-&C&-&N&-&C&-&C&- \\
&\mid&&\mid&&\|&&\mid&&\mid&&\|&&\mid&&\mid&&\|&&\mid&&\mid&&\| \\
&H&&R_1&&O&&H&&R_2&&O&&H&&R_3&&O&&H&&R_4&&O
\end{array}
$$

where R_1, R_2, R_3, R_4, . . . may be the same or different.

A protein molecule consists of a large number of monomer units, bonded together as shown above. Typically, the number is 100 or more and can be as large as 10^5. Remember also that about 20 different monomer units (different R groups) are possible. As you can imagine, there are many, many, many, *many* different proteins. That is why we say that proteins have more complex structures than carbohydrates or fats.

EXAMPLE 16.8

Give the structure of the polymer formed by condensing molecules of glycine, aspartic acid, and glycine (in that order, from left to right).

SOLUTION

Referring to Table 16.2, we see that R is a H atom in glycine and a —CH_2—COOH group in aspartic acid. The structure is:

```
             H           H          H
             |           |          |
     H—N—C—C—N—C—C—N—C—C—OH
        |   |   ||   |   |    ||  |   |   ||
        H   H   O   H  CH₂O  H   H   O
                           |
                         COOH
     glycine      aspartic acid     glycine
       unit            unit           unit
```

The richest sources of proteins are eggs, meats, poultry, and fish. Milk and milk products such as cheese also contain proteins. They are also found in some vegetables and grains. Soy beans (16% protein) are one of the cheapest sources. Soybean protein is sometimes added to hamburger and other meat products.

We now know in detail the structures of many proteins.

Proteins are required to build and maintain body tissues. Every living cell contains at least some protein molecules. They are present in blood, muscles, the brain, and all body organs. Skin and hair are nearly 100% protein. There is even some protein in bones and teeth. Normally, the amount of protein we eat exceeds that required for growth or repair of tissue. Excess protein is oxidized to produce energy, or is eliminated as waste.

16.6 SUMMARY

A polymer is a large molecule formed when many small molecules (monomers) combine with one another. A polymer may be derived from a single monomer. In other cases, two or more different monomers are involved. Combination may occur by:

1. *Addition* Here, a single monomer is usually used. It must contain at least one double bond per molecule. In polymerization a C═C bond per monomer unit is converted to a C—C bond.

2. *Condensation* Two monomers are usually involved. Both monomers have a functional group (—COOH, —OH) at each end of the molecule. Upon polymerization, a small molecule such as H_2O is split out. Two common types of condensation polymers are polyesters and polyamides. In these, the monomer units are joined through ester and amide groups.

A wide variety of polymeric materials are found in living organisms. Among these are the carbohydrates starch and cellulose. They are derived from the monomer glucose, a simple sugar. Another class of natural polymers is formed by the proteins. These are complex condensation polymers of alpha amino acids.

NEW TERMS

addition polymer
amide group
amino acid
carbohydrate
condensation polymer
copolymer
monomer
polyamide
polyester
polymer
protein

QUESTIONS

1. Explain what is meant by a polymer; a monomer. Which of the two generally has the higher molecular mass?

2. A certain polyethylene molecule contains 2400 $—CH_2—$ groups. How many ethylene molecules combined to form it?

3. Distinguish between an addition and a condensation polymer. Which type has the same chemical composition as the monomer from which it is formed?

4. Draw the structural formula of the monomer used to make:

   a. polyethylene b. polyvinyl chloride c. Teflon

5. Give at least one commercial use for each of the polymers listed in Question 4.

6. Draw the structure of the monomer from which natural rubber is derived.

7. Which of the following polymers contain double bonds?

   a. polyethylene b. neoprene c. Dacron

8. What is meant by a copolymer? Which of the following are copolymers?

   a. polyethylene b. neoprene c. Dacron

9. A synthetic rubber can be made using a mixture of butadiene and styrene as the starting materials. Draw the structures of these two monomers. Which one, by itself, would you expect to polymerize to give a rubber-like material?

10. Explain why condensation polymers are formed by:

    a. oxalic acid, $HOOC—COOH$, but not formic acid.
    b. ethylene glycol, $HO—CH_2—CH_2—OH$, but not methyl alcohol.
    c. urea, $NH_2—CO—NH_2$, but not ammonia.

11. Draw structural formulas of the monomers used to make Dacron and Nylon.

12. Show the structure of an:

    a. ester group in a polyester b. amide group in a polyamide

13. Which two of the following compounds have the same molecular formula?

    a. sucrose b. glucose c. fructose d. starch

14. A compound with the formula $C_5H_{10}O_5$ is most likely:

    a. an amino acid b. a protein c. a fat d. a carbohydrate

15. Which of the following has the smallest molecular mass? the largest?

    a. sucrose b. starch c. glucose d. water

16. What is the general structure of an alpha amino acid?

17. Using Table 16.2, give the structural formula of:

    a. methionine b. aspartic acid c. phenylalanine

18. In the text, we showed one way in which a molecule of glycine can condense with a molecule of alanine. There is another way this can occur, giving an isomer of the compound shown in the text. Draw the structural formula of this isomer.

19. Explain, in terms of structure, why there are so many proteins.

PROBLEMS

1. *Addition Polymers (Example 16.1)* A certain type of Teflon has a molecular mass of about 1.0×10^5.

    a. How many $—CF_2—$ groups are there per Teflon molecule?
    b. How many C_2F_4 molecules are required to make the polymer molecule?
    c. How many carbon atoms are there per polymer molecule?

2. *Addition Polymers (Example 16.2)* Identify the monomer from which the following polymer is made.

$$\begin{array}{cccccccccccccc} & \mathrm{H} & & \mathrm{H} & & \mathrm{H} & & \mathrm{H} & & \mathrm{H} & & \mathrm{H} & \\ & | & & | & & | & & | & & | & & | & \\ — & \mathrm{C} & — & \mathrm{C} & — & \mathrm{C} & — & \mathrm{C} & — & \mathrm{C} & — & \mathrm{C} & — \\ & | & & | & & | & & | & & | & & | & \\ & \mathrm{H} & & \mathrm{CN} & & \mathrm{H} & & \mathrm{CN} & & \mathrm{H} & & \mathrm{CN} & \end{array}$$

3. *Addition Polymers (Example 16.3)* Sketch a portion of the addition polymer formed by:

    a. propylene b. C_2F_4 c. styrene

4. *Addition Polymers (Example 16.4)* Give the percentages by mass of the elements in

    a. ethylene b. polyethylene c. polypropylene

5. *Synthetic Rubber (Example 16.5)* Sketch a portion of a synthetic rubber molecule made by polymerizing butadiene as the only monomer.

6. *Condensation Polymers (Example 16.6)* Which of the following would react to form a polyester?

    a. $HOOC—CH_2—COOH$ and $HO—CH_2—CH_2—OH$
    b. $CH_3—COOH$ and $HO—CH_2—CH_2—OH$
    c. $HOOC—CH_2—COOH$ and

$$\begin{array}{ccccc} & & & \mathrm{CH_3} & \\ & & & | & \\ \mathrm{HO}—\mathrm{CH_2} & — & & \mathrm{C} & —\mathrm{OH} \\ & & & | & \\ & & & \mathrm{H} & \end{array}$$

7. *Condensation Polymers (Example 16.6)* Sketch a portion of any condensation polymers formed in Problem 6.

8. *Condensation Polymers (Example 16.7)* Sketch a portion of the polyamide formed by: $HOOC—(CH_2)—COOH$ and $NH_2—CO—NH_2$

9. *Protein Structures (Example 16.8)* Give the structure of the polymer formed by condensing molecules of alanine, glycine, and phenylalanine (in that order, from left to right).

10. *Protein Structures (Example 16.8)* What amino acids were condensed to make:

```
        H           H
        |           |
  H—N—C—C—N—C—C—OH
     |   |   ||  |   |   ||
     H  CH2 O   H  CH2 O
         |            |
        OH          COOH
```

* * * * *

11. Identify the monomer used to make the following addition polymer:

```
   CH3 H   CH3 H   CH3 H   CH3 H
    |   |    |   |    |   |    |   |
  —C—C—C—C—C—C—C—C—
    |   |    |   |    |   |    |   |
    H  Cl   H  Cl   H  Cl   H  Cl
```

12. Identify the monomer used to make the following synthetic rubber.

```
        CH3 CH3                  CH3 CH3                  CH3 CH3
         |   |                     |   |                     |   |
—CH2—C=C—CH2—CH2—C=C—CH2—CH2—C=C—CH2—
```

13. Identify the monomers used to make the following condensation polymer:

```
 —C—(CH2)5—C—N—C—N—C—(CH2)5—C—N—C—N—
   ||            ||  |  ||  |  ||            ||  |  ||  |
   O             O  H  O  H  O             O  H  O  H
```

14. Sketch a portion of the synthetic rubber made by using a 1:1 mol ratio of isoprene and chloroprene.

15. Identify the two monomers used to make the following polyamide.

```
      CH3                              CH3
       |                                 |
 —C—C—C—N—CH2—N—C—C—C—N—CH2—N—
   ||  |  ||  |             |  ||  |  ||  |          |
   O  H  O  H             H  O  H  O  H          H
```

16. The molecular mass of a certain polymer made from ethylene is 5.6×10^4. How many molecules of ethylene are required to form one polymer molecule?

17. What is the mass in grams of a polymer made from 6.0×10^{26} molecules of vinyl chloride?

18. What are the percentages by mass of C, H, and Cl in polyvinyl chloride?

19. Draw the structural formula of a portion of the polyester formed from the two monomers:

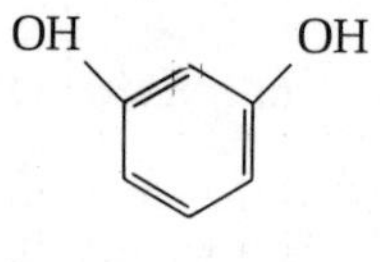

OH OH (1,3-dihydroxybenzene) and COOH COOH (1,2-benzenedicarboxylic acid)

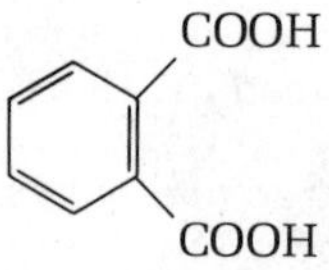

20. Consider the amino acid phenylalanine (Table 16.2).

a. What is the molecular formula of phenylalanine?
b. What is its molecular mass?
c. What are the mass percentages of the elements in phenylalanine?

21. What are the molecular masses of:

a. glycine b. serine
c. the dimer formed by combining a molecule of glycine with one of serine.

22. Draw the structure of the polymer: serine—alanine—glycine—aspartic acid.

*23. In making the synthetic rubber SBR referred to on p. 403, how many grams of styrene should be used per gram of butadiene?

*24. What are the percentages by mass of C, H, O, and N in nylon, whose structure is shown on p. 407?

*25. A protein molecule consists of 30 molecules of glycine and 60 of alanine. What is its molecular mass? Remember that H_2O molecules are condensed out in forming the protein.

The chemical reactions that supply the energy for lift-off must take place in a fraction of a second. *NASA.*

17

RATE OF REACTION

An important feature of any reaction is the rate at which it takes place. Frequently, we want a reaction to occur as quickly as possible. This is usually the case with reactions that you carry out in the laboratory. They must occur rapidly if you are to conduct the reaction, make observations, and draw conclusions, all in a single lab period. In industry as well, chemical processes must occur rather rapidly. A manufacturer cannot wait years to get a product for which there is a large market today.

Sometimes we want to slow down a reaction, making it occur less rapidly. Chemicals are added to bread and other foods to slow down the growth of molds and harmful bacteria. Biochemists are searching for ways to slow down aging, a chemical process taking place in the body.

Many factors influence the rate of a reaction. One of the most important is the nature of the reactants. Some reactions are naturally rapid. A freshly-cut piece of sodium reacts quickly with oxygen and water in the air. It tarnishes instantly and later crumbles to a white powder. Other reactions are slow. Copper reacts slowly with air under ordinary conditions. For that reason, deposits of the element are still found on the earth's surface.

Still other reactions take place at intermediate speeds. They are difficult to classify as "slow" or "fast". One such reaction is that between iron and oxygen. A steel bar exposed to dry air reacts very slowly. Months or even years may be required to form a coating of rust. In contrast, if a piece of steel wool is heated and placed in a bottle of pure oxygen, it reacts almost instantly. The steel wool bursts into flame, shooting off sparks of red-hot iron and iron oxide. This effect is shown in Color Plate 7C, in the center of the book.

Reactions such as the one just described can be speeded up or slowed down by adjusting the conditions of reaction. Factors such as temperature, concentration, and surface area affect reaction rate. Later in this chapter we will see why these factors are so important. First, though, we had best explain what is meant by reaction rate.

17.1 MEANING OF REACTION RATE

All of us are familiar with the concept of rate. A teacher may tell you that your reading rate is 200 words per minute. You may be told by a policeman that you were driving at a rate of 60 miles per hour. In both cases, rate describes how position (on the page of a book or on the highway) changes with time. The rate of any process can be expressed as a ratio. In this ratio, time appears in the denominator:

Rate always equals a change in some quantity, divided by the time it takes for the change to occur.

$$200\,\frac{\text{words}}{\text{minute}},\ 60\,\frac{\text{miles}}{\text{hour}}$$

The rate of a reaction describes how the "position" or extent of reaction changes with time. The quantity which changes is the concentration of reactant or product. Consider a simple reaction in which a reactant A yields a single product, B:

$$A(g) \rightarrow B(g) \qquad (17.1)$$

We can take the rate of this reaction to be the change in concentration of B in unit time. That is:

$$\text{rate} = \frac{\text{change in concentration of B}}{\text{time interval}} = \frac{\Delta\ \text{conc. B}}{\Delta t} \qquad (17.2)$$

Here, as always, the symbol Δ means "final" minus "original". Thus if the concentration of B increases from 0 to 0.10 mol/ℓ in one minute:

$$\text{rate} = \frac{(0.10 - 0)\ \text{mol}/\ell}{(1 - 0)\ \text{min}} = 0.10\,\frac{\text{mol}/\ell}{\text{min}}$$

If, under different conditions, the concentration of B increases from 0.12 to 0.24 mol/ℓ in four minutes, the rate is:

$$\frac{(0.24 - 0.12)\ \text{mol}/\ell}{(4 - 0)\ \text{min}} = 0.030\ \frac{\text{mol}/\ell}{\text{min}}$$

In expressing the rate of reaction, we will always use concentrations in moles per liter (mol/ℓ). Time, however, may be expressed in many different units. For example, we might use hours or seconds rather than minutes. Reaction rate in one time unit is readily converted to a different unit (Example 17.1).

EXAMPLE 17.1

Consider a reaction where the rate is $0.10\ \frac{\text{mol}/\ell}{\text{min}}$. Express this rate in $\frac{\text{mol}/\ell}{\text{hr}}$.

SOLUTION

The conversion factor required is obtained from the relation:

$$60\ \text{min} = 1\ \text{hr}$$

$$\text{rate} = 0.10\ \frac{\text{mol}/\ell}{\text{min}} \times \frac{60\ \text{min}}{1\ \text{hr}} = 6.0\ \frac{\text{mol}/\ell}{\text{hr}}$$

To check this answer, note that rate should be larger when time is expressed in hours rather than minutes. If you drive a car at the rate of one mile per *minute*, your speedometer will read 60 miles per *hour*.

Determination of Reaction Rate

Suppose you were driving from New York to Boston, a distance of about 400 km. If you wanted to determine your driving rate over a certain time period, it could be done very easily. You need only record distance travelled as a function of time. Using the following data:

Distance from New York (km)	0	83	161	245	320	400
Time	1 P.M.	2 P.M.	3 P.M.	4 P.M.	5 P.M.	6 P.M.

it is clear that your rate between 1 and 2 P.M. is:

Is this driver exceeding the speed limit of 55 miles/hr?

$$\frac{(83 - 0)\ \text{km}}{1\ \text{hr}} = 83\ \text{km/hr}$$

The rate between 2 and 3 P.M. is:

$$\frac{(161 - 83)\ \text{km}}{1\ \text{hr}} = 78\ \text{km/hr}$$

and so on.

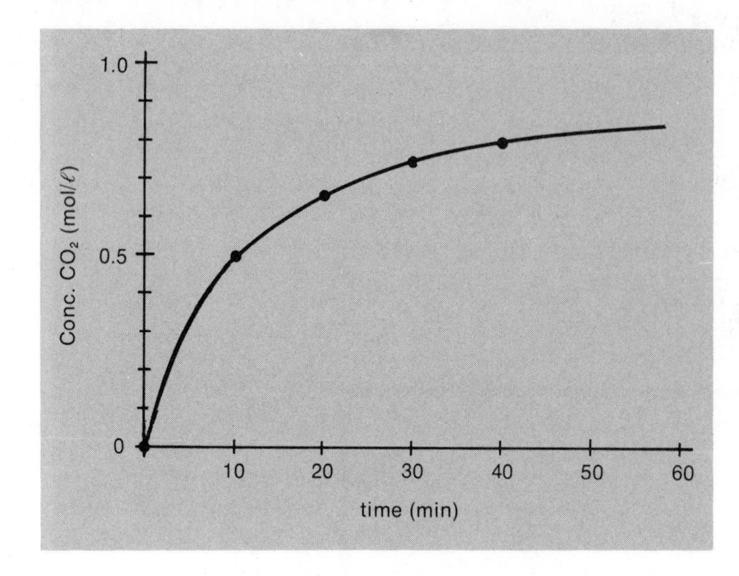

FIGURE 17.1

When CO reacts with NO_2, the concentration of the product, CO_2, increases as the reaction proceeds. The amount of change in concentration of CO_2 in a given period of time goes down as the reaction goes on, so the rate decreases with time.

Reaction rates can be obtained in much the same way. Consider the reaction between carbon monoxide and nitrogen dioxide:

$$CO(g) + NO_2(g) \rightarrow CO_2(g) + NO(g) \tag{17.3}$$

Here, we measure the concentration of one substance as a function of time. We might choose carbon dioxide. Since CO_2 is a product, its concentration starts at zero and increases as time passes. Perhaps we would obtain the data shown in Figure 17.1 and Table 17.1.

From Table 17.1, we can calculate the rate over any ten-minute time interval. We take the rate to be the change in concentration of CO_2 with time.

$$\text{rate} = \frac{\Delta \text{ conc. } CO_2}{\Delta t}$$

Thus we have:

$$\text{rate from 0 to 10 min} = \frac{(0.50 - 0.00) \text{ mol}/\ell}{(10 - 0) \text{ min}} = 0.050 \frac{\text{mol}/\ell}{\text{min}}$$

We conclude that over this time interval the concentration of CO_2 is increasing at the rate of 0.050 mol/ℓ per minute. The reaction rate during other time intervals is calculated in a similar way (Example 17.2).

TABLE 17.1 CHANGE OF CO_2 CONCENTRATION WITH TIME (REACTION 17.3 AT 300°C)

Original Concentrations: CO = NO_2 = 1.00 mol/ℓ					
CONC. CO_2 (mol/ℓ)	0.00	0.50	0.67	0.75	0.80
TIME (min)	0	10	20	30	40

EXAMPLE 17.2

Using Table 17.1, calculate the rate of Reaction 17.3 between 10 and 20 min.

SOLUTION

$$\text{rate} = \frac{\Delta \text{ conc. } CO_2}{\Delta t} = \frac{(0.67 - 0.50) \text{ mol}/\ell}{(20 - 10) \text{ min}}$$

$$= \frac{0.17 \text{ mol}/\ell}{10 \text{ min}} = 0.017 \frac{\text{mol}/\ell}{\text{min}}$$

As you can see from these calculations, reaction rate decreases as time passes. Between 0 and 10 min, the rate is 0.050 mol/ℓ per minute. Between 10 and 20 min, it is only 0.017 mol/ℓ per minute. The same effect is evident from Figure 17.1. The concentration of CO_2 builds up rapidly in the early stages of the reaction. Then it starts to level off. The slope of the curve in Figure 17.1, which is a measure of the rate, decreases steadily with time.

Actually, these are average rates over the time interval.

This behavior is typical of most reactions. They start out rapidly and then slow down. To understand why this happens, we must consider the effect upon rate of the concentration of reactants.

17.2 EFFECT OF CONCENTRATION UPON RATE

We ordinarily find that the rate of reaction is directly related to the concentration of reactants. That is:

—*an increase in the concentration of reactants increases the rate.*
—*a decrease in the concentration of reactants decreases the rate.*

This tells us why reactions slow down as time passes. At first, the concentrations of reactants are high and the rate is rapid. As reaction proceeds, reactants are consumed. Their concentrations decrease and the reaction slows down.

To explain the effect of concentration upon rate, we consider the behavior of reactant molecules. For reaction to occur, two molecules must collide with each other. At low concentrations, the molecules are far apart. They have to move a long way to meet each other. Hence, there are few collisions in a given time and reaction occurs slowly. At higher concentrations, the molecules are closer to one another. Collisions occur more frequently and the rate of reaction increases.

Most reactions that take place between gases or in water solution are speeded up by increasing the concentration of reactants. For example:

1. Reaction of oxygen with metals such as iron occurs more rapidly in pure O_2 than in air. As pointed out earlier, a piece of steel wool heated

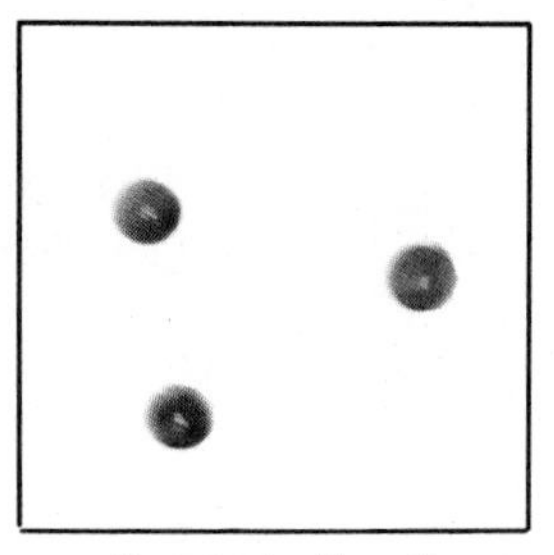
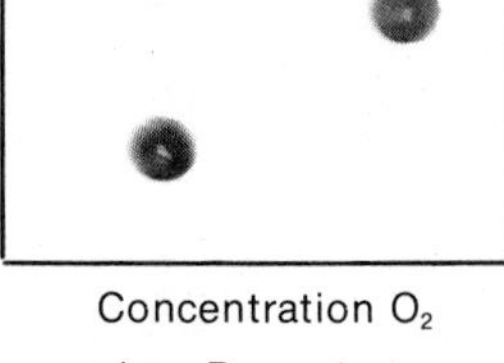

Concentration O_2
when P_{air} = 1 atm

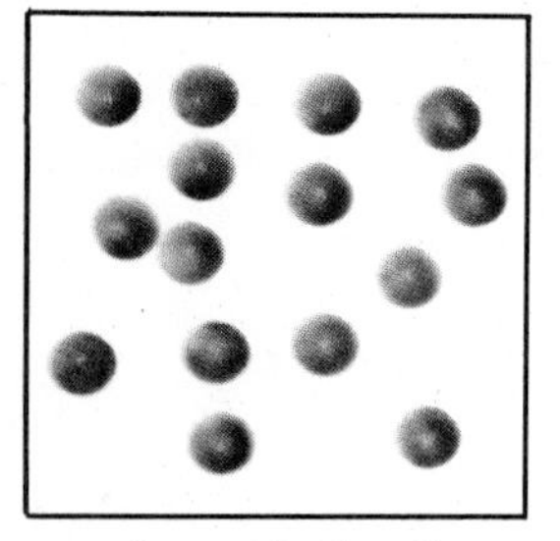

Concentration O_2
when P_{O_2} = 1 atm

FIGURE 17.2

In air at 1 atm the concentration of O_2 molecules is only about 1/5 that in a sample of pure O_2 at 1 atm. Reactions involving oxygen go much faster in pure O_2 than they do in air.

in air bursts into flame if exposed to pure oxygen. The explanation is a simple one (Figure 17.2). In air, only about one molecule in five is an O_2 molecule. The others are mostly N_2 molecules. In pure oxygen, there are five times as many O_2 molecules per unit volume. Thus their concentration is greater than that in air by a factor of five. The rate of reaction increases as a result of this concentration effect.

2. Metals ordinarily react more rapidly with acid when the concentration of H^+ ions is raised. To generate hydrogen by the reaction of zinc with acid:

$$Zn(s) + 2\,H^+(aq) \rightarrow Zn^{2+}(aq) + H_2(g) \qquad (17.4)$$

you use 6 M HCl rather than 1 M HCl. The higher concentration of H^+ makes the reaction go faster. As time passes, the concentration of H^+ drops and the reaction slows down. When this happens, you add more acid to restore the original rate.

Reactions Which Occur in a Single Step

As we have seen, reactant molecules must collide in order to react. Sometimes reaction occurs in a single step: a collision between two different molecules. One such reaction is that between carbon monoxide and nitrogen dioxide:

$$CO(g) + NO_2(g) \rightarrow CO_2(g) + NO(g)$$

This reaction involves a collision between CO and NO_2 molecules. Product molecules form directly as a result of this collision. No further steps are required to complete the reaction.

For reactions of this type, there is a simple relation between rate and concentration. To obtain this relation, we start by asking a question. What will happen to the number of collisions, and hence the rate, if we double the concentration of CO? Looking at Figure 17.3, the effect should be clear. The rate of reaction will be doubled. With twice as many CO molecules in the same space, a molecule of NO_2 will collide with a molecule of CO twice as often. The same argument applies to the other reactant. Suppose we keep the concentration of CO constant but double that of NO_2. Once again, there will be twice as many collisions in a given time. When we double the number of collisions we double

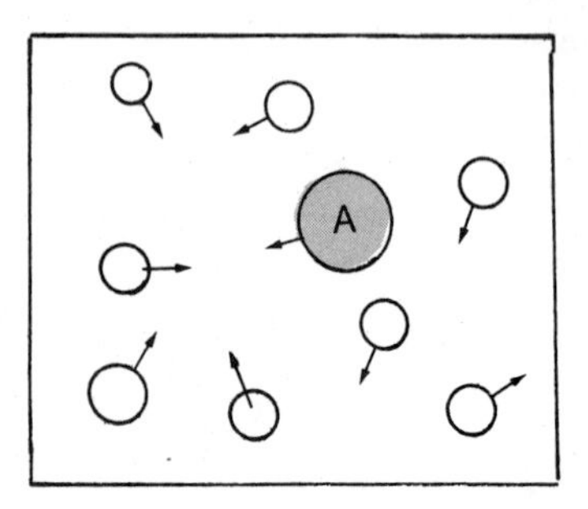

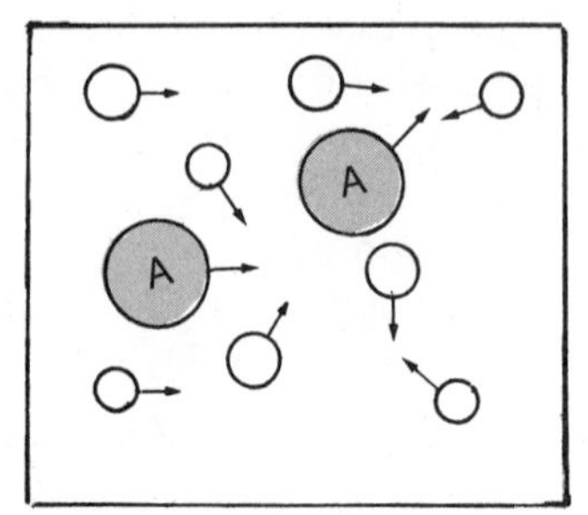

FIGURE 17.3

Effect of concentration of a reactant, A, on collision rate. When the concentration of A is doubled, there are twice as many collisions with the other reactant.

the rate. Thus we expect that doubling the concentration of either CO or NO_2 will double the rate.

Summarizing the discussion we have just gone through, we can say that the rate of the reaction:

$$CO(g) + NO_2(g) \rightarrow CO_2(g) + NO(g)$$

is:

—directly proportional to the concentration of CO (constant concentration NO_2)

—directly proportional to the concentration of NO_2 (constant concentration CO)

Putting these two statements together, it follows that:

The rate of reaction is directly proportional to the product of the concentrations of CO and NO_2. That is:

$$\text{rate} = k\ (\text{conc. CO}) \times (\text{conc. } NO_2) \qquad (17.5)$$

where k is the proportionality constant. It is called a *rate constant.* An equation such as 17.5, which tells us how the rate depends upon concentrations of reactants is called a *rate equation.*

EXAMPLE 17.3

What happens to the rate of the reaction between CO and NO_2 if:

a. conc. CO increases from 0.40 mol/ℓ to 0.80 mol/ℓ (conc. NO_2 constant)

b. conc. NO_2 decreases from 0.20 mol/ℓ to 0.10 mol/ℓ (conc. CO constant)

SOLUTION

In general, we can say that for two sets of conditions 1 and 2:

Given an equation like 17.5, you can always use this approach to relate rates under two sets of conditions.

$$\text{rate}_2 = k(\text{conc. CO})_2(\text{conc. } NO_2)_2$$

$$\text{rate}_1 = k(\text{conc. CO})_1(\text{conc. } NO_2)_1$$

Dividing the first equation by the second:

$$\frac{\text{rate}_2}{\text{rate}_1} = \frac{(\text{conc. CO})_2(\text{conc. } NO_2)_2}{(\text{conc. CO})_1(\text{conc. } NO_2)_1}$$

a. Here, conc. NO_2 is constant, so it cancels out and we have:

$$\frac{\text{rate}_2}{\text{rate}_1} = \frac{(\text{conc. CO})_2}{(\text{conc. CO})_1} = \frac{0.80\ \text{mol}/\ell}{0.40\ \text{mol}/\ell} = 2.0$$

We conclude that the rate *doubles.*

b. In this case conc. CO is constant so:

$$\frac{\text{rate}_2}{\text{rate}_1} = \frac{(\text{conc. } NO_2)_2}{(\text{conc. } NO_2)_1} = \frac{0.10 \text{ mol}/\ell}{0.20 \text{ mol}/\ell} = 0.50$$

The rate is only one-half the original value.

Several-Step Reactions

Most reactions do not occur by the simple process just described. They require more than a single collision between reactant molecules. Consider, for example, the reaction:

$$4\ NH_3(g) + 5\ O_2(g) \rightarrow 4\ NO(g) + 6\ H_2O(l) \qquad (17.6)$$

If you want to see an accident like this, go to Istanbul.

This reaction could hardly occur through a simultaneous collision between four NH_3 and five O_2 molecules. The odds against such an event are enormous. It is about as unlikely as a simultaneous collision between nine automobiles, coming from nine different directions.

Some reactions which might appear to occur in a single step do not actually do so. The equation for the reaction

$$CH_4(g) + Cl_2(g) \rightarrow CH_3Cl(g) + HCl(g) \qquad (17.7)$$

looks very much like that for the reaction between CO and NO_2. We could imagine a single collision between a CH_4 molecule and a Cl_2 molecule bringing about reaction. However, experimental evidence suggests that this does not happen. Instead, reaction occurs through a series of three steps.

(1) The dissociation of a Cl_2 molecule into atoms:

$$Cl_2 \rightarrow Cl + Cl$$

This step occurs very rapidly in the presence of ultraviolet radiation. This furnishes the energy to break the Cl—Cl bond.

(2) A collision between a Cl atom and a CH_4 molecule:

$$Cl + CH_4 \rightarrow CH_3Cl + H$$

This is the slow step in the process and hence controls the overall rate of the reaction. It involves breaking a C—H bond in the CH_4 molecule, which absorbs a large amount of energy.

(3) A collision between a Cl atom and an H atom:

$$H + Cl \rightarrow HCl$$

This step occurs very rapidly. No bonds are broken. As soon as an H atom is produced by step (2), it reacts in this step.

This type of reaction path is typical of most chemical reactions. They take place in a series of steps. Each step is a simple one. It may involve the dissociation of a molecule into atoms (step 1 above). Other steps may involve collisions between molecules or atoms (steps 2 and 3 above). Frequently, one step is much slower than all the others. It governs the overall rate at which products are formed.

In any stepwise reaction, the equations for the individual steps must add to give the overall reaction. For example, for the reaction of Cl_2 with CH_4:

$$Cl_2 \rightarrow Cl + Cl$$

$$Cl + CH_4 \rightarrow CH_3Cl + H$$

$$H + Cl \rightarrow HCl$$

$$Cl_2(g) + CH_4(g) \rightarrow CH_3Cl(g) + HCl(g) \qquad (17.7)$$

The three individual steps written above comprise the path or *mechanism* of the reaction. The overall equation, Equation 17.7, tells us the end result of the reaction. In the three-step mechanism, the net effect is to convert a molecule of Cl_2 and a molecule of CH_4 to a molecule of CH_3Cl and a molecule of HCl.

Rate Equation and Order of Reaction The equation relating reaction rate to concentration is called a rate equation. For a reaction which occurs in a single step, such as:

$$CO(g) + NO_2(g) \rightarrow CO_2(g) + NO(g)$$

the rate equation, as we have seen, is a simple one:

$$\text{rate} = k\ (\text{conc. CO}) \times (\text{conc. } NO_2)$$

For stepwise reactions, the rate equation is more complex. For the reaction between Cl_2 and CH_4:

$$Cl_2(g) + CH_4(g) \rightarrow CH_3Cl(g) + HCl(g)$$

the rate equation is found to be:

$$\text{rate} = k\ (\text{conc. } CH_4) \times (\text{conc. } Cl_2)^{1/2}$$

This equation tells us that the reaction rate is directly proportional to

1. the concentration of CH_4.
2. the *square root* of the concentration of Cl_2.

In another case, for the reaction:

$$2\ H_2(g) + 2\ NO(g) \rightarrow N_2(g) + 2\ H_2O(g) \qquad (17.8)$$

the rate equation is:

$$\text{rate} = k\ (\text{conc. } H_2) \times (\text{conc. NO})^2 \qquad (17.9)$$

Here, the rate is directly proportional to:

1. the concentration of H_2.
2. the *square* of the concentration of NO.

In general, for any reaction

$$A + B \rightarrow \text{products}$$

it is safe to say that the rate equation is:

$$\text{rate} = k\ (\text{conc. A})^m \times (\text{conc. B})^n \qquad (17.10)$$

where m and n are exponents. They may be whole numbers (1, 2, . . . or, in some cases, 0). For other reactions, the exponents are fractional ($^1/_2$, $^3/_2$, . . .). In any case, these exponents define what we call the *order* of the reaction. We say that:

m = order of reaction with respect to A

n = order of reaction with respect to B

For the reaction of H_2 with NO, we see that:

m = 1; the reaction is first order with respect to H_2.
n = 2; the reaction is second order with respect to NO.

17.3 EFFECT OF SURFACE AREA ON REACTION RATE

The rates of reactions involving solids are usually increased by an increase in surface area. To get a wood fire going, we start with kindling or shavings rather than a large block of wood. The increased surface area makes reaction with oxygen of the air occur more rapidly. If we go further and pulverize the solid, reaction may go faster than we wish. The sawdust produced in lumber mills can be ignited by a spark or small flame. The same is true of other combustible solids such as coal dust or grain. Tragic fires and explosions in grain elevators result from the rapid reaction of oxygen of the air with finely divided particles of grain.

The effect of surface area on reaction rate is explained quite simply. Reactions involving a solid must take place at its surface. Only surface molecules (or atoms or ions) are exposed to the other reactant. Ordinarily, the fraction of molecules at the surface is small. In a cube one centimeter on an edge, only about 1 out of 10^8 molecules are at the surface. The others are buried inside the solid, unavailable for reaction.

If a solid is subdivided, the total surface area is increased. Suppose, for example, we divide a 1-cm cube into eight smaller cubes. Each of these will now be ½ cm on an edge. As shown in Figure 17.5, this doubles the surface area. Twice as many molecules are now exposed. We would expect the reaction rate to double. We might continue this

FIGURE 17.4

A grain elevator burning after an explosion near New Orleans. Finely divided grain and oxygen of the air form an explosive mixture. *UPI photo.*

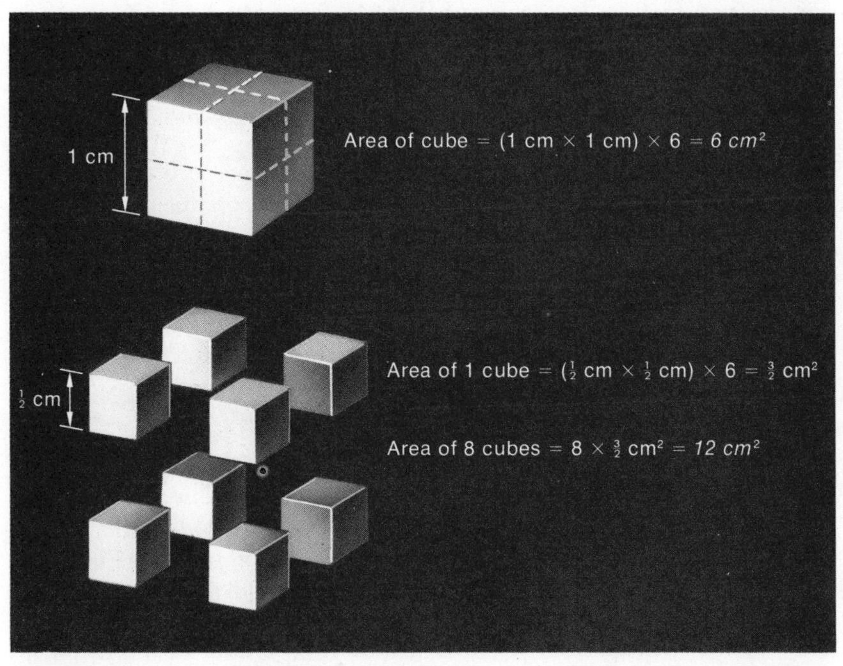

FIGURE 17.5

Effect of particle size on area. Decreasing the size of the particles in a solid by a factor of 2 doubles the total area of the particles. There will be twice as many atoms on the total surface of the particles, so the rate of reaction at the surface would be expected to double.

process of subdivision until we reach molecular dimensions, about 10^{-8} cm. Now reaction should occur very rapidly, perhaps instantaneously.

Most reactions between dissolved ions of opposite charge occur very rapidly.

To illustrate this effect, consider the reaction between lead nitrate, $Pb(NO_3)_2$, and potassium iodide, KI. Both of these compounds are white solids. They react to form a yellow solid, lead iodide, PbI_2. If large crystals of $Pb(NO_3)_2$ and KI are mixed, considerable time passes before we notice the yellow color. This color develops more quickly if the two solids are ground together in a mortar. (See Color Plate 7D. in the center of the book.) If the solid mixture is dissolved in water, the ions are set free. All the Pb^{2+} and I^- ions become available for reaction, which occurs immediately.

$$Pb^{2+}(aq) + 2\ I^-(aq) \rightarrow \underset{\text{yellow}}{PbI_2(s)} \qquad (17.11)$$

17.4 EFFECT OF TEMPERATURE ON REACTION RATE

An increase in temperature almost always makes a reaction go faster. Your mother takes advantage of this when she uses a pressure cooker. The reactions involved in cooking are speeded up by raising the temperature, perhaps from 100°C to 110°C. To prevent foods from spoiling, they are kept cold in a refrigerator at about 5°C. To slow down the process even further, you can put them in a freezer at −15°C.

One reaction whose rate is strongly affected by temperature is:

$$N_2(g) + O_2(g) \rightarrow 2\ NO(g) \qquad (17.12)$$

This reaction occurs quite rapidly at the high temperature reached in an automobile engine. As the nitric oxide passes out the exhaust, it cools to air temperature. In principle, it should decompose back to the elements at low temperatures.

For this reaction at 25°C, $\Delta H = -43.2$ kcal

$$2\ NO(g) \rightarrow N_2(g) + O_2(g)$$

However, this reaction takes place very slowly at 25°C. The NO stays around long enough to be a serious air pollutant.

An interesting example of the effect of temperature upon reaction rate involves animal metabolism. Warm-blooded animals maintain a constant body temperature, just as we do. Their metabolism rate is nearly independent of air temperature. When certain animals hibernate in winter, this situation changes. With a groundhog, body temperature may drop by as much as 30°C. All body processes, including metabolism, slow down. In this way, the groundhog is able to live through the winter without eating. Body fat is slowly consumed to keep life processes going at a reduced rate.

To estimate the effect of temperature upon reaction rate, we often use a simple rule. *Rate is approximately doubled by an increase in temperature of 10°C*. In other words, the time required for a given amount of product to form is about cut in half by raising the temperature 10°C. Using this rule, we can predict that:

1. cooking time should be cut in half by using a pressure cooker to raise the temperature from 100 to 110°C.
2. milk stored in a refrigerator at 5°C would stay fresh four times as long as at room temperature, 25°C. Two 10°C intervals are involved here: $2 \times 2 = 4$.

This rule, although useful, is only approximate. Depending upon the reaction involved, the factor for a 10°C increase varies considerably. It can be as low as 1 (no increase in rate) or as high as 4 (four-fold increase). The factor of 2 is an average value.

We have seen that temperature has a pronounced effect on reaction rate. It is by no means obvious why this should be the case. To be sure, raising the temperature makes molecules move faster. This results in an increased number of collisions. However, this factor by itself is quite small. Typically, raising the temperature 10°C increases the average speed of molecules by 1 or 2%.

Clearly we cannot explain the effect of temperature upon reaction rate in terms of an increased number of collisions. If this were the only factor involved, a 10° increase in temperature would have a very small effect. The rate might be a few percent greater at the higher temperature. Certainly, the reaction would not be twice as fast. To understand the temperature effect, we must take a closer look at the nature of molecular collisions. In particular, we must consider the energies involved in these collisions.

17.5 ACTIVATION ENERGY

As we saw in Section 17.2, reactions take place through collisions between reactant molecules. However, under ordinary conditions, **only a small fraction of collisions result in reaction.** Consider, for example, the reaction between CO and NO_2:

$$CO(g) + NO_2(g) \rightarrow CO_2(g) + NO(g)$$

The rate of collision between CO and NO_2 molecules can be calculated from the kinetic theory of gases. In a one-liter vessel containing one mole each of CO and NO_2 at 300°C, about 10^{37} collisions occur each second. If every collision led to reaction, the reaction would be over in a fraction of a second. In practice, this doesn't happen. Even after an hour, there are appreciable amounts of CO and NO_2 left unreacted. Clearly, the rate at which molecules react is much smaller than the rate at which they collide.

There would also be an explosion.

It is not too difficult to see why collisions between molecules seldom result in reaction. Stable molecules are held together by strong chemical bonds. These bonds must be broken if products are to form. Only those molecules which are moving very rapidly will collide with enough force to break apart. Most of the molecules are moving too slowly

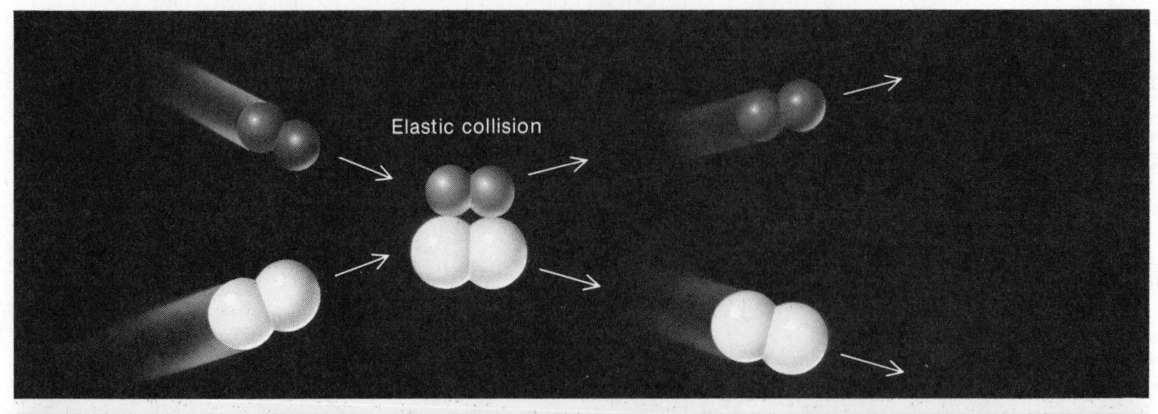

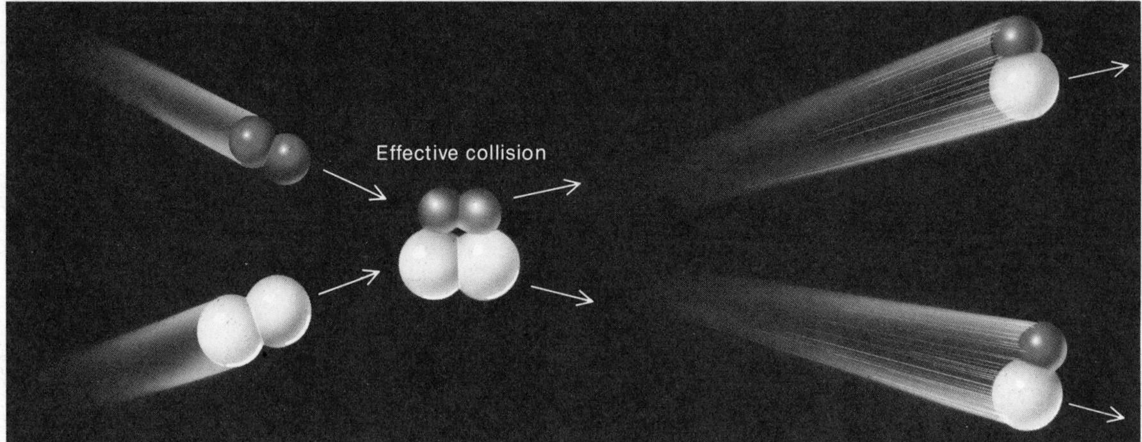

FIGURE 17.6

If two molecules collide, they usually just bounce away from each other unchanged, in a so-called elastic collision. If the energy of the collision is high enough, E_a, then the atoms in the colliding molecules can interact to produce products. Such a collision would be called effective.

for this to happen. They simply bounce off one another without reacting (Figure 17.6). Collisions must involve a certain minimum amount of energy if they are to result in reaction.

The energy which colliding molecules must have if they are to react is called the *activation energy*. It is given the symbol E_a and expressed in kilocalories per mole of reacting molecules. For every reaction, E_a has a definite value. Typically, the activation energy falls in the range 5–50 kcal. For the reaction between CO and NO_2, it is 32 kcal.

$$CO(g) + NO_2(g) \rightarrow CO_2(g) + NO(g); E_a = 32 \text{ kcal}$$

This means that molecules of CO and NO_2 must have a combined energy of at least 32 kcal (per mole) if they are to react upon collision. If their total energy is less than this, nothing will happen when they collide.

Activation Energy and Kinetic Energy

Ordinarily, activation energy comes from the kinetic energy of colliding molecules. You may recall (Chapter 5) that molecules in a

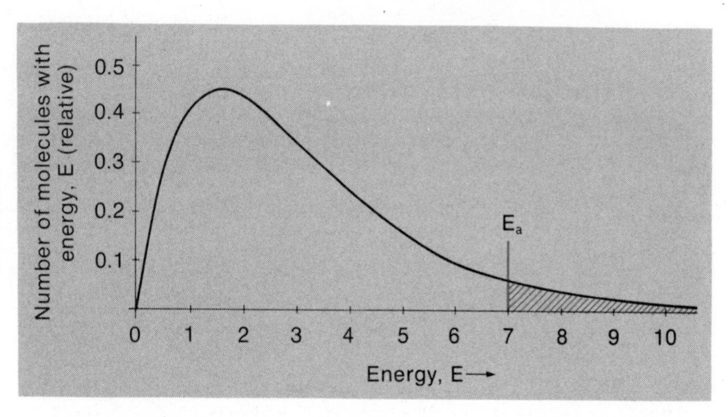

FIGURE 17.7

Graph showing number of molecules having a given energy. Most of the molecules have an energy near the peak in the curve. Only a small fraction, those in the shaded area, have an energy E_a or greater. Only the high energy molecules can undergo reaction.

gas sample have a distribution of kinetic energies. Some of them are moving very slowly, with low energy. Most have an energy close to the average. Only a few are moving very rapidly, with very high kinetic energy. Ordinarily, *the activation energy corresponds to a high kinetic energy*. It is possessed by only a small fraction of the molecules.

This situation is shown graphically in Figure 17.7. Here, we have plotted the fraction of molecules having a certain kinetic energy vs that energy. The vertical line on the graph corresponds to a typical activation energy, E_a. Notice that this line is quite far out (to the right) on the x axis. Only those molecules having kinetic energies equal to or greater than E_a will react when they collide. These molecules are included in the shaded area at the right of the graph. As you can see, the shaded area is very small compared to the total area. Perhaps only 1 in 100 molecules fall within the shaded area. If so, only about 1 in 100 collisions will result in reaction.

Molecules need energy in order to react because most reactions start by breaking a reactant bond. That takes energy.

Clearly the fraction of molecules having enough energy to react upon collision depends upon the value of E_a. If, for a particular reaction, E_a were zero, all collisions would be effective. Reaction would take place in a fraction of a second. If E_a lies farther to the right on the kinetic energy curve, fewer molecules will have the activation energy. In general, the greater the value of E_a, the smaller the number of effective collisions. Other things being equal, we expect an inverse relation between activation energy and reaction rate. That is:

1. Reactions with large activation energies should occur relatively slowly.
2. Reactions with small activation energies should occur relatively rapidly.

Activation Energy and the Effect of Temperature Upon Rate

The effect of temperature upon reaction rate can be explained in terms of activation energy. As we have seen, E_a is usually so high that only a small fraction of the molecules have this energy. However, consider what happens when temperature is raised (Fig. 17.8, p. 436). *A small increase in temperature greatly increases the fraction of high-*

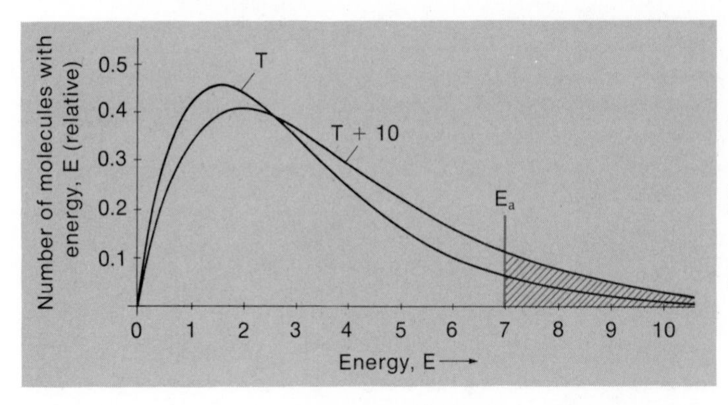

FIGURE 17.8

Graph showing effect of temperature on the energy of gas molecules. At high temperatures the fraction of molecules with high energy increases dramatically. This increase is responsible for the large increase in reaction rate with temperature.

energy molecules. This means that the fraction of molecules with energy equal to or greater than E_a increases sharply. Hence, the reaction occurs much more rapidly at the higher temperature.

From Figure 17.8, it appears that the shaded area to the right of E_a is about twice as large at the higher temperature. This means that twice as many molecules have enough energy to react when they collide. Hence, the reaction rate doubles when the temperature increases by 10°C. As pointed out in Section 17.4, this behavior is typical of many reactions.

Energy Diagrams of Reactions

In Chapter 4, we discussed energy effects in reactions in terms of ΔH. This quantity is the difference in heat content between products and reactants. Exothermic reactions are ones for which ΔH is negative. That is, the products have a lower heat content than the reactants. An endothermic reaction corresponds to a positive value of ΔH. In this case, the heat content of the products is greater than that of the reactants.

You might wonder whether the activation energy for a reaction, E_a, is related to ΔH. As a matter of fact, there is no simple relation between these two quantities. All reactions, exothermic as well as endothermic, have a positive activation energy. Consider, for example,

$$CO(g) + NO_2(g) \rightarrow CO_2(g) + NO(g)$$

ΔH measures the energy difference between reactants and products. E_a measures the energy difference between reactants and the activated complex.

This is an exothermic reaction; $\Delta H = -56$ kcal. The activation energy is a positive quantity, +32 kcal.

This situation is shown in the energy diagram of Figure 17.9. The horizontal line at the left represents the energy of the reactants (1 mol CO + 1 mol NO_2). The line at the right indicates the energy of the products (1 mol CO_2 + 1 mol NO). It lies 56 kcal below that for the reactants. This corresponds to a ΔH of −56 kcal. The peak in the middle represents a species often called an "*activated complex*". It is an intermediate state between reactants and products. Its energy is some 32 kcal above that of the reactants. The value of E_a, 32 kcal, is the energy that must be absorbed to form the activated complex. Once this species is formed, it can decompose to give products.

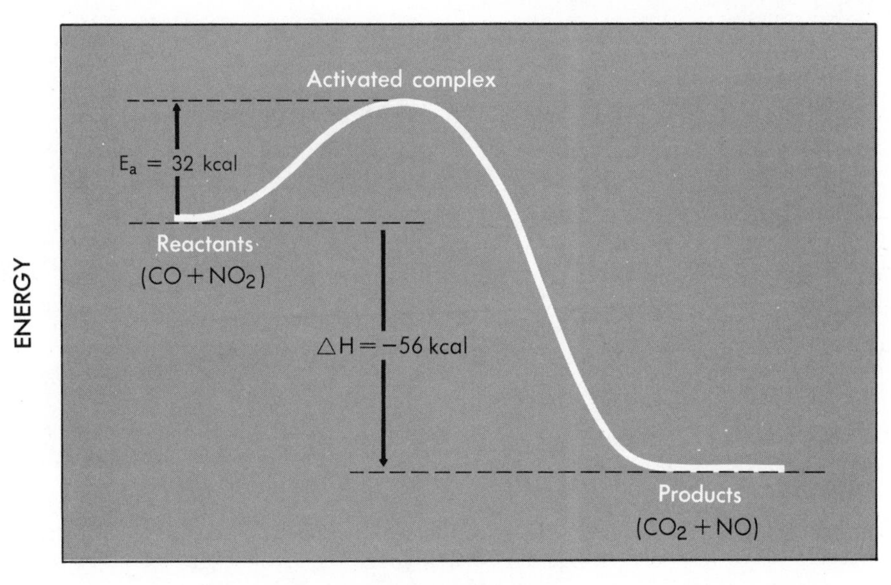

FIGURE 17.9

In a reaction, the reactants follow the path shown on the graph. The reactants first must increase their energy by E_a, at which point they can become an activated complex. Then they can react to form products. In the CO-NO_2 reaction, the reactants absorb 32 kcal/mole in forming the activated complex. When the complex breaks down to form products, 88 kcal of energy are given off. The net change in energy in going from reactants to products is −56 kcal/mole.

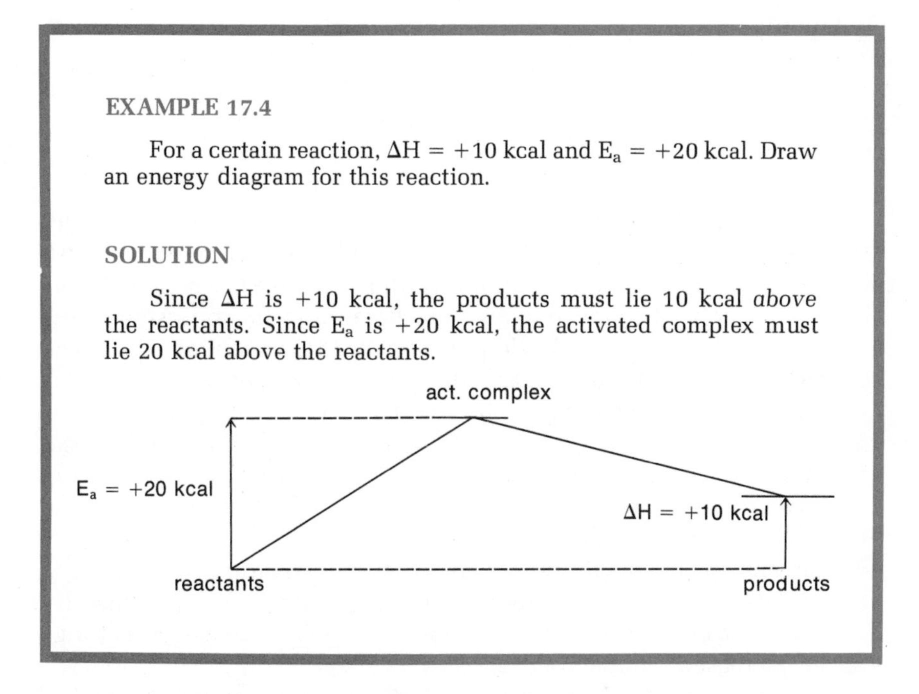

EXAMPLE 17.4

For a certain reaction, $\Delta H = +10$ kcal and $E_a = +20$ kcal. Draw an energy diagram for this reaction.

SOLUTION

Since ΔH is +10 kcal, the products must lie 10 kcal *above* the reactants. Since E_a is +20 kcal, the activated complex must lie 20 kcal above the reactants.

Energy diagrams for reactions resemble in a way those that mountain climbers experience. To draw the analogy, see Figure 17.10, p. 438. The "energies" here are heights above sea level. Suppose you climb the mountain from the left, starting at an elevation of 1000 m. Even-

FIGURE 17.10

A hiker going from a height of 1000 meters to sea level doesn't necessarily have an easy time of it. Similarly, simply because ΔH for a reaction is negative does not mean the reaction will occur rapidly.

tually, you arrive at the chalet on the right, at sea level. Overall, you have gone downhill ($\Delta H = -1000$ m). However, as your aching muscles will tell you, the trip wasn't that easy. First you had to go up the mountain (elevation 5000 m). This means that you had to climb 4000 m ($E_a = 5000 \text{ m} - 1000 \text{ m} = 4000 \text{ m}$). Only then could you start to go downhill.

17.6 CATALYSIS

A catalyst, as pointed out earlier, is a substance which changes the rate of a reaction without being consumed by it. Often the simplest, least expensive way to speed up a reaction is to find the right catalyst. This is easier said than done. We are beginning to understand how catalysts work. Still, the selection of a catalyst for a reaction is more an art than a science. Usually many different substances are tested. The one which is most effective is chosen. The results can be impressive. Catalysts are used in the production of such common materials as gasoline, synthetic rubber, and plastics. In all, 100 billion dollars worth of chemicals is produced each year in the United States by processes that involve catalysts.

Most catalysts are finely divided solids.

In Chapter 15, we described some of the catalytic processes used in organic chemistry. Solid catalysts are used in the petroleum industry for many purposes. The amount of hydrocarbons in the gasoline range can be increased by cracking:

$$C_{16}H_{34}(l) \xrightarrow[Al_2O_3]{SiO_2} C_8H_{18}(l) + C_8H_{16}(l) \tag{17.13}$$

Other catalysts are used to produce branched-chain from straight-chain hydrocarbons:

$$CH_3{-}CH_2{-}CH_2{-}CH_2{-}CH_3 \xrightarrow{AlCl_3} CH_3{-}\underset{\displaystyle CH_3}{\overset{\displaystyle CH_3}{\underset{|}{\overset{|}{C}}}}{-}CH_3 \tag{17.14}$$

This approach is used to raise the octane number of gasoline.

Many important inorganic chemicals are made by processes that require a catalyst. One such compound is ammonia, whose synthesis will be discussed in Chapter 18. Another is sulfuric acid, which is made from sulfur by a three-step process. The slow step is the conversion of SO_2 to SO_3. This is carried out with a catalyst of platinum or vanadium pentoxide:

$$2\ SO_2(g) + O_2(g) \xrightarrow[V_2O_5]{\text{Pt or}} 2\ SO_3(g) \tag{17.15}$$

Activation Energy and Catalysis

A catalyst acts by lowering the activation energy for a reaction. As we saw in Section 17.5, this makes the reaction go more rapidly. The effect is shown in Figure 17.11. The catalyst lowers the activation energy but has no effect on the energies of reactants and products. In other words, *it changes E_a but not ΔH.*

Finding a catalyst for a reaction is a little like finding a pass through a mountain range. The objective is to make it easier to get over the mountain from either side. The relative positions of the valleys on both sides of the mountain remain unchanged. As with mountain passes, it is often not easy to find catalysts.

As you might guess, some chemists are also backpackers.

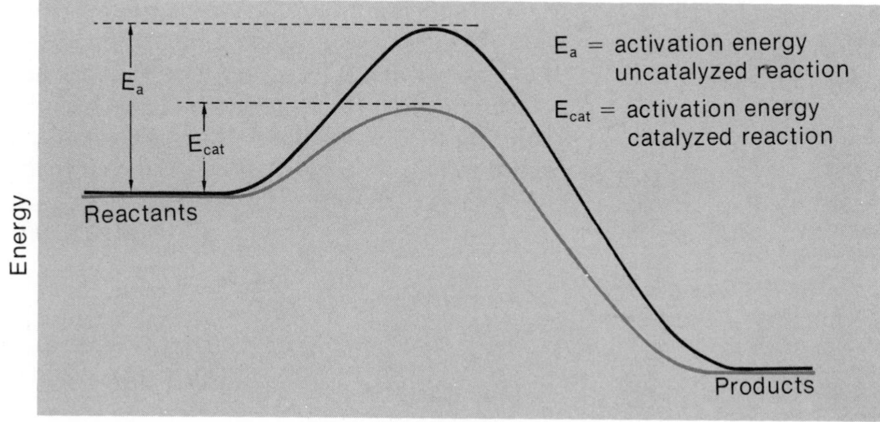

FIGURE 17.11

A catalyst changes the activation energy for a reaction by changing the path by which the reaction occurs. Usually the catalyst lowers the activation energy at the same time and so speeds up the reaction.

Mechanism of Catalysis

To change the activation energy for a reaction, a catalyst must change the reaction path or mechanism. In most cases, we do not know exactly how it does this. In a few cases, the mechanism of catalysis is known. Consider, for example, the decomposition of hydrogen peroxide:

$$2\ H_2O_2(aq) \rightarrow 2\ H_2O + O_2(g) \qquad (17.16)$$

The direct reaction occurs very slowly. The water solution of H_2O_2 that you buy in a drugstore is stable for several months. However, if you add iodide ions to the solution, reaction occurs immediately. You can see bubbles of oxygen gas form in the solution. With I^- ions, the reaction follows a two-step path:

$$(1)\quad H_2O_2(aq) + I^-(aq) \rightarrow H_2O + IO^-(aq)$$

$$(2)\quad H_2O_2(aq) + IO^-(aq) \rightarrow H_2O + O_2(g) + I^-(aq)$$

$$2\ H_2O_2(aq) \rightarrow 2\ H_2O + O_2(g) \qquad (17.16)$$

Notice that the end result is the same as in the direct reaction. The I^- ion is a true catalyst. It is not consumed in the reaction. For every I^- ion consumed in the first step, one is produced in the second step. The activation energy for this two-step path is much smaller than for the uncatalyzed reaction.

Solid catalysts operate in a quite different way. The catalyst usually *adsorbs* one of the reactants on its surface. The bonds within that reactant are weakened, perhaps even broken, when it is attached to the surface. This lowers the activation energy for the reaction. This happens in the reaction of SO_2 with O_2 using a platinum catalyst:

Pt is a good catalyst because it adsorbs a reactant strongly enough to weaken a bond, but not so strongly that it won't let go when the reaction is complete.

$$2\ SO_2(g) + O_2(g) \xrightarrow{Pt} 2\ SO_3(g)$$

Here it appears to be O_2 molecules that are adsorbed. The attractive forces between O_2 molecules and Pt atoms are strong enough to weaken the bond within O_2. This leads to the formation of free oxygen atoms. These are much more reactive than O_2 molecules. An SO_2 molecule colliding with an O atom at the surface of the catalyst is quickly converted to SO_3. The path of the catalyzed reaction is:

$$O_2(g) \rightarrow 2\ O(g)$$

$$SO_2(g) + O(g) \rightarrow SO_3(g)$$

$$SO_2(g) + O(g) \rightarrow SO_3(g)$$

$$2\ SO_2(g) + O_2(g) \rightarrow 2\ SO_3(g)$$

The platinum catalyst in the catalytic converter of an automobile probably operates in much the same way. In this case, unburned carbon monoxide is converted to carbon dioxide:

$$2\ CO(g) + O_2(g) \rightarrow 2\ CO_2(g) \qquad (17.17)$$

At the same time, any unburned hydrocarbons are converted to CO_2 and H_2O.

Inhibitors

Frequently, small amounts of foreign substances slow down a reaction. These are referred to as *inhibitors* or "negative catalysts". They may be added deliberately to decrease the rate of an undesirable reaction. In other cases, they come from natural sources. Here the problem is to remove them so a desired reaction can occur more rapidly.

Usually an inhibitor works in one or more of the following ways.

1. *It prevents the reactants from coming in contact with each other.* Inhibitors in antifreeze work this way. They form a coating on the surface of the metal which prevents dissolved air and water from reaching it. In essence, they have the same effect as a coat of paint.

2. *It reacts preferentially with one of the reactants.* The "antioxidants" added to foods have this effect. They react with the oxygen of the air. In this way, fats in the foods are protected from attack by oxygen. The shelf life of such foods as potato chips or peanut butter is greatly increased by this type of inhibitor.

3. *It destroys a catalyst or makes it ineffective.* Solid catalysts are vulnerable to "poisoning" of this type. Typically, an impurity in the reaction mixture is adsorbed on the catalyst. This means that the catalyst is no longer effective. The reaction slows down.

Platinum catalysts are easily poisoned. In forming SO_3 from SO_2 with a Pt catalyst, all traces of arsenic compounds must be removed. Otherwise, the surface of the metal becomes coated and the reaction stops. This is one reason why V_2O_5 is usually preferred to platinum as a catalyst in this reaction.

17.7 SUMMARY

The rate of a reaction tells us how concentration changes with time. Rate has the units of moles per liter per unit time. In general, reaction rate depends upon:

—the nature of the reactants
—the concentration of reactants (in the gas phase or in water solution)
—surface area (for reactions occurring at solid surfaces)
—the temperature
—the presence of a catalyst or inhibitor

In the simplest case, reaction takes place as the result of a collision between two high-energy molecules:

$$A + B \rightarrow \text{products (single step)}$$

The two molecules must have enough kinetic energy to supply the activation energy needed for reaction. To speed up such a reaction, we can:

—increase the concentrations of A and B. This increases the rate of collisions (rate = k × conc. A × conc. B).

—increase the temperature. This increases the fraction of molecules which have high enough kinetic energies to react when they collide.

—add a catalyst which lowers the activation energy.

In practice, most reactions do not occur in this simple way. Instead, they take place in a series of steps. These steps add to give the overall equation for the reaction. The rate expressions for such reactions are often complex. However, the effects of concentration and temperature are usually in the direction indicated above.

NEW TERMS

activated complex
activation energy
inhibitor
mechanism
rate constant
rate equation
rate of reaction

QUESTIONS

1. What is meant by "rate of reaction"? What quantities would you measure to determine the rate of the reaction:

$$NO(g) + O_3(g) \rightarrow NO_2(g) + O_2(g)$$

2. What units would you use to express the rate at which:

   a. an airplane flies from San Francisco to New York?
   b. a child becomes taller?
   c. a person on a diet loses mass?
   d. a "sparkler" burns on the 4th of July?

3. How would you measure each of the rates referred to in Question 2?

4. Consider the reaction: $CO(g) + NO_2(g) \rightarrow CO_2(g) + NO(g)$. As reaction takes place, how does each of the following change (increase or decrease)?

   a. concentration of CO b. concentration of NO c. rate of reaction

5. Suppose that, in drawing Figure 17.1, the concentration of CO rather than that of CO_2 were plotted against time. What would the curve look like?

6. Explain, in terms of molecular collisions, why an increase in concentration of reactants increases reaction rate.

7. The reaction: $NO(g) + O_3(g) \rightarrow NO_2(g) + O_2(g)$ takes place in a single step. Write an expression, similar to Equation 17.5, for the rate of this reaction.

8. A certain reaction occurs in three steps:

$$I_2 \rightarrow 2\ I$$

$$I + H_2 \rightarrow H_2I$$

$$H_2I + I \rightarrow 2\ HI$$

What is the overall equation for the reaction?

9. The reaction: $2\ O_3(g) \rightarrow 3\ O_2(g)$ occurs in two steps. The first step is:

$$O_3 \rightarrow O_2 + O$$

What is the second step?

10. Why does coal dust burn more rapidly than a lump of coal?

11. Why is meat which is not to be cooked for some time stored in a freezer rather than a refrigerator?

12. What happens to rates of reactions taking place in your body when you run a fever?

13. Which of the following best explains why an increase in temperature increases reaction rate?

    a. the average molecular speed increases, so collisions are more frequent.
    b. the activation energy decreases.
    c. pressure increases, so collisions are more frequent.
    d. the fraction of high energy molecules increases.

14. The rate of a certain reaction doubles when the temperature rises by 10°C. What happens to the rate when the temperature changes from 20°C to:

    a. 30°C b. 10°C c. 40°C

15. Explain, in terms of Figure 17.7, why reactions with a low activation energy tend to occur rapidly.

16. Choose the words or phrases which correctly complete the following statements. A catalyst (increases, decreases, has no effect upon) E_a. At the same time, the catalyst (increases, decreases, has no effect upon) ΔH.

17. What is an inhibitor? Discuss the ways in which inhibitors act to slow down a reaction.

PROBLEMS

1. *Units of Reaction Rate (Example 17.1)* The rate of a certain reaction is 0.50 mol/ℓ·min. Express this rate in mol/ℓ·hr.

2. *Determination of Reaction Rate (Example 17.2)* In a certain reaction, the concentration of product increases from 0.10 to 0.26 mol/ℓ in 15 min. What is the rate?

3. *Determination of Reaction Rate (Example 17.2)* Using Table 17.1, determine the rate of reaction:

    a. between 20 and 30 min b. between 30 and 40 min

4. *Effect of Concentration Upon Rate (Example 17.3)* For the reaction:

$$NO(g) + O_3(g) \rightarrow NO_2(g) + O_2(g)$$

the rate equation is: rate = k(conc. NO) × (conc. O_3)

What happens to the rate of reaction if:

a. conc. NO increases from 0.10 M to 0.30 M (conc. O_3 constant)
b. conc. O_3 decreases from 0.10 M to 0.020 M (conc. NO constant)

5. *Activation Energy (Example 17.4)* For a certain reaction, $\Delta H = -18$ kcal, $E_a = +22$ kcal. Draw an energy diagram similar to that shown in Example 17.4.

6. *Activation Energy (Example 17.4)* By adding a catalyst to the reaction in Problem 5, the activation energy is lowered to +12 kcal. Draw an energy diagram for the catalyzed reaction.

* * * * *

7. The rate of a certain reaction is 0.50 mol/ℓ·s. What is the rate expressed in mol/ℓ·min?

8. For the reaction in Problem 4, what happens to the rate if the concentration of NO is doubled and that of O_3 is cut in half?

9. For the reaction: A + B → C, the following data are obtained:

conc. C (mol/ℓ)	0.00	0.50	0.75	0.88
t (hr)	0	1	2	3

Calculate:

a. the rate in mol/ℓ·hr between 1 and 2 hr.
b. the rate in mol/ℓ·min, between 0 and 1 hr.
c. the rate in mol/ℓ·s between 2 and 3 hr.

10. For a reaction, $\Delta H = +10$ kcal. The activation energy is +30 kcal. Upon addition of a catalyst, E_a is reduced to +20 kcal. Draw an energy diagram similar to Figure 17.11 for this reaction.

11. Consider the following energy diagram for a reaction:

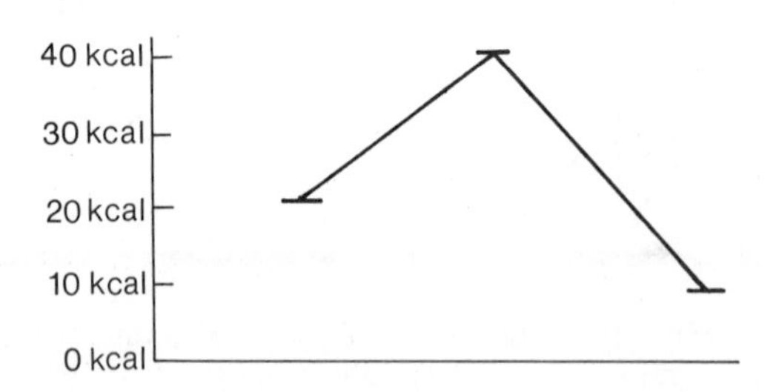

What is the value of ΔH? of E_a?

*12. For the reaction: $NO(g) + O_3(g) \rightarrow NO_2(g) + O_2(g)$, the following data are obtained:

conc. NO_2 (mol/ℓ)	0.000	0.004	0.006
time (s)	0	10	20

Estimate the rate at t = 10 s.

*13. For the reaction: $2\,HI(g) \rightarrow H_2(g) + I_2(g)$, the concentration of HI decreases from 0.42 to 0.24 mol/ℓ in a 20 min interval. What is the rate of the reaction, expressed as: rate $= \frac{\Delta \text{ conc. } H_2}{\Delta t}$?

*14. For a reaction between X and Y, the following data are obtained:

rate (mol/ℓ·min)	*conc.* X (mol/ℓ)	*conc.* Y (mol/ℓ)
0.020	0.10	0.10
0.020	0.10	0.20
0.080	0.20	0.10

What is the order of the reaction with respect to X? Y?

Photo by Cary Wolinski/Stock Boston.

18 CHEMICAL EQUILIBRIUM

Most reactions that you are familiar with appear to go to completion. When natural gas is burned in excess air

$$CH_4(g) + 2\ O_2(g) \rightarrow CO_2(g) + 2\ H_2O(l)$$

you do not notice any unreacted methane in the products. The same is true for the combustion of gasoline in an automobile engine. Ordinarily, you don't see any unburned gasoline dripping out of the exhaust pipe. Even reactions that take place slowly seem to go to completion. Given enough time, cement "sets" and iron rusts, apparently completely.

There are, however, many reactions which behave quite differently. They do not go to completion. Instead, they stop at a certain point, with some of the starting materials unreacted. A case in point is the reaction between iodine and hydrogen.

$$I_2(g) + H_2(g) \rightarrow 2\ HI(g)$$

Suppose we put a mixture of 0.10 mol I_2 and 0.10 mol H_2 in a one-liter container at 150°C. Originally, the mixture has a deep violet color because of the iodine present. As time passes, the purple color fades. Iodine is being consumed by the reaction above. However, no matter how long we wait, the purple color does not disappear completely. Some of the I_2 (and some of the H_2) remains unreacted.

This behavior is typical of many reactions in water solution. Consider, for example, what happens when acetic acid, CH_3COOH, is added to water. Some of it ionizes:

$$CH_3COOH(aq) \rightarrow H^+(aq) + CH_3COO^-(aq)$$

However, not all of the acetic acid reacts. If you add 0.10 mol of acetic acid to a liter of water, only about 1% of it forms ions. The other 99% remains as CH_3COOH molecules.

The ions that are present have important properties, so they cannot be ignored.

These two reactions are by no means unusual. Indeed, most of the reactions that we carry out in the laboratory do not go 100% to completion. The same is true of many important industrial processes. Instead of obtaining a pure product, we get an *equilibrium mixture* of products and reactants.

It is possible to control the amount of product obtained in such reactions. This can be done by adjusting such factors as concentration of reactants, temperature or pressure. To proceed in a logical manner, we need to understand the principles that govern *chemical equilibrium*. In this chapter, we will develop these principles. They are valid for all states of matter, including water solutions. Here, we will apply equilibrium principles to reactions involving gases. We start with a rather simple system involving three gases: PCl_5, PCl_3 and Cl_2.

18.1 THE PCl_5–PCl_3–Cl_2 SYSTEM

Suppose we heat a sample of phosphorus pentachloride, PCl_5, in a closed container to 250°C. Almost immediately, it starts to decompose. The following reaction occurs:

$$PCl_5(g) \rightarrow PCl_3(g) + Cl_2(g)$$

The products are phosphorus trichloride, PCl_3, and chlorine, Cl_2.

At first, this is the only reaction taking place. However, as the concentrations of PCl_3 and Cl_2 build up, the reverse reaction begins:

$$PCl_3(g) + Cl_2(g) \rightarrow PCl_5(g)$$

As time passes, this reaction speeds up while the forward reaction slows down. Eventually, their rates become equal. When this happens, we

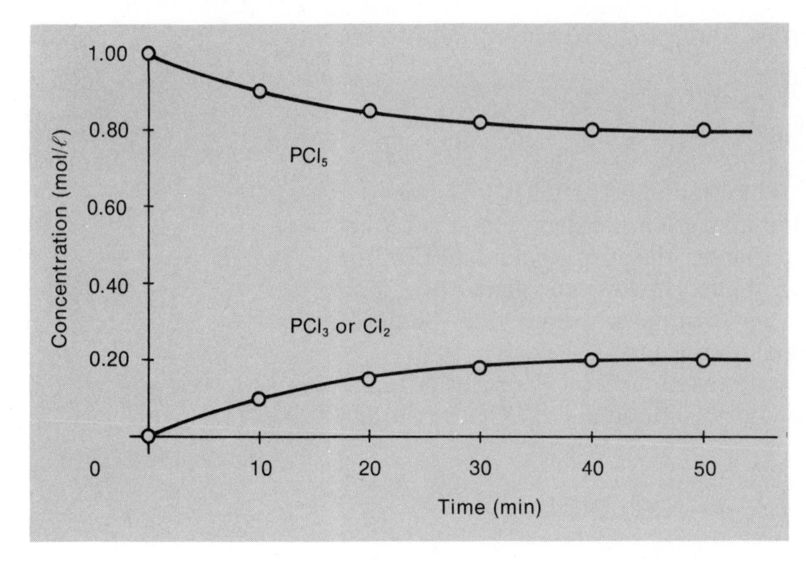

FIGURE 18.1

When PCl_5 is put into a container at high temperatures, some, but not all, of it decomposes to PCl_3 and Cl_2. The number of moles of PCl_5 that break down will equal the number of moles of PCl_3 formed or the number of moles of Cl_2 formed. After a certain period of time the concentrations of all three species will stop changing. At that point the system is in a state of chemical equilibrium.

reach a position of *equilibrium*. The concentrations of PCl_5, PCl_3, and Cl_2 no longer change with time (Figure 18.1). To an outside observer, it appears that nothing is happening. We describe this situation by writing the equation:

In a system at equilibrium, there is a balance between the tendency of reactants to form products and that of products to reform reactants.

$$PCl_5(g) \rightleftharpoons PCl_3(g) + Cl_2(g) \quad (18.1)$$

The double arrow implies that the equilibrium is "dynamic". That is, the forward and reverse reactions take place at the same rate.

There are many different ways in which we can establish this equilibrium at 250°C. We can, as indicated above, start with pure PCl_5. Another possibility is to start with "products", PCl_3 and Cl_2. If we do this, they partially react to form PCl_5. Regardless of where we start, we arrive at equilibrium. Moreover, we find that, at equilibrium, there is a simple relation between the concentrations of PCl_5, PCl_3, and Cl_2. To discover this relation, we conduct a series of carefully controlled experiments, as described below.

Relation Between $[PCl_5]$, $[PCl_3]$, and $[Cl_2]$

We can study this system at 250°C using the apparatus shown in Figure 18.2. This is a closed container with a volume of one liter. In this flask, we place a known amount of reactant (PCl_5) or products (PCl_3 and Cl_2). Reaction takes place to approach equilibrium. From time to time, we draw off samples of the gas mixture. These samples are analyzed to determine the concentrations of PCl_5, PCl_3, and Cl_2. We continue in this way until two successive samples give identical results. When this happens, the system is at equilibrium. The concentrations measured then must be equilibrium values.

The results of four different experiments of this type are shown in Table 18.1. The experiments are similar except that we start at different points.

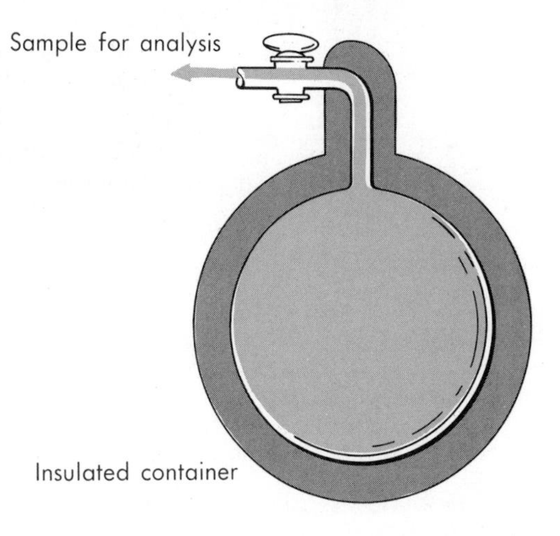

FIGURE 18.2

Apparatus that might be used to study the PCl_5-PCl_3-Cl_2 equilibrium. Samples of the gas present would be removed and analyzed to establish that equilibrium existed and the equilibrium concentrations of each of the species.

Expt. 1 1.00 mol PCl_5 is placed in the 1.00 ℓ container. As time passes, the PCl_5 decomposes to PCl_3 and Cl_2. After perhaps 40 min, equilibrium is reached. At that point, 0.80 mol PCl_5 remains. The other 0.20 mol has decomposed, forming 0.20 mol PCl_3 and 0.20 mol Cl_2.

Expt. 2 This is similar to (1), except that we start with only 0.60 mol of PCl_5. At equilibrium, 0.15 mol of PCl_3 and 0.15 mol of Cl_2 are present. This means that 0.15 mol of PCl_5 has reacted. So, its final concentration is:

To make 0.15 mol of PCl_3 and of Cl_2 takes 0.15 mol of PCl_5

$$0.60 \frac{\text{mol}}{\ell} - 0.15 \frac{\text{mol}}{\ell} = 0.45 \frac{\text{mol}}{\ell}$$

Expt. 3 Here, we start from the "other side" of the system. That is, we put 1.00 mol of PCl_3 and 1.00 mol of Cl_2 in the 1.00 ℓ flask. They react, by the reverse of Reaction 18.1, to form PCl_5. The concentrations at equilibrium are the same as in Expt. 1.

Expt. 4 This is similar to (3) in that the starting materials are PCl_3 and Cl_2. This time, we begin with 0.60 mol PCl_3 and 0.50 mol Cl_2. During the experiment, 0.40 mol PCl_5 is formed. The equilibrium concentrations are:

PCl_5: 0.40 mol/ℓ
PCl_3: 0.60 mol/ℓ − 0.40 mol/ℓ = 0.20 mol/ℓ
Cl_2: 0.50 mol/ℓ − 0.40 mol/ℓ = 0.10 mol/ℓ

TABLE 18.1 ORIGINAL AND EQUILIBRIUM CONCENTRATIONS (mol/ℓ) IN EXPT. 1–4

	Expt. 1		Expt. 2		Expt. 3		Expt. 4	
	Orig.	Equil.	Orig.	Equil.	Orig.	Equil.	Orig.	Equil.
Conc. PCl_5	1.00	0.80	0.60	0.45	0.00	0.80	0.00	0.40
Conc. PCl_3	0.00	0.20	0.00	0.15	1.00	0.20	0.60	0.20
Conc. Cl_2	0.00	0.20	0.00	0.15	1.00	0.20	0.50	0.10

Looking at the data in Table 18.1, we might ask a question. Is there any relation between the equilibrium concentrations of PCl_5, PCl_3, and Cl_2 that holds for all four experiments? As a matter of fact, there is. It is a rather simple mathematical relation. For all the experiments, the equilibrium concentrations are related by the equation:

$$\frac{(\text{equil. conc. } PCl_3) \times (\text{equil. conc. } Cl_2)}{(\text{equil. conc. } PCl_5)} = 0.050$$

To check that this is the case, let's substitute numbers:

The fact that equilibrium concentrations behave this way is surprising but true.

$$\text{Expt. 1:} \quad \frac{0.20 \times 0.20}{0.80} = \frac{0.040}{0.80} = 0.050$$

$$\text{Expt. 2:} \quad \frac{0.15 \times 0.15}{0.45} = \frac{0.0225}{0.45} = 0.050$$

$$\text{Expt. 3:} \quad \frac{0.20 \times 0.20}{0.80} = \frac{0.040}{0.80} = 0.050$$

$$\text{Expt. 4:} \quad \frac{0.20 \times 0.10}{0.40} = \frac{0.020}{0.40} = 0.050$$

Notice that this relationship is valid only for *equilibrium* concentrations. It would not, for example, hold for the original concentrations listed in Table 18.1. Neither would it apply to data obtained on the way to equilibrium. To emphasize this, we write equilibrium concentrations in a special way. We enclose them in square brackets. For this system at 250°C:

$$\frac{[PCl_3] \times [Cl_2]}{[PCl_5]} = 0.050$$

where **the symbol [ ] stands for equilibrium concentrations in moles per liter.** We will follow this practice consistently from now on. When we deal with non-equilibrium concentrations, we will write out "conc. PCl_5", and so on.

The Equilibrium Constant, K

The number just obtained, 0.050, is an example of what is called an equilibrium constant. Equilibrium constants are given the general symbol K. Thus for the reaction system:

$$PCl_5(g) \rightleftharpoons PCl_3(g) + Cl_2(g)$$

$$K = \frac{[PCl_3] \times [Cl_2]}{[PCl_5]} = 0.050 \text{ at } 250°C \qquad (18.2)$$

In words, Equation 18.2 tells us that:

a. At equilibrium, the product of the concentrations of PCl_3 and Cl_2, divided by that of PCl_5, has a constant value, regardless of what we start with.

b. At 250°C, this constant, K, is 0.050.

Throughout the rest of this chapter, we will deal with equilibrium constants for many different chemical systems. The expressions we write for K may look quite different. The numerical values of K will certainly differ from one system to another. However, in all cases:

1. *The value of K is independent of the individual concentrations of reactants and products.* In Equation 18.2, $[PCl_3]$, $[Cl_2]$, and $[PCl_5]$ separately can have any values whatsoever. The only requirement is that the quotient, $[PCl_3] \times [Cl_2]/[PCl_5]$ be 0.050. To cite an analogy, suppose we write the equation:

$$xy = 6$$

Here, it is understood that x can have any value you wish (1, 2, 3, . . .). However, once you choose a value for x, y is fixed. If x = 1, y = 6; if x = 2, y = 3, and so on. In this case, the product, xy, is a constant, 6. In Equation 18.2, it is a quotient that is constant, 0.050. There, we might find:

$[PCl_3] = 0.050$ M, $[Cl_2] = 1.0$ M, in which case $[PCl_5] = 1.0$ M

or: $[PCl_3] = 1.0$ M, $[Cl_2] = 0.050$ M, in which case $[PCl_5] = 1.0$ M

or: $[PCl_3] = 0.10$ M, $[Cl_2] = 1.0$ M, in which case $[PCl_5] = 2.0$ M

and so on.

We can set the value of any two of the equilibrium concentrations anywhere we want, but not all three.

EXAMPLE 18.1

In a certain experiment, it is found that the equilibrium concentrations of PCl_3 and PCl_5 are 0.10 and 0.20 mol/ℓ respectively. At 250°C, what must be the equilibrium concentration of Cl_2 (K = 0.050)?

SOLUTION

The equation is: $\frac{[PCl_3] \times [Cl_2]}{[PCl_5]} = 0.050$

Solving for $[Cl_2]$: $[Cl_2] = 0.050 \times \frac{[PCl_5]}{[PCl_3]}$

Substituting numbers: $[Cl_2] = 0.050 \times \frac{0.20}{0.10} = 0.10$ mol/ℓ

2. *The value of K is independent of how the system is set up originally.* We can start with any amount of PCl_5 we please. Or, we can start with a mixture of PCl_3 and Cl_2 in any proportions. Regardless of what we do, it will still be true that at equilibrium:

$$\frac{[PCl_3] \times [Cl_2]}{[PCl_5]} = 0.050 \text{ at } 250°\text{C}$$

Reaction will proceed in one direction or the other to reach this value.

3. *The value of K is independent of the volume of the container.* For the PCl_5-PCl_3-Cl_2 system, K = 0.050 whether the container has a volume of one liter or 10 ℓ or 1 cm^3. The relative amounts of PCl_5, PCl_3, and Cl_2 may be quite different in containers of different sizes. However, the concentration quotient at equilibrium will be the same, 0.050.

4. *The value of K changes with temperature* (see Section 18.4). For this system, K is 0.050 at only one temperature, 250°C. At a different temperature, it will have some other value. That value might be 0.010, 100, or almost any other number. However, it will still be true that:

$$\frac{[PCl_3] \times [Cl_2]}{[PCl_5]} = \text{a constant (number) at each temperature}$$

18.2 GENERAL EXPRESSION FOR K

We have pointed out that for the system:

$$PCl_5(g) \rightleftharpoons PCl_3(g) + Cl_2(g)$$

the expression for the equilibrium constant is:

$$K = \frac{[PCl_3] \times [Cl_2]}{[PCl_5]} \qquad (18.2)$$

For other systems, the form of the expression for K may be quite different. In general, it is related in a simple way to the chemical equation for the equilibrium system. Perhaps you can see what this relation is from a couple more examples.

$$2\ SO_2(g) + O_2(g) \rightleftharpoons 2\ SO_3(g);\ K = \frac{[SO_3]^2}{[SO_2]^2 \times [O_2]} \qquad (18.3)$$

$$N_2(g) + 3\ H_2(g) \rightleftharpoons 2\ NH_3(g);\ K = \frac{[NH_3]^2}{[N_2] \times [H_2]^3} \qquad (18.4)$$

Looking at these three systems, we see that in the expression for K:

Learn these rules for setting up the expression for K.

1. **The equilibrium concentrations of products (right side of equation) appear in the numerator of K.**

2. **The equilibrium concentrations of reactants (left side of equation) appear in the denominator of K.**

3. **Each of these concentrations is raised to a power (1, 2, 3, . . .) equal to its coefficient in the balanced equation.**

4. **Where there is more than one product** (18.2), **the concentration terms are multiplied by each other** ($[PCl_3] \times [Cl_2]$). **Similarly, if there is more than one reactant** (18.3, 18.4), **their concentration terms are multiplied** ($[SO_2]^2 \times [O_2]$; $[N_2] \times [H_2]^3$).

EXAMPLE 18.2

Write the expression for the equilibrium constant corresponding to the equation:

$$H_2(g) + I_2(g) \rightleftharpoons 2\ HI(g)$$

SOLUTION

Since HI appears on the right of the equation, its concentration should be in the numerator. Since HI has a coefficient of 2, its concentration should be squared.

H_2 and I_2 are on the left of the equation. Their coefficients are understood to be one. Hence the concentrations of H_2 and I_2 should be in the denominator, each raised to the first power. The two concentrations are multiplied together.

$$K = \frac{[HI]^2}{[H_2] \times [I_2]}$$

The equilibrium constants we have written here have the same properties as that for the PCl_5-PCl_3-Cl_2 system (Section 18.1). The value of K is:

A mixture that is not at equilibrium will tend to react until it is. At that point, the concentration quotient will equal K.

—independent of the individual concentrations of reactants and products.
—independent of the original amounts of reactants and products.
—independent of the volume of the container.
—dependent upon temperature.

These constants can be calculated from equilibrium concentrations in the usual way (Example 18.3).

EXAMPLE 18.3

Consider the reaction between nitrogen and hydrogen to form ammonia:

$$N_2(g) + 3\ H_2(g) \rightleftharpoons 2\ NH_3(g)$$

At a certain temperature, the equilibrium concentrations of NH_3, N_2 and H_2 are found to be 0.20, 0.18 and 0.10 mol/ℓ respectively. Calculate K.

SOLUTION

The expression for K is:

$$K = \frac{[NH_3]^2}{[N_2] \times [H_2]^3}$$

Substituting numbers:

$$K = \frac{(0.20)^2}{(0.18) \times (0.10)^3} = \frac{4.0 \times 10^{-2}}{(1.8 \times 10^{-1}) \times (1.0 \times 10^{-3})} = \frac{4.0 \times 10^{-2}}{1.8 \times 10^{-4}}$$

$$= 2.2 \times 10^2$$

Systems Involving Solids or Liquids

The position of an equilibrium involving gases is not affected by the amount of a liquid or solid present. For example, the position of the equilibrium:

$$CaCO_3(s) \rightleftharpoons CaO(s) + CO_2(g)$$

is independent of how much $CaCO_3$ or CaO is present. Provided there is at least a small amount of both solids, the concentration of CO_2 is constant. Adding or removing CaO or $CaCO_3$ does not change the concentration of CO_2.

This situation is reflected in the equilibrium constant expression. Concentration terms for solids or liquids do not appear in the expression for K. For example, for the equilibrium just described:

Try to state in words what Eqn. 18.5 means if K = 0.1.

$$K = [CO_2] \quad (18.5)$$

Again, for the system:

$$2\ H_2O(l) \rightleftharpoons 2\ H_2(g) + O_2(g)$$

the expression for K is:

$$K = [H_2]^2 \times [O_2] \quad (18.6)$$

Notice that terms for $CaCO_3(s)$, CaO(s), and $H_2O(l)$ do not appear in these expressions.

EXAMPLE 18.4

Write expressions for K corresponding to:

a. $4\ NH_3(g) + 5\ O_2(g) \rightleftharpoons 4\ NO(g) + 6\ H_2O(l)$ (at 25°C)

b. $CuO(s) + H_2(g) \rightleftharpoons Cu(s) + H_2O(g)$ (at 500°C)

SOLUTION

a. $K = \dfrac{[NO]^4}{[NH_3]^4 \times [O_2]^5}$

b. $K = \dfrac{[H_2O]}{[H_2]}$

A term for water does not appear in (a). It is present as a liquid at 25°C. At 500°C, water is a gas. So, its concentration appears in (b).

18.3 THE EQUILIBRIUM CONSTANT AND THE EXTENT OF REACTION

When we carry out a reaction in the laboratory, we are concerned with its yield. That is, we want to know how much product we can expect to get from a given amount of reactant. Ideally, we would like to get 100%. In reality, if the system goes to equilibrium, we may well have to settle for less.

We can predict in advance the extent to which a reaction will occur. All we need to know is the equilibrium constant. Using K, we can determine the fraction of reactants that will be converted to products. It is possible to carry out exact calculations of this sort. Here, however, we will consider only some simple ways of using K to estimate the extent of reaction. We will attempt to answer an important question. How can we use the magnitude of K to predict whether an equilibrium mixture will contain:

1. mostly reactants, with very little product?
2. mostly products, with very little reactant left?
3. appreciable amounts of both reactants and products?

K Very Small

If the equilibrium constant is very small, very little product will be formed. As an example, consider the system:

$$N_2(g) + O_2(g) \rightleftharpoons 2\ NO(g)$$

for which:

$$K = \frac{[NO]^2}{[N_2] \times [O_2]} = 1 \times 10^{-30} \text{ at } 25°C \qquad (18.7)$$

Since K is so small, the equilibrium concentration of NO will be tiny compared to those of N_2 and O_2. Suppose, for example $[N_2]$ and $[O_2]$ are both 1 M. Under these conditions, [NO] is only 1×10^{-15} M. That is:

This is necessary because the expression for K must be satisfied.

$$[NO]^2 = 1 \times 10^{-30} \times 1 \times 1 = 1 \times 10^{-30}$$

$$[NO] = (1 \times 10^{-30})^{1/2} = 1 \times 10^{-15}$$

This is a very small concentration indeed. Clearly, the forward reaction does not proceed to any noticeable extent at 25°C.

K Very Large

If the equilibrium constant is very large, we expect a reaction to go virtually to completion. For the reaction:

$$2\ CO(g) + O_2(g) \rightleftharpoons 2\ CO_2(g)$$

$$K = \frac{[CO_2]^2}{[CO]^2 \times [O_2]} = 4 \times 10^{30} \text{ at } 500°C \qquad (18.8)$$

Since K is so large, the equilibrium concentration of CO_2 must be huge compared to those of CO and O_2. Carbon monoxide exposed to air at 500°C should be almost completely converted to CO_2. This is indeed the case. This reaction takes place in the catalytic converter of an automobile. Any CO present in the exhaust is converted to CO_2 by this reaction.

There is one restriction on the prediction we have just made. Sometimes reactions for which K is very large occur slowly. We may have to wait forever for the reaction to take place if the activation energy is large. Consider, for example, the reaction between CO and O_2 at room temperature:

$$2\ CO(g) + O_2(g) \rightleftharpoons 2\ CO_2(g);\ K = 2 \times 10^{91} \text{ at } 25°C$$

Notice that K is even larger at 25°C than it is at 500°C (2×10^{91} vs. 4×10^{30}). However, reaction occurs much more slowly at the lower temperature. Carbon monoxide exposed to air at 25°C may stay around for hours without reacting. In general, to be sure that a reaction will give a high yield of products in a reasonable time, two conditions must be met. First, K must be very large. Equally important, the rate must be great enough so that equilibrium is reached quickly.

Since reaction rates increase with temperature, it is easier to get to equilibrium at high temperatures than at low.

Intermediate Value of K

We have seen what happens at the extremes, where K is very small or very large. Suppose, however, that K has an "in-between" value. Perhaps it falls between 10^{-5} and 10^5, just to pick a couple of numbers. What happens now? Quite simply, we obtain an equilibrium mixture with appreciable amounts of both products and reactants. Consider, for example, the reaction between N_2 and O_2 at 2000°C:

$$N_2(g) + O_2(g) \rightleftharpoons 2\ NO(g)$$

$$K = \frac{[NO]^2}{[N_2] \times [O_2]} = 0.10 \text{ at } 2000°C$$

Here, the situation is quite different from that at 25°C. At 2000°C, K is neither very large nor very small. Hence, we expect to find appreciable amounts of both product and reactants in the equilibrium mixture. This is indeed the case. Suppose, for instance, $[N_2] = [O_2] = 1$ M. Now:

$$[NO]^2 = 0.10 \times 1 \times 1 = 0.10$$

$$[NO] = (0.10)^{1/2} = 0.32\ M$$

We see that the equilibrium concentration of NO is nearly ⅓ that of N_2 or O_2. This explains why appreciable amounts of NO are formed in an automobile engine. At the high temperatures involved, the equilibrium constant is much larger than at 25°C. Hence N_2 and O_2 of the air can combine to form some NO.

EXAMPLE 18.5

Consider the three systems, all at 500°C.

a. $H_2(g) + Cl_2(g) \rightleftharpoons 2\ HCl(g)$; $K = 1 \times 10^{14}$

b. $H_2(g) + Br_2(g) \rightleftharpoons 2\ HBr(g)$; $K = 1 \times 10^{8}$

c. $H_2(g) + I_2(g) \rightleftharpoons 2\ HI(g)$; $K = 60$

Starting with the same concentration of reactants, compare the extent of reaction in the three cases.

SOLUTION

The extent of reaction would be greatest in (a), since K is the largest. H_2 and Cl_2 are converted virtually completely to HCl at 500°C. The extent of reaction would be somewhat less in (b). However, K is still large enough to give a nearly 100% yield of HBr. In (c), K has an intermediate value. We expect to get some HI, along with unreacted H_2 and I_2. The yield is considerably less than in (a) or (b).

Equilibrium principles are most useful for reactions where K has an intermediate value. When K is very small, we can ignore the possibility of reaction. If K is very large, all we need worry about is reaction rate. With intermediate values of K, we must be more careful. In particular, we must consider the factors that can influence the position of the equilibrium. This is the topic of Section 18.4.

18.4 CHANGES IN EQUILIBRIUM SYSTEMS: LE CHATELIER'S PRINCIPLE

We now consider a very practical question related to equilibrium systems. Suppose we are dealing with a gaseous equilibrium where K is neither very large nor very small. Suppose further that the system has come to equilibrium. Is there anything we can do to change the relative amounts of products and reactants? In other words, can we disturb the equilibrium so as to get more of the substance we want?

Sometimes, when we disturb a quiet scene, the effect we get is not the one we want.

There are, it turns out, a few things that we can do to change the composition of an equilibrium system involving gases. We can:

1. Add or remove a gaseous reactant or product.
2. Compress or expand the system, thereby changing the total pressure.
3. Change the temperature by heating or cooling the system.

The effect of these changes can be predicted by applying a rather simple principle. It was first suggested by the French chemist Le Chatelier (1850–1936). The principle states that:

When a system at equilibrium is disturbed by a change in conditions, the system shifts so as to partially counteract the change.

This means that the final equilibrium state will lie somewhere between

—the original equilibrium state, and
—the temporary, nonequilibrium state produced by the change.

Le Chatelier's Principle in Economics

Before applying Le Chatelier's Principle to chemical systems, it may be useful to see how it operates in a different area. Let's apply it to the stock market. To be specific, let's consider how changes in the price of a stock come about. Suppose that the stock of the XYZ company is selling at $50 a share. This price reflects a balance or equilibrium between opposing forces. One group of people, who own the stock, think it is worth more than $50. Another group, who don't own it, feel it is worth somewhat less than $50. So, nobody buys or sells, and the price is stable at $50 a share.

Now, suddenly, the equilibrium is disturbed by a new development. Perhaps the XYZ company announces that it has developed a process for converting waste paper into gasoline. More people want to buy it rather than sell it. The price rises to $60 a share.

When this happens, investors start to have second thoughts. Those who own the stock may decide to take their profit and run. Prospective buyers are put off by the higher price. As a result, the price drops to a new equilibrium position of perhaps $55 a share. It stays there until some new factor enters to bring about further change.

This situation is shown graphically in Figure 18.3. As we shall see, this behavior is very similar to that of a chemical system at equilibrium. The effect of the change is *partially* counteracted by a reaction taking place after the change. With the stock, the new equilibrium price is greater than its original price. However, it is less than it was shortly after the favorable news was released.

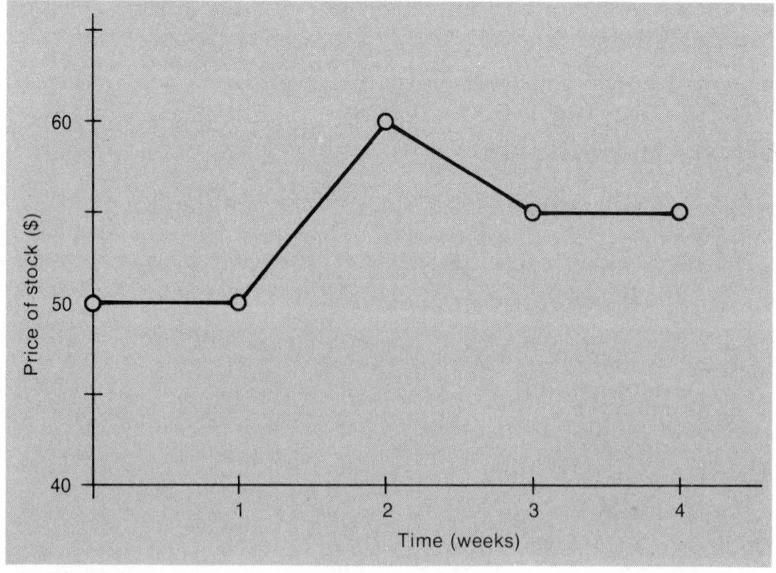

FIGURE 18.3

The price of a stock on the stock market varies depending on the pressures to buy and sell. A change in opinion favorable to the stock will tend to initially increase its value, but after a time the value will usually go down to a point intermediate between the value before the change occurred and the value just after the change.

Adding or Removing a Reactant or Product

Let us consider again the chemical equilibrium:

$$PCl_5(g) \rightleftharpoons PCl_3(g) + Cl_2(g)$$

Suppose now that to this equilibrium system we add some PCl_5. According to Le Chatelier's principle, we expect part of the added PCl_5 to decompose to PCl_3 and Cl_2. When equilibrium is re-established, the amount of PCl_5 will be:

—somewhat greater than it was originally.
—somewhat less than it was immediately after the PCl_5 was added.

The decomposition of part of the added PCl_5 produces some PCl_3 and Cl_2. Hence in the final state there will be more PCl_3 and Cl_2 than there was originally.

In Table 18.2, this reasoning is illustrated with some numbers. In Experiment 1, we first establish equilibrium in a one-liter container at 250°C. At this stage we have 0.80 mol PCl_5, 0.20 mol PCl_3, and 0.20 mol Cl_2. Now we add 0.20 mol PCl_5 to the equilibrium mixture. Temporarily, this raises the amount of PCl_5 to 1.00 mol. Then, the forward reaction:

Remember, K equals 0.05, so the concentrations adjust themselves until they satisfy Eqn. 18.2.

$$PCl_5(g) \rightarrow PCl_3(g) + Cl_2(g)$$

takes place. This partially counteracts the change. About 0.02 mol PCl_5 is consumed. Its amount drops to 0.98 mol. At the same time, 0.02 mol PCl_3 and 0.02 mol Cl_2 are formed. Hence, their amounts increase to a final value of 0.22 mol.

TABLE 18.2 EFFECT OF ADDING OR REMOVING PCl_5 ON THE SYSTEM:
$PCl_5(g) \rightleftharpoons PCl_3(g) + Cl_2(g)$; K = 0.050
(250°C IN A 1.00 ℓ CONTAINER)

Expt. 1 Adding 0.20 mol PCl_5

	Original Concentration (Equilibrium)	Temporary Concentration (Non-equilibrium)	Final Concentration (Equilibrium)
PCl_5	0.80	1.00	0.98
PCl_3	0.20	0.20	0.22
Cl_2	0.20	0.20	0.22

Expt. 2 Removing 0.20 mol PCl_5

	Original Concentration (Equilibrium)	Temporary Concentration (Non-equilibrium)	Final Concentration (Equilibrium)
PCl_5	0.80	0.60	0.62
PCl_3	0.20	0.20	0.18
Cl_2	0.20	0.20	0.18

At the bottom of Table 18.2, we show the results of a second experiment. This time, we remove 0.20 mol PCl_5 instead of adding it. Here, the reverse reaction occurs:

$$PCl_5(g) \leftarrow PCl_3(g) + Cl_2(g)$$

This restores part of the PCl_5 that was removed. To be exact, it produces 0.02 mol PCl_5. This is about 10% of the amount removed. The final amounts of PCl_3 and Cl_2 are somewhat less than they were originally.

We can apply this reasoning to any equilibrium system involving gases. In general, we can do one of two things.

1. *Add a gaseous substance to a system at equilibrium.* In this case, reaction occurs in such a direction as to consume part of what we added. If the substance added is a "reactant" (PCl_5), reaction occurs in the forward direction (left to right). If it is a "product" (PCl_3, Cl_2), the reverse reaction occurs (right to left).

2. *Remove a gaseous substance from a system at equilibrium.* In this case, reaction occurs in such a direction as to restore part of what we removed. If the substance removed is a reactant, reaction occurs in the reverse direction. If it is a product, the forward reaction occurs.

Do you think public opinion obeys Le Chatelier's principle?

Notice that these statements conform to Le Chatelier's principle. The "change" in this case is the addition or removal of a reactant or product. The system "shifts" because a reaction takes place. It may go in either the forward or reverse direction. The reaction that occurs is the one that opposes the change. In this way, the result of the change is partially counteracted.

EXAMPLE 18.6

Consider the equilibrium system:

$$N_2(g) + 3\ H_2(g) \rightleftharpoons 2\ NH_3(g)$$

What will happen if you:

a. add some N_2? b. remove some NH_3?

SOLUTION

a. Reaction will occur in the forward direction. Part of the added N_2 will be consumed. You will end up with less H_2 and more NH_3 than you had originally.

b. Again, the forward reaction will occur. Part of the NH_3 removed will be restored. You will end up with less N_2 and H_2 than you had originally.

Expansion or Compression

Consider once again the equilibrium system:

$$PCl_5(g) \rightleftharpoons PCl_3(g) + Cl_2(g)$$

There is another way in which we might "disturb" this system. Suppose, as in Figure 18.4, that we expand the system by increasing the volume. By doing this, we decrease the concentration of gas molecules. The same number of molecules now occupy a larger space. The number of molecules per unit volume decreases.

According to Le Chatelier's principle, the system will shift so as to oppose this change. The question is, how can this happen? How can the system shift to increase the number of molecules per unit volume?

The answer to this question may not be obvious. However, consider what happens if the forward reaction occurs:

$$PCl_5(g) \rightarrow PCl_3(g) + Cl_2(g)$$

The effect is to increase the number of molecules. Two molecules (one PCl_3, one Cl_2) are formed from one (PCl_5). In this way, the number of molecules per unit volume increases. This partially counteracts the expansion, which decreased the concentration of gas molecules.

By expanding the system, we lower the total concentration of all species. Reaction occurs so as to increase that total concentration.

This reasoning is illustrated by the data in Table 18.3, p. 462. In Expt. 1, we start once again with 0.80 mol PCl_5, 0.20 mol PCl_3, and 0.20 mol Cl_2. This represents an equilibrium mixture in a one-liter container at 250°C. Now we expand the system to a volume of 2.00 ℓ. The immediate effect is to reduce all the concentrations to one half their original value. However, the forward reaction occurs to partially counteract this. To be

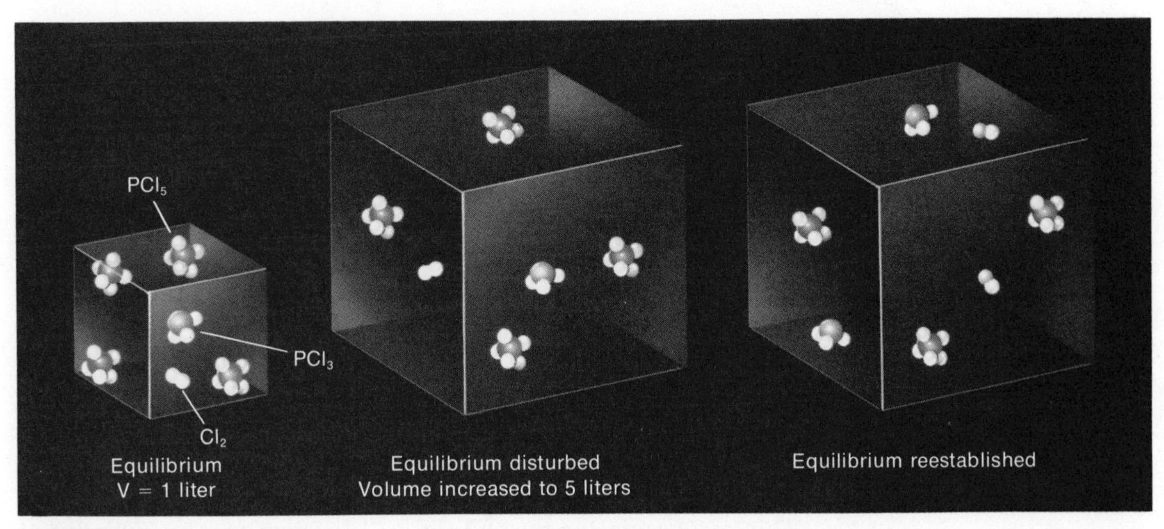

FIGURE 18.4

Diagram showing the effect of an expansion on an equilibrium mixture of PCl_5, PCl_3, and Cl_2. The equilibrium system will tend to shift toward PCl_3 and Cl_2 to counteract the decrease in concentrations caused by the expansion.

TABLE 18.3 EFFECT OF EXPANSION AND COMPRESSION ON THE EQUILIBRIUM SYSTEM: $PCl_5(g) \rightleftharpoons PCl_3(g) + Cl_2(g)$ at 250°C; K = 0.050

Expt. 1 Expansion from 1.00 ℓ to 2.00 ℓ

	Original Concentration (Equilibrium)	Temporary Concentration (Non-Equilibrium)	Final Concentration (Equilibrium)
PCl_5	0.80	0.40	0.36
PCl_3	0.20	0.10	0.14
Cl_2	0.20	0.10	0.14
total	1.20	0.60	0.64

Expt. 2 Compression from 1.00 ℓ to 0.50 ℓ

	Original Concentration (Equilibrium)	Temporary Concentration (Non-Equilibrium)	Final Concentration (Equilibrium)
PCl_5	0.80	1.60	1.71
PCl_3	0.20	0.40	0.29
Cl_2	0.20	0.40	0.29
total	1.20	2.40	2.29

exact, 0.04 mol PCl_5 decomposes, forming 0.04 mol PCl_3 and 0.04 mol Cl_2. The net effect is an increase in the total number of moles.

$$\text{change in no. moles} = 0.04 \text{ mol } PCl_3 + 0.04 \text{ mol } Cl_2 - 0.04 \text{ mol } PCl_5$$
$$= +0.04 \text{ mol}$$

The changes occur so as to make the equilibrium concentrations satisfy Eqn. 18.2.

Notice what happens to the "total" concentration. Originally, it is 1.20 mol/ℓ. Upon expansion, it drops to half that value, 0.60 mol/ℓ. As equilibrium is restored, it bounces back to an intermediate value, 0.64 mol/ℓ.

Consider now what happens if we compress the system shown in Figure 18.4. To do this, we *decrease* the volume. The immediate effect is to increase the number of molecules per unit volume. To counteract this, the reverse reaction occurs:

$$PCl_5(g) \leftarrow PCl_3(g) + Cl_2(g)$$

Here, one molecule forms at the expense of two. Data obtained upon compression are shown at the bottom of Table 18.3. Notice that the total concentration doubles (2.40 mol/ℓ vs 1.20 mol/ℓ) upon compression. However, as the reverse reaction occurs, we return to an intermediate value, 2.29 mol/ℓ. Again, Le Chatelier's principle applies.

This discussion for the PCl_5-PCl_3-Cl_2 system can be extended to all gaseous equilibria. There is a simple rule for predicting the effect of expansion or compression upon an equilibrium system. In stating it, we usually refer to expansion as a "decrease in pressure". Compression is called an "increase in pressure". The rules are

1. **When the pressure on an equilibrium system is decreased, reaction occurs so as to increase the number of moles of gas.** Examples include:

$PCl_5(g) \rightarrow PCl_3(g) + Cl_2(g)$ 1 mol gas → 2 mol gas

$2\ NH_3(g) \rightarrow N_2(g) + 3\ H_2(g)$ 2 mol gas → 4 mol gas

The system in both cases reacts so as to partially counteract the change in pressure.

2. **When the pressure on an equilibrium system is increased, reaction occurs so as to decrease the number of moles of gas.** Examples include:

$PCl_3(g) + Cl_2(g) \rightarrow PCl_5(g)$ 2 mol gas → 1 mol gas

$N_2(g) + 3\ H_2(g) \rightarrow 2\ NH_3(g)$ 4 mol gas → 2 mol gas

EXAMPLE 18.7

Consider the following systems:

a. $2\ SO_2(g) + O_2(g) \rightleftharpoons 2\ SO_3(g)$ b. $2\ CO_2(g) \rightleftharpoons 2\ CO(g) + O_2(g)$

c. $H_2(g) + I_2(g) \rightleftharpoons 2\ HI(g)$

If the object is to obtain more products, would you increase or decrease the pressure?

SOLUTION

a. The forward reaction forms two moles of gas (SO_3) at the expense of three ($2\ SO_2 + 1\ O_2$). Hence, it will occur if the pressure is increased.

b. In the forward reaction, three moles of gas are formed from two. Decrease the pressure to increase the yield.

c. There is no change in the number of moles of gas. If you want to get more HI, you will have to change some other factor, not the pressure.

Changing the Temperature

Still another way to disturb an equilibrium system is to heat it or cool it. In this way, the temperature is changed. As usual, we can apply Le Chatelier's principle to predict what will happen when we do this. Consider, one last time, the system:

$$PCl_5(g) \rightleftharpoons PCl_3(g) + Cl_2(g)$$

Let's suppose we have established equilibrium at 250°C. We now supply heat, perhaps with a Bunsen burner. The immediate effect is to raise the temperature. According to Le Chatelier's principle, the system will shift to counteract this temperature increase. How can it do this? Very simply, it turns out. The forward reaction is *endothermic:*

$$PCl_5(g) \rightarrow PCl_3(g) + Cl_2(g); \qquad \Delta H = +22.2 \text{ kcal}$$

So, this reaction occurs, absorbing part of the heat supplied by the burner. As a result, the temperature increase will be less than expected. When equilibrium is established again, there will be more products (PCl_3, Cl_2) and less reactant (PCl_5).

No matter how you try to change it, the system will always fight you.

Now suppose we carry out the same experiment with the system:

$$2\ SO_2(g) + O_2(g) \rightleftharpoons 2\ SO_3(g)$$

Here, the temperature effect is quite different. For the forward reaction, ΔH is negative:

$$2\ SO_2(g) + O_2(g) \rightarrow 2\ SO_3(g); \qquad \Delta H = -47.0 \text{ kcal}$$

This time it is the reverse reaction which is *endothermic* ($\Delta H = +47.0$ kcal). When we supply heat, the reverse reaction takes place. Some SO_3 decomposes, forming SO_2 and O_2. At the higher temperature, there will be less product (SO_3) and more reactants (SO_2, O_2).

In general, we can say that:

1. **When the temperature of an equilibrium system is increased, the reaction which is endothermic ($\Delta H > 0$) takes place.** If the forward reaction is endothermic, more product is formed. If the reverse reaction is endothermic, there will be less product at the higher temperature.

Applying the same reasoning:

2. **When the temperature of an equilibrium system is decreased, the reaction which is exothermic ($\Delta H < 0$) takes place.**

EXAMPLE 18.8

Consider the equilibrium:

$$N_2O_4(g) \rightleftharpoons 2\ NO_2(g); \qquad \Delta H = +13.9 \text{ kcal}$$

What will happen if you increase the temperature? decrease the temperature?

SOLUTION

The forward reaction is endothermic (The quoted value of ΔH always applies to the reaction from left to right). Consequently, if the temperature is increased, the forward reaction takes place. N_2O_4 is converted to NO_2. If you reduce the temperature, the exothermic reaction will occur. This is the reverse reaction: $2\ NO_2(g) \rightarrow N_2O_4(g)$. For this reaction, ΔH is -13.9 kcal. Some N_2O_4 will be formed at the expense of NO_2.

These effects are shown in Color Plate 8A, in the center of the book. This shows the equilibrium system at two different temperatures, 25 and 100°C. The colored species is NO_2. Note that the color is much deeper at the higher temperature.

TABLE 18.4 EFFECT OF TEMPERATURE UPON THE EQUILIBRIUM CONSTANT

$PCl_5(g) \rightleftharpoons PCl_3(g) + Cl_2(g)$; $\Delta H = +22.2$ kcal

t (°C)	200	250	300	400	500
K	0.0049	0.050	0.25	3.7	28

$2\ SO_2(g) + O_2(g) \rightleftharpoons 2\ SO_3(g)$; $\Delta H = -47.0$ kcal

t (°C)	500	600	700	800	900	1000
K	150,000	5,000	350	40	6.6	1.5

You will recall (Section 18.1) that changing the temperature changes the equilibrium constant. We sometimes describe the effect of temperature in terms of its effect upon K. The general rule is that an increase in temperature:

—increases K if the forward reaction is endothermic ($\Delta H > 0$).
—decreases K if the forward reaction is exothermic ($\Delta H < 0$).

This rule is illustrated in Table 18.4. Note that the effect of temperature upon K is often quite large. For the SO_2-O_2-SO_3 system, K decreases by a factor of 100,000 when the temperature increases from 500 to 1000°C. This means that temperature is a key factor to consider if you want to change the extent of reaction.

18.5 THE SYNTHESIS OF AMMONIA

A very important equilibrium system involves the synthesis of ammonia from the elements:

$$N_2(g) + 3\ H_2(g) \rightleftharpoons 2\ NH_3(g); \qquad \Delta H = -22.0 \text{ kcal} \qquad (18.4)$$

About 2×10^{10} kg of ammonia are made each year in the United States by this reaction. The process was first developed by a German chemist, Fritz Haber, just prior to World War I. The Germans needed ammonia to make nitric acid (Chap. 23), which in turn was used to make explosives. Today, NH_3 is used mainly to make compounds used in fertilizers.

Sad to say, the Haber process was in part responsible for World War I.

In carrying out the Haber process, it is important to get as high a yield of ammonia as possible. We must also be concerned with the rate of reaction. It will do no good to get a high yield of NH_3 if it takes forever to do it. With these two factors in mind, let us consider the conditions under which Reaction 18.4 should be carried out.

Effect of Pressure

You will note from Equation 18.4 that 4 mol of reactants (1 N_2, 3 H_2) give 2 mol NH_3. By compressing the system, we make the forward reaction occur, producing more ammonia. Compressing the system also increases the rate of reaction. The reactant molecules are confined to a smaller volume. Hence, they collide more often.

From both standpoints, a high pressure is desirable. We arrive at equilibrium faster and get a higher yield of ammonia. The pressure commonly used is 400 atm. Presumably, a still higher pressure would be even better. However, there is a practical problem. Equipment designed to withstand very high pressures (1000 atm or more) is expensive and subject to leaks.

Effect of Temperature

Notice that Reaction 18.4 is exothermic. This means that an increase in temperature decreases the yield of ammonia. So far as the position of the equilibrium is concerned, a low temperature is best. However, we must be concerned with rate as well. As we saw in Chapter 17, reactions go faster at higher temperatures. From the standpoint of rate, we prefer a high temperature.

Like most industrial processes, NH_3 synthesis depends on a proper choice of temperature, pressure, and catalyst.

In this case, the two factors operate in opposite directions. So, we compromise, using a "moderate" temperature. Commonly, this is about 400°C. At this temperature, K for Reaction 18.4 is about 0.5. At 400 atm, this gives about a 50% yield of NH_3. We could do better by going to lower temperatures (K = 650 at 200°C). However, if we did this, the reaction would be too slow.

Effect of a Catalyst

A catalyst has no effect whatsoever on the position of this or any other equilibrium. The yield of NH_3 is exactly the same whether we use a catalyst or not. However, with the proper catalyst, the rate at which we reach equilibrium can be greatly increased. A solid catalyst is used

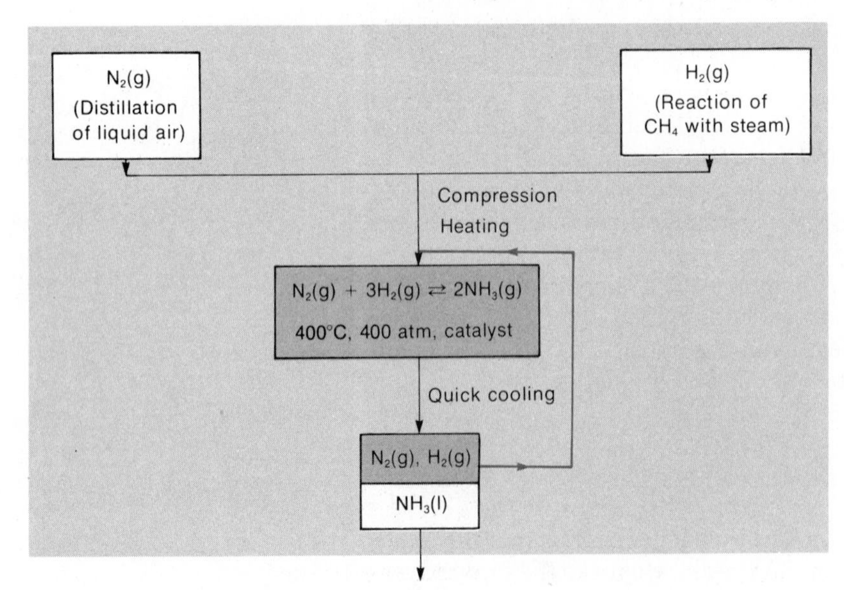

FIGURE 18.5

The Haber process. This process is carried out at high pressure, to favor formation of product NH_3, and at about 400°C, to give a good rate of reaction. The catalyst helps to increase the rate, but does not affect the equilibrium state of the system. Any N_2 and H_2 that does not react is recycled following removal of the NH_3 by condensation following cooling of the equilibrium mixture.

in the Haber process. It consists largely of iron with small amounts of potassium oxide, K_2O, and aluminum oxide, Al_2O_3.

Figure 18.5 shows a flow chart for the Haber process. Notice, among other things, that ammonia is condensed out as a liquid (bp = −33°C). Unreacted nitrogen and hydrogen are recycled to increase the yield of ammonia.

EXAMPLE 18.9

Consider the equilibrium:

$$N_2(g) + O_2(g) \rightleftharpoons 2\ NO(g); \qquad \Delta H = +43.2\ kcal$$

Suppose you want to make NO from N_2 and O_2. Discuss the effect of each of the following upon both the position of the equilibrium and the rate at which it is reached.

a. pressure b. temperature c. catalyst

SOLUTION

a. Pressure has no effect on the yield of NO. There is no change in the number of moles of gas in the reaction (2 mol reactants → 2 mol NO). However, a high pressure will increase the reaction rate. This is generally true for any reaction. By compressing the system, we increase the concentration of reactants and hence the rate of collision. Thus we reach equilibrium more quickly.

b. Since the reaction is endothermic, an increase in temperature will increase the yield of NO. It will also make the reaction go faster. From both standpoints, we should use as high a temperature as possible.

c. By choosing the proper catalyst, we can increase the rate of reaction. No catalyst will affect the position of this or any other equilibrium.

18.6 SUMMARY

In this chapter, we have considered gas-phase equilibria. For each reaction, there is an equilibrium constant, K. This constant relates the equilibrium concentrations of products and reactants. For a given reaction, the magnitude of K depends only upon temperature. It does *not* depend upon the pressure, the volume of the container, the presence of a catalyst, or anything else. Reactions for which K is very large tend to go virtually to completion. If K is very small, essentially no reaction occurs. Most of the reactions we have considered have intermediate values of K. When K falls between about 10^5 and 10^{-5}, there are appreciable quantities of both products and reactants at equilibrium.

The form of K depends upon the equation for the reaction. It may be very simple:

$$CaCO_3(s) \rightleftharpoons CaO(s) + CO_2(g);\ K = [CO_2]$$

or more complex:

$$N_2(g) + 3\ H_2(g) \rightleftharpoons 2\ NH_3(g);\ K = \frac{[NH_3]^2}{[N_2] \times [H_2]^3}$$

The square brackets refer to equilibrium concentrations in moles per liter. Remember that:

1. Only gases appear in the expression for K.
2. Products are in the numerator, reactants in the denominator.
3. The concentration of each gas is raised to a power equal to its coefficient in the balanced equation.

Once equilibrium is reached, it can be disturbed in any of three ways. In each case, the result can be predicted by Le Chatelier's principle. Thus, if we:

1. *add a substance* (reactant or product), reaction will occur in such a direction as to consume part of it.
2. *compress the system*, reaction will occur in a direction so as to reduce the number of moles of gas.
3. *heat the system* (raise the temperature), the reaction which absorbs heat will occur.

NEW TERMS

chemical equilibrium
equilibrium constant (K)
Haber process
Le Chatelier's principle

QUESTIONS

1. Suppose a sample of Cl_2 is placed in a closed container. It partially dissociates to form Cl atoms.

$$Cl_2(g) \rightleftharpoons 2\ Cl(g)$$

How do the rates of forward and reverse reactions compare:

a. at the beginning of the experiment?
b. when equilibrium is reached?

2. In Section 18.1, we found that at 250°C:

$$\frac{[PCl_3] \times [Cl_2]}{[PCl_5]} = 0.050$$

Would this ratio change if we changed

a. the volume of the container? b. the temperature?
c. the original concentrations?

3. Consider the reaction: $PCl_5(g) \rightarrow PCl_3(g) + Cl_2(g)$. If 0.60 mol PCl_5 reacts:

a. How many moles of PCl_3 are formed?
b. How many moles of Cl_2 are formed?

4. Consider the reaction: $H_2(g) + I_2(g) \rightarrow 2\ HI(g)$. If 0.036 mol H_2 reacts:

   a. How many moles of I_2 are consumed?
   b. How many moles of HI are formed?

5. For which of the following systems would there be a term for N_2 in the numerator of K?

   a. $N_2(g) + 3\ H_2(g) \rightleftharpoons 2\ NH_3(g)$
   b. $2\ NCl_3(g) \rightleftharpoons N_2(g) + 3\ Cl_2(g)$
   c. $N_2(g) + O_2(g) \rightleftharpoons 2\ NO(g)$
   d. $N_2H_4(g) \rightleftharpoons N_2(g) + 2\ H_2(g)$
   e. $2\ MgN_3(s) \rightleftharpoons 2\ Mg(s) + 3\ N_2(g)$

6. Referring to the systems in Question 5, in which ones would:

   a. the equilibrium concentration of N_2 be raised to a power other than 1?
   b. the equilibrium concentration of N_2 be multiplied by that of another substance (to the proper power)?

7. For which of the following systems would a term for H_2O appear in the expression for K?

   a. $2\ H_2O_2(l) \rightleftharpoons 2\ H_2O(l) + O_2(g)$
   b. $2\ H_2(g) + O_2(g) \rightleftharpoons 2\ H_2O(l)$
   c. $SnO_2(s) + 2\ H_2(g) \rightleftharpoons Sn(s) + 2\ H_2O(g)$

8. For a certain reaction, K is 10^{20}. This means that, under ordinary conditions:

   a. the reaction is very rapid
   b. very little product is formed
   c. the product is formed in high yield

9. For the system: $CO(g) + H_2O(g) \rightleftharpoons CO_2(g) + H_2(g)$ at a certain temperature, K = 1. At equilibrium you would expect to find:

   a. only CO_2 and H_2
   b. mostly CO_2 and H_2
   c. about the same amount of products as reactants
   d. mostly CO and H_2O
   e. only CO and H_2O

10. At another temperature, K for the reaction in Question 9 is 10^{-7}. Under these conditions, which of the answers in Question 9 would be correct?

11. State Le Chatelier's principle. Discuss briefly how it applies in chemistry.

12. List three different ways in which a chemical system at equilibrium can be "disturbed".

13. Explain why, in terms of Le Chatelier's principle, adding a reactant causes the forward reaction to occur.

14. Why, in terms of Le Chatelier's principle, does an increase in temperature favor an endothermic reaction?

15. Criticize the following statement: If a system at equilibrium is disturbed, it eventually returns to the same equilibrium state.

16. For the system: $A_2(g) + B_2(g) \rightleftharpoons 2\ AB(g)$; $\Delta H = 0$. The position of this equilibrium could be shifted to favor the formation of AB by

    a. compression
    b. expansion
    c. raising T
    d. lowering T
    e. none of these

17. Consider the data in Table 18.2. Show that in the final equilibrium state:

$$\frac{[PCl_3] \times [Cl_2]}{[PCl_5]} = 0.05$$

for both experiments.

18. In the Haber process for the synthesis of NH_3, what would happen to the yield of ammonia and the rate at which it is formed if:

    a. the pressure was decreased to 100 atm?
    b. the temperature was lowered to 100°C?
    c. the catalyst was not used?

PROBLEMS

1. *Expression for K (Example 18.2)* Write the expression for the equilibrium constant for:

    a. $2\ SO_2(g) + O_2(g) \rightleftharpoons 2\ SO_3(g)$
    b. $H_2(g) + Br_2(g) \rightleftharpoons 2\ HBr(g)$
    c. $N_2H_4(g) \rightleftharpoons N_2(g) + 2\ H_2(g)$

2. *Expression for K (Example 18.4)* Write the expression for the equilibrium constant for:

    a. $Ca(OH)_2(s) \rightleftharpoons CaO(s) + H_2O(g)$
    b. $CO(g) + H_2O(l) \rightleftharpoons CO_2(g) + H_2(g)$
    c. $Fe_2O_3(s) + 3\ H_2(g) \rightleftharpoons 2\ Fe(s) + 3\ H_2O(g)$

3. *Calculations Involving K (Example 18.1)* For the equilibrium:

$$H_2(g) + I_2(g) \rightleftharpoons 2\ HI(g); \qquad K = 60$$

Calculate [HI] if $[H_2] = [I_2] = 0.10$ M.

4. *Calculations Involving K (Example 18.1)* If, in Problem 3, [HI] = 0.60 M and $[H_2] = 0.10$ M, calculate $[I_2]$.

5. *Calculation of K (Example 18.3)* For the equilibrium:

$$N_2(g) + 3\ H_2(g) \rightleftharpoons 2\ NH_3(g)$$

at a certain temperature, $[NH_3] = 0.10$ M, $[N_2] = 0.10$ M, $[H_2] = 0.20$ M. Calculate K.

6. *Magnitude of K (Example 18.5)* For the equilibrium:

$$N_2(g) + 3\ H_2(g) \rightleftharpoons 2\ NH_3(g)$$

K is 650 at 200°C and 0.50 at 400°C. At which temperature would you expect to get the better yield of ammonia, with the same concentrations of N_2 and H_2? Explain your reasoning.

7. *Effect of Adding or Removing Substance on Position of Equilibrium (Example 18.6)* Consider the equilibrium: $N_2(g) + 3\ H_2(g) \rightleftharpoons 2\ NH_3(g)$. Which way would reaction proceed if you were to:

    a. add H_2? b. remove N_2? c. remove NH_3? d. add NH_3?

8. *Effect of Pressure on Position of Equilibrium (Example 18.7)* To increase the yield of product(s) in each of the following reactions, would you expand or compress the system?

    a. $N_2(g) + 3\ Cl_2(g) \rightleftharpoons 2\ NCl_3(g)$
    b. $2\ CO_2(g) \rightleftharpoons 2\ CO(g) + O_2(g)$
    c. $CO_2(g) + H_2(g) \rightleftharpoons CO(g) + H_2O(g)$

9. *Effect of Temperature Upon Position of Equilibrium (Example 18.8)*
For the system: $2\ H_2(g) + O_2(g) \rightleftharpoons 2\ H_2O(g)$; $\Delta H = -116$ kcal
What would happen to the yield of H_2O if the temperature were increased? decreased?

10. *Effect of Conditions on Rate and Position of Equilibrium (Example 18.9)*
For the system: $PCl_5(g) \rightleftharpoons PCl_3(g) + Cl_2(g)$; $\Delta H = +22$ kcal
State the effect upon the rate and yield of products if:

a. the pressure is increased b. the temperature is increased
c. a catalyst is used

* * * * *

11. For the equilibrium: $BaCl_2 \cdot 2H_2O(s) \rightleftharpoons BaCl_2(s) + 2\ H_2O(g)$ the concentration of water vapor at a certain temperature is 0.16 M. Calculate K.

12. Consider the system:

$$2\ SO_3(g) \rightleftharpoons 2\ SO_2(g) + O_2(g)$$

a. Write the expression for K.
b. At a certain temperature, all three concentrations at equilibrium are 0.10 M. Calculate K.

13. For the system: $2\ CO(g) + O_2(g) \rightleftharpoons 2\ CO_2(g)$; $K = 2.0 \times 10^5$

a. Write the expression for K.
b. Calculate $[O_2]$ when $[CO] = [CO_2] = 1$ M

14. At a certain temperature, K for the equilibrium: $Cl_2(g) \rightleftharpoons 2\ Cl(g)$ is 1×10^{-12}. How would you expect the concentration of Cl atoms at equilibrium to compare to that of Cl_2 molecules?

15. At a different temperature, K for the system in Problem 13 is 200. Would you expect the yield of CO_2 from CO and O_2 to be larger or smaller at this temperature as compared to that in Problem 13?

16. For the system: $2\ I(g) \rightleftharpoons I_2(g)$ at a certain temperature, $K = 3 \times 10^9$. Would you expect the equilibrium system to contain mostly I or mostly I_2?

17. Consider the equilibrium: $H_2(g) + I_2(g) \rightleftharpoons 2\ HI(g)$. To convert as much I_2 as possible to HI, should you add or remove H_2?

18. Which way does each of the following equilibria shift if the volume of the container is increased?

a. $PCl_5(g) \rightleftharpoons PCl_3(g) + Cl_2(g)$
b. $2\ NO(g) + O_2(g) \rightleftharpoons 2\ NO_2(g)$
c. $3\ O_2(g) \rightleftharpoons 2\ O_3(g)$
d. $H_2(g) + F_2(g) \rightleftharpoons 2\ HF(g)$

19. For the system: $N_2(g) + 2\ O_2(g) \rightleftharpoons 2\ NO_2(g)$; $\Delta H = +16$ kcal.
Would K for this system increase or decrease if the temperature were raised?

20. Consider the system: $N_2O_4(g) \rightleftharpoons 2\ NO_2(g)$; $\Delta H = +14$ kcal
Describe three ways in which you could increase the rate of reaching equilibrium and two ways in which you could get a higher yield of NO_2 from a given amount of N_2O_4.

21. Consider the equilibrium: $2\ NO(g) + O_2(g) \rightleftharpoons 2\ NO_2(g)$; $\Delta H = -27$ kcal
List three different ways in which you could increase the yield of NO_2, starting with the same amount of NO.

22. What effect would an increase in pressure or temperature have on the position of the equilibrium: $N_2O_4(g) \rightleftharpoons 2\ NO_2(g)$; $\Delta H = +14$ kcal

23. Consider the system: $2\ SO_2(g) + O_2(g) \rightleftharpoons 2\ SO_3(g)$; $\Delta H = -47$ kcal
List three different ways in which you could increase the rate of this reaction. What effect would each of these changes have on the yield of SO_3?

24. For the system: $PCl_5(g) \rightleftharpoons PCl_3(g) + Cl_2(g)$, K is 0.050 at 250°C. Which of the following represent equilibrium concentrations in this system?

	a.	b.	c.
conc. PCl_3 (mol/ℓ)	0.50	0.25	0.050
conc. Cl_2 (mol/ℓ)	0.10	0.25	0.10
conc. PCl_5 (mol/ℓ)	1.00	1.00	0.10

25. For the system: $H_2(g) + I_2(g) \rightleftharpoons 2\ HI(g)$ at a certain temperature, K is 60. Which of the following represent equilibrium concentrations for this system?

	a.	b.	c.
conc. HI (mol/ℓ)	6.0	60.0	1.00
conc. H_2 (mol/ℓ)	0.60	1.0	0.10
conc. I_2 (mol/ℓ)	1.00	1.0	0.17

*26. Consider the system: $PCl_5(g) \rightleftharpoons PCl_3(g) + Cl_2(g)$. At a certain temperature, it is found that when one starts with 0.20 mol of PCl_5 in a 4.0 ℓ container, 0.10 mol is left at equilibrium. Calculate:

a. the equilibrium concentration of PCl_5.
b. the equilibrium concentrations of PCl_3 and Cl_2.
c. K for the system.

*27. For the system: $CaCO_3(s) \rightleftharpoons CaO(s) + CO_2(g)$, the equilibrium pressure of CO_2 is found to be 1.00 atm at 850°C.

a. Using the gas laws (Chapter 5), calculate the molar volume of CO_2 at 1 atm and 850°C.
b. Calculate the equilibrium concentration, in moles per liter, of CO_2 at 850°C.
c. What is K for this system at 850°C?

*28. Consider the system: $CO(g) + H_2O(l) \rightleftharpoons CO_2(g) + H_2(g)$. If this system is compressed, what happens to the position of the equilibrium (note that H_2O is present as a liquid). Explain your reasoning.

*29. The equilibrium constant for a certain system is:

$$K = \frac{[H_2O]}{[H_2]}$$

Write chemical equations for at least two different systems which would give this expression for K. (Remember that terms for liquids or solids do not appear in K.)

Some of these foods are acidic; others are basic. Can you tell which is which? *Photo by Amy Shapiro.*

19 ACIDS AND BASES

This is the first of several chapters dealing with reactions taking place in water solution. Here, we will consider two types of compounds, acids and bases. These have been mentioned in previous chapters. Now we are ready to take a closer look at their structures and chemical properties in water solution.

The concept of acids and bases is an old one. Indeed, these compounds were known long before chemistry became a science. The earliest acids were organic compounds such as acetic acid. They were isolated from natural materials such as vinegar. The inorganic acids became available later. The early alchemists worked with acids such as HCl, HNO_3 and H_2SO_4. They used water solutions of these compounds to dissolve metals and minerals. Other compounds, known as "alkalis" or bases, were obtained by leaching ashes with water. The alchemists found that bases were able to "neutralize" acids. That is, they destroyed the characteristic acidic properties.

19.1 ACIDIC AND BASIC WATER SOLUTIONS

All water solutions can be classified as acidic, basic, or neutral. A neutral solution is one that shows neither acidic nor basic properties. We will now consider the properties of acidic and basic solutions.

Properties of Acidic Solutions. The H^+ Ion.

All acidic water solutions have certain properties in common. They:

1. *Have a sour taste.* The sour taste of many foods is due to the presence of organic acids. These include acetic acid (vinegar), citric acid (lemon juice), and oxalic acid (rhubarb). Solutions of inorganic acids such as HCl, HNO_3, and H_2SO_4 are also reported to taste sour. However, you wouldn't want to taste these solutions because they can be extremely corrosive.

2. *Change the color of certain organic dyes known as indicators.* In the laboratory, the dye known as litmus is often used to test for acids. Strips of paper treated with litmus turn red in acidic solution. Acids can also cause a color change in many natural products including grape juice. The color of certain flowers depends upon the acidity of the soil in which they are grown. Azaleas can be fiery red, orange, or yellow, depending upon soil acidity.

Actually, most plants are sensitive to the acidity of the soil, and will not grow well in too acidic or too basic a soil.

3. *React with active metals such as zinc to give hydrogen gas.*

4. *React with basic solutions to neutralize their properties* (Section 19.4).

These properties of acidic water solutions are due to a particular ion, the **H^+ ion (proton).** This ion causes the sour taste and change in color of indicators. It is also the H^+ ion which reacts with zinc.

$$Zn(s) + 2\ H^+(aq) \rightarrow Zn^{2+}(aq) + H_2(g) \qquad (19.1)$$

We should point out that the H^+ ion, like most ions in water solution, is hydrated. That is, it is closely surrounded by water molecules. To emphasize this fact, it is sometimes written as

$$H^+(H_2O)$$

or, more frequently:

$$H_3O^+$$

The acidic properties of water solutions are sometimes ascribed to the H_3O^+ *(hydronium)* ion. In this text, we will keep things simple by using H^+ rather than H_3O^+ in all the equations we write.

Properties of Basic Water Solutions. The OH^- Ion.

All basic water solutions have certain properties in common. They:

1. *Have a slippery feeling.* You may notice this property when you use a strong detergent or laundry soap. These products produce a basic water solution. Lye, (NaOH) often used to unplug drains, has a similar but more drastic effect. It should not be handled without gloves, since it can burn the skin.

2. *Change the colors of indicators.* Litmus, which is red in acidic solution, turns blue in basic solution. Many other indicators show similar color changes.

3. *React with solutions containing* Mg^{2+} *to give a white precipitate of magnesium hydroxide,* $Mg(OH)_2$.

4. *React with acidic solutions to neutralize their properties* (Section 19.4).

What simple experiment could you do to find out if a solution is acidic or basic?

The properties of basic water solutions are due to a particular ion, the **OH^- ion.** It is this ion which turns red litmus blue. It is also the OH^- ion which reacts with Mg^{2+} to form a precipitate:

$$Mg^{2+}(aq) + 2\ OH^-(aq) \rightarrow Mg(OH)_2(s) \qquad (19.2)$$

Relation Between $[H^+]$ and $[OH^-]$

We have pointed out that H^+ ions make a water solution acidic. The properties common to basic solutions are those of the OH^- ion. With this background, you may be surprised to learn that *all* water solutions contain *both* of these ions. They are present because water itself is slightly ionized:

$$H_2O(l) \rightleftharpoons H^+(aq) + OH^-(aq) \qquad (19.3)$$

We can write an equilibrium constant for Reaction 19.3. Following the rule cited in Chapter 18, we omit $H_2O(l)$ from the expression for K. We are left with:

$$K = [H^+] \times [OH^-]$$

Here the square brackets, as always, indicate equilibrium concentrations in moles per liter.

The equilibrium constant for Reaction 19.3 is a very small number. At 25°C, it is 1.0×10^{-14}. Hence we can write the following equation, valid for all water solutions:

$$[H^+] \times [OH^-] = 1.0 \times 10^{-14} \text{ (at 25°C)} \qquad (19.4)$$

This relationship is plotted in Fig. 19.1, p. 476. Values of $[OH^-]$ are shown on the vertical axis, $[H^+]$ on the horizontal axis. Notice that the two quantities are inversely proportional. Figure 19.1 looks like a

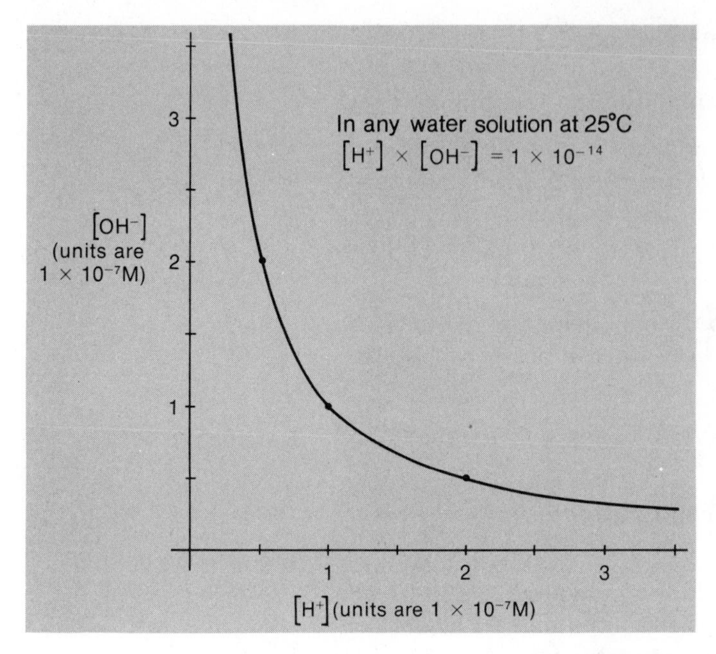

FIGURE 19.1

In water solutions the concentrations of hydrogen and hydroxide ions are related as shown by the graph. Given the concentration of one of these ions, the concentration of the other can be found from the graph or from the fact that the product of the two concentrations must be 1×10^{-14}.

Boyle's Law plot (PV = constant). When the concentration of OH^- is high (left side of the curve), that of H^+ is low. At the right of the curve, where $[H^+]$ is large, $[OH^-]$ is small.

Using Equation 19.4, we can divide all water solutions into three types.

1. *In pure water, or in a neutral water solution, there are equal numbers of H^+ and OH^- ions.* These ions come from the reaction:

$$H_2O \rightleftharpoons H^+(aq) + OH^-(aq)$$

In pure water, one H^+ ion is formed for every OH^- ion. Hence, under these conditions, the two ions are present in equal amounts.

From Equation 19.4, we can calculate $[H^+]$ and $[OH^-]$ in pure water. Since there are equal numbers of these two ions:

$$[H^+] = [OH^-]$$

Substituting in Equation 19.4:

$$[H^+] \times [H^+] = 1.0 \times 10^{-14}$$

Taking the square root of both sides:

$$[H^+] = 1.0 \times 10^{-7}\text{ M} = [OH^-]$$

In pure water, there are relatively few H^+ and OH^- ions.

These concentrations are very small indeed. In pure water, only about two out of every billion water molecules are ionized. No wonder pure water is a nonconductor of electricity!

2. *In an acidic water solution, there are more H^+ ions than OH^- ions.* To make a water solution acidic, we add a substance such as HCl

which furnishes H^+ ions. This makes the concentration of H^+ greater than its original value, 1×10^{-7} M. Since $[OH^-]$ is inversely proportional to $[H^+]$, it must decrease when $[H^+]$ is increased. Hence the concentration of OH^- ions in the acidic solution must be less than that in pure water, 1×10^{-7} M.

To be specific, suppose we add enough H^+ to make its concentration 1×10^{-5} M. That is:

We could do this by adding a solution of HCl.

$$[H^+] = 1 \times 10^{-5}\ M$$

According to Equation 19.4:

$$[OH^-] = \frac{1.0 \times 10^{-14}}{1 \times 10^{-5}} = 1 \times 10^{-9}\ M$$

Notice that the concentration of H^+ is now 10,000 times that of OH^-. That is:

$$\frac{[H^+]}{[OH^-]} = \frac{1 \times 10^{-5}\ M}{1 \times 10^{-9}\ M} = 1 \times 10^4 = 10{,}000$$

In this particular acidic solution, there are 10,000 H^+ ions for every OH^- ion. It is the excess H^+ ions that make the solution acidic.

3. *In a basic water solution, there are more OH^- ions than H^+ ions.* The reasoning here is similar to that just gone through. Suppose we add a compound such as NaOH to water. This increases $[OH^-]$. At the same time, it decreases $[H^+]$. To be specific, suppose we add enough NaOH to make:

$$[OH^-] = 1 \times 10^{-5}\ M$$

From Equation 19.4 we see that:

$$[H^+] = \frac{1.0 \times 10^{-14}}{1 \times 10^{-5}} = 1 \times 10^{-9}\ M$$

In a basic solution, $[H^+]$ is very, very low.

In this basic solution, the concentration of OH^- is 10,000 times that of H^+. In other words, OH^- ions outnumber H^+ ions by 10,000 to 1. Hence, the solution shows basic properties.

Table 19.1, p. 478 shows values of $[H^+]$ and $[OH^-]$ in a series of different water solutions. The three solutions at the left are all acidic. In each of these solutions:

$$[H^+] > [OH^-]$$
$$[H^+] > 10^{-7}\ M;\ [OH^-] < 10^{-7}\ M \qquad \text{acidic solutions (1, 2, 3)}$$

The solution in the center column of the Table is neutral.

$$[H^+] = [OH^-] = 10^{-7}\ M \qquad \text{neutral solution (4)}$$

TABLE 19.1 RELATION BETWEEN $[H^+]$, $[OH^-]$, AND PH IN WATER SOLUTIONS

Solution	1	2	3	4	5	6	7
$[H^+]$	10^{-1}	10^{-3}	10^{-5}	10^{-7}	10^{-9}	10^{-11}	10^{-13}
$[OH^-]$	10^{-13}	10^{-11}	10^{-9}	10^{-7}	10^{-5}	10^{-3}	10^{-1}
pH	1	3	5	7	9	11	13

The three solutions at the right are all basic.

$$[OH^-] > [H^+]$$

$$[OH^-] > 10^{-7}\ M;\ [H^+] < 10^{-7}\ M$$

basic solutions (5, 6, 7)

EXAMPLE 19.1

In a certain water solution, $[OH^-] = 2.0 \times 10^{-7}$ M.

a. What is $[H^+]$?

b. Is this solution acidic, basic, or neutral?

SOLUTION

a. From Equation 19.4:

$$[H^+] = \frac{1.0 \times 10^{-14}}{[OH^-]} = \frac{1.0 \times 10^{-14}}{2.0 \times 10^{-7}} = 0.50 \times 10^{-7} \text{ or } 5.0 \times 10^{-8}\ M$$

Given $[H^+]$ or $[OH^-]$, it is always possible to find the concentration of the other ion by Eqn. 19.4.

b. Clearly, $[OH^-]$ is greater than $[H^+]$. 2.0×10^{-7} is a larger number than 0.50×10^{-7}. Hence the solution must be basic. As in all basic solutions:

$$[OH^-] > 10^{-7}\ M;\ [H^+] < 10^{-7}\ M$$

pH

The acidity or basicity of a water solution can always be described by giving the concentration of H^+. In most cases, this turns out to be an exponential number, such as 1×10^{-5} or 5×10^{-8}. Unfortunately, many people are unfamiliar with the meaning of exponents. If you tell a friend that the H^+ ion concentration in his well water is 5×10^{-8} M, he may not know what you're talking about. (Unless, of course, he's had a course in chemistry!).

There is a way to describe acidity or basicity which does not require the use of exponents. It involves a quantity known as pH. The pH of a solution is defined as:

$$pH = -\log [H^+] \qquad (19.5)$$

Using Equation 19.5, we can assign a pH to any water solution of known $[H^+]$. When this is done, the pH usually (but not always) turns out to be a number between 0 and 14. It might, for example, be 6 or 7.2 or 10.5. As we shall see, these numbers are readily interpreted as a measure of acidity or basicity.

To show how Equation 19.5 is used, consider the first solution listed in Table 19.1. Here:

$$[H^+] = 10^{-1}\ M$$

To obtain the pH of this solution, we must find the logarithm of 10^{-1}. This is readily done if you recall that:

The common logarithm of a number is the power to which 10 must be raised to give that number.

In this case, the power is -1. That is, 10 raised to the -1 power is, by definition, 10^{-1}. So, we have:

$$\log 10^{-1} = -1$$

From Equation 19.5, we see that:

$$pH = -\log [H^+] = -\log 10^{-1} = -(-1) = 1$$

Once you know how to find a logarithm, getting the pH is easy.

In other words, this solution has a pH of 1.

The pH's of all the other solutions in Table 19.1 can be calculated in the same way. We can also use Equation 19.5 to convert a known pH to the corresponding value of $[H^+]$.

EXAMPLE 19.2

Calculate:

a. the pH of pure water, in which $[H^+] = 10^{-7}$ M.

b. the pH of a solution made by adding 0.01 mol of H^+ to one liter of water.

c. $[H^+]$ in a solution of pH 8.

SOLUTION

a. First we obtain the logarithm of the hydrogen ion concentration:

$$\log [H^+] = \log 10^{-7} = -7$$

Then we change the sign to get the pH:

$$pH = -\log [H^+] = -(-7) = 7$$

b. $[H^+] = \dfrac{0.01\ \text{mol}}{1\ \ell} = 0.01\ M = 10^{-2}\ M$

$$\log [H^+] = \log 10^{-2} = -2$$

$$pH = -\log [H^+] = -(-2) = 2$$

c. From the definition: $pH = -\log [H^+]$ we see that:

$$8 = -\log [H^+] \quad \text{or} \quad \log [H^+] = -8$$

Here we need to find the number whose logarithm is −8. That number is 10^{-8}. That is:

$$\log 10^{-8} = -8$$

So: $[H^+] = 10^{-8}$ M

In this example and throughout this text, we will keep conversions between $[H^+]$ and pH simple by dealing only with cases where:

$[H^+]$ is expressed as a simple exponent such as 10^{-2}, 10^{-7}, . . .
pH is expressed as a whole number such as 2, 7, . . .

In the laboratory, things are seldom quite so simple. You may find that the concentration of H^+ in a solution is 2.5×10^{-7} or that its pH is 6.60. Conversions between numbers of these types require that you have either:

—a table of logarithms (and know how to use it)
—a calculator which allows you to take logarithms and antilogarithms.

Let's take a look now at the pH's in Table 19.1, p. 478. Notice that:

Chemists use pH a lot, simply because its easier to say the pH than it is to state $[H^+]$ when $[H^+]$ is low.

1. *Acidic solutions have a pH less than 7.* Here, the pH's are 1, 3, and 5 in the three acidic solutions. In other acidic solutions, the pH might be 2.4, 6.7, 0, or any other number less than 7.

2. *Neutral solutions have a pH of 7.* This is the pH of pure water.

3. *Basic solutions have a pH greater than 7.* The basic solutions listed in the Table have pH's of 9, 11, and 13. Solutions having a pH of 7.2, 9.6, or 15 would also be basic.

We might express these rules in a slightly different way. The more acidic the solution, the lower its pH. A solution of pH 3 has a higher concentration of H^+ ions than one of pH 5, and is more acidic. The more basic the solution, the higher its pH. A solution of pH 9 has a higher $[OH^-]$ than one of pH 7.

EXAMPLE 19.3

Solutions A, B, and C have pH's of 3.0, 12.2, and 1.6 respectively.

a. Classify each solution as acidic, basic, or neutral.

b. In which solution is $[H^+]$ the highest?

c. Which solution contains the highest concentration of OH^-?

SOLUTION

a. A and C are acidic ($pH < 7$). B is basic ($pH > 7$).

b. Solution C (lowest pH)

c. Solution B (highest pH)

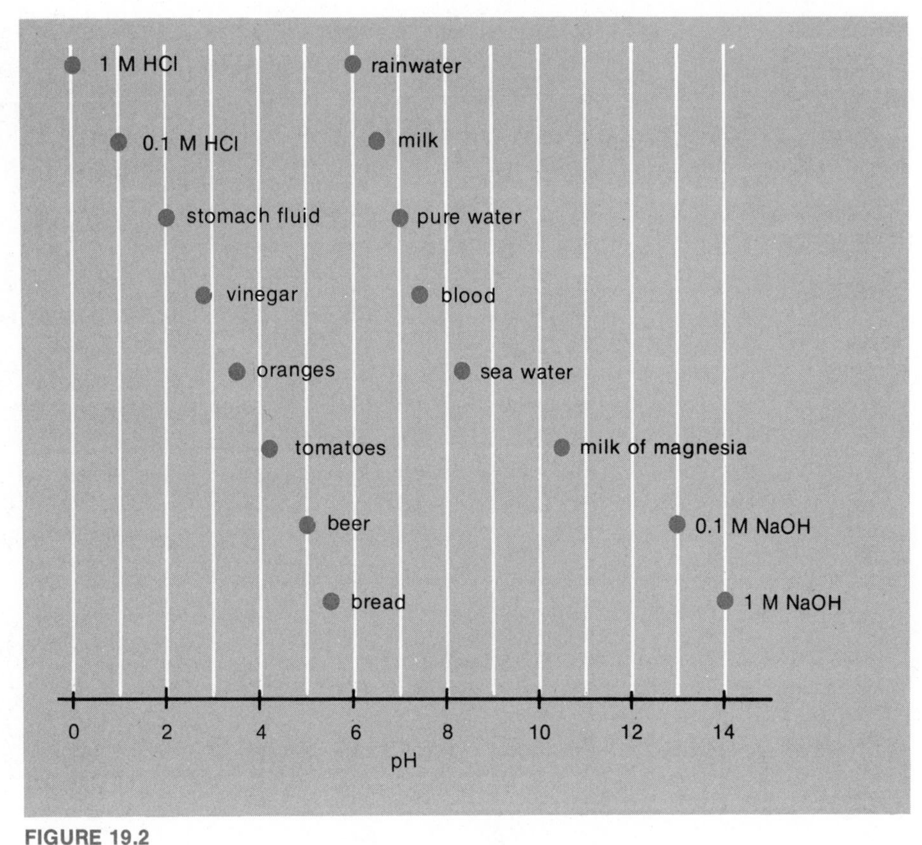

FIGURE 19.2

Most water solutions are not neutral since most solutes have acidic or basic properties.

Figure 19.2 shows the approximate pH of several common materials. Notice that most foods are at least slightly acidic. Many of these, especially fruits and certain vegetables, contain organic acids. Digestive fluids also tend to be acidic. The low pH of stomach fluids is due to the presence of HCl, about 0.01 M. Sometimes the pH in the digestive tract becomes too low, leading to an upset stomach. To relieve this condition, you take a product that is slightly basic. One possibility is sodium bicarbonate, $NaHCO_3$ (pH = 8.5). Another is milk of magnesia, $Mg(OH)_2$ (pH = 10.5).

Too much pizza might cause this problem.

19.2 STRONG AND WEAK ACIDS

For our purposes, an acid can be defined as a species which, upon addition to water, increases the concentration of H^+ ions. Most commonly, an acid is a molecular substance which contains at least one hydrogen atom. We might represent its formula by HX. When added to water, it forms H^+ and X^- ions in solution. We distinguish between strong acids, which are completely ionized in dilute water solution:

$$HX(aq) \rightarrow H^+(aq) + X^-(aq) \tag{19.6}$$

FIGURE 19.3

The conductance of a 1 M solution of a weak acid like HF is much less than that of a 1 M solution of a strong acid like HCl. The difference is due to the fact that HF is only slightly ionized, whereas HCl is completely ionized.

and weak acids, where ionization is incomplete:

$$HX(aq) \rightleftharpoons H^+(aq) + X^-(aq) \qquad (19.7)$$

In a solution of a strong acid, we expect to find no HX molecules. All of them have been converted to H^+ and X^- ions by Reaction 19.6. Putting it another way, the equilibrium constant for Reaction 19.6 is extremely large, approaching infinity. Hence the forward reaction, for all practical purposes, goes to completion.

The situation is quite different with a weak acid. Here, we expect to find an equilibrium mixture in water solution. This mixture contains measurable amounts of HX molecules, H^+ ions, and X^- ions. In one case, we might find that 10% of the weak acid was converted to ions. If this were so, we would say that the weak acid was 10% ionized.

One way to determine whether an acid is strong or weak is to measure the electrical conductivity of its water solution (Figure 19.3). Solutions of strong acids are good conductors because they contain high concentrations of ions. For example, 1 M HCl, which is completely ionized, contains 1 mol/ℓ of H^+ and 1 mol/ℓ of Cl^-. These ions carry the electrical current. The behavior of a weak acid such as HF is quite different. In 1 M HF, only about 3% of the HF molecules are converted to H^+ and F^- ions. Hence the concentrations of H^+ and F^- are only 0.03 M. This explains why 1 M HF is a poor conductor.

In 1 M HF, [HF] = 0.97 M.

Common Strong Acids

There are only a few acids which are completely ionized in water. In fact, there are only six common strong acids. They are:

HCl	hydrochloric acid
HBr	hydrobromic acid
HI	hydriodic acid
HNO_3	nitric acid
$HClO_4$	perchloric acid
H_2SO_4	sulfuric acid

When these acids are added to water, the acid molecule dissociates completely. Two ions are formed. One is a positive ion, H^+. The other is the negative ion derived from the strong acid. Thus we have:

$$HCl(aq) \rightarrow H^+(aq) + Cl^-(aq) \qquad (19.8)$$

In 0.5 M HCl and in 0.5 M HNO_3, $[H^+] = 0.5$ M.

$$HNO_3(aq) \rightarrow H^+(aq) + NO_3^-(aq) \qquad (19.9)$$

EXAMPLE 19.4

Consider the strong acid $HClO_4$.

a. Write a balanced equation for its ionization in water.

b. If 0.1 mol $HClO_4$ is added to one liter of water, what is $[H^+]$ in this solution?

SOLUTION

a. $HClO_4(aq) \rightarrow H^+(aq) + ClO_4^-(aq)$

b. Since perchloric acid is a strong acid, it is completely ionized. 0.1 mol $HClO_4$ forms 0.1 mol H^+. Hence:

$$[H^+] = \frac{0.1 \text{ mol}}{1 \ \ell} = 0.1 \text{ M} = 10^{-1} \text{ M}$$

In the sulfuric acid molecule, there are two hydrogen atoms. The first one ionizes completely.

$$H_2SO_4(aq) \rightarrow H^+(aq) + HSO_4^-(aq) \qquad (19.10)$$

For this reason, we call sulfuric acid a strong acid. The second ionization is incomplete:

$$HSO_4^-(aq) \rightleftharpoons H^+(aq) + SO_4^{2-}(aq) \qquad (19.11)$$

Of the six strong acids, you are likely to find only three in the laboratory. These are HCl, HNO_3, and H_2SO_4. They are most often used as the dilute solutions:

dilute HCl	6 mol/ℓ
dilute HNO_3	6 mol/ℓ
dilute H_2SO_4	3 mol/ℓ

If you buy these acids from a supply house, you get the concentrated solutions.

Concentrated solutions of these acids are also available (12 M HCl, 16 M HNO_3, 18 M H_2SO_4). They are used for special purposes but are generally unpleasant to work with. Concentrated HCl has a pungent odor due to HCl gas over the solution. Concentrated HNO_3 is both colorless and odorless when pure. However, it often contains NO_2 gas, which has a brown color and a choking odor. The concentrated acid burns the skin and causes brown stains on clothing.

Concentrated sulfuric acid is particularly dangerous to work with. Its reaction with water (Equations 19.10 and 19.11) gives off a great deal of heat. Some water may actually be converted to steam by this reaction. If you ever have to mix concentrated sulfuric acid with water, always slowly *add the acid to water,* never the reverse.

Weak Acids

There are thousands of weak acids, far too many to list here. We will consider only a few of the more common weak acids. All of these are molecules which contain at least one ionizable hydrogen atom. In water solution, some of the acid molecules dissociate, forming an H^+ ion and a negative ion. A simple example is hydrogen fluoride.

$$HF(aq) \rightleftharpoons H^+(aq) + F^-(aq) \qquad (19.12)$$

All of the common organic acids are weak. Perhaps the most familiar example is acetic acid, CH_3COOH, found in vinegar. In water, acetic acid undergoes the following reversible reaction:

$$CH_3COOH(aq) \rightleftharpoons H^+(aq) + CH_3COO^-(aq) \qquad (19.13)$$

The products are a proton, H^+, and the acetate ion, CH_3COO^-.

EXAMPLE 19.5

Nitrous acid, HNO_2, is a weak acid.

a. Write a balanced equation for its (partial) ionization in water.

b. When 0.1 mol of HNO_2 is added to one liter of water, 10% of it ionizes. Calculate $[H^+]$ in this solution.

SOLUTION

a. $HNO_2(aq) \rightleftharpoons H^+(aq) + NO_2^-(aq)$

b. Since 10% or 0.10 of the HNO_2 is ionized:

$$[H^+] = 0.10 \times 0.1 \frac{mol}{\ell} = 0.01 \frac{mol}{\ell} = 10^{-2}\ M$$

Notice the contrast with Example 19.4. In 0.1 M $HClO_4$, which is 100% ionized, $[H^+] = 10^{-1}$ M. In 0.1 M HNO_2, which is only 10% ionized, $[H^+] = 10^{-2}$ M.

Several weak acids contain more than one ionizable hydrogen atom. They ionize in successive steps, forming one H^+ at a time. We will consider two such species.

1. *Carbonic Acid,* H_2CO_3. When carbon dioxide gas is dissolved in water, the following reversible reaction occurs:

$$CO_2(g) + H_2O \rightleftharpoons H_2CO_3(aq)$$

The product, H_2CO_3, is a weak acid with two ionizable hydrogen atoms. It ionizes in two steps. The first step produces a proton, H^+, and a *hydrogen carbonate* ion, HCO_3^-.

Sometimes the HCO_3^- ion is called bicarbonate.

$$H_2CO_3(aq) \rightleftharpoons H^+(aq) + HCO_3^-(aq) \qquad (19.14)$$

This reaction explains why distilled water is slightly acidic (pH about 6). In the distillation process, the water becomes saturated with air. Air contains a small amount of CO_2 which reacts with the water to form H_2CO_3.

The second step in the ionization involves the HCO_3^- ion, produced in the first step, as a reactant. It ionizes to form a H^+ ion and the *carbonate* ion, CO_3^{2-}.

$$HCO_3^-(aq) \rightleftharpoons H^+(aq) + CO_3^{2-}(aq) \qquad (19.15)$$

This reaction, like 19.4, is reversible. However, the extent of ionization is much smaller than in the first step. Virtually all the H^+ ions in a solution of CO_2 come from the first ionization of H_2CO_3.

2. *Phosphoric Acid,* H_3PO_4. As its formula implies, phosphoric acid contains three ionizable hydrogen atoms. It ionizes in three steps:

$$H_3PO_4(aq) \rightleftharpoons H^+(aq) + H_2PO_4^-(aq) \qquad (19.16)$$

phosphoric acid — dihydrogen phosphate ion

$$H_2PO_4^-(aq) \rightleftharpoons H^+(aq) + HPO_4^{2-}(aq) \qquad (19.17)$$

monohydrogen phosphate ion

$$HPO_4^{2-}(aq) \rightleftharpoons H^+(aq) + PO_4^{3-}(aq) \qquad (19.18)$$

phosphate ion

Pure phosphoric acid is a low-melting solid (mp = 42°C). It is extremely soluble in water. Commercially, it is available as an 82% water solution. Compounds containing the $H_2PO_4^-$, HPO_4^{2-}, and PO_4^{3-} ions have a variety of uses. Calcium dihydrogen phosphate, $Ca(H_2PO_4)_2$, is the effective component in fertilizers known as "superphosphates". Calcium monohydrogen phosphate, $CaHPO_4$, is used as a polishing agent in toothpaste. Sodium phosphate, Na_3PO_4, is used in water softeners and detergents. This compound also acts as a fertilizer. Indeed, its use has been curtailed because it has been held responsible for promoting the growth of algae in lakes.

Ionization Constants of Weak Acids

As we have seen, a weak acid HX in water solution is in equilibrium with its ions, H^+ and X^-:

$$HX(aq) \rightleftharpoons H^+(aq) + X^-(aq)$$

We can write an equilibrium constant for this reaction, following the rules given in Chapter 18.

In any solution of a weak acid, this condition will be obeyed.

$$K_a = \frac{[H^+] \times [X^-]}{[HX]} \tag{19.19}$$

Here, as always, the square brackets refer to equilibrium concentrations in moles per liter. The products, H^+ and X^-, appear in the numerator. The reactant, HX, is in the denominator. The equilibrium constant for this type of reaction is given a special symbol, K_a. It is called the *ionization constant of the weak acid.*

The ionization constant of a weak acid can be determined in the laboratory. One method involves adding a known amount of the acid to a given volume of water. We then measure the concentration of H^+ ion (or the pH) in the water solution. Using the equation for the ionization of the weak acid, we can then calculate K_a (Example 19.6).

EXAMPLE 19.6

To determine K_a of a certain weak acid, HA, a student adds 0.50 mol of HA to a liter of water. The following reaction occurs:

$$HA(aq) \rightleftharpoons H^+(aq) + A^-(aq)$$

She measures the concentration of H^+ in the solution and finds it to be 0.020 M.

a. Write the expression for K_a.

b. Calculate K_a.

SOLUTION

a. $K_a = \frac{[H^+] \times [A^-]}{[HA]}$

b. For every mole of H^+ formed, one mole of A^- is formed at the same time. Hence:

$$[A^-] = [H^+] = 0.020\ M$$

For every mole of H^+ formed, one mole of HA is consumed. We started with 0.50 mol HA and formed 0.02 mol of H^+, all in one liter of solution. Hence:

$$[HA] = 0.50\ M - 0.02\ M = 0.48\ M$$

Substituting in the expression for K_a:

$$K_a = \frac{(0.020) \times (0.020)}{(0.48)} = \frac{4.0 \times 10^{-4}}{0.48} = 8.3 \times 10^{-4}$$

What would hàppen to K_a if we diluted the solution of HA with water? Answer: Nothing!

Ionization constants are small numbers, ordinarily much less than one. Depending on the weak acid, they may be as large as 10^{-2} or as small as 10^{-12}. For the two weak acids HF and CH_3COOH:

$$HF(aq) \rightleftharpoons H^+(aq) + F^-(aq); \qquad K_a = \frac{[H^+] \times [F^-]}{[HF]} = 7.0 \times 10^{-4}$$

$$CH_3COOH(aq) \rightleftharpoons H^+(aq) + CH_3COO^-(aq); \ K_a = \frac{[H^+] \times [CH_3COO^-]}{[CH_3COOH]} = 1.8 \times 10^{-5}$$

The value of K_a is a measure of the extent to which a weak acid ionizes in water. The smaller K_a is, the fewer H^+ ions there will be

TABLE 19.2 IONIZATION CONSTANTS OF WEAK ACIDS

Acid	Ionization	K_a
Acetic Acid	$CH_3COOH(aq) \rightleftharpoons H^+(aq) + CH_3COO^-(aq)$	1.8×10^{-5}
Carbonic Acid (1)	$H_2CO_3(aq) \rightleftharpoons H^+(aq) + HCO_3^-(aq)$	4.2×10^{-7}
(2)	$HCO_3^-(aq) \rightleftharpoons H^+(aq) + CO_3^{2-}(aq)$	4.8×10^{-11}
Formic Acid	$HCOOH(aq) \rightleftharpoons H^+(aq) + HCOO^-(aq)$	1.8×10^{-4}
Hydrogen Cyanide	$HCN(aq) \rightleftharpoons H^+(aq) + CN^-(aq)$	4.0×10^{-10}
Hydrogen Fluoride	$HF(aq) \rightleftharpoons H^+(aq) + F^-(aq)$	7.0×10^{-4}
Hypochlorous Acid	$HClO(aq) \rightleftharpoons H^+(aq) + ClO^-(aq)$	3.2×10^{-8}
Phosphoric Acid (1)	$H_3PO_4(aq) \rightleftharpoons H^+(aq) + H_2PO_4^-(aq)$	7.5×10^{-3}
(2)	$H_2PO_4^-(aq) \rightleftharpoons H^+(aq) + HPO_4^{2-}(aq)$	6.2×10^{-8}
(3)	$HPO_4^{2-}(aq) \rightleftharpoons H^+(aq) + PO_4^{3-}(aq)$	1.7×10^{-12}

at a given concentration of weak acid. Compare, for example, HF ($K_a = 7.0 \times 10^{-4}$) and CH_3COOH ($K_a = 1.8 \times 10^{-5}$). In a 1 M solution of HF, $[H^+]$ is about 0.03 M. In 1 M acetic acid, $[H^+]$ is considerably smaller, about 0.004 M. We describe this situation by saying that CH_3COOH is a weaker acid than HF. Both of these acids are, of course, weaker than the strong acid HCl. In 1 M HCl, the concentration of H^+ is 1 M.

Given a solution of an acid, how could you tell if the acid was weak or strong?

EXAMPLE 19.7

With the aid of Table 19.2, arrange the three acids

$$HClO_4, HCOOH, HClO$$

in order of decreasing strength.

SOLUTION

$HClO_4$ is a strong acid, completely ionized in water. Both HCOOH and HClO are weak acids, listed in Table 19.2. Of these, HClO is the weaker, since it has the smaller ionization constant (3.2×10^{-8} is smaller than 1.8×10^{-4}).

$$HClO_4 > HCOOH > HClO$$

19.3 STRONG AND WEAK BASES

We can define a base as a species which, upon addition to water, increases the concentration of OH^- ions. As with acids, we distinguish between strong and weak bases. A strong base is completely converted to OH^- ions in water. A weak base is only partially ionized in water to form OH^- ions.

Strong Bases

You will recall that there are only a few strong acids. Similarly, there are few strong bases. The common ones are:

—*the hydroxides of the Group 1 metals:* LiOH, NaOH, KOH, RbOH, CsOH

—*the hydroxides of the heavier Group 2 metals:* $Ca(OH)_2$, $Sr(OH)_2$, $Ba(OH)_2$

These compounds consist of ions when in the solid state. When they dissolve in water, the ions are set free. Thus we have:

In 1 M NaOH, there are no NaOH molecules. $[OH^-] = [Na^+] = 1$ M

$$NaOH(s) \rightarrow Na^+(aq) + OH^-(aq) \quad (19.20)$$

$$Ca(OH)_2(s) \rightarrow Ca^{2+}(aq) + 2\ OH^-(aq) \quad (19.21)$$

These "reactions", for all practical purposes, go 100% to completion.

EXAMPLE 19.8

0.10 mol of barium hydroxide, $Ba(OH)_2$, is added to one liter of water.

a. Write an equation for the solution process in water.

b. What are the concentrations of Ba^{2+} and OH^- in the solution?

SOLUTION

a. $Ba(OH)_2(s) \rightarrow Ba^{2+}(aq) + 2\ OH^-(aq)$

b. Note from the equation that *one* mole of $Ba(OH)_2$ forms *one* mole of Ba^{2+} and *two* moles of OH^-. Hence:

$$[Ba^{2+}] = 0.10\ \frac{mol}{\ell} = 0.10\ M$$

$$[OH^-] = 2 \times 0.10\ \frac{mol}{\ell} = 0.20\ M$$

There is only one strong base which you are likely to use in the laboratory. This is sodium hydroxide, NaOH. It is relatively cheap and very soluble in water (20 M at 25°C). Concentrated solutions of NaOH are very caustic. You should be careful not to get any on your skin. Calcium hydroxide is somewhat cheaper than sodium hydroxide. However, it is much less soluble in water (0.02 M). $Ca(OH)_2$ is often used in industry as a strong base in processes where high solubility is not necessary.

The stock solution is usually 6 M or 1 M NaOH.

Weak Bases

There are a great many species which act as weak bases in water. Of these, we will consider only one. This is the ammonia molecule, NH_3. When ammonia is added to water, the following reaction occurs:

$$NH_3(aq) + H_2O(aq) \rightleftharpoons NH_4^+(aq) + OH^-(aq) \qquad (19.22)$$

Notice that the NH_3 molecule has picked up a proton from an H_2O molecule. In this way, it is converted to an NH_4^+ ion (Fig. 19.4, p. 490). At the same time, an OH^- ion is formed. This makes the solution basic.

It may be helpful to consider the reaction of ammonia with water as occurring in two steps:

(a) the dissociation of a water molecule, H_2O, to H^+ and OH^- ions:

$$H_2O \rightleftharpoons H^+(aq) + OH^-(aq)$$

(b) the reaction of the ammonia molecule, NH_3, with a H^+ ion to form an ammonium ion, NH_4^+:

$$NH_3(aq) + H^+(aq) \rightarrow NH_4^+(aq)$$

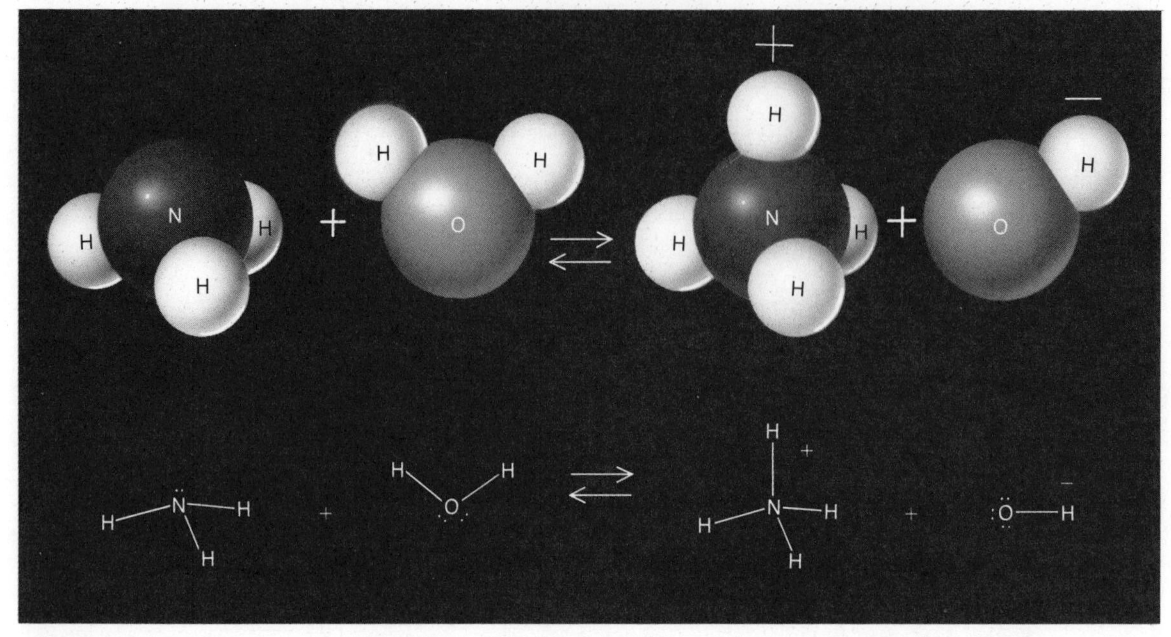

FIGURE 19.4

If NH_3 molecules are dissolved in water, they tend to remove hydrogen ions from H_2O molecules, forming NH_4^+ ions. This leaves OH^- ions in the solution and makes it basic.

Adding these two equations, we obtain the overall equation:

$$
\begin{array}{l}
H_2O \rightleftharpoons H^+(aq) + OH^-(aq) \\
NH_3(aq) + H^+(aq) \rightarrow NH_4^+(aq) \\
\hline
NH_3(aq) + H_2O \rightleftharpoons NH_4^+(aq) + OH^-(aq)
\end{array}
\tag{19.22}
$$

In a sense, the NH_3 molecule competes for a proton with an OH^- ion. In fact, it doesn't compete very well. The equilibrium constant for Reaction 19.22 is small, about 1.8×10^{-5}. Only a small fraction of the NH_3 molecules is converted to NH_4^+ ions. Relatively few OH^- ions are present in solution. If we add one mole of NH_3 to a liter of water, only about 0.004 mol of OH^- is produced. Contrast this to the situation with the strong base NaOH. Adding one mole of NaOH to a liter of water gives one mole of OH^-.

NH_4OH is called ammonium hydroxide. It's just NH_3 in water.

In the laboratory, you will find two different water solutions of ammonia. Both of these, for historical reasons, may be labelled "NH_4OH". The more dilute solution is 6 M in NH_3. "Concentrated NH_3" contains 15 mol/ℓ of NH_3. It, like household ammonia, has a strong odor due to the NH_3 gas present. You should work under the hood and protect your eyes in any experiment involving 15 M NH_3.

Acid-Base Properties of Ions

The most common weak base in the laboratory is ammonia, NH_3. However, many negative ions also act as weak bases. A typical example is the fluoride ion, F^-. It reacts with water in much the same way as the NH_3 molecule. Following the same two-step path, we have:

$$
\begin{array}{l}
H_2O \rightleftharpoons H^+(aq) + OH^-(aq) \\
F^-(aq) + H^+(aq) \rightarrow HF(aq) \\
\hline
F^-(aq) + H_2O \rightleftharpoons HF(aq) + OH^-(aq)
\end{array}
$$

The products are the weak acid molecule, HF, and the OH^- ion. This reaction occurs to only a small extent ($K = 1.4 \times 10^{-11}$). However, enough OH^- ions are formed to make the solution slightly basic. A 0.1 M solution of sodium fluoride, NaF, has a pH of about 8.

All the negative ions derived from weak acids behave like the F^- ion. They undergo a reaction with water to produce OH^- ions. This explains why solutions of sodium acetate (CH_3COO^- ion) and sodium carbonate (CO_3^{2-} ion) are slightly basic. (See Color Plate 8B, in the center of the book.) In contrast, negative ions derived from strong acids (Cl^-, NO_3^-, . . .) do not react with water. To do so, they would have to form molecules such as HCl or HNO_3. As we saw earlier, these molecules are not stable in water. Solutions of NaCl or $NaNO_3$ are neutral (pH 7) rather than basic.

Certain positive ions act as weak acids in water. The simplest example is the NH_4^+ ion. In solution, it ionizes to NH_3 molecules and H^+ ions:

$$NH_4^+(aq) \rightleftharpoons H^+(aq) + NH_3(aq) \tag{19.23}$$

The K_a of the NH_4^+ ion is very small, about 5.6×10^{-10}. Even so, enough H^+ ions are formed to make the solution slightly acidic. A 0.1 M solution of NH_4Cl or NH_4NO_3 has a pH of about 5.

19.4 ACID-BASE REACTIONS

Acid-base reactions are very common, and very important, in chemistry.

When a basic water solution is added to an acidic solution, a reaction occurs. Perhaps the most direct evidence of reaction is the heat flow. When a solution of sodium hydroxide is added to hydrochloric acid, the temperature rises. Heat is given off by the acid-base reaction. There is also a change in pH. As NaOH is added, the H^+ concentration decreases. Hence, the pH increases.

There are three major types of acid-base reactions. All of these have very large equilibrium constants. They go virtually to completion.

1. Strong acid-strong base. Example: HCl–NaOH
2. Strong acid-weak base. Example: HCl–NH_3
3. Weak acid-strong base. Example: HF–NaOH

(The other possible combination, weak acid-weak base, is of little practical importance. It does not go to completion. Instead it produces an equilibrium mixture of products and reactants.)

We will now consider each of these reactions. We will be interested in the nature of the reaction and the equation written for it. Later, in Chapters 21 and 22, we will discuss some of the practical uses that chemists make of acid-base reactions.

Strong Acid-Strong Base

Suppose we add a strong base such as NaOH to a strong acid such as HCl. A simple reaction occurs. The OH^- ions of the basic solution react with the H^+ ions of the acid.

$$H^+(aq) + OH^-(aq) \rightarrow H_2O \tag{19.24}$$

This reaction is called **neutralization.** It is the reverse of Reaction 19.3, the ionization of water into H^+ and OH^- ions. Recalling that water is only ionized to a very small extent ($K = 1 \times 10^{-14}$), you might expect the reverse reaction to go virtually to completion. Indeed it does; K for Reaction 19.24 is 1×10^{14}.

In writing this equation we do not include ions such as Na^+ and Cl^-. Granted, these ions are present in a solution made by neutralizing HCl with NaOH. However, they take no part in the reaction. So, they are not included in the equation. **Equation 19.24 applies to the reaction of any strong acid with any strong base.** The reaction and hence the equation is the same regardless of the acid or base used. It makes no difference whether NaOH reacts with HCl or $Ca(OH)_2$ with HNO_3, or . . .

Strong Acid-Weak Base

Even though NH_3 is a weak base, it takes 1 mole of H^+ ion to neutralize 1 mole of NH_3.

When ammonia is added to a solution of a strong acid, the following reaction occurs:

$$H^+(aq) + NH_3(aq) \rightarrow NH_4^+(aq) \quad (19.25)$$

The H^+ ions from the strong acid react with NH_3 molecules in the ammonia solution. The product is the ammonium ion, NH_4^+. The equilibrium constant for this reaction is very large, about 2×10^9. Hence the reaction goes virtually to completion. If we add an excess of strong acid to a solution of ammonia, nearly all of the NH_3 molecules are converted to NH_4^+ ions.

The reaction with ammonia is the same regardless of what strong acid is used. Addition of hydrochloric acid, HCl, to ammonia gives a solution of ammonium chloride, NH_4Cl. If nitric acid, HNO_3, is used, a solution of NH_4NO_3 is obtained. The negative ions Cl^- or NO_3^- are not included in the equation. They are present as free ions in solution before and after reaction. In the equation, we include only those species (H^+, NH_3, NH_4^+) which take part in the reaction.

Weak Acid-Strong Base

Suppose we add a strong base such as NaOH to a solution of the weak acid, HF. The following reaction occurs:

$$HF(aq) + OH^-(aq) \rightarrow F^-(aq) + H_2O \quad (19.26)$$

The OH^- ion from the strong base reacts with the weak acid molecule, HF. A proton (H^+) is removed from the HF molecule. The products are an H_2O molecule and a fluoride ion, F^- (Figure 19.5).

We might think of Reaction 19.26 as occurring in two steps:

(a) the ionization of an HF molecule:

$$HF(aq) \rightleftharpoons H^+(aq) + F^-(aq)$$

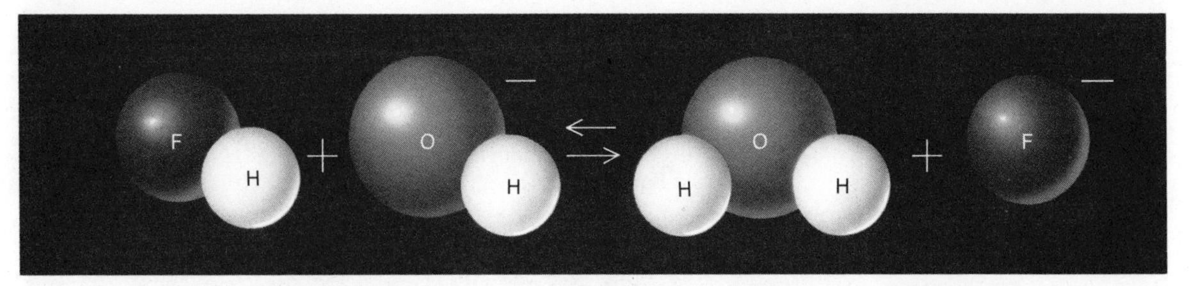

FIGURE 19.5

When solutions containing HF molecules and OH^- ions are mixed, the reaction to form H_2O molecules and F^- ions goes essentially to completion. This is a typical weak acid-strong base reaction.

(b) the neutralization of the H^+ ion formed by the OH^- ion of the strong base:

$$H^+(aq) + OH^-(aq) \rightarrow H_2O$$

The sum of these two equations gives Equation 19.26.

Eqn. 19.25 can also be written as occurring in two steps. Can you write the steps?

$$HF(aq) \rightleftharpoons H^+(aq) + F^-(aq)$$
$$H^+(aq) + OH^-(aq) \rightarrow H_2O$$
$$HF(aq) + OH^-(aq) \rightarrow F^-(aq) + H_2O \quad (19.26)$$

In a sense, the OH^- ion is competing for a proton with the F^- ion. Looking at it this way, the OH^- ion is a winner. The equilibrium constant for Reaction 19.26 is a large number, about 7×10^{10}. The reaction goes virtually to completion.

Other weak acids behave much like HF. If we add a solution of NaOH (or $Ca(OH)_2$, or any strong base) to a solution of acetic acid, CH_3COOH, the reaction is:

$$CH_3COOH(aq) + OH^-(aq) \rightarrow CH_3COO^-(aq) + H_2O \quad (19.27)$$

The general equation for the reaction of any strong base with a weak acid HX is:

$$HX(aq) + OH^-(aq) \rightarrow X^-(aq) + H_2O \quad (19.28)$$

EXAMPLE 19.9

Write an equation for the reaction between:

a. solutions of HNO_3 and $Ca(OH)_2$.

b. solutions of HNO_3 and NH_3.

c. solutions of the weak acid HClO and NaOH.

SOLUTION

a. HNO_3 is a strong acid, $Ca(OH)_2$ a strong base. The reaction is a simple neutralization of the H^+ ion of HNO_3 with the OH^- ion of $Ca(OH)_2$. The equation is identical to 19.24:

$$H^+(aq) + OH^-(aq) \rightarrow H_2O$$

b. The H^+ ions of HNO_3 react with NH_3 molecules (Reaction 19.25).

$$H^+(aq) + NH_3(aq) \rightarrow NH_4^+(aq)$$

c. The OH^- ion of the strong base reacts with the weak acid molecule, HClO. The products are a ClO^- ion and an H_2O molecule. Compare Equation 19.28.

$$HClO(aq) + OH^-(aq) \rightarrow ClO^-(aq) + H_2O$$

Bronsted-Lowry Acids and Bases

In this chapter, we have considered an acid to be a substance which produces H^+ ions when added to water. Similarly, we have taken a base to be a substance which forms OH^- ions in water solution. This is the most common way of defining acids and bases. It is also the most useful for our purposes. It relates directly to the prediction as to whether a given solution will be acidic, neutral, or basic.

This concept of acids and bases was first proposed by the Swedish physical chemist Svante Arrhenius nearly a century ago. Over the years, many other, more general, models of acids and bases have been proposed. One of the most popular was suggested independently by Bronsted (Denmark) and Lowry (England) in 1923. This model concentrates upon the nature of acid-base reactions. It considers that any such reaction involves the transfer of a proton (H^+ ion) from one species to another. The species that gives up ("donates") the proton is called an acid. The species that takes on ("accepts") the proton is called a base.

Perhaps the simplest example of a proton transfer reaction is that involving a strong acid and a strong base:

$$\underset{\text{acid}}{H^+(aq)} + \underset{\text{base}}{OH^-(aq)} \rightarrow H_2O$$

Another simple proton transfer occurs in the reaction of a strong acid with ammonia:

$$\underset{\text{acid}}{H^+(aq)} + \underset{\text{base}}{NH_3(aq)} \rightarrow NH_4^+(aq)$$

In these reactions, the OH^- ion or the NH_3 molecule act as Bronsted-Lowry bases. They accept a proton. In the reaction between the weak acid HF and a strong base:

$$\underset{\text{acid}}{HF(aq)} + \underset{\text{base}}{OH^-(aq)} \rightarrow H_2O + F^-(aq)$$

HF acts as a Bronsted-Lowry acid because it donates a proton (to the OH^- ion). The OH^- ion acts as a base because it accepts a proton (from the HF molecule).

You will recall that the three reactions just considered were discussed earlier in this section. They were described as acid-base reactions. In these cases, the Bronsted-Lowry model leads to the same conclusion as the simple model we have used. However, there are some cases where the Bronsted-Lowry model expands the scope of acid-base reactions. Consider, for example, the reaction of ammonia with water:

$$NH_3(aq) + H_2O \rightarrow NH_4^+(aq) + OH^-(aq)$$

When we discussed this reaction earlier (p. 489), we did not think of it as an acid-base reaction. However, according to the Bronsted-Lowry model, it is. The NH_3 molecule acts as a base, accepting an H^+ ion to form the NH_4^+ ion. At the same time, the H_2O molecule behaves as a Bronsted-Lowry acid. By donating a proton to ammonia, it is converted to the OH^- ion.

At first glance, it may seem strange to think of water as an acid. However, by the Bronsted-Lowry model, it is quite reasonable. Since the H_2O molecule can be converted to the OH^- ion, it can be a proton donor. Indeed, this model goes one step further. It says that in certain reactions H_2O can accept a proton and in that sense act as a base. To see how this is possible, consider what happens when HCl is added to water. We have written the reaction as a simple ionization:

$$HCl(aq) \rightarrow H^+(aq) + Cl^-(aq)$$

According to the Bronsted-Lowry model, this process is a bit more complex. It considers that a proton transfer is involved. An HCl molecule transfers a proton to an H_2O molecule:

$$HCl(aq) + H_2O \rightarrow H_3O^+(aq) + Cl^-(aq)$$

The products are the hydronium ion, H_3O^+, and the Cl^- ion. From this point of view, HCl is acting as an acid while H_2O acts as a base.

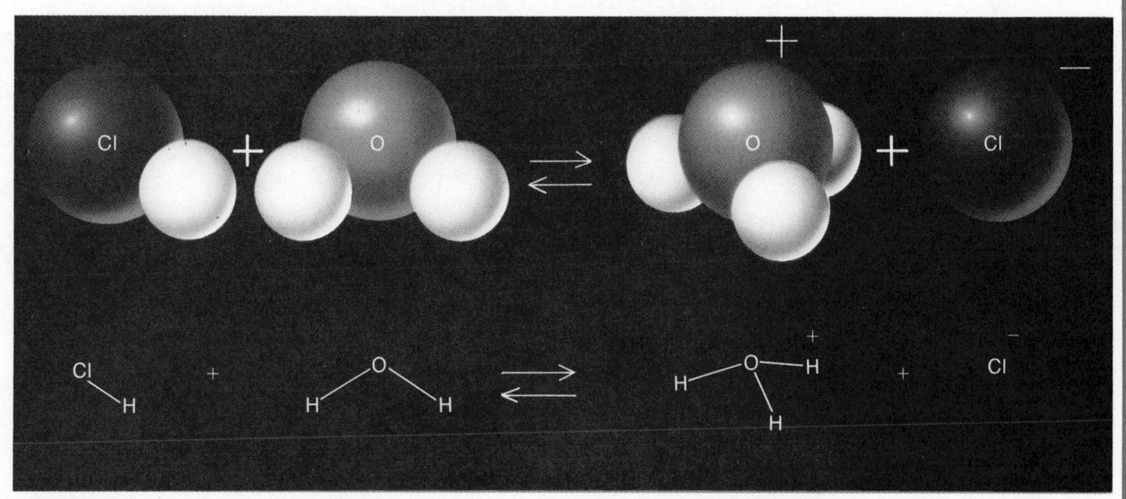

FIGURE 19.6

When HCl molecules are dissolved in water, hydrogen ions are transferred to the water, forming H_3O^+ ions and Cl^- ions. This reaction goes to completion, and explains why HCl "ionizes" in water. In this text we write the formula of the hydrogen ion as H^+, rather than as the hydrate H_3O^+, just to keep things as simple as possible.

19.5 SUMMARY

An acidic water solution is one which contains an excess of H^+ ions. A basic solution contains excess OH^- ions. More exactly:

neutral solution:	$[H^+] = [OH^-]$;	$[H^+] = 10^{-7}$ M;	pH = 7
acidic solution:	$[H^+] > [OH^-]$;	$[H^+] > 10^{-7}$ M;	pH < 7
basic solution:	$[OH^-] > [H^+]$;	$[OH^-] > 10^{-7}$ M;	pH > 7

In all cases: $[H^+] \times [OH^-] = 1.0 \times 10^{-14}$ (at 25°C)

$$pH = -\log [H^+]$$

A strong acid is one which is completely converted to H^+ ions in water. There are six common strong acids: HCl, HBr, HI, HNO_3, $HClO_4$, H_2SO_4. All other acids are weak. In water, they form relatively few H^+ ions, but enough to make $[H^+] > 10^{-7}$ M. A strong base is completely converted to OH^- ions in water. The only common strong bases are the hydroxides of Group 1 and the heavier Group 2 metals (Ca, Sr, Ba). There are many weak bases, of which we considered only one, NH_3. In water, ammonia forms relatively few OH^- ions, but enough to make $[OH^-] > 10^{-7}$ M.

The strength of a weak acid is related to the value of its ionization constant.

$$HX(aq) \rightleftharpoons H^+(aq) + X^-(aq); \qquad K_a = \frac{[H^+] \times [X^-]}{[HX]}$$

The smaller the value of K_a, the weaker the acid. In general, K_a is a small number, considerably less than one.

In this chapter, we have written several equations. They can be classified in five groups.

1. Ionization of a strong acid:

$$HX(aq) \rightarrow H^+(aq) + X^-(aq)$$

2. Ionization of a strong base:

$$MOH(s) \rightarrow M^+(aq) + OH^-(aq) \qquad M = \text{Group 1 metal}$$

$$M(OH)_2(s) \rightarrow M^{2+}(aq) + 2\,OH^-(aq) \qquad M = \text{Group 2 metal}$$

3. Neutralization of a strong acid by a strong base:

$$H^+(aq) + OH^-(aq) \rightarrow H_2O$$

4. Reaction of a weak acid.

   a. ionization: $HX(aq) \rightleftharpoons H^+(aq) + X^-(aq)$

   b. with a strong base: $HX(aq) + OH^-(aq) \rightarrow X^-(aq) + H_2O$

5. Reaction of ammonia, a weak base.

   a. ionization: $NH_3(aq) + H_2O \rightleftharpoons NH_4^+(aq) + OH^-(aq)$

   b. with a strong acid: $NH_3(aq) + H^+(aq) \rightarrow NH_4^+(aq)$

All of these reactions except 4a and 5a have large equilibrium constants and go virtually to completion.

NEW TERMS

acid
acid-base indicator
acid-base reaction
acidic solution
base
basic solution
dihydrogen phosphate ion
ionization constant (K_a)
monohydrogen phosphate ion
neutralization
neutral solution
pH
strong acid
strong base
weak acid
weak base

QUESTIONS

1. A certain water solution turns blue litmus red. Would you expect this solution to *react* with:

   a. Zn? b. H^+ ions? c. OH^- ions? d. Mg^{2+} ions?

2. Suppose the solution in Question 1 turned red litmus blue. Which of the species listed in Question 1 would you expect this solution to react with?

3. Classify each of the following solutions as acidic, basic, or neutral. A solution which:

   a. does not react with any of the species listed in Question 1.
   b. has a sour taste and turns blue litmus red.
   c. has a pH of 7.
   d. has a $[H^+]$ of 1×10^{-5}.
   e. has a slippery feeling and forms a precipitate with Mg^{2+}.

4. Which of the following solutions are basic?

   a. pH = 9
   b. pH = 4.6
   c. $[H^+] = 3 \times 10^{-12}$
   d. $[OH^-] = 2 \times 10^{-2}$

5. Solutions A, B, and C have pH's of 2, 7, and 9 in that order. For each solution, state which is larger, $[H^+]$ or $[OH^-]$.

6. Classify each of the following household solutions as acidic or basic.

   a. vinegar b. lye c. grapefruit juice d. ammonia

7. Criticize each of the following statements concerning a strong acid.

   a. It produces a high concentration of H^+ in water.
   b. It reacts violently with water.

c. It eats holes in clothing.
d. It is completely ionized to form OH^- ions in solution.

8. There are six common strong acids. Without referring back to the text, give their formulas.

9. State whether each of the following acids is strong or weak.

   a. HCl b. HF c. HNO_2 d. HNO_3 e. HCOOH

10. Give the formulas of:

    a. sodium phosphate b. sodium carbonate
    c. sodium monohydrogen phosphate d. sodium hydrogen carbonate
    e. sodium dihydrogen phosphate

11. Write the general expression for the first ionization constant of H_2CO_3.

12. Write the chemical equation for which the expression for K_a is:

$$K_a = \frac{[H^+] \times [HCOO^-]}{[HCOOH]}$$

13. The ionization constant of the strong acid HNO_3 would be:

    a. 0 b. 1 c. a very small number d. a very large number

14. Using Table 19.2, arrange HCl, HCN and CH_3COOH in order of decreasing strength as acids.

15. Give the formulas of the three strongest acids in Table 19.2.

16. Classify each of the following as a strong acid, weak acid, strong base, or weak base.

    a. NaOH b. $HClO_4$ c. HClO d. HNO_2 e. $Ca(OH)_2$ f. NH_3

17. One mole of a certain substance is added to one liter of water. The concentration of OH^- in the solution formed is about 0.004 M. Which one of the following would behave in this way?

    a. NH_3 b. CH_3COOH c. $Ba(OH)_2$ d. HNO_3

18. To obtain a solution with a pH of 0, which of the following substances would you add to water?

    a. HF b. HCl c. KOH d. NH_3
    e. none of these; a pH of 0 is impossible

19. Arrange 1 M water solutions of the following in order of increasing pH.

    a. HBr b. CH_3COOH c. NaOH d. HNO_3 e. KOH f. NH_3
    (some of these solutions may have the same pH).

20. What reagent would you add to a water solution of ammonia to form a water solution of:

    a. NH_4Cl b. $(NH_4)_2SO_4$ c. NH_4NO_3

PROBLEMS

1. *Relation between [H^+] and [OH^-] (Example 19.1)* Calculate [OH^-] in a solution where [H^+] is:

   a. 10^{-6} M b. 2×10^{-4} M c. 4.0×10^{-12} M

2. *Relation between [H^+] and [OH^-] (Example 19.1)* Calculate [H^+] in a solution where [OH^-] is:

   a. 10^{-4} M b. 5×10^{-10} M c. 3.6×10^{-7} M

3. *Relation between [H^+] and pH (Example 19.2)* Determine the pH of solutions in which [H^+] is:

   a. 10^{-8} M b. 10^{-1} M c. 0.001 M

4. *Relation between [H^+] and pH (Example 19.2)* Determine [H^+] in a solution with a pH of:

   a. 5 b. 9 c. −1

5. *Ionization of Strong Acids (Example 19.4)* Write a balanced equation for the ionization of HBr in water. What would [H^+] be in a 0.20 M solution of HBr?

6. *Ionization of Weak Acids (Example 19.5)* Write a balanced equation for the (partial) ionization of HNO_2 in water. In 0.20 M solution, HNO_2 is 5% ionized. What is the concentration of H^+ in this solution?

7. *Ionization Constant (Example 19.6)* When 0.40 mol of a certain weak acid HB is added to a liter of water, 0.040 mol of H^+ is formed. Calculate:

   a. [H^+] b. [B^-] c. [HB], after ionization d. K_a

8. *Strong Bases (Example 19.8)* Write a balanced equation for the dissolving of $Sr(OH)_2$ in water. Calculate [OH^-] in a solution prepared by dissolving 0.20 mol of $Sr(OH)_2$ in 4.0 ℓ of water.

9. *Acid-Base Reactions (Example 19.9)* Write balanced equations for the reactions between solutions of:

   a. NaOH and HCl b. LiOH and HNO_3 c. $Ba(OH)_2$ and HBr

10. *Acid-Base Reactions (Example 19.9)* Write a balanced equation for the reaction of NH_3 with a solution of:

    a. HCl b. HNO_3 c. HI

11. *Acid-Base Reactions (Example 19.9)* Write balanced equations for the reaction of OH^- ions with the weak acids:

    a. CH_3COOH b. HNO_2 c. H_2CO_3 (two equations)

* * * * *

12. What would be [H^+], [OH^-], and the pH of a solution prepared by dissolving 0.01 mol of HNO_3 in a liter of water?

13. What would be [OH^-], [H^+], and the pH of a solution prepared by adding 0.30 mol of KOH to 3.0 ℓ of water?

14. Calculate [OH^-] for each solution in Problem 4.

15. The [OH^-] in a certain solution is 10^{-12} M. What is the pH?

16. What are the pH's of solutions made by adding, to one liter of water:

    a. 0.1 mol HCl b. 0.1 mol NaOH

17. For acetic acid, K_a is 1.8×10^{-5}.

    a. Write the expression for K_a.
    b. Calculate $[H^+]$ when $[CH_3COOH] = 0.10$ M, $[CH_3COO^-] = 1.0$ M.

18. When 0.10 mol of a certain weak acid is added to 500 cm^3 of water, 10% of it ionizes. Calculate $[H^+]$ and K_a for the weak acid.

19. As the temperature is raised, K for Reaction 19.3 increases. At a certain temperature: $[H^+] \times [OH^-] = 1.0 \times 10^{-12}$. What is the pH of pure water at this temperature?

20. Write balanced equations for the reaction of:

    a. NH_3 with H_2O.
    b. NH_3 with a solution of H_2SO_4.
    c. solutions of NaOH and HF.

21. Write balanced equations for the ionization in water of:

    a. HCl b. HCOOH c. RbOH d. $Ca(OH)_2$

22. Using Table 19.2, calculate $[H^+]$ in a solution in which:

    a. $[HCO_3^-] = 0.10$ M, $[H_2CO_3] = 0.20$ M.
    b. $[HCO_3^-] = 1.0 \times 10^{-5}$ M, $[H_2CO_3] = 2.0 \times 10^{-6}$ M.

23. Write balanced equations for the reaction of OH^- ions with:

    a. HClO b. CH_3CH_2COOH c. HCO_3^-

*24. What is the pH of a solution in which:

    a. $[H^+] = 2 \times 10^{-4}$ b. $[OH^-] = 2.5 \times 10^{-6}$

*25. The pH of a solution prepared by dissolving 0.10 mol of HF in one liter of water is 2.1. Calculate $[H^+]$ and K_a of HF.

*26. Write balanced equations to explain why the following ions act as weak bases.

    a. F^- b. CH_3COO^- c. NO_2^-

*27. Classify each species in the following reactions as either a Bronsted acid or base.

    a. $HF(aq) + H_2O \rightleftharpoons F^-(aq) + H_3O^+(aq)$
    b. $F^-(aq) + H_2O \rightleftharpoons HF(aq) + OH^-(aq)$
    c. $HCO_3^-(aq) + HCOOH(aq) \rightleftharpoons H_2CO_3(aq) + HCOO^-(aq)$

Stalagmites and stalactites are formed by the precipitation of $CaCO_3$ from water solution over a period of centuries. *National Park Service.*

20

PRECIPITATION REACTIONS

In the last section of Chapter 19, we discussed one type of reaction in water solution. This was the reaction of an acid with a base. In this chapter, we consider another type of reaction in water solution. Frequently, when water solutions of two ionic compounds are mixed, a solid separates. We refer to this type of reaction as precipitation.

To show how precipitation occurs, consider the following experiment. (See Color Plate 8C, in the center of the book.) We start with water solutions of two ionic compounds. One of these solutions, colored blue, contains copper(II) chloride, $CuCl_2$. The other solution, which is colorless, contains sodium hydroxide, NaOH. When the two solutions are mixed, a light blue solid precipitates from the solution.

We can readily decide what happened in the reaction just described. To start with, we list the ions present in the two solutions.

	positive ion	*negative ion*
solution of $CuCl_2$	Cu^{2+}	Cl^-
solution of NaOH	Na^+	OH^-

The precipitate must have come from the reaction of a positive ion in one solution with the negative ion in the other solution. There are two possibilities:

—$Cu(OH)_2$, formed from the reaction of Cu^{2+} ions with OH^- ions.
—NaCl, formed from the reaction of Na^+ ions with Cl^- ions.

Experience allows us to choose between these. As pointed out above, a blue precipitate forms. Sodium chloride (table salt) is neither blue nor insoluble in water. By elimination, the solid formed must be copper(II) hydroxide, $Cu(OH)_2$. There are several ways to confirm this. We could go to the storeroom and ask to see a sample of $Cu(OH)_2$. Sure enough, we find that $Cu(OH)_2$ is both blue and very insoluble in water.

There is another way we might identify the precipitate in this reaction. Suppose we mix solutions of:

$Cu(NO_3)_2$ and NaOH
$CuSO_4$ and KOH
$CuCl_2$ and $Ca(OH)_2$

If solutions containing the + and − ions present in an insoluble substance are mixed, the substance will precipitate.

In each case we obtain a blue precipitate. Whenever Cu^{2+} and OH^- ions come together in water solution, they react to form insoluble $Cu(OH)_2$.

The equation for this precipitation reaction is easily written. We work backwards from the formula of the precipitate:

$$? + ? \rightarrow Cu(OH)_2(s)$$

Clearly, one mole of Cu^{2+} and two moles of OH^- ions are required to form one mole of $Cu(OH)_2$. So, the balanced equation is:

$$Cu^{2+}(aq) + 2\ OH^-(aq) \rightarrow Cu(OH)_2(s) \qquad (20.1)$$

In this equation, as always, we include only those species which take part in the reaction. Ions such as Cl^- and Na^+ are not included. They are present in solution to begin with and are still there at the end.

We have gone through this precipitation reaction in some detail. In principle, we can always deduce the nature of the reaction by this process. However, there is a simple way to predict in advance the results of precipitation reactions. We will now consider how this can be done.

20.1 SOLUBILITY RULES

Suppose we want to predict whether or not a precipitate will form when two solutions are mixed. To do this, we must know whether the possible products are insoluble in water. It would take a table longer than this chapter to list the water solubilities of all ionic compounds. However, a shorter table will meet our needs. The solubility rules given in Table 20.1 will be sufficient for all the ionic compounds you are likely to work with in the laboratory.

The solubility rules are organized according to the negative ion

TABLE 20.1 SOLUBILITY RULES FOR IONIC COMPOUNDS

NO_3^-	All nitrates are soluble.
Cl^-	The compounds AgCl, $PbCl_2$ and Hg_2Cl_2 are insoluble. All other chlorides are soluble.
SO_4^{2-}	The following sulfates are insoluble: Ag_2SO_4, $PbSO_4$, Hg_2SO_4; $CaSO_4$, $SrSO_4$, $BaSO_4$. All other sulfates are soluble.
CO_3^{2-}	The carbonates of the Group 1 metals (Li_2CO_3, Na_2CO_3, K_2CO_3, . . .) and $(NH_4)_2CO_3$ are soluble. All other carbonates are insoluble.
OH^-	The hydroxides of the Group 1 metals (LiOH, NaOH, KOH, . . .) and the heavier Group 2 metals $Sr(OH)_2$, $Ba(OH)_2$ are soluble. All other hydroxides are insoluble.
S^{2-}	The sulfides of the Group 1 metals, the Group 2 metals, and $(NH_4)_2S$ are soluble. All other sulfides are insoluble.

present. Under "chlorides", we include all common ionic compounds containing the Cl^- ion. All compounds containing the SO_4^{2-} ion are included under sulfates. The entries in the table are made as brief as possible. For example, there are only three common insoluble chlorides. These are listed (AgCl, $PbCl_2$, Hg_2Cl_2). By elimination, all other chlorides are soluble, including NaCl, $CaCl_2$, $ZnCl_2$, and so on.

EXAMPLE 20.1

Using Table 20.1, decide whether the following compounds are soluble or insoluble in water.

a. $Pb(NO_3)_2$ b. $PbCl_2$ c. $PbCO_3$

SOLUTION

a. Since all nitrates are soluble, $Pb(NO_3)_2$ must be soluble.

b. $PbCl_2$ is stated to be insoluble.

c. Since lead is not a Group 1 metal, $PbCO_3$ must be insoluble.

The words "soluble" and "insoluble" are not very precise. Ionic compounds vary in solubility from perhaps 20 M to as low as 10^{-26} M. Somewhere, we have to draw a line. In Table 20.1 the line is drawn at a convenient place.

A compound is "insoluble" if it precipitates when equal volumes of solutions 0.1 M in the corresponding ions are mixed. Otherwise, it is "soluble".

Note, for example, that AgCl is listed as being insoluble. This means that it precipitates when two solutions, one 0.1 M in Ag^+, the other 0.1 M in Cl^- are mixed. On the other hand, $Ba(OH)_2$ is soluble. If we mix a solution 0.1 M in Ba^{2+} with another solution 0.1 M in OH^-, no precipitate forms.

Table 20.1 is easily used to make predictions about precipitation reactions. To illustrate its use, consider the following problem. Suppose we want to know what will happen if 0.1 M solutions of $Cd(NO_3)_2$ and Na_2S are mixed. To decide, we follow a logical, stepwise procedure.

1. *Write down the ions present in the two solutions.*
   $Cd(NO_3)_2$ solution: Cd^{2+}, NO_3^-
   Na_2S solution: Na^+, S^{2-}
2. *Identify the two possible precipitates.*
   Cd^{2+} ions could combine with S^{2-} ions to form CdS(s).
   Na^+ ions could combine with NO_3^- ions to form $NaNO_3$(s).
3. *Use Table 20.1 to decide whether a precipitate actually forms.*
   Since all nitrates are soluble, $NaNO_3$ cannot precipitate.
   CdS will precipitate, since it is insoluble. (Cadmium is not a Group 1 or Group 2 metal!)
4. *Write an equation to describe the precipitation reaction.*
   The product is CdS(s). The equation is:

$$Cd^{2+}(aq) + S^{2-}(aq) \rightarrow CdS(s) \quad (20.2)$$

Color Plate 8D, in the center of the book, tells us that our reasoning is correct. When 0.1 M solutions of $Cd(NO_3)_2$ and Na_2S are mixed, a yellow precipitate of CdS forms.

EXAMPLE 20.2

Equal volumes of 0.1 M solutions of the following compounds are mixed. Predict whether or not a precipitate will form in each case. If it does, write an equation to describe the reaction.

a. $AgNO_3$ and K_2CO_3 b. NaCl and KNO_3

SOLUTION

a. Ions present: Ag^+, NO_3^-; K^+, CO_3^{2-}. Possible precipitates: Ag_2CO_3, KNO_3. From Table 20.1, we deduce that KNO_3 is soluble. On the other hand, Ag_2CO_3 is insoluble. Hence, Ag_2CO_3 must precipitate. The equation is:

$$2\ Ag^+(aq) + CO_3^{2-}(aq) \rightarrow Ag_2CO_3(s)$$

Note that 2 mol of Ag^+ are required to react with 1 mol of CO_3^{2-} tp form 1 mol of Ag_2CO_3.

b. Ions present: Na^+, Cl^-; K^+, NO_3^-. Possible precipitates: $NaNO_3$, KCl. From Table 20.1, we conclude that both of these compounds are soluble. No reaction will occur (Figure 20.1).

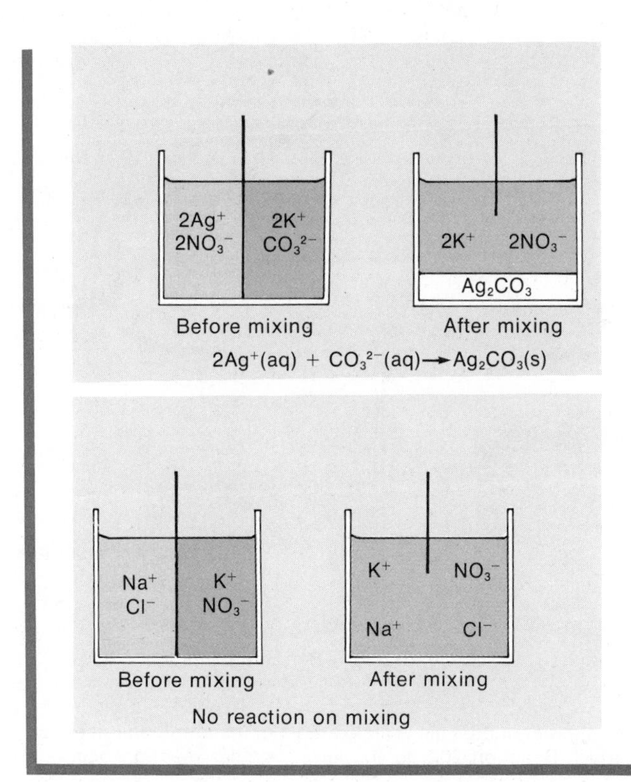

FIGURE 20.1

If a solution of $AgNO_3$ is mixed with a solution of K_2CO_3, a precipitate of Ag_2CO_3 forms because Ag_2CO_3 is insoluble. If solutions of NaCl and KNO_3 are mixed, there is no reaction because none of the possible products are insoluble.

20.2 SOLUBILITY EQUILIBRIUM

When enough ionic solid dissolves in water, it reaches equilibrium with its ions in solution. If the compound is very soluble, as with NaCl, the equilibrium concentration of ions is high:

$$NaCl(s) \rightleftharpoons Na^+(aq) + Cl^-(aq)$$

The solubility of sodium chloride is 6.2 mol/ℓ at 25°C. In the saturated solution, the concentrations of Na^+ and Cl^- are 6.2 M. The situation is quite different with AgCl. Silver chloride has a very low solubility, about 1.3×10^{-5} mol/ℓ at 25°C.

$$AgCl(s) \rightleftharpoons Ag^+(aq) + Cl^-(aq)$$

Here, at equilibrium, the concentrations of Ag^+ and Cl^- are only 1.3×10^{-5} M.

Solubility equilibria were referred to briefly in Chapter 13. There, we considered only the effect of temperature. Here, we will discuss concentration effects. We want to know what happens to the equilibrium system when the concentration of one of the ions in solution is changed.

Common Ion Effect

Let us suppose that we form a saturated water solution of NaCl at 25°C. We do this by shaking an excess of the solid with water until no

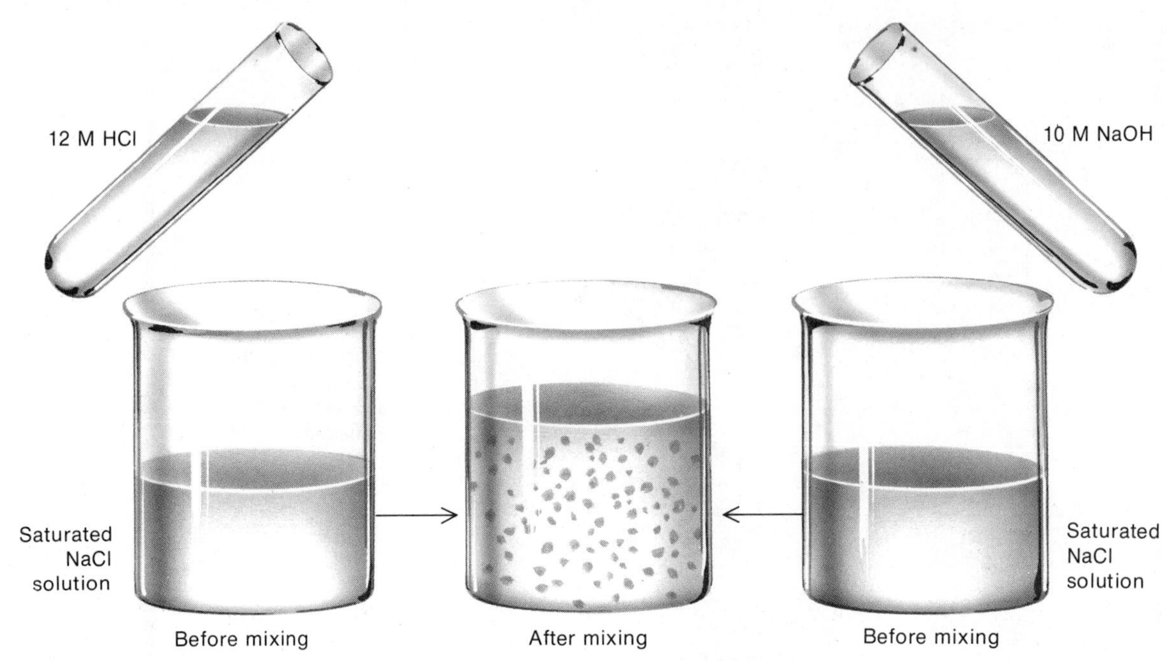

FIGURE 20.2

If 12 M HCl is added to a saturated solution of NaCl, NaCl will precipitate. This is an example of the common ion effect. Adding a common ion, Cl^-, to the saturated solution drives Reaction 20.3 to the left. A similar reaction occurs if 10 M NaOH is added; in this case Na^+ is the common ion.

more dissolves. The equilibrium:

$$NaCl(s) \rightleftharpoons Na^+(aq) + Cl^-(aq) \qquad (20.3)$$

Raising the concentration of a product tends to drive a reaction to the left.

is established. As pointed out above, the concentrations of $Na^+(aq)$ and $Cl^-(aq)$ are 6.2 M. Now suppose we add concentrated HCl to this solution. This reagent contains H^+ and Cl^- ions, both at a concentration of 12 M. In effect, we have increased the Cl^- ion concentration in solution. This disturbs the equilibrium described by Equation 20.3. Following Le Chatelier's principle, the reverse reaction occurs. The added Cl^- ions react with Na^+ ions in the solution. Some solid sodium chloride precipitates (Figure 20.2).

At the right of Figure 20.2, we show another way of precipitating NaCl from its saturated solution. This time, we add a concentrated solution of NaOH, perhaps 10 M. This again disturbs the equilibrium. The concentration of Na^+ ions is now higher than it was to begin with. Some of these Na^+ ions react with Cl^- ions in the solution to form solid NaCl.

The experiments just described illustrate the *common ion* effect. The "common ion" is one that was already present in the saturated solution. Here, the common ion is Cl^- in the first case, Na^+ in the second. By increasing the concentration of either of these ions, we disturb the solubility equilibrium. As a result, a solid precipitates from the solution. The equilibrium shifts in such a way as to relieve the disturbing force.

EXAMPLE 20.3

Consider $Ca(OH)_2$, which has a rather low water solubility. A small amount of $Ca(OH)_2$ is shaken with water to form a saturated solution. Name at least two reagents which you could add to precipitate calcium hydroxide from this solution.

SOLUTION

The equilibrium involved is:

$$Ca(OH)_2(s) \rightleftharpoons Ca^{2+}(aq) + 2\ OH^-(aq)$$

One way to precipitate $Ca(OH)_2$ would be to increase the concentration of Ca^{2+} ions. This could be done by adding a solution of calcium chloride, $CaCl_2$. Any other solution containing a high concentration of Ca^{2+} ions would work as well. For example, calcium nitrate solution, $Ca(NO_3)_2$, could be used. A different approach would be to add OH^- ions. This could be done with a solution of sodium hydroxide, NaOH, barium hydroxide, $Ba(OH)_2$, . . .

The common ion effect is often stated in a somewhat different way. We say that the solubility of an ionic compound is less in a solution containing a common ion than it is in pure water. For example:

NaCl is less soluble in a solution of NaOH or HCl than it is in pure water.

$Ca(OH)_2$ is less soluble in a solution of $CaCl_2$ or NaOH than it is in water.

AgCl is less soluble in a solution of $AgNO_3$ or HCl than it is in water.

Solubility Product Constant, AgCl

We have seen how the position of a solubility equilibrium can be shifted by adding a common ion. We have said nothing, however, about the extent to which this shift occurs. That is, we have not made any calculations of equilibrium concentrations of ions in solution. This can be done by using a type of equilibrium constant known as the solubility product constant. It can be applied to all very slightly soluble ionic compounds. Typical of these is silver chloride, AgCl.

When solid silver chloride is shaken with water, the following equilibrium is established.

$$AgCl(s) \rightleftharpoons Ag^+(aq) + Cl^-(aq) \qquad (20.4)$$

We can write an equilibrium constant for Reaction 20.4. Recall from Chapter 18 that terms for solids do not appear in K. So, we have:

$$K = [Ag^+] \times [Cl^-]$$

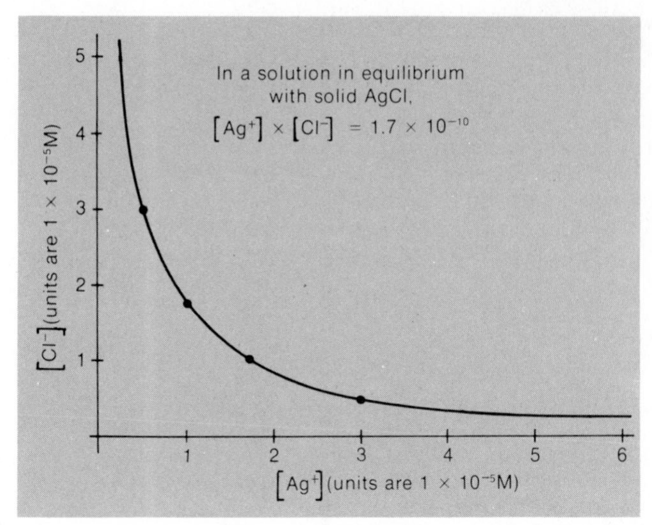

FIGURE 20.3

Although the concentrations of Ag^+ or Cl^- ions can be changed in a solution in equilibrium with AgCl(s), the product of the concentrations cannot be changed. It will remain constant and equal to the solubility product, K_{sp}, for AgCl, which is 1.7×10^{-10}.

This type of equilibrium constant is given a special name. It is called the *solubility product constant.* It is also given a special symbol, $\mathbf{K_{sp}}$. For AgCl, as for all very slightly soluble compounds, K_{sp} is a very small number.

$$K_{sp} \text{ of AgCl} = [Ag^+] \times [Cl^-] = 1.7 \times 10^{-10} \text{ (at 25°C)} \qquad (20.5)$$

Equation 20.5 is valid for all water solutions in equilibrium with solid silver chloride. It tells us that, in such solutions, the product of the concentrations of Ag^+ and Cl^- ions is constant. This means that if we add Ag^+ ions, thereby increasing $[Ag^+]$, the concentration of Cl^- must decrease. Again, if $[Cl^-]$ increases, $[Ag^+]$ must decrease.

This relation is plotted in Figure 20.3. Note that Figure 20.3 looks very much like Figure 19.1, p. 476, where we plotted $[H^+]$ *versus* $[OH^-]$. In both cases, the concentrations of the two ions are inversely proportional to each other.

We can use Equation 20.5 to calculate the concentration of one of these ions, knowing that of the other. Suppose, for example, we have a 0.10 M solution of $AgNO_3$ which is saturated with AgCl. There will be a few Cl^- ions in that solution. To calculate the concentration of Cl^-, we start by solving Equation 20.5 for $[Cl^-]$.

$$[Ag^+] \times [Cl^-] = 1.7 \times 10^{-10}$$

Dividing both sides of this equation by $[Ag^+]$, we have:

If you added AgCl to 1 liter 0.1 M $AgNO_3$, how many moles of AgCl would dissolve? Ans.: 1.7×10^{-9}

$$[Cl^-] = \frac{1.7 \times 10^{-10}}{[Ag^+]}$$

Substituting

$$[Ag^+] = 0.10 \text{ M} = 1.0 \times 10^{-1} \text{ M}:$$

$$[Cl^-] = \frac{1.7 \times 10^{-10}}{1.0 \times 10^{-1}} = 1.7 \times 10^{-9} \text{ M}$$

Notice that this is a very low concentration of Cl^-:

$$1.7 \times 10^{-9} = 0.000\ 000\ 001\ 7$$

In this case, the concentration of Ag^+ is rather high, 1.0×10^{-1} M. In Example 20.4, we consider the reverse situation: $[Cl^-]$ is considerably higher than $[Ag^+]$.

EXAMPLE 20.4

In seawater, the concentration of Cl^- is about 0.53 M. Assuming that seawater is saturated with AgCl, calculate $[Ag^+]$.

SOLUTION

Here, we solve Equation 20.5 for $[Ag^+]$.

$$[Ag^+] \times [Cl^-] = 1.7 \times 10^{-10}$$

$$[Ag^+] = \frac{1.7 \times 10^{-10}}{[Cl^-]}$$

Substituting

$[Cl^-] = 0.53$ M:

$$[Ag^+] = \frac{1.7 \times 10^{-10}}{0.53} = 3.2 \times 10^{-10} \text{ M}$$

Could you have $[Cl^-]$ and $[Ag^+]$ both equal to 0.53 M in the same solution?

This is indeed a very low concentration of Ag^+ ion. It would probably not be worth your while to try to extract silver from seawater.

You will recall from Chapter 19 that in pure water, $[H^+] = [OH^-]$. A similar relation exists between $[Ag^+]$ and $[Cl^-]$ when AgCl is dissolved in pure water.

$$AgCl(s) \rightarrow Ag^+(aq) + Cl^-(aq)$$

Every Ag^+ ion entering the solution must be balanced by a Cl^- ion. Hence, the concentrations of the two ions must be equal. We can use Equation 20.5 to calculate these concentrations (Example 20.5).

EXAMPLE 20.5

Calculate $[Ag^+]$ and $[Cl^-]$ in a solution prepared by saturating pure water with AgCl.

SOLUTION

As pointed out above, the two concentrations must be equal under these conditions. That is:

$$[Ag^+] = [Cl^-]$$

But, from Equation 20.5: $[Ag^+] \times [Cl^-] = 1.7 \times 10^{-10}$

Substituting for $[Cl^-]$: $[Ag^+] \times [Ag^+] = [Ag^+]^2 = 1.7 \times 10^{-10}$

To solve for $[Ag^+]$, we take the square root of both sides of this equation:

$$[Ag^+] = (1.7 \times 10^{-10})^{1/2} = (1.7)^{1/2} \times (10^{-10})^{1/2}$$
$$= (1.7)^{1/2} \times 10^{-5}$$

The square root of 1.7 is about 1.3. Hence:

$$[Ag^+] = 1.3 \times 10^{-5}\ M$$

This is also the concentration of Cl^-, since the two quantities are equal.

In the solution $[Ag^+]$ and $[Cl^-]$ would equal 1.3×10^{-5} M. By Ex. 20.5, the solution is saturated.

The concentration calculated in Example 20.5, 1.3×10^{-5} M, is an important one. It is the solubility, in moles per liter, of AgCl in pure water. In other words, if we shake AgCl with water, 1.3×10^{-5} mol of the solid dissolves per liter. Notice that this quantity is the square root of the solubility product, K_{sp}. This relation holds for any solid in which the two ions are present in a 1:1 ratio (1 Ag^+ for 1 Cl^-).

$$\text{solubility (mol}/\ell) = (K_{sp})^{1/2} \qquad (20.6)$$

General Expression for K_{sp}

We can write an expression for the solubility product constant of any slightly soluble ionic compound. The form of K_{sp} will differ, depending upon the type of compound. Its numerical value depends upon the nature of the compound and the temperature. Values of K_{sp} are ordinarily given at 25°C.

The K_{sp} expression is readily written, using the solubility equation. It is always a product of two concentration terms, one for each ion. Each concentration is raised to a power equal to its coefficient in the equation. Thus we have:

$$CaCO_3(s) \rightleftharpoons Ca^{2+}(aq) + CO_3^{2-}(aq)$$

$$K_{sp} \text{ of } CaCO_3 = [Ca^{2+}] \times [CO_3^{2-}] = 5 \times 10^{-9} \qquad (20.7)$$

$$Mg(OH)_2(s) \rightleftharpoons Mg^{2+}(aq) + 2\ OH^-(aq)$$

$$K_{sp} \text{ of } Mg(OH)_2 = [Mg^{2+}] \times [OH^-]^2 = 1 \times 10^{-11} \qquad (20.8)$$

Notice that:

a. The expression for K_{sp} of $CaCO_3$ is similar to that for AgCl. The concentrations of the two ions in equilibrium with the solid are inversely proportional to one another. A plot of $[Ca^{2+}]$ *versus* $[CO_3^{2-}]$ would look very much like Figure 20.3. The only difference is in the magnitude of K_{sp}: 5×10^{-9} for $CaCO_3$ *versus* 1.7×10^{-10} for AgCl.

b. The expression for K_{sp} of $Mg(OH)_2$ is a bit more complex than that for AgCl or $CaCO_3$. Here, the concentrations of the two ions are not inversely *proportional* to each other. However, they are inversely related. As $[Mg^{2+}]$ increases, $[OH^-]$ decreases, and vice versa.

EXAMPLE 20.6

Write expressions for K_{sp} for:

a. Ag_2CrO_4 b. $Al(OH)_3$

SOLUTION

a. The solubility equation is $Ag_2CrO_4(s) \rightleftharpoons 2\ Ag^+(aq) + CrO_4^{2-}(aq)$

so $K_{sp} = [Ag^+]^2 \times [CrO_4^{2-}]$

b. $Al(OH)_3(s) \rightleftharpoons Al^{3+}(aq) + 3\ OH^-(aq)$

$K_{sp} = [Al^{3+}] \times [OH^-]^3$

Why doesn't the concentration of the solid appear in the expression for K_{sp}?

The solubility product constant, K_{sp}, can always be used to calculate the concentration of one ion, knowing that of the other. In our calculations, we will consider only compounds such as AgCl or $CaCO_3$, where the ions are present in a 1:1 ratio. Here, the arithmetic is rather simple. Two such calculations were carried out for AgCl in Examples 20.4 and 20.5. In Example 20.7, we show a similar calculation for calcium carbonate.

EXAMPLE 20.7

K_{sp} of $CaCO_3$ is 5×10^{-9}. Calculate:

a. $[Ca^{2+}]$ in a solution 0.1 M in CO_3^{2-} in equilibrium with $CaCO_3$.

b. the solubility, in moles per liter, of $CaCO_3$ in water.

SOLUTION

a. Since: $[Ca^{2+}] \times [CO_3^{2-}] = 5 \times 10^{-9}$

$$[Ca^{2+}] = \frac{5 \times 10^{-9}}{[CO_3^{2-}]}$$

Substituting $[CO_3^{2-}] = 0.1\ M = 1 \times 10^{-1}\ M$

$$[Ca^{2+}] = \frac{5 \times 10^{-9}}{1 \times 10^{-1}} = 5 \times 10^{-8}\ M$$

b. Here, as with AgCl, Equation 20.6 applies:

$$\text{solubility (mol}/\ell) = K_{sp}^{1/2} = (5 \times 10^{-9})^{1/2}$$

To take the square root of 5×10^{-9}, it is convenient to make the exponent an even number. To do this, we multiply the coefficient (5) by 10 and divide the exponential term (10^{-9}) by 10:

$$5 \times 10^{-9} = 50 \times 10^{-10}$$

Now we have:

$$\text{solubility (mol}/\ell) = (50 \times 10^{-10})^{1/2} = (50)^{1/2} \times (10^{-10})^{1/2}$$

The square root of 10^{-10} is 10^{-5}. That of 50 is about 7 (1 significant figure). So:

$$\text{solubility (mol}/\ell) = 7 \times 10^{-5}\ \text{M}$$

Note that the water solubility of $CaCO_3$ is somewhat larger than that of AgCl (7×10^{-5} M *vs.* 1.3×10^{-5} M). This follows, since K_{sp} is larger for $CaCO_3$ ($5 \times 10^{-9} > 1.7 \times 10^{-10}$).

TABLE 20.2 SOLUBILITY PRODUCT CONSTANTS

Compound	Formula	K_{sp}
Barium carbonate	$BaCO_3$	2×10^{-9}
Barium sulfate	$BaSO_4$	1×10^{-10}
Cadmium sulfide	CdS	1×10^{-27}
Calcium carbonate	$CaCO_3$	5×10^{-9}
Calcium sulfate	$CaSO_4$	3×10^{-5}
Cobalt(II) sulfide	CoS	1×10^{-21}
Copper(II) sulfide	CuS	1×10^{-36}
Iron(II) sulfide	FeS	2×10^{-18}
Lead carbonate	$PbCO_3$	1×10^{-13}
Lead chromate	$PbCrO_4$	2×10^{-14}
Lead sulfate	$PbSO_4$	1×10^{-8}
Lead sulfide	PbS	1×10^{-28}
Magnesium carbonate	$MgCO_3$	2×10^{-8}
Manganese(II) sulfide	MnS	1×10^{-13}
Mercury(II) sulfide	HgS	1×10^{-52}
Nickel sulfide	NiS	1×10^{-21}
Silver bromide	AgBr	5×10^{-13}
Silver chloride	AgCl	1.7×10^{-10}
Silver iodide	AgI	1×10^{-16}
Zinc sulfide	ZnS	1×10^{-23}

Other Applications of K_{sp}

The solubility product calculations we have carried out are limited to solids where the ions are present in a 1:1 ratio. Such solids include all those listed in Table 20.2. With these compounds, we have shown how to obtain:

—the concentration of one ion, knowing that of the other (Examples 20.4, 20.7).

—the water solubility in moles per liter (Examples 20.5, 20.7).

Values of K_{sp} can be used for many purposes other than those we have described. For example, they can be used to determine whether or not a precipitate will form when two solutions are mixed. You will recall that we can use the solubility rules for this purpose (Table 20.1). However, for that approach to be valid, the ions must be at concentrations close to 0.1 M. With K_{sp}, we can make a more exact prediction at any concentration. The rule is a simple one:

For a precipitate to form when two ions are mixed, the concentration product must exceed K_{sp}. If it is less than K_{sp}, no precipitate will form.

To show how this rule is applied, consider $CaCO_3$:

$$K_{sp} = 5 \times 10^{-9} = [Ca^{2+}] \times [CO_3^{2-}]$$

Suppose now that we add CO_3^{2-} ions to a solution 1×10^{-4} M in Ca^{2+}. If we add enough CO_3^{2-} to make its concentration 1×10^{-5} M:

$$(\text{conc. } Ca^{2+}) \times (\text{conc. } CO_3^{2-}) = (1 \times 10^{-4}) \times (1 \times 10^{-5}) = 1 \times 10^{-9}$$

Since the concentration product (1×10^{-9}) is less than K_{sp} (5×10^{-9}), no precipitate forms. However, if we continue to add CO_3^{2-} until its concentration becomes 1×10^{-4} M:

$$(\text{conc. } Ca^{2+}) \times (\text{conc. } CO_3^{2-}) = (1 \times 10^{-4}) \times (1 \times 10^{-4}) = 1 \times 10^{-8}$$

Now the concentration product (1×10^{-8}) exceeds K_{sp} (5×10^{-9}, or 0.5×10^{-8}). We deduce that a precipitate should form. It does. Moreover, it continues to form, using up Ca^{2+} and CO_3^{2-} ions, until the concentration product drops to the equilibrium value given by K_{sp}.

20.3 DISSOLVING PRECIPITATES

Consider an ionic compound which is "insoluble" in water. That is, its water solubility is less than about 0.1 M. No matter how much water we add, we cannot bring very much of it into solution. Yet in many cases we need to get at least one of the ions in the precipitate into solution. This is particularly important in qualitative analysis (Chapter 22). Here, we frequently separate an ion from a mixture by precipitating it. Once the ion has been separated, we need to test for its presence. To do this, it is usually necessary to get it back into solution.

To dissolve a solid which is insoluble in water, we must carry out a chemical reaction. We add a reagent which reacts with one or the other of the ions in the precipitate. In certain cases, it is the negative ion of the solid that reacts. In others, the reagent attacks the positive ion. We will consider both types of reactions in this section.

Strong Acids (H^+)

When solid magnesium hydroxide is shaken with water, equilibrium is established:

$$Mg(OH)_2(s) \rightleftharpoons Mg^{2+}(aq) + 2\ OH^-(aq)$$

The concentrations of Mg^{2+} and OH^- in solution are very low. Magnesium hydroxide is quite insoluble in water (1.3×10^{-4} M).

We now ask an important question. How can we shift the position of this equilibrium to bring more $Mg(OH)_2$ into solution? One way to do this would be to lower the concentration of $OH^-(aq)$. According to Le Chatelier's principle, this should cause the forward reaction to occur. How might we lower the concentration of OH^-? Recall from Chapter 19 that the neutralization reaction

$$H^+(aq) + OH^-(aq) \rightarrow H_2O$$

goes virtually to completion. We could accomplish our purpose by adding a strong acid. The H^+ ions of the acid will react with OH^- ions, lowering their concentration. This will shift the equilibrium, bringing more $Mg(OH)_2$ into solution. If we add enough strong acid, we should be able to dissolve all the solid.

We can force many reactions to occur by disturbing the equilibrium system in such a way as to lower the concentration of a product.

Experiment confirms the prediction just made (Figure 20.4). By adding an excess of a strong acid such as HCl, we can dissolve $Mg(OH)_2$. The solution process involves a chemical reaction. The H^+ ions of the acid react with the OH^- ions of the solid. We might consider the reaction as taking place in two steps:

$$Mg(OH)_2(s) \rightleftharpoons Mg^{2+}(aq) + 2\ OH^-(aq)$$
$$2\ H^+(aq) + 2\ OH^-(aq) \rightarrow 2\ H_2O$$
$$\overline{Mg(OH)_2(s) + 2\ H^+(aq) \rightarrow Mg^{2+}(aq) + 2\ H_2O} \quad (20.9)$$

In a sense, the H^+ and Mg^{2+} ions compete for OH^- ions. The H^+ ion wins.

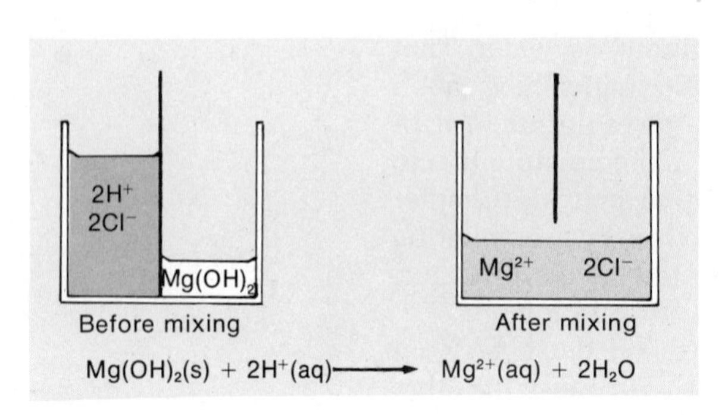

FIGURE 20.4

$Mg(OH)_2$ will dissolve in 1M HCl very easily because $[OH^-]$ ions in an acidic solution is very low. $Mg(OH)_2$ will tend to dissolve until $[Mg^{2+}] \times [OH^-]^2$ equals K_{sp} for $Mg(OH)_2$; in an acidic solution this equality is never achieved, so all the $Mg(OH)_2$ dissolves.

The equilibrium constant for Reaction 20.9 is very large, about 1×10^{17}. This means that the reaction goes virtually to completion.

This procedure can be used with any water-insoluble hydroxide. Addition of excess strong acid brings it into solution. For example, aluminum hydroxide is dissolved by an acid such as HCl or HNO_3. The reaction is similar to 20.9.

$$Al(OH)_3(s) \rightleftharpoons Al^{3+}(aq) + 3\ OH^-(aq)$$

$$3\ H^+(aq) + 3\ OH^-(aq) \rightarrow 3\ H_2O$$

$$Al(OH)_3(s) + 3\ H^+(aq) \rightarrow Al^{3+}(aq) + 3\ H_2O \qquad (20.10)$$

Strong acids can also be used to dissolve metal carbonates. Consider, for example, calcium carbonate. When this compound is added to water, an equilibrium is set up:

$$CaCO_3(s) \rightleftharpoons Ca^{2+}(aq) + CO_3^{2-}(aq)$$

As we saw earlier (Example 20.7), very little calcium carbonate dissolves in water. Its solubility is only about 7×10^{-5} M. However, if we add a strong acid, the situation changes. The H^+ ions react with the CO_3^{2-} ions. They form the weak acid H_2CO_3, which then decomposes to CO_2 and H_2O:

$$2\ H^+(aq) + CO_3^{2-}(aq) \rightarrow [H_2CO_3(aq)] \rightarrow CO_2(g) + H_2O \qquad (20.11)$$

This reaction lowers the concentration of CO_3^{2-} ions in solution. Following Le Chatelier's principle, more $CaCO_3$ dissolves in an attempt to re-establish equilibrium. If an excess of strong acid is used, all the solid dissolves. The overall equation is the sum of the two just written.

$$CaCO_3(s) \rightleftharpoons Ca^{2+}(aq) + CO_3^{2-}(aq)$$

$$2\ H^+(aq) + CO_3^{2-}(aq) \rightarrow CO_2(g) + H_2O$$

$$CaCO_3(s) + 2\ H^+(aq) \rightarrow Ca^{2+}(aq) + CO_2(g) + H_2O \qquad (20.12)$$

Here again, there is a competition, this time for the CO_3^{2-} ion. The two contestants are the Ca^{2+} ion and the H^+ ion. The proton wins; K for Reaction 20.12 is 2×10^8.

H^+ ion competes very effectively for OH^-, CO_3^{2-}, and most negative ions obtained from weak acids.

All metal carbonates, like all hydroxides, are soluble in strong acid. The equations are similar to those written above (Example 20.8).

EXAMPLE 20.8

Write balanced equations to explain why the following solids, which are insoluble in water, dissolve in strong acid:

a. $Fe(OH)_3$ b. $ZnCO_3$

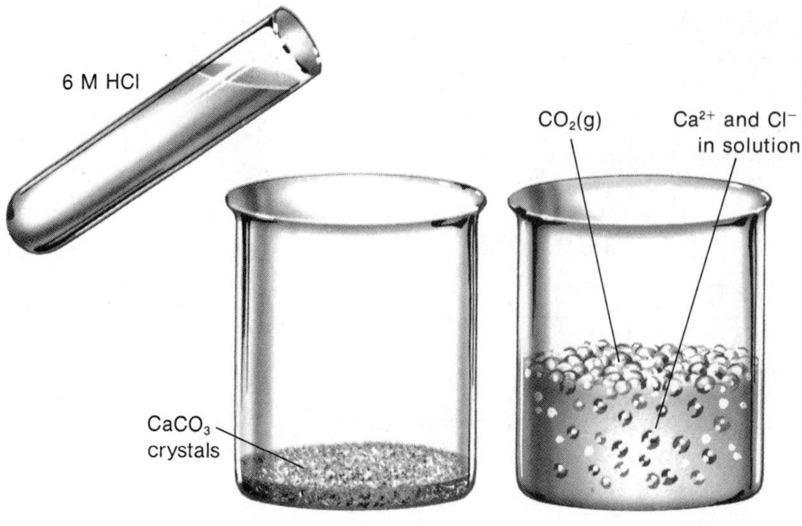

FIGURE 20.5

Calcium carbonate dissolves in solutions of strong acids, because in such solutions $[CO_3^{2-}]$ will be very small due to reaction of CO_3^{2-} with H^+ ions. In excess acid it is not possible for $[Ca^{2+}] \times [CO_3^{2-}]$ to become as large as K_{sp} for $CaCO_3$, and all the solid dissolves.

SOLUTION

a. Here, the H^+ ions of the acid react with OH^- ions to form water. The overall reaction is:

$$Fe(OH)_3(s) + 3\ H^+(aq) \rightarrow Fe^{3+}(aq) + 3\ H_2O$$

(Compare Equation 20.10.)

b. The H^+ ions react with CO_3^{2-} ions to form $CO_2(g)$ and H_2O:

$$ZnCO_3(s) + 2\ H^+(aq) \rightarrow Zn^{2+}(aq) + CO_2(g) + H_2O$$

(Compare Equation 20.12.)

Looking at the equations we have written, it is clear that chemical reactions are involved in bringing hydroxides and carbonates into solution. The solid "dissolves" in strong acid in the sense that:

1. The positive ion (Mg^{2+}, Ca^{2+}, . . .) goes into solution.
2. The negative ion (OH^- or CO_3^{2-}) is destroyed. The OH^- ion is converted to an H_2O molecule. The CO_3^{2-} ion is converted to a CO_2 molecule.

Complexing Agents (NH_3, OH^-)

We have just considered one way in which a water-insoluble ionic compound can be brought into solution. It involves the reaction of the negative ion of the solid (OH^-, CO_3^{2-}, . . .) with H^+. We can often bring about solution by adding a reagent which instead reacts with the *positive* ion of the compound. The species used is called a *complexing agent.* It reacts with the positive ion of the solid to form a *complex ion.*

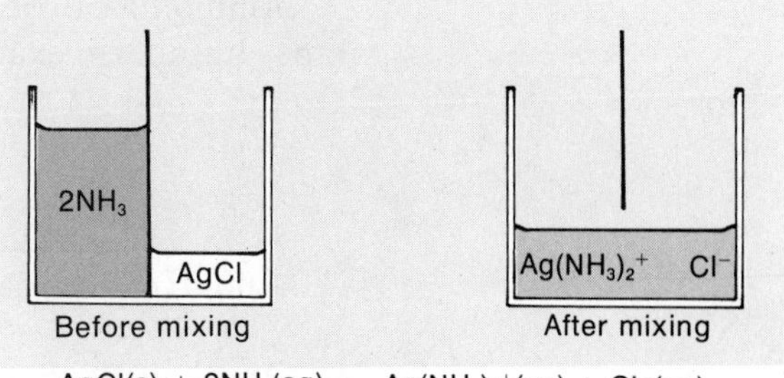

FIGURE 20.6

When AgCl is treated with 6M NH_3, the concentration of Ag^+ ion in the solution becomes very small, due to formation of $Ag(NH_3)_2^+$. AgCl will dissolve until $[Ag^+] \times [Cl^-]$ becomes equal to K_{sp} for AgCl. If $[Ag^+]$ is kept sufficiently low by the formation of $Ag(NH_3)_2^+$, all the AgCl will dissolve.

To illustrate this process, consider silver chloride:

$$AgCl(s) \rightleftharpoons Ag^+(aq) + Cl^-(aq)$$

This compound has a low water solubility, about 1.3×10^{-5} M. However, if we add 6 M NH_3 to silver chloride, it dissolves (Figure 20.6). The ammonia molecules react with Ag^+ ions to form a complex ion. This complex ion has the formula $Ag(NH_3)_2^+$. The NH_3 molecules in the complex ion are bonded to Ag^+ through the nitrogen atom. The structural formula is

$$\left[\begin{array}{ccccc} H & & & & H \\ & \diagdown & & \diagup & \\ H-&N-Ag-N&&-H \\ & \diagup & & \diagdown & \\ H & & & & H \end{array}\right]^+$$

With excess ammonia, virtually all of the Ag^+ ions in solution can be converted to $Ag(NH_3)_2^+$. The concentration of free Ag^+ ions in equilibrium with AgCl(s) drops sharply. More AgCl dissolves in an attempt to restore equilibrium.

We can consider that the reaction of AgCl with NH_3 occurs in two steps:

Even though $[Ag^+]$ is small in equilibrium with AgCl, it is much smaller in equilibrium with $Ag(NH_3)_2^+$.

$$AgCl(s) \rightleftharpoons Ag^+(aq) + Cl^-(aq)$$

$$Ag^+(aq) + 2\ NH_3(aq) \rightarrow Ag(NH_3)_2^+(aq)$$

The overall equation is obtained by adding these two equations:

$$AgCl(s) + 2\ NH_3(aq) \rightarrow Ag(NH_3)_2^+(aq) + Cl^-(aq) \qquad (20.13)$$

Looking at the two-step process, we see that NH_3 molecules and Cl^- ions compete for Ag^+ ions. Ammonia forms a complex with Ag^+; Cl^- ions form a precipitate. The complex $Ag(NH_3)_2^+(aq)$ is stable enough to bring the precipitate, AgCl, into solution.

Ammonia forms stable complex ions with many transition metal ions. With Cu^{2+}, it forms the $Cu(NH_3)_4^{2+}$ ion. Here, there are four ammonia molecules bonded through nitrogen to the central Cu^{2+} ion.

$$\left[\begin{array}{ccc} NH_3 & & NH_3 \\ & \diagdown \ \diagup & \\ & Cu & \\ & \diagup \ \diagdown & \\ NH_3 & & NH_3 \end{array}\right]^{2+}$$

Ammonia can be used to bring water-insoluble compounds of Cu^{2+} into solution. An example is $Cu(OH)_2$.

$$Cu(OH)_2(s) \rightleftharpoons Cu^{2+}(aq) + 2\ OH^-(aq)$$

$$Cu^{2+}(aq) + 4\ NH_3(aq) \rightarrow Cu(NH_3)_4^{2+}(aq)$$

$$Cu(OH)_2(s) + 4\ NH_3(aq) \rightarrow Cu(NH_3)_4^{2+}(aq) + 2\ OH^-(aq) \qquad (20.14)$$

Another effective complexing agent is the hydroxide ion, OH^-. This ion, like NH_3, forms stable complex ions with many transition metal ions. In addition, the OH^- ion forms a complex ion with Al^{3+}. This complex ion has the formula $Al(OH)_4^-$. Its stability explains why aluminum hydroxide dissolves in excess 6 M NaOH solution.

$$Al(OH)_3(s) + OH^-(aq) \rightarrow Al(OH)_4^-(aq) \qquad (20.15)$$

In Table 20.3, we list some of the water-insoluble compounds that dissolve in 6 M NH_3 or NaOH. In each case, solution comes about because of complex ion formation. In the column at the right the formulas of the complex ions are listed.

EXAMPLE 20.9

Using Table 20.3 write balanced equations to explain why:

a. $Zn(OH)_2$ dissolves in concentrated NaOH solution.

b. $Zn(OH)_2$ dissolves in NH_3.

TABLE 20.3 WATER-INSOLUBLE COMPOUNDS DISSOLVED BY NH_3 OR OH^-

	NH_3	
Positive Ion	Compounds	Complex Ion Formed
Ag^+	AgCl, AgBr	$Ag(NH_3)_2^+$
Cd^{2+}	$Cd(OH)_2$, $CdCO_3$	$Cd(NH_3)_4^{2+}$
Cu^{2+}	$Cu(OH)_2$, $CuCO_3$	$Cu(NH_3)_4^{2+}$
Ni^{2+}	$Ni(OH)_2$, $NiCO_3$	$Ni(NH_3)_6^{2+}$
Zn^{2+}	$Zn(OH)_2$, $ZnCO_3$	$Zn(NH_3)_4^{2+}$
	OH^-	
Positive Ion	Compounds	Complex Ion Formed
Al^{3+}	$Al(OH)_3$	$Al(OH)_4^-$
Cr^{3+}	$Cr(OH)_3$, $Cr_2(CO_3)_3$	$Cr(OH)_6^{3-}$
Zn^{2+}	$Zn(OH)_2$, $ZnCO_3$	$Zn(OH)_4^{2-}$

SOLUTION

a. From Table 20.3, we see that the product of the reaction is the complex ion $Zn(OH)_4^{2-}$. The reactants are solid $Zn(OH)_2$ and OH^- ions in aqueous solution. The balanced equation is:

$$Zn(OH)_2(s) + 2\ OH^-(aq) \rightarrow Zn(OH)_4^{2-}(aq)$$

(Compare Equation 20.15.)

b. The complex ion formed is $Zn(NH_3)_4^{2+}$. Clearly, 4 mol of NH_3 are required to react with 1 mol of solid $Zn(OH)_2$. Two moles of OH^- enter the water solution.

$$Zn(OH)_2(s) + 4\ NH_3(aq) \rightarrow Zn(NH_3)_4^{2+}(aq) + 2\ OH^-(aq)$$

(Compare Equation 20.14)

20.4 SUMMARY

The solubility rules (Table 20.1) allow us to predict whether a precipitate will form when two solutions are mixed. For example, the Table tells us that aluminum hydroxide is insoluble in water. This means that when solutions of $Al(NO_3)_3$ and NaOH are mixed, the following reaction occurs:

$$Al^{3+}(aq) + 3\ OH^-(aq) \rightarrow Al(OH)_3(s)$$

The equilibrium between a slightly soluble compound and its ions is described by the solubility product constant, K_{sp}. For AgCl, we have

$$AgCl(s) \rightleftharpoons Ag^+(aq) + Cl^-(aq); \qquad K_{sp} = [Ag^+] \times [Cl^-] = 1.7 \times 10^{-10}$$

Knowing K_{sp}, we can calculate the concentration of one ion, given that of the other. If in a solution in equilibrium with AgCl(s), $[Cl^-] = 1.0 \times 10^{-2}$ M, then:

$$[Ag^+] = \frac{1.7 \times 10^{-10}}{1.0 \times 10^{-2}} = 1.7 \times 10^{-8}\ M$$

If AgCl is shaken with pure water, $[Ag^+] = [Cl^-]$ and

$$[Ag^+]^2 = 1.7 \times 10^{-10};\ [Ag^+] = 1.3 \times 10^{-5}\ M$$

This means that the solubility of AgCl in water is 1.3×10^{-5} M.

To dissolve a precipitate, we carry out a reaction with either the positive or the negative ion of the solid. In the case of $Al(OH)_3$, we could add a strong acid, H^+. This reacts with the OH^- ions of the solid:

$$Al(OH)_3(s) + 3\ H^+(aq) \rightarrow Al^{3+}(aq) + 3\ H_2O$$

Another possibility is to add a strong base, OH^-. The OH^- ion reacts with Al^{3+} to form a complex ion:

$$Al(OH)_3(s) + OH^-(aq) \rightarrow Al(OH)_4^-(aq)$$

In both cases, the solid, $Al(OH)_3$, is brought into solution.

NEW TERMS

common ion effect
complexing agent
complex ion
precipitation
solubility product constant (K_{sp})
solubility rule

QUESTIONS

1. When solutions of $NiCl_2$ (green) and KOH (colorless) are mixed, a green precipitate forms. Give the formulas of the two possible precipitates. Which one do you think it is? Why?

2. What ions are present in solutions of:

   a. KBr b. $Pb(NO_3)_2$ c. $(NH_4)_2CO_3$ d. $CuCl_2$ e. $Al_2(SO_4)_3$

3. Referring to Table 20.1, state which of the following are soluble in water.

   a. $PbCl_2$ b. $NaNO_3$ c. $CuCO_3$ d. CaS

4. Using Table 20.1, give the formulas of all the soluble sulfides.

5. Using Table 20.1, give the formulas of five insoluble hydroxides.

6. Using Table 20.1, complete the following table on a separate sheet of paper. Enter "S" for soluble or "I" for insoluble.

	NO_3^-	Cl^-	SO_4^{2-}	CO_3^{2-}	OH^-	S^{2-}
Pb^{2+}	S	I	–	–	–	–
Ca^{2+}	S	–	–	–	–	–
Ag^+	S	–	–	–	–	–

7. You are given a saturated water solution of $CaSO_4$. Name at least two reagents which would precipitate $CaSO_4$ from this solution by the common ion effect.

8. Explain, in your own words, why silver iodide is less soluble in silver nitrate solution than it is in water.

9. For $Cu(OH)_2$, K_{sp} is given by the expression:

   a. $[Cu^{2+}] \times [OH^-]$ b. $[Cu^{2+}] \times [OH^-]^2$ c. $[Cu^{2+}]^2 \times [OH^-]$

   d. $\dfrac{[Cu^{2+}] \times [OH^-]^2}{[Cu(OH)_2]}$

10. The solubility product of AgCl is 1.7×10^{-10} at 25°C. At higher temperatures, AgCl becomes more soluble in water. At 100°C, a reasonable value for K_{sp} might be:

    a. 1.7×10^{-12} b. 1.7×10^{-10} c. 1.7×10^{-8} d. zero

11. In a certain solution, $[Ag^+]$ in equilibrium with AgCl is twice what it is when AgCl is shaken with water. How does $[Cl^-]$ in this solution compare to that in water saturated with AgCl?

    a. half as large b. the same c. twice as large

12. What property do the equations:

$$PV = \text{constant}; \quad K = [H^+] \times [OH^-]; \quad K_{sp} = [Ag^+] \times [Cl^-]$$

    have in common?

13. When $CaCO_3$ is dissolved in water, $[Ca^{2+}] = [CO_3^{2-}]$. Why must this be true?

14. Each of the following compounds is dissolved in water. In which cases are the concentrations of two ions in solution equal to each other?

    a. $CaSO_4$ b. K_2SO_4 c. $Ba(OH)_2$ d. AgI

15. Looking at Table 20.2:

    a. Which one of the compounds listed has the highest water solubility?
    b. Which has the lowest water solubility?

16. Which of the following compounds would you expect to dissolve in strong acid but not in water?

    a. AgCl b. $BaCO_3$ c. NaCl d. $Mg(OH)_2$

    Use Table 20.1 if necessary.

17. When a strong acid is added to $CuCO_3$, two different ions "compete" for CO_3^{2-}. What are these ions?

18. When ammonia is added to $Cu(OH)_2(s)$, what ion is being competed for? What species are competing for it?

19. Consider the compound $Zn(OH)_2$. Would you expect it to dissolve in:

    a. 6 M HCl b. 6 M NaOH c. 6 M NH_3

    Use Table 20.3 where needed.

20. What is meant by a complex ion? a complexing agent? Give examples of both.

PROBLEMS

1. *Solubility Rules (Example 20.2)* For each of the following pairs of solutions, state what ions are present and give the formulas of the two possible precipitates.

    a. $AgNO_3$ and Na_2S b. $CuSO_4$ and $BaCl_2$ c. $NiCl_2$ and $Ca(OH)_2$
    d. $Cu(NO_3)_2$ and $ZnCl_2$

2. *Solubility Rules (Example 20.2)* For each pair of solutions in Problem 1, identify the precipitate that forms, if any, and write a balanced equation for its formation.

3. *Solubility Product Constant (Examples 20.4, 20.7)* Given that K_{sp} for AgBr is 1×10^{-13}, calculate:

    a. $[Ag^+]$ when $[Br^-] = 5 \times 10^{-4}$ M b. $[Br^-]$ when $[Ag^+] = 0.10$ M

4. *Solubility Product Constant (Examples 20.5, 20.7)* Using Table 20.2, calculate the water solubility, in moles per liter, of:

   a. AgI b. MnS

5. *Solubility Product Constant (Example 20.6)* Write expressions for K_{sp} for:

   a. $PbCl_2$ b. $Cr(OH)_3$ c. As_2S_3 d. $BaCrO_4$

6. *Dissolving Precipitates (Example 20.8)* Write balanced equations for the reactions by which the following hydroxides dissolve in $H^+(aq)$.

   a. $Zn(OH)_2$ b. AgOH c. $Al(OH)_3$

7. *Dissolving Precipitates (Example 20.8)* Write a balanced equation for the reaction of $BaCO_3(s)$ with strong acid.

8. *Dissolving Precipitates (Example 20.9)* Write balanced equations to explain why the following compounds dissolve in NH_3 (see Table 20.3):

   a. $Cd(OH)_2$ b. $CdCO_3$ c. AgBr

9. *Dissolving Precipitates (Example 20.9)* Write a balanced equation to explain why $Cr(OH)_3$ dissolves in excess OH^-.

* * * * *

10. In a solution in equilibrium with $BaSO_4$, $[Ba^{2+}] = 5.0 \times 10^{-4}$, $[SO_4^{2-}] = 3.0 \times 10^{-6}$. Calculate K_{sp} for $BaSO_4$.

11. Write balanced equations for the precipitation reactions that occur when the following 0.1 M solutions are mixed. If no reaction occurs, write "NR."

    a. $BaCl_2$ and K_2SO_4 b. $FeCl_2$ and NaOH c. $Cr(NO_3)_3$ and KOH
    d. $AgNO_3$ and $SrCl_2$ e. $CaCl_2$ and $Zn(NO_3)_2$

12. Which of the compounds precipitated in Problem 11 would dissolve in strong acid? Write equations for the reactions of these compounds with H^+.

13. Which of the compounds precipitated in Problem 11 would dissolve in NH_3 (see Table 20.3)? Write equations for the reactions of these compounds with NH_3.

14. Which of the compounds precipitated in Problem 11 would dissolve in OH^-? Write equations for the reactions of these compounds with OH^-.

15. In a certain solution saturated with $Mg(OH)_2$, $[Mg^{2+}] = 1 \times 10^{-3}$, $[OH^-] = 1 \times 10^{-4}$ M. Calculate K_{sp} of $Mg(OH)_2$.

16. What is the formula of iron(III) sulfide? Write an expression for its solubility product constant?

17. Using Table 20.2, calculate the solubility in water (mol/ℓ) of:

    a. CdS b. ZnS c. $CaSO_4$

18. Using Table 20.2, calculate $[Pb^{2+}]$ in a solution saturated with PbS in which $[S^{2-}] = 1 \times 10^{-4}$ M.

19. Using Table 20.2, calculate $[CrO_4^{2-}]$ in a solution saturated with $PbCrO_4$ in which $[Pb^{2+}] = 5 \times 10^{-7}$ M.

20. K_{sp} for HgS is 1×10^{-52}. Calculate the water solubility of HgS in:

    a. moles per liter b. grams per liter

21. Write balanced equations for the reaction with strong acid of:

    a. $CoCO_3$ b. $Al_2(CO_3)_3$

22. The concentration of Mg^{2+} in seawater is 0.0520 mol/ℓ. How many liters of seawater must be processed to obtain:

    a. 1.00 mol of Mg.
    b. 1.00 kg of Mg.

*23. Show that, for $PbCl_2$, the relation between solubility, S, (mol/ℓ) and K_{sp} is: $4\ S^3 = K_{sp}$. Hint: start by writing down the equation for the process by which $PbCl_2$ dissolves in water and the expression for its K_{sp}.

*24. Will a precipitate of $CaSO_4$ form when conc. $Ca^{2+} = 2 \times 10^{-3}$ M, conc. $SO_4^{2-} = 1 \times 10^{-2}$ M?

*25. The solubility product constant of AgBr is 5×10^{-13}. How many grams of AgBr will dissolve in 100 cm^3 of water?

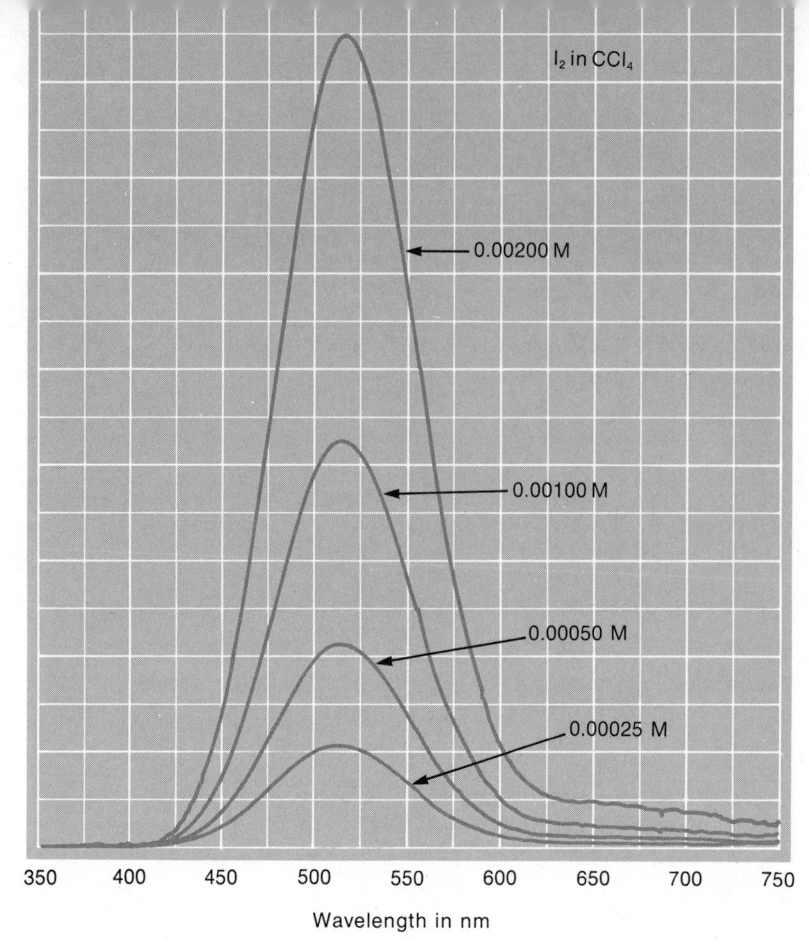

A curve showing the absorbance of I_2 dissolved in CCl_4 as a function of wavelength. The height of the curve at its maximum is directly proportional to the concentration of I_2.

21 QUANTITATIVE ANALYSIS

An important task of the chemist is to determine the chemical composition of mixtures. To do this, he or she uses the methods of chemical analysis. Analytical chemistry is divided into:

1. *Qualitative Analysis* (Chapter 22), where the substances present are identified. For example, we might use qualitative analysis to show that a U.S. quarter coin contains the two metals copper and nickel.
2. *Quantitative Analysis,* where the amounts of the substances present are determined. By quantitative analysis, we can show that a quarter contains 75% Cu and 25% Ni.

Today, chemists carry out a wide variety of quantitative analyses. Large cities employ chemists to analyze drinking water for such ele-

Reaction: $Ag^+(aq) + Cl^-(aq) \rightarrow AgCl(s)$

Procedure: Mix 0.1 M $AgNO_3$ with 0.1 M HCl

100 ml 0.100 M $AgNO_3$ containing 0.0100 mole Ag^+, 1.08 g Ag^+	+	100 ml 0.100 M HCl containing 0.0100 mole Cl^-, 0.36 g Cl^-

→ 0.0100 mole AgCl, 1.44 g AgCl

The Ag^+ ion in a solution can be precipitated quantitatively as AgCl by addition of an equal number of moles of Cl^- ion

FIGURE 21.1

A reaction that goes essentially to completion. If a solution containing 1.08g Ag^+ is mixed with one containing 0.36g Cl^-, a precipitate of AgCl weighing 1.44g is obtained. (There will be a tiny amount of Ag^+ and Cl^- remaining in solution, since the system must be in equilibrium, but the amounts are negligible as compared to the amount of precipitate.)

ments as calcium or chlorine. The Environmental Protection Agency (EPA) is interested in analyzing polluted air for such compounds as sulfur dioxide and carbon monoxide. An analyst at a drug company may be asked to determine the percentage of streptomycin in an antibiotic.

In this chapter, we will describe some of the methods used in quantitative analysis. First (Sections 21.1, 21.2) we will deal with analyses based on chemical reactions. Many of these reactions are ones considered in the last few chapters. They include acid-base reactions (Chapter 19) and precipitation reactions (Chapter 20). These reactions, you will recall, go to completion. That is, one or both of the reacting species is completely converted to products. This is essential in quantitative analysis. The "product", whatever it is, must be formed in 100% yield (Fig. 21.1).

K for these reactions is very large, so essentially all of at least one reactant is used up.

There is one drawback to using chemical reactions in analysis. The substance being analyzed is converted to another substance. In that sense, it is destroyed. For that reason among others, chemists often prefer another type of analysis. Here, the physical properties of a species are used. Analyses of this type are usually carried out with an instrument which measures a particular property. One such instrument will be described in Section 21.3.

21.1 GRAVIMETRIC ANALYSIS

One method available to the analytical chemist is gravimetric analysis. This method is based on the masses of reactants and products. Most analyses, at one point or another, require weighing a sample. In gravimetric analysis, the composition of the sample is determined solely from mass measurements. The general procedure is straightforward. To illustrate the principle involved, we will describe a few typical analyses of this type.

Determination of Silver in a Coin

Let us suppose that you are given a "silver" coin and asked to determine the mass percent of silver. After weighing the coin, you would need to separate out the silver and determine its mass. There are many different ways to do this. We will describe one of the simpler procedures in some detail. It is typical of the approach used in gravimetric analysis.

Step 1. Weigh the coin on an analytical balance.

The nitric acid should be present in excess.

Step 2. Dissolve the coin in 10 cm^3 dilute nitric acid, 6M HNO_3, in a small beaker, heating gently with a Bunsen burner. If necessary, add more nitric acid to complete the reaction. Any silver in the coin will go into solution as Ag^+ ion, any copper as Cu^{2+}. Any residue will not contain silver and can be discarded.

So should the hydrochloric acid.

Step 3. To the solution, add 10 cm^3 dilute hydrochloric acid, 6M HCl, and stir well. A white precipitate of AgCl will form, removing the Ag^+ ion from solution. To make sure all the Ag^+ has precipitated, let the AgCl settle and add a drop of 6M HCl. Stir and check again, adding HCl until no further precipitation occurs. Add 20 cm^3 distilled water.

Essentially, no silver chloride will dissolve.

Step 4. Pour the beaker contents into a funnel in which there is a weighed filter paper and filter out the AgCl. Make sure all the solid is transferred to the paper. Use distilled water to aid with the transfer. Wash the solid well with distilled water. In this way the AgCl can be removed from the other ions formed when the coin dissolved and from the HNO_3 and the HCl that were used.

This will remove all the water from the sample.

Step 5. Carefully remove the filter paper from the funnel and put it on a watch glass. Put the watch glass and its contents in an oven for one hour at 110°C to dry the AgCl and the paper.

Step 6. Take the watch glass and its contents out of the oven and let them cool to room temperature. Weigh the filter paper and the AgCl on an analytical balance.

Often the product is a compound containing the species to be determined.

Step 7. Calculate the mass of AgCl formed in the reaction. From its formula, find the amount of silver in the AgCl. This is the same as the amount in the coin. From the mass of silver and the mass of the coin, calculate the percentage of silver in the coin.

EXAMPLE 21.1

A 1929 U.S. dime was found by a student to weigh 2.8357 g. She dissolved the dime in dilute HNO_3 and then added dilute HCl in excess. She filtered the AgCl on a filter paper weighing 0.7942 g. She washed the solid well with water, then dried the paper and AgCl in an oven. She found the mass of the AgCl plus paper to be 4.1860 g. What percent of the coin was silver?

SOLUTION

Since all the silver in the coin is precipitated as AgCl by the reaction:

$$Ag^+(aq) + Cl^-(aq) \rightarrow AgCl(s)$$

we need to calculate the amount of silver in the recovered silver chloride. The formula mass of AgCl is 143.32 (AM Ag = 107.87, AM Cl = 35.45). One mole of AgCl weighs 143.32 g and contains 107.87 g Ag. In the sample of AgCl:

$$\text{mass AgCl} = (\text{mass of AgCl} + \text{paper}) - \text{mass of paper}$$

$$= 4.1860 \text{ g} - 0.7942 \text{ g} = 3.3918 \text{ g}$$

$$\text{mass of Ag} = \text{mass of AgCl} \times \frac{\text{mass of Ag in one mole AgCl}}{\text{mass of one mole of AgCl}}$$

$$= 3.3918 \text{ g} \times \frac{107.87 \text{ g}}{143.32 \text{ g}} = 2.5528 \text{ g}$$

$$\%\ \text{silver} = \frac{\text{mass of Ag}}{\text{mass of coin}} \times 100 = \frac{2.5528 \text{ g}}{2.8357 \text{ g}} \times 100 = 90.02$$

(A modern U.S. dime contains no silver at all. In the 1960's, the value of the silver in the coin became equal to the value of the coin itself. At that point, U.S. "silver" coins were changed to a Cu-Ni alloy.)

There are several points about this analysis that should be mentioned. They apply to all gravimetric analyses.

1. You must know a great deal about the chemical properties of the substances involved. Most metals, including silver, go into solution in warm 6M HNO_3, forming positive ions.

2. A pure substance, containing the species being analyzed for, must be separated from the rest of the sample. This can be the species itself or one of its compounds. If excess HCl is added to a solution of Ag^+, all of the silver precipitates as AgCl.

3. All key steps in the procedure must be quantitative. In Step 2, all the silver in the coin must dissolve. In Step 3, HCl must be present in excess. This is necessary to make sure all the Ag^+ ions precipitate as AgCl. In Step 4, all the AgCl must be transferred to the filter paper. In Step 5, all the water must be driven off.

FIGURE 21.2

Purposes of the steps in the gravimetric analysis of a silver coin.

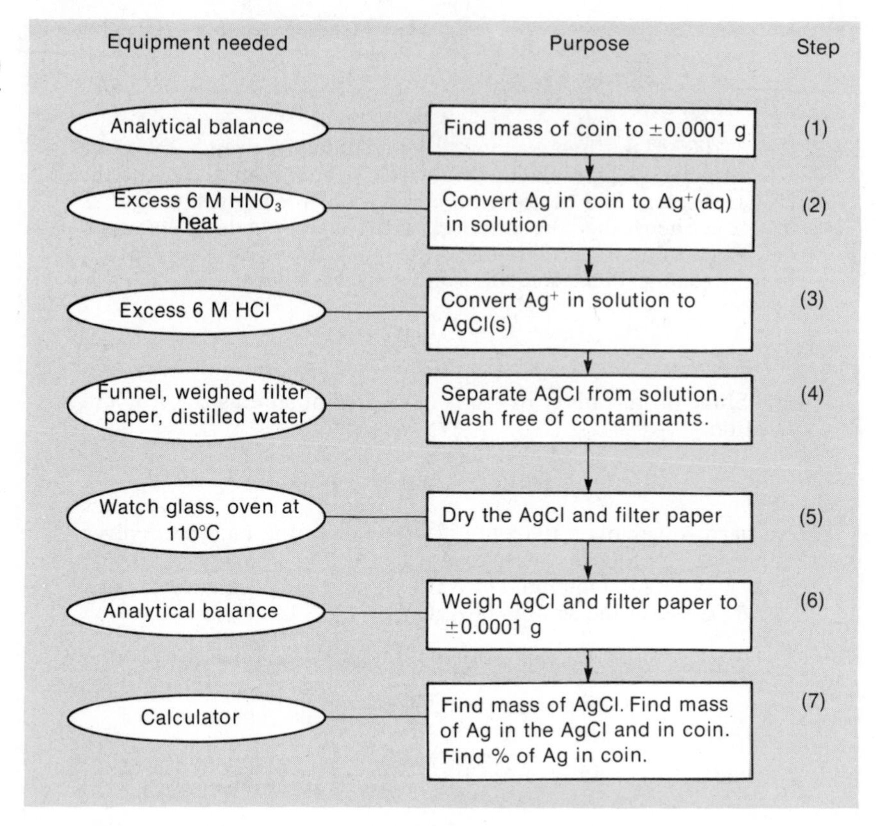

Good gravimetric analysis requires a truly quantitative procedure. It also requires good technique. Good results in gravimetric analysis are no accident.

Other Analyses

Gravimetric procedures can be devised for a great many different elements and compounds. We will cite only two more examples.

1. The percentage of Cl^- in a sample can be determined by:
   a. dissolving a weighed sample in water.
   b. adding excess Ag^+ ions to the water solution to precipitate AgCl.
   c. weighing the AgCl produced.

2. The percentage of carbon in an organic sample can be determined by:
   a. burning a weighed sample in excess oxygen.
   b. collecting and weighing the CO_2 produced.

We will not attempt to describe these procedures in detail. It may help, however, to give one more example of the calculations involved.

EXAMPLE 21.2

You are given a solid known to contain sodium chloride. To determine the percentage of NaCl, you weigh out a 1.500 g sample. This is dissolved in water and treated with an excess of $AgNO_3$

solution. You separate the AgCl formed and find that it weighs 1.200 g. What is the percentage of NaCl in the original sample?

SOLUTION

You need to know how many grams of NaCl are required to form 1.200 g of AgCl. Clearly, one mole of NaCl gives one mole of AgCl, since they both contain one mole of Cl^- ions. One mole of NaCl weighs: 22.99 g + 35.45 g = 58.44 g. One mole of AgCl weighs: 107.87 g + 35.45 g = 143.32 g. Hence:

$$1 \text{ mol NaCl} \simeq 1 \text{ mol AgCl}$$

$$58.44 \text{ g NaCl} \simeq 143.32 \text{ g AgCl}$$

Using the conversion factor approach:

An analysis like this can be accurate to about 0.1%.

$$\text{mass of NaCl} = 1.200 \text{ g AgCl} \times \frac{58.44 \text{ g NaCl}}{143.32 \text{ g AgCl}} = 0.4893 \text{ g NaCl}$$

Recall that the original sample weighed 1.500 g. Hence:

$$\% \text{ NaCl} = \frac{0.4893 \text{ g}}{1.500 \text{ g}} \times 100 = 32.62\%$$

21.2 VOLUMETRIC ANALYSIS

As we have seen, gravimetric analysis depends entirely upon mass measurements. This is not the case in other areas of quantitative analysis. Although they may require a weighing at some stage, other measurements are involved as well. In many procedures, the determination of the volume of a solution is important. Such analyses are called volumetric. In volumetric analysis, a solution reacts with another solution or with a solid. Calculations are based on the masses and volumes involved.

Acid-Base Titrations

A very common method of volumetric analysis uses acid-base reactions. It involves what is called an acid-base titration. Commonly, acid-base titrations are used to determine either:

1. The concentration of an acidic or basic water solution, or:
2. The percentage of an acid or base in a solid mixture.

To illustrate the method, suppose you want to determine the molarity of an NaOH solution. You are given an HCl solution of known concentration, let us say 0.1000 M. You might proceed as follows.

Step 1 Transfer **25.00 cm³** of the **0.1000 M HCl** solution to a flask, using a pipet. The flask should be clean but need not be dry.

Step 2 Add to the flask a few drops of an acid-base indicator. A good choice here is phenolphthalein. This compound is colorless in acid but turns pink in basic solution.

Step 3 Fill a buret with the NaOH solution whose concentration is to be determined. The buret should be clean and *dry* (Figure 21.3); otherwise, the NaOH solution will be diluted. Run out a little NaOH solution through the stopcock so that the level of solution drops to the zero mark, near the top of the buret.

Step 4 Slowly run the NaOH solution into the flask containing the HCl. Swirl the flask as you add the NaOH, to insure good mixing. At first you will not observe any color change, since the solution is highly acidic. At this stage, you can add the NaOH steadily in a thin stream.

The reaction between H^+ and OH^- ions occurs very rapidly.

Step 5 As the titration proceeds, you will reach a point at which the solution in the flask turns pink when the NaOH strikes it. The pink color will quickly disappear when you swirl the flask (see Color Plate 9A, in the center of the book.) You are now close to the "end point". That is, the HCl solution is nearly neutralized. Close the stopcock, then open it slightly so the NaOH solution enters the flask drop by drop.

Step 6 Sooner or later, you will find that a drop of NaOH solution produces a permanent pink color. That is, the color persists when you swirl the flask for perhaps ten seconds. This means that all the HCl has been neutralized. The acid-base reaction is com-

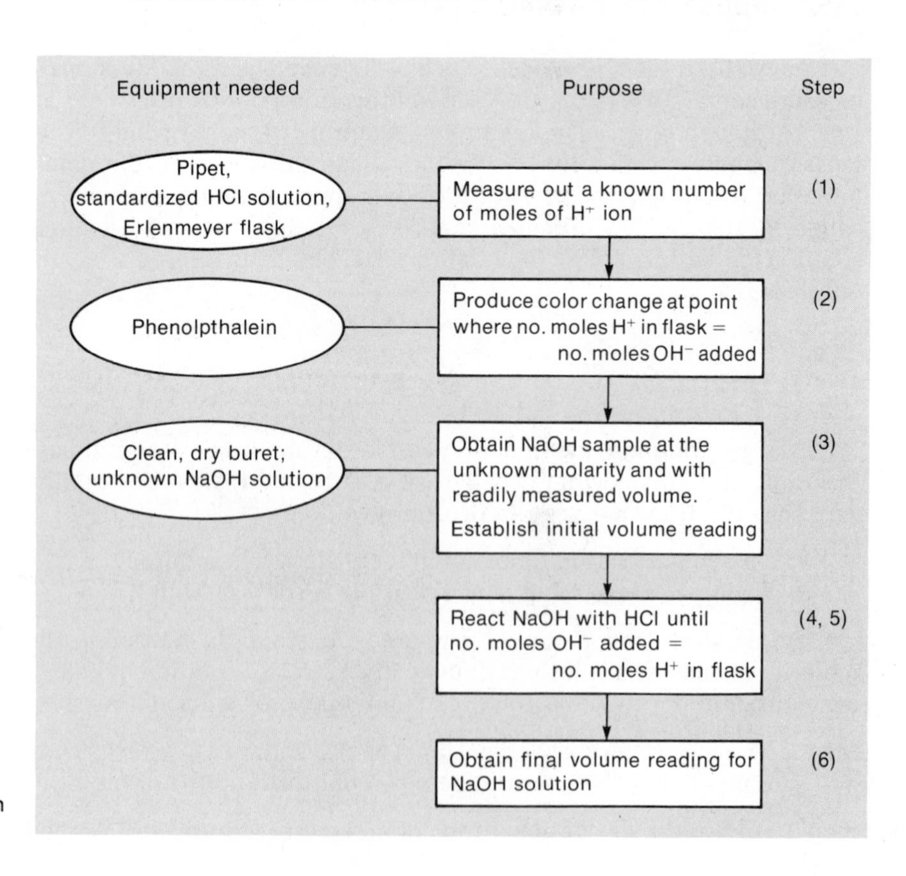

FIGURE 21.3

Purposes of the steps in an acid-base titration.

plete. Close the stopcock and carefully read the level of the NaOH solution in the buret. Let us suppose that your reading is **20.80 cm³**.

From data obtained in this experiment, you can calculate the concentration of the NaOH solution. The reaction is:

$$H^+(aq) + OH^-(aq) \rightarrow H_2O \qquad (21.1)$$

From this equation, we see that when reaction is complete:

$$\text{no. moles } H^+ \text{ originally present} = \text{no. moles } OH^- \text{ added} \qquad (21.2)$$

But one mole of HCl furnishes one mole of H^+; one mole of NaOH yields one mole of OH^-. Hence:

$$\text{no. moles HCl originally present} = \text{no. moles NaOH added} \qquad (21.3)$$

For any solute:

$$\text{molarity} = \frac{\text{no. moles solute}}{\text{volume solution in liters}};\ M = \frac{n}{V};\ \text{or, } n = M \times V \qquad (21.4)$$

Substituting this relation in Equation 21.3, we obtain:

$$(\text{M HCl}) \times (\text{V HCl}) = (\text{M NaOH}) \times (\text{V NaOH}) \qquad (21.5)$$

This is the relation we use to calculate the molarity of the NaOH solution. To do this, recall that:

$$\text{M HCl} = 0.1000\ \text{mol}/\ell$$
$$\text{V HCl} = 25.00\ \text{cm}^3 = 0.02500\ \ell$$
$$\text{V NaOH} = 20.80\ \text{cm}^3 = 0.02080\ \ell$$

So we have:

$$(0.1000\ \text{mol}/\ell) \times (0.02500\ \ell) = (\text{M NaOH}) \times (0.02080\ \ell)$$

Solving: $$\text{M NaOH} = 0.1000\ \text{mol}/\ell \times \frac{0.02500\ \ell}{0.02080\ \ell} = 0.1202\ \text{mol}/\ell$$

Properly done, a titration like this is accurate to within about 0.1%.

The procedure we have described can be applied equally well to determine the concentration of an "unknown" acid (Example 21.3). It can also be used to determine the percentage of an acid or base in a solid mixture (Example 21.4).

EXAMPLE 21.3

To determine the concentration of acetic acid in vinegar, a student titrates the vinegar with 0.2000 M NaOH solution. He finds that 21.60 cm³ of this NaOH solution is required to titrate a

25.00 cm^3 sample of vinegar. The reaction is:

$$CH_3COOH(aq) + OH^-(aq) \rightarrow CH_3COO^-(aq) + H_2O$$

What is the molarity of the acetic acid?

SOLUTION

We use an indicator in acid-base titrations so we can tell when we have added equivalent amounts of acid and base.

From the chemical equation, we see that:

$$\text{no. moles } CH_3COOH = \text{no. moles } OH^-$$

But since 1 mol NaOH yields 1 mol OH^-:

$$\text{no. moles } CH_3COOH = \text{no. moles NaOH}$$

Using the relation: $M = \frac{n}{V}$; $n = M \times V$

$$(M\ CH_3COOH) \times (V\ CH_3COOH) = (M\ NaOH) \times (V\ NaOH)$$

Substituting: M NaOH = 0.2000 mol/ℓ; V NaOH = 0.02160 ℓ; V CH_3COOH = 0.02500 ℓ

$$(M\ CH_3COOH) \times (0.02500\ \ell) = (0.2000\ mol/\ell) \times (0.02160\ \ell)$$

Solving: $M\ CH_3COOH = 0.2000 \frac{mol}{\ell} \times \frac{0.02160\ \ell}{0.02500\ \ell} = 0.1728 \frac{mol}{\ell}$

EXAMPLE 21.4

A student determines the percentage of NaOH in a solid mixture with NaCl. She finds that 20.0 cm^3 of 0.100 M HCl is required to titrate a 0.100 g sample of the solid. The reaction is:

$$H^+(aq) + OH^-(aq) \rightarrow H_2O$$

The H^+ ions come from the HCl and the OH^- ions from the NaOH. The NaCl takes no part in the reaction. What is the percentage of NaOH in the mixture?

SOLUTION

A logical path to follow is to calculate, in order:

Can you see why this path is a logical one?

(1) The number of moles of HCl added, using the relation:

$$n\ HCl = M\ HCl \times V\ HCl$$

(2) The number of moles of NaOH present. Since one mole of NaOH (1 mol OH^-) reacts with one mole of HCl (1 mol H^+):

$$n\ NaOH = n\ HCl$$

(3) The number of grams of NaOH (1 mol NaOH = 40.0 g NaOH)

(4) The percent of NaOH in the sample:

$$\% \text{ NaOH} = \frac{\text{mass NaOH}}{\text{mass sample}} \times 100$$

Carrying out the calculations:

(1) $\text{n HCl} = 0.100 \frac{\text{mol}}{\ell} \times 0.0200\ \ell = 0.00200 \text{ mol}$

(2) $\text{n NaOH} = 0.00200 \text{ mol}$

(3) $\text{mass NaOH} = 0.00200 \text{ mol} \times 40.0 \frac{\text{g}}{\text{mol}} = 0.0800 \text{ g}$

(4) $\% \text{ NaOH} = \frac{0.0800 \text{ g}}{0.100 \text{ g}} \times 100 = 80.0\%$

Other Types of Titrations

Acid-base titrations are the most common type of volumetric analysis. However, many other kinds of reactions are possible. There are only three requirements.

1. The reaction must go to completion. This means that the equilibrium constant for the forward reaction must be a large number. This is true for acid-base reactions (Chapter 19) and precipitation reactions (Chapter 20).

2. A suitable indicator must be available to tell us when reaction is complete. With acid-base reactions, this is no problem. For other types of reactions, it is sometimes difficult to find a good indicator.

3. It must be possible to prepare a "standard solution" of one reagent used in the titration. Its concentration must be known accurately. For example, to titrate a base with HCl, we must know the concentration of the HCl solution.

Among the reactions used in volumetric analysis is that of precipitation. Suppose, for example, we want to know the percent of NaCl in a solid sample. We can do this by volumetric analysis, using the set-up shown in the Color Plate 9B, in the center of the book. Here, we measure the volume of standard $AgNO_3$ solution required to react with a weighed sample. The reaction is:

$$Ag^+(aq) + Cl^-(aq) \rightarrow AgCl(s) \quad (21.6)$$

The indicator used is potassium chromate, K_2CrO_4. The CrO_4^{2-} ions give the solution a yellow color. So long as there are Cl^- ions present, Reaction 21.6 occurs and a white precipitate of AgCl forms. When the reaction is complete, Ag^+ ions start to react with CrO_4^{2-} ions. A red precipitate, Ag_2CrO_4, forms. This color change is your cue to stop adding $AgNO_3$ solution and do the calculations (Example 21.5).

Equation 21.6 gives us this mole ratio.

EXAMPLE 21.5

A student finds that 25.0 cm^3 of 0.100 M $AgNO_3$ solution is required to titrate a sample which weighs 0.200 g and contains NaCl. Calculate the percentage of NaCl in the sample.

SOLUTION

A logical path would be to calculate:

(1) the number of moles of $AgNO_3$ added, using the relation:

$$n\ AgNO_3 = M\ AgNO_3 \times V\ AgNO_3$$

(2) the number of moles of NaCl in the sample. Since one mole of $AgNO_3$ (1 mol Ag^+) reacts with one mole of NaCl (1 mol Cl^-):

$$n\ NaCl = n\ AgNO_3$$

(3) the number of grams of NaCl in the sample:

$$1\ mol\ NaCl = 22.99\ g + 35.45\ g = 58.44\ g$$

(4) the % of NaCl in the sample: $\%\ NaCl = \frac{mass\ NaCl}{mass\ sample} \times 100$

Substituting numbers:

(1) $n\ AgNO_3 = (0.100\ mol/\ell) \times (0.0250\ \ell) = 0.00250\ mol$

(2) $n\ NaCl = 0.00250\ mol$

(3) $mass\ NaCl = 0.00250\ mol \times \frac{58.44\ g}{1\ mol} = 0.146\ g$

(4) $\%\ NaCl = \frac{0.146\ g}{0.200\ g} \times 100 = 73.0\%$

You will recall from Section 21.1 that the percentage of chloride in a sample can also be determined by gravimetric analysis. The calculations were discussed in Example 21.2. Comparing Examples 21.2 and 21.5, you will note one important difference. In the gravimetric determination, the AgCl formed must be separated, dried, and weighed. In the volumetric analysis, we need only measure the volume of $AgNO_3$ added. This is much simpler and less tedious.

21.3 INSTRUMENTAL METHODS OF ANALYSIS

In the 19th century, gravimetric and volumetric methods dominated analytical chemistry. However, other methods, which involved physical, rather than chemical, properties were sometimes used. These methods were gradually improved, starting in about 1945. Now there is a wide variety of instrumental methods. They are used for most analyses in industrial laboratories. They have certain advantages over the two

TABLE 21.1 WAVELENGTHS OF THE COLORS IN THE VISIBLE SPECTRUM

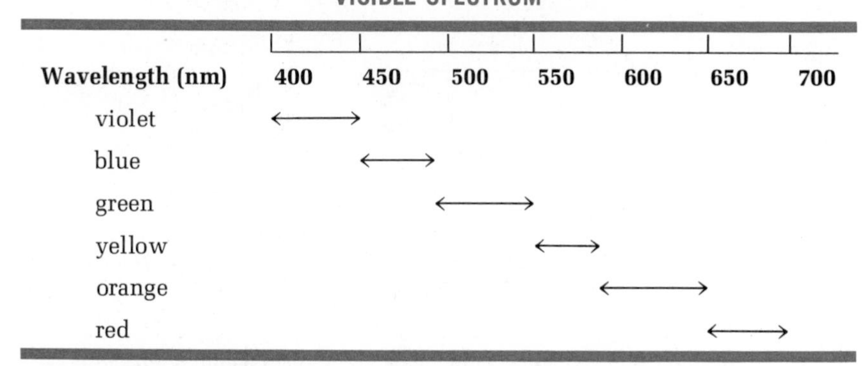

Wavelengths less than 400 nm (1 nm = 10^{-9} m) are in the ultraviolet region. Wavelengths greater than 700 nm are in the infrared.

approaches we have discussed. For one thing, they tend to be rapid. They can handle complex mixtures and analyze for several components at the same time. Instrumental methods can also be used for components present at very low concentrations. Furthermore, they can be carried out without destroying a sample.

To show how instruments are used in quantitative analysis, we will consider one such device, commonly called a colorimeter, which sells for about $500. It is available in many high school chemistry laboratories.

The colorimeter is used to determine the concentrations of colored species in solution. As you may know, visible light consists of a mixture of radiation at different wavelengths. (See Table 21.1 which shows the wavelength range for various colors.) A colored substance absorbs certain wavelengths and allows others to pass through. The color that we see is that of the light transmitted. Consider, for example, a substance which absorbs light in the blue, green, and yellow regions. A solution containing this substance transmits violet and red light. It appears purple, a mixture of violet and red. A solution of potassium permanganate, $KMnO_4$, behaves this way. (See Color Plate 6C, in the center of the book.)

What colors would be absorbed by a colorless liquid like water?

Figure 21.4 p. 536 shows the essential parts of a colorimeter. The solution to be studied is contained in a small test tube which fits into the sample holder. The wavelength of the light passing through this solution can be adjusted. By turning a dial, we can select any wavelength range between 340 and 625 nm. On a scale, we read either:

1. The percent transmittance (% T). This is the percent of incident light which passes through the solution without being absorbed.
2. The **absorbance,** which is defined by the relation:

$$\text{absorbance} = \log \frac{100}{(\%\ \text{T})}$$

In quantitative analysis, we read the absorbance. The concentration, c, of the colored species in solution is directly proportional to the absorbance. That is:

$$c = k\ (\text{absorbance}) \qquad (21.7)$$

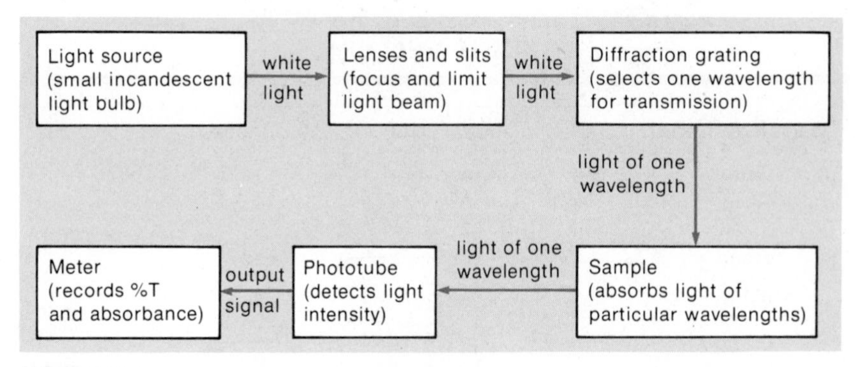

FIGURE 21.4

Schematic diagram showing the components of a colorimeter. Light from an incandescent bulb is broken into its component wavelengths by the diffraction grating. Light at a given wavelength is first passed through a sample of pure water, and the instrument is adjusted to read 100% transmission. The light is then sent through the sample and either the % transmission or the absorbance is read directly off the scale.

At a given wavelength, with a given species in the test tube, k is a constant.

To show how the colorimeter is used, consider the determination of chromate ion, CrO_4^{2-}. As you may know, water solutions containing this ion are colored yellow. The chromate ion absorbs in the violet and ultraviolet regions of the spectrum. To analyze for CrO_4^{2-}, we set the wavelength dial at a fixed value, usually 370 nm.

All of the solutions in Table 21.2 would look yellow, but the color would be fainter at the lower concentrations.

The first step in the analysis involves the determination of k in Equation 21.7. To do this, we start by preparing a series of standard solutions, containing known concentrations of CrO_4^{2-} (Table 21.2). This could be done by weighing out samples of pure K_2CrO_4 and dissolving them in enough water to give a specified volume. To prepare a solution 3.00×10^{-4} M in CrO_4^{2-}, we might dissolve:

$$3.00 \times 10^{-4} \text{ mol} \times 194 \frac{\text{g}}{\text{mol}} = 0.0582 \text{ g}$$

of K_2CrO_4 to form a liter of solution.

We then measure, with the colorimeter, the absorbances of these standard solutions. In this way, we obtain the data in Table 21.2. These data are plotted in Figure 21.5. Notice that the graph of concentration versus absorbance is a straight line through the origin. This must be the

TABLE 21.2 ABSORBANCE OF STANDARD SOLUTIONS OF CrO_4^{2-}

Conc. of CrO_4^{2-} (mol/ℓ)	Absorbance
3.00×10^{-4}	1.20
2.00×10^{-4}	0.80
1.00×10^{-4}	0.40
0.50×10^{-4}	0.20

FIGURE 21.5

The concentration of an absorbing solute is directly proportional to its absorbance, c equals k × absorbance. Since the concentration of CrO_4^{2-} is 2.0×10^{-4} M when the absorbance is 0.80, we can say that 2.0×10^{-4} M equals k × 0.80. Therefore, k must equal 2.5×10^{-4} M.

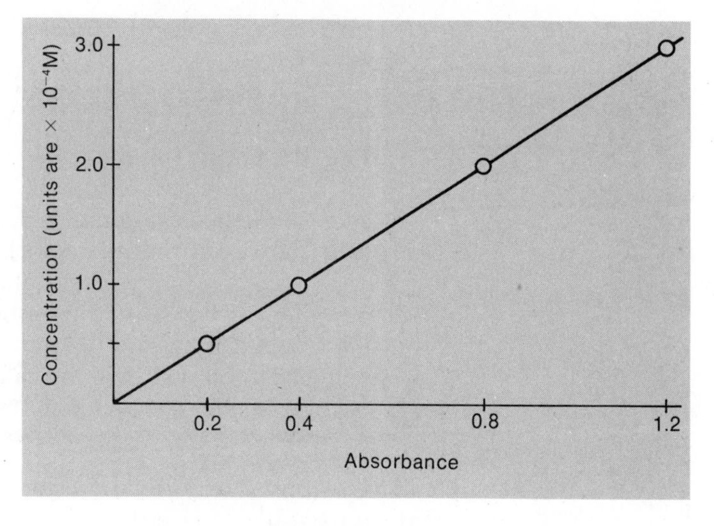

case since the two quantities are directly proportional to each other. To obtain the value of k in Equation 21.7, we take the slope of this line. As you can see from Figure 21.5:

$$k = 2.5 \times 10^{-4} \text{ M}$$

Once we know the value of k, we can readily determine the concentration of CrO_4^{2-} in an unknown solution. Suppose, for example, that we read an absorbance of 0.64. For that solution:

$$c = (2.5 \times 10^{-4}) \times 0.64 = 1.6 \times 10^{-4} \text{ M}$$

In other words, the concentration of chromate ion in this solution is 1.6×10^{-4} mol/ℓ.

We could also use the graph to obtain this concentration.

EXAMPLE 21.6

It is possible, with somewhat less accuracy, to obtain the value of k in Equation 21.7 with a single standard solution. Suppose that, at a different wavelength, you find that a solution 2.00×10^{-4} M in CrO_4^{2-} has an absorbance of 0.40.

a. What is the value of k?

b. What is the concentration of CrO_4^{2-} in a solution which has an absorbance of 0.83?

SOLUTION

a. Solving Equation 21.7 for k:

$$k = \frac{\text{conc. } CrO_4^{2-}}{\text{absorbance}} = \frac{2.00 \times 10^{-4} \text{ M}}{0.40} = 5.0 \times 10^{-4} \text{ M}$$

b. conc. $CrO_4^{2-} = (5.0 \times 10^{-4} \text{ M}) \times 0.83 = 4.2 \times 10^{-4} \text{ M}$

This example illustrates two points that are worth commenting upon.

1. In using a colorimeter, the wavelength setting must be the same for standard and unknown solutions. If the wavelength changed, say from 370 nm to 400 nm, k would change.

2. Colored species can be determined at very low concentrations. Here, the concentration of CrO_4^{2-} was around 10^{-4} M. This does not mean, however, that the colorimeter is restricted to this low concentration range. More concentrated solutions can be used. Suppose, for example, you had a CrO_4^{2-} solution which was about 10^{-2} M. You would first dilute this solution 1:100. That is, dilute 10 cm^3 to one liter. You then read the absorbance and calculate c for the diluted solution. Finally, you multiply by 100 to obtain the concentration of the original solution.

Absorption of Light and Energy Levels

You may wonder why some substances, like I_2 and CrO_4^{2-}, are colored, while others, like H_2O and NaCl, are not. The reason has to do with the energy levels available to electrons. These levels have different spacings in colored as opposed to colorless substances. A colored species has one or more "excited" levels which can be reached by absorbing visible light. In colorless substances, these upper levels are farther above the lowest level. They can be reached only by the absorption of ultraviolet light. When a substance absorbs light, electrons jump from the lowest to a higher level. If the levels are fairly close to one another, visible absorption occurs. If they are far apart, we get ultraviolet absorption. Water and NaCl both absorb ultraviolet light of the proper wavelengths.

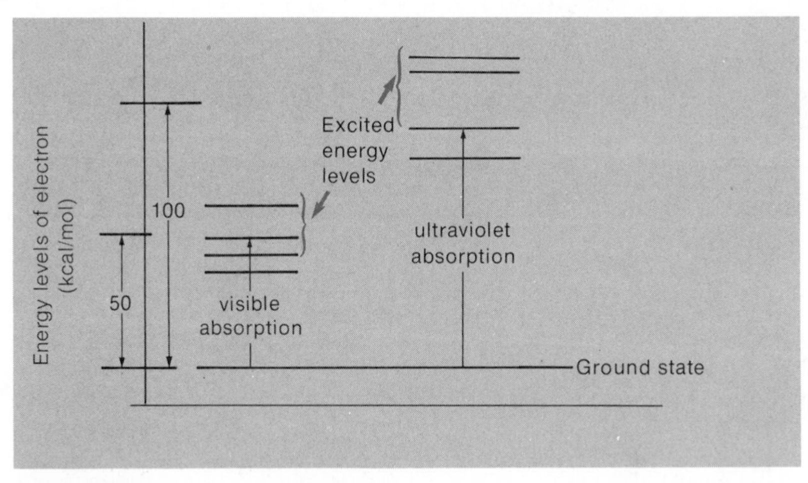

FIGURE 21.6

In substances which absorb visible light, there are electron energy levels that lie about 50 kcal/mol above the lowest energy level (ground state). Substances which absorb in the ultraviolet region have electronic energy levels about 100 kcal/mol above the ground state. A substance which absorbs only ultraviolet light will appear colorless or white to the eye.

Absorption in the infrared occurs for a different reason. When a molecule absorbs in the infrared, its vibrational energy increases. All molecules containing more than one atom undergo vibrations. The spacing of energy levels for vibrations is small. Low-energy, infrared light will excite these vibrations. Large molecules have more vibrations than small ones. Hence, their infrared spectra tend to be more complex.

21.4 SUMMARY

Quantitative analysis deals with the determination of the amounts of substances present in a sample. Sometimes we determine the percentage of a substance in a mixture (Examples 21.1, 21.2, 21.4, 21.5). In other cases, we determine the concentration of a species in solution (Examples 21.3, 21.6). To analyze a sample, we make use of the chemical or physical properties of the substances present. Two methods involving chemical properties are:

1. gravimetric analysis: based on masses of sample and of a pure product.
2. volumetric analysis: based on measurements of the volumes of solutions.

Usually the component to be analyzed for in volumetric analysis is in one of the solutions. The other solution contains a known concentration of a species which reacts with the unknown. An indicator is used to determine the point at which reaction is complete.

Instrumental methods are used in analyses based on physical properties of substances. There are many instrumental methods. The colorimeter discussed here measures the absorption of light by a sample. It requires the use of standard samples to determine k in the equation:

$$\text{concentration} = k\ (\text{absorbance})$$

Once k is known, the concentration of a colored species is easily calculated from its measured absorbance.

NEW TERMS

absorbance
colorimeter
end point
gravimetric analysis
instrumental analysis
quantitative analysis
standard solution
titration
volumetric analysis

QUESTIONS

1. Which of the following are in the area of gravimetric analysis? volumetric analysis? instrumental analysis?
   a. Determination of concentration of Ag^+ by titration with a standard solution of Cl^-.
   b. Analysis for Cl^- by precipitation with Ag^+ followed by weighing AgCl.
   c. Determination of I_2 using a colorimeter.
   d. Analysis of an antacid tablet for $NaHCO_3$ by titration with HCl.

2. In the analysis of a coin for silver, described on p. 526, what is the purpose of adding HNO_3? HCl?

3. Suppose that, in analyzing a coin for silver, a student makes the following errors. How will they affect the value reported for the percentage of silver (too high or too low)?

   a. Too little Cl^- is added to precipitate all the Ag^+ in Step 3.
   b. In Step 4, some of the AgCl is lost.
   c. In Step 5, the AgCl is not completely dried.

4. Which of the following compounds would be suitable products for the determination of SO_4^{2-} in gravimetric analysis (see Table 20.1, p. 503).

   a. Na_2SO_4 b. $BaSO_4$ c. $PbSO_4$ d. $CuSO_4$

5. Describe briefly the gravimetric analysis for determination of:

   a. % of Cl^- in a solid b. % C in an organic compound

6. What is an acid-base titration? What is the purpose of adding an indicator?

7. As NaOH is added to HCl, does the pH increase or decrease?

8. In the acid-base titration described on p. 530, why is it important that the buret used for the NaOH solution be dry?

9. In an acid-base titration, how can you tell when you are getting close to the end point?

10. In the titration of NaOH with HCl, a student makes the following errors. What effect will they have on the value reported for the concentration of NaOH? (too high or too low)

    a. The pipet used in Step 1 is not allowed to drain completely.
    b. Step 2 is omitted.
    c. There is 1 cm^3 of water in a buret to which NaOH is added.
    d. NaOH is added past the point where the indicator changes color.

11. A student titrates HCl with $Ca(OH)_2$. Which of the following statements apply?

    a. no. moles OH^- added at end point = no. moles H^+ present
    b. no. moles $Ca(OH)_2$ added = no. moles H^+ present
    c. no. moles $Ca(OH)_2$ = no. moles OH^-
    d. no. moles HCl = no. moles H^+

12. What is the color of light with the following wavelengths?

    a. 350 nm b. 500 nm c. 600 nm d. 750 nm

13. What would be the color of a substance that absorbed between 400 and 450 nm and between 580 and 700 nm?

14. In the text, a method was discussed for preparing a solution 3.00×10^{-4} M in CrO_4^{2-}. Describe how you could prepare from this, by dilution with water, a solution 1.00×10^{-4} M in CrO_4^{2-}. How would you obtain 30 cm^3 of the more dilute solution, starting with the more concentrated solution?

15. Which of the following equations could be used to describe the relation between concentration (c) and absorbance (A)? (k_1, k_2, k_3, and k_4 are constants).

    a. $c = k_1A$ b. $c = k_2/A$ c. $A = k_3c$ d. $A = k_4/c$

16. In using a colorimeter, it is important that the wavelength setting not be changed between measurements. Why?

17. Describe briefly how you would use a colorimeter to determine CrO_4^{2-} in a solution where its concentration is about 0.1 M. Note that if the solution were put directly into the instrument the absorbance would be much too high to read accurately.

PROBLEMS

1. *Gravimetric Analysis (Example 21.1)* "Silver" jewelry is made from an alloy of silver and copper. A piece of jewelry weighing 1.063 g is dissolved in HNO_3 and treated with HCl. All the silver is precipitated as AgCl, which weighs 1.271 g. What is the percent of silver present?

2. *Gravimetric Analysis (Example 21.2)* A sample containing KCl weighs 1.000 g. It is dissolved in water and treated with excess $AgNO_3$. The AgCl formed weighs 1.250 g. What is the percentage of KCl in the sample?

3. *Volumetric Analysis (Example 21.3)* A student finds that 43.60 cm^3 of 0.1050 M HCl is required to titrate 28.40 cm^3 of a KOH solution. What is the molarity of the KOH?

4. *Volumetric Analysis (Example 21.4)* An organic solid contains benzoic acid, C_6H_5COOH. A sample weighing 1.202 g reacts with 40.0 cm^3 of 0.150 M NaOH. The reaction is:

$$C_6H_5COOH(s) + OH^-(aq) \rightarrow C_6H_5COO^-(aq) + H_2O$$

What is the percentage of benzoic acid in the sample?

5. *Volumetric Analysis (Example 21.5)* A student analyzes a sample for NH_4Cl by titrating it with $AgNO_3$ solution. He finds that a sample weighing 1.542 g requires 26.2 cm^3 of 0.200 M $AgNO_3$ to react with it. What is the percentage of NH_4Cl in the sample?

6. *Instrumental Analysis (Example 21.6)* In a colorimeter, a solution 1.00×10^{-4} M in CrO_4^{2-} has an absorbance of 1.12. What is the concentration of CrO_4^{2-} in a solution which has an absorbance of 0.42?

* * * * *

7. An alloy of Cu and Zn is treated with hydrochloric acid, which reacts only with the zinc. A sample weighing 0.642 g is treated with excess HCl. The metal remaining is dried and found to weigh 0.414 g. What is the percentage of copper in the sample? the percentage of zinc?

8. A sample containing KBr weighs 1.602 g. It is dissolved in water and treated with excess $AgNO_3$ solution. The AgBr formed weighs 1.261 g. What is the percentage of KBr in the sample?

9. A sample containing KI weighs 1.528 g. It is dissolved in water and titrated with 0.200 M $AgNO_3$. The volume of this solution required is 31.8 cm^3. What is the percentage of KI in the sample?

10. What mass of $BaSO_4$ is formed from a 1.000 g sample Epsom salts, $MgSO_4 \cdot 7H_2O$?

11. A sample of Epsom salts weighing 1.444 g is heated to drive off all the water. What mass of $MgSO_4$ remains?

12. A solution of $Ca(OH)_2$ is titrated with HCl. 15.45 cm^3 of 0.1000 M HCl is required to react with 20.00 cm^3 of the $Ca(OH)_2$ solution. What is the molarity of $Ca(OH)_2$?

13. An antacid tablet weighing 1.24 g contains $NaHCO_3$. 26.0 cm^3 of 0.250 M HCl is required to react with the tablet. The reaction is:

$$HCO_3^-(aq) + H^+(aq) \rightarrow CO_2(g) + H_2O$$

What is the percentage of $NaHCO_3$ in the tablet?

14. A solution of ammonia is titrated with HCl. 24.16 cm^3 of the NH_3 solution requires 31.62 cm^3 of 0.1098 M HCl. The reaction is:

$$NH_3(aq) + H^+(aq) \rightarrow NH_4^+(aq)$$

What is the molarity of NH_3?

15. At a certain wavelength, the following data are obtained for CrO_4^{2-} solutions.

conc. CrO_4^{2-} *(M)*	*absorbance*
2.00×10^{-4}	0.50
1.00×10^{-4}	0.24
0.40×10^{-4}	0.10

Using this data, draw a graph to determine the value of k in the equation:

$$\text{concentration} = k\ (\text{absorbance})$$

16. Using the value of k calculated in Problem 15, determine the concentration of CrO_4^{2-} in a solution which has an absorbance of 0.62.

*17. A certain solution of K_2CrO_4 is diluted by adding 2.0 cm^3 to 98.0 cm^3 of water. The dilute solution has an absorbance of 1.00. Another solution, known to be 1.00×10^{-4} M in CrO_4^{2-}, has an absorbance of 0.40. What is the molarity of CrO_4^{2-} in the original solution?

*18. A sample containing $AlCl_3$ weighs 1.400 g. It requires 50.0 cm^3 of 0.100 M $Pb(NO_3)_2$ to precipitate the chloride as $PbCl_2$. What is the percent of $AlCl_3$ in the sample?

*19. A certain solid is a mixture of NaCl and KCl. A sample weighing 1.000 g gives 2.100 g of AgCl. What is the percent of NaCl in the sample?

The centrifuge is a useful tool in qualitative analysis. *Witco Chemical Corporation.*

22

QUALITATIVE ANALYSIS

In the last chapter, we discussed some of the common methods of quantitative analysis. There we were interested in finding out how much of a given substance was present in a sample. In quantitative analysis you know which species are present. The problem is to determine their amount or concentration.

In qualitative analysis, the problem is quite different. Here you want to know what species are present in a sample. If you are dealing with a pure substance, the task is quite simple. All you need do is to identify the substance. More often in qualitative analysis you are working with a mixture. In that case, you need to separate and identify all the components of the mixture.

In this chapter we will consider procedures for the qualitative analysis of water solutions containing positive ions. This type of analysis can be done by examining the physical properties of these ions. This is how such problems probably would be handled in an industrial laboratory. However in this chapter we will emphasize a method of analysis which depends upon chemical properties. This method can be used in a high school laboratory. Our purpose here is twofold. First, to show how the properties of the positive ions allow us to establish their presence in solution. Second, to review and apply some of the reactions that were discussed in Chapters 19 and 20. By studying such reactions you can become familiar with some common chemical reagents and their properties.

22.1 ANALYSIS FOR A SINGLE POSITIVE ION

Different ions have different properties.

We start with a rather simple problem which often arises in qualitative analysis. You are given a water solution known to contain only one positive ion. Your task is to identify that ion. Ordinarily, you are given a list of ions that might be present. Let us suppose that the positive ion must be one of the following:

$$Ag^+, Cu^{2+}, Zn^{2+}, Ni^{2+}, Al^{3+}, Mg^{2+}, Na^+, \text{ or } K^+$$

To decide which positive ion is present, you carry out chemical tests on the solution. The results of these tests are compared to the known properties of the eight ions. They should match the properties of one and only one ion. That must be the positive ion present in the solution.

For such an analysis, solubility tests are particularly useful. You can readily determine which compounds of a given positive ion are soluble in water. To do this, you add a solution containing the appropriate negative ion. You might, for example, add a solution of NaOH. If the positive ion forms an insoluble hydroxide, it will precipitate upon addition of OH^- ions. If no precipitate forms, you conclude that the hydroxide of the positive ion must be soluble in water.

A different kind of solubility test is often useful. Here you start with a precipitate formed by the procedure just described. You test its solubility in various reagents. The most useful reagents here are 6 M HCl, 6 M NaOH, and 6 M NH_3. Some, but not all, precipitates dissolve in these reagents. In this way, you can often distinguish one positive ion from another.

Solubility Table

By conducting several solubility tests, it is usually possible to identify a single positive ion present in a solution. To do this, you need to know how all the possible positive ions would behave under the conditions of your tests. The solubility behavior of the eight positive ions listed at the beginning of this section is summarized in Table 22.1. To make sure that you understand how to read the Table, let us consider it in some detail.

TABLE 22.1 SOLUBILITY TABLE FOR SOME COMMON POSITIVE IONS

1. Water Solubility

	NO_3^-	Cl^-	OH^-	S^{2-}
Ag^+	S	I	D	I (black)
Cu^{2+} (blue)	S	S	I (blue)	I (black)
Zn^{2+}	S	S	I	I
Ni^{2+} (green)	S	S	I (green)	I (black)
Al^{3+}	S	S	I	D
Mg^{2+}	S	S	I	D
Na^+	S	S	S	S
K^+	S	S	S	S

2. Compounds Insoluble in Water but Soluble in 6 M H^+

Hydroxides (form H_2O): $Cu(OH)_2$, $Zn(OH)_2$, $Ni(OH)_2$, $Al(OH)_3$, $Mg(OH)_2$

Sulfide (forms H_2S): ZnS

3. Compounds Insoluble in Water but Soluble in 6 M OH^-

$Zn(OH)_2$; forms $Zn(OH)_4^{2-}$

$Al(OH)_3$; forms $Al(OH)_4^-$

4. Compounds Insoluble in Water but Soluble in 6 M NH_3

AgCl; forms $Ag(NH_3)_2^+$

$Cu(OH)_2$; forms $Cu(NH_3)_4^{2+}$ (dark blue)

$Zn(OH)_2$; forms $Zn(NH_3)_4^{2+}$

$Ni(OH)_2$; forms $Ni(NH_3)_6^{2+}$ (blue)

5. Additional Tests

Na^+ gives a yellow flame test; K^+ gives a violet flame test

Ni^{2+} forms a red precipitate upon addition of dimethylglyoxime, an organic reagent

1. Water Solubility

Here we show what happens when a given negative ion (NO_3^-, Cl^-, OH^-, or S^{2-}) is added to a solution containing a particular positive ion. If a precipitate forms, the compound produced must be insoluble in water **(I)**. Compounds in this category include AgCl and Ag_2S made by mixing solutions containing Ag^+ with ones containing Cl^- or S^{2-}. If no precipitate forms, the compound is soluble in water **(S)**. Silver nitrate, $AgNO_3$, is the only compound of Ag^+ in this category. Those compounds which decompose in contact with water are listed in the Table as "D". We will not consider these compounds further.

This portion of the Table also gives information about colors. Note, for example, that solutions containing the Cu^{2+} ion are blue. (See Color Plate 9C, in the center of the book.) Among the compounds of Cu^{2+}, $Cu(OH)_2$ is blue while CuS is black. *All positive ions whose colors are not listed are colorless in solution. All compounds whose colors are not listed are white.*

2. *Solubility in H^+*

Here we list those compounds that are insoluble in water but dissolve in a strong acid. These include hydroxides and sulfides. The reactions involved were discussed in Chapter 20. The equations with $Zn(OH)_2$ and ZnS are:

You may recall that all hydroxides are soluble in solutions of strong acids.

$$Zn(OH)_2(s) + 2\ H^+(aq) \rightarrow Zn^{2+}(aq) + 2\ H_2O \qquad (22.1)$$

$$ZnS(s) + 2\ H^+(s) \rightarrow Zn^{2+}(aq) + H_2S(g) \qquad (22.2)$$

The acid used here is usually 6 M HCl. It is added in small portions to the solid being tested. Frequent stirring insures that the solid has a chance to dissolve (Figure 22.1).

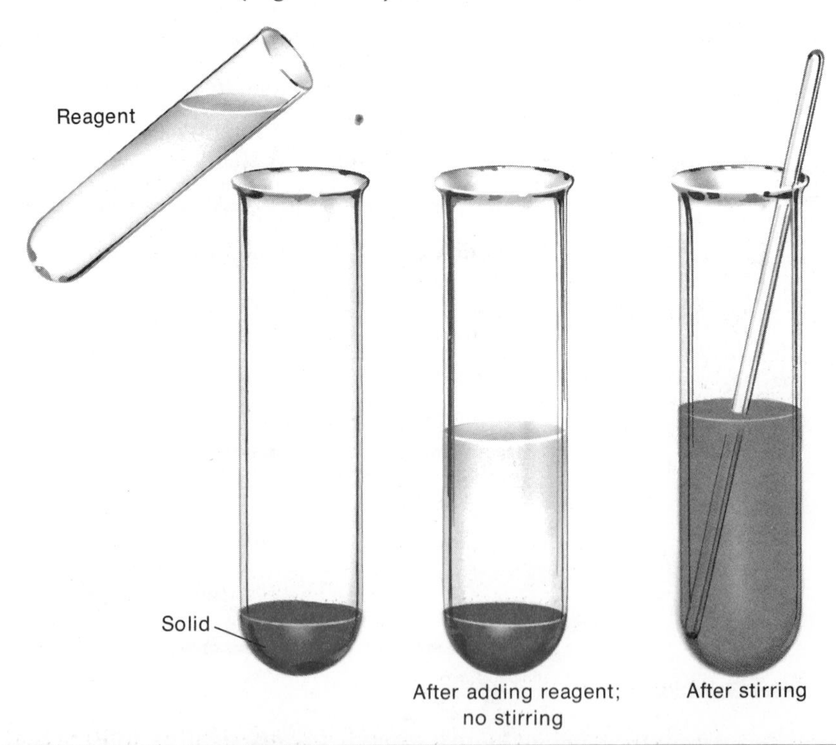

FIGURE 22.1

In order to dissolve a precipitate it is usually necessary to stir the reagent into the solid. Without stirring, the solution process may be very slow.

EXAMPLE 22.1

Write balanced equations for the reactions, if any, that occur when:

a. OH^- and S^{2-} ions are added to two portions of a solution containing Cu^{2+} ions.

b. H^+ ions are added to any precipitates formed in (a).

SOLUTION

a. Table 22.1 tells us that $Cu(OH)_2$ and CuS are both insoluble in water. In each case a precipitate forms. Following the rules discussed in Chapter 20, we write the equations:

$$Cu^{2+}(aq) + 2\ OH^-(aq) \rightarrow Cu(OH)_2(s)$$
$$Cu^{2+}(aq) + S^{2-}(aq) \rightarrow CuS(s)$$

These equations both follow from the information in Table 22.1.

b. From Table 22.1, we see that $Cu(OH)_2$ is soluble in acid. The equation is similar to Equation 22.1:

$$Cu(OH)_2(s) + 2\ H^+(aq) \rightarrow Cu^{2+}(aq) + 2\ H_2O$$

Copper sulfide, CuS, does not dissolve in acid (Table 22.1). So, there is no equation to write.

3. Solubility in OH^-

Here we list the water-insoluble compounds that are dissolved by OH^- ions. Solution comes about because the positive ion forms a stable complex with OH^-. The formulas of the complex ions formed are listed at the right in the Table. A typical reaction is:

$$Zn(OH)_2(s) + 2\ OH^-(aq) \rightarrow Zn(OH)_4^{2-}(aq) \quad (22.3)$$

Relatively few hydroxides dissolve in 6 M NaOH.

The reagent used here is 6 M NaOH. It is added, with stirring, to the solid being tested. Sometimes, 6 M NaOH is used to first form a precipitate and then dissolve it. Suppose, for example, we add this reagent to a solution containing Zn^{2+} ions. The first product is $Zn(OH)_2$, which is insoluble in water:

$$Zn^{2+}(aq) + 2\ OH^-(aq) \rightarrow Zn(OH)_2(s) \quad (22.4)$$

When more NaOH is added, the concentration of OH^- increases. This brings the solid into solution through Reaction 22.3 (Figure 22.2).

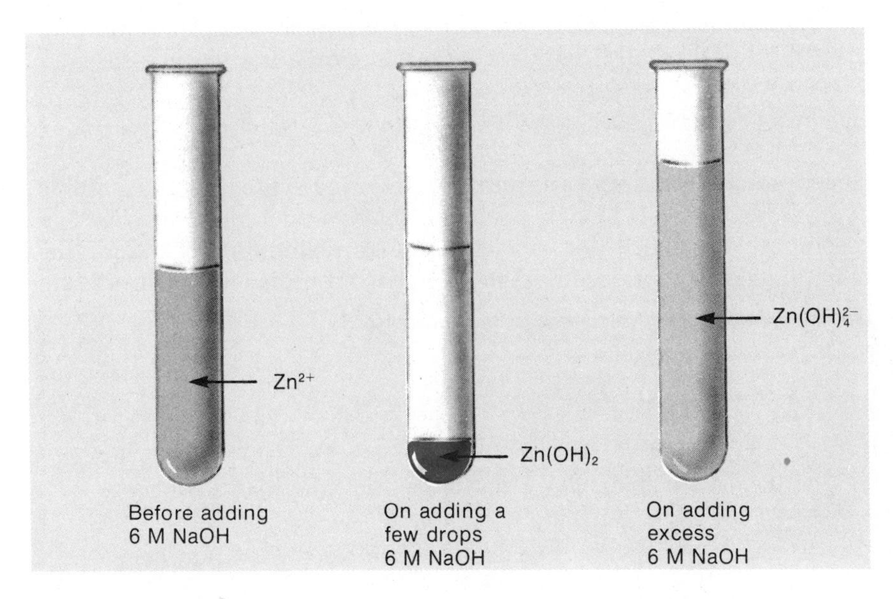

FIGURE 22.2

Reaction of Zn^{2+} ion with 6 M NaOH. A few drops of the NaOH cause $Zn(OH)_2$ to precipitate. In excess NaOH the $Zn(OH)_2$ dissolves as the $Zn(OH)_4^{2-}$ ion is formed.

EXAMPLE 22.2

Write balanced equations for all reactions that occur when 6 M NaOH is added to a solution containing:

a. Ni^{2+} b. Al^{3+}

SOLUTION

a. From Table 22.1, we see that $Ni(OH)_2$ is insoluble in water and does not dissolve in NaOH. Only one reaction occurs. A precipitate forms:

$$Ni^{2+}(aq) + 2\ OH^-(aq) \rightarrow Ni(OH)_2(s)$$

b. $Al(OH)_3$ is insoluble in water. It dissolves in excess 6 M NaOH to form the $Al(OH)_4^-$ ion. Addition of 6 M NaOH to a solution of Al^{3+} gives two successive reactions. First, a precipitate forms:

$$Al^{3+}(aq) + 3\ OH^-(aq) \rightarrow Al(OH)_3(s)$$

Then, as more NaOH is added, the precipitate dissolves:

$$Al(OH)_3(s) + OH^-(aq) \rightarrow Al(OH)_4^-(aq)$$

4. Solubility in NH_3

Here we list the water-insoluble compounds that dissolve in ammonia. In each case, the positive ion forms a stable complex with NH_3 (Chapter 20). The formulas of these complexes are listed at the right in Table 22.1. With silver chloride, the reaction is:

$$AgCl(s) + 2\ NH_3(aq) \rightarrow Ag(NH_3)_2^+(aq) + Cl^-(aq) \qquad (22.5)$$

Ammonia, like NaOH, can form precipitates as well as dissolve them. You will recall (Chapter 19) that ammonia is a weak base:

$$NH_3(aq) + H_2O \rightleftharpoons NH_4^+(aq) + OH^-(aq) \qquad (22.6)$$

If you get a precipitate on addition of NH_3 solution, it is a hydroxide.

Hence ammonia can act as a source of OH^- ions to precipitate insoluble hydroxides. When 6 M NH_3 is added to a solution containing Zn^{2+}, a precipitate of $Zn(OH)_2$ forms at first. As more ammonia is added, this precipitate dissolves, forming the $Zn(NH_3)_4^{2+}$ complex ion (Figure 22.3).

EXAMPLE 22.3

Write balanced equations to show:

a. the formation of a precipitate of $Zn(OH)_2$ when NH_3 is added to a solution containing Zn^{2+}.

b. the dissolving of this precipitate when more NH_3 is added.

FIGURE 22.3

The effect of 6 M NH_3 on a solution of Zn^{2+} ion. To each of the test tubes about 15 cm^3 0.1 M $Zn(NO_3)_2$ were added. Then, starting with the tube on the left, the following volumes in cm^3 of 6 M NH_3 were added: 0, 0.5, 1.0, 1.5, 2.0. Initially, a precipitate of $Zn(OH)_2$ formed, since 6 M NH_3 is basic. In excess NH_3 the precipitate dissolved due to formation of the $Zn(NH_3)_4^{2+}$ complex ion.

SOLUTION

a. To form one mole of $Zn(OH)_2$, we need two moles of OH^- ions. This requires that two moles of NH_3 react by Equation 22.6. The overall reaction is:

$$Zn^{2+}(aq) + 2\ NH_3(aq) + 2\ H_2O \rightarrow Zn(OH)_2(s) + 2\ NH_4^+(aq)$$

b. Here the product is the complex ion $Zn(NH_3)_4^{2+}$. Four moles of NH_3 must react with one mole of $Zn(OH)_2$.

$$Zn(OH)_2(s) + 4\ NH_3(aq) \rightarrow Zn(NH_3)_4^{2+}(aq) + 2\ OH^-(aq)$$

Whenever an initial precipitate dissolves in excess reagent, it is because a complex ion forms.

At the bottom of Table 22.1 are some additional tests used to identify particular ions. Two of these are flame tests, described in Chapter 8. They are carried out as indicated in Figure 22.4 p. 550. Such tests are useful for ions such as Na^+ or K^+, which do not form precipitates with common negative ions. Another test referred to here is that for Ni^{2+}. It involves adding an organic reagent called dimethylglyoxime. This gives a bright red precipitate with Ni^{2+}. (See Color Plate 9D, in the center of the book.) No other positive ion behaves in this way with dimethylglyoxime.

Identification of a Positive Ion

Let us now return to the problem posed at the beginning of this section. You are told that the positive ion present in an "unknown" solution is one of the eight listed in Table 22.1. Your problem is to find out which ion it is. If you're lucky, this problem may be solved very easily. Suppose your solution has a bright green color. From Table 22.1, it is clear that Ni^{2+} must be present. It is the only positive ion with a green

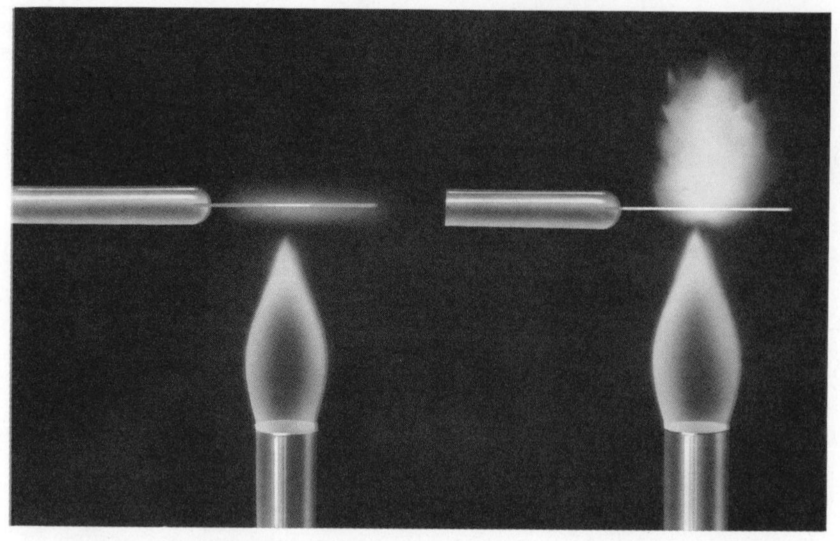

On heating a clean platinum wire

On heating the same wire after dipping it in a solution containing Na^+

FIGURE 22.4

To carry out a flame test, one heats a piece of platinum or chromel wire in a Bunsen flame until any ions from previous tests are vaporized off and the wire glows but gives off no flame. The wire is then dipped in the solution to be tested and put back into the flame. If the solution contains Na^+ ions a yellow flame is produced. If K^+ ion is present in the solution a violet flame is observed.

color in solution. Just to be sure, you might confirm the presence of Ni^{2+} by a simple test. Add dimethylglyoxime. You should obtain a bright red precipitate. If you don't, something is very, very wrong.

If your solution is colorless, the problem is more difficult. Six of the eight positive ions listed in Table 22.1 are colorless. However, it may still be possible to identify the ion by a simple solubility test (Example 22.4).

EXAMPLE 22.4

A solution containing one of the eight positive ions in Table 22.1 is colorless. It gives a precipitate upon addition of HCl. What is the cation present?

SOLUTION

The precipitate formed by adding HCl must contain the Cl^- ion. From Table 22.1, we see that the only positive ion that forms an insoluble chloride is Ag^+. That must be the positive ion present.

Ordinarily, things don't work out quite this simply. Suppose your solution were colorless and did *not* give a precipitate upon addition of HCl. At this stage, any one of five positive ions might be present.

$$Zn^{2+}, Al^{3+}, Mg^{2+}, Na^{+} \text{ or } K^{+}$$

To proceed further, you might add OH^- ions until the solution is just basic to litmus. Suppose that when you do this, you get a white precipitate. Referring to Table 22.1, we see that this eliminates two ions, Na^+ and K^+. Both of these positive ions have soluble hydroxides. You are now down to three possibilities.

$$Zn^{2+}, Al^{3+}, \text{ or } Mg^{2+}$$

At this point, you might test the solubility of your hydroxide in different reagents. From Table 22.1, we see that a good choice here would be 6 M NaOH. If the solid does not dissolve in excess OH^-, it must be $Mg(OH)_2$, which is insoluble in this reagent. Suppose, however, that the hydroxide goes into solution in 6 M NaOH. This would happen with either $Zn(OH)_2$ or $Al(OH)_3$. So, the ion originally present could be:

The identification of the positive ion simply requires a logical interpretation of your experimental observations.

$$Zn^{2+} \text{ or } Al^{3+}$$

To complete the analysis, only one more step is required. Take another portion of your water-insoluble hydroxide and add ammonia. From Table 22.1, we see that $Zn(OH)_2$ is soluble in NH_3; $Al(OH)_3$ is not. So, if the hydroxide dissolves in NH_3 the positive ion originally present must have been Zn^{2+}. If it doesn't dissolve, the positive ion was Al^{3+}. Either way, your problem is solved.

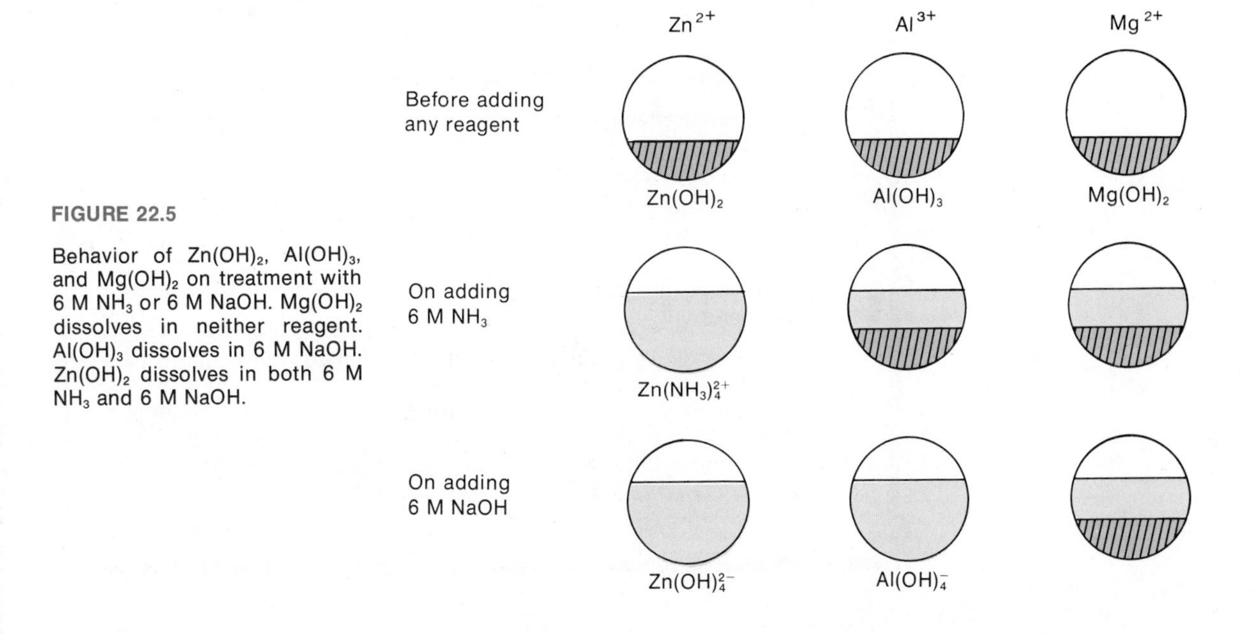

FIGURE 22.5

Behavior of $Zn(OH)_2$, $Al(OH)_3$, and $Mg(OH)_2$ on treatment with 6 M NH_3 or 6 M NaOH. $Mg(OH)_2$ dissolves in neither reagent. $Al(OH)_3$ dissolves in 6 M NaOH. $Zn(OH)_2$ dissolves in both 6 M NH_3 and 6 M NaOH.

EXAMPLE 22.5

A solution which contains one and only one of the positive ions listed in Table 22.1 is colorless. It does not give a precipitate with Cl^- ions. However, addition of OH^- ions does give a precipitate. The solid formed is insoluble in both excess OH^- and NH_3. What is the positive ion?

SOLUTION

The positive ion *cannot* be:

Cu^{2+} or Ni^{2+}	(both are colored in solution)
Ag^+	(would precipitate AgCl)
Na^+ or K^+	(would not precipitate with OH^-)
Al^{3+}	($Al(OH)_3$ is soluble in excess OH^-)
Zn^{2+}	($Zn(OH)_2$ is soluble in NH_3)

It must be Mg^{2+}; $Mg(OH)_2$ is insoluble in water, excess OH^-, or NH_3.

Many other schemes of analysis are possible. One of these is illustrated in Example 22.6.

EXAMPLE 22.6

You are given a solution and asked to determine the positive ion present. You add S^{2-} ions and observe that a black precipitate forms. Referring to Table 22.1:

a. Which positive ions might be present?

b. What further tests would you use to identify the positive ion?

SOLUTION

a. Only three of these ions give black, insoluble sulfides: Ag^+, Cu^{2+}, and Ni^{2+}.

b. Simply looking at the color of the original solution should be enough. The Ag^+ ion is colorless in solution; Cu^{2+} is blue, Ni^{2+} is green. If you want to be entirely sure, you could conduct chemical tests. Chloride ions would give a precipitate with Ag^+, but not with Cu^{2+} or Ni^{2+}. Dimethylglyoxime would give a red precipitate with Ni^{2+} but not with Ag^+ or Cu^{2+}.

22.2 IDENTIFICATION OF A MIXTURE OF IONS

We now turn to a somewhat more difficult problem in qualitative analysis. Suppose you are given a solution which may contain *several* different positive ions. Your task is to identify all the positive ions present. In other words, you cannot stop when you find a single positive ion. You must continue with the analysis, looking for other positive ions that may be present. There is one other complication. Frequently, one ion in solution will interfere with a test for another ion. That factor has to be taken into account in devising a scheme of analysis.

In this type of analysis, the general approach is as follows:

1. Separate the ions, one by one from the solution.
2. Once an ion has been separated, identify it by appropriate tests.

We separate them by precipitation reactions.

This approach works well where the mixture can contain only a few positive ions. This will be the case for all the mixtures considered in this section. We will choose two or three of the positive ions listed in Table 22.1. We will then develop a simple scheme to find out which of those ions are present in an unknown solution.

Ag^+, Mg^{2+}

Suppose you are given a solution which may contain Ag^+ and Mg^{2+} but no other positive ions. You must devise a scheme of analysis to show whether each of these ions is present or absent. There are several ways you could do this. We will consider one simple approach, based on Table 22.1.

The first step might be to add HCl to a test tube containing the solution. This precipitates any Ag^+ ions present as AgCl.

$$Ag^+(aq) + Cl^-(aq) \rightarrow AgCl(s)$$

The HCl should be in excess. Why?

To remove the AgCl, you use a centrifuge (p. 543). After the solid has settled, the clear solution above it is poured off into another test tube. To test for Mg^{2+}, you could add NaOH to this solution. At first, the OH^- ions are consumed in neutralizing the H^+ ions from the HCl. As more NaOH is added, the solution becomes basic to litmus. At this point, any Mg^{2+} ions precipitate as $Mg(OH)_2$:

$$Mg^{2+}(aq) + 2\ OH^-(aq) \rightarrow Mg(OH)_2(s) \qquad (22.7)$$

If a precipitate does not form here, Mg^{2+} must be absent. Similarly, if AgCl does not form in the first step, Ag^+ must be absent.

A procedure such as that just described can be summarized by a diagram called a flow chart. In a flow chart, we show the species present at each point in the analysis. We also show the reagents used in each step. Fig. 22.6, p. 554 is a flow chart for the analysis of a solution that may contain Ag^+ and Mg^{2+}.

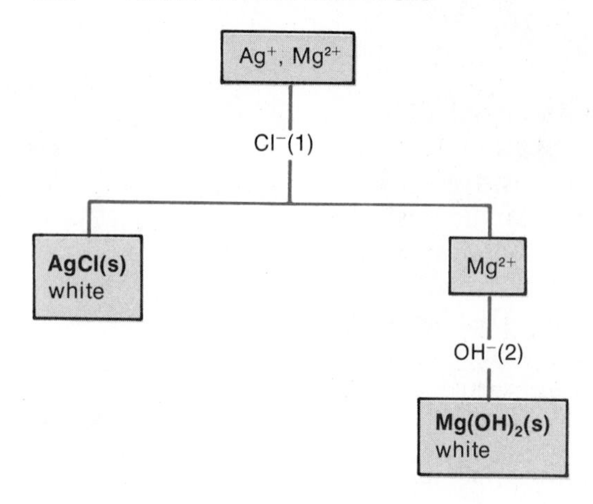

FIGURE 22.6

Flow Chart for Analysis of Ag^+ and Mg^{2+}.

Ag^+, Al^{3+}, K^+

If we work with a solution containing three cations rather than two, the analysis is more difficult. More steps are involved. For an unknown that could contain Ag^+, Al^{3+}, and K^+, we might proceed as follows.

1. Add HCl to precipitate any Ag^+ ions as AgCl. The solid is removed from the solution, which may now contain Al^{3+} and K^+ ions.

Why not use a solution of NaOH?

2. To this solution, add an excess of NH_3. As the solution becomes basic, any Al^{3+} ions precipitate as $Al(OH)_3$.

$$Al^{3+}(aq) + 3\ NH_3(aq) + 3\ H_2O \rightarrow Al(OH)_3(s) + 3\ NH_4^+(aq) \qquad (22.8)$$

The $Al(OH)_3$ is removed by centrifuging. At this stage the solution can contain only K^+ of the three positive ions we started with.

3. As you will note from Table 22.1, K^+ ions do not readily precipitate from solution. To test for K^+, carry out a flame test. A violet color indicates the presence of K^+.

This procedure is summarized in the flow chart shown in Figure 22.7. Given this flow chart, you can easily determine if a solution contains Ag^+, Al^{3+}, or K^+ (Example 22.7).

EXAMPLE 22.7

A solution may contain Ag^+, Al^{3+}, and K^+ but no other positive ions. Addition of HCl does not give a precipitate. Ammonia is then added until the solution is basic to litmus. A white precipitate forms. The solution remaining is tested in a flame. The flame takes on a violet color. Which of the positive ions are present?

SOLUTION

Since no precipitate is obtained with HCl, Ag^+ is not present. The precipitate formed with NH_3 is $Al(OH)_3$. Hence, Al^{3+} must be present. The positive flame test indicates that K^+ is present.

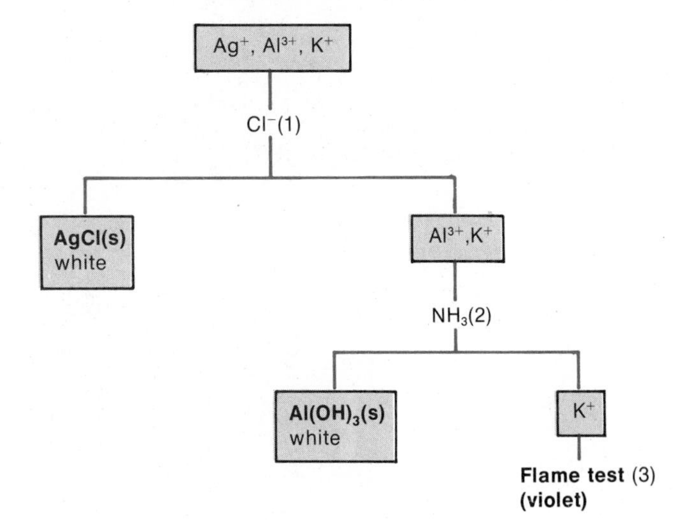

FIGURE 22.7

Flow Chart for Analysis of Ag^+, Al^{3+}, and K^+.

Other Mixtures

Using Table 22.1, you should be able to devise flow charts for any combination of two or three positive ions. Example 22.8 shows how this is done in one case.

EXAMPLE 22.8

A solution may contain Al^{3+}, Mg^{2+}, and Na^+ and no other positive ions. Devise a scheme of analysis which would separate and identify these three ions. Summarize your scheme by constructing a flow chart.

SOLUTION

To devise a scheme of analysis, we need to find ways in which the three ions differ from each other. Note from Table 22.1 that:

—Al^{3+} forms a hydroxide, $Al(OH)_3$, which is soluble in excess OH^-.

—Mg^{2+} forms a hydroxide, $Mg(OH)_2$, which is insoluble in excess OH^-.

—Na^+ gives a yellow flame test; its hydroxide is water-soluble.

We can use these differences to analyze for the three ions. One reasonable procedure is described below.

1. Run a flame test on a portion of the solution. A yellow color indicates that Na^+ is present.
2. To another portion of the solution add excess NaOH. If Mg^{2+} is present, it will precipitate as $Mg(OH)_2$. Any Al^{3+} ions present will precipitate at first as $Al(OH)_3$. However, with excess NaOH, the precipitate will dissolve to form the complex ion $Al(OH)_4^-$.
3. To the solution which may contain $Al(OH)_4^-$, add acid (HCl) until acidic to litmus. The H^+ ions of the acid destroy the $Al(OH)_4^-$ ion, converting it back to Al^{3+}.
4. To test the solution from (3) for Al^{3+}, add NH_3. Any Al^{3+} present will precipitate as $Al(OH)_3$, which is insoluble in NH_3.

The flow chart is shown in Figure 22.8, p. 556.

It's best to test for Na^+ first. Can you see why?

How would you know if Mg^{2+} ion was in the unknown?

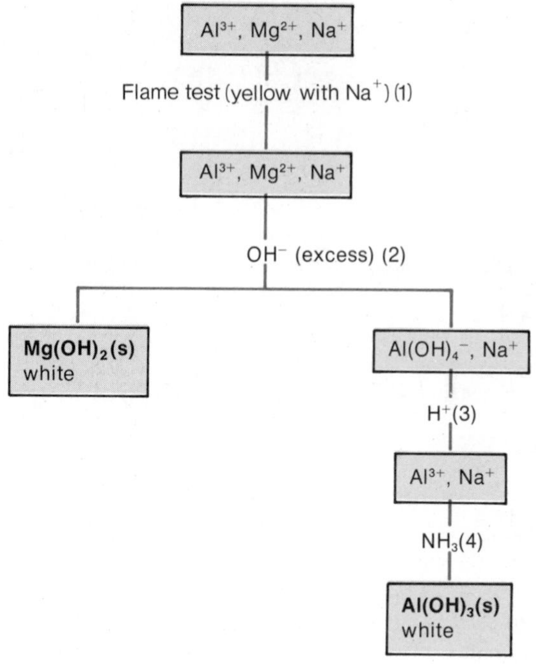

FIGURE 22.8

Flow Chart for Analysis of Al^{3+}, Mg^{2+}, Na^+.

One point concerning Example 22.8 is worth noting. In the third step of the scheme, we added acid to convert $Al(OH)_4^-$ to Al^{3+}. The reaction which takes place is:

$$Al(OH)_4^-(aq) + 4\ H^+(aq) \rightarrow Al^{3+}(aq) + 4\ H_2O \qquad (22.9)$$

This reaction is a general one. Any complex ion formed by adding OH^- ions, such as $Al(OH)_4^-$ or $Zn(OH)_4^{2-}$, can be destroyed by adding H^+ ions. The product is the simple ion (Al^{3+} or Zn^{2+}).

A similar reaction can be used to destroy complex ions formed by NH_3. Addition of acid gives the simple ion. The reaction with $Zn(NH_3)_4^{2+}$ is typical.

$$Zn(NH_3)_4^{2+}(aq) + 4\ H^+(aq) \rightarrow Zn^{2+}(aq) + 4\ NH_4^+(aq) \qquad (22.10)$$

Reactions of this type are often useful in qualitative analysis. Frequently we form a complex ion such as $Zn(OH)_4^{2-}$ or $Zn(NH_3)_4^{2+}$ in one step of a scheme. Later in the analysis, we want to get rid of the complex ion. To do this we add a strong acid, usually HCl, to get back to the simple positive ion.

22.3 THE STANDARD SCHEME FOR QUALITATIVE ANALYSIS

Simple tests of the type we have described are usually sufficient for mixtures containing only two or three positive ions. However, as the

TABLE 22.2 GENERAL SEPARATION SCHEME FOR GROUPS OF POSITIVE IONS

Group	Ions	Conditions for Separation
I	Ag^+, Pb^{2+}, Hg_2^{2+}	Precipitated as chlorides in 1 M HCl solution
II	Cu^{2+}, Cd^{2+}, Bi^{3+}, Sn^{2+}, Sn^{4+}, Sb^{3+}, Sb^{5+}, Hg^{2+}	Precipitated as sulfides at pH 0.5 by 0.1 M H_2S solution
III	Co^{2+}, Ni^{2+}, Cr^{3+}, Fe^{2+}, Fe^{3+}, Mn^{2+}, Al^{3+}, Zn^{2+}	Precipitated as sulfides or hydroxides at pH 9 by 0.1 M H_2S solution
IV	Ba^{2+}, Ca^{2+}, Mg^{2+}, NH_4^+, Na^+, K^+	First three ions are precipitated as carbonates by 0.2 M $(NH_4)_2CO_3$ at pH 10

number of possible ions increases, the scheme of analysis becomes more complex. Interferences between ions become more common. It becomes more and more difficult to devise a scheme that will cleanly separate and identify each ion.

This situation applies with "general unknowns" that may contain *any* common positive ion. There are at least 25 such ions. These are listed in Table 22.2. To work out a scheme of analysis for a solution containing all these ions would be a difficult task.

Fortunately, a detailed scheme of analysis has been developed for a general unknown of this type. It was worked out during the late 19*th* century, but is still being modified and improved. The standard scheme divides the positive ions into four Groups (I, II, III, and IV). These groups are removed, one after the other, from the mixture. Each Group is then analyzed by a well-developed procedure. This procedure separates and identifies the ions within a particular group.

We will not attempt to describe the standard scheme of analysis in detail. As you can imagine, it is rather lengthy. It may be helpful, however, to explain how the four Groups are separated from each other. Later, in Section 22.4, we will consider in some detail how to test for the individual ions in Group I.

Group I

This Group contains two ions with which you are familiar, Ag^+ and Pb^{2+}. The other ion, Hg_2^{2+} is a diatomic positive ion of mercury. It is held together by a covalent bond:

$$(Hg\text{—}Hg)^{2+}$$

These are the only common positive ions that form insoluble chlorides. They are separated from the other Groups by adding 6 M HCl.

The precipitation reactions that occur are:

$$Ag^+(aq) + Cl^-(aq) \rightarrow AgCl(s) \quad (22.11)$$

$$Pb^{2+}(aq) + 2\ Cl^-(aq) \rightarrow PbCl_2(s) \quad (22.12)$$

$$Hg_2{}^{2+}(aq) + 2\ Cl^-(aq) \rightarrow Hg_2Cl_2(s) \quad (22.13)$$

Group II

This Group contains a total of eight positive ions. Included are the common positive ions derived from copper (Cu^{2+}), cadmium (Cd^{2+}), bismuth (Bi^{3+}), tin (Sn^{2+}, Sn^{4+}), and antimony (Sb^{3+}, Sb^{5+}). Finally, another positive ion of mercury, Hg^{2+}, is in this Group.

The positive ions in Group II have one property in common. They form extremely insoluble sulfides. For example, the solubility product constant of CuS is a very small number, 1×10^{-36}. That of HgS is even smaller, 1×10^{-52}. Indeed, HgS is one of the least soluble of all known compounds.

The Group II ions are separated by treating with H_2S in acidic solution, at pH = 0.5. Under these conditions, the concentration of S^{2-} ions, produced by the reaction:

$$H_2S(aq) \rightleftharpoons 2\ H^+(aq) + S^{2-}(aq) \quad (22.14)$$

is very low. However, it is still high enough to precipitate the Group II sulfides. Typical precipitation reactions include:

The equations involve H_2S, since H_2S, not S^{2-}, is present at appreciable concentration.

$$Cu^{2+}(aq) + H_2S(aq) \rightarrow CuS(s) + 2\ H^+(aq) \quad (22.15)$$

$$2\ Bi^{3+}(aq) + 3\ H_2S(aq) \rightarrow Bi_2S_3(s) + 6\ H^+(aq) \quad (22.16)$$

Several of the Group II sulfides have characteristic colors. (See Color Plate 10, in the center of the book.)

Group III

The positive ions in this Group include those of several familiar transition metals. Among these are the ions derived from chromium (Cr^{3+}), manganese (Mn^{2+}), iron (Fe^{2+}, Fe^{3+}), cobalt (Co^{2+}), nickel (Ni^{2+}), and zinc (Zn^{2+}). The aluminum ion, (Al^{3+}), is also in this Group.

As you might suppose, the chlorides of these ions are soluble in water. Hence, they do not precipitate with the Group I ions. Again, the positive ions in Group III are not precipitated by H_2S in acidic solution. Their sulfides are considerably more soluble than those of the Group II ions. Compare, for example, MnS ($K_{sp} = 1 \times 10^{-13}$) to CuS ($K_{sp} = 1 \times 10^{-36}$).

To separate this Group from a general unknown, the solution is saturated with H_2S in basic solution (pH = 9). The concentration of H^+ is low enough to drive the equilibrium in Reaction 22.14 far to the right. Under these conditions, enough sulfide ions are formed to precipitate the following Group III sulfides:

CoS	*NiS*	*MnS*	*ZnS*	*FeS*
black	black	pink	white	black

Would you expect these sulfides to dissolve in 6 M HCl?

The two ions Cr^{3+} and Al^{3+} behave differently. The OH^- ions present in the basic solution cause them to precipitate as hydroxides:

$Al(OH)_3$	*$Cr(OH)_3$*
white	green

Group IV

This Group contains

—three ions derived from metals in Group 2 of the Periodic Table (Mg^{2+}, Ca^{2+}, Ba^{2+})
—the ions of the two most common metals in Group 1 of the Periodic Table (Na^+, K^+)
—the ammonium ion, NH_4^+

These six positive ions form fewer insoluble compounds than those in the first three Groups. They do not precipitate with 6 M HCl (Group I), H_2S at pH = 0.5 (Group II), or H_2S at pH = 9 (Group III). However, three positive ions in Group IV can be precipitated with CO_3^{2-} ions:

$$Mg^{2+}(aq) + CO_3^{2-}(aq) \rightarrow MgCO_3(s) \quad (22.17)$$

Would these carbonates dissolve in 6 M HCl?

$$Ca^{2+}(aq) + CO_3^{2-}(aq) \rightarrow CaCO_3(s) \quad (22.18)$$

$$Ba^{2+}(aq) + CO_3^{2-}(aq) \rightarrow BaCO_3(s) \quad (22.19)$$

The other three ions are usually tested for individually. No attempt is made to separate them from the mixture. Flame tests can be used for Na^+ and K^+. To test for NH_4^+, the solution is heated with OH^- ions:

$$NH_4^+(aq) + OH^-(aq) \rightarrow NH_3(g) + H_2O \quad (22.20)$$

The ammonia produced is easily detected (Figure 22.9 p. 560). It is the only common gas that turns red litmus blue.

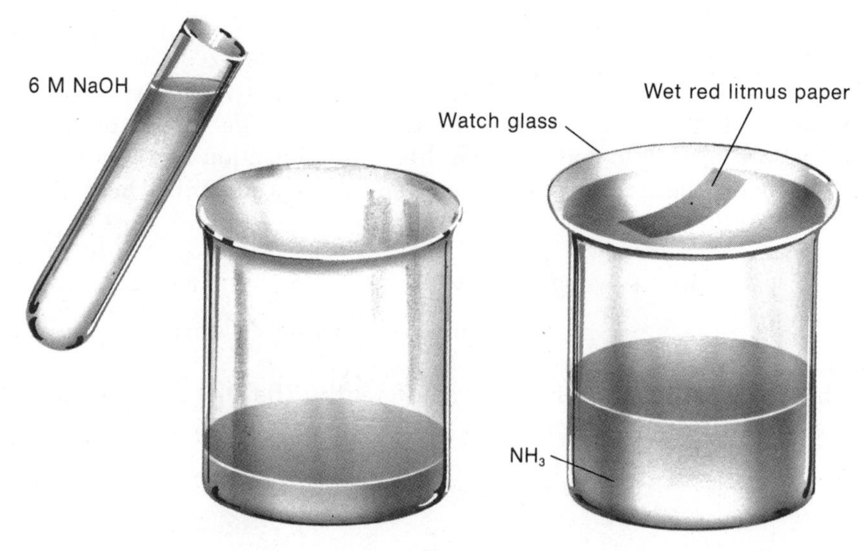

FIGURE 22.9

Procedure for testing for the presence of NH_4^+ ion.

22.4 ANALYSIS OF THE GROUP I IONS: Ag^+, Pb^{2+}, Hg_2^{2+}

As we have pointed out, the Group I ions are separated from a general unknown by adding HCl. They precipitate as the chlorides AgCl, $PbCl_2$, Hg_2Cl_2. This precipitate must then be treated so as to separate and identify the positive ions.

The first step in the analysis of Group I is to heat the chloride precipitate with water. This has no effect upon AgCl or Hg_2Cl_2. However, $PbCl_2$ dissolves in hot water:

If a simple way to separate ions exists, we might as well use it.

$$PbCl_2(s) \rightarrow Pb^{2+}(aq) + 2\ Cl^-(aq) \tag{22.21}$$

In this way, Pb^{2+} is separated from the other two ions and brought into solution. To test the hot solution for Pb^{2+}, CrO_4^{2-} ions are added. This gives a bright yellow precipitate of lead chromate if Pb^{2+} is present:

$$Pb^{2+}(aq) + CrO_4^{2-}(aq) \rightarrow PbCrO_4(s) \tag{22.22}$$

At this stage, the original chloride precipitate can contain only AgCl and Hg_2Cl_2. To distinguish between these compounds, the precipitate is treated with NH_3. Ammonia reacts with Hg_2Cl_2 in a rather unusual way. Two insoluble substances are produced. One of these is finely divided mercury, Hg, which is black. The other is a white compound whose formula is $HgNH_2Cl$. The reaction is:

$$\underset{\text{white}}{Hg_2Cl_2(s)} + 2\ NH_3(aq) \rightarrow \underset{\text{black}}{Hg(l)} + \underset{\text{white}}{HgNH_2Cl(s)} + NH_4^+(aq) + Cl^-(aq) \tag{22.23}$$

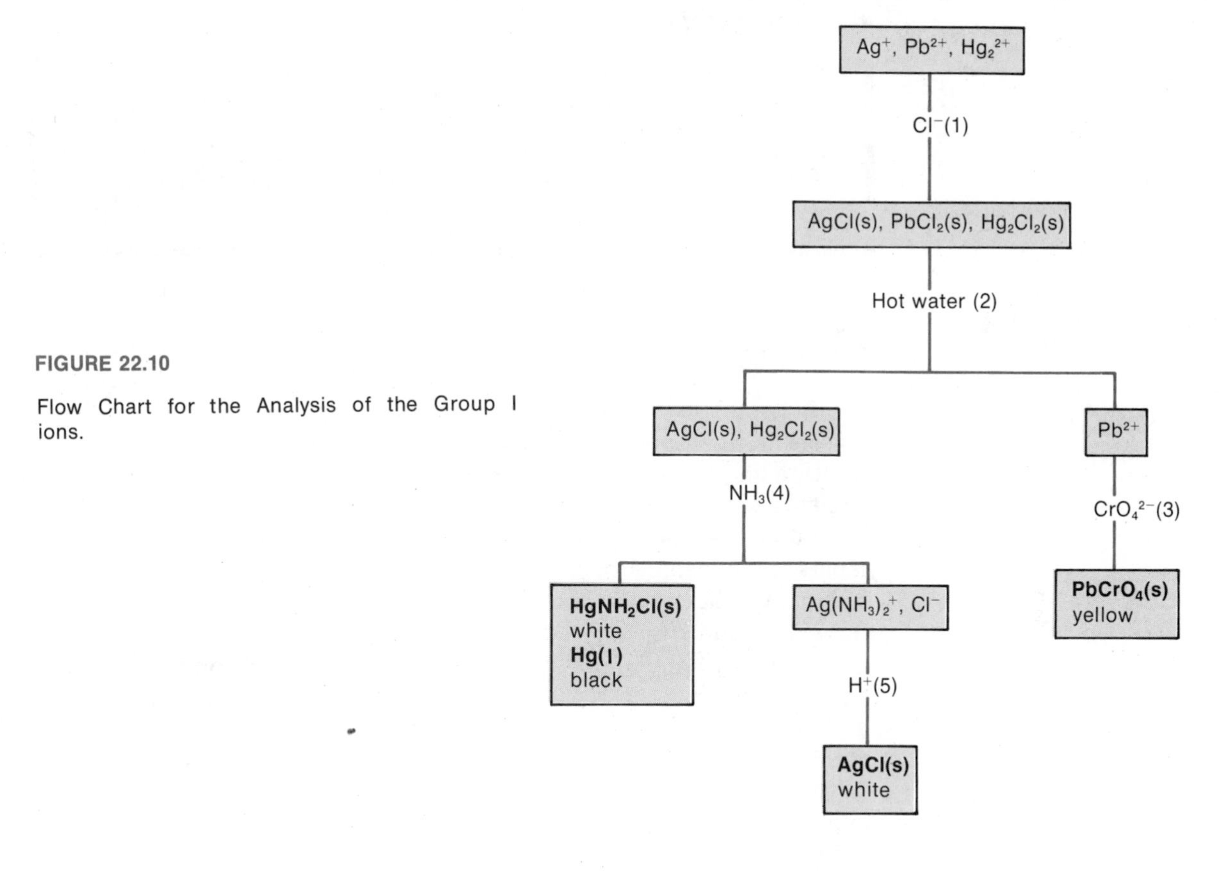

FIGURE 22.10

Flow Chart for the Analysis of the Group I ions.

The formation of this mixed precipitate upon addition of NH_3 shows the presence of Hg_2^{2+} ions.

The behavior of AgCl with NH_3 is quite different. It goes into solution, forming the $Ag(NH_3)_2^+$ complex ion:

$$AgCl(s) + 2\ NH_3(aq) \rightarrow Ag(NH_3)_2^+(aq) + Cl^-(aq) \qquad (22.24)$$

To test this solution for the presence of silver, we add acid. This destroys the $Ag(NH_3)_2^+$ ion, forming free Ag^+. This ion then reacts with Cl^- ions in the solution to precipitate white, insoluble AgCl. The reaction is:

The Group I procedure illustrates most of the kinds of reactions that are used in qualitative analysis.

$$Ag(NH_3)_2^+(aq) + 2\ H^+(aq) \rightarrow Ag^+(aq) + 2\ NH_4^+(aq)$$
$$Ag^+(aq) + Cl^-(aq) \rightarrow AgCl(s)$$
$$\overline{Ag(NH_3)_2^+(aq) + 2\ H^+(aq) + Cl^-(aq) \rightarrow AgCl(s) + 2\ NH_4^+(aq)} \qquad (22.25)$$

EXAMPLE 22.9

A solution may contain Ag^+, Pb^{2+}, Hg_2^{2+} but no other positive ions. When the solution is treated with 6 M HCl, a white precipitate forms. The precipitate is unaffected by hot water but completely soluble in 6 M NH_3. Which ions are present, which are absent, and which are still undetermined?

SOLUTION

Since the precipitate is completely insoluble in hot water, Pb^{2+} must be missing. $PbCl_2$ would tend to dissolve when treated with hot water. The fact that the precipitate dissolves in NH_3 tells us that the unknown contains Ag^+ and that Hg_2^{2+} is absent. AgCl dissolves in 6 M NH_3 and Hg_2Cl_2 turns black. No ions are still in doubt.

22.5 SUMMARY

In qualitative analysis, the goal is to find out what species are present in a sample. The qualitative analysis for positive ions in water solution may be carried out using their chemical properties. The method depends on separating the ions into Groups. Groups I through IV are separated one after the other from the unknown solution. Following separation of each Group precipitate, it is analyzed by a standard procedure. We have considered one such procedure, for the analysis of Ag^+, Pb^{2+}, and Hg_2^{2+} in Group I.

Schemes for the analysis of mixtures containing only a few ions can easily be devised. Here, we use the information in solubility tables. In using such schemes, reactions involving H^+ ions, OH^- ions, or NH_3 molecules are important. Many ionic compounds which are insoluble in water dissolve in acidic solution. Others can be brought into solution by OH^- or NH_3, which form complex ions with the positive ion of the compound.

NEW TERMS

flow chart   qualitative analysis

QUESTIONS

1. Referring to Table 22.1, which negative ions (one or more):
   a. do not give a precipitate with any of the eight positive ions?
   b. give a precipitate with only one positive ion?
   c. give white precipitates with Zn^{2+}?
   d. give black precipitates with certain positive ions?
   e. do not give a precipitate with Na^+?

2. Referring to Table 22.1, which positive ions (one or more):
   a. form insoluble chlorides?
   b. are colored in solution?
   c. give colored flames?

3. Referring to Table 22.1, give the formulas of all the water-insoluble hydroxides that:
   a. dissolve in 6 M H^+.
   b. do not dissolve in 6 M OH^-.
   c. do not dissolve in 6 M NH_3.
   d. do not dissolve in either 6 M OH^- or 6 M NH_3.

4. Referring to Table 22.1, state which reagents, 6 M H^+, 6 M OH^-, or 6 M NH_3, will dissolve:

   a. ZnS b. $Zn(OH)_2$ c. $Ni(OH)_2$ d. CuS

5. Which of the following reagents bring precipitates into solution by forming complex ions?

   a. H^+ b. NH_3 c. OH^- d. hot water

6. Describe what you would observe when a solution of NaOH is added, drop by drop, to a solution containing Al^{3+}. (Two successive reactions occur.)

7. When a solution of ammonia is added to a solution containing Mg^{2+}, a precipitate forms.

   a. What is its formula?
   b. Where do the negative ions in the precipitate come from?

8. When ammonia is added to a solution containing Zn^{2+}, a white precipitate forms at first. As more ammonia is added, the precipitate dissolves. Explain what is happening, using balanced equations.

9. The analysis for a mixture containing Ag^+, Al^{3+} and K^+ is described on p. 554.

   a. In the first step, why not add KCl instead of HCl?
   b. In the second step, why not add excess NaOH instead of NH_3?
   c. In the third step, why not precipitate K^+ instead of using a flame test?

10. You are given a solution which may contain Ag^+ and Cu^{2+} and no other positive ions. You have thirty seconds to determine whether each ion is present or absent. What would you do?

11. A solution contains the $Zn(OH)_4^{2-}$ complex ion. It is made acidic by adding H^+ ions. Write a balanced equation for the reaction that occurs.

12. You have a solution containing the $Cu(NH_3)_4^{2+}$ complex ion. What reagent would you use to obtain the simple Cu^{2+} ion from the complex ion? Write a balanced equation for the reaction involved.

13. Brass is an alloy of copper and zinc. It can be dissolved by heating with nitric acid. In the solution formed, would you expect to find ions in:

    a. Group I? b. Group II? c. Group III? d. Group IV?

14. Of the four groups of positive ions in the standard scheme, which ones contain:

    a. transition metal ions?
    b. ions derived from mercury?
    c. ions of the metals in Groups 1 and 2 of the Periodic Table?

15. Write balanced equations for the reactions that occur in the standard scheme when H_2S is used to precipitate:

    a. Cu^{2+} b. Bi^{3+} c. Mn^{2+} d. Sb^{3+}

16. Write a balanced equation for the reaction that occurs when hydroxide ions are used to precipitate:

    a. Cr^{3+} b. Al^{3+}

17. In separating Group II from a general unknown, a student adds H_2S in basic solution. To his surprise, he finds that his precipitate contains several ions which do not belong in Group II. Where did he go wrong?

18. How would you test for NH_4^+ in a Group IV unknown?

19. What property do the ions in Group I have in common?

20. Of the Group I chlorides, which ones are soluble in

    a. cold water b. hot water c. NH_3

21. Write a balanced equation for the reaction that occurs when:

    a. NH_3 is added to a precipitate of Hg_2Cl_2.
    b. CrO_4^{2-} ions are added to a solution containing Pb^{2+}.

PROBLEMS

1. *Writing Equations (Example 22.1)* Write balanced equations for the reactions, if any that occur when:

   a. OH^- and S^{2-} ions are added to portions of a solution containing Zn^{2+}.
   b. H^+ ions are added to any precipitates formed in (a).

2. *Writing Equations (Example 22.2)* Write balanced equations for all reactions that occur when 6 M NaOH is added to a solution that contains:

   a. Mg^{2+} b. Zn^{2+}

3. *Writing Equations (Example 22.3)* Write balanced equations for the two successive reactions that occur when 6 M NH_3 is added to a solution containing Cu^{2+}.

4. *Analysis for a Single Ion (Examples 22.4, 22.5, 22.6)* A colored solution containing one of the positive ions in Table 22.1 gives a red precipitate with dimethylglyoxime. Identify the positive ion.

5. *Analysis for a Single Ion (Example 22.4, 22.5, 22.6)* A colorless solution containing one of the positive ions in Table 22.1 does not give a precipitate with Cl^-. It does give a precipitate with OH^- which is soluble in either 6 M NH_3 or 6 M NaOH. What is the positive ion?

6. *Analysis for a Single Ion (Examples 22.4, 22.5, 22.6)* A solution containing one of the positive ions in Table 22.1 does not give a precipitate with Cl^- or OH^-.

   a. What ions might be present?
   b. How would you decide which one of the ions in (a) is present?

7. *Analysis of a Mixture (Example 22.7)* A solution may contain Ag^+, Al^{3+}, and K^+ but no other positive ions. Addition of HCl gives a precipitate. Addition of excess NH_3 to the solution remaining does not give a precipitate. Classify each positive ion as present, absent, or undetermined.

8. *Analysis of a Mixture (Example 22.8)* A solution may contain Cu^{2+}, Zn^{2+} and K^+ but no other positive ion. Devise a scheme of analysis for such a mixture and summarize it in a flow chart.

9. *Analysis of Group I (Example 22.9)* The chloride precipitate of Group I is heated with water. The solution formed gives a yellow precipitate when CrO_4^{2-} ions are added. The precipitate remaining is completely soluble in NH_3. Which ions are present? absent?

* * * * *

10. On a separate sheet of paper, complete and balance the following equations, if a reaction occurs (see Table 22.1).

a. $Ni(OH)_2(s) + H^+(aq) \rightarrow ?$ b. $ZnS(s) + H^+(aq) \rightarrow ?$
c. $CuS(s) + H^+(aq) \rightarrow ?$ d. $AgCl(s) + H^+(aq) \rightarrow ?$

11. Complete and balance the following equations, if a reaction occurs (see Table 22.1).

a. $Zn(OH)_2(s) + OH^-(aq) \rightarrow ?$ b. $Al(OH)_3(s) + OH^-(aq) \rightarrow ?$
c. $Ni(OH)_2(s) + OH^-(aq) \rightarrow ?$

12. Complete and balance the following equations, if a reaction occurs (see Table 22.1).

a. $Ni(OH)_2(s) + NH_3(aq) \rightarrow ?$ b. $CuS(s) + NH_3(aq) \rightarrow ?$
c. $AgCl(s) + NH_3(aq) \rightarrow ?$

13. A solution which contains one and only one of the positive ions in Table 22.1 gives a colored precipitate with OH^-.
a. What positive ions might be present?
b. How would you determine which one of the ions in (a) is present?

14. Draw a flow chart for a mixture which may contain Mg^{2+} and Al^{3+} but no other positive ions.

15. Draw a flow chart for a mixture which may contain Ag^+, Zn^{2+} and Na^+ but no other positive ions.

16. A solution may contain Cu^{2+} and Al^{3+} but no other positive ions. It is treated with NH_3. A blue precipitate forms at first but completely dissolves as more NH_3 is added. Referring to Table 22.1, what ions are present? absent?

17. A solution may contain Cu^{2+}, Zn^{2+}, and Mg^{2+} but no other positive ions. Ammonia is added to the solution. A precipitate forms at first. This precipitate partially dissolves as more NH_3 is added. Which ions are present? definitely absent? doubtful? What further tests would you make to identify the ions in the solution?

18. A Group I precipitate is completely soluble in hot water. Which ions are present? absent?

19. A Group I precipitate turns gray on addition of NH_3. What ion is definitely present? How would you test for the other two ions?

*20. For the reaction: $H_2S(aq) \rightleftharpoons 2\,H^+(aq) + S^{2-}(aq)$; $K = 1 \times 10^{-22}$.
When $[H_2S] = 0.1$ M, what is $[S^{2-}]$ if:

a. $[H^+] = 0.3$ M b. $[H^+] = 10^{-9}$

*21. What reagent would you use to convert:

a. $Ag(NH_3)_2{}^+$ to Ag^+ b. $Ag(NH_3)_2{}^+$ to $AgCl$
c. ZnS to Zn^{2+} d. $Zn(OH)_4{}^{2-}$ to $Zn(NH_3)_4{}^{2+}$ (two reagents)

*22. How would you distinguish between:

a. $AlCl_3$ and $Al(OH)_3$ b. $Al(OH)_3$ and $Zn(OH)_2$
c. $MgCl_2$ and $ZnCl_2$ d. CuS and $Cu(OH)_2$

*23. Using an advanced textbook or other reference source, find out what happens when the following ions are mixed:
a. Ag^+ and OH^- b. Al^{3+} and S^{2-} c. Mg^{2+} and S^{2-}
(These are the entries marked "D" in Table 22.1)

The Lunar Excursion Module above the surface of the moon—see page 588. *Photo courtesy of NASA.*

23

OXIDATION-REDUCTION REACTIONS

An oxidation-reduction reaction involves the transfer of electrons from one species to another. Many of the reactions considered in previous chapters have been of this type. Some of these involve pure substances. An example is the reaction between the two elements, sodium and chlorine:

$$2\ Na(s) + Cl_2(g) \rightarrow 2\ NaCl(s)$$

Others, such as the reaction of zinc with acid:

$$Zn(s) + 2\ H^+(aq) \rightarrow Zn^{2+}(aq) + H_2(g)$$

take place in water solution.

In this chapter, we will examine how oxidation-reduction (*"redox"*) reactions take place (Section 23.1). We will consider how they are translated into balanced chemical equations (Section 23.2). Finally, in Section 23.3, we will look at the chemistry of some of the substances that take part in these reactions.

23.1 OXIDATION AND REDUCTION

Consider what happens when a strip of magnesium is heated with sulfur:

$$Mg(s) + S(s) \rightarrow MgS(s) \qquad (23.1)$$

The reactants, magnesium and sulfur, consist of neutral atoms. The product, magnesium sulfide, is ionic. It is made up of equal numbers of Mg^{2+} and S^{2-} ions.

To better understand Reaction 23.1, we break it down into two "half-reactions". One of these involves the loss of two electrons by a Mg atom:

$$Mg \rightarrow Mg^{2+} + 2\ e^- \qquad (23.1a)$$

Reactions 23.1a and 23.1b occur simultaneously.

In the other half-reaction, a sulfur atom gains two electrons:

$$S + 2\ e^- \rightarrow S^{2-} \qquad (23.1b)$$

The overall reaction (23.1) is simply the sum of these two half-reactions.

A half-reaction such as 23.1a, in which a species **loses electrons,** is referred to as **oxidation.** Other examples of oxidation half-reactions are:

$$Na \rightarrow Na^+ + e^-$$

$$Ca \rightarrow Ca^{2+} + 2\ e^-$$

$$Al \rightarrow Al^{3+} + 3\ e^-$$

A half-reaction such as 23.1b, in which a species **gains electrons,** is referred to as **reduction.** Other examples of reduction half-reactions are:

$$Cl_2 + 2\ e^- \rightarrow 2\ Cl^-$$

$$F_2 + 2\ e^- \rightarrow 2\ F^-$$

$$\tfrac{1}{2}\ O_2 + 2\ e^- \rightarrow O^{2-}$$

The two processes, oxidation and reduction, cannot occur separately. Any oxidation-reduction reaction is the sum of two half-reactions. Consider, for example, the reaction between sodium and chlorine.

$$2\ Na(s) \rightarrow 2\ Na^+ + 2\ e^- \quad \text{oxidation}$$
$$Cl_2(g) + 2\ e^- \rightarrow 2\ Cl^- \quad \text{reduction}$$
$$\overline{2\ Na(s) + Cl_2(g) \rightarrow 2\ NaCl(s)} \quad \text{oxidation-reduction} \tag{23.2}$$

The reaction between zinc and H^+ ions can be analyzed in the same way.

$$Zn(s) \rightarrow Zn^{2+}(aq) + 2\ e^- \quad \text{oxidation}$$
$$2\ H^+(aq) + 2\ e^- \rightarrow H_2(g) \quad \text{reduction}$$
$$\overline{Zn(s) + 2\ H^+(aq) \rightarrow Zn^{2+}(aq) + H_2(g)} \quad \text{oxidation-reduction} \tag{23.3}$$

Looking at Equations 23.2 and 23.3, we see an important property of redox reactions. *The electron gain exactly balances the electron loss.* In 23.3, the two electrons lost by the Zn atom are picked up by the two H^+ ions. Again, in the reaction of aluminum with fluorine

$$2\ Al(s) + 3\ F_2(g) \rightarrow 2\ AlF_3(s) \tag{23.4}$$

there is no net change in the number of electrons. Six electrons are lost in the oxidation half-reaction:

$$2\ Al(s) \rightarrow 2\ Al^{3+} + 6\ e^-$$

The same number of electrons, 6, is gained in the reduction half-reaction:

$$3\ F_2(g) + 6\ e^- \rightarrow 6\ F^-$$

This means that electrons are not created, destroyed, or mislaid during the chemical reaction.

Notice that none of the oxidation-reduction reactions we have considered (23.1–23.4) involve oxygen. Despite its name, "oxidation" does not require reaction with oxygen. A species is oxidized whenever it loses electrons. Sometimes this comes about by reaction with oxygen:

$$2\ Mg(s) + O_2(g) \rightarrow 2\ MgO(s) \tag{23.5}$$

(Mg oxidized to Mg^{2+}; O_2 reduced to O^{2-})

Anything that burns in air is oxidized.

More frequently, it does not involve oxygen:

$$Mg(s) + Cl_2(g) \rightarrow MgCl_2(s) \tag{23.6}$$

(Mg oxidized to Mg^{2+}; Cl_2 reduced to Cl^-)

$$Mg(s) + F_2(g) \rightarrow MgF_2(s) \tag{23.7}$$

(Mg oxidized to Mg^{2+}; F_2 reduced to F^-)

All the reactions considered so far are relatively simple ones. They involve atoms (Mg, Al, . . .), simple molecules (Cl_2, O_2, . . .), or monatomic ions (Mg^{2+}, Al^{3+}, Cl^-, O^{2-}, . . .). Here, we can readily see which species is losing electrons and which one is gaining them. However, many redox reactions are more complex. They may involve molecules such as HNO_3 or polyatomic ions such as CrO_4^{2-}. For such reactions, it is helpful to refine the concepts of oxidation and reduction. To do this, we introduce the quantity known as oxidation number.

Oxidation Number

Consider the reaction between H_2 and Cl_2:

$$H_2(g) + Cl_2(g) \rightarrow 2\ HCl(g) \quad (23.8)$$

In many ways, this resembles the reaction between sodium and chlorine. The major difference is that the product, HCl, is molecular rather than ionic. It has the Lewis structure:

$$H{-}\ddot{\underset{..}{Cl}}:$$

However, the bonding electrons in HCl are not equally shared. Instead, they are displaced toward the more electronegative chlorine atom. So far as "electron bookkeeping" is concerned, we could assign these electrons to the chlorine atom.

$$H \mid :\ddot{\underset{..}{Cl}}:$$

In doing this, in effect we give the chlorine atom a −1 charge. It now controls 8 valence electrons as compared to 7 in the neutral atom. The hydrogen atom now has no electrons. It originally had 1. Hence, we have assigned it a +1 charge.

The process we have just gone through leads to the concept of oxidation number. In a molecule or polyatomic ion, the oxidation number of an atom is the charge it would have if all the bonding electrons were assigned to the more electronegative atom. In the HCl molecule:

Oxidation numbers are calculated rather than measured experimentally.

the oxidation number of Cl is −1.
the oxidation number of H is +1.

In principle, we could assign oxidation numbers in any species by the procedure used with HCl. In practice, we almost never do it this way. For one thing, it would require writing Lewis structures for each species. There is a much simpler approach which gives the same results. It uses a set of rules listed below.

1. **The oxidation number of an atom in any elementary substance is 0.** For example, the oxidation number of chlorine in the Cl_2 molecule is 0.

2. **The oxidation number of an atom in a simple, monatomic ion is the charge of that ion.** The oxidation number of sodium in Na^+ is +1. That of sulfur in the S^{2-} ion is −2.

3. **Certain elements have the same oxidation number in all or nearly all of their compounds.** The most important cases are those listed below.

a. *The Group 1 metals have an oxidation number of +1 in all their compounds.* Thus sodium in NaCl has an oxidation number of +1 (it is present as a +1 ion). Again, potassium in K_2SO_4 has an oxidation number of +1.

b. *The Group 2 metals have an oxidation number of +2 in all their compounds.* For example, calcium in $Ca(NO_3)_2$ has an oxidation number of +2.

If you learn these rules, you will be able to easily find oxidation numbers.

c. *Hydrogen, in all of its compounds with nonmetals, has an oxidation number of +1.* This is the case in such common molecules as H_2O, NH_3, HCl, and HNO_3.

d. *Oxygen, with a few exceptions, has an oxidation number of −2 in its compounds.* This is true in molecules such as H_2O and HNO_3 and polyatomic ions such as OH^- and CrO_4^{2-}

4. **The sum of the oxidation numbers of all the atoms in any neutral species is 0.** For example:

$$\text{in } H_2O,\ 2 \times \text{oxid. no. H} + \text{oxid. no. O} = 0$$
$$\text{in } HNO_3,\ \text{oxid. no. H} + \text{oxid. no. N} + 3 \times \text{oxid. no. O} = 0$$

5. **The sum of the oxidation numbers of all the atoms in a polyatomic ion is the charge of the ion.** For example, in the CrO_4^{2-} ion:

$$\text{oxid. no. Cr} + 4 \times \text{oxid. no. O} = -2$$

Using these rules, it is possible to assign oxidation numbers to atoms in many different species. Example 23.1 illustrates how this is done.

EXAMPLE 23.1

Assign oxidation numbers to

a. N in HNO_3 b. Cr in CrO_4^{2-} c. N in $NaNO_2$

SOLUTION

a. By rule 4: oxid. no. H + oxid. no. N + 3 × oxid. no. O = 0
By rule 3: oxid. no. H = +1; oxid. no. O = −2
So: +1 + oxid. no. N + 3(−2) = 0
Solving: oxid. no. N = 3(2) − 1 = +5

b. By rule 5: oxid. no. Cr + 4 × oxid. no. O = −2
By rule 3: oxid. no. O = −2
So: oxid. no. Cr + 4(−2) = −2
Solving: oxid. no. Cr = −2 + 8 = +6

c. By rule 4: oxid. no. Na + oxid. no. N + 2 × oxid. no. O = 0
By rule 3: oxid. no. Na = +1; oxid. no. O = −2
So: +1 + oxid. no. N + 2(−2) = 0
Solving: oxid. no. N = 4 − 1 = +3

The concept of oxidation number leads directly to a working definition of oxidation and reduction:

Oxidation = Increase in Oxidation Number
Reduction = Decrease in Oxidation Number

This definition agrees with our earlier one in terms of the loss and gain of electrons. Thus, for the half-reaction:

$$Al \rightarrow Al^{3+} + 3\ e^-$$

we say that aluminum is oxidized. It loses electrons *and* increases in oxidation number, from 0 to +3. Similarly, for the half-reaction:

When the oxidation number of an atom changes, its properties change dramatically.

$$Cl_2 + 2\ e^- \rightarrow 2\ Cl^-$$

we conclude, from either point of view, that Cl_2 is reduced. It gains electrons *and* decreases in oxidation number, from 0 to −1.

With more complex half-reactions, the definitions in terms of oxidation number are more easily applied. Example 23.2 illustrates how this is done.

EXAMPLE 23.2

State whether nitrogen is being oxidized or reduced in each of the following processes.

a. $N_2 \rightarrow HNO_3$ b. $N_2 \rightarrow NH_3$ c. $NaNO_2 \rightarrow HNO_2$

SOLUTION

a. The oxidation number of N in N_2 is 0 (rule 1). In HNO_3, it is +5 (Example 23.1 a). Hence, nitrogen is increasing in oxidation number from 0 to +5. It is oxidized.

b. To find the oxidation number of nitrogen in NH_3, we start by assigning hydrogen an oxidation number of +1 (Rule 3). Applying rule 4:

$$\text{oxid. no. N} + 3(+1) = 0;\ \text{oxid. no. N in } NH_3 = -3$$

Nitrogen is decreasing in oxidation number, from 0 to −3. So, it is being reduced.

Oxidation numbers have both signs and magnitudes.

c. In Example 23.1 c, we found the oxidation number of N in $NaNO_2$ to be +3. In HNO_2:

$$\text{oxid. no. H} + \text{oxid. no. N} + 2 \times \text{oxid. no. O} = 0$$

$$+1 + \text{oxid. no. N} + 2(-2) = 0$$

$$\text{oxid. no. N} = +4 - 1 = +3$$

The oxidation number of nitrogen remains constant. It is neither oxidized nor reduced.

Oxidizing and Reducing Agents

In describing redox reactions, we make use of the terms "oxidizing agent" and "reducing agent". The oxidizing agent is the species which brings about oxidation. Similarly, the reducing agent is responsible for reduction.

To illustrate the use of these terms, consider the reaction:

$$2\ Mg(s) + O_2(g) \rightarrow 2\ MgO(s)$$

For this reaction, we can say that:

Mg is oxidized (oxidation number increases from 0 to +2)
O_2 is reduced (oxidation number decreases from 0 to −2)
Mg is the reducing agent (it reduces O_2)
O_2 is the oxidizing agent (it oxidizes Mg)

You will notice from this example that:

1. *A species which acts as a reducing agent is itself oxidized.* It shows an increase in oxidation number. Mg, in reducing O_2, is itself oxidized to Mg^{2+}.

2. *A species which acts as an oxidizing agent is itself reduced.* It shows a decrease in oxidation number. O_2, in oxidizing Mg, is itself reduced to O^{2-}.

Oxygen is a good oxidizing agent.

23.2 BALANCING OXIDATION-REDUCTION EQUATIONS

Many redox reactions lead to simple equations. These can be balanced by the "trial-and-error" method described in Chapter 3. Most reactions between pure substances are of this type. Equations such as:

$$Mg(s) + Cl_2(g) \rightarrow MgCl_2(s)$$

$$2\ Na(s) + Cl_2(g) \rightarrow 2\ NaCl(s)$$

are easy to balance.

However, certain oxidation-reduction reactions in water solution are a bit more complex. For such reactions, we need a general approach to balancing equations. The approach used here is the **half-equation method.** It involves dividing the equation into two parts, oxidation and reduction. The half-equations for these processes are balanced separately. They are then combined to give a single balanced equation for the overall reaction.

To illustrate the half-equation approach, let us start with a simple example. Consider the oxidation-reduction reaction:

$$Al(s) + Cl_2(g) \rightarrow Al^{3+}(aq) + Cl^-(aq)$$

To balance this equation, we follow a three-step approach:

1. **Divide the equation into two half-equations, an oxidation and a reduction.** To do this, we first establish the oxidation numbers in each species.

$$\text{Al: 0 in Al(s)} \rightarrow +3 \text{ in } Al^{3+}(aq)$$

$$\text{Cl: 0 in } Cl_2(g) \rightarrow -1 \text{ in } Cl^{-}(aq)$$

We see that Al is increasing in oxidation number. Cl is decreasing. Hence:

$$\text{oxidation half-equation: } Al(s) \rightarrow Al^{3+}(aq)$$

$$\text{reduction half-equation: } Cl_2(g) \rightarrow Cl^{-}(aq)$$

The two half-equations are usually easy to write.

2. **Balance the half-equations, first with respect to number of atoms and then with respect to charge.**

In the oxidation half-equation, there is the same number of atoms of Al on both sides, one. However, charge is not balanced. We have zero charge on the left as opposed to +3 on the right. *To balance charge, electrons are added to the proper side of the equation.* Here we add three electrons to the right:

$$\mathbf{Al(s) \rightarrow Al^{3+}(aq) + 3\ e^{-}}$$

Looking at the reduction, we see two chlorine atoms on the left and only one on the right. To balance atoms, we write a coefficient of 2 in front of Cl^{-}.

$$Cl_2(g) \rightarrow 2\ Cl^{-}(aq)$$

This half-equation is still unbalanced with respect to charge. We have zero charge on the left, −2 on the right. To balance, we add two electrons to the left. This gives a charge of −2 on both sides.

$$\mathbf{Cl_2(g) + 2\ e^{-} \rightarrow 2\ Cl^{-}(aq)}$$

3. **Combine the balanced half-equations so that electrons "cancel", leaving none on either side.**

Remember, electrons are neither produced nor destroyed in chemical reactions.

There are 3 electrons on the right of the oxidation half-equation. In the reduction half-equation, there are 2 electrons on the left. To make these cancel, we multiply the first half-equation by 2 and the second half-equation by 3.

$$2\ Al(s) \rightarrow 2\ Al^{3+}(aq) + 6\ e^{-}$$

$$3\ Cl_2(g) + 6\ e^{-} \rightarrow 6\ Cl^{-}(aq)$$

We now add these two equations. This gets rid of the electrons, 6 on both sides. The final balanced equation for the overall reaction is:

$$\mathbf{2\ Al(s) + 3\ Cl_2(g) \rightarrow 2\ Al^{3+}(aq) + 6\ Cl^{-}(aq)} \qquad (23.9)$$

In the redox equation just balanced, only two elements were involved. One of those, aluminum, was oxidized. The other, chlorine, was reduced. Frequently, in water solution, the situation is a bit more complex. Other elements, in addition to those oxidized or reduced, may be involved. Most often, these elements are:

—hydrogen, which keeps a +1 oxidation number throughout the reaction.

—oxygen, which keeps a −2 oxidation number throughout the reaction.

A reaction of this type occurs when copper is treated with 16 M nitric acid. (See Color Plate 11A, in the center of the book.) The unbalanced equation for the reaction is:

$$Cu(s) + H^+(aq) + NO_3^-(aq) \rightarrow Cu^{2+}(aq) + NO_2(g) + H_2O$$

In this reaction, copper is oxidized. It goes from an oxidation number of 0 to +2. Nitrogen is reduced. Its oxidation number drops from +5 in the NO_3^- ion to +4 in NO_2. The other two elements, hydrogen and oxygen, do not change oxidation number. They are, however, involved in the reaction and must be included in the equation.

Equations of this type are readily balanced by the three-step process we have described. The only difference is that the half-equations are a little more difficult to balance. Let's apply the method to the reaction of Cu with NO_3^-.

Step 1

$$\text{oxidation: } Cu(s) \rightarrow Cu^{2+}(aq)$$

$$\text{reduction: } NO_3^-(aq) \rightarrow NO_2(g)$$

Step 2

The oxidation half-equation is readily balanced. We add 2 electrons to the right to give zero charge on both sides:

$$\mathbf{Cu(s) \rightarrow Cu^{2+}(aq) + 2\ e^-}$$

The reduction half-equation is not as simple. We proceed as follows.

a. *Balance the number of atoms of the element whose oxidation number is changing.* Here, this is already done. We have one N atom on both sides.

It's easier to balance a half-equation than a whole equation.

b. *Balance oxygen by adding H_2O molecules.* We have 2 oxygen atoms on the right and 3 on the left. To balance, we add 1 H_2O to the right.

$$NO_3^-(aq) \rightarrow NO_2(g) + H_2O$$

c. *Balance hydrogen by adding H^+ ions.* We have two hydrogen atoms on the right (1 H_2O molecule). There are no hydrogen atoms on the left. So, we must add 2 H^+ ions to the left.

$$NO_3^-(aq) + 2\ H^+(aq) \rightarrow NO_2(g) + H_2O$$

d. *Balance charge by adding electrons.* At the moment:

charge on left = −1 + 2 = +1

charge on right = 0

We must add one electron on the left. This gives zero charge on both sides. The balanced half-equation for the reduction is:

$$\mathbf{NO_3^-(aq) + 2\ H^+(aq) + e^- \rightarrow NO_2(g) + H_2O}$$

Step 3

As before, we combine the two half-equations so as to cancel electrons. To do this, we multiply the reduction equation by 2 and add to the oxidation equation.

$$Cu(s) \rightarrow Cu^{2+}(aq) + 2\ e^-$$

$$2\ NO_3^-(aq) + 4\ H^+(aq) + 2\ e^- \rightarrow 2\ NO_2(g) + 2\ H_2O \quad (23.10)$$

$$\mathbf{Cu(s) + 2\ NO_3^-(aq) + 4\ H^+(aq) \rightarrow Cu^{2+}(aq) + 2\ NO_2(g) + 2\ H_2O}$$

At this stage, it is well to check the equation to make sure that it is balanced. We should check both atom balance and charge balance.

	reactants	products
no. atoms Cu	1	1
no. atoms N	2	2
no. atoms O	6	4 + 2 = 6
no. atoms H	4	4
charge	+2	+2

You can always use this kind of check to make sure you balanced the equation properly.

The equation is balanced.

Reviewing the equation just balanced, we see that the only complication came in the second step. To balance a half-equation involving oxygen and hydrogen atoms in water solution, we proceed in the order shown. That is:

First, balance the element oxidized or reduced.

Next, balance oxygen by adding H_2O molecules. These are, of course, readily available in water solution.

Then, balance hydrogen by adding H^+ ions. These are available in acidic solution, as in a solution of HNO_3.

Finally, balance charge by adding electrons.

EXAMPLE 23.3

Chlorine gas can be prepared by adding hydrochloric acid to a water solution of potassium chromate, K_2CrO_4. The unbalanced equation for the reaction is:

$$CrO_4^{2-}(aq) + H^+(aq) + Cl^-(aq) \rightarrow Cr^{3+}(aq) + Cl_2(g) + H_2O$$

Balance this equation.

SOLUTION

We must first decide what is being oxidized and reduced:

Cr: oxid. no. = +6 in CrO_4^{2-}, +3 in Cr^{3+} (reduced)
O: oxid. no. = −2 in CrO_4^{2-}, −2 in H_2O
H: oxid. no. = +1 in H^+, +1 in H_2O
Cl: oxid. no. = −1 in Cl^-, 0 in Cl_2 (oxidized)

The unbalanced half-equations are:

$$\text{oxidation: } Cl^-(aq) \rightarrow Cl_2(g)$$
$$\text{reduction: } CrO_4^{2-}(aq) \rightarrow Cr^{3+}(aq)$$

To balance the oxidation half-equation, we give Cl^- a coefficient of 2. We then add 2 electrons to the right.

$$2\ Cl^-(aq) \rightarrow Cl_2(g) + 2\ e^- \qquad \text{(a)}$$

To balance the reduction half-equation, we:

—note that chromium is already balanced. There is one Cr atom on both sides.

—balance oxygen by adding 4 H_2O molecules to the right:

When balancing equations like this, we can use H_2O and H^+ as either reactants or products because water is the solvent and the solution is acidic.

$$CrO_4^{2-}(aq) \rightarrow Cr^{3+}(aq) + 4\ H_2O$$

—balance hydrogen by adding 8 H^+ ions to the left:

$$CrO_4^{2-}(aq) + 8\ H^+(aq) \rightarrow Cr^{3+}(aq) + 4\ H_2O$$

—balance charge by adding 3 e^- to the left. This gives a net charge of +3 on both sides. The balanced half-equation for the reduction is:

$$CrO_4^{2-}(aq) + 8\ H^+(aq) + 3\ e^- \rightarrow Cr^{3+}(aq) + 4\ H_2O \qquad \text{(b)}$$

We now combine the balanced half-equations, (a) and (b). To eliminate electrons, we multiply (a) by 3, (b) by 2, and add:

$$6\ Cl^-(aq) \rightarrow 3\ Cl_2(g) + 6\ e^-$$
$$2\ CrO_4^{2-}(aq) + 16\ H^+(aq) + 6\ e^- \rightarrow 2\ Cr^{3+}(aq) + 8\ H_2O$$
$$6\ Cl^-(aq) + 2\ CrO_4^{2-}(aq) + 16\ H^+(aq) \rightarrow 3\ Cl_2(g) + 2\ Cr^{3+}(aq) + 8\ H_2O$$

Balanced redox equations can be used like any balanced equation. In particular, they allow us to determine mass relations in chemical reactions. The calculations are very similar to those considered in Chapter 3.

EXAMPLE 23.4

Using the balanced equation obtained in Example 23.3, calculate

a. the number of moles of $Cl_2(g)$ formed from 1.20 mol of CrO_4^{2-}.

b. the number of grams of Cr^{3+} formed in (a).

SOLUTION

a. The mole relationship we need is given by the coefficients of the balanced equation:

$$2 \text{ mol } CrO_4^{2-} \simeq 3 \text{ mol } Cl_2$$

Hence:

$$\text{moles } Cl_2 = 1.20 \text{ mol } CrO_4^{2-} \times \frac{3 \text{ mol } Cl_2}{2 \text{ mol } CrO_4^{2-}} = 1.80 \text{ mol } Cl_2$$

b. Here, we need a relation between moles of CrO_4^{2-} and grams of Cr^{3+}. From the coefficients of the balanced equation:

$$2 \text{ mol } CrO_4^{2-} \simeq 2 \text{ mol } Cr^{3+}$$

or:

$$1 \text{ mol } CrO_4^{2-} \simeq 1 \text{ mol } Cr^{3+}$$

But, since the atomic mass of Cr is 52.0, 1 mol of Cr^{3+} weighs 52.0 g. So, the relation we need is:

$$1 \text{ mol } CrO_4^{2-} \simeq 52.0 \text{ g } Cr^{3+}$$

$$\text{mass } Cr^{3+} = 1.20 \text{ mol } CrO_4^{2-} \times \frac{52.0 \text{ g } Cr^{3+}}{1 \text{ mol } CrO_4^{2-}} = 62.4 \text{ g } Cr^{3+}$$

23.3 COMMON OXIDATION NUMBERS OF THE ELEMENTS

In Figure 23.1, p. 578, we list the oxidation numbers of elements in their stable compounds. Notice that the great majority of elements show more than one oxidation number. The principal exceptions are the metals at the far left of the Periodic Table. As previously mentioned, the metals in Groups 1 and 2 always have oxidation numbers of +1 and +2 respectively. In all of their compounds, they exist as +1 and +2 ions.

Multiple oxidation numbers are the rule rather than the exception among:

—the transition metals, located near the center of the Periodic Table.

—the nonmetals, located at the far right of the Periodic Table.

Elements with multiple oxidation numbers tend to have interesting chemical properties.

We will now examine the oxidation numbers shown by some of these elements.

Transition Metals

Most of the transition metals show at least two different oxidation numbers. Commonly, these are +1, +2, or +3. They correspond to the charges of monatomic ions. For example, iron forms two different series of compounds. In one series (iron(II) compounds), it is present as a +2

1	2			Transition Metals									3	4	5	6	7	8
1 H +1 −1																	1 H +1 −1	2 He
3 Li +1	4 Be +2												5 B +3	6 C +4 +2 −4	7 N +5 +4 +3 +2 +1 −3	8 O −1 −2	9 F −1	10 Ne
11 Na +1	12 Mg +2												13 Al +3	14 Si +4 −4	15 P +5 +3 −3	16 S +6 +4 +2 −2	17 Cl +7 +5 +3 +1 −1	18 Ar
19 K +1	20 Ca +2	21 Sc +3		22 Ti +4 +3 +2	23 V +5 +4 +3 +2	24 Cr +6 +3 +2	25 Mn +7 +6 +4 +3 +2	26 Fe +3 +2	27 Co +3 +2	28 Ni +2	29 Cu +2 +1	30 Zn +2	31 Ga +3	32 Ge +4 −4	33 As +5 +3 −3	34 Se +6 +4 −2	35 Br +5 +1 −1	36 Kr +4 +2
37 Rb +1	38 Sr +2	39 Y +3		40 Zr +4	41 Nb +5 +4	42 Mo +6 +4 +3	43 Tc +7 +6 +4	44 Ru +8 +6 +4 +3	45 Rh +4 +3 +2	46 Pd +4 +2	47 Ag +1	48 Cd +2	49 In +3	50 Sn +4 +2	51 Sb +5 +3 −3	52 Te +6 +4 −2	53 I +7 +5 +1 −1	54 Xe +6 +4 +2
55 Cs +1	56 Ba +2	57 La +3	58 Ce → 71 Lu +3	72 Hf +4	73 Ta +5	74 W +6 +4	75 Re +7 +6 +4	76 Os +8 +4	77 Ir +4 +3	78 Pt +4 +2	79 Au +3 +1	80 Hg +2 +1	81 Tl +3 +1	82 Pb +4 +2	83 Bi +5 +3	84 Po +2	85 At −1	86 Rn
87 Fr +1	88 Ra +2	89 Ac +3	90 Th → 103 Lr	104	105													

FIGURE 23.1

The oxidation states of the elements. The most common or stable states are shown in bold type.

ion, Fe^{2+}. In the other series (iron(III) compounds), it exists as the +3 ion, Fe^{3+}.

$$Fe^{2+}: FeCl_2, FeSO_4$$

$$Fe^{3+}: FeCl_3, Fe_2(SO_4)_3, Fe_2O_3$$

The Fe^{2+} ion has an interesting property. It can go up or down in oxidation number, losing or gaining electrons.

$$Fe^{2+}(aq) \rightarrow Fe^{3+}(aq) + e^-$$

$$Fe^{2+}(aq) + 2\ e^- \rightarrow Fe(s)$$

When Fe^{2+} is oxidized to Fe^{3+}, it acts as a reducing agent. The electrons produced in this half-reaction can reduce another species. On the other hand, when Fe^{2+} is reduced to Fe(s), it acts as an oxidizing agent. Here, the Fe^{2+} ion takes electrons away from some other species.

In contrast, the Fe^{3+} ion can act only as an oxidizing agent. It can only be reduced, either to Fe^{2+} or Fe(s).

$$Fe^{3+}(aq) + e^- \rightarrow Fe^{2+}(aq)$$

$$Fe^{3+}(aq) + 3\ e^- \rightarrow Fe(s)$$

EXAMPLE 23.5

Copper shows oxidation numbers of +1 and +2. These correspond to the ions Cu^+ and Cu^{2+}.

a. Give the formulas of the two oxides of copper.

b. Name these oxides.

c. Can Cu^+ act as an oxidizing agent? a reducing agent?

d. Can Cu^{2+} act as an oxidizing agent? a reducing agent?

SOLUTION

a. Since oxygen has an oxidation number of −2, the formulas are:

$$Cu_2O, CuO$$

b. Recall the rules given on p. 251, Chapter 10.

Cu_2O: copper(I) oxide

CuO: copper(II) oxide

c. Cu^+ can act as an oxidizing agent, in which case it is reduced to copper metal.

$$Cu^+(aq) + e^- \rightarrow Cu(s)$$

It can also act as a reducing agent. When this happens, it is oxidized to Cu^{2+}.

$$Cu^+(aq) \rightarrow Cu^{2+}(aq) + e^-$$

d. Cu^{2+} can act as an oxidizing agent. When it does, it may be reduced to either Cu^+ or Cu.

$$Cu^{2+}(aq) + e^- \rightarrow Cu^+(aq)$$

$$Cu^{2+}(aq) + 2\ e^- \rightarrow Cu(s)$$

It cannot act as a reducing agent. To do so, it would have to be oxidized. Copper does not ordinarily show an oxidation number greater than +2.

As you can see, the oxidation numbers of an element tell us quite a lot about its chemical properties.

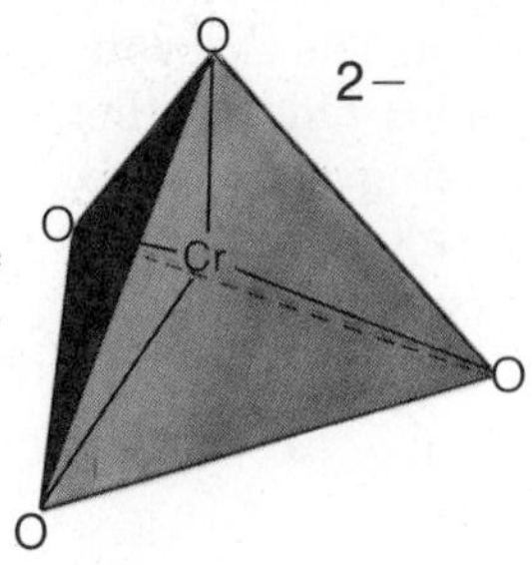

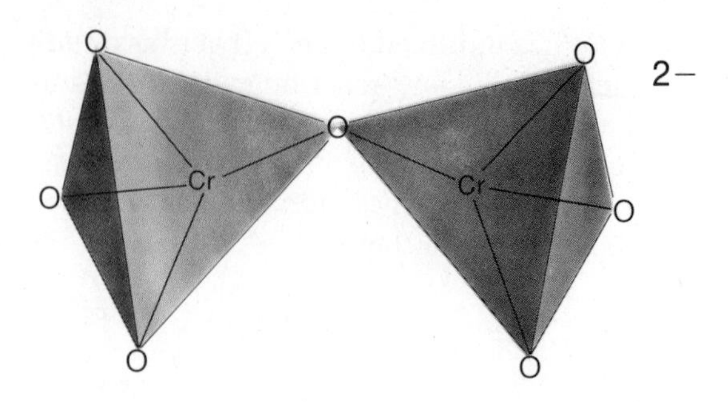

FIGURE 23.2

Structures and geometries of the chromate and dichromate ions. In both ions the Cr atom is at the center of a tetrahedron, on the corners of which are O atoms.

Several transition metals have oxidation numbers higher than +3. When this happens, the metal is *not* present as a simple, monatomic ion. Instead, it is covalently bonded to a nonmetal atom, most often oxygen. As an example, consider chromium. It shows an oxidation number of +6 in two polyatomic ions:

In any polyatomic ion, the atoms are covalently bonded.

—CrO_4^{2-}, found in such compounds as Na_2CrO_4 and $BaCrO_4$. Within the ion, the bonding is covalent (Figure 23.2).

—$Cr_2O_7^{2-}$ ($K_2Cr_2O_7$, etc.). Here, as in CrO_4^{2-}, the bonding between chromium and oxygen is covalent.

Polyatomic ions in which a transition metal is in its highest oxidation state are often strong oxidizing agents. A solution of potassium dichromate, $K_2Cr_2O_7$, in sulfuric acid is used to clean laboratory glassware. It does this by oxidizing grease, oil, and other organic materials. The dichromate ion is reduced to Cr^{3+}, a green species. (See Color Plate 11B, in the middle of the book.)

$$Cr_2O_7^{2-}(aq) + 14\ H^+(aq) + 6\ e^- \rightarrow 2\ Cr^{3+}(aq) + 7\ H_2O$$

Chlorine

The element chlorine shows a variety of oxidation numbers in its compounds. These range from +7 to −1. In its positive oxidation states, chlorine is most often bonded to oxygen (Table 23.1). All of the oxides of chlorine are unstable. They decompose, often violently, upon heating.

TABLE 23.1 OXIDATION NUMBERS OF CHLORINE IN ITS COMPOUNDS

	Acid	Negative Ion	Oxide
+7	$HClO_4$	ClO_4^-	Cl_2O_7
+5	$HClO_3$	ClO_3^-	
+4			ClO_2
+3	$HClO_2$	ClO_2^-	
+1	HClO	ClO^-	Cl_2O
−1	HCl	Cl^-	

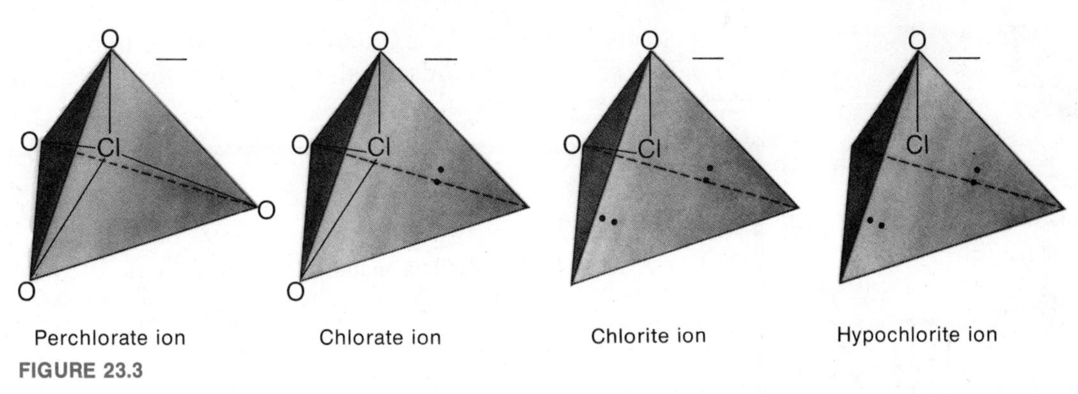

Perchlorate ion Chlorate ion Chlorite ion Hypochlorite ion

FIGURE 23.3

Structures and geometries of the oxyanions of chlorine. In each ion there are four electron pairs arranged tetrahedrally around the Cl atom. These electron pairs form the covalent bonds to O atoms or are unshared.

The Lewis structures of the four oxygen acids formed by chlorine are:

$$\mathrm{H{-}O{-}\underset{:\ddot{O}:}{\overset{:\ddot{O}:}{Cl}}{-}\ddot{O}:} \qquad \mathrm{H{-}O{-}\overset{:\ddot{O}:}{Cl}{-}\ddot{O}:} \qquad \mathrm{H{-}\ddot{O}{-}\ddot{Cl}{-}\ddot{O}} \qquad \mathrm{H{-}\ddot{O}{-}\ddot{Cl}:}$$

$HClO_4$	$HClO_3$	$HClO_2$	$HClO$
perchloric acid	chloric acid	chlorous acid	hypochlorous acid

Notice that in each case the hydrogen atom is bonded to oxygen rather than chlorine. The geometries of the corresponding negative ions are shown in Figure 23.3. The names of the acids and their ions were discussed in Chapters 10 and 11, pages 251 and 279.

Hypochlorous Acid and the Hypochlorite Ion

When chlorine gas is bubbled through water, the following reaction takes place:

How would you describe the geometries of the four ions in Figure 23.3?

$$Cl_2(g) + H_2O \rightleftharpoons HClO(aq) + H^+(aq) + Cl^-(aq) \qquad (23.11)$$

The products are the weak acid HClO ($K_a = 3.2 \times 10^{-8}$) and the strong acid, HCl. Notice that this is an oxidation-reduction reaction. Chlorine starts off in the zero oxidation state in Cl_2. Half of it is oxidized to the +1 state in HClO. The other half is reduced to the −1 state in Cl^-.

The equilibrium in Reaction 23.11 is driven to the right by OH^- ions. They react with the H^+ ions to form water. In sodium hydroxide solution, the products are ClO^- ions, Cl^- ions, and H_2O molecules.

$$Cl_2(g) + 2\ OH^-(aq) \rightarrow ClO^-(aq) + Cl^-(aq) + H_2O \qquad (23.12)$$

The solution prepared by this reaction is sold under various trade names (Fig. 23.4, p. 582). It is used in the home as a bleach and disinfectant.

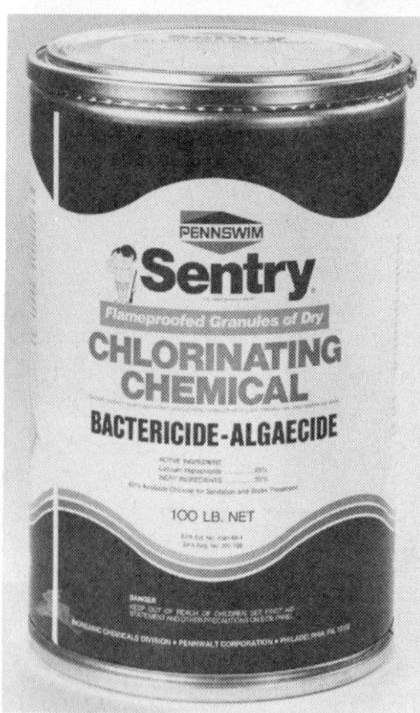

FIGURE 23.4

These and similar products used as household bleaches are water solutions containing Na^+, Cl^-, and ClO^- ions. The active ingredient is the hypochlorite ion, which is a strong oxidizing agent. *top;* courtesy of the Clorox Corporation. *bottom;* courtesy of Pennwalt Corporation.

These effects are due to the oxidizing power of the hypochlorite ion, ClO^-. The solid compounds used to disinfect swimming pools or drinking water also contain this ion.

EXAMPLE 23.6

A household bleach is made by Reaction 23.12. It contains 5.25 mass percent of NaClO. How many moles of Cl_2 are required to form one liter of this bleach (density = 1.05 g/cm³)?

SOLUTION

This is a straightforward problem involving mass relations in chemical reactions. It is solved by the methods used in Chapter 3. First, we calculate the mass in grams of NaClO, then convert to moles. Finally, we use the coefficients of the balanced equation to obtain the number of moles of Cl_2.

The mass of one liter of the bleach is:

$$1000\ \text{cm}^3 \times 1.05\ \frac{\text{g}}{\text{cm}^3} = 1050\ \text{g}$$

Since it contains 5.25% by mass of NaClO:

$$\text{mass NaClO} = 1050\ \text{g} \times 0.0525 = 55.1\ \text{g NaClO}$$

The formula mass of NaClO is: 23.0 + 35.5 + 16.0 = 74.5. So:

$$\text{moles NaClO} = 55.1\ \text{g} \times \frac{1\ \text{mol}}{74.5\ \text{g}} = 0.740\ \text{mol NaClO}$$

According to Equation 23.12, one mole of Cl_2 gives one mole of ClO^-, or one mole of NaClO. So, we require 0.740 mol Cl_2.

Sulfur

The common oxidation states of sulfur are indicated in Table 23.2 at the bottom of p. 583. They range from +6 to −2. The Lewis structures of the two oxygen acids are:

```
        :Ö:
         |
   H—O—S—O—H          H—Ö—S̈—Ö—H
         |                  |
        :Ö:                :Ö:

       H2SO4               H2SO3
   sulfuric acid       sulfurous acid
```

Figure 23.5 shows the geometry of the sulfate (SO_4^{2-}), sulfite (SO_3^{2-}), and thiosulfate ($S_2O_3^{2-}$) ions. Note that the $S_2O_3^{2-}$ ion has the same structure as the SO_4^{2-} ion. The only difference is that one of the oxygen atoms has been replaced by sulfur.

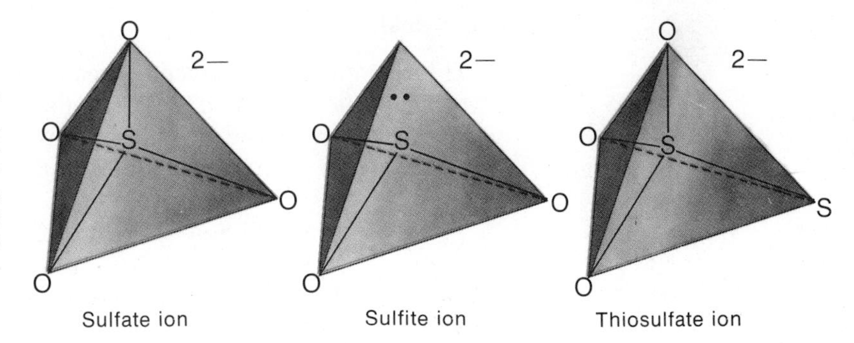

FIGURE 23.5

Structures and geometries of the SO_4^{2-}, $S_2O_3^{2-}$, and SO_3^{2-} ions. In each ion there are four electron pairs arranged tetrahedrally around the central S atom. These pairs may form the covalent bonds to O atoms or to the S atom in $S_2O_3^{2-}$, or may be unshared, as in the SO_3^{2-} ion.

Hydrogen Sulfide

Hydrogen sulfide is a poisonous gas. It is easily detected by its foul odor (rotten eggs). It can be prepared by adding a strong acid such as HCl to a solution containing sulfide ions.

$$S^{2-}(aq) + 2\ H^+(aq) \rightarrow H_2S(g) \quad (23.13)$$

or with a solid metal sulfide such as FeS (Figure 23.6):

$$FeS(s) + 2\ H^+(aq) \rightarrow Fe^{2+}(aq) + H_2S(aq) \quad (23.14)$$

In water solution, H_2S can act:

1. as a *weak acid.* It ionizes in two steps:

$$H_2S(aq) \rightleftharpoons H^+(aq) + HS^-(aq); \quad K_a = 1 \times 10^{-7} \quad (23.15)$$

$$HS^-(aq) \rightleftharpoons H^+(aq) + S^{2-}(aq); \quad K_a = 1 \times 10^{-15} \quad (23.16)$$

2. as a *precipitating agent.* All of the sulfides of the transition metals are insoluble in water. Many transition metal ions can be precipitated from water solution by adding H_2S. The reaction with Cu^{2+} is typical:

$$Cu^{2+}(aq) + H_2S(aq) \rightarrow CuS(s) + 2\ H^+(aq) \quad (23.17)$$

You will recall (Chapter 22) that this reaction was used to precipitate Cu^{2+} in qualitative analysis.

3. as a *reducing agent.* Hydrogen sulfide is readily oxidized, most often to sulfur. Water solutions of H_2S slowly turn cloudy due to the formation of colloidal sulfur. The oxidizing agent is dissolved oxygen:

$$2\ H_2S(aq) + O_2(g) \rightarrow 2\ S(s) + 2\ H_2O \quad (23.18)$$

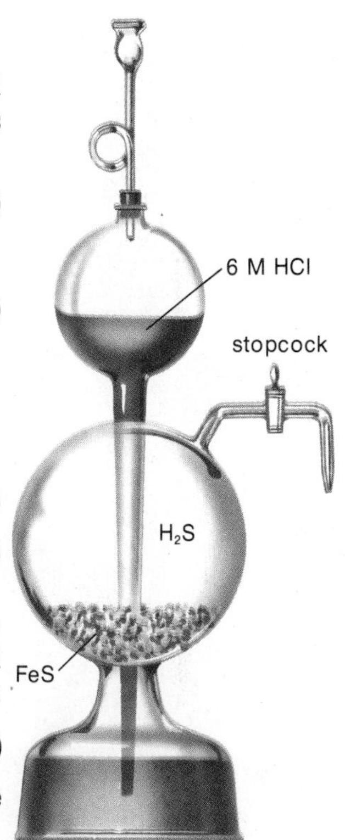

FIGURE 23.6

Kipp generator, used for preparation of H_2S. Opening the stopcock allows the HCl solution to come into contact with the solid FeS, producing H_2S. Closing the stockcock causes the gas pressure inside the generator to rise, forcing the acid back up into the reservoir and stopping the formation of H_2S.

TABLE 23.2 OXIDATION NUMBERS SHOWN BY SULFUR IN ITS COMPOUNDS

	Acid	Negative Ion	Oxide
+6	H_2SO_4	SO_4^{2-}, HSO_4^-	SO_3
+4	H_2SO_3	SO_3^{2-}, HSO_3^-	SO_2
+2		$S_2O_3^{2-}$	
−2	H_2S	S^{2-}, HS^-	

Sulfuric Acid

This chemical is produced in greater amounts than any other (about 3×10^{10} kg/yr in the United States; that's more than 100 kg per person!). Sulfuric acid is made by a three-step process:

(a) Sulfur is burned in air to form SO_2:

$$S(s) + O_2(g) \rightarrow SO_2(g) \quad (23.19)$$

(b) Sulfur dioxide is converted to SO_3 by the reaction discussed in Chapter 17:

$$SO_2(g) + \tfrac{1}{2} O_2(g) \rightarrow SO_3(g) \quad (23.20)$$

Large amounts of H_2SO_4 are used to solubilize $Ca_3(PO_4)_2$ for fertilizers, to clean oxide scale from steel, and to remove impurities from petroleum.

(c) Sulfur trioxide reacts with water:

$$SO_3(g) + H_2O(l) \rightarrow H_2SO_4(aq) \quad (23.21)$$

This reaction is carried out by adding SO_3 to dilute sulfuric acid. If SO_3 is passed directly into water, H_2SO_4 is formed as a fog of tiny particles.

In dilute water solution, H_2SO_4 is a strong acid. The *concentrated* acid (18 M) is also:

1. a *dehydrating agent.* The reaction of concentrated sulfuric acid with water is strongly exothermic. Small amounts of water can be removed from organic liquids such as gasoline by shaking with sulfuric acid. Sometimes, it is even possible to remove the elements of water from a compound by treating with 18 M sulfuric acid. Treatment of sugar with concentrated H_2SO_4 leaves a black char which is mostly carbon (Figure 23.7):

$$C_{12}H_{22}O_{11}(s) \xrightarrow{H_2SO_4} 12\ C(s) + 11\ H_2O(l) \quad (23.22)$$

FIGURE 23.7

Effect of adding concentrated H_2SO_4 to sugar. The black, gummy solid formed is mostly carbon.

2. an *oxidizing agent*. Hot concentrated H_2SO_4 reacts with copper, oxidizing it to Cu^{2+}. The sulfuric acid is reduced to $SO_2(g)$. The balanced equation for the redox reaction is:

$$Cu(s) + 4\ H^+(aq) + SO_4^{2-}(aq) \rightarrow Cu^{2+}(aq) + SO_2(g) + 2\ H_2O \quad (23.23)$$

Concentrated H_2SO_4 is not a reagent to be taken lightly. If you have to work with it, use due caution.

Nitrogen

Nitrogen shows more different oxidation numbers than any other element. A few typical species in various oxidation states are shown in Table 23.3. The Lewis structures of the two principal oxygen acids of nitrogen are:

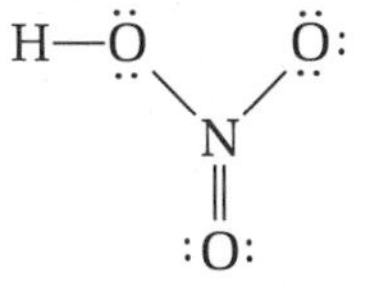

HNO_3
nitric acid

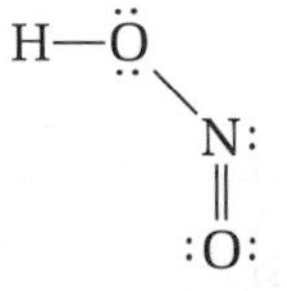

HNO_2
nitrous acid

Here, as in all oxygen acids, the acidic hydrogen is bonded to oxygen.

TABLE 23.3 OXIDATION NUMBERS SHOWN BY NITROGEN IN ITS COMPOUNDS

	Acid	Negative Ion	Oxide	Hydride	Positive Ion
+5	HNO_3	NO_3^-	N_2O_5		
+4			NO_2, N_2O_4		
+3	HNO_2	NO_2^-	N_2O_3		
+2			NO		
+1			N_2O		
−2				N_2H_4	$N_2H_5^+$
−3				NH_3	NH_4^+

Ammonia

The preparation and uses of ammonia, NH_3, were discussed in Chapter 18. Ammonia in water behaves as a weak base (Chapter 19). It can also act as a complexing agent, forming complex ions such as $Ag(NH_3)_2^+$ (Chapter 20). Finally, ammonia can act as a reducing agent. When it does this, it may be oxidized to elementary nitrogen, N_2

$$NH_3(g) \rightarrow \tfrac{1}{2}\ N_2(g) + 3\ H^+(aq) + 3\ e^- \qquad \text{(oxid. no. N: } -3 \text{ to } 0)$$

or to a species in which nitrogen has a positive oxidation number.

EXAMPLE 23.7

Write a balanced half-equation for the oxidation of NH_3 to NO. Use H^+ ions and H_2O molecules to balance hydrogen and oxygen.

SOLUTION

We start with the equation:

$$NH_3(g) \rightarrow NO(g)$$

To balance oxygen, we add one water molecule to the left:

$$NH_3(g) + H_2O \rightarrow NO(g)$$

We now have 5 H atoms on the left and none on the right. To balance hydrogen, we must add 5 H^+ ions to the right:

$$NH_3(g) + H_2O \rightarrow NO(g) + 5\ H^+(aq)$$

Finally, to balance the charge, we add 5 electrons to the right:

$$NH_3(g) + H_2O \rightarrow NO(g) + 5\ H^+(aq) + 5\ e^-$$

Nitric Acid

This compound is made from ammonia by the Ostwald process. This involves three separate steps.

All of our nitric acid is made by this process.

(1) Ammonia is burned in air at about 1000°C. With a platinum catalyst, the product is NO:

$$4\ NH_3(g) + 5\ O_2(g) \rightarrow 4\ NO(g) + 6\ H_2O(l) \qquad (23.24)$$

(2) The NO formed is cooled in air to convert it to NO_2:

$$2\ NO(g) + O_2(g) \rightarrow 2\ NO_2(g) \qquad (23.25)$$

(3) The nitrogen dioxide is added to warm water. An oxidation-reduction reaction takes place. Part of the NO_2 is oxidized to nitric acid. The rest of it is reduced back to NO.

$$3\ NO_2(g) + H_2O(l) \rightarrow 2\ HNO_3(aq) + NO(g) \qquad (23.26)$$

The NO is recycled to produce more NO_2. The aqueous solution is distilled to produce concentrated nitric acid (16 M).

EXAMPLE 23.8

How many grams of NO_2 must be used to produce one liter of 16 M HNO_3 by Reaction 23.26?

SOLUTION

According to the equation, 3 mol of NO_2 are required to form 2 mol of HNO_3 (2 mol H^+, 2 mol NO_3^-).

$$3 \text{ mol } NO_2 \simeq 2 \text{ mol } HNO_3$$

We need 16 mol of HNO_3. Hence:

$$\text{no. moles } NO_2 \text{ required} = 16 \text{ mol } HNO_3 \times \frac{3 \text{ mol } NO_2}{2 \text{ mol } HNO_3} = 24 \text{ mol } NO_2$$

One mole of NO_2 weighs: 14.0 g + 2(16.0 g) = 46.0 g

$$\text{mass } NO_2 \text{ required} = 24 \text{ mol } NO_2 \times \frac{46.0 \text{ g } NO_2}{1 \text{ mol } NO_2} = 1.1 \times 10^3 \text{ g } NO_2$$

As pointed out in Chapter 19, HNO_3 is a strong acid. It is completely ionized to H^+ and NO_3^- in dilute water solution. Nitric acid is also a strong oxidizing agent. It can be reduced to any of the species listed below it in Table 23.3. With concentrated HNO_3, the reduction product is most often $NO_2(g)$. You will recall (Section 23.2) that copper reacts with 16 M HNO_3 to give Cu^{2+} ions and brown fumes of NO_2:

Nitric acid is the most commonly used oxidizing agent in the laboratory.

$$Cu(s) + 4\ H^+(aq) + 2\ NO_3^-(aq) \rightarrow Cu^{2+}(aq) + 2\ NO_2(g) + 2\ H_2O$$

Oxides of Nitrogen

Nitrogen forms a total of six different oxides, more than any other element. Their physical properties are listed in Table 23.4. Of these, N_2O_3 and N_2O_5 are stable only at low temperatures. Two others, NO and NO_2, are discussed in other chapters (Chapters 18, 26) in connection with air pollution.

TABLE 23.4 PROPERTIES OF OXIDES OF NITROGEN

	mp (°C)	bp (°C)	Remarks
N_2O_5	41	–	Decomposes to NO_2 and O_2 above the melting point
N_2O_4	−9	21	
NO_2	–	–	Exists in equilibrium with N_2O_4
N_2O_3	−102	3	Decomposes to NO and NO_2 at −20°C
NO	−161	−151	
N_2O	−103	−91	

Dinitrogen tetroxide, N_2O_4, is a colorless gas. It exists in equilibrium with NO_2:

$$2\ NO_2(g) \rightleftharpoons N_2O_4(g); \qquad \Delta H = -14 \text{ kcal} \qquad (23.27)$$

As the temperature drops, the system shifts to form N_2O_4. At the boiling point, 21°C, the mixture is mostly N_2O_4.

Dinitrogen tetroxide was one of the chemicals used in the rockets of the Lunar

Excursion Module. The other substance used was dimethylhydrazine, $(CH_3)_2N_2H_2$. These two compounds were stored in separate tanks under pressure. When a valve connecting the two tanks was opened, the gaseous mixture ignited. The reaction between N_2O_4 and $(CH_3)_2N_2H_2$ is exothermic and produces large volumes of gases.

$$2\ N_2O_4(g) + (CH_3)_2N_2H_2(g) \rightarrow 3\ N_2(g) + 4\ H_2O(g) + 2\ CO_2(g) \qquad (23.28)$$

The hot gases propelled the LEM from the surface of the moon, sending the astronauts on their way back to earth (chapter opening photo).

Dinitrogen oxide, N_2O, is made by *carefully* heating small amounts of ammonium nitrate, NH_4NO_3, below 250°C:

$$NH_4NO_3(s) \rightarrow N_2O(g) + 2\ H_2O(g) \qquad (23.29)$$

It is a colorless gas with a pleasant odor and a slightly sweet taste. Because of the effect N_2O has on some people who inhale it, it is often referred to as "laughing gas". Its anaesthetic properties were discovered nearly 200 years ago. It was the first anaesthetic used in surgery, in 1837. Today, it is still used for this purpose, particularly in dentistry.

Dinitrogen oxide is also used as the pressurizing gas in whipped cream cans. It is quite soluble in fats under pressure. When the pressure is released, the N_2O comes out of solution. In doing so, it produces a foam of "whipped" cream.

23.4 SUMMARY

In any oxidation-reduction reaction, two different processes are occurring. One species is oxidized. That is, it shows an increase in oxidation number (loses electrons). The other species is reduced. It shows a decrease in oxidation number (gains electrons). The species which brings about the oxidation is called the oxidizing agent. It is itself reduced. The other species is the reducing agent. It is oxidized. Thus we have:

$$2\ Na(s) + Cl_2(g) \rightarrow 2\ NaCl(s)$$

Na is oxidized (oxid. no. $0 \rightarrow +1$). It is the reducing agent.

Cl_2 is reduced (oxid. no. $0 \rightarrow -1$). It is the oxidizing agent.

Oxidation numbers are assigned according to a consistent set of rules, given on p. 569. A few metals, notably those in Groups 1 and 2, have only a single oxidation number in all their compounds. Most transition metals have two or more possible oxidation numbers (Figure 23.1). Among nonmetals, multiple oxidation numbers are very common. We considered three such elements in this chapter: chlorine, sulfur, and nitrogen. They show negative oxidation numbers in their monatomic ions (Cl^-, S^{2-}, N^{3-}) and in their hydrides (HCl, H_2S, NH_3). Where they have positive oxidation numbers, they are most often covalently bonded to oxygen. This is the case for example, with:

—the oxyacids ($HClO_4$, H_2SO_4, HNO_3 . . .)

—negative ions derived from these acids (ClO_4^-, SO_4^{2-}, NO_3^- . . .)
—nonmetal oxides (Cl_2O_7, SO_3, N_2O_5 . . .).

To write a balanced equation for a redox reaction, we start by splitting it into two half-equations. Each of these is balanced in a systematic way as described on p. 573. The two half-equations are then combined so as to make the electron gain cancel the electron loss. The balanced equation obtained is used in the ordinary way to calculate mass relationships (Examples 23.4, 23.6, 23.8).

NEW TERMS

half-equation
Ostwald process
oxidation
oxidation number
oxidizing agent
reducing agent
reduction

QUESTIONS

1. In the reaction: $2\ Ca(s) + O_2(g) \rightarrow 2\ CaO(s)$, identify:

   a. the species that is oxidized
   b. the species that is reduced
   c. the oxidizing agent
   d. the reducing agent

2. Answer parts a through d of Question 1 for the reaction:

$$Fe^{2+}(aq) + Ag^+(aq) \rightarrow Fe^{3+}(aq) + Ag(s)$$

3. Consider the reaction: $2\ Al(s) + 3\ F_2(g) \rightarrow 2\ AlF_3(s)$

   a. How many electrons are lost by the two aluminum atoms?
   b. How many electrons are gained by the three F_2 molecules?
   c. What is the oxidizing agent?
   d. What is the reducing agent?

4. Which of the following statements apply to the reducing agent in a reaction?

   a. It loses electrons
   b. It gains electrons
   c. Its oxidation number increases
   d. Its oxidation number decreases
   e. It is reduced
   f. It is oxidized.

5. Which of the statements (a–f) in Question 4 apply to an oxidizing agent?

6. What is the most common oxidation number shown by the following elements in their compounds?

   a. Na b. Mg c. Ba d. K e. H f. O

7. What are the oxidation numbers of the elements in the following ions?

   a. Ca^{2+} b. F^- c. Fe^{3+} d. S^{2-}

8. In oxidation-reduction equations, H_2O molecules are used to balance oxygen. Why not use O_2 molecules instead?

9. Cobalt forms two ions, Co^{2+} and Co^{3+}. For each ion, state whether it can act as a reducing agent; an oxidizing agent.

10. Give the formulas and names of the two chlorides of cobalt, formed by the Co^{2+} and Co^{3+} ions. Do the same for the two sulfates of cobalt.

11. The highest oxidation number of Mn is +7, shown in the MnO_4^- ion. Would you expect this ion to be:

    a. a reducing agent? b. a weak oxidizing agent?
    c. a strong oxidizing agent?

12. Consider the ions listed in Table 23.1. Which of these could act as reducing agents; oxidizing agents?

13. Give the Lewis structures of:

    a. HCl b. HClO c. perchloric acid

14. How would you make a solution containing hypochlorous acid, starting with chlorine gas?

15. Write a balanced equation for the reaction that occurs when chlorine is added to a solution containing OH^- ions.

16. Give the Lewis structures of:

    a. H_2S b. H_2SO_3 c. H_2SO_4

17. Describe in words how hydrogen sulfide can be prepared in the laboratory.

18. Write balanced equations for a reaction in which H_2S

    a. acts as a weak acid.
    b. acts as a reducing agent.
    c. is used to precipitate a transition metal ion from water solution.

19. The compound H_2SO_4 can act as which of the following?

    a. an acid b. a base
    c. an oxidizing agent d. a reducing agent
    e. a dehydrating agent f. a foreign agent

20. Ammonia undergoes the following reactions. In each case, state whether it is acting as a base, a reducing agent, or a complexing agent.

    a. $NH_3(aq) + H_2O \rightleftharpoons NH_4^+(aq) + OH^-(aq)$
    b. $4\ NH_3(g) + 5\ O_2(g) \rightarrow 4\ NO(g) + 6\ H_2O(l)$
    c. $Ag^+(aq) + 2\ NH_3(aq) \rightarrow Ag(NH_3)_2^+(aq)$
    d. $NH_3(aq) + H^+(aq) \rightarrow NH_4^+(aq)$

21. How is nitric acid produced commercially? What are the basic starting materials? What product is formed in the first step? the second step?

22. Give the formulas of four different acids, described in this chapter, in which the acidic hydrogen atom is bonded to oxygen.

PROBLEMS

1. *Oxidation Number (Example 23.1)* Assign oxidation numbers to chromium in:

    a. CrO_3 b. Cr_2O_3 c. Na_2CrO_4 d. $Cr_2O_7^{2-}$

2. *Oxidation Number (Example 23.1)* Assign oxidation numbers to all the elements in:

   a. PH_3 b. SiO_2 c. Na_2MnO_4 d. HPO_4^{2-}

3. *Oxidation Number (Example 23.2)* Classify each of the following as oxidations or reductions:

   a. $NH_3 \rightarrow N_2$ b. $N_2 \rightarrow NO$ c. $NO_2 \rightarrow NO$ d. $NO \rightarrow NO_3^-$

4. *Oxidation Number (Example 23.2)* Classify each of the following as oxidation, reduction, or neither:

   a. $CrO_4^{2-} \rightarrow Cr^{3+}$ b. $Cr_2O_7^{2-} \rightarrow CrO_4^{2-}$ c. $Fe^{2+} \rightarrow Fe^{3+}$
   d. $Al^{3+} \rightarrow Al(OH)_3$ e. $NH_3 \rightarrow NH_4^+$ f. $Al \rightarrow Al_2O_3$

5. *Half-Equations (Example 23.7)* Write balanced half-equations for each of the reductions and oxidations in Problem 3. Use H^+ ions and H_2O molecules to balance hydrogen and oxygen.

6. *Balancing Equations (Example 23.3)* Balance the following oxidation-reduction equation: $Ag(s) + NO_3^-(aq) + H^+(aq) \rightarrow Ag^+(aq) + NO_2(g) + H_2O$.

7. *Balancing Equations (Example 23.3)* Balance the following oxidation-reduction equation: $MnO_4^-(aq) + I^-(aq) + H^+(aq) \rightarrow MnO_2(s) + I_2(s) + H_2O$

8. *Balancing Equations (Example 23.3)* Balance the following oxidation-reduction equation:

   $$CuS(s) + NO_3^-(aq) + H^+(aq) \rightarrow Cu^{2+}(aq) + S(s) + NO_2(g) + H_2O$$

9. *Mass Relations in Balanced Equations (Examples 23.4, 23.6, 23.8)* Refer to the equation balanced in Example 23.3. Starting with 1.20 mol of Cl^-, calculate:

   a. the number of moles of Cl_2 formed.
   b. the number of moles of Cr^{3+} formed.

10. *Mass Relations in Balanced Equations (Examples 23.4, 23.6, 23.8)* Referring to the equation you balanced in Problem 6:

    a. How many moles of NO_3^- are required to oxidize 1.00 g of Ag?
    b. How many grams of NO_2 are formed, using 1.00 g of Ag?

* * * * *

11. Complete and balance the following half-equations:

    a. $Cu(s) \rightarrow Cu^{2+}(aq)$ b. $NO_3^-(aq) \rightarrow N_2(g)$ c. $SO_2(g) \rightarrow S(s)$
    d. $Al(s) \rightarrow Al^{3+}(aq)$

12. Write balanced oxidation-reduction reactions combining:

    a. (a) and (b) of Problem 11 b. (a) and (c) of Problem 11
    c. (b) and (d) of Problem 11 d. (c) and (d) of Problem 11

13. Balance the equation: $NO_3^-(aq) + H_2S(aq) \rightarrow N_2(g) + SO_2(g)$
    Use H^+ ions and H_2O molecules where needed.

14. Hydrazine, used as a rocket fuel, is made by the reaction:

    $$2\ NH_3(aq) + ClO^-(aq) \rightarrow N_2H_4(g) + Cl^-(aq) + H_2O$$

To make one kilogram of hydrazine, N_2H_4, how many grams should you use of:

a. NH_3 b. ClO^- c. $NaClO$

15. Consider Reaction 23.11. Calculate:

    a. the number of moles of Cl_2 required to form 1.00 kg of HClO.
    b. the number of grams of HClO produced from 1.00 kg of Cl_2.

16. "Aqua regia" is a mixture of concentrated nitric and hydrochloric acids. It dissolves gold by the following (unbalanced) reaction:

    $$Au(s) + H^+(aq) + NO_3^-(aq) + Cl^-(aq) \rightarrow AuCl_4^-(aq) + NO_2(g) + H_2O$$

    Taking the oxidation number of Cl to be -1 throughout:

    a. Assign oxidation numbers to all the other elements.
    b. Write balanced half-equations for the reduction and oxidation.
    c. Write a balanced equation for the reaction.

17. The unbalanced equation for the reaction of tin with nitric acid is:

    $$Sn(s) + H^+(aq) + NO_3^-(aq) \rightarrow SnO_2(s) + NO_2(g) + H_2O$$

    a. Balance this equation.
    b. How many moles of NO_3^- are required to react with 1.00 g of Sn?

18. For Reaction 23.18, what volume of O_2 at 0°C and 1 atm is required to react with 10.0 g of H_2S (molar volume of a gas at STP = 22.4 ℓ).

19. What is the oxidation number of carbon in:

    a. CO_2 b. CO_3^{2-} c. CH_3COOH

20. Referring to Figure 23.1, state which of the following can act as both oxidizing and reducing agents. (Remember that a positive ion can always be reduced to the metal.)

    a. Cr^{3+} b. Ni^{2+} c. Zn^{2+} d. Sn^{2+} e. Pb^{2+}

*21. Balance the following equation: $ClO_3^- \rightarrow ClO_4^- + Cl^-$

*22. Nitric acid (H^+, NO_3^- ions) reacts with bismuth sulfide, Bi_2S_3, an insoluble solid. The products include $NO_2(g)$, solid sulfur, and Bi^{3+} ions in solution. Write a balanced equation for the reaction.

*23. Consider Reaction 23.28. What volume of N_2O_4 at 1.00 atm and 25°C is required to react with 2.40 ℓ of $(CH_3)_2N_2H_2$ at 25°C and

    a. 1.00 atm b. 10.0 atm

*24. Consider Reaction 23.21. Suppose you want to make one liter of 18 M H_2SO_4. What volume of $SO_3(g)$ at STP will be required?

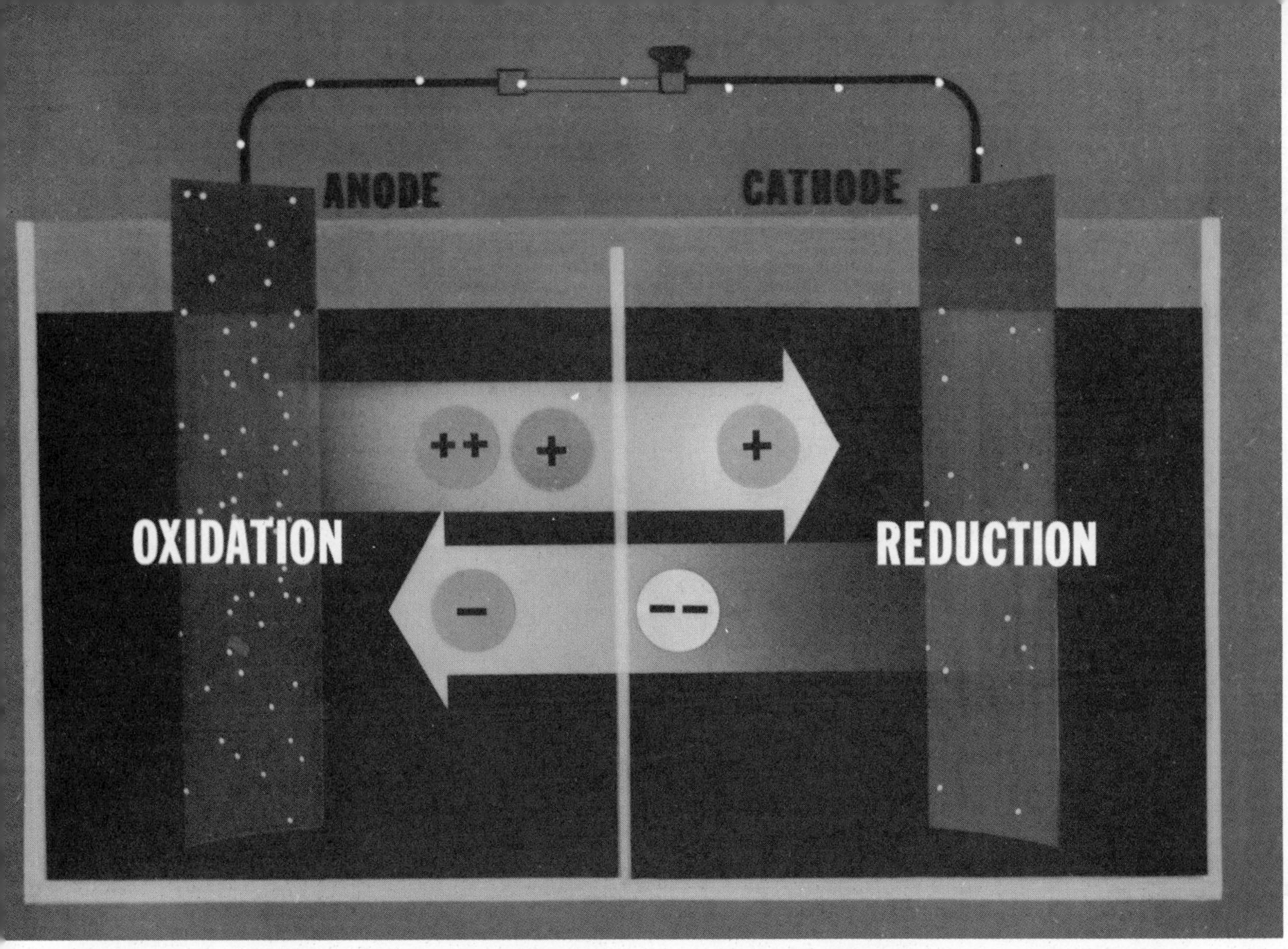

Anions move to the anode; cations to the cathode. Oxidation occurs at the anode, reduction at the cathode. These relations hold for all types of electrical cells. *From the CHEM Study Film:* Electrochemical Cells.

24 ELECTROCHEMISTRY

In Chapter 23 we discussed oxidation-reduction reactions. They involve a transfer of electrons from one species to another. One substance gives up electrons and is oxidized. Another substance acquires these electrons and is reduced.

When an oxidation-reduction reaction is carried out in the laboratory, the reactants come in direct contact with each other. Electron transfer takes place directly and spontaneously. Consider, for example, what happens when a piece of zinc is added to a solution containing Cu^{2+} ions. The following reaction occurs. (See Color Plate 11C, in the center of the book.)

$$Zn(s) + Cu^{2+}(aq) \rightarrow Zn^{2+}(aq) + Cu(s) \qquad (24.1)$$

The electrons given up by zinc atoms are acquired by Cu^{2+} ions. Electron transfer takes place at the surface of the zinc.

Reaction 24.1 occurs spontaneously because the energy of the products is less than that of the reactants. By carrying out the reaction in a somewhat different way, we can use this energy difference to produce electrical work. For this to happen, electron transfer must occur indirectly. That is, the electrons given up by the zinc must pass through an electrical circuit before they reach the Cu^{2+} ions. The device used for this purpose is called a **voltaic cell.** In a voltaic cell, a spontaneous reaction such as 24.1 produces electrical energy. This energy can be used to operate a flashlight, start an automobile, or for many other purposes.

A voltaic cell furnishes electrical energy. An electrolytic cell consumes it.

The operation of a voltaic cell will be discussed in some detail in Section 24.1. Later, in Section 24.3, we will discuss **electrolytic cells.** In an electrolytic cell, electrical energy is supplied from an outside source. This energy is used to make a non-spontaneous reaction occur in the cell. It might, for example, be used to reverse Reaction 24.1:

$$Zn^{2+}(aq) + Cu(s) \rightarrow Zn(s) + Cu^{2+}(aq)$$

More commonly, it is used to produce nonmetals such as Cl_2 or H_2 or to plate out metals such as copper or silver.

As we have indicated, there is a basic difference between these two types of cells. However, they have certain features in common. In both types of cells:

1. There are two *electrodes*. Ordinarily, these are wires or strips of metal. At any rate, they are conducting materials, through which electrons can move.

2. At one electrode, an **oxidation** half-reaction occurs. This electrode is referred to as the **anode.** At the other electrode, **reduction** takes place. This electrode is called the **cathode.**

3. The two electrodes are physically separated from one another. Between them is a liquid through which ions can move. Most commonly, this is a water solution. Positive ions **(cations)** move through the cell to the cathode. Negative ions **(anions)** move in the opposite direction. They migrate to the anode of the cell.

You may notice that we have said nothing about the polarity of electrodes. This subject can be confusing, particularly since different systems are used around the world. However, if you are hooking up an electrical circuit, it helps to know what is meant by a "positive" or "negative" electrode. In the United States, the system used is the following.

1. The terminal at which electrons come out of a voltaic cell is considered to be negative (−). An electrode connected to that terminal is referred to as a "negative" electrode.

2. The terminal at which electrons enter a voltaic cell is considered to be positive (+). An electrode connected to that terminal is called a "positive" electrode.

This system seems very logical. However, it turns out that the reverse system would be just as logical. That occurred to the English, among others. We will not mention the subject again and would just as soon you didn't worry about it.

24.1 VOLTAIC CELLS

The Zinc-Copper Cell

As we have seen, the oxidation-reduction reaction

$$Zn(s) + Cu^{2+}(aq) \rightarrow Zn^{2+}(aq) + Cu(s) \quad (24.1)$$

is spontaneous. If the reactants are placed in a flask, the process takes place directly. Two half-reactions occur, both at the surface of the zinc. Zinc atoms are oxidized:

$$Zn(s) \rightarrow Zn^{2+}(aq) + 2\ e^-$$

The electrons given up by the zinc atoms are acquired by Cu^{2+} ions. They are reduced to copper atoms:

$$Cu^{2+}(aq) + 2\ e^- \rightarrow Cu(s)$$

Carried out a different way, Reaction 24.1 can serve as a source of energy in a voltaic cell. A simple cell using this reaction is shown in Figure 24.1. This voltaic cell consists of:

—a porous cup, containing a solution of $ZnSO_4$. Immersed in this solution is a strip of zinc metal. The cup sits inside:

—a beaker containing a solution of $CuSO_4$. A strip of copper metal is immersed in this solution. The zinc and copper strips are connected through wires to:

—an electrical circuit. As shown, this is a simple light bulb. It could be a more complex circuit.

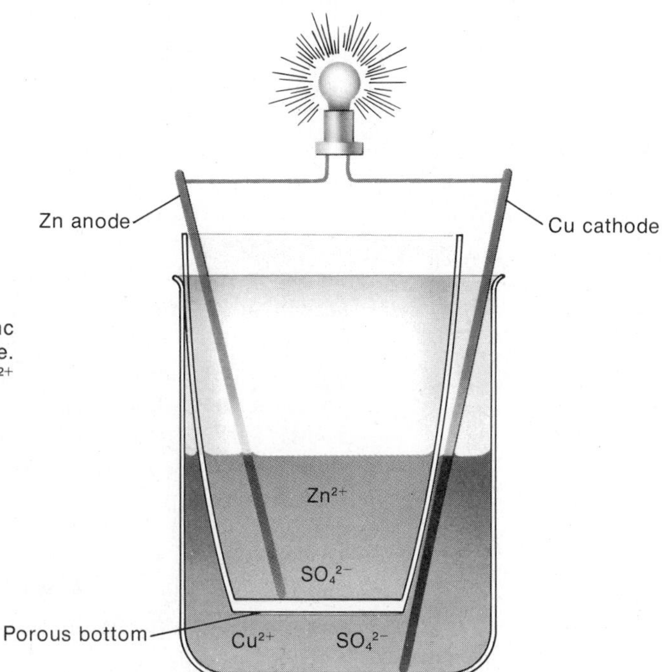

FIGURE 24.1

A Zn-Cu^{2+} voltaic cell. Electrons flow from the zinc anode through the light bulb to the copper cathode. The zinc electrode is oxidized to Zn^{2+}. The Cu^{2+} ion is reduced to Cu at the copper electrode.

To understand how Reaction 24.1 takes place in this cell, consider what happens in the various regions.

(1) Within the cup, the following half-reaction occurs:

$$\text{ANODE:} \quad Zn(s) \rightarrow Zn^{2+}(aq) + 2\,e^-; \quad \text{OXIDATION} \qquad (24.1\text{ A})$$

Since this is an oxidation half-reaction, the zinc electrode is referred to as the anode. As time passes, the mass of the zinc anode decreases.

(2) The electrons given up by the zinc pass through the circuit in the direction indicated. As they flow through the circuit, they cause the bulb to light up. In this way, the energy given off by the reaction is used to do useful electrical work.

(3) After passing through the circuit, the electrons enter the copper strip. Moving through this strip, they reach the $CuSO_4$ solution in the beaker. Here, the following half-reaction occurs:

$$\text{CATHODE:} \quad Cu^{2+}(aq) + 2\,e^- \rightarrow Cu(s); \quad \text{REDUCTION} \qquad (24.1\text{ C})$$

Since this is a reduction half-reaction, the copper electrode is the cathode. As time passes, the blue color of the solution, due to Cu^{2+} ions, fades. The copper cathode increases in mass as copper metal plates out on it.

(4) As the cell operates, ions move through it. Positive ions (cations) move toward the copper cathode. Negative ions (anions) move toward the zinc anode. In this way, there is a flow of electrical charge through the cell itself. This completes the electrical circuit.

The overall reaction in the cell is obtained by adding the two half-reactions 24.1 A and 24.1 C:

$$Zn(s) + Cu^{2+}(aq) \rightarrow Zn^{2+}(aq) + Cu(s) \qquad (24.1)$$

In a voltaic cell the reaction can occur only if the electron transfer occurs through the external circuit.

This is the same reaction that occurs directly when the reactants are mixed in a flask. Notice, however, an important difference. In the cell shown in Figure 24.1, Zn and Cu^{2+} ions never come in contact. If they did, the cell would "short-circuit". The light would go out and the energy would be wasted as heat.

The need for keeping the reactants separate explains why the cell shown in Figure 24.1 is separated into two compartments. By placing $CuSO_4$ solution in the beaker, we keep Cu^{2+} ions away from the zinc strip in the cup.

However, it is vital that the walls of the cup be porous. Ions must be able to move through these walls. Otherwise the cell would not operate. Reaction 24.1 A would build up a positive (+) charge in the cup as Zn^{2+} ions form. At the same time, Reaction 24.1 C, by using up Cu^{2+}

ions, would create a negative (−) charge in the beaker. Under these conditions, the electrons would refuse to move from the zinc strip to the copper. There would be an "open circuit" and no current would flow.

Migration of ions through the cell prevents this buildup of charge. Zn^{2+} ions move out of the cup into the beaker. At the same time, SO_4^{2-} ions move in the opposite direction. This way we avoid forming a positive charge in the cup or a negative charge in the beaker. Electrical neutrality exists throughout the cell.

We see then that the use of the porous cup prevents two "disasters". These are a short circuit (direct contact of reactants) and an open circuit (no current flow).

Other Simple Voltaic Cells

In principle, any spontaneous oxidation-reduction reaction can serve as a source of energy in a voltaic cell. For reactions in water solution, a cell such as that shown in Figure 24.1 will work. Its design depends upon the particular reaction used (Example 24.1).

EXAMPLE 24.1

Consider the spontaneous reaction:

$$2\ Ag^+(aq) + Cu(s) \rightarrow 2\ Ag(s) + Cu^{2+}(aq)$$

Describe a voltaic cell which uses this reaction as a source of electrical energy. What is the half-reaction at the anode? at the cathode? What materials would be in the anode compartment? the cathode compartment? In what direction would electrons move through the electrical circuit? cations through the cell? anions through the cell?

SOLUTION

The half-reactions are:

ANODE:	$Cu(s) \rightarrow Cu^{2+}(aq) + 2\ e^-$;	OXIDATION
CATHODE:	$2\ Ag^+(aq) + 2\ e^- \rightarrow 2\ Ag(s)$;	REDUCTION

The anode compartment might consist of a strip of copper dipping into a solution of $Cu(NO_3)_2$ (Cu^{2+} ions). This could be placed in a porous cup. Outside the cup would be a beaker. This might contain a silver wire dipping into a solution of $AgNO_3$ (Ag^+ ions).

Once you get the idea, it's easy to design cells like this.

Electrons are produced at the copper electrode. They move through the circuit from copper to silver. At the silver electrode, they reduce Ag^+ ions to Ag(s). Within the cell, cations (Ag^+ and Cu^{2+}) move to the silver cathode. Anions move to the copper anode. A diagram of this cell is shown in Figure 24.2, p. 598.

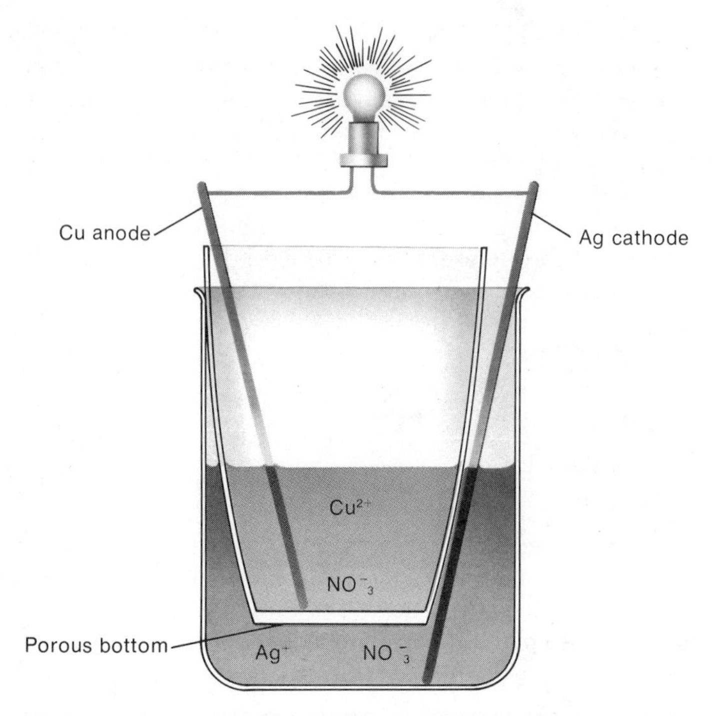

FIGURE 24.2

A Cu-Ag^+ voltaic cell. The electron flow is from the copper anode to the silver cathode. In the solutions, cations migrate toward the cathode, anions toward the anode.

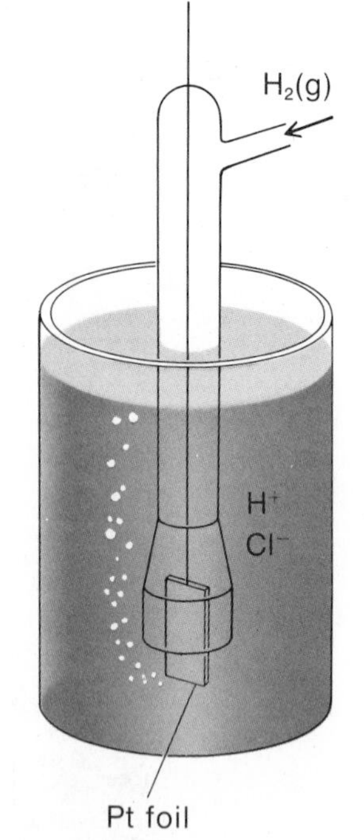

FIGURE 24.3

An H_2-H^+ half cell. Platinum serves as the electron conductor but does not participate in the chemical reaction. When the cell operates, H_2 gas is slowly bubbled past the Pt foil. Hydrochloric acid is usually the source of the H^+ ion.

In some cells, one of the half-reactions does not involve a metal. An example is the Zn-H^+ cell.

$$Zn(s) + 2\ H^+(aq) \rightarrow Zn^{2+}(aq) + H_2(g)$$

The half-reactions are:

$$\text{ANODE:} \quad Zn(s) \rightarrow Zn^{2+}(aq) + 2\ e^-; \quad \text{OXIDATION}$$

$$\text{CATHODE:} \quad 2\ H^+(aq) + 2\ e^- \rightarrow H_2(g); \quad \text{REDUCTION}$$

Notice that the cathode reaction involves an ion, H^+, and a nonmetal, H_2.

Here we have to use an inert electrode for the H^+-H_2 half cell. This electrode can be made of any material through which electrons can flow. However, it must be inert in the sense that it does not react with molecules or ions around it. One possibility is a platinum wire. Platinum is a good conductor of electricity and is very unreactive. The wire would be surrounded by a solution containing H^+ ions. This might be a solution of hydrochloric acid. Hydrogen gas is bubbled over the platinum electrode (Figure 24.3).

EXAMPLE 24.2

The following oxidation-reduction reaction occurs in a voltaic cell:

$$Cl_2(g) + 2\ Br^-(aq) \rightarrow 2\ Cl^-(aq) + Br_2(l)$$

Write equations for the reactions at the anode and cathode. What materials would you use in the anode compartment? the cathode compartment?

SOLUTION

The half-reactions are:

ANODE: $2\ Br^-(aq) \rightarrow Br_2(l) + 2\ e^-$; OXIDATION

CATHODE: $Cl_2(g) + 2\ e^- \rightarrow 2\ Cl^-(aq)$; REDUCTION

Neither reaction involves a metal, so two inert electrodes would be used.

In the anode compartment, we might use a platinum wire. This would be surrounded by a solution containing both Br_2 molecules and Br^- ions. This could be prepared by shaking a water solution of NaBr with liquid bromine.

Platinum is a good electrode material, but it is very expensive.

In the cathode compartment, we could also use a platinum electrode. It would dip into a water solution containing both Cl_2 molecules and Cl^- ions. To prepare this solution, chlorine gas could be bubbled through a solution of NaCl.

Commercial Voltaic Cells

The type of voltaic cell just discussed is easy to set up. Also, as we will see later, it gives valuable information about the redox reaction taking place within it. However, it is not a practical source of electrical energy, at least for regular use, since it cannot deliver much power.

A variety of cells have been developed for commercial use. They operate flashlights, portable radios, and automobiles. Here we will see how two of the most common commercial cells operate.

Dry Cell

The ordinary "dry cell" used in flashlights is shown in Fig. 24.4, p. 600. The outside case of the cell is made of zinc, which is the anode. Through the center of the cell there is a graphite rod which acts as the cathode. The two electrodes are separated by a moist paste. This contains manganese dioxide, MnO_2, ammonium chloride, NH_4Cl, and finely divided carbon.

The anode reaction in the dry cell is a simple one. Zinc atoms are oxidized to Zn^{2+} ions.

$$\text{ANODE:} \quad Zn(s) \rightarrow Zn^{2+}(aq) + 2\ e^-$$

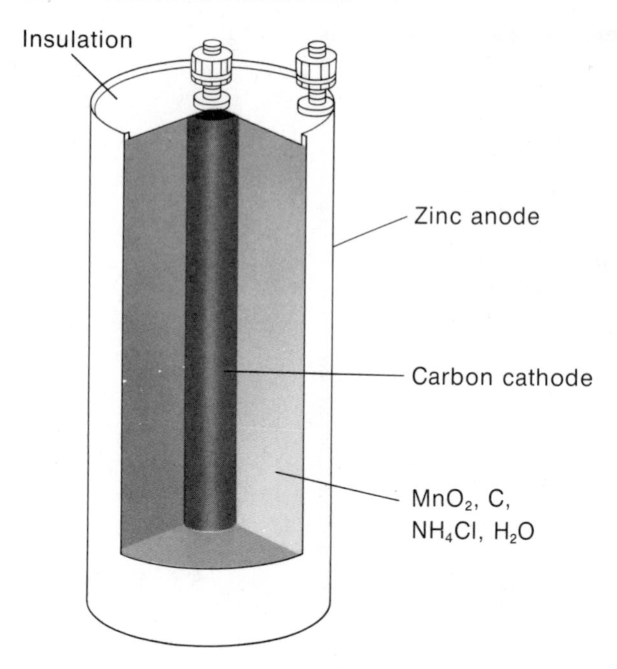

FIGURE 24.4

A "dry cell". In this cell Zn serves as the anode. MnO_2 is reduced at the carbon cathode. This cell has a potential of 1.5 volts and is not rechargeable. It is a relatively cheap source of small amounts of electrical energy.

A dry cell can only furnish small amounts of energy.

The cathode reaction is more complex. The graphite electrode is inert and takes no part in the reaction. Manganese is reduced, from the +4 to the +3 state. The half-reaction may be written:

$$\text{CATHODE:}\ MnO_2(s) + 4\,NH_4^+(aq) + e^- \rightarrow Mn^{3+}(aq) + 4\,NH_3(aq) + 2\,H_2O$$

The overall cell reaction is obtained by combining the two half-reactions.

$$Zn(s) + 2\ MnO_2(s) + 8\ NH_4^+(aq) \rightarrow Zn^{2+}(aq) + 2\ Mn^{3+}(aq) + 8\ NH_3(aq) + 4\ H_2O \qquad (24.2)$$

A dry cell produces a voltage of about 1.5 V. If current is drawn too quickly, gaseous ammonia builds up around the cathode. This cuts off the flow of ions and the cell ceases to operate. Upon standing, the ammonia dissolves and the cell can be used again. However, over a period of a few months, the cell becomes "dead". This happens because Reaction 24.2 occurs even when the cell is not being used to furnish energy. The reaction is slow, but eventually uses up the reactants. There is no practical way to recharge an ordinary dry cell. That is, Reaction 24.2 cannot be reversed so as to return the cell to its original condition.

Lead Storage Battery

The 12-volt battery used in automobiles consists of six voltaic cells of the type shown in Figure 24.5. Each of these cells gives a voltage of 2 V. The electrodes are both built of lead in the form of grids. In the

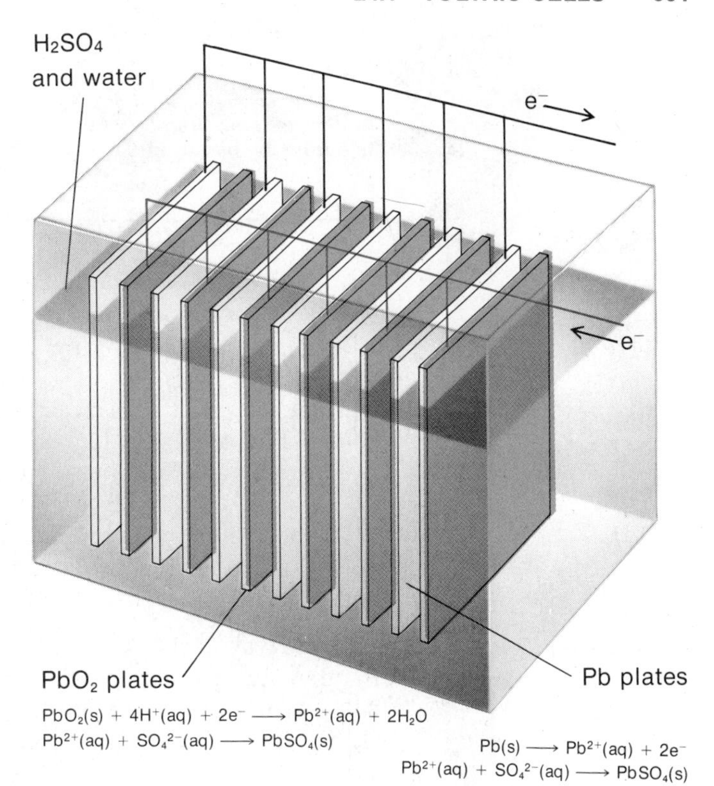

FIGURE 24.5

A lead storage battery. A battery like this has a voltage of two volts and can deliver a large amount of electrical energy for a short time. Another advantage of this battery is that it can be recharged. A disadvantage is that it is very heavy.

anode, the grids are filled with compressed, "spongy" lead, which has a grey appearance. The grids on the cathode are filled with brown lead(IV) oxide, PbO_2. The two types of electrodes are immersed in a dilute solution of sulfuric acid, H_2SO_4.

When the cell operates, a spontaneous oxidation-reduction reaction occurs. At the anode, lead metal is oxidized to the +2 state. At the cathode, lead in the form of PbO_2 is reduced from the +4 to the +2 state. In both cases, the Pb^{2+} ions formed are converted to insoluble $PbSO_4$. This occurs when they come in contact with the SO_4^{2-} ions in the sulfuric acid solution.

A storage battery can deliver about 300 amperes when starting a car.

ANODE:

$$Pb(s) \rightarrow Pb^{2+}(aq) + 2\ e^-$$

$$Pb^{2+}(aq) + SO_4^{2-}(aq) \rightarrow PbSO_4(s)$$

$$Pb(s) + SO_4^{2-}(aq) \rightarrow PbSO_4(s) + 2\ e^- \quad (24.3\ A)$$

CATHODE:

$$PbO_2(s) + 4\ H^+(aq) + 2\ e^- \rightarrow Pb^{2+}(aq) + 2\ H_2O$$

$$Pb^{2+}(aq) + SO_4^{2-}(aq) \rightarrow PbSO_4(s)$$

$$PbO_2(s) + 4\ H^+(aq) + 2\ e^- + SO_4^{2-}(aq) \rightarrow PbSO_4(s) + 2\ H_2O \quad (24.3\ C)$$

The net reactions at the electrodes are given by 24.3 A and 24.3 C.

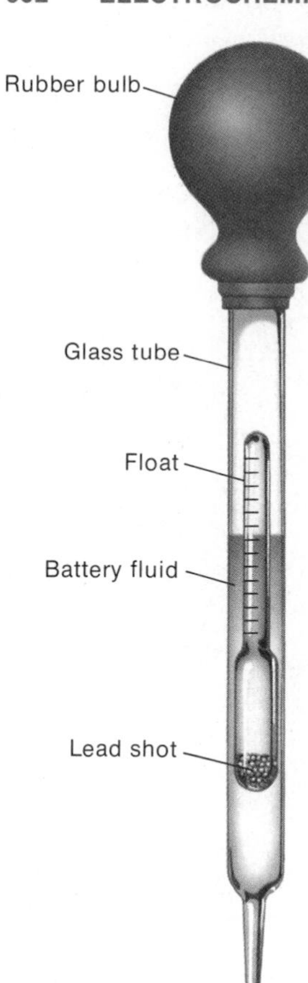

FIGURE 24.6

Hydrometer for checking density of battery fluid. The float inside the hydrometer is calibrated so that it will float high in fluid from a fully charged battery and low when the fluid is from a run-down battery. The suction bulb allows the serviceman to draw fluid from the battery into the float chamber.

The overall reaction in the lead storage battery is obtained by combining 24.3 A and 24.3 C.

$$Pb(s) + SO_4^{2-}(aq) \rightarrow PbSO_4(s) + 2\ e^-$$

$$PbO_2(s) + 4\ H^+(aq) + 2\ e^- + SO_4^{2-}(aq) \rightarrow PbSO_4(s) + 2\ H_2O$$

$$Pb(s) + PbO_2(s) + 4\ H^+(aq) + 2\ SO_4^{2-}(aq) \rightarrow 2\ PbSO_4(s) + 2\ H_2O \quad (24.3)$$

Notice that as the battery supplies electrical energy by Reaction 24.3:

—both electrodes become coated with lead sulfate.

—sulfuric acid is consumed. One mole of H_2SO_4 (2 H^+, 1 SO_4^{2-}) is used up for every mole of $PbSO_4$ formed.

One way to check the condition of a lead storage battery is to check the density of the sulfuric acid solution, using a hydrometer (Figure 24.6). When fully charged, the solution contains about 40 mass % of H_2SO_4. It has a density of about 1.30 g/cm^3. As the cell reaction consumes H_2SO_4, the density drops. A density of 1.15 g/cm^3 means that about half of the sulfuric acid has been consumed. At this point it is necessary to recharge the battery. Otherwise, both the voltage and the conductivity will drop off sharply with further use.

To recharge a storage battery, Reaction 24.3 must be reversed. This is possible because the reaction product, $PbSO_4$, is formed directly on the electrodes. So, it can be converted back to Pb on one electrode, to PbO_2 on the other. In an automobile, recharging occurs automatically when the car is running. In older-model cars, the electrical energy required is furnished by a direct current generator run off the motor. With modern cars, an alternator equipped with a rectifier is used.

In principle, a lead storage battery should last indefinitely. Unfortunately, as you well know, this is not the case. One problem is that both lead and lead sulfate tend to flake off the electrodes. They collect as a sludge at the bottom of the battery. When this happens, Reaction 24.3 occurs directly and the cell short-circuits.

Other Commercial Cells

Many different kinds of voltaic cells have been developed for special purposes. One of the most popular nowadays is the nickel-cadmium cell. It is used to power hand calculators and flash attachments on cameras. The anode in this cell is made of cadmium metal. The cathode contains nickel(IV) oxide, NiO_2. The electrodes are surrounded by a solution of potassium hydroxide, KOH. The cell reactions are:

ANODE: $Cd(s) + 2\ OH^-(aq) \rightarrow Cd(OH)_2(s) + 2\ e^-$ OXIDATION

CATHODE: $NiO_2(s) + 2\ e^- + 2\ H_2O \rightarrow Ni(OH)_2(s) + 2\ OH^-(aq)$ REDUCTION

$$Cd(s) + NiO_2(s) + 2\ H_2O \rightarrow Cd(OH)_2(s) + Ni(OH)_2(s) \quad (24.4)$$

The products of the cell reaction, $Cd(OH)_2$ and $Ni(OH)_2$, are formed on the electrodes. This means that, unlike a dry cell, a nickel-cadmium cell can be recharged. Hence, it can be used over and over again, like a lead storage battery. In contrast to a storage battery, nickel-cadmium cells can be scaled down to a small size. This makes them ideal as a power source for small, portable, electronic equipment.

In recent years, there has been renewed interest in electrically-driven automobiles. Such cars would get their energy from a voltaic cell instead of the combustion of gasoline. As of now, there is no practical cell for this purpose. A lead storage battery is too heavy for the amount of energy it produces. A car using such batteries as its only power source would weigh twice as much as those now on the road. Otherwise, it would have to stop every few miles for recharging.

Many oxidation-reduction reactions produce more energy per unit mass than a lead storage battery. The most promising are those that involve a light, reactive metal such as sodium (d = 0.97 g/cm^3) as the anode. Recently, a cell has been developed that uses the reaction between sodium and sulfur to produce electrical energy:

$$\text{ANODE:}\quad 2\,Na(l) \rightarrow 2\,Na^+ + 2\,e^- \quad \text{OXIDATION}$$

$$\text{CATHODE:}\quad S(l) + 2\,e^- \rightarrow S^{2-} \quad \text{REDUCTION}$$

$$2\,Na(l) + S(l) \rightarrow Na_2S(s) \qquad (24.5)$$

This would be an ideal source of electrical energy for an automobile if it were not for one factor. Sodium metal reacts violently with water, so the reaction can not be carried out in water solution. At present, the cell has to operate at 200°C, using liquid sodium.

24.2 STANDARD VOLTAGES

As we have pointed out, oxidation-reduction reactions taking place in voltaic cells are spontaneous. Such a reaction produces a positive voltage. This voltage is a measure of the driving force behind the reaction. The magnitude of the voltage depends upon:

—the nature of the reaction;
—the concentrations of reactants and products;
—the temperature.

Throughout our discussion, we will deal with "standard" concentrations. We will assume that all ions or molecules in aqueous solution are at a concentration of 1 M. Further, we assume that any gas taking part in the reaction is at a pressure of 1 atm. Finally, we will take the temperature to be 25°C. We refer to the voltage measured under these conditions as the **standard voltage.**

Assignment of Half-Cell Voltages

Let us consider the voltaic cell shown in Fig. 24.7, p. 604. Here, both Zn^{2+} and H^+ ions are at a concentration of 1 M. The H_2 gas is at a pressure of 1 atm. The temperature is 25°C. Under these conditions, the voltage read on the meter shown at the top of the figure is the standard

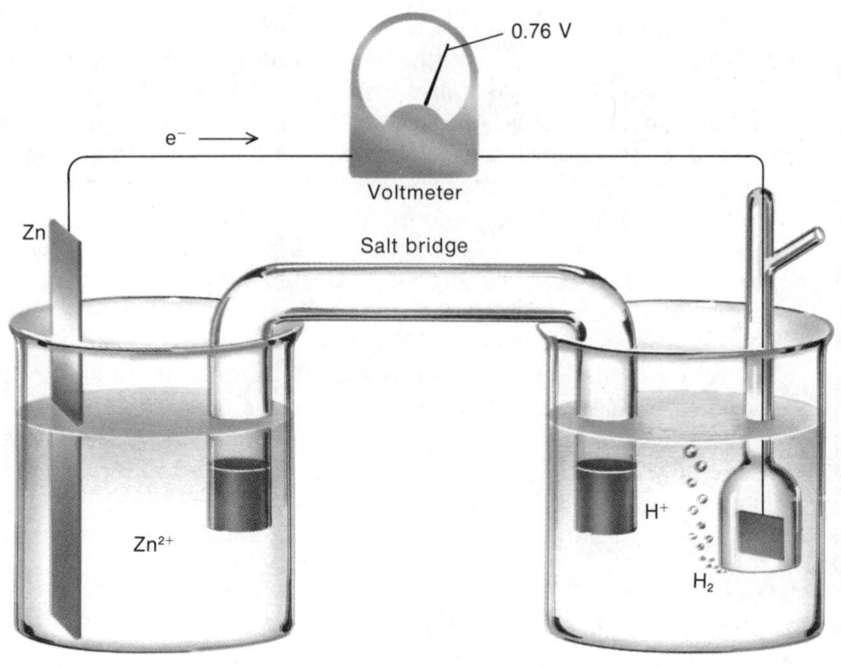

FIGURE 24.7

A Zn-H^+ voltaic cell. In this cell Zn is the anode. The H^+ ion is reduced to H_2 gas at the platinum cathode. If H^+ and Zn^{2+} are both 1 M, and P_{H_2} is 1 atm, the voltage of the cell is 0.76 V. The "salt bridge" is filled with a solution of KNO_3. Its function is similar to that of the porous cup shown in Fig. 24.1.

voltage. This is given the symbol **E⁰**. We see that for the Zn-H^+ cell, $E^0 = +0.76$ V.

The standard voltage of a cell is associated with the reaction taking place in the cell. Here we can say that for the reaction:

$$Zn(s) + 2\,H^+(1\,M) \rightarrow Zn^{2+}(aq) + H_2(g,\,1\,atm); \qquad E^0 = +0.76\text{ V}$$

When you use a salt bridge, it's easier to see the species involved in the two half reactions.

As we know, any oxidation-reduction reaction can be broken down into two half-reactions. One of these is an oxidation, the other a reduction. In the same way, we can think of E^0 as being made up of two parts. One of these is a standard half-cell voltage for the oxidation half-reaction. We give this the symbol E^0_{ox}. The other is the standard half-cell voltage for the reduction. This has the symbol E^0_{red}. For any reaction:

$$E^0 = E^0_{ox} + E^0_{red} \qquad (24.6)$$

To illustrate the use of Equation 24.6, consider the reaction between zinc and H^+ ions. We can say that:

$$\begin{array}{ll} Zn(s) \rightarrow Zn^{2+}(1\,M) + 2\,e^- & E^0_{ox}\ Zn \rightarrow Zn^{2+} \\ 2\,H^+(1\,M) + 2\,e^- \rightarrow H_2(g,\,1\,atm) & E^0_{red}\ 2H^+ \rightarrow H_2 \\ \hline Zn(s) + 2\,H^+(1\,M) \rightarrow Zn^{2+}(1\,M) + H_2(g,\,1\,atm) & E^0 = +0.76\text{ V} \end{array}$$

That is:

$$E^0 = +0.76\text{ V} = E^0_{ox}\ Zn \rightarrow Zn^{2+} + E^0_{red}\ 2H^+ \rightarrow H_2 \qquad (24.7)$$

or, in words:

The standard cell voltage, E^0, is the sum of the standard voltage for the oxidation half-reaction ($Zn \rightarrow Zn^{2+}$) and that for the reduction half-reaction ($2\ H^+ \rightarrow H_2$).

Every redox reaction leads to an equation like 24.7. Each such equation contains three terms. One of these, the standard cell voltage, E^0, can be measured. The other two quantities, E^0_{ox} and E^0_{red}, cannot be measured. We cannot set up a voltaic cell with a single electrode or carry out a half-reaction by itself. Oxidation and reduction must occur together.

However, it is possible to establish relative values for standard half-cell voltages. To do this, *we arbitrarily take the standard voltage for the reduction of H^+ ions to H_2 to be zero.* That is:

$$E^0_{red}\ 2H^+ \rightarrow H_2 = 0.00\ V \qquad (24.8)$$

All standard voltages are based on this assumption.

Now, going back to Equation 24.7, we can obtain the standard voltage for the oxidation of Zn to Zn^{2+} ions:

$$+0.76\ V = E^0_{ox}\ Zn \rightarrow Zn^{2+} + 0.00\ V$$

$$E^0_{ox}\ Zn \rightarrow Zn^{2+} = +0.76\ V$$

Once we have found a few half-reaction voltages, it is easy to obtain others. Suppose, for example, we want to know the standard voltage for the reduction of Cu^{2+} to Cu. To do this, we use the cell shown in Figure 24.1, p. 595. At standard concentrations, we measure the voltage and find it to be +1.10 V. That is:

$$Zn(s) + Cu^{2+}(1\ M) \rightarrow Zn^{2+}(1\ M) + Cu(s); \qquad E^0 = +1.10\ V$$

But, E^0 for this cell is the sum of the two half-reaction voltages:

$$+1.10\ V = E^0_{ox}\ Zn \rightarrow Zn^{2+} + E^0_{red}\ Cu^{2+} \rightarrow Cu \qquad (24.9)$$

By Eqn. 24.6.

Having found $E^0_{ox}\ Zn \rightarrow Zn^{2+}$ to be +0.76 V, we have:

$$+1.10\ V = +0.76\ V + E^0_{red}\ Cu^{2+} \rightarrow Cu$$

or: $$E^0_{red}\ Cu^{2+} \rightarrow Cu = +1.10\ V - 0.76\ V = +0.34\ V$$

Continuing in this way, we can find E^0_{ox} or E^0_{red} for any half-reaction. Table 24.1, p. 606, lists several such values. Standard voltages for reduction half-reactions are listed in the column at the far right. They apply to the forward reaction (left to right). For example:

$$Li^+(aq) + e^- \rightarrow Li(s) \qquad E^0_{red} = -3.05\ V$$

Standard voltages for oxidation half-reactions are listed in the column at the far left. They apply to the reverse reaction (right to left). That is:

$$Li(s) \rightarrow Li^+(aq) + e^- \qquad E^0_{ox} = +3.05\ V$$

TABLE 24.1 STANDARD HALF-CELL VOLTAGES

E^0_{ox} (volts)	Half Reaction	E^0_{red} (volts)
	OXIDIZING AGENT REDUCING AGENT	
+3.05	$Li^+(aq) + e^- \rightleftharpoons Li(s)$	−3.05
+2.93	$K^+(aq) + e^- \rightleftharpoons K(s)$	−2.93
+2.87	$Ca^{2+}(aq) + 2\ e^- \rightleftharpoons Ca(s)$	−2.87
+2.71	$Na^+(aq) + e^- \rightleftharpoons Na(s)$	−2.71
+2.37	$Mg^{2+}(aq) + 2\ e^- \rightleftharpoons Mg(s)$	−2.37
+1.66	$Al^{3+}(aq) + 3\ e^- \rightleftharpoons Al(s)$	−1.66
+0.76	$Zn^{2+}(aq) + 2\ e^- \rightleftharpoons Zn(s)$	−0.76
+0.44	$Fe^{2+}(aq) + 2\ e^- \rightleftharpoons Fe(s)$	−0.44
+0.25	$Ni^{2+}(aq) + 2\ e^- \rightleftharpoons Ni(s)$	−0.25
+0.14	$Sn^{2+}(aq) + 2\ e^- \rightleftharpoons Sn(s)$	−0.14
+0.13	$Pb^{2+}(aq) + 2\ e^- \rightleftharpoons Pb(s)$	−0.13
0.00	$2\ H^+(aq) + 2\ e^- \rightleftharpoons H_2(g)$	0.00
−0.14	$S(s) + 2\ H^+(aq) + 2\ e^- \rightleftharpoons H_2S(g)$	+0.14
−0.15	$Sn^{4+}(aq) + 2\ e^- \rightleftharpoons Sn^{2+}(aq)$	+0.15
−0.20	$SO_4^{2-}(aq) + 4\ H^+(aq) + 2\ e^- \rightleftharpoons SO_2(g) + 2\ H_2O$	+0.20
−0.34	$Cu^{2+}(aq) + 2\ e^- \rightleftharpoons Cu(s)$	+0.34
−0.53	$I_2(s) + 2\ e^- \rightleftharpoons 2\ I^-(aq)$	+0.53
−0.77	$Fe^{3+}(aq) + e^- \rightleftharpoons Fe^{2+}(aq)$	+0.77
−0.80	$Ag^+(aq) + e^- \rightleftharpoons Ag(s)$	+0.80
−0.96	$NO_3^-(aq) + 4\ H^+(aq) + 3\ e^- \rightleftharpoons NO(g) + 2\ H_2O$	+0.96
−1.07	$Br_2(l) + 2\ e^- \rightleftharpoons 2\ Br^-(aq)$	+1.07
−1.23	$O_2(g) + 4\ H^+(aq) + 4\ e^- \rightleftharpoons 2\ H_2O$	+1.23
−1.23	$MnO_2(s) + 4\ H^+(aq) + 2\ e^- \rightleftharpoons Mn^{2+}(aq) + 2\ H_2O$	+1.23
−1.33	$Cr_2O_7^{2-}(aq) + 14\ H^+(aq) + 6\ e^- \rightleftharpoons 2\ Cr^{3+}(aq) + 7\ H_2O$	+1.33
−1.36	$Cl_2(g) + 2\ e^- \rightleftharpoons 2\ Cl^-(aq)$	+1.36
−1.50	$Au^{3+}(aq) + 3\ e^- \rightleftharpoons Au(s)$	+1.50
−1.52	$MnO_4^-(aq) + 8\ H^+(aq) + 5\ e^- \rightleftharpoons Mn^{2+}(aq) + 4\ H_2O$	+1.52
−2.87	$F_2(g) + 2\ e^- \rightleftharpoons 2\ F^-(aq)$	+2.87

Looking at Table 24.1, you will note that:

1. E^0_{red} values become more positive as you move down the Table. We start with −3.05 V ($Li^+ \rightarrow Li$), pass through zero ($2\ H^+ \rightarrow H_2$), and end, at the bottom of the column, with +2.87 V ($F_2 \rightarrow 2\ F^-$).

2. E^0_{ox} values become more negative as you move down the Table. We start with +3.05 V ($Li \rightarrow Li^+$), pass through zero ($H_2 \rightarrow 2\ H^+$), and end, at the bottom of the column, with −2.87 V ($2\ F^- \rightarrow F_2$).

3. **The standard voltages for forward and reverse half-reactions (reduction and oxidation) are equal in magnitude but opposite in sign.** For example:

$E^0_{red}\ Li^+ \rightarrow Li = -3.05\ V$; $E^0_{ox}\ Li \rightarrow Li^+ = +3.05\ V$
$E^0_{red}\ 2H^+ \rightarrow H_2 = 0.00\ V$; $E^0_{ox}\ H_2 \rightarrow 2H^+ = 0.00\ V$
$E^0_{red}\ F_2 \rightarrow 2F^- = +2.87\ V$; $E^0_{ox}\ 2F^- \rightarrow F_2 = -2.87\ V$

Standard half-cell voltages are useful in many ways. We will consider three of their uses:

—to calculate cell voltages.
—to compare the strengths of different oxidizing and reducing agents.
—to decide whether a given redox reaction is likely to occur in the laboratory.

Calculation of Cell Voltages

Half-cell voltages can be used to calculate voltages for many different cells. To do this, we apply Equation 24.6:

$$E^0 = E^0_{ox} + E^0_{red} \qquad (24.6)$$

We select the values of E^0_{ox} and E^0_{red} from Table 24.1. These are then added to obtain E^0.

EXAMPLE 24.3

Calculate E^0 for cells in which the reaction is:

a. $Ni(s) + 2\ Ag^+(aq) \rightarrow Ni^{2+}(aq) + 2\ Ag(s)$

b. $Cl_2(g) + 2\ Br^-(aq) \rightarrow 2\ Cl^-(aq) + Br_2(l)$

SOLUTION

In each case, we first split the reaction into two half-reactions. Then we look up the values of E^0_{ox} and E^0_{red} in Table 24.1. Finally, we add the two half-cell voltages to obtain E^0.

Cell voltages are usually small, of the order of a volt or so.

a. OXIDATION: $Ni(s) \rightarrow Ni^{2+}(aq) + 2\ e^-$; $E^0_{ox} = +0.25\ V$

REDUCTION: $2\ Ag^+(aq) + 2\ e^- \rightarrow 2\ Ag(s)$; $E^0_{red} = +0.80\ V$

$Ni(s) + 2\ Ag^+(aq) \rightarrow Ni^{2+}(aq) + 2\ Ag(s)$ $E^0 = +1.05\ V$

Notice that the coefficients for the reduction half-equation are twice those listed in the Table. However, we do *not* multiply E^0_{red} by two. Voltage is a measure of the driving force behind the reaction. It does not depend upon the number of electrons transferred.

b. OXIDATION: $2\ Br^-(aq) \rightarrow Br_2(l) + 2\ e^-$ $E^0_{ox} = -1.07\ V$

REDUCTION: $Cl_2(g) + 2\ e^- \rightarrow 2\ Cl^-(aq)$ $E^0_{red} = +1.36\ V$

$Cl_2(g) + 2\ Br^-(aq) \rightarrow 2\ Cl^-(aq) + Br_2(l)$ $E^0 = +0.29\ V$

Strength of Oxidizing and Reducing Agents

At least in principle, all the species in the left column of Table 24.1 can act as oxidizing agents. By picking up electrons they can oxidize another substance. Their strengths as oxidizing agents depend upon how readily they are reduced. This in turn depends upon the E^0 value for their reduction. *The more positive its E^0_{red} value, the stronger the oxidizing agent.*

If E^0_{red} is large and positive the reduction half reaction will tend to go.

Applying this principle, we see that the strongest oxidizing agents are located near the bottom of the left column of Table 24.1. These include F_2 ($E^0_{red} = +2.87$ V) and MnO_4^- ($E^0_{red} = +1.52$ V). Both of these are strong enough to oxidize water, producing oxygen gas. In the case of fluorine, this reaction takes place immediately when the gas comes in contact with water.

$$F_2(g) + H_2O \rightarrow 2\ F^-(aq) + 2\ H^+(aq) + \tfrac{1}{2}\ O_2(g) \qquad (24.10)$$

As we move up the left column, oxidizing strength decreases. Chlorine ($E^0_{red} = +1.36$ V) is a powerful enough oxidizing agent to produce Br_2 and I_2 from their ions.

$$Cl_2(g) + 2\ Br^-(aq) \rightarrow 2\ Cl^-(aq) + Br_2(l) \qquad (24.11)$$

$$Cl_2(g) + 2\ I^-(aq) \rightarrow 2\ Cl^-(aq) + I_2(s) \qquad (24.12)$$

The H^+ ion ($E^0_{red} = 0.00$ V) is a much weaker oxidizing agent. It does, however, oxidize active metals such as zinc or aluminum:

$$Zn(s) + 2\ H^+(aq) \rightarrow Zn^{2+}(aq) + H_2(g) \qquad (24.13)$$

$$Al(s) + 3\ H^+(aq) \rightarrow Al^{3+}(aq) + \tfrac{3}{2}\ H_2(g) \qquad (24.14)$$

Finally, as we approach the top of the left column, the oxidizing agents become very weak. Indeed, such species as Ca^{2+}, K^+ and Li^+ never act as oxidizing agents in water solution. Their E^0_{red} values are large negative numbers (−2.87 V, −2.93 V, −3.05 V).

If E^0_{ox} is large and positive, the oxidation half-reaction will tend to go.

We can apply the same argument to reducing agents. Their strength depends upon how readily they are oxidized. This in turn depends upon the E^0 value for their oxidation. *The more positive its E^0_{ox} value, the stronger the reducing agent.*

Looking at Table 24.1, we see that the strongest reducing agents are at the top of the right column. These are the metals in Groups 1 and 2, all of which lose electrons readily. Their E^0_{ox} values range from +3.05 V (Li) to +2.37 V (Mg). All of these are strong enough reducing agents to liberate hydrogen from water. (Magnesium reacts very slowly at room temperature.) The reactions of potassium and calcium are typical:

$$2\ K(s) + 2\ H_2O \rightarrow 2\ K^+(aq) + H_2(g) + 2\ OH^-(aq) \qquad (24.15)$$

$$Ca(s) + 2\ H_2O \rightarrow Ca^{2+}(aq) + H_2(g) + 2\ OH^-(aq) \qquad (24.16)$$

As we move down the right column, reducing strength decreases. Zinc (E^0_{ox} = +0.76 V) is strong enough to reduce H^+ ions (Reaction 24.13). Hydrogen (E^0_{ox} = 0.00 V) is less effective. It will, however, reduce non-metals such as oxygen or chlorine.

$$2\ H_2(g) + O_2(g) \rightarrow 2\ H_2O \qquad (24.17)$$

$$H_2(g) + Cl_2(g) \rightarrow 2\ H^+(aq) + 2\ Cl^-(aq) \qquad (24.18)$$

Metals with negative values of E^0_{ox} such as copper (−0.34 V) and silver (−0.80 V) are very weak reducing agents. Finally, the F^- ion, at the bottom of the right column, never acts as a reducing agent in water solution.

EXAMPLE 24.4

Consider the following species:

$$Mg^{2+},\ Fe^{2+},\ Ni,\ O_2,\ Au$$

Using Table 24.1:

a. Classify each of these as possible oxidizing or reducing agents.

b. Arrange those which are oxidizing agents in order of decreasing strength.

c. Arrange those which are reducing agents in order of decreasing strength.

SOLUTION

a. In Table 24.1, we locate Mg^{2+}, Fe^{2+} and O_2 in the left column. At least in principle, they can all act as oxidizing agents. In the right column, we find Ni, Fe^{2+} and Au. They are all potential reducing agents. (Notice that Fe^{2+} can either be reduced to Fe(s) or oxidized to Fe^{3+}. So, it can act as either an oxidizing agent or reducing agent.)

How can you tell if a species can behave as an oxidizing agent?

b. For the oxidizing agents:

Mg^{2+} $E^0_{red} = -2.37$ V
Fe^{2+} $E^0_{red} = -0.44$ V
O_2 $E^0_{red} = +1.23$ V

Hence O_2 is the strongest oxidizing agent and Mg^{2+} the weakest. The order is:

$$O_2 > Fe^{2+} > Mg^{2+}$$

c. For the three reducing agents:

Ni $E^0_{ox} = +0.25$ V
Fe^{2+} $E^0_{ox} = -0.77$ V
Au $E^0_{ox} = -1.50$ V

The order is: $Ni > Fe^{2+} > Au$

Spontaneity of Oxidation-Reduction Reactions

Certain reactions are spontaneous. That is, they take place by themselves when the reagents are mixed in the laboratory. Reactants are nearly 100% converted to products. Other reactions are nonspontaneous. They do not occur to any measurable extent when the reactants are mixed.

Most redox reactions occur fairly rapidly, so when E^0 is positive the reaction will probably occur.

Standard voltages can be used to predict whether a given redox reaction will be spontaneous. The principle is a simple one. **If the calculated value of E^0 is positive we expect the reaction to "go". If, on the other hand, we calculate a negative value for E^0, the reaction is unlikely to take place.** In other words:

if E^0 (i.e., $E^0_{ox} + E^0_{red}$) is +, reaction is spontaneous

if E^0 (i.e., $E^0_{ox} + E^0_{red}$) is −, reaction is nonspontaneous

To apply this principle, consider the reaction:

$$Cl_2(g) + 2\ Br^-(aq) \rightarrow 2\ Cl^-(aq) + Br_2(l)$$

Splitting this into two half-reactions, we have:

REDUCTION:	$Cl_2(g) + 2\ e^- \rightarrow 2\ Cl^-(aq)$	$E^0_{red} = +1.36\ V$
OXIDATION:	$2\ Br^-(aq) \rightarrow Br_2(l) + 2\ e^-$	$E^0_{ox} = -1.07\ V$
	$Cl_2(g) + 2\ Br^-(aq) \rightarrow 2\ Cl^-(aq) + Br_2(l)$	$E^0 = +0.29\ V$

Since the calculated standard voltage is positive, +0.29 V, we decide that the reaction should take place. Indeed it does. If chlorine gas is bubbled through a solution containing Br^- ions, a reaction occurs. The products are elementary bromine and chloride ions.

In contrast, consider the reaction between Cl_2 and F^- ions:

REDUCTION:	$Cl_2(g) + 2\ e^- \rightarrow 2\ Cl^-(aq)$	$E^0_{red} = +1.36\ V$
OXIDATION:	$2\ F^-(aq) \rightarrow F_2(g) + 2\ e^-$	$E^0_{ox} = -2.87\ V$
	$Cl_2(g) + 2\ F^-(aq) \rightarrow 2\ Cl^-(aq) + F_2(g)$	$E^0 = -1.51\ V$

Since the calculated E^0 is negative, −1.51 V, we conclude that the reaction will not go. Experiment confirms this prediction. You can bubble Cl_2 gas through a solution of NaF forever. No elementary fluorine is formed.

EXAMPLE 24.5

Would you expect a redox reaction to take place when:

a. Copper metal is added to a solution containing Ag^+ ions?

b. Copper metal is added to a solution containing H^+ ions?

SOLUTION

a. If a reaction occurs, it must involve the oxidation of Cu (to Cu^{2+} ions) and the reduction of Ag^+ ions (to silver metal). Taking the standard half-cell voltages from Table 24.1:

OXIDATION:	$Cu(s) \rightarrow Cu^{2+}(aq) + 2\,e^-$	$E^0_{ox} = -0.34$ V
REDUCTION:	$2\,Ag^+(aq) + 2\,e^- \rightarrow 2\,Ag(s)$	$E^0_{red} = +0.80$ V
	$Cu(s) + 2\,Ag^+(aq) \rightarrow Cu^{2+}(aq) + 2\,Ag(s)$	$E^0 = +0.46$ V

We conclude that the reaction should go. It does. (See Color Plate 11D, in the center of the book.)

b. Applying the same reasoning:

OXIDATION:	$Cu(s) \rightarrow Cu^{2+}(aq) + 2\,e^-$	$E^0_{ox} = -0.34$ V
REDUCTION:	$2\,H^+(aq) + 2\,e^- \rightarrow H_2(g)$	$E^0_{red} = 0.00$ V
	$Cu(s) + 2\,H^+(aq) \rightarrow Cu^{2+}(aq) + H_2(g)$	$E^0 = -0.34$ V

The reaction should not take place. It doesn't. Copper cannot be oxidized by H^+ ions.

Effect of Concentration upon Voltage

As pointed out earlier, standard voltages apply exactly only when the species involved are at unit concentrations. Changes in concentration affect the half-cell voltages and hence the cell voltage, E. We can easily predict the direction in which the cell voltage will change.

1. The voltage will increase ($E > E^0$) if the concentration of a reactant is increased or that of a product is decreased. Both of these changes make the forward reaction more spontaneous. Hence, they increase the cell voltage.

2. The voltage will decrease ($E < E^0$) if the concentration of a reactant is decreased or that of a product is increased. These changes favor the reverse reaction. They make the forward reaction less spontaneous and hence decrease the cell voltage.

Applying these principles to the reaction:

$$Zn(s) + Cu^{2+}(aq) \rightarrow Zn^{2+}(aq) + Cu(s); \qquad E^0 = 1.10\text{ V}$$

the voltage will be:

1. greater than 1.10 V if the concentration of Cu^{2+} is greater than 1 M or that of Zn^{2+} is less than 1 M.
2. less than 1.10 V if the concentration of Cu^{2+} is less than 1 M or that of Zn^{2+} is greater than 1 M.

Ordinarily, the effect of concentration on voltage is small. Suppose, for example, that in the Zn-Cu^{2+} cell, we decrease the concentration of Cu^{2+} from 1 M to 0.1 M. The voltage drops only slightly, to 1.07 V. However, a drastic change in concentration can have a major effect. If we allow the Zn-Cu^{2+} cell to discharge for a long time, the Cu^{2+} ions are virtually used up. To be exact, the concentration of Cu^{2+} drops to 10^{-37} M, a very low value indeed. When this happens, the voltage drops to zero and the cell is "dead".

By changing concentrations, it is sometimes possible to make a reaction with a negative E^0 value take place. Consider the reaction:

$$MnO_2(s) + 4\ H^+(aq) + 2\ Cl^-(aq) \rightarrow Mn^{2+}(aq) + Cl_2(g) + 2\ H_2O \qquad (24.19)$$

From Table 24.1, we see that:

$$E^0 = E^0_{red} + E^0_{ox} = +1.23\ V - 1.36\ V = -0.13\ V$$

So, according to the rules given on p. 610, we would not expect MnO_2 to react with hydrochloric acid (H^+ and Cl^- ions). However, by increasing the concentration of HCl, we can make E more positive and the reaction more spontaneous. With 12 M HCl, the reaction goes to produce chlorine gas, particularly if we raise the temperature.

Situations such as this arise only when E^0 is close to zero. If E^0 is more negative than -0.20 V, there is very little we can do to make the reaction go. Increasing the concentration of reactants doesn't help very much. The cell voltage, E, remains negative and very little product, if any, is formed.

24.3 ELECTROLYTIC CELLS

As pointed out earlier, an electrolytic cell is one in which electrical energy is used to make an oxidation-reduction reaction take place. Ordinarily, the reaction is nonspontaneous. This is the case in the electrolysis of water.

$E^0 < 0$ for this reaction.

$$2\ H_2O(l) \rightarrow 2\ H_2(g) + O_2(g)$$

This reaction does not go by itself. The energy of the products is greater than that of the reactant, water. Electrical energy has to be supplied to decompose H_2O molecules into the elements hydrogen and oxygen.

In order to carry out an electrolysis, ions must be present to carry the current. To electrolyze water, we add a small amount of an electrolyte such as H_2SO_4. It provides the necessary ions. In many water solutions, the ions themselves take part in the electrode reactions. In this section, we will examine a couple of processes of this type. They involve the electrolysis of water solutions of $CuCl_2$ and NaCl.

We would supply the electrical energy through Pt electrodes.

$CuCl_2$ Solution

The reaction:

$$Cu^{2+}(aq) + 2\ Cl^-(aq) \rightarrow Cu(s) + Cl_2(g) \qquad (24.20)$$

does not occur by itself. That is, a water solution of $CuCl_2$ does not spontaneously decompose to the elements, copper and chlorine. However, the reaction can be made to take place in the electrolytic cell shown in Figure 24.8. This cell consists of a beaker filled with a solution of $CuCl_2$. Dipping into this solution are two platinum electrodes. They are attached to the terminals of a lead storage battery.

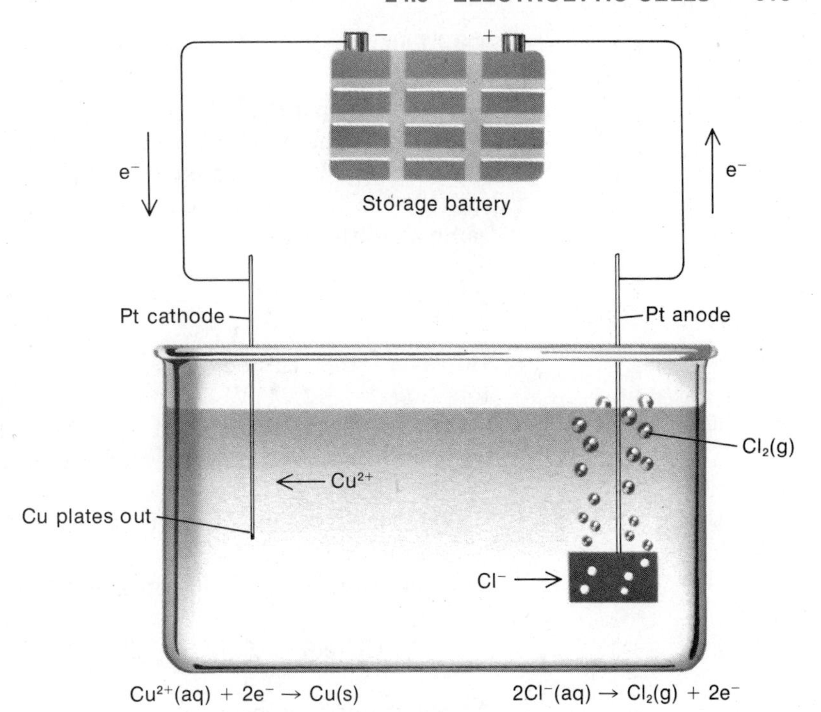

FIGURE 24.8

Electrolysis of $CuCl_2$ solution. In this cell the electrical energy is furnished by a storage battery. At the cathode, copper metal plates out on the surface of the platinum. At the anode oxidation of Cl^- ions produces Cl_2 gas.

The lead storage battery acts as an "electron pump". That is, it tends to push electrons in a particular direction. As shown in Figure 24.8, the battery tends to pump electrons into the electrode at the left. At this electrode, the cathode, electrons reduce Cu^{2+} ions:

$$\text{CATHODE:} \quad Cu^{2+}(aq) + 2\ e^- \rightarrow Cu(s); \quad \text{REDUCTION} \quad (24.20\ C)$$

At the same time, the storage battery pulls electrons from the electrode at the right. These electrons come from Cl^- ions, which are oxidized at that electrode, the anode.

$$\text{ANODE:} \quad 2\ Cl^-(aq) \rightarrow Cl_2(g) + 2\ e^- \quad \text{OXIDATION} \quad (24.20\ A)$$

The overall reaction in the electrolytic cell is the sum of 24.20 A and 24.20 C:

$$\begin{array}{r} Cu^{2+}(aq) + 2\ e^- \rightarrow Cu(s) \\ 2\ Cl^-(aq) \rightarrow Cl_2(g) + 2\ e^- \\ \hline Cu^{2+}(aq) + 2\ Cl^-(aq) \rightarrow Cu(s) + Cl_2(g) \end{array} \quad (24.20)$$

As electrolysis proceeds, the concentrations of Cu^{2+} and Cl^- ions drop. Bubbles of chlorine gas form at the anode. A deposit of reddish-brown copper forms on the cathode. Current is carried through the cell by the movement of ions. Cations (Cu^{2+} ions) move toward the cathode. Anions (Cl^- ions) move in the opposite direction, toward the anode.

NaCl Solution

The set-up for the electrolysis of a water solution of sodium chloride is similar to that for $CuCl_2$. A storage battery serves as a source of electrical energy. Wires connected to the battery act as electrodes. They dip into the NaCl solution (Figure 24.9). The half-reaction at the anode is the same as with $CuCl_2$:

$$\text{ANODE:} \quad 2\,Cl^-(aq) \rightarrow Cl_2(g) + 2\,e^-; \quad \text{OXIDATION} \quad (24.21\ A)$$

If Na metal formed, it would react with the water, producing $H_2(g)$ and OH^- ions.

However, the cathode half-reaction is quite different. The Na^+ ion ($E^0_{red} = -2.71$ V) is much more difficult to reduce than Cu^{2+} ($E^0_{red} = +0.34$ V). Indeed, it is more difficult to reduce than an H_2O molecule. So, the following half-reaction occurs at the cathode:

$$\text{CATHODE:} \quad 2\,H_2O + 2\,e^- \rightarrow H_2(g) + 2\,OH^-(aq); \quad \text{REDUCTION} \quad (24.21\ C)$$

The overall reaction is obtained by summing 24.21 A and 24.21 C:

$$2\,Cl^-(aq) + 2\,H_2O \rightarrow Cl_2(g) + H_2(g) + 2\,OH^-(aq) \qquad (24.21)$$

As electrolysis proceeds, bubbles of chlorine gas form at the anode. At the cathode, hydrogen gas is given off. At the same time, the solution around the cathode becomes basic because of the OH^- ions formed. As a result of electrolysis, the original solution of NaCl is converted to a solution of NaOH.

From a commercial standpoint, Reaction 24.21 is the most important of all electrolyses carried out in water solution. Most of the chlorine

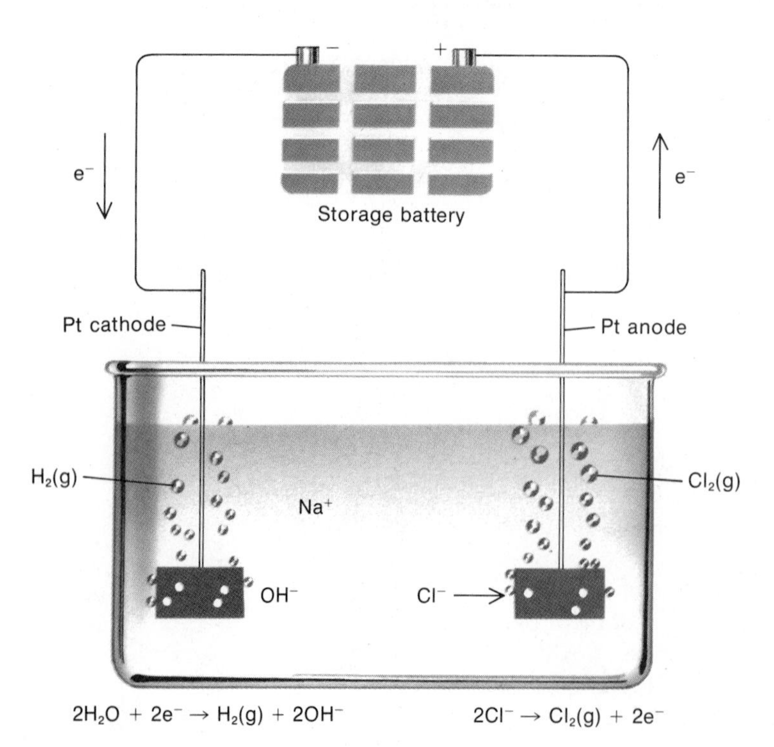

FIGURE 24.9

Electrolysis of a NaCl solution. At the anode, Cl^- ions are oxidized to $Cl_2(g)$. At the cathode, H_2O molecules are reduced to $H_2(g)$ and OH^- ions.

used in the United States is formed this way. This amounts to some 1×10^{10} kg each year. At the same time, about 1×10^{10} kg of sodium hydroxide, NaOH, are produced. Hydrogen gas is an important by-product.

EXAMPLE 24.6

The electrolysis of a KI solution is similar to that of NaCl. The K^+ ion ($E^0_{red} = -2.93$ V), like the Na^+ ion, is difficult to reduce. So, water molecules are reduced instead. The I^- ion is oxidized to the element iodine, I_2. For the electrolysis of KI, write balanced equations for:

a. the anode half-reaction.

b. the cathode half-reaction.

c. the overall reaction.

In electrolysis, the reaction that can occur most easily is the one that does occur.

SOLUTION

a. I^- ions are oxidized to I_2:

$$2\ I^-(aq) \rightarrow I_2(s) + 2\ e^- \qquad \text{(compare 24.21 A)}$$

b. H_2O molecules are reduced to H_2 molecules and OH^- ions:

$$2\ H_2O + 2\ e^- \rightarrow H_2(g) + 2\ OH^-(aq) \qquad \text{(compare 24.21 C)}$$

c. Summing the two half-equations:

$$2\ I^-(aq) + 2\ H_2O \rightarrow I_2(s) + H_2(g) + 2\ OH^-(aq) \qquad \text{(compare 24.21)}$$

24.4 SUMMARY

There are two kinds of electrochemical cells. In a voltaic cell, a spontaneous oxidation-reduction reaction produces electrical energy. In an electrolytic cell, the reverse is true. Electrical energy is supplied from the outside, often from a voltaic cell. This energy is used to make a non-spontaneous reaction occur. In both types of cells, oxidation takes place at the anode. Reduction occurs at the cathode. Current is carried through the cell by ions. Anions (negative ions) move to the anode. Cations (positive ions) move to the cathode.

Every voltaic cell has a standard voltage. This is the measured voltage when all ions in solution are at a concentration of 1 M and all gases are at a pressure of 1 atm. Table 24.1 can be used to calculate standard cell voltages. The relation used is:

$$E^0 = E^0_{ox} + E^0_{red}$$

where E^0 is the standard cell voltage. E^0_{ox} is the standard voltage for the oxidation half-reaction. E^0_{red} is the standard voltage for the reduction. If the E^0 calculated this way is positive, the reaction is spontaneous, at least at standard concentrations. That is, the reaction will take place when the reactants are mixed with each other.

Values of E^0_{ox} and E^0_{red} can be used to compare the strengths of different oxidizing or reducing agents. Strong reducing agents such as Li have large, positive values of E^0_{ox} (+3.05 V). Strong oxidizing agents such as F_2 have large, positive values of E^0_{red} (+2.87 V).

When a water solution of an ionic compound is electrolyzed, there are several possible half-reactions. At the cathode, a cation such as Cu^{2+} may be reduced to the metal. This occurs in the electrolysis of a $CuCl_2$ solution. With ions, such as Na^+, which are very difficult to reduce, water molecules are reduced at the cathode. At the anode, an anion such as Cl^- or I^- may be oxidized, giving a nonmetal (Cl_2, I_2). Other half-reactions are possible.

NEW TERMS

anion
anode
cathode
cation
dry cell
electrode
electrolytic cell
inert electrode
lead storage battery
standard oxidation voltage (E^0_{ox})
standard reduction voltage (E^0_{red})
standard cell voltage (E^0)
voltaic cell

QUESTIONS

1. In what kind of cell does a spontaneous oxidation-reduction reaction produce electrical energy? In what kind of cell is electrical energy used to make a nonspontaneous reaction occur?

2. Which of the following statements apply to the cathode? It is the electrode:

   a. at which reduction occurs
   b. at which oxidation occurs
   c. toward which positive ions move
   d. toward which negative ions move

3. Which of the statements in Question 2 apply to the anode?

4. In the Zn-Cu^{2+} voltaic cell, using Zn and Cu electrodes:

   a. At which electrode do electrons enter the cell?
   b. At which electrode do electrons leave the cell?
   c. Toward which electrode do Zn^{2+} and Cu^{2+} ions move?
   d. Toward which electrode do negative ions move?

5. Explain why the anode and cathode compartments in the Zn-Cu^{2+} cell are separated from each other. What would happen if the "porous cup" in Figure 24.1 were replaced by a glass beaker? What would happen if it were removed completely?

6. What is meant by an "inert" electrode? Why must inert electrodes be used in the Cl_2-Br^- cell?

7. For an ordinary dry cell, give the equation for:

   a. the anode reaction   b. the cathode reaction
   c. the overall reaction

8. In a lead storage battery, what is the anode made of? What is the formula of the oxide of lead in the cathode? What is the electrolyte in the water solution?

9. For the lead storage battery, give the equation for the:

   a. anode reaction   b. cathode reaction   c. overall reaction

10. Suppose, instead of taking E^0_{red} $2\ H^+ \rightarrow H_2 = 0.00$ V, we were to make it +0.20 V. What would be the effect upon:

    a. E^0_{ox} $Zn \rightarrow Zn^{2+}$ b. E^0_{ox} $H_2 \rightarrow 2\ H^+$ c. E^0 for the Zn-H^+ cell?

11. For the reaction: $A(s) \rightarrow A^{2+}(aq) + 2\ e^-$, $E^0_{ox} = 2.00$ V. Is A a

    a. strong oxidizing agent?   b. strong reducing agent?
    c. weak oxidizing agent?   d. weak reducing agent?

12. Referring to Question 11, what is E^0_{red} for $A^{2+}(aq) + 2\ e^- \rightarrow A(s)$? Which of the descriptions in Question 11 apply to A^{2+}?

13. Referring to Table 24.1, find:

    a. E^0_{ox} for $Cu \rightarrow Cu^{2+}$ b. E^0_{red} for $Sn^{4+} \rightarrow Sn^{2+}$ c. E^0_{ox} for $Sn \rightarrow Sn^{2+}$

14. Consider Table 24.1. Where are the strong oxidizing agents located?

    a. upper left   b. lower left   c. upper right   d. lower right

    Where are the strong reducing agents located?

15. To estimate the strength of a reducing agent, do you look at the value of E^0_{ox} or E^0_{red}? Is the strength of an oxidizing agent related to its value of E^0_{ox} or E^0_{red}?

16. A reaction has a calculated E^0 value of −1.61 V. Which of the following statements apply to this reaction?

    a. It is spontaneous.   b. It is not spontaneous.
    c. It can occur in a voltaic cell.
    d. It can be carried out in an electrolytic cell.

17. When $CuCl_2$ solution is electrolyzed, at which electrode is copper produced? What is the other electrode called? What element is produced there?

18. When a water solution of NaCl is electrolyzed, sodium is not formed. Why? What are the products at the cathode?

19. When a water solution of $Al(NO_3)_3$ is electrolyzed, the mass of the aluminum cathode does not change. However, the solution around the cathode becomes basic and a gas is given off. Explain what is happening.

PROBLEMS

1. *Voltaic Cells (Example 24.1)* Draw a diagram of a voltaic cell in which the reaction is: $Ni(s) + 2\ Ag^+(aq) \rightarrow Ni^{2+}(aq) + 2\ Ag(s)$. Label the anode and cathode. Indicate the ions present in both compartments. In which direction do electrons move through the electrical circuit?

2. *Voltaic Cells (Example 24.2)* Draw a diagram of a voltaic cell in which the reaction is: $Cl_2(g) + H_2(g) \rightarrow 2\ Cl^-(aq) + 2\ H^+(aq)$. Label anode and cathode (both Pt wires). In which compartment would you put H^+ ions? Cl^- ions? In which direction do H^+ ions move through the cell? Cl^- ions?

3. *Standard Voltages (Example 24.3)* Using Table 24.1, calculate E^0 for the reaction in Problem 1; Problem 2.

4. *Standard Voltages (Example 24.3)* Calculate E^0 for the following reaction:

$$MnO_2(s) + 4\ H^+(aq) + 2\ I^-(aq) \rightarrow Mn^{2+}(aq) + I_2(s) + 2\ H_2O$$

5. *Strength of Reducing Agents (Example 24.4)* Arrange the following reducing agents in order of decreasing strength: Cu, Pb, Cl^-, H_2.

6. *Strength of Oxidizing Agents (Example 24.4)* Of the following oxidizing agents, which is the weakest? the strongest? I_2, H^+, Au^{3+}, O_2

7. *Spontaneous Reactions (Example 24.5)* Using your answers to Problem 3 and 4, state which of the reactions involved are spontaneous.

8. *Electrolytic Cells (Example 24.6)* The electrolysis of $CaCl_2$ is similar to that of NaCl. Indeed, the products are the same in the two cases. Write equations for the reactions at each electrode and for the overall reaction.

* * * * *

9. Using Table 24.1, predict which of the following will react spontaneously with H^+ ions.

   a. Al b. Cu c. Au d. Mg e. Ni

10. The electrolysis of $NiBr_2$ is similar to that of $CuCl_2$. Write equations for:

    a. the cathode half-reaction b. the anode half-reaction
    c. the overall reaction

11. Draw a diagram of a voltaic cell in which the following reaction occurs:

$$Pb(s) + 2\ H^+(aq) \rightarrow Pb^{2+}(aq) + H_2(g)$$

Label anode and cathode. Show the flow of electrons outside the cell and ions inside the cell. What is E^0 for the cell?

12. Follow the directions of Problem 11 for the cell:

$$Zn(s) + 2\ Fe^{3+}(aq) \rightarrow Zn^{2+}(aq) + 2\ Fe^{2+}(aq)$$

13. Draw a diagram of a cell which could be used to electrolyze a solution of $SnCl_2$. Label anode and cathode. Write half-equations for the reactions at the two electrodes (the products are tin and chlorine). What is the overall cell reaction?

14. Follow the directions of Problem 13 for a solution of KBr. Water is reduced at the cathode; Br^- ions are oxidized at the anode.

15. Using Table 24.1, decide which of the following will react spontaneously with Cu^{2+} ions:

    a. Fe^{2+} b. Ni c. H_2 d. Br^-

16. Calculate the standard voltage of a cell in which sodium reacts with fluorine. Despite the high voltage, this cell has never been developed commercially. Can you suggest why?

17. Using Table 24.1, calculate E^0 for cells in which I^- ions are oxidized to I_2 by:

    a. Cl_2 b. Ni^{2+} c. Zn^{2+} d. Au^{3+}

18. Which of the reactions in Problem 17 are spontaneous?

19. Using Table 24.1, decide which of the following will reduce Pb^{2+} ions to lead.

    a. Ni b. Cu c. Ag d. Sn

20. Refer to Table 24.1. Of those listed,

    a. how many reducing agents are stronger than $H_2(g)$?
    b. how many oxidizing agents are stronger than $Cl_2(g)$?

*21. Using Table 24.1, decide whether copper metal will react with nitric acid (H^+, NO_3^- ions).

*22. How many electrons must pass through an electroplating cell to convert 1.00 g of Cu^{2+} to copper ($1 \text{ mol } e^- = 6.02 \times 10^{23} \text{ } e^-$)?

*23. Explain, in terms of standard voltages, why:

    a. iron(II) compounds are slowly oxidized to iron(III) in water saturated with air.
    b. Ag^+ and Fe^{2+} ions are never found in the same solution.
    c. Solutions of Sn^{2+} compounds are protected from oxidation to Sn^{4+} by adding tin metal.

*24. Consider a cell using the reaction: $Cl_2(g) + 2\,Br^-(aq) \rightarrow 2\,Cl^-(aq) + Br_2(l)$ List three ways in which you could increase the voltage of this cell by changing concentrations. How could you reduce the cell voltage?

From the standpoint of the chemist, gold is the easiest metal to separate from its ores. This prospector probably wouldn't agree. *Library, The State Historical Society of Colorado.*

METALS AND THEIR ORES

Metals were first discussed in Chapter 7 in connection with the Periodic Table. In Chapter 8, we looked at the chemistry of the metals in Groups 1 and 2. The properties of other metals and their compounds were referred to in several later chapters. Here, we will look more closely at the most important chemical process involving metals. This is the extraction of metals from their **ores.** An ore, quite simply, is a mineral from which a metal can be extracted profitably.

Table 25.1 gives an overview of the production of metals from their ores. There are several interesting features of this table.

1. The annual production of metals (column 1) amounts to a total of about 4.4×10^{11} kg. Of this, about 90% is accounted for by a single

TABLE 25.1 PRODUCTION OF METALS FROM THEIR ORES
(All masses quoted are in billions of kilograms, i.e., 10^9 kg)

Metal	Production (Annual, Global)	Ore Reserves Global	%, US	Other Major Sources	Total Available (Earth's Crust)
	(1)	(2)	(3)	(4)	(5)
Iron	400	100 000	10	USSR, Canada	3×10^{11}
Aluminum	12	1200	4	Australia, Jamaica	5×10^{11}
Copper	8.6	310	28	Chile	4×10^{8}
Manganese	8	800	0	So. Africa, USSR	6×10^{9}
Zinc	5.3	120	27	Canada	8×10^{8}
Lead	3.5	91	39	Australia, USSR	1×10^{8}
Chromium	1.8	780	0	Rhodesia, So. Africa	1×10^{9}
Nickel	0.35	67	4	Cuba, Canada	5×10^{8}
Tin	0.26	4.4	0	Thailand, Malaysia	6×10^{6}
Magnesium	0.22	10^{9}		seawater	1×10^{11}
Molybdenum	0.062	4.9	58	USSR	5×10^{7}
Tungsten	0.032	1.3	5	China	2×10^{8}
Sodium	0.025	10^{10}		seawater	2×10^{11}
Cobalt	0.020	2.2	7	Congo, Zambia	1×10^{8}
Silver	0.011	0.17	24	Canada, USSR	6×10^{5}
Mercury	0.0088	0.12	3	Spain, Italy	2×10^{6}
Gold	0.0010	0.011	5	So. Africa	3×10^{4}

metal, iron. Next in order are aluminum, copper, manganese, zinc, lead, and chromium. All the other metals listed are produced in relatively small amounts. Together, they account for less than 0.2% of the total. Yet many of these metals, notably gold, silver, and mercury, have unique properties and are essential to our economy.

It's not just oil that will be in short supply by the year 2000.

2. The known reserves of metal ores (column 2) exceed their annual production by factors that range from 10 to many thousands. Certain metals are in very short supply. Among these are gold, silver, and mercury. At the current rate, ores of these metals will be exhausted in 10 to 20 years. At the other extreme, a few metals will be available for centuries. An example is magnesium, obtained from seawater.

3. The United States has adequate supplies of the ores of most of the common metals (column 3). It is, however, deficient in a few key metals. These include manganese, chromium, and tin. To obtain these metals, the United States is dependent upon imports from the countries listed in column 4.

4. The total amount of a metal available in the earth's crust (column 5) ordinarily far exceeds known reserves. Consider, for example,

aluminum. The common ore of aluminum is bauxite, Al_2O_3. This accounts for only a tiny fraction:

$$\frac{1.2 \times 10^{12}\ \text{kg}}{5 \times 10^{20}\ \text{kg}} = 2 \times 10^{-9}$$

of the total amount of aluminum in the earth's crust. Most of our aluminum is tied up in such rocks as granite and feldspar. In principle, it could be obtained from these sources. However, that would require spending a tremendous amount of energy. Another interesting case is gold. Here again, the fraction which is commercially available is very small:

$$\frac{1.1 \times 10^{7}\ \text{kg}}{3 \times 10^{13}\ \text{kg}} = 4 \times 10^{-7}$$

25.1 SOURCES OF METALS

Metals occur in nature in many different chemical forms (Figure 25.1). Most of the ores of metals fall into one or more of the following groups.

1. *Native ores* A few of the less active transition metals occur as the free element. Gold is the most familiar example. It is found occasionally as nuggets which are almost pure. (See Color plate 12, in the center of the book.) More often, gold occurs in veins of quartz or other rocky material.

Would you say that the nature of the ores of the metals are in line with the trends in the Periodic Table?

2. *Oxides* The most familiar ores of this type are those of aluminum and iron, Al_2O_3 and Fe_2O_3. Several other transition metals occur as oxides. They include chromium (Cr_2O_3) and manganese (MnO_2). The principal ore of tin is the +4 oxide, SnO_2.

3. *Sulfides* These are perhaps more common than any other type of ore. Most of the transition metals toward the right of the Periodic Table occur as sulfides. Among these are cobalt (CoS), nickel (NiS), copper (Cu_2S), zinc (ZnS), cadmium (CdS), and mercury (HgS). The metalloids arsenic and antimony are extracted from their sulfides (As_2S_3, Sb_2S_3). So are bismuth and lead (Bi_2S_3, PbS).

4. *Chlorides* The alkali metals are extracted from chloride ores ($NaCl$, KCl). So, in a sense, is magnesium. It is obtained from seawater where it occurs as the Mg^{2+} ion. Of the many negative ions in seawater, Cl^- is the most abundant.

5. *Carbonates* The carbonates are the most common compounds of the alkaline earth elements. Calcium, strontium, and barium are obtained from their carbonates ($CaCO_3$, $SrCO_3$, $BaCO_3$).

The ores of metals, whether elements or compounds, rarely occur pure. Instead they are usually mixed with large amounts of rocky material. The first step in extracting a metal is to *concentrate* its ore. One way or another, the ore is separated from most of the impurities. If the

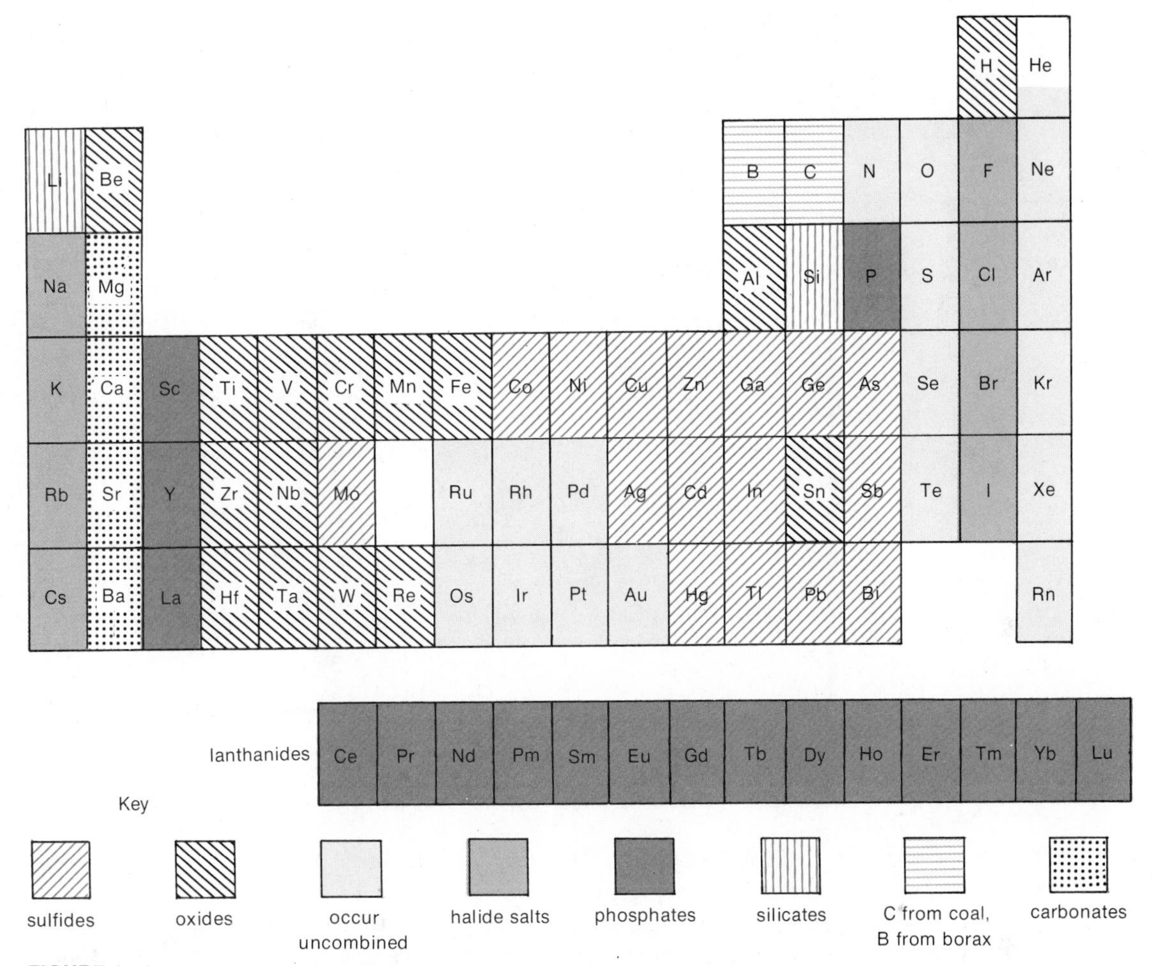

FIGURE 25.1

Natural sources of the elements. The most common ores are oxides and sulfides, but significant numbers of elements occur uncombined or as halides or carbonates.

ore is a chemical compound, it must then be *reduced* to the free metal. The final step in the process ordinarily consists of *refining* the crude metal. This frees it from traces of remaining impurities.

As we discuss the extraction of various metals, we will see examples of all of these processes: concentration, reduction, and refining. We will not attempt to survey all of the more than 60 metals that are found in nature. Instead, we will focus on a few of the more important and more interesting metals. These include:

—Na and Mg, obtained from their chlorides, NaCl and $MgCl_2$.

—Al and Fe, obtained from their oxides, Al_2O_3 and Fe_2O_3.

—Cu, Zn, and Hg, obtained from their sulfides, Cu_2S, ZnS, and HgS.

EXAMPLE 25.1

How much metal can be obtained from one kilogram of each of the following ores?

a. $MgCl_2$ b. Al_2O_3 c. Cu_2S

SOLUTION

a. One mole of $MgCl_2$ weighs: 24.30 g + 2(35.45 g) = 95.20 g. It contains one mole of Mg, 24.30 g. Hence:

$$24.30 \text{ g Mg} \simeq 95.20 \text{ g MgCl}_2$$

$$\text{mass Mg} = 1000 \text{ g MgCl}_2 \times \frac{24.30 \text{ g Mg}}{95.20 \text{ g MgCl}_2} = 255.3 \text{ g Mg}$$

b. Similarly:

one mole Al_2O_3 weighs: 2(26.98 g) + 3(16.00 g) = 101.96 g

It contains two moles of Al, 53.96 g. So:

$$\text{mass Al} = 1000 \text{ g Al}_2\text{O}_3 \times \frac{53.96 \text{ g Al}}{101.96 \text{ g Al}_2\text{O}_3} = 529.2 \text{ g Al}$$

c. 2 mol Cu $\simeq$ 1 mol Cu_2S

$$127.1 \text{ g Cu} \simeq 159.2 \text{ g Cu}_2\text{S}$$

$$\text{mass Cu} = 1000 \text{ g Cu}_2\text{S} \times \frac{127.1 \text{ g Cu}}{159.2 \text{ g Cu}_2\text{S}} = 798.4 \text{ g Cu}$$

25.2 SODIUM AND MAGNESIUM

Both sodium and magnesium are extremely reactive metals. Their ions, Na^+ and Mg^{2+}, are difficult to reduce (E^0_{red} $Na^+ = -2.71$ V; E^0_{red} $Mg^{2+} = -2.37$ V). Reduction to the free metals is carried out by electrolysis. However, electrolysis cannot be done in water solution. There, water molecules are reduced instead of Na^+ or Mg^{2+} ions. Electrolysis has to be carried out using the molten chlorides. High temperatures are required (mp NaCl = 800°C, $MgCl_2$ = 712°C).

The reactive metals are made by electrolysis of their molten salts.

Sodium

The cell used to electrolyze sodium chloride is shown in Figure 25.2. A mixture of about 40% NaCl and 60% $CaCl_2$ is put into the cell. The calcium chloride lowers the melting point of NaCl in much the same way that it lowers the melting point of ice. The mixture melts at about 600°C as compared to 800°C for pure NaCl. Consequently, less energy has to be used to keep the mixture molten.

The half reactions in the Downs cell are simple ones. At the iron cathode, Na^+ ions are reduced to Na atoms.

$$\text{CATHODE:} \quad Na^+ + e^- \rightarrow Na \qquad (25.1\ C)$$

At the graphite anode, Cl^- ions are oxidized to Cl_2 molecules:

$$\text{ANODE:} \quad Cl^- \rightarrow \tfrac{1}{2} Cl_2 + e^- \qquad (25.1\ A)$$

FIGURE 25.2

Electrolysis cell for production of sodium metal. Molten NaCl is the material electrolyzed. Liquid Na is produced at the iron cathode; chlorine gas is formed at the carbon anode. All of our sodium metal is made by this process.

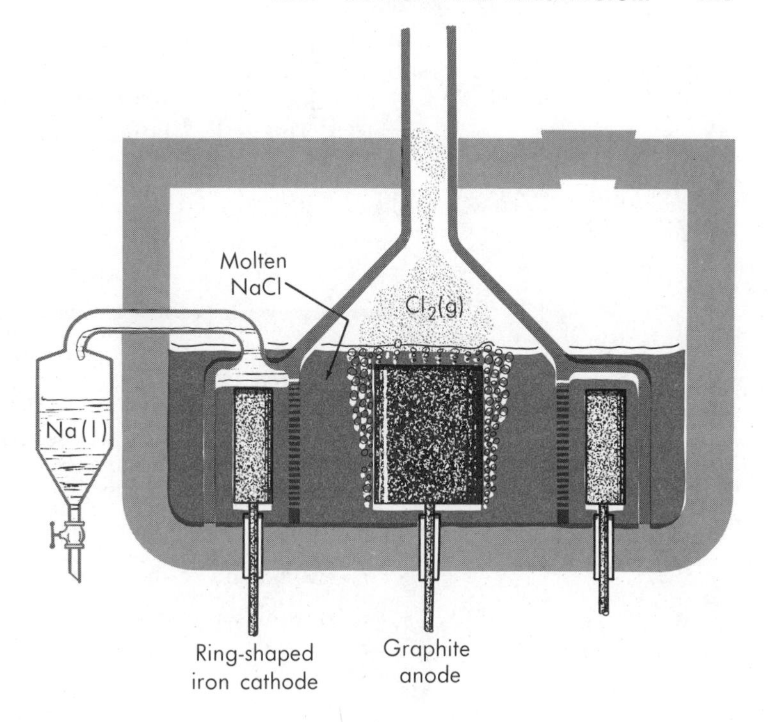

The overall cell reaction is:

$$NaCl(l) \rightarrow Na(l) + \frac{1}{2} Cl_2(g) \qquad (25.1)$$

Liquid sodium, which has a lower density than the $NaCl$-$CaCl_2$ mixture, rises to the top of the cathode compartment. The sodium is drawn off and allowed to cool. At 97°C it solidifies. Chlorine is collected as a gas over the anode compartment.

The two compartments in the Downs cell are separated by an iron screen. This prevents the sodium and chlorine from coming in contact with each other. If they did, they would react by the reverse of Reaction 25.1:

$$Na(l) + \frac{1}{2} Cl_2(g) \rightarrow NaCl(l)$$

If this happened, all the electrical energy required to electrolyze NaCl would be wasted.

We'd also have an explosion.

Magnesium

More than 60% of the magnesium we use is obtained from seawater. The Mg^{2+} ion is the second most abundant positive ion, after Na^+, in the oceans. To remove Mg^{2+}, a solution of $Ca(OH)_2$ is added. Magnesium hydroxide precipitates.

$$Mg^{2+}(aq) + 2OH^-(aq) \rightarrow Mg(OH)_2(s)$$

The solid is filtered off and dried. It is then treated with hydrochloric acid to give a water solution of magnesium chloride. Evaporation gives solid $MgCl_2$.

Magnesium chloride is electrolyzed in a cell similar to that shown for NaCl. The cell reactions are:

$$\text{CATHODE:} \quad Mg^{2+} + 2\,e^- \rightarrow Mg(l) \qquad (25.2\ C)$$

$$\text{ANODE:} \quad 2\,Cl^- \rightarrow Cl_2(g) + 2\,e^- \qquad (25.2\ A)$$

$$MgCl_2(l) \rightarrow Mg(l) + Cl_2(g) \qquad (25.2)$$

EXAMPLE 25.2

Calcium metal is produced in much the same way as magnesium. The major difference is that the starting material is $CaCO_3$ rather than $MgCl_2$. Write balanced equations for the:

a. reaction of $CaCO_3(s)$ with hydrochloric acid.

b. evaporation of the solution obtained in (a) to obtain $CaCl_2(s)$.

c. electrolysis of molten calcium chloride.

Magnesium is used in lightweight alloys. Calcium has few industrial uses.

SOLUTION

a. Recall from Chapter 20 that metal carbonates react with acid (H^+ ions) to form carbon dioxide.

$$CaCO_3(s) + 2\,H^+(aq) \rightarrow Ca^{2+}(aq) + CO_2(g) + H_2O$$

With hydrochloric acid, there are Cl^- ions in the solution, two for every Ca^{2+} ion.

b. $Ca^{2+}(aq) + 2\,Cl^-(aq) \rightarrow CaCl_2(s)$

c. $CaCl_2(l) \rightarrow Ca(l) + Cl_2(g)$ (compare Equation 25.2)

25.3 ALUMINUM

Aluminum is a very reactive metal. The Al^{3+} ion, like Na^+ and Mg^{2+}, is difficult to reduce ($E^0_{red} = -1.66$ V). The only practical way to carry out the reduction to aluminum metal is to use an electrolytic cell. Here again electrolysis cannot be carried out in water solution. Any aluminum produced would immediately react with water. Instead, electrolysis is carried out in a molten state.

The only important aluminum ore is bauxite, which contains about 50% Al_2O_3. The principal impurities are iron(III) oxide, Fe_2O_3, and silicon dioxide, SiO_2. To concentrate the ore, it is treated with a solution of sodium hydroxide. Since aluminum hydroxide dissolves in base, the following reaction occurs:

$$Al_2O_3(s) + 3\,H_2O \rightarrow 2\,Al(OH)_3(s)$$

$$2\,Al(OH)_3(s) + 2\,OH^-(aq) \rightarrow 2\,Al(OH)_4^-(aq)$$

$$Al_2O_3(s) + 3\,H_2O + 2\,OH^-(aq) \rightarrow 2\,Al(OH)_4^-(aq) \qquad (25.3)$$

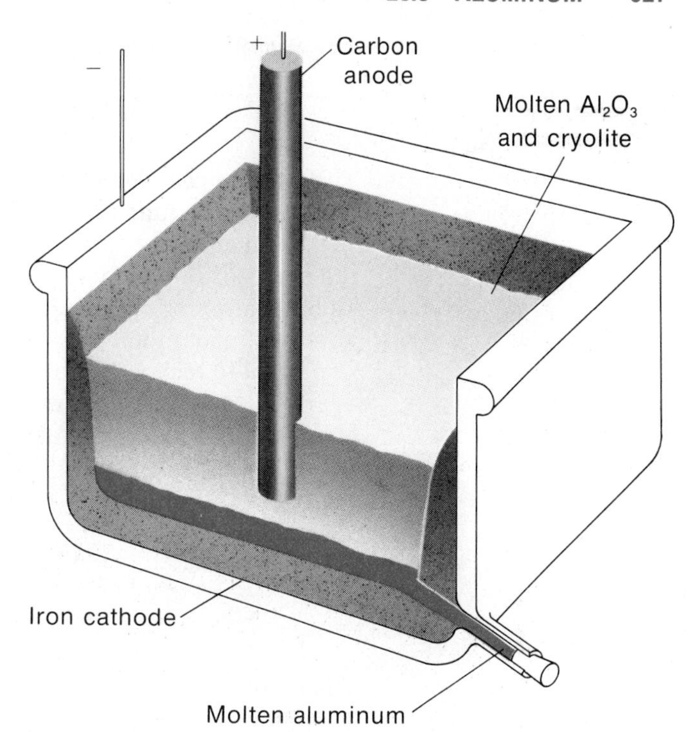

FIGURE 25.3

Electrolysis cell for production of aluminum metal. The molten Al_2O_3-cryolite mixture is at 1000°C. In the process liquid Al is formed at the iron cathode and remains below the molten mixture.

In this way, Al^{3+} goes into solution as the $Al(OH)_4^-$ complex ion. The impurities, Fe_2O_3 and SiO_2, do not dissolve. They are filtered off. The solution is then acidified to precipitate $Al(OH)_3$:

$$Al(OH)_4^-(aq) + H^+(aq) \rightarrow Al(OH)_3(s) + H_2O \qquad (25.4)$$

Finally, the aluminum hydroxide is heated to 1000°C to drive off water and give Al_2O_3:

$$2\ Al(OH)_3(s) \rightarrow Al_2O_3(s) + 3\ H_2O(g) \qquad (25.5)$$

The aluminum oxide produced by this three-step process is nearly 100% pure.

Aluminum oxide is converted to aluminum in the electrolytic cell shown in Figure 25.3. Cryolite, Na_3AlF_6, is added to the Al_2O_3. This reduces the melting point to about 1000°C (mp pure Al_2O_3 = 2000°C). The cryolite does not interfere in the electrolysis.

Charles Martin Hall discovered the process for making aluminum while he was a graduate student at Oberlin College.

The iron wall of the cell acts as the cathode. Here, Al^{3+} ions are reduced:

$$\text{CATHODE:} \quad 2\ Al^{3+} + 6\ e^- \rightarrow 2\ Al(l) \qquad (25.6\ C)$$

Molten aluminum, which is more dense than the Al_2O_3-Na_3AlF_6 mixture, sinks to the bottom of the cell. The anodes are carbon rods, at which oxygen is formed:

$$\text{ANODE:} \quad 3\ O^{2-} \rightarrow \tfrac{3}{2}\ O_2(g) + 6\ e^- \qquad (25.6\ A)$$

The overall cell reaction is:

$$Al_2O_3(l) \rightarrow 2\ Al(l) + {}^3/_2\ O_2(g) \quad (25.6)$$

The oxygen gas attacks the anodes, gradually converting them to CO and CO_2. From time to time, new carbon rods are inserted into the cell.

From an energy standpoint, the preparation of aluminum is one of the most costly chemical processes. In the United States, about 3×10^9 kg of Al are produced each year. This requires about 6×10^{10} kilowatt hours, 5% of our total demand for electrical energy. You can see why it is important to recycle aluminum cans. It would also be helpful to devise a process for making aluminum that requires less energy. One which is being studied involves treating aluminum oxide with chlorine:

That's a lot of electricity for just one product.

$$2\ Al_2O_3(s) + 6\ Cl_2(g) \rightarrow 4\ AlCl_3(s) + 3\ O_2(g) \quad (25.7)$$

The aluminum chloride formed in this step is electrolyzed:

$$2\ AlCl_3(l) \rightarrow 2\ Al(s) + 3\ Cl_2(g) \quad (25.8)$$

The chlorine produced by electrolysis is recycled to form more $AlCl_3$. This process uses 30% less energy than the direct electrolysis of Al_2O_3.

More aluminum is produced than any other metal except iron. Its major use is in aircraft and automobiles. Here its low density (2.70 g/cm^3) is an important advantage. Automobile makers are using more and more aluminum these days to save mass and improve gasoline mileage. Aluminum is also an excellent conductor of heat and electricity. This explains its use in cooking utensils and electrical wiring. On a smaller scale, it is used as a mirror in reflecting telescopes because its surface reflects light so well.

Even though aluminum is a very reactive metal, it does not corrode readily. Freshly cut aluminum quickly picks up a very thin coating of the oxide, Al_2O_3. This adheres tightly to the surface, preventing further corrosion.

Recycling of Metals

The recycling of cans, frozen food trays, and other articles made of aluminum is becoming a major industry. About 20% of the aluminum we use comes from this source. The objective here is to save electrical energy. It "costs" about 20 kcal of energy to produce a gram of aluminum from Al_2O_3. In contrast, only about 0.2 kcal of heat is required to produce a gram of aluminum by melting down aluminum scrap.

Recycling of other metals has been going on for centuries. Ancient peoples recovered gold and copper by melting down objects made of these metals. In colonial times, lead bullets were salvaged and remolded. During World War II, the United States was cut off from its supply of tin ore. Tin cans and foil were collected and processed to recover the metal. Today, with many metals becoming scarce and expensive, recycling is on the increase. About 40% of lead, 30% of copper, iron, and tin, and 10% of zinc is obtained this way.

Objects containing more than one metal are often difficult to recycle. Junk automobiles are a prime example. There are 20 million of these scattered across the United

FIGURE 25.4

Scrap steel recovered from an automobile recycling plant in Cleveland is shredded and then loaded into railroad cars for shipment to a steel mill. *Courtesy of Luria Brothers and Co., Inc., a division of the Ogden Corporation.*

States. The problem here is one of separating the steel from other metals that are present. Copper, used in the electrical wiring of cars, is most objectionable. More than 0.1% of copper weakens steel to the point that it can not be reused.

Several methods are used to remove copper from automobile scrap. All of them are costly. One involves shredding the metal into pieces small enough (Figure 25.4) that the iron can be separated by a magnet. Another approach is to immerse the metal in a molten salt bath. The copper melts (mp = 1083°C) and flows away from the iron (mp = 1535°C). In the long run, the simplest approach may be to use aluminum rather than copper wiring. Aluminum is a good electrical conductor and is easily separated from steel.

25.4 IRON

The process used to obtain iron from its principal ore, Fe_2O_3, is unusual in that the end product is not the pure metal. Instead it is steel, an alloy of iron with other elements, principally carbon. The properties of steel make it superior to pure iron as a structural metal.

The conversion from iron ore to steel takes place in two steps. The first of these is carried out in a blast furnace. Here, the ore is converted to a product called "pig iron". This contains too much carbon to be useful. Most of the other impurities in the iron are removed in the blast furnace. Then, in a separate operation, the pig iron is converted to steel. The percentage of carbon is lowered and remaining impurities removed.

Blast Furnace

A model of a modern blast furnace is shown in Figure 25.5. Typically, the furnace is about 30 m high and perhaps 10 m in diameter. The solid "charge", admitted at the top, is a mixture of three materials. These are:

1. *Iron ore,* which is mostly Fe_2O_3. The major impurity is silicon dioxide, SiO_2.

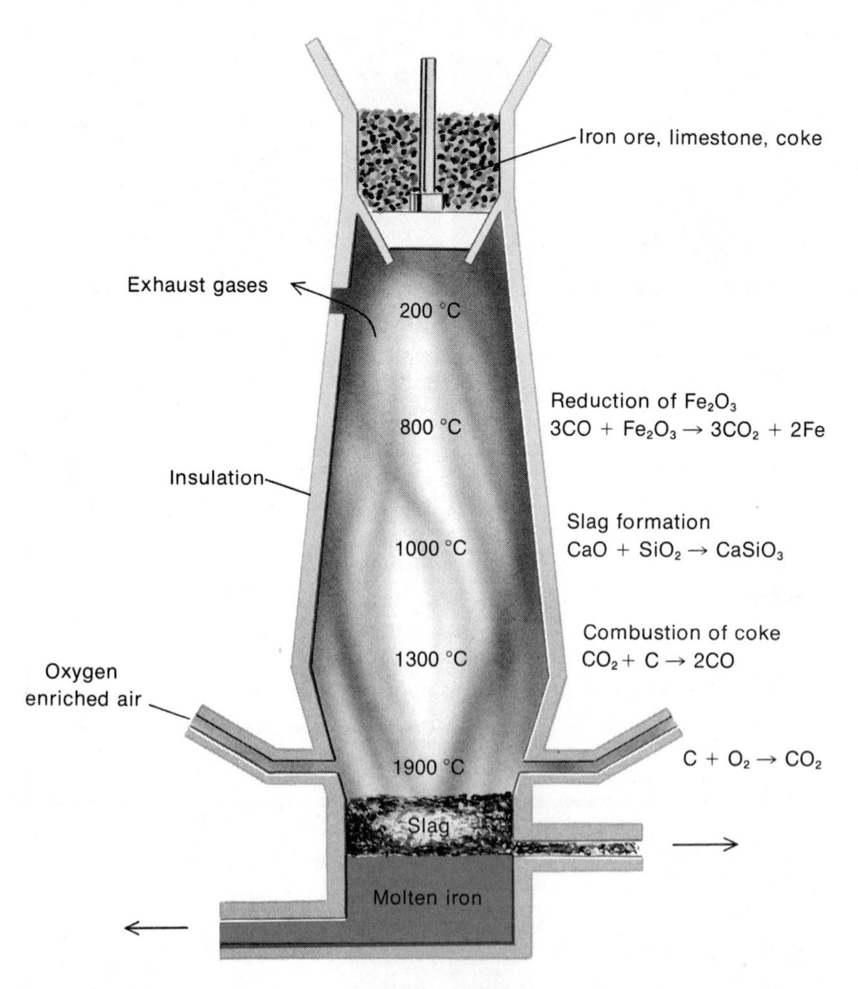

FIGURE 25.5

Blast furnace used for reduction of iron oxide. Heat is furnished by the combustion of coke to CO_2. The main reducing agent is CO. Limestone serves to convert sand in the ore to silicate slag. Molten iron falls to the bottom of the furnace where it is periodically drawn off.

2. *Coke*, which is made by heating coal in the absence of air. Coke is nearly pure carbon.
3. *Limestone*, $CaCO_3$.

Hot air (500°C) under pressure is blown into the furnace through nozzles located near the bottom.

Several different chemical reactions occur in the blast furnace. Three of these are most important.

(1) *Conversion of carbon to carbon monoxide.* As the hot air comes in contact with the coke, it burns part of it to carbon dioxide:

$$C(s) + O_2(g) \rightarrow CO_2(g); \qquad \Delta H = -94.1 \text{ kcal}$$

This reaction gives off enough heat to raise the temperature to about 1500°C. Under these conditions, a further, endothermic, reaction occurs:

$$C(s) + CO_2(g) \rightarrow 2\ CO(g); \qquad \Delta H = +41.3 \text{ kcal}$$

The overall reaction is the sum of these two reactions:

$$2\ C(s) + O_2(g) \rightarrow 2\ CO(g); \qquad \Delta H = -52.8 \text{ kcal} \qquad (25.9)$$

(2) *Reduction of Fe_2O_3 to iron by CO.* The CO produced by Reaction 25.9 rises through the solid mixture in the furnace. It reacts with the Fe_2O_3 in the iron ore.

Carbon can reduce Fe_2O_3, but the CO gas is what makes best contact with the ore.

$$Fe_2O_3(s) + 3\ CO(g) \rightarrow 2\ Fe(l) + 3\ CO_2(g) \qquad (25.10)$$

The molten iron produced collects at the bottom of the furnace. Four or five times a day, it is drawn off. The daily production of iron from a single furnace is about 1×10^6 kg, enough iron to make 20 railroad freight cars.

(3) *Formation of slag.* The limestone added to the furnace decomposes in the lower regions. At about 800°C, it reacts as follows:

$$CaCO_3(s) \rightarrow CaO(s) + CO_2(g) \qquad (25.11)$$

Carbon dioxide gas escapes. The calcium oxide reacts with impurities in the iron ore to form a glassy material called slag. The principal reaction is the formation of calcium silicate, $CaSiO_3$:

$$CaO(s) + SiO_2(s) \rightarrow CaSiO_3(s) \qquad (25.12)$$

The slag is less dense than molten iron. It forms a layer on the surface of the metal. The slag is drawn off through an opening located above that used for the iron. It is used to make cement and as a base in road construction.

EXAMPLE 25.3

Suppose a sample of iron ore contains 95% Fe_2O_3 and 5% SiO_2. From one kilogram of this ore:

a. how much iron is obtained (Reaction 25.10)?

b. how much $CaCO_3$ should be added to react with all the SiO_2 (Reactions 25.11 and 25.12)?

SOLUTION

a. From Eqn. 25.10, we see that:

We use mass relationships in all of chemistry, even in the steel mills.

$$1 \text{ mol } Fe_2O_3 \simeq 2 \text{ mol Fe}$$
$$159.7 \text{ g } Fe_2O_3 \simeq 111.7 \text{ g Fe}$$

We start with 950 g of Fe_2O_3 (95% of 1000 g). Hence:

$$\text{mass Fe} = 950 \text{ g } Fe_2O_3 \times \frac{111.7 \text{ g Fe}}{159.7 \text{ g } Fe_2O_3} = 660 \text{ g Fe}$$

b. From Eqns. 25.11 and 25.12:

$$1 \text{ mol } CaCO_3 \simeq 1 \text{ mol } SiO_2$$
$$100.1 \text{ g } CaCO_3 \simeq 60.1 \text{ g } SiO_2$$

We start with 50 g of SiO_2 (5% of 1000 g). Hence:

$$\text{mass } CaCO_3 = 50 \text{ g } SiO_2 \times \frac{100.1 \text{ g } CaCO_3}{60.1 \text{ g } SiO_2} = 80 \text{ g } CaCO_3 \text{ (1 sig. fig.)}$$

Manufacture of Steel

Can you suggest how pig iron got its name?

The pig iron that comes out of the blast furnace is highly impure. Typically it contains about 4% of carbon. This comes from the coke used to reduce the ore. Other impurities, present in smaller amounts, include silicon, manganese, and phosphorus. To make steel from pig iron, we need to do two things.

(1) *Remove the Si, Mn, and P.* This is done by oxidizing these elements:

$$Si(s) + O_2(g) \rightarrow SiO_2(s) \quad (25.13)$$

$$2\ Mn(s) + O_2(g) \rightarrow 2\ MnO(s) \quad (25.14)$$

$$P_4(s) + 5\ O_2(g) \rightarrow P_4O_{10}(s) \quad (25.15)$$

The oxides formed are removed by adding $CaCO_3$ to form a slag, as in the blast furnace.

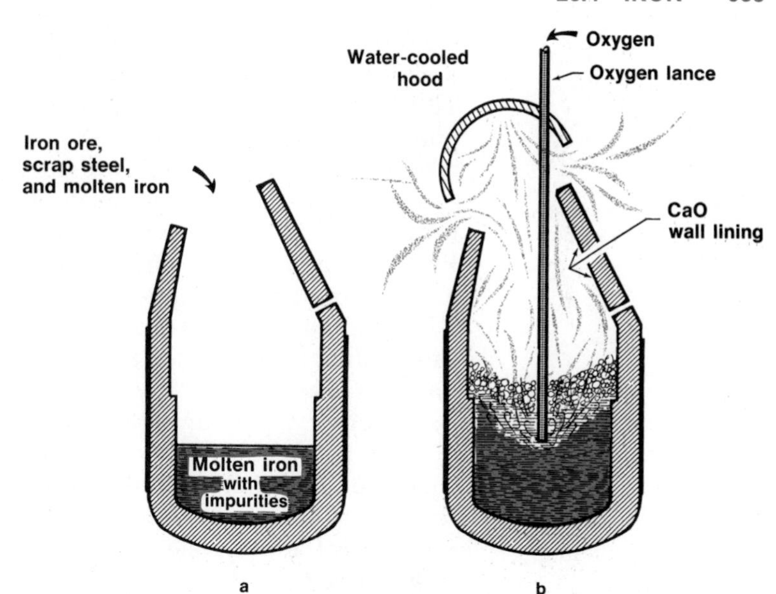

FIGURE 25.6

Converter used to make steel from iron from the blast furnace. Pure oxygen under pressure is blown into the molten metal, oxidizing impurities such as silicon and manganese and lowering the carbon content of the iron from 4 per cent to well below 2 per cent.

(2) *Lower the carbon content below 2%.* This is necessary to produce a strong, ductile metal that can be heat-treated. Most of the carbon is burned to carbon dioxide:

$$C(s) + O_2(g) \rightarrow CO_2(g) \qquad (25.16)$$

It is possible to adjust the carbon content of the steel within very narrow limits.

Most of the steel produced in the world today is made by the "basic oxygen" process. This is the only process used in Japan and western Europe. It accounts for about 60% of steel production in the United States. A diagram of the "converter" used is shown in Figure 25.6. This is filled with a mixture of about 70% pig iron and 30% scrap iron or steel. The pig iron is brought over from the blast furnace while it is still molten. Most of the scrap comes from the steel plant itself. Some is obtained from recycling of steel products such as automobiles.

Pure oxygen under a pressure of about 10 atm is admitted to the converter through the water-cooled "lance" at the top. When it reaches the surface of the molten metal, it causes vigorous stirring. Reactions 25.13 through 25.16 occur rapidly. Their progress is followed by an automatic, computerized system of chemical analysis. When the carbon content drops to the desired level, the supply of oxygen is cut off. (Too much oxygen would oxidize the iron, a reaction we don't want.) At this stage, the steel is ready to be poured. The whole process takes from 30 minutes to an hour. It yields about 200,000 kg of steel in a single "blow".

This process consumes tremendous amounts of pure oxygen.

Properties and Uses of Steel

The properties of steel depend to a large extent upon the amount of carbon present. "Mild" steels, containing less than 0.2% carbon, are very ductile. They are used in making wire, pipe, and sheet for automobile bodies. Medium steels (0.2–0.6% C) are less ductile but stronger. They are used in railroad rails, machine parts, girders, and bridge supports. High-carbon steels (0.6–1.5% C) are hard and brittle. Springs, razor blades, and surgical instruments are made from this type of steel.

Metallurgical methods are the result of both theory and experience.

The properties of steel can also be varied by heat treatment. If high-carbon steel is heated to redness and slowly cooled, it becomes relatively soft and tough. However, if it is "quenched" by plunging into an oil bath at room temperature, its properties are quite different. It is hard and brittle. By suitable heat treatment it is possible to produce a hard surface over a softer, tougher interior. This is desirable for making knives, scissors, and screws.

Another way to adjust the properties of steel is to add small amounts of other metals. In this way, an *alloy* is produced. Stainless steel is one such alloy. It contains about 15% chromium and 8% nickel and is very resistant to corrosion. Many other alloy steels are used for special purposes. Steel armor plate contains 10% or more of manganese, which makes it very hard and tough. High-speed cutting tools are made by adding tungsten or molybdenum. They remain sharp even when used for long periods at high temperatures.

25.5 ZINC AND MERCURY

The metals zinc and mercury are in the same vertical column of the Periodic Table. Yet they differ greatly in their physical and chemical properties. Zinc is a solid at room temperature (mp = 420°C, bp = 907°C). It is one of the more reactive transition metals (E^0_{ox} = +0.76 V). It is readily oxidized to Zn^{2+} ions by weak oxidizing agents such as the H^+ ion in dilute acids. Mercury is the only metal that is a liquid at room temperature (mp = −39°C, bp = 357°C). It is very unreactive (E^0_{ox} = −0.79 V). It does not react with dilute acids or with oxygen of the air.

Both zinc and mercury occur in nature as sulfides, ZnS and HgS. The processes used to extract the metals differ in many respects. These differences reflect the quite different properties of the metals themselves.

Zinc

The two most important steps in the extraction of zinc from ZnS are:

(1) *"Roasting" the sulfide ore.* Finely divided zinc sulfide is heated in air (Figure 25.7). This converts the sulfide to the oxide, which can more readily be reduced to the metal. The balanced equation for the reaction of zinc sulfide with oxygen of the air is:

$$2\ ZnS(s) + 3\ O_2(g) \rightarrow 2\ ZnO(s) + 2\ SO_2(g) \qquad (25.17)$$

(2) *Reduction of ZnO to Zn.* The zinc oxide is heated with carbon

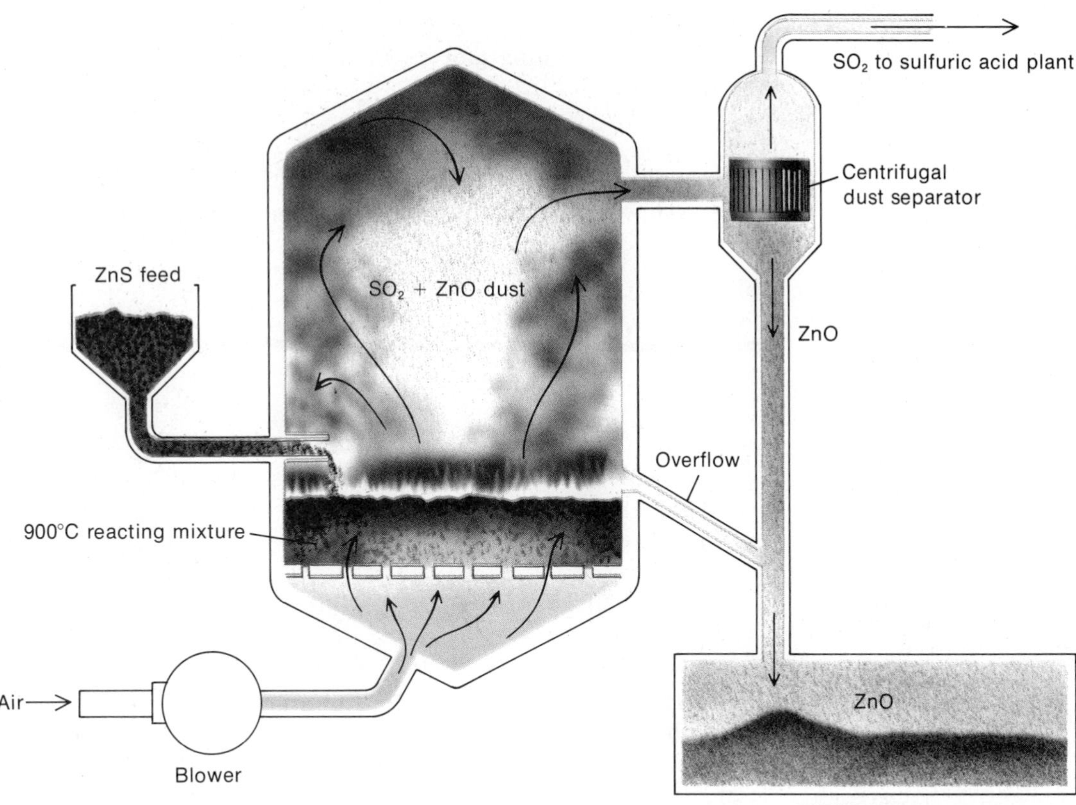

FIGURE 25.7

Roasting zinc ore. The hot ore is treated with compressed air, which converts the ZnS to ZnO dust and SO_2 gas. The dust is separated and collected, and the SO_2 is used to make sulfuric acid.

in the form of coal or coke. The reactions involved are similar to those with iron(III) oxide. The reducing agent is carbon monoxide.

$$ZnO(s) + CO(g) \rightarrow Zn(g) + CO_2(g) \quad (25.18)$$

Zinc is produced as a vapor at a temperature of 1200°C. If the vapor is cooled quickly "zinc dust" (finely divided zinc metal) is formed. More commonly, the zinc is allowed to cool until it condenses to a liquid at 907°C. If the liquid is quenched by pouring into water, "mossy zinc" is formed. These irregular pieces of solid zinc are frequently used as a laboratory reagent.

The main use of zinc is as a protective coating for steel. The zinc quickly picks up a thin film of the oxide or carbonate which prevents further corrosion. Zinc is also a major component of two familiar alloys:

—brass (10–30% Zn, 70–90% Cu)

—bronze (1–25% Zn, 70–95% Cu, 1–18% Sn)

EXAMPLE 25.4

Bismuth is obtained from its sulfide ore, Bi_2S_3. The process used resembles that for zinc. The sulfide ore is first roasted to the oxide, Bi_2O_3. The oxide is then reduced to the metal by heating with coke. Write balanced equations similar to 25.17 and 25.18 for the two reactions involved.

SOLUTION

In the roasting process, the reactants are Bi_2S_3 and O_2. The products are Bi_2O_3 and SO_2. The balanced equation for the reaction involved is:

$$2\ Bi_2S_3(s) + 9\ O_2(g) \rightarrow 2\ Bi_2O_3(s) + 6\ SO_2(g)$$

As with ZnO and Fe_2O_3, the reducing agent in the second step is CO:

$$Bi_2O_3(s) + 3\ CO(g) \rightarrow 2\ Bi(s) + 3\ CO_2(g)$$

Mercury

When the mineral cinnabar, HgS, is roasted in air, the following reaction occurs:

$$HgS(s) + O_2(g) \rightarrow Hg(g) + SO_2(g) \qquad (25.19)$$

Notice the difference between this reaction and that for the roasting of ZnS (Eqn. 25.17). Here, the product is the free element rather than the oxide. This happens because mercury is a much less active metal than zinc. At temperatures of 400°C or above, mercuric oxide, HgO, is unstable. So, any HgO formed in roasting decomposes immediately to the elements.

Mercury was once used to make felt for hats. The mad hatter got that way by inhaling mercury vapor.

The mercury vapor produced by Reaction 25.19 is condensed to a bright, shiny liquid. Liquid mercury can be purified by distillation. However, this is dangerous because mercury vapor is extremely toxic. In the laboratory, dirty mercury is often purified by shaking with dilute nitric acid. The HNO_3 oxidizes and dissolves out the impurities, but does not affect the very inactive metal.

Mercury is the liquid found in laboratory barometers and thermometers. Its use in barometers reflects its high density, 13.5 g/cm^3. A column of mercury 76.0 cm high exerts the same pressure as a column of water

$$76.0\ cm \times 13.5 = 1030\ cm$$

high. To use water in a barometer, the tube would have to be more than 10 m long. Mercury is useful in thermometers because it expands at a uniform rate when the temperature is raised. A mercury column heated from 0 to 50°C expands by almost exactly the same amount as when heated from 50 to 100°C.

Mercury is a relatively poor electrical conductor compared to other metals. Its conductivity is only about 1% that of silver. Yet, because it is a liquid metal, it is almost indispensable in many electrical applications. It is used, for example, in noiseless light switches and in many types of electrical cells.

The common fluorescent lamp contains a very small amount of mercury. When Hg atoms absorb electrical energy, they give off ultraviolet light at 365 and 366 nm. This light strikes a thin coating of fluorescent powder on the inside of the lamp. The powder in turn gives off light in the visible region. Overall, about 30% of the electrical energy is converted to visible light. This compares favorably with an efficiency of 5% for an ordinary light bulb. Fluorescent lights save both energy and money.

25.6 COPPER

Copper is one of the less reactive transition metals ($E^0_{ox} = -0.34$ V). It is slowly oxidized by air to black copper(II) oxide, CuO. The metal does not react with dilute acids (H^+). It is, however, oxidized to Cu^{2+} ions by concentrated nitric acid, HNO_3.

The extraction of copper from its ores is perhaps more complex than that of any other metal. The basic problem is that the percentage of copper in its ores is small. Typically, they contain less than 10% of copper, usually in the form of copper(I) sulfide, Cu_2S. Some deposits contain as little as 0.25% copper. A major impurity is iron(II) sulfide, FeS. Even larger amounts of rocky materials (granite, feldspar, clay) are present.

Concentration of Copper Ore

Copper ore is first treated to remove the rocky impurities. This is done by a process called flotation (Figure 25.8 p. 638). The ore is pulverized and treated with a mixture of oil, water, and detergent. Compressed air is then blown through the mixture to form a froth. The sulfide particles, which are wet by the oil, rise to the surface. The rocky material sinks to the bottom of the tank.

At this stage, the remaining ore consists largely of FeS mixed with Cu_2S. These two compounds are present in roughly equal amounts. To remove the FeS, the ore is roasted in a limited amount of air. This converts iron(II) sulfide to the oxide:

$$2\ FeS(s) + 3\ O_2(g) \rightarrow 2\ FeO(s) + 2\ SO_2(g) \qquad (25.20)$$

It does not, however, affect the less reactive Cu_2S. Sand, largely SiO_2, is added to react with the iron(II) oxide:

Otherwise the FeO would remain as a solid.

$$FeO(s) + SiO_2(s) \rightarrow FeSiO_3(l) \qquad (25.21)$$

The product is a liquid slag similar to that formed in the blast furnace. This is drawn off, leaving a product that consists mostly of Cu_2S.

Conversion of Cu_2S to Cu

With most of the impurities removed, copper(I) sulfide is then converted to copper metal. Hot compressed air is blown through the

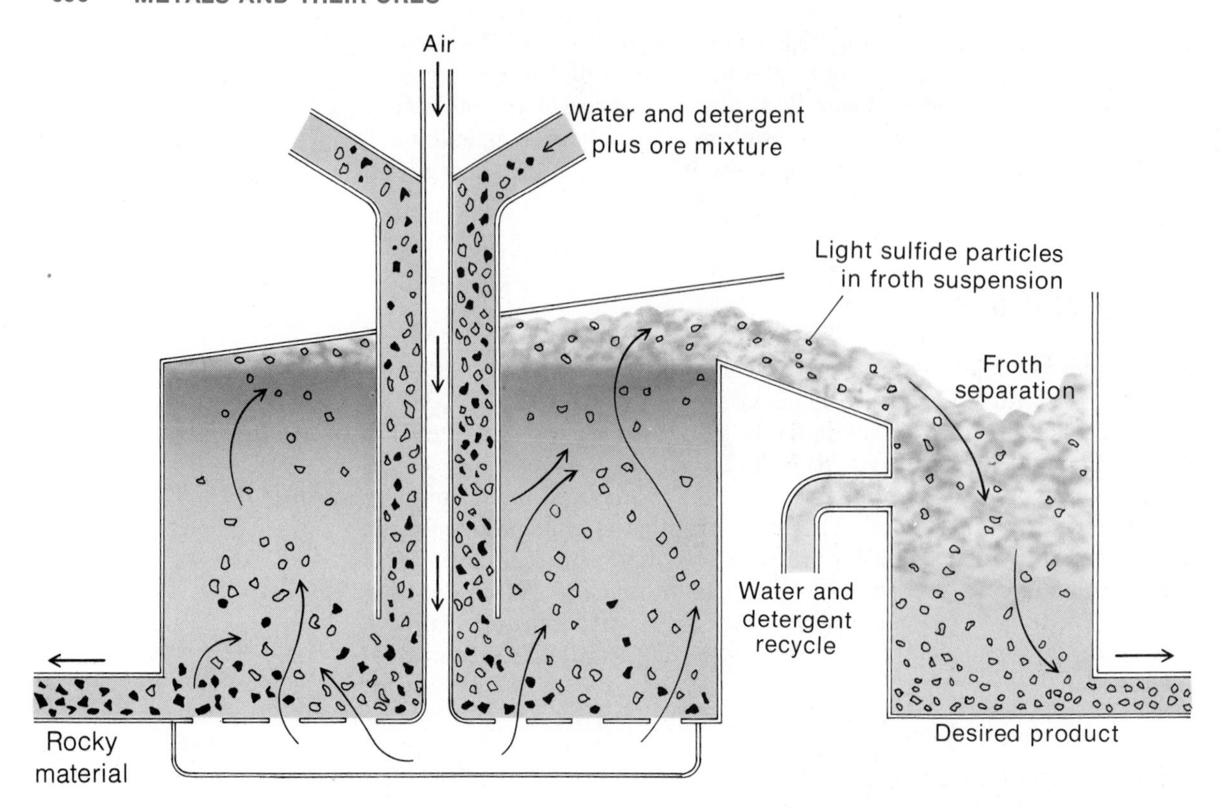

FIGURE 25.8

Flotation process for enrichment of copper sulfide ore. The relatively light sulfide particles are put into suspension in the water-oil-detergent mixture and collected as a froth. The denser material sinks to the bottom of the container.

sulfide. At 1500°C, the following reaction occurs:

$$Cu_2S(s) + O_2(g) \rightarrow 2\ Cu(l) + SO_2(g) \qquad (25.22)$$

Here, as with mercury, the product is the free metal. Any oxide formed decomposes at the high temperature.

The product of Reaction 25.22 is called "blister" copper. It has a rough, irregular appearance. This is largely due to air bubbles trapped when the molten metal solidifies. Blister copper still contains about 1% of impurities, all of them metals. These include iron, zinc, and small amounts of silver, gold, and platinum. For its major use, in electrical wiring, copper must be 99.9% pure. So, the blister copper is refined to remove these metals.

Refining of Cu

Copper is refined by electrolysis. The impure, blister copper serves as the anode of the cell. It is oxidized to Cu^{2+} ions. These ions are reduced to copper atoms at the cathode, which consists of a sheet of pure copper. The cell contains a water solution of $CuSO_4$. The concentration of Cu^{2+} stays constant at about 1 M throughout the electrolysis.

$$\text{ANODE:} \quad Cu(s, impure) \rightarrow Cu^{2+}(aq) + 2\ e^-$$

$$\text{CATHODE:} \quad Cu^{2+}(aq) + 2\ e^- \rightarrow Cu(s, pure)$$

$$Cu(s, impure) \rightarrow Cu(s, pure) \qquad (25.23)$$

The energy required for this electrolysis is small. Why?

As electrolysis proceeds, more active metals are oxidized at the anode. Iron and zinc are converted to Fe^{2+} and Zn^{2+} ions. These are difficult to reduce (E^0_{red} $Fe^{2+} = -0.44$ V, $Zn^{2+} = -0.76$ V vs. $Cu^{2+} = +0.34$ V). Moreover, they are present at very low concentrations, below 0.01 M. Consequently, they do not plate out at the cathode. From time to time, the solution in the cell is drawn off and replaced with fresh $CuSO_4$.

Less active metals in the blister copper are not oxidized. Instead, they drop off and collect at the bottom of the cell. The so-called "anode mud" that forms contains silver, gold, and platinum. These metals are recovered, separated from one another, and purified. Their value exceeds the cost of the energy used in the electrolysis.

25.7 SUMMARY

In this chapter, we have considered the extraction of several different metals from their ores. In most cases, the metal is present in the ore as a positive ion, associated with Cl^-, O^{2-}, or S^{2-} ions. To obtain the metal, it is necessary to reduce the positive ion. With chlorides (NaCl, $MgCl_2$), this is ordinarily done by electrolysis.

$$2\ NaCl(l) \xrightarrow{\text{electr.}} 2\ Na(l) + Cl_2(g)$$

$$MgCl_2(l) \xrightarrow{\text{electr.}} Mg(l) + Cl_2(g)$$

With oxides (Al_2O_3, Fe_2O_3), reduction can be brought about by electrolysis:

$$2\ Al_2O_3(l) \xrightarrow{\text{electr.}} 4\ Al(l) + 3\ O_2(g)$$

or with carbon monoxide:

$$Fe_2O_3(s) + 3\ CO(g) \rightarrow 2\ Fe(l) + 3\ CO_2(g)$$

With sulfides (HgS, Cu_2S), roasting in air sometimes gives the metal:

$$HgS(s) + O_2(g) \rightarrow Hg(g) + SO_2(g)$$

$$Cu_2S(s) + O_2(g) \rightarrow 2\ Cu(l) + SO_2(g)$$

With more active metals such as zinc, roasting gives the oxide (ZnO) rather than the metal itself. This must then be reduced, usually by CO, to the metal.

Usually, an ore has to be freed from impurities before the metal can be profitably extracted. Sometimes this involves a simple physical process such as flotation (Figure 25.8). More frequently, it involves a chemical reaction. The reaction may be with the ore itself or with the impurities. Thus:

—with Al, the ore is dissolved by complex ion formation ($Al(OH)_4^-$). The impurities stay behind as solids.

—with Fe and Cu, the impurities are removed by forming a glassy slag. This material is readily separated from the metal (Reaction 25.12) or its ore (Reaction 25.21).

Small amounts of impurities often have a pronounced effect on the properties of a metal. These impurities may be removed by electrolysis, as in the refining of copper. In other cases, it is more practical to adjust the concentration of foreign elements in the final product. This is the case with iron. Steel, containing up to 2% of carbon, is a much more useful material than pure iron.

NEW TERMS

basic oxygen process
blast furnace
flotation
ore
pig iron
refining
roasting
slag

QUESTIONS

1. What is meant by an ore? The minerals bauxite and granite both contain aluminum. Why is bauxite considered to be an ore of aluminum while granite is not?

2. Give the chemical formula of the principal ore of:

   a. Al b. Fe c. Cu d. Zn e. Hg

3. Referring to Table 25.1, list, in decreasing order, the three most abundant metals in the earth's crust.

4. Referring to Table 25.1, determine the number of kilograms of copper in:

   a. global reserves b. U.S. reserves c. the earth's crust

5. Of the 17 metals listed in Table 25.1, how many are transition metals?

6. It is cheaper to electrolyze a water solution than a molten salt. Why not produce sodium by electrolyzing a water solution of NaCl?

7. Give balanced equations for the reactions involved in the electrolysis of:

   a. NaCl b. $MgCl_2$ c. Al_2O_3

8. In the electrolysis of sodium chloride, what would happen if the products of the reaction came in contact with each other?

9. Why, in the process used to obtain aluminum, is:

    a. the bauxite ore treated with NaOH?
    b. cryolite, Na_3AlF_6, added to the electrolysis cell?
    c. the electrolysis not carried out in water solution?

10. Why is aluminum used in:

    a. automobiles (instead of steel)? b. frying pans?

11. What are the purposes of the coke and limestone added to the blast furnace?

12. Give the formula of the reducing agent used in the blast furnace. What compound does it reduce?

13. Why is oxygen used in making steel?

14. Describe two different ways to vary the properties of steel.

15. Which would you say is generally more reactive (toward O_2, for example)

    a. Na or Al? b. Zn or Hg? c. Cu or Au?

16. What is meant by "roasting" an ore? For what type of ore is this process used?

17. Why does roasting HgS yield the free metal while roasting ZnS gives an oxide?

18. What property of mercury makes it useful in:

    a. barometers b. thermometers c. electrical switches

19. Describe briefly how the process of flotation works.

20. Write equations for the chemical reactions involved in the separation of FeS from Cu_2S.

21. How is "blister" copper purified?

PROBLEMS

1. *Percentages of Elements (Example 25.1)* How many grams of metal can be obtained from one kilogram of:

    a. Fe_2O_3 b. ZnS c. Bi_2S_3

2. *Mass Relations in Equations (Example 25.3)* In the reduction of ZnO to Zn (Equation 25.18) how many grams of:

    a. CO are required to react with 1.00 g of ZnO?
    b. C are required to form the CO needed to reduce 1.00 g of ZnO?

3. *Mass Relations in Equations (Example 25.3)* A sample of copper ore contains 6.7% Cu_2S and 13.8% FeS. For one kilogram of this ore, how much O_2 is required to react with the:

    a. FeS (Equation 25.20) b. Cu_2S (Equation 25.22)

4. *Writing Equations (Example 25.2)* Write equations for the electrolysis of:

    a. $AlCl_3$ b. $BaCl_2$ c. KCl

5. *Writing Equations (Example 25.4)* Write equations for the reactions that occur when the following ores are roasted.

   a. ZnS (products: ZnO, SO_2) b. Cu_2S (products: Cu, SO_2)
   c. CoS (products: CoO, SO_2) d. As_2S_3 (products: As_2O_3, SO_2)

6. *Writing Equations (Example 25.4)* Write equations for the reduction of the following metal oxides by CO.

   a. Fe_2O_3 b. ZnO c. SnO_2 d. Cr_2O_3

* * * * *

7. Which one of the following compounds contains the largest mass percent of copper?

   a. Cu_2S b. CuS c. CuO d. $CuSO_4$

8. In the electrolysis of Al_2O_3 (Equation 25.6), how many grams of O_2 are formed for every gram of Al produced?

9. The principal ore of manganese is pyrolusite, MnO_2.

   a. Write a balanced equation for the reduction of MnO_2 by CO.
   b. How many grams of manganese can be obtained from one kilogram of MnO_2?

10. The principal ore of antimony is stibnite, Sb_2S_3.

    a. Write a balanced equation for the roasting of Sb_2S_3 to give the corresponding oxide and SO_2.
    b. Write a balanced equation for the reduction of antimony(III) oxide by CO.
    c. How many grams of Sb_2S_3 are required to produce a kilogram of Sb?

11. A certain ore contains 5.2 mass percent of TiO_2. Starting with 120 g of this ore:

    a. how much TiO_2 can be obtained? b. how much Ti can be obtained?

12. A bauxite ore contains 32.0% by mass of Al_2O_3. To dissolve all the Al_2O_3 in one kilogram of this ore:

    a. how many moles of OH^- are required? (Reaction 25.3)
    b. how many grams of NaOH are required?

13. Referring to Table 25.1, how long would it take, at the current rate of production, to consume the reserves of zinc? How many years would it take to consume all the zinc in the earth's crust?

14. About one kilowatt hour of electrical energy must be used to electrolyze one mole of Al_2O_3. How many kilowatt hours must be used to obtain the 3×10^9 kg of Al produced annually in the United States?

*15. Blister copper contains 99% Cu, 0.2% Ag, 0.04% Au, and 0.01% Pt. (The rest is Zn and Fe.) Take the prices of copper, silver, gold, and platinum to be $1.30, $150, $5000, and $5000 per kilogram in that order. Calculate the value of each of the metals recovered from one kilogram of blister copper.

*16. Tungsten is recovered from its oxide, WO_3, by treatment with hydrogen gas.

    a. Write a balanced equation for the reaction of WO_3 with H_2.
    b. What volume of H_2, at STP, is required to produce one gram of tungsten?

A smoggy vs. a clear day in New York City. The two photographs were taken at the same hour on successive days. *N.Y. Daily News photo.*

26 CHEMISTRY OF THE ENVIRONMENT

To an increasing extent, humanity is changing the nature of its environment. The air we breathe, the water we drink, and the soil we cultivate, contain foreign substances. These "chemicals" were unknown to our ancestors of 100 years ago. Some of them, from all the evidence we have, are beneficial. Others appear to be harmless. A few are clearly harmful. They have an adverse effect on our environment, sometimes upon our health. These are the substances that cause pollution.

In some cases, the addition of chemicals is deliberate. Chemical fertilizers are added to soil to increase the yield of crops. Chlorine is added to drinking water to kill organisms that cause disease. No one can argue that "additives" of this type should be abandoned. Without fertilizers, much of the world's population would starve to death. Untreated drinking water has been the source of epidemics of cholera and typhoid fever. Other additives are more controversial. The balance of evidence supports fluoridation of drinking water to prevent tooth decay. Yet there are many reasonable people who argue otherwise. In other areas, chemical additives are questionable, to say the least. This applies to many of the substances added, for no good reason, to our food.

Most pollutants enter the environment by the back door, so to speak. They arise as a side-effect of man's activities. When you drive a car, it is not your intention to pollute the air with noxious gases. Yet it is quite possible for that to happen. The operators of nuclear power plants do not set out to stockpile radioactive isotopes. Yet these dangerous materials are formed as by-products of the generation of power.

In this chapter, we will look at the chemistry of some of the substances that are being added to the environment. In each case, we will be looking for answers to three questions:

1. Where do they come from?
2. What effects do they have, harmful or otherwise?
3. How can we get rid of them, and what will happen if we do?

As you can imagine, these questions, particularly (2) and especially (3), do not always have simple answers. Indeed, in some cases, there are no answers, only more questions.

26.1 GASEOUS POLLUTANTS

"Pure air", as you know, is a mixture. It consists largely of two gases, nitrogen and oxygen. Smaller amounts of argon, water vapor, and carbon dioxide are present. Together these five gases, all of which are harmless, make up at least 99.99% of the air we breathe. In this section, we will be concerned with a few of the gases that account for the remaining 0.01%. Even though present in very small amounts, they can affect our health and comfort.

There are many different air pollutants. Suspended particles of dirt and soot can be a serious problem. The same is true of unburned hydrocarbons from the incomplete combustion of gasoline. Here, however, we will concentrate upon three major types of air pollutants. These are:

1. The oxides of sulfur, SO_2 and SO_3.
2. Carbon monoxide, CO.
3. Two oxides of nitrogen, NO and NO_2.

Sulfur Oxides (SO_2, SO_3)

In the early part of this century, when coal was used in home furnaces, air pollution was often worse than it is now.

Sulfur oxides come mostly from the burning of coal. This fuel contains 1 to 3% sulfur, mostly in the form of metal sulfides. One of these is pyrite, FeS_2. When coal containing FeS_2 is burned, SO_2 is produced by reaction with the oxygen of the air.

$$4\ FeS_2(s) + 11\ O_2(g) \rightarrow 8\ SO_2(g) + 2\ Fe_2O_3(s) \qquad (26.1)$$

Some of the sulfur dioxide produced is further oxidized to SO_3:

$$2\ SO_2(g) + O_2(g) \rightarrow 2\ SO_3(g) \qquad (26.2)$$

Ordinarily, less than 1% of the SO_2 in the air undergoes this reaction. However, the formation of SO_3 is catalyzed by suspended particles such as those in coal smoke. In the gas coming out of the chimney of a coal-burning power plant, 5% or more of the sulfur may be in the form of SO_3.

FIGURE 26.1

Marble statues often suffer greatly when attacked by the oxides of sulfur. *From Wagner, R. H.,* Environment and Man, *W. W. Norton and Co., New York, 1971.*

Most of the adverse effects of sulfur oxides come from the sulfuric acid produced when SO_3 reacts with water:

$$SO_3(g) + H_2O(l) \rightarrow H_2SO_4(l) \quad (26.3)$$

The pH of rainfall in polluted areas may be as low as 2 or 3 because of this reaction. At this pH, metals such as steel and aluminum corrode rapidly. They react with the H^+ ions in the dilute sulfuric acid solution. Buildings made of limestone and marble statues are also affected (Figure 26.1). The calcium carbonate present dissolves through the reaction:

$$CaCO_3(s) + 2\ H^+(aq) \rightarrow Ca^{2+}(aq) + CO_2(g) + H_2O \quad (26.4)$$

People also suffer from exposure to oxides of sulfur. As little as 1×10^{-4} mol percent of SO_2 can be fatal to patients with lung diseases. The London "killer smog" of 1952 was responsible for over 4000 deaths. Many people developed symptoms of bronchitis or asthma for the first time. The culprit here was sulfur dioxide and the sulfuric acid produced from it. These chemicals were trapped in a stagnant air mass that hung over the city for several days.

During the 1950's, sulfur oxide emissions were lowered largely by switching from coal to other fuels. Petroleum contains little sulfur. Natural gas is virtually free of sulfur. Now, with these fuels in short supply, power plants are converting back to coal. By law, these plants

FIGURE 26.2

"Scrubbers" are used in coal-burning electric plants to prevent air pollution. In this case, the sulfur dioxide formed by combustion of high-sulfur coal is removed by treating with limestone. Courtesy of Pullman-Kellogg.

have to keep the emissions of sulfur oxides at a low level. One way to do this is to treat the gases coming out of the smokestack with a substance such as magnesium oxide. This reacts with the SO_2 present, converting it to magnesium sulfate.

$$MgO(s) + SO_2(g) + \tfrac{1}{2} O_2(g) \rightarrow MgSO_4(s) \qquad (26.5)$$

The equipment used to carry out this reaction (Figure 26.2) is expensive to build and operate. If the electrical energy you use comes from the burning of coal, SO_2 removal may add as much as 10% to your utility bill.

EXAMPLE 26.1

A power plant burns coal containing 1.2% of combined sulfur.

a. How many grams of SO_2 are formed from one kilogram of this coal?

b. How many grams of MgO are required to remove the SO_2 produced?

SOLUTION

a. In one kilogram (1000 g) of coal:

$$\text{mass S} = 0.012 \times 1000 \text{ g} = 12 \text{ g S}$$

One mole of S, 32.1 g, produces one mole of SO_2, 64.1 g. Hence:

$$\text{mass } SO_2 = 12 \text{ g S} \times \frac{64.1 \text{ g } SO_2}{32.1 \text{ g S}} = 24 \text{ g } SO_2$$

b. According to Equation 26.5, one mole of MgO, 40.3 g, reacts with one mole of SO_2, 64.1 g. So:

$$\text{mass MgO} = 24 \text{ g } SO_2 \times \frac{40.3 \text{ g MgO}}{64.1 \text{ g } SO_2} = 15 \text{ g MgO}$$

Carbon Monoxide, CO

The source of carbon monoxide in air differs from that of the oxides of sulfur. About 80% of the CO in the air over our cities comes from automobiles. It is produced by the incomplete combustion of gasoline. Reactions such as the following are responsible:

$$2\ C_8H_{18}(l) + 17\ O_2(g) \rightarrow 16\ CO(g) + 18\ H_2O(l) \qquad (26.6)$$

Figure 26.3 shows how the concentration of CO in New York City relates to time of day. Notice that it rises rapidly in the early morning rush hour. This is when commuter traffic reaches its peak. In the early evening, as the number of cars on the road decreases, so does the level of CO.

It's not a good idea to go jogging in New York at rush hour.

The toxic effects of carbon monoxide come from its reaction with hemoglobin of the blood. Hemoglobin is a large organic molecule which acts as an oxygen carrier. It forms a stable complex with O_2. We will represent this complex simply as $Hem \cdot O_2$. In this form, oxygen is carried from the lungs to the tissues where it is needed. Carbon monoxide also forms a complex with hemoglobin, $Hem \cdot CO$. Unfortunately, this complex is more stable than that with oxygen. For the reaction:

$$CO(g) + Hem \cdot O_2(aq) \rightleftarrows O_2(g) + Hem \cdot CO(aq) \qquad (26.7)$$

$$K = \frac{[O_2] \times [Hem \cdot CO]}{[CO] \times [Hem \cdot O_2]} = 200$$

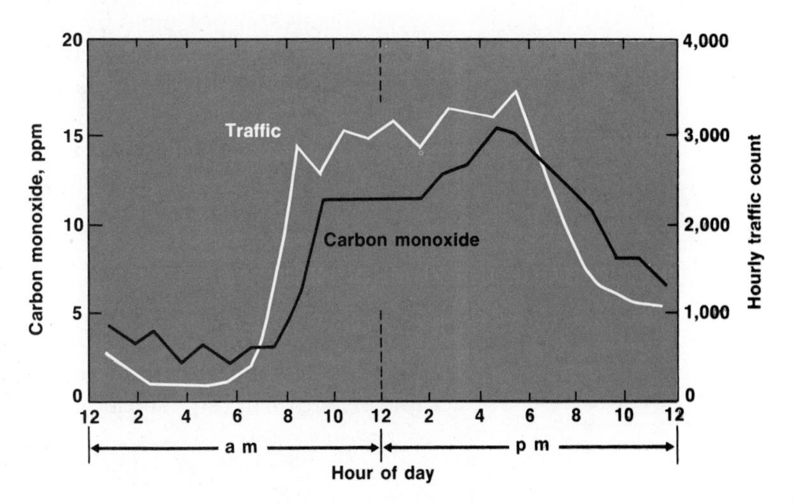

FIGURE 26.3

Variation of carbon monoxide concentration with time of day in New York City. *From Johnson, K. L., Dworetzky, L. H., and Heller, A. N., Carbon Monoxide and Air Pollution from Automobile Emissions in New York City,* Science, *Vol. 160, No. 3823, p. 67 (1968).*

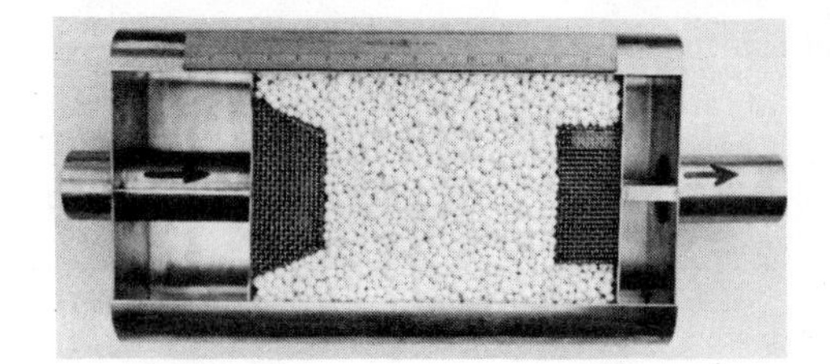

FIGURE 26.4

Cross section of a catalytic converter showing catalyst pellets. Such a converter contains a few grams of platinum and palladium on the surface of the pellets. *From Turk, A. et al,* Environmental Science, *W. B. Saunders Co., Philadelphia, 1974.*

Since the equilibrium constant is large, a small concentration of CO can tie up much of the hemoglobin. Suppose, for example, that $[CO] = 1/200\ [O_2]$. At this point:

$$\frac{[\text{Hem}\cdot\text{CO}]}{[\text{Hem}\cdot\text{O}_2]} = 200 \times \frac{[\text{CO}]}{[\text{O}_2]} = 200 \times \frac{1}{200} = 1.$$

This means that half of the hemoglobin will be in the form of the CO complex. Too little of the $Hem \cdot O_2$ complex is left to supply the body with the oxygen it needs. Indeed, if as little as 20% of the Hem·CO complex is present, the results can be fatal.

To reduce CO emissions, we must go to their source, the automobile engine. Before 1975, car manufacturers approached the problem by minor changes in engine design. Among other things, they increased the air-to-fuel ratio to get more complete combustion. Changes of this type had several side effects. They reduced gas mileage and made starting more difficult.

Since 1975, most cars made in the United States have used a catalytic converter to get rid of CO (and unburned hydrocarbons as well). A diagram of the converter is shown in Figure 26.4. It contains from 1 to 3 g of two metals, platinum and palladium. They serve as a catalyst to oxidize carbon monoxide to CO_2:

$$2\ \text{CO}(g) + \text{O}_2(g) \xrightarrow{\text{Pt, Pd}} 2\ \text{CO}_2(g) \qquad (26.8)$$

We've made progress in reducing air pollution by cars but we're still seeking better methods.

Like all methods of reducing pollution, the catalytic converter has created new problems. It adds about $300 to the cost of a car. Moreover, the catalyst is poisoned by gasoline additives. Cars using catalytic converters must burn unleaded gasoline. Producing a high-octane gasoline without lead additives is not easy. It requires changing the blend of hydrocarbons to use more aromatics. This raises the price by perhaps 10%. It may also cause a health hazard. Some aromatic hydrocarbons are known to cause cancer.

Nitrogen Oxides (NO, NO_2)

These compounds, like all air pollutants, are byproducts of the combustion of fuels. At high temperatures, some of the nitrogen in air combines with oxygen. The first product formed is nitric oxide, NO:

$$\text{N}_2(g) + \text{O}_2(g) \rightarrow 2\ \text{NO}(g) \qquad (26.9)$$

Upon cooling, further reaction occurs to form nitrogen dioxide, NO_2:

$$2\ NO(g) + O_2(g) \rightarrow 2\ NO_2(g) \qquad (26.10)$$

About 35% of the NO and NO_2 in air is produced by automobiles. The high temperatures in an automobile engine favor Reaction 26.9. Another 50% is produced by the combustion of fuels in power plants and industry. Natural gas, a "clean" fuel in all other respects, is a problem here. Its combustion temperature is higher than that of other fuels, so more NO is formed.

So far as we know, nitric oxide in air is not itself harmful. Nitrogen dioxide, however, is a different story. It ends up as nitric acid, formed by reactions such as:

$$3\ NO_2(g) + H_2O(l) \rightarrow 2\ HNO_3(l) + NO(g) \qquad (26.11)$$

Nitric acid in the environment is no more pleasant to live with than sulfuric acid. Moreover, NO_2 is involved in forming photochemical smog. This is a major problem in many cities (see Chapter opening photo).

Unfortunately there are other problems as well.

Smog formation starts when the NO_2 molecule breaks down in sunlight:

$$NO_2(g) \rightarrow NO(g) + O(g) \qquad (26.12)$$

The free oxygen atoms produced react with O_2 molecules. The product is ozone, O_3.

$$O_2(g) + O(g) \rightarrow O_3(g) \qquad (26.13)$$

The three species NO_2, O, and O_3 are responsible for the effects of smog. The brown or yellow color comes mostly from NO_2. Eye irritation is caused by O atoms or O_3 molecules. Both of these are strong oxidizing agents.

Reduction of NO-NO_2 emissions is more difficult than for any other air pollutant. Reaction 26.9 can be prevented by lowering the combustion temperature. This generally means incomplete combustion, which gives more CO. However, this is the approach now used in automobiles. About 10 to 20% of the exhaust gas is recirculated through the engine. This lowers the temperature enough to suppress the formation of NO. It also reduces gasoline economy.

A better approach to getting rid of NO and NO_2 is needed. Methods now under study attempt to decompose NO to nitrogen, N_2. In a power plant, this can be done by injecting a small amount of ammonia into the gases produced.

$$4\ NH_3(g) + 6\ NO(g) \rightarrow 6\ H_2O(g) + 5\ N_2(g) \qquad (26.14)$$

For automobiles, a special catalyst containing rhodium metal is being developed. With a Pt-Pd-Rh catalyst, NO is reduced to N_2 by the CO in the exhaust:

$$2\ CO(g) + 2\ NO(g) \rightarrow 2\ CO_2(g) + N_2(g) \qquad (26.15)$$

26.2 TRACE METALS

Several metals are needed by the body. Most of these are required only in very small amounts. Among the essential trace metals are:

chromium	manganese
cobalt	molybdenum
copper	tin
iron	vanadium
lithium	zinc

On the other hand, a few metals are toxic at low concentrations. Two of these, mercury and lead, are discussed below. Other toxic trace metals include:

—*beryllium*. Fortunately, this metal is quite rare and is not widely used. It is extremely poisonous. As little as 1×10^{-5} g of Be in the lungs causes the disease known as beryllosis. The symptoms are similar to those of tuberculosis or lung cancer. They show up many years after exposure. The disease is often fatal; no cure is known.

If you plan to work with unfamiliar chemicals, check their toxicities first.

—*cadmium*. This metal occurs in nature with zinc. Most objects made of zinc contain small amounts of cadmium as an impurity. At low levels, Cd^{2+} ions can cause high blood pressure and heart disorders. Higher concentrations affect the bones and joints.

Concentrations of trace metals in soil, water or other materials are often expressed in *parts per million* **(ppm)**.

$$\begin{aligned} 1 \text{ ppm of X in Y} &= 1 \text{ g of X in } 10^6 \text{ g of Y} \\ &= 1 \times 10^{-6} \text{ g of X per gram of Y} \end{aligned} \qquad (26.16)$$

To show how this unit is used, consider the standard set for Pb^{2+} ions in drinking water. The highest concentration allowed is 0.05 ppm. This means that there can be:

$$0.05 \times 10^{-6} \text{ g} = 5 \times 10^{-8} \text{ g of } Pb^{2+} \text{ per gram of water}$$

Mercury

There are two main sources of mercury poisoning.

1. *Mercury vapor*. This is a hazard in many laboratories. Drops of the liquid are difficult to pick up. They tend to collect in crevices of floors, sinks, and benches. Even though the vapor pressure is low (1.6×10^{-6} atm at 20°C, 3.7×10^{-6} atm at 30°C), enough vapor can form to be harmful. Taken into the lungs, it quickly diffuses into the blood. Over a period of time, mercury concentrates in the brain. Only 5 ppm can cause severe damage to the brain and nervous system.

Needless to say, if you spill mercury, you should report it to your instructor. He or she will know how to collect and remove it (Figure 26.5). This applies even to a broken thermometer.

2. *Organic Mercury Compounds*. One compound that is particularly

dangerous is dimethyl mercury, $(CH_3)_2Hg$. Certain microorganisms have the ability to convert elementary mercury into dimethyl mercury. This explains why plants growing in lakes or rivers where mercury has been discarded often contain $(CH_3)_2Hg$.

The toxic effects of dimethyl mercury were discovered tragically in Minamata, Japan in 1953. Discharge of mercury from a nearby chemical plant led to concentrations in fish of 10 ppm of Hg, in the form of $(CH_3)_2Hg$. Of the people who ate these fish, 43 died and 68 more became seriously ill. Later, similar cases of poisoning by dimethyl mercury occurred in other parts of the world.

In 1970, there was a "mercury scare" in the United States and Canada. It was found that pike, bass, and salmon caught in Lake Erie contained as much as 2 ppm of Hg. Here, the culprit was again $(CH_3)_2Hg$. This came from mercury discarded by industries on the lake shore. Fish in unpolluted waters normally contain 0.2 ppm or less of mercury. The "safe" limit set by the FDA is 0.5 ppm.

FIGURE 26.5

Mercury collector. This device, used at Macalester College, consists of an evacuated flask attached through a stopcock to a tube with a fine orifice. Mercury droplets from a spill are readily sucked into the trap in the tube when the stopcock is opened.

EXAMPLE 26.2

Fish containing more than 0.5 ppm of Hg cannot be sold. How many grams of mercury are allowed in:

a. one gram of fish?

b. a fish weighing 4 kg (about 10 lb)?

SOLUTION

a. 0.5×10^{-6} g $= 5 \times 10^{-7}$ g Hg

b. Noting that 4 kg $= 4 \times 10^3$ g, and setting up a simple conversion:

$$4 \times 10^3 \text{ g of fish} \times \frac{5 \times 10^{-7} \text{ g Hg}}{\text{g fish}} = 20 \times 10^{-4} \text{ g} = 2 \times 10^{-3} \text{ g Hg}$$

This is a very small amount of mercury. Indeed, many chemists question whether Hg at such a low level can be determined accurately.

Lead

Lead poisoning is due to the Pb^{2+} ion. In a normal adult, the concentration of Pb^{2+} in the blood is about 0.4 ppm. Symptoms of lead poisoning show up when this concentration reaches 0.8 ppm. Above 1.2 ppm, lead can cause brain damage. As you can see, there is not much margin for error here. Small increases in the level of Pb^{2+} can be dangerous.

There are two main sources of lead compounds in the environment.

1. *Tetraethyl lead, $(C_2H_5)_4Pb$* This compound is added to gaso-

FIGURE 26.6

Lead pigments have not been used in interior paints for many years. However, they are still a cause of lead poisoning. Small children eat flakes of paint peeling from walls and woodwork of old houses. *Courtesy of the Food and Drug Administration.*

line, along with $C_2H_4Br_2$, to increase the octane number. "High test" gasoline contains about 0.5 g/ℓ of tetraethyl lead. Combustion produces compounds such as $PbBr_2$ and PbO. These come out of the exhaust and end up in the soil or water. The dust in city streets can contain as much as 1.5% by mass of lead.

As more cars are made with catalytic converters, the use of tetraethyl lead is dropping. We can hope that within a few years, Pb^{2+} ions from this source will no longer be a problem.

It took a long time to recognize the cause of the illness we now call lead poisoning.

2. *Lead pigments in paints* These include:
   white lead, a mixture of $Pb(OH)_2$ and $PbCO_3$
   red lead, Pb_3O_4
   chrome yellow, $PbCrO_4$

Hundreds of children die of lead poisoning from this source each year. The victims are mostly between the ages of 1 and 3. They eat chips of paint, each of which may contain as much as 0.1 g of lead. Paints with lead pigments have not been sold for interior use since 1940. However, they are still found on the walls of many older houses and apartment buildings (Figure 26.6).

26.3 RADIOACTIVITY

The process of radioactivity was discovered, almost by accident, in 1896. A French scientist, Henri Becquerel, was carrying out experiments with a uranium salt. He placed crystals of this salt in contact with a photographic plate. To his surprise, Becquerel found that the plate was darkened by radiation from the uranium. Later, Marie and Pierre Curie isolated two new elements from uranium ore. These elements, polonium and radium, were strongly radioactive. They gave off radiation much more intense than that from uranium.

For many years, the hazard of radiation was not recognized. Scientists studying radioactive processes took few if any precautions. The results were tragic. Nuclear radiation was in part responsible for the death of Madame Curie in 1934. We now know that this high-energy radiation can have two adverse effects. Direct exposure can lead to leukemia and other forms of cancer. Often, cancer develops many years after exposure. Radiation also has a genetic effect. It causes changes in the genes which bring about defects in the children of those exposed to radiation.

Later in this section we will consider the sources of radiation in the environment. First, though, we need to examine the nature of radioactive processes. Here we will be interested in the types of radiation and the rates at which they are emitted.

Types of Radiation

Radiation comes from the decomposition of unstable nuclei. These nuclei contain either too many neutrons or, less often, too many protons. They are said to be *radioactive*. Radioactive nuclei "decay" spontaneously to other, more stable nuclei. In doing so, they emit high-energy rays. The properties of these rays can be studied in an electrical field (Figure 26.7). We can distinguish three different types of rays: **alpha** (α), **beta** (β), and **gamma** (γ).

We can't see or feel this radiation, so researchers tended to ignore the risks involved in studying it.

1. Alpha rays consist of a stream of positively charged particles called **alpha particles.** An alpha particle is made up of two protons and two neutrons. It has a charge of +2 and a mass of 2 + 2 = 4 on the atomic mass scale. An alpha particle is identical with the nucleus of an ordinary helium atom (at. no. = 2, mass no. = 4). Its nuclear symbol is:

$$^{4}_{2}He$$

When a nucleus emits an alpha particle, a new nucleus is formed.

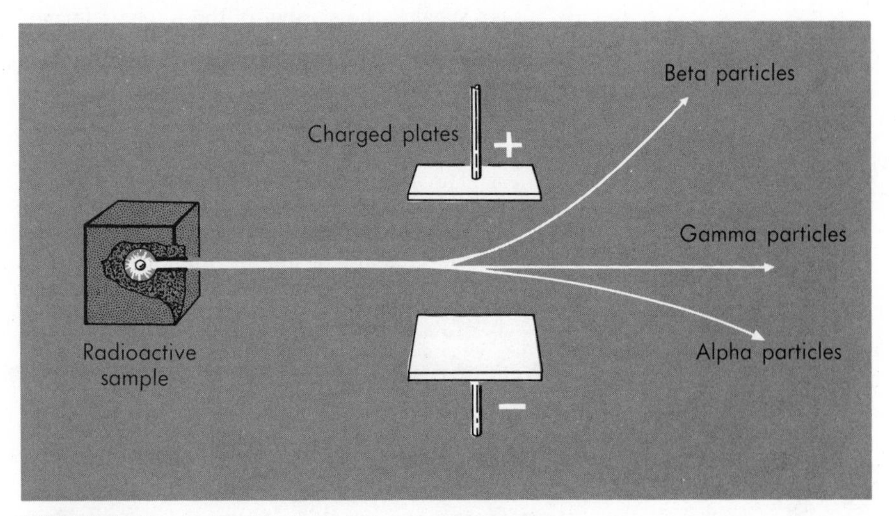

FIGURE 26.7

A beam of radiation from a radioactive source can be resolved into three components by an electric field. These components, called α, β, and γ rays, consist of helium nuclei, electrons, and short wavelength X-rays respectively.

This nucleus corresponds to a different element. It has:

an atomic number 2 units smaller than the original nucleus
a mass number 4 units smaller than the original nucleus

To illustrate α-decay, let us consider the most common isotope of uranium. This has an atomic number of 92 and a mass number of 238. Its nuclear symbol is $^{238}_{92}U$. It is radioactive, decaying by emitting an alpha particle, $^{4}_{2}He$. The new nucleus formed has:

an atomic number of 92 − 2 = 90
a mass number of 238 − 4 = 234

This is an isotope of the element of atomic number 90, thorium. Its nuclear symbol is $^{234}_{90}Th$.

This reaction can be described by a nuclear equation. In such an equation, we use nuclear symbols for each species. For the α-decay of $^{238}_{92}U$, the equation is:

$$^{238}_{92}U \rightarrow {}^{4}_{2}He + {}^{234}_{90}Th \qquad (26.17)$$

In this and all other nuclear equations, there is a balance of both atomic number and mass number (nuclear charge and mass). Here, we have:

Mass and nuclear charge are conserved in nuclear reactions.

	reactants	*products*
at. no.	92	90 + 2 = 92
mass no.	238	234 + 4 = 238

EXAMPLE 26.3

Another radioactive isotope that decays by giving off an alpha particle is radium-226, which has the nuclear symbol $^{226}_{88}Ra$.

a. What is the symbol of the nucleus formed by α-decay?

b. Write a balanced equation for the nuclear reaction.

SOLUTION

a. The atomic number decreases by 2, the mass number by 4. The nucleus formed has an atomic number of 88 − 2 = 86. Its mass number is 226 − 4 = 222. From the Periodic Table, we see that the element of atomic number 86 is radon (Rn). The nuclear symbol of this isotope of radon is $^{222}_{86}Rn$.

b. $^{226}_{88}Ra \rightarrow {}^{222}_{86}Rn + {}^{4}_{2}He$

Alpha particles given off by radioactive nuclei have very high energies. As they pass through matter, they give off this energy to molecules with which they collide. By absorbing this energy, a molecule may dissociate, breaking apart into atoms. In other cases, it may ionize, losing an electron to become a positive ion. Either way, the molecule is destroyed. This accounts for the radiation damage caused by alpha particles and by other types of radiation as well.

Since alpha particles are relatively heavy, they move rather slowly. Most of their energy is given up in collisions within a short distance of their source. A piece of aluminum foil or even a piece of paper can protect you from α-rays. However, if a sample which emits alpha particles is taken into the body, the danger is much greater. The alpha particles can burn and destroy tissue with which they come in contact.

2. Beta rays consist of a stream of negatively charged particles called **beta particles.** A beta particle is simply an electron. An electron, you will recall, has a charge of -1. It has a very small mass, nearly zero on the atomic mass scale. Its symbol is:

β particles are electrons.

$$_{-1}^{0}e$$

Beta particles are formed by a reaction within the nucleus. A neutron (at. no. = 0, mass no. = 1) decomposes to emit an electron and leave behind a proton (at. no. = 1, mass no. = 1). Hence, when an electron is emitted, the nucleus formed has:

an atomic number 1 unit larger than the original nucleus
the same mass number as the original nucleus

Consider, for example, the isotope formed by Reaction 26.17, $^{234}_{90}Th$. This isotope of thorium is radioactive. It decays by emitting an electron, $_{-1}^{0}e$. The new nucleus formed has:

A radioactive isotope decays spontaneously.

an atomic number of 90 + 1 = 91
a mass number of 234

The element of atomic number 91 is protoactinium, Pa. Hence, the nuclear equation is:

$$^{234}_{90}Th \rightarrow {}_{-1}^{0}e + {}^{234}_{91}Pa \quad (26.18)$$

Here, as always, there is a balance of atomic number and mass number:

	reactants	*products*
at. no.	90	91 − 1 = 90
mass no.	234	234 + 0 = 234

EXAMPLE 26.4

Small amounts of a radioactive isotope of carbon, $^{14}_{6}C$, are found in the atmosphere. It decays by emitting an electron. Write a balanced nuclear equation for the decay.

SOLUTION

The new nucleus formed has the same mass number as $^{14}_{6}C$, but an atomic number 1 unit greater. It is an isotope of nitrogen, $^{14}_{7}N$. The equation is:

$$^{14}_{6}C \rightarrow {}_{-1}^{0}e + {}^{14}_{7}N$$

Beta particles interact with matter in much the same way as alpha particles. However, since their size is much smaller, they have greater penetrating power. In air, β-particles typically travel a meter or more before giving up their energy. They readily pass through the walls of glass test tubes or beakers. They can move up to 1 cm through human tissue.

3. Gamma rays are a form of high energy radiation. They are similar to X-rays but have even shorter wavelengths. Gamma rays usually accompany the emission of alpha and beta particles. They have no mass or charge. Hence, emission of a γ-ray does not change atomic number or mass number. For this reason, we will not include gamma radiation in writing nuclear equations.

Radiation-caused cancer is usually the result of gamma rays.

Of the three types of nuclear radiation, gamma rays are by far the most penetrating. In that sense, they are the most dangerous. In human tissue, they can cause damage over a range 10 to 20 times as great as beta particles. Heavy lead shielding is used to protect against gamma rays.

Rate of Decay

An important property of a radioactive isotope is the rate at which it decays. This is usually described in terms of its **half-life.** The half-life is the time required for one half of a radioactive sample to decay. To illustrate, suppose that a certain isotope has a half-life of 10 years. This means that after 10 years half of it will be gone. The other half will remain, still giving off radiation.

Every radioactive species has a characteristic half-life. This half-life remains constant, regardless of how much of the species is present. Consider, for example, a radioactive isotope of iodine, $^{131}_{53}I$. This has a half-life of 8 days. Suppose we start with a 1.00 g sample of $^{131}_{53}I$. After 8 days, there will be left:

$$\tfrac{1}{2}(1.00\ g) = 0.500\ g$$

During the next 8 days, one half of the remaining $^{131}_{53}I$ will decay. This means that after 16 days we will have:

$$\tfrac{1}{2}(0.500\ g) = 0.250\ g$$

The same fraction of $^{131}_{53}I$ will decay in the next half-life period. So, after 24 days, there will be left:

$$\tfrac{1}{2}(0.250\ g) = 0.125\ g$$

How long is it before all the $^{131}_{53}I$ is gone?

This process continues through successive half-lives. The amount of $^{131}_{53}I$ drops off with time, approaching zero after a large number of half-lives.

Our discussion of the decay of $^{131}_{53}I$ is summarized in Table 26.1. Looking at the last two columns, you can see that there is a simple rela-

TABLE 26.1 RATE OF DECAY OF $^{131}_{53}I$ (HALF-LIFE = 8 DAYS)

Time (Days)	Amount Left (g)	Fraction Left	Number of Half-Lives
0	1.00		
8	0.500	$1/2$	1
16	0.250	$(1/2)^2 = 1/4$	2
24	0.125	$(1/2)^3 = 1/8$	3
32	0.0625	$(1/2)^4 = 1/16$	4
⋮		⋮	⋮
8n		$(1/2)^n$	n

tion between the fraction left and the number of half-lives. This relation is:

$$\text{fraction left} = (\tfrac{1}{2})^n \qquad (26.19)$$

where n is the number of half-lives. We can apply Equation 26.19 to the decomposition of any radioactive species (Example 26.6).

EXAMPLE 26.5

The isotope $^{90}_{38}Sr$ is radioactive. It decays with a half-life of 29 years. If we start with a sample weighing 0.120 g, after 87 years:

a. What fraction of the sample will be left?

b. How many grams of $^{90}_{38}Sr$ will remain?

SOLUTION

a. 87 years is three half-lives ($87/29 = 3$). So, the fraction left will be:

$$(1/2)^3 = 1/8$$

b. $\frac{1}{8}(0.120 \text{ g}) = 0.0150$ g

Half-lives of different isotopes vary widely. Uranium-238, $^{238}_{92}U$, decomposes very slowly. It has a half-life of 4.5×10^9 yr. As you can imagine, the intensity of radiation from $^{238}_{92}U$ is very low. Alpha particles are given off very slowly. At the opposite extreme, $^{234}_{91}Pa$ decays very rapidly. It has a half-life of only about 1 min. If you are exposed to this isotope at the instant it is formed, you will be subject to very intense radiation. However, within an hour at the most the hazard will be over. Virtually all of the $^{234}_{91}Pa$ will be gone.

About half the $^{238}_{92}U$ present when the earth was formed is still present.

The most dangerous isotopes are those with intermediate half-lives. Consider, for example, $^{90}_{38}Sr$, with a half-life of 29 yr. This is short enough to give a high intensity of radiation. At the same time, you cannot wait until it disappears. As we saw in Example 26.5, 1/8 of it will remain after 87 yr. This is more than enough to be dangerous, even if you start with a small amount of strontium-90.

Sources of Radiation

Isotopes with short half lives tend to emit large amounts of energy per gram.

Table 26.2 lists the sources and properties of several radioactive isotopes found in the environment. As you can see, many of them occur naturally. Typically, these isotopes have long half-lives. They have to, if they are still around after billions of years. Taken together, radiation from natural sources accounts for about 70% of the total to which we are exposed.

Some radioactive nuclei come from cosmic radiation. One such nucleus is $^{14}_{6}C$. It is produced by the reaction of $^{14}_{7}N$ nuclei in the air with neutrons.

$$^{14}_{7}N + ^{1}_{0}n \rightarrow ^{14}_{6}C + ^{1}_{1}H \quad (26.20)$$

The neutrons are formed by nuclear reactions taking place in outer space. The intensity of cosmic radiation varies with altitude. It is higher in Denver than at sea level. Again, your exposure is greater in a jet plane at 10 km than it is on the ground.

An important source of radiation is the fall-out from the testing of nuclear weapons. As we will see in Chapter 27, the fission reaction produces large amounts of radioactive isotopes. Many of these have half-

TABLE 26.2 SOURCES AND PROPERTIES OF RADIOACTIVE ISOTOPES IN THE ENVIRONMENT

Isotope	Half-life	Type of Decay	Source
$^{87}_{37}Rb$	4.7×10^{10} yr	beta	natural
$^{232}_{90}Th$	1.4×10^{10} yr	alpha (gamma)	natural
$^{238}_{92}U$	4.5×10^{9} yr	alpha	natural
$^{40}_{19}K$	1.3×10^{9} yr	beta (gamma)	natural
$^{14}_{6}C$	5.8×10^{3} yr	beta (gamma)	cosmic, fall-out
$^{226}_{88}Ra$	1.6×10^{3} yr	alpha (gamma)	natural
$^{137}_{55}Cs$	30 yr	beta (gamma)	fall-out
$^{90}_{38}Sr$	29 yr	beta	fall-out
$^{3}_{1}H$	12 yr	beta	cosmic
$^{7}_{4}Be$	53 d	*	cosmic
$^{131}_{53}I$	8 d	beta (gamma)	fall-out

*This isotope decays by electron capture. One of the electrons in the first energy level falls into the nucleus, converting a proton to a neutron. A gamma ray is emitted.

lives of several years. They stay around in the environment for a long time. One of the most dangerous is strontium-90, $^{90}_{38}Sr$. Recall that strontium is in the same group of the Periodic Table as calcium. If strontium-90 is taken into the body, it substitutes for calcium in the bones. There it slowly decays, giving off intense radiation for years.

For many years, Russia, the United States, and other countries carried out an extensive program of testing nuclear weapons. The concentration of dangerous isotopes such as $^{90}_{38}Sr$ rose rapidly in the atmosphere. Public concern led to the first Nuclear Test Ban Treaty, signed in 1963. Since then, the level of radioactivity in the atmosphere has been falling slowly. Now, many people are worried about radiation from a different source. This involves the radioactive isotopes produced by fission reactors in power plants. The safe disposal of these highly active waste products is a problem we will consider in Chapter 27.

26.4 SUMMARY

In this chapter, we have looked at a few of the substances that humanity has added to its environment. Virtually all of these are present in very small amounts. Yet many of them raise hazards to our health and well-being. The air we breathe contains certain toxic gases. The ones discussed here are all non-metal oxides. One of these, CO, comes almost entirely from automobiles. Two others, SO_2 and SO_3, come mainly from the burning of coal. The oxides of nitrogen, NO and NO_2, are byproducts of all sorts of combustion processes.

Trace metals in the environment can have both good and bad effects. Many of them are needed for life processes. Others can be extremely toxic. Mercury and lead are particularly dangerous. At the part per million level, they cause serious damage to the brain and nervous system.

The hazards of nuclear radiation are well-established. This radiation comes from radioactive isotopes in the environment. Most of these arise from natural processes which we cannot control. Some of the most dangerous, however, result from man's activities. We have looked in some detail at the types and rates of decay of these isotopes.

Throughout this chapter, we have emphasized ways of reducing pollution. Sometimes this is relatively easy. Industry can stop dumping mercury into lakes and rivers. Food additives suspected to cause cancer can be eliminated. Usually, however, there are substantial costs involved in reducing pollution. Removing SO_2 from stack gases or CO from automobile exhaust costs money. Often, the cost of removing pollutants cannot be measured in dollars alone. When anti-pollution devices make combustion less efficient, they contribute to the energy crisis. Here, as in most areas of society, there is no "free ride".

NEW TERMS

alpha particle
beta particle
gamma radiation
half-life
nuclear equation
parts per million
pollutant
radioactive decay
radioactivity

QUESTIONS

1. Give the formulas of five different gases that are air pollutants.

2. Of the pollutants you listed in Question 1, choose:

   a. two that are produced directly in an automobile engine.
   b. one that is produced directly in the combustion of coal.
   c. two that are produced by the reaction of O_2 with other pollutants.

3. Rainfall near industrial areas is often acidic. It may contain H_2SO_4 or HNO_3 or both. Explain, by means of chemical equations, how these two acids are formed in polluted air.

4. One way to remove SO_2 from the gas coming out of smokestacks is to add MgO (Reaction 26.5). Another is to add calcium carbonate. The $CaCO_3$ first decomposes to form calcium oxide. This in turn reacts with SO_2. Write balanced equations for:

   a. the decomposition of $CaCO_3$ to calcium oxide.
   b. the reaction of calcium oxide with SO_2 (similar to 26.5).

5. Carbon monoxide is formed by incomplete combustion of any hydrocarbon. Write a balanced equation for the combustion of $CH_4(g)$ to give CO(g) and $H_2O(l)$.

6. Explain why carbon monoxide is toxic, even at very low concentrations.

7. Write a balanced equation for the reaction that occurs in a catalytic converter to remove CO.

8. Photochemical smog contains O_3 molecules and O atoms. Explain, using balanced equations, where these species come from (NO_2 is involved in the process).

9. Give the symbols of four trace metals that are poisonous.

10. Mercury spills are particularly dangerous near radiators or other sources of heat. Can you suggest why this should be the case?

11. Give the formula of a toxic organic compound of

    a. Hg b. Pb

12. In the text, two sources of lead poisoning were discussed. The number of cases of poisoning from both sources is expected to decrease in the future. Why?

13. What is an alpha particle made up of? It is identical to the nucleus of what kind of atom? To what is a beta particle identical?

14. Explain why alpha, beta, and gamma radiation behave differently in an electrical field.

15. Which of the following processes result in an increase in atomic number? a decrease? no change?

    a. α-emission b. β-emission c. γ-emission

16. Which of the processes listed in Question 15 increase the mass number? decrease the mass number? do not change the mass number?

17. Arrange the three types of radiation, alpha, beta, and gamma, in order of decreasing penetrating power.

18. When a beta particle is emitted, what happens to the number of protons in the nucleus? the number of neutrons?

19. Two isotopes, A and B, decay with half-lives of 10 yr and 10 min in that order. If you start off with equal amounts of the two isotopes:

    a. Which one gives off the more intense radiation?
    b. Which one lasts longer?

20. We pointed out in Chapter 17 that reaction rate ordinarily decreases as time passes. Is this true of radioactive decay? (Remember that reaction rate describes how the amount of reactant changes in unit time.)

21. What are the three common sources of nuclear radiation in the environment?

PROBLEMS

1. *Mass Relations in Equations (Example 26.1)* Consider Reaction 26.1. Starting with one kilogram of coal containing 2.5% of FeS_2:

   a. How many moles of FeS_2 are present?
   b. How many moles of SO_2 are formed?
   c. How many grams of SO_2 are formed?

2. *Parts per Million (Example 26.2)* The maximum allowed concentration of mercury in drinking water is 0.005 ppm. How many grams of mercury are allowed in:

   a. one gram of water b. one liter of water (d = 1.0 g/cm^3)

3. *Parts per Million (Example 26.2)* The concentration of lead in the dust around the highways in New York City is estimated to be 1.0 mass percent.

   a. How many grams of lead are there in one gram of dust?
   b. What is the concentration of lead in parts per million?

4. *Nuclear Equations (Example 26.3)* Write a balanced nuclear equation for alpha emission by:

   a. $^{235}_{92}U$ b. $^{232}_{90}Th$

5. *Nuclear Equations (Example 26.4)* Write a balanced equation for beta emission by:

   a. $^{137}_{55}Cs$ b. $^{90}_{38}Sr$

6. *Rate of Decay (Example 26.5)* Consider tritium, $^{3}_{1}H$, which has a half-life of 12 yr. Starting with 2.40 g of tritium, how much will be left after:

   a. 12 yr b. 36 yr

7. *Rate of Decay (Example 26.5)* A certain radioactive isotope of beryllium, $^{7}_{4}Be$, has a half-life of 53 days. What fraction of a sample of this isotope will be left after:

   a. 53 days? b. 530 days?

* * * * *

8. Using Table 26.2, as it applies to potassium-40:

   a. Write a balanced nuclear equation for its decay.
   b. Calculate the fraction left after 5.2×10^9 yr.

9. On your worksheet, complete and balance the following nuclear equation.

$$^{28}_{13}Al \rightarrow ? + ^{28}_{14}Si$$

   What type of decay is involved here?

10. On your worksheet, complete and balance the following nuclear equation.

$$^{27}_{13}Al + ^{1}_{0}n \rightarrow ^{1}_{1}H + ?$$

11. A certain radioactive isotope has a half-life of 20 min. What fraction of it *has decayed* after:

    a. 40 min b. 60 min

12. The concentration of the radioactive isotope of potassium, $^{40}_{19}K$, in the body is about 3 ppm. How many grams of this isotope are there in a person weighing 50 kg?

13. Coal from Montana typically contains about 0.5 mass percent sulfur. What is the concentration of the sulfur in the coal in parts per million?

14. How many grams of CO would be formed from the combustion of 1.00 g of C_8H_{18} by Reaction 26.6?

*15. Consider the equilibrium shown by Equation 26.7. What is the ratio $[CO]/[O_2]$ when 20% of the hemoglobin is in the form of the CO complex? (This is the level at which death occurs from carbon monoxide poisoning.)

*16. A certain fish weighing 1.3 kg is found to contain 0.0024 g of dimethyl mercury, $(CH_3)_2Hg$.

    a. How many grams of Hg are present?
    b. What is the concentration of Hg, in parts per million?

*17. Plot the data in Table 26.1 (amount left on the Y axis, time on the X axis). Draw a smooth curve through the points. From the curve, estimate how much iodine-131 is left after 10 days.

*18. A child eats a piece of paint weighing 11 g. The paint contains 5.0% by mass of red lead, Pb_3O_4.

    a. How many grams of lead are there in this chip of paint?
    b. If the child weighs 10 kg, by how much will this increase the concentration of Pb^{2+} in his body, in parts per million? (Actually, much of the lead is eliminated in the urine, so the effect is not as great as this calculation suggests.)

One energy source not discussed in this chapter is geothermal energy. Steam from deep beneath the earth supplies the power for this electrical generating plant near San Francisco. *Department of Energy Photo.*

27

ENERGY RESOURCES

Since prehistoric times, fuels have been used as a source of energy. Combustion of materials such as wood and coal gives off energy in the form of heat. Prior to 1700, almost all of this heat was used directly to keep people warm and cook their food. The mechanical energy needed for agriculture and transportation came from a different source. Fields were cleared, homes were built, and carriages were drawn by the muscles of people and their domestic animals.

In 1705, Thomas Newcomen built the first steam engine. This opened a new era in the use of energy. For the first time, heat could be converted to mechanical energy on a large scale. Over the years, many different kinds of engines have been developed for this purpose. They furnish the energy required to run an automobile or generate electricity. Modern civilization requires cheap, abundant energy sources. You use about 100 times as much energy as your ancestor of 200 years ago.

The Industrial Revolution really began in 1705.

Virtually all of our energy today comes from the combustion of *fossil fuels.* These include coal, natural gas, and petroleum. All of these react with oxygen to give off heat. Typical reactions are:

$$C(s) + O_2(g) \rightarrow CO_2(g) \quad ; \quad \Delta H = -94 \text{ kcal} \tag{27.1}$$

$$CH_4(g) + 2\ O_2(g) \rightarrow CO_2(g) + 2\ H_2O(l) \quad ; \quad \Delta H = -213 \text{ kcal} \tag{27.2}$$

$$C_8H_{18}(l) + \tfrac{25}{2}\ O_2(g) \rightarrow 8\ CO_2(g) + 9\ H_2O(l); \quad \Delta H = -1308 \text{ kcal} \tag{27.3}$$

For 200 years, we have used fossil fuels at an ever-increasing rate. Until recently, we took this source of energy for granted. Only within the past decade have we become aware of the "energy crisis". Prices have jumped as we have become more and more dependent upon foreign imports of these fuels. This is a problem that will not go away. There are no easy solutions. If your children are to enjoy a standard of living comparable to yours, new sources of energy must be found. Moreover, we must make better use of the sources we now have.

27.1 ENERGY SOURCES; CONSUMPTION AND CONSERVATION

When we use a fuel, we convert energy from one form to another.

We hear a great deal nowadays about how rapidly we are "consuming energy". Strictly speaking, no one ever consumes energy. According to the Law of Conservation of Energy, that is impossible. The energy released when you burn gasoline in your car doesn't disappear. Sooner or later, it appears in the surroundings as heat. In a practical sense, however, this energy is "lost". You can't use it to refill your tank with gasoline. Perhaps it would be better to say that it is the *energy source,* gasoline, which is consumed. "Conserving energy" has a similar meaning. When you turn down the thermostat at night, you conserve the *energy source,* heating oil, natural gas, or whatever.

TABLE 27.1 ENERGY USAGE FOR SELECTED NATIONS (1970)

	Total (kcal/yr)	Population	Per Person (kcal/yr)
United States	17.1×10^{15}	0.203×10^{9}	84×10^{6}
Canada	1.3	0.022	60
Great Britain	2.9	0.054	54
West Germany	2.4	0.059	40
U.S.S.R.	9.4	0.240	39
Japan	1.7	0.112	15
Mexico	0.6	0.056	11
India	1.1	0.540	2
World	49×10^{15}	3.61×10^{9}	14×10^{6}

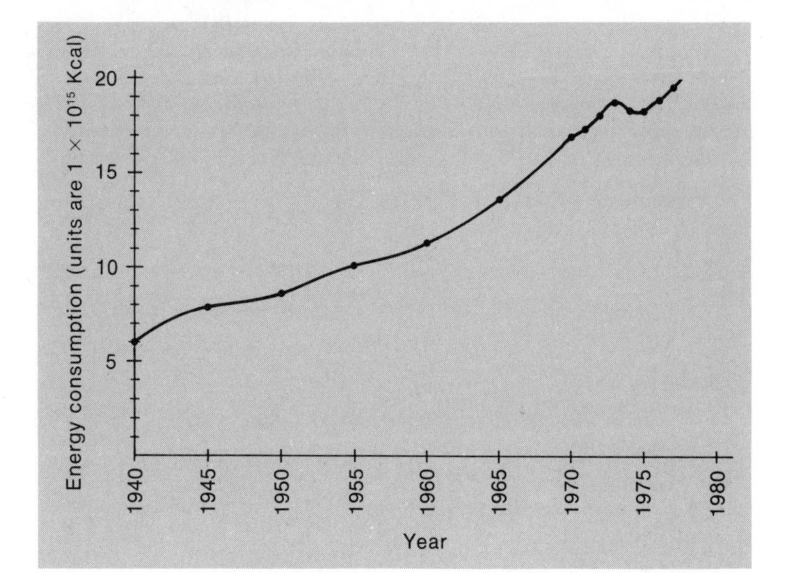

FIGURE 27.1

Graph showing energy consumption in the U. S. since 1940. In 1972 the consumption was three times that in 1940. Energy usage has levelled off considerably since the oil embargo in 1973.

In 1970, the world consumed its energy sources at the rate of 49×10^{15} kcal/yr. The United States, with only 6% of the world's population, accounted for about 35% of this consumption. Table 27.1 gives energy usage figures for various countries. Notice that, on the average, each person in the United States uses 84×10^6 kcal of energy each year. This compares with a worldwide average of 14×10^6 kcal/yr.

As you might expect, other countries resent this.

Figure 27.1 shows how energy usage in the United States has increased. Our demand for energy doubles in about 20 years. This amounts to an annual growth rate of about 4%. In other words, we use about 4% more energy each year than we did the year before. Notice that there was a slight dip in the curve in 1974, the year after the oil embargo. However, energy usage in the United States soon started to increase again, to a higher level than before. In contrast, Japan and West Germany are using *less* energy than they did in 1973. Clearly these countries and many others have been much more successful in "conserving energy" than the United States.

Table 27.2, p. 666 shows the purposes for which energy is used in the United States. Nearly 25%, about 4.2×10^{15} kcal/yr, goes for transportation. Of this, about ⅔ is used to operate automobiles. As you can see from Figure 27.2, p. 666 the typical automobile, carrying one person, is one of the least efficient means of transportation. Carpooling or the use of buses would save a considerable amount of energy. Another way to "conserve energy" is to improve the fuel economy of automobiles. If all cars driven in the United States today got 30 miles to the gallon instead of an average of 15, we could cut gasoline consumption by 50%.

Someday our cars may well do better than 30 mpg.

Another area in which we can easily "conserve energy" is in space heating. About 11% of all the energy used in the United States goes to heat our homes. Most of this heat comes from burning oil or natural gas. Turning the thermostat down saves fuel and in that sense conserves energy. Better insulation has the same effect. Space heating requirements could be reduced by one third if all buildings were prop-

TABLE 27.2 USES OF ENERGY IN THE UNITED STATES (1970)

Item	Energy Used (10^{15} kcal/yr)		% of Total	
Transportation	4.22		24.7	
Autos		2.75		16.1
Others		1.47		8.6
Space Heating	3.03		17.7	
Homes		1.88		11.0
Other		1.15		6.7
Industry	7.23		42.3	
Production of Steam		2.80		16.4
Direct heat		1.88		11.0
Electric drive		1.38		8.1
Fuels as Raw Materials		0.96		5.6
Electrolysis		0.21		1.2
Water Heating	0.68		4.0	
Air Conditioning	0.50		2.9	
Refrigeration	0.39		2.3	
Cooking	0.21		1.2	
Others*	0.84		4.9	
	17.1			

*includes clothes drying, small appliances, lighting, . . .

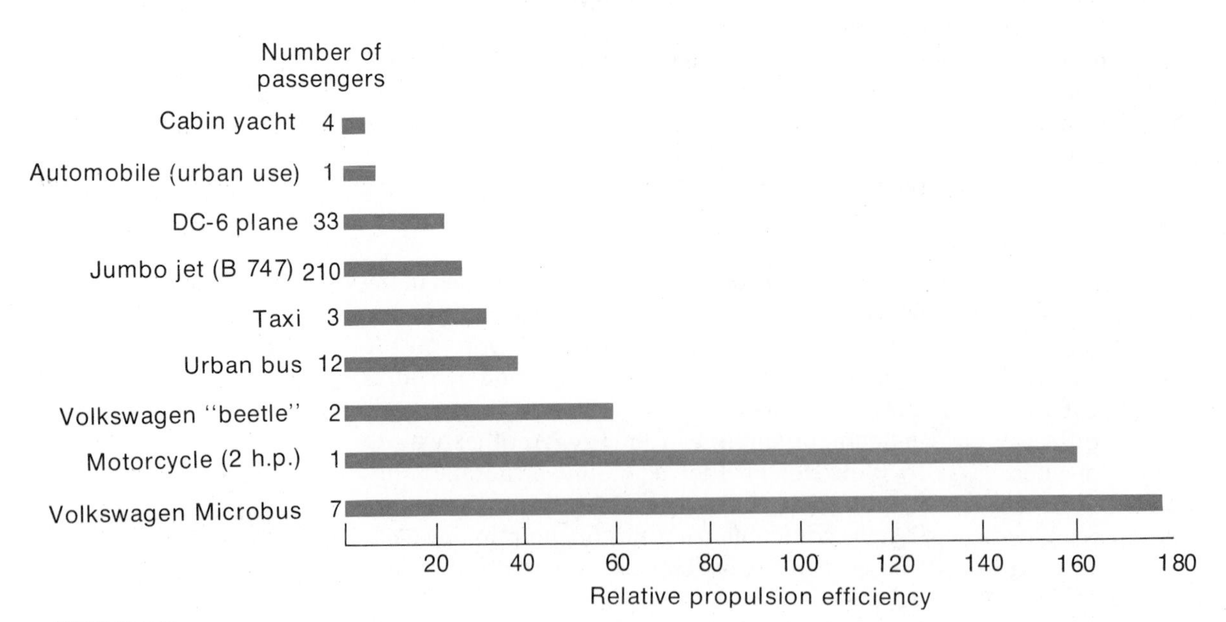

FIGURE 27.2

Relative efficiencies of different kinds of motor vehicles, listed in terms of passenger mile per gallon of gasoline. The efficiency goes up dramatically as more passengers are accommodated. Adapted from *Technology Review*, Jan. 1972, p. 34.

erly insulated. This would also reduce the use of energy for air conditioning. The fraction of total energy used for this purpose is rather small, about 2.9% in 1970. However, it is growing very rapidly. In 1960, the figure was only about 1.5%.

27.2 PETROLEUM AND NATURAL GAS

Figure 27.3 shows how energy sources used in the United States have changed since 1900. As the years have passed, we have become more and more dependent upon petroleum and natural gas. In 1900 only 5% of the energy used in the United States came from these sources. Today that fraction is above 70%.

There are several reasons for the trend shown in Figure 27.3.

1. Liquid and gaseous fuels are easier and cheaper to transport than solid fuels such as wood or coal. Petroleum and natural gas can be shipped by tankers or pipeline. Coal is ordinarily shipped by railroad car.

2. The automobile engine uses petroleum (gasoline) as a fuel. Other sources of energy are possible (Figure 27.4 p. 668). However, so far, steam driven or electrical cars have not proved practical.

3. For space heating, petroleum (heating oil) and natural gas are more convenient than solid fuels. If you have ever used a woodstove, you probably know how much labor is involved. Your parents, or perhaps your grandparents, can tell you about the "joys" of operating a coal furnace. Shoveling in the coal and taking out the ashes were no fun.

4. At least in the period between 1950 and 1970, petroleum and natural gas were by far the cheapest fuels. Recently, this situation has changed. In recent years, prices of gasoline and heating oil have increased fivefold.

The wood stove business is booming in Minnesota.

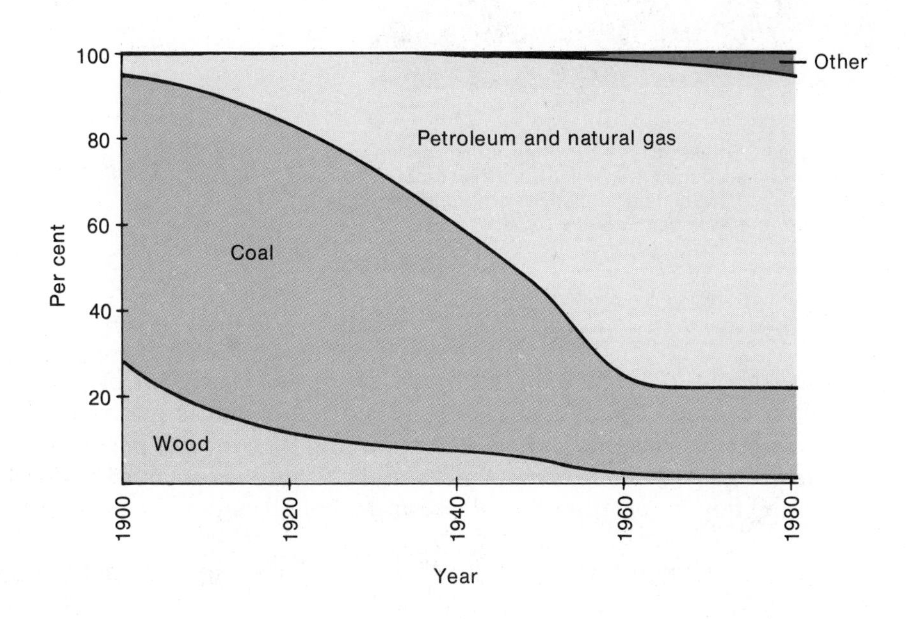

FIGURE 27.3

Relative importance of different fuels in the U. S. since 1900. Since 1950 more than half the fuel consumed has been either natural gas or petroleum. With the current fuel shortage, coal is likely to become much more important in the next few years.

FIGURE 27.4

Picture *A,* taken in 1897, shows Francis and Freelan Stanley in their steam driven automobile (Stanley Steamer). The steam was produced in a boiler from the combustion of coal. *B,* an electrically driven car, is a test vehicle designed for urban driving. *A: courtesy of the Motor Vehicle Manufacturers Association of the United States, Inc. B: AiResearch Manufacturing Company.*

Until about 1950, the United States produced nearly all of the petroleum it used. Today, nearly 50% of our petroleum is imported. For economic reasons, the United States is trying to reduce its dependence on foreign petroleum. So far, the results have not been encouraging. Despite the flow of oil from the Alaska pipeline, oil imports continue to increase.

There is a simple reason why the United States is importing petroleum. We are running out of this fuel, and natural gas as well. It is estimated that, without imports, our reserves would last only about 20 years. Actually, this figure is optimistic. It is based on the current rate of consumption. Unfortunately, this rate has been increasing by about 3% each year.

There is a further problem so far as petroleum and natural gas are concerned. World reserves of these fuels are limited. They amount to about 400×10^{16} kcal. Each year, about 5×10^{16} kcal are used. Simple division gives a time limit of 80 years:

$$\frac{400 \times 10^{16}\ \text{kcal}}{5 \times 10^{16}\ \text{kcal/yr}} = 80\ \text{yr}$$

This does not allow for future increases in consumption. Everything considered, 50 years is probably more realistic. Within your lifetime, the world will effectively run out of these fuels. In the remainder of this chapter, we will consider how this crisis may be met.

Things must change in the area of energy and soon.

27.3 COAL

As reserves of petroleum and natural gas dwindle, we might turn to other fossil fuels. One such fuel, coal, has been used for centuries. Almost certainly, we will burn more of it in the future than we do today.

Most of the coal mined today is used to produce electrical energy. More than half of the power plants in the United States burn coal as a fuel. Most of the remainder operate with oil or gas. We could reduce consumption of these scarce fuels by 10% if all power plants converted to coal. This would be desirable from the standpoint of conserving natural resources. The energy equivalent of coal reserves in the United States is 33 times that of petroleum and natural gas.

Unfortunately, coal has certain disadvantages as a fuel. Most coal deposits in the eastern United States have a high sulfur content (~2 to 3%). When this coal is burned, large amounts of sulfur dioxide are produced. As we saw in Chapter 26, SO_2 is a dangerous air pollutant.

Mining coal can also have a bad effect upon the environment. In the coal-mining areas of Pennsylvania and West Virginia, there are huge piles of waste. These have been building for 100 years. Besides being unsightly, they are a fire hazard. Low-grade coal in the piles can catch fire and smoulder for long periods of time. Strip mining of coal can also leave ugly scars (Figure 27.5, p. 670). Coal companies are now required to restore the land affected. The results, as you can see from the figure, are impressive. The cost adds only a few cents per ton to the price of the coal.

Good restoration programs require adequate rainfall.

Gaseous and Liquid Fuels from Coal

Since coal is a solid fuel, it is an unlikely energy source for transportation. For this use, it needs to be converted to a liquid or gaseous fuel. In principle at least, this can be done. If solid carbon is heated

FIGURE 27.5

Photograph *A* shows a strip-mined area 20 years after grading and seeding, completely reforested. Photograph *B* shows the original condition of the land. Note the landmark poplar, just to the right of center; everything else has changed. *Courtesy of the National Coal Association.*

with steam at 700°C, the following reaction occurs:

$$C(s) + H_2O(g) \rightarrow CO(g) + H_2(g) \quad (27.4)$$

The product is a mixture of carbon monoxide and hydrogen called "water gas". It is a reasonably good fuel with a heating value somewhat less than natural gas. If coal is used instead of carbon, a similar but more complex reaction occurs. The gas produced, called *synthesis gas,* contains about:

40 mol % H_2, 15 mol % CO, 15 mol % CH_4, 30 mol % CO_2

Synthesis gas can serve as a fuel. Upon burning, it produces water and carbon dioxide.

$$H_2(g) + \tfrac{1}{2} O_2(g) \rightarrow H_2O(l) \qquad \Delta H = -68.3 \text{ kcal} \qquad (27.5)$$

$$CO(g) + \tfrac{1}{2} O_2(g) \rightarrow CO_2(g) \qquad \Delta H = -67.7 \text{ kcal} \qquad (27.6)$$

$$CH_4(g) + 2\ O_2(g) \rightarrow CO_2(g) + 2\ H_2O(l) \qquad \Delta H = -212.8 \text{ kcal} \qquad (27.7)$$

EXAMPLE 27.1

Using the equations just written and the composition given above, calculate the amount of heat given off when one mole of synthesis gas burns.

SOLUTION

In one mole of this gaseous fuel, there is:

0.40 mol H_2, 0.15 mol CO, 0.15 mol CH_4, 0.30 mol CO_2

For home heating, synthesis gas may be used instead of natural gas.

heat evolved H_2: $0.40 \text{ mol} \times 68.3 \dfrac{\text{kcal}}{\text{mol}} = 27.3 \text{ kcal}$

heat evolved CO: $0.15 \text{ mol} \times 67.7 \dfrac{\text{kcal}}{\text{mol}} = 10.2 \text{ kcal}$

heat evolved CH_4: $0.15 \text{ mol} \times 212.8 \dfrac{\text{kcal}}{\text{mol}} = 31.9 \text{ kcal}$

The CO_2 does not burn and so makes no contribution. The total heat evolved is:

$$27.3 \text{ kcal} + 10.2 \text{ kcal} + 31.9 \text{ kcal} = 69.4 \text{ kcal}$$

This is only about ⅓ as much heat as that evolved when a mole of methane burns. On a mole basis, synthesis gas is not as good a fuel as natural gas, which is mostly methane. On the other hand, we are running out of natural gas. We have enough coal for a long time to come.

For the future, the greatest value of synthesis gas may be to make liquid fuels. If CO and H_2 are heated with the proper catalyst, they form liquid hydrocarbons. These are similar to those found in gasoline. A typical reaction is:

$$6\ CO(g) + 13\ H_2(g) \rightarrow C_6H_{14}(l) + 6\ H_2O(l) \qquad (27.8)$$

This process was used in Germany during World War II to make a low-grade gasoline. In general, the product is more expensive than ordinary gasoline. However, as the price of petroleum rises, this process begins to look more attractive. Quite possibly, it will be a major source of liquid fuels 20 years from now. Pilot plants are already in operation.

27.4 NUCLEAR ENERGY

Today, about 10% of the electricity produced in the United States comes from nuclear energy. This is the energy given off in certain nuclear reactions. Such reactions were considered briefly in Chapter 27. You will recall that they involve rearrangements within the nucleus. Protons and neutrons, rather than electrons, are involved.

Nuclear reactions differ from ordinary chemical reactions in several ways. For one thing, the energy change is much larger. The amount of energy evolved per gram of reactant ranges from 100,000 to 100,000,000 kcal. This is many times greater than the energy given off when fossil fuels burn. Let us now consider where this energy comes from.

Mass-Energy Relation

Energy changes in any reaction can be interpreted in terms of mass relationships. Modern physics tells us that there is a change in mass in a reaction. The products have a slightly different mass than the reactants. In exothermic reactions, the products weigh slightly less than the reactants. In a sense, mass is "converted" to energy in such reactions.

The energy change in a reaction, ΔE, is directly proportional to the change in mass, Δm. That is:

$$\Delta E = \text{constant} \times \Delta m$$

The constant in this equation can be expressed in many different units. To relate ΔE in kilocalories to Δm in grams, we express the constant in kcal/g. In these units it has the value 2.15×10^{10} kcal/g.

We can get a lot of energy from a small amount of mass.

$$\Delta E = 2.15 \times 10^{10} \frac{\text{kcal}}{\text{g}} \times \Delta m \qquad (27.9)$$

The use of this equation is shown in Example 27.2.

EXAMPLE 27.2

Suppose that in a certain reaction, the products weigh one microgram (10^{-6} g) less than the reactants. Calculate:

a. Δm in grams b. ΔE in kilocalories

SOLUTION

a. Since the products weigh *less* than the reactants, Δm is *negative:*

$$\Delta m = \text{mass products} - \text{mass reactants}$$
$$= -1.00 \times 10^{-6} \text{ g}$$

b. $\Delta E = 2.15 \times 10^{10} \frac{\text{kcal}}{\text{g}} \times (-1.00 \times 10^{-6}\ \text{g}) = -2.15 \times 10^{4}\ \text{kcal}$

Since ΔE is *negative*, energy is *evolved*. To be exact, 21,500 kcal of energy is evolved.

In an "ordinary" chemical reaction, Δm is extremely small. Indeed, it is too small to be detected with the most sensitive balance (Example 27.3). On the other hand, in a nuclear reaction, Δm is significant. As much as one milligram of mass may be "lost" per gram of starting material. For this reason, the energy change in nuclear reactions is much greater than that in ordinary reactions.

EXAMPLE 27.3

Using Equation 27.9, calculate Δm for:

a. a chemical reaction in which $\Delta E = -10$ kcal (this is about equal to the energy change for the combustion of a gram of fossil fuel).

b. a nuclear reaction in which $\Delta E = -1.0 \times 10^{7}$ kcal.

This is enough energy to heat a house for 100 years.

SOLUTION

Solving Equation 27.9 for Δm:

$$\Delta m = \frac{\Delta E}{2.15 \times 10^{10}\ \text{kcal/g}}$$

a. $\Delta m = \frac{-10\ \text{kcal}}{2.15 \times 10^{10}\ \text{kcal/g}} = -4.7 \times 10^{-10}\ \text{g}$

This change is much too small to be detected. The most sensitive balance cannot weigh to better than 1×10^{-6} g.

b. $\Delta m = \frac{-1.0 \times 10^{7}\ \text{kcal}}{2.15 \times 10^{10}\ \text{kcal/g}} = -4.7 \times 10^{-4}\ \text{g}$

Here, the change in mass is much greater. If we started with one gram of reactant, the products would weigh:

$$1.00000\ \text{g} - 0.00047\ \text{g} = 0.99953\ \text{g}$$

In principle, any spontaneous nuclear reaction can serve as a source of energy. In practice, only two such reactions are suitable for large-scale use. These are:

—*nuclear fission*, in which a heavy nucleus splits into two smaller nuclei.

—*nuclear fusion*, in which two very small nuclei combine with each other.

We will now consider these two processes.

The Fission Process

Until about 40 years ago, only one type of nuclear reaction was known. This was radioactivity, discussed in Chapter 26. Here, you will recall, an unstable nucleus emits a small particle (electron, alpha particle). The product nucleus differs only slightly from that of the reactant. The number of protons in the nucleus changes by only one or two units.

This discovery had many unforeseen consequences.

In 1938, two German scientists, Hahn and Strassman, discovered a new kind of nuclear reaction. They exposed uranium (atomic number 92) to neutrons. Among the products was barium (atomic number 56). At first, they were reluctant to believe that this had really happened. A colleague, Lisa Meitner, interpreted their results. She pointed out that there could be only one explanation. Uranium atoms had undergone fission. That is, they had split into at least two fragments.

The discovery of nuclear fission came on the eve of World War II (1939–1945). This process became the object of intense research in Germany, Great Britain, Canada, and the United States. Very quickly, several facts were established.

1. Nuclear fission produces enormous amounts of energy. There is a significant decrease in mass when a very heavy atom splits into lighter atoms (Figure 27.6). In the fission of one gram of uranium, there is a decrease in mass of about one milligram. That is, the total mass of the fission products is about 0.999 g. According to Equation 27.9, a change in mass of one milligram (10^{-3} g) corresponds to about 2×10^7 kcal of energy.

2. Only one isotope of uranium, $^{235}_{92}U$, splits apart when struck by neutrons. This isotope makes up only 0.7% of natural uranium. The more abundant isotope, $^{238}_{92}U$, does not take part in the reaction.

3. The fission of $^{235}_{92}U$ gives many different products. One atom may split to isotopes of barium (at. no. = 56) and krypton (at. no. = 36). Another may yield isotopes of cesium (at. no. = 55) and rubidium (at. no. = 37). These reactions can be written as:

These reactions are both fast.

$$^{235}_{92}U + ^{1}_{0}n \nearrow ^{146}_{56}Ba + ^{87}_{36}Kr + 3\,^{1}_{0}n \quad (27.10)$$

$$^{235}_{92}U + ^{1}_{0}n \searrow ^{144}_{55}Cs + ^{90}_{37}Rb + 2\,^{1}_{0}n \quad (27.11)$$

Many other reactions are possible. More than 35 different elements have been detected in the fission products.

4. Fission produces excess neutrons ($^{1}_{0}n$). Two or three neutrons are formed for every one required to get the reaction started. Once a few $^{235}_{92}U$ nuclei split, enough neutrons are produced to split many more. This keeps the reaction going. Indeed, unless some of the neutrons are removed, the reaction gets out of control. This, of course, is what happens in an atomic bomb such as that exploded over Hiroshima in 1945.

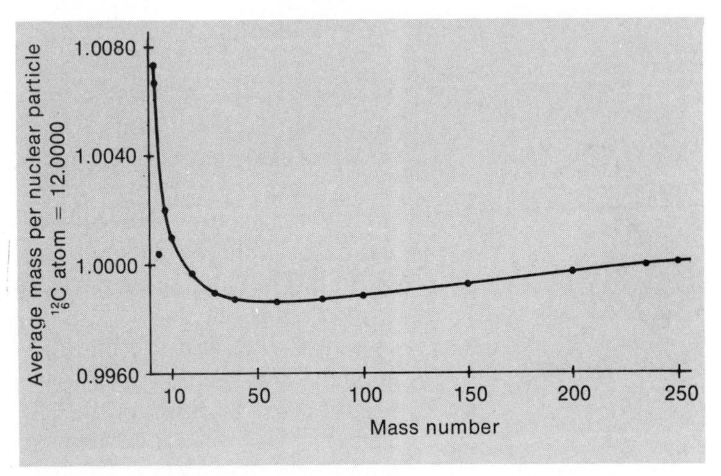

FIGURE 27.6

Graph showing average mass per nuclear particle as a function of mass number. Nuclei with masses between 50 and 100 have the lowest average masses per nuclear particle. This means that when heavy nuclei like uranium undergo fission to form nuclei with lower masses, there is a net decrease in mass. This loss is associated with the large amount of energy that is released in the process.

EXAMPLE 27.4

Two other possible fission reactions are:

a. $^{235}_{92}U + ^{1}_{0}n \rightarrow ^{146}_{57}La + ^{87}_{35}Br + ?\ ^{1}_{0}n$

b. $^{235}_{92}U + ^{1}_{0}n \rightarrow ^{160}_{62}Sm + ? + 4\ ^{1}_{0}n$

Balance these nuclear equations by filling in the blanks on the right.

SOLUTION

Both atomic number and mass number must be balanced.

a. Here, atomic number is already balanced.

left: 92 right: 57 + 35 = 92

However, mass number is not balanced:

left: 235 + 1 = 236 right: 146 + 87 = 233

Three neutrons are needed to balance this equation:

$$^{235}_{92}U + ^{1}_{0}n \rightarrow ^{146}_{57}La + ^{87}_{35}Br + 3\ ^{1}_{0}n$$

b. In this case, neither atomic number nor mass is balanced. The atomic number of uranium, 92, is 30 units greater than that of samarium, 62. Clearly we need an isotope of atomic number 30 on the right. Referring to the Periodic Table, we see that zinc has an atomic number of 30. At this stage we have:

$$^{235}_{92}U + ^{1}_{0}n \rightarrow ^{160}_{62}Sm + ^{?}_{30}Zn + 4\ ^{1}_{0}n$$

To find the mass number of the zinc isotope, we add mass numbers on the two sides:

left: 235 + 1 = 236 right: 160 + 4(1) = 164

Since: 236 − 164 = 72, we conclude that the mass number of the zinc isotope must be 72. The final balanced equation is:

$$^{235}_{92}U + ^{1}_{0}n \rightarrow ^{160}_{62}Sm + ^{72}_{30}Zn + 4\ ^{1}_{0}n$$

Fission Reactors

At the present time, there are about 70 nuclear power plants operating in the United States. These plants are often referred to as fission reactors. They convert the heat given off by the fission of $^{235}_{92}U$ into electrical energy. The heat is used to vaporize water to steam. The steam in turn drives an electrical generator. The reactor is designed so that the fission rate remains constant. In this way, the possibility of an explosion is avoided. Energy is produced at a constant rate until all of the $^{235}_{92}U$ fuel has reacted.

Figure 27.7 is a diagram of a common type of reactor. The fuel elements consist of stainless steel tubes. These tubes contain pellets of uranium oxide enriched in $^{235}_{92}U$. Control rods are located side-by-side with the fuel elements. They are made of cadmium or boron, both of which absorb neutrons. The position of the control rods can be adjusted. In an emergency, they can be dropped to absorb more neutrons and stop the reaction.

If the neutrons are absorbed, they can't cause fission.

The heat generated by fission is transferred to water under pressure. This water, which becomes radioactive, circulates through a closed loop. The heat is given off to another stream of water (right side

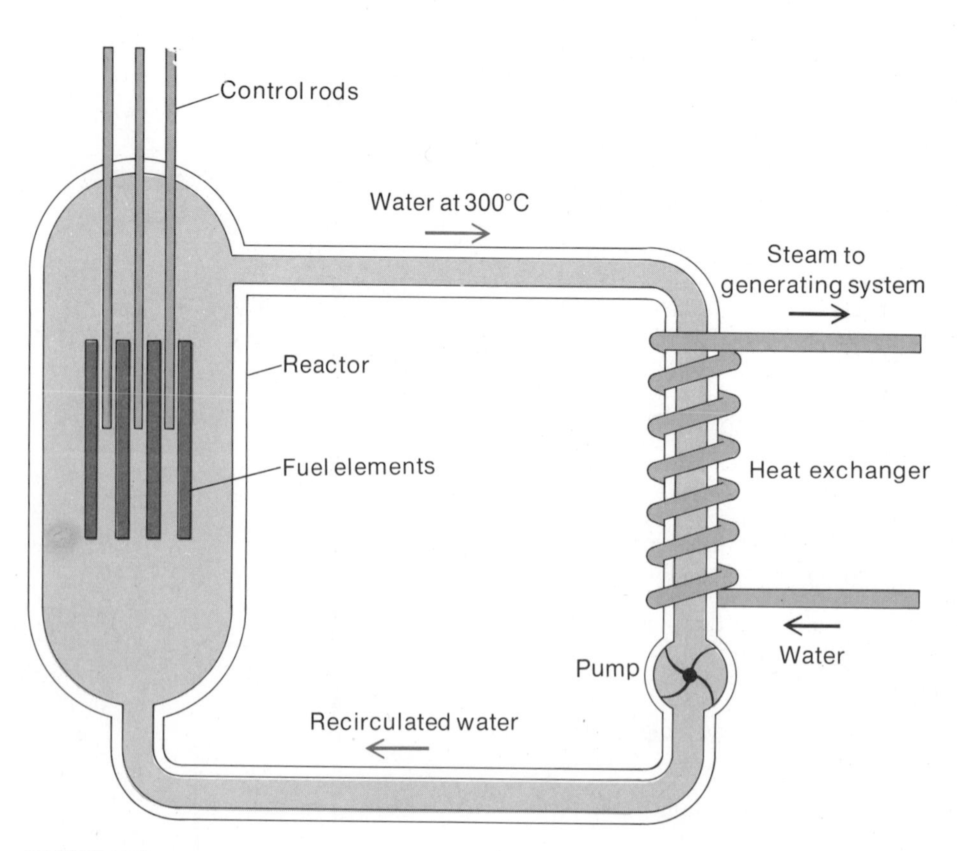

FIGURE 27.7

A nuclear reactor of the pressurized water type. Enough neutrons are emitted by uranium-235 nuclei undergoing fission to maintain a chain reaction. The control rods absorb just enough neutrons to keep the process self-sustaining.

of Figure 27.7). This water is converted to steam. The entire reactor is surrounded by lead blocks or other shielding material.

Nuclear power plants have several advantages. The more important of these are listed below.

1. Very little fuel is required. One gram of $^{235}_{92}U$ produces as much energy as 3 tons of coal or 2 tons of heating oil. This means that fuel cost is a relatively small factor in nuclear power plants.

2. The level of air pollution is greatly reduced. Fossil fuels produce appreciable amounts of oxides of sulfur and ash particles. A nuclear plant emits virtually nothing to the atmosphere.

3. Increased use of fission reactors is a possible way for the United States to reduce consumption of oil and natural gas. Conversion to coal will take many years. Other energy sources, to be discussed later in this chapter, are still in the development stage.

Balanced against these advantages of fission energy are certain drawbacks. All of these are related to a simple fact. The products of nuclear fission are highly radioactive. Consider, for example, the $^{90}_{37}Rb$ produced by Reaction 27.11. It decomposes by a series of steps. Two of these are:

$$^{90}_{37}Rb \rightarrow ^{90}_{38}Sr + ^{0}_{-1}e; \text{ half-life} = 2.8 \text{ min} \qquad (27.12)$$

$$^{90}_{38}Sr \rightarrow ^{90}_{39}Y + ^{0}_{-1}e; \text{ half-life} = 29 \text{ yrs} \qquad (27.13)$$

The isotope $^{90}_{38}Sr$ ("strontium-90") is particularly dangerous. In the form of $SrCO_3$, it is readily taken into the bones, replacing $CaCO_3$. Since it has a half-life of 29 years, it stays around for a long time. A child exposed to strontium-90 could still be suffering from its effects 60 years later. About ¼ of this isotope would remain.

The radiation given off by fission products poses a hazard for several reasons.

1. There is always a chance of an accident. A fire or earthquake at a nuclear reactor could release huge amounts of radioactive material. In the 20 years that fission reactors have been operating, this has not happened. However, no one can guarantee that it never will.

The accident at Three Mile Island was an example of what can occur. No one was killed, but the cooling system did fail and the reactor had to be shut down for major overhaul.

2. Terrorists or saboteurs might be able to obtain radioactive material. The greatest danger comes when spent fuel elements are being transported.

3. The radioactive wastes must be stored for long periods of time. At present, they are dissolved in water and buried in underground tanks. Eventually, it is planned to evaporate the water to obtain solids. These will be stored in abandoned salt mines, perhaps a mile below the earth. Let's hope our descendants know where these deposits are. The residues will still be radioactive hundreds of years from now.

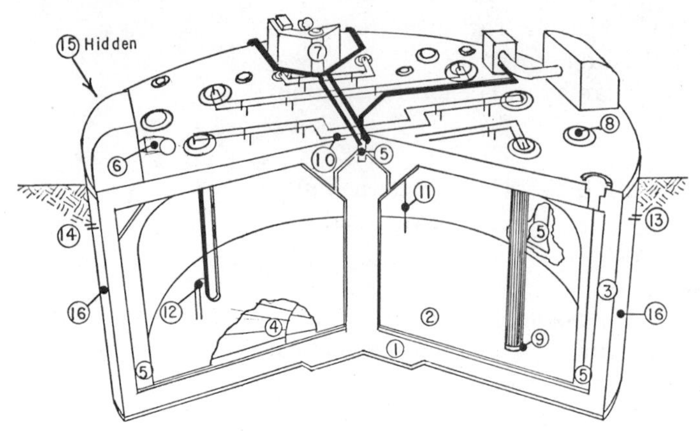

HIGH LEVEL WASTE STORAGE TANK

CAPACITY 1,300,000 GALLONS

TOP & INSIDE

1. CONCRETE
2. PRIMARY TANK
3. SECONDARY TANK
4. INSULATING REFRACTORY
5. ANNULUS AIR SUPPLY
6. ANNULUS EXHAUST
7. TANK EXHAUST
8. REMOVABLE PLUGS
9. PUMP OUT JET
10. RECIRCULATING COOLING WATER
11. INSTRUMENT PROBE
12. PUMP OUT JET

OUTSIDE

13. INLET PIPE
14. INLET-OUTLET PIPE
15. STEAM PIPE
16. EARTH

FIGURE 27.8

These tanks are used to store radioactive waste at the U.S. Department of Energy's Savannah River plant. They hold about 4000 cubic meters of liquid or solid waste. The tanks are equipped with cooling coils to remove heat produced by radioactive decay of fission products. E. I. duPont de Nemours & Co.

Breeder Reactors

The fission reactors now in use rely upon the rare isotope of uranium, ${}^{235}_{92}U$. If we depend upon fission to meet increasing demands for energy, sources of this isotope will soon be exhausted. This could happen by the year 2000, perhaps sooner. For fission to make a long-range contribution, it will be necessary to use $^{238}_{92}U$. There is about 150 times as much of this isotope as there is of $^{235}_{92}U$.

The heavy isotope of uranium does not undergo fission directly. However, some of it is converted to an isotope of plutonium (at. no. = 94) by neutron bombardment.

$$^{238}_{92}U + ^{1}_{0}n \rightarrow ^{239}_{94}Pu + 2\ ^{0}_{-1}e \qquad (27.14)$$

This isotope of plutonium can undergo fission. A typical reaction is:

$$^{239}_{94}Pu + ^{1}_{0}n \rightarrow ^{147}_{56}Ba + ^{90}_{38}Sr + 3\ ^{1}_{0}n \qquad (27.15)$$

Reaction 27.15, like 27.10 and 27.11, produces an excess of neutrons. These can react with $^{238}_{92}U$ to give $^{239}_{94}Pu$. In a sense, this process "breeds" fissionable material. Reactors using this process are called breeder reactors.

At first sight, breeder reactors are a very attractive energy source.

With breeder reactors, we could satisfy our energy needs for at least 100 years. Unfortunately, there are problems. In the first place, these devices have had a relatively poor safety record. In an early model (1955), the fuel elements melted and the reactor was destroyed. A similar but less serious accident occurred in 1966.

There are no breeder reactors producing electrical energy on a large scale at the present time. One is being built at Clinch River, Tennessee. There is considerable debate as to whether it should be completed. In part, this is because of the problems already described. There are other difficulties. Plutonium is probably the most toxic of all known materials. In microgram (10^{-6} g) amounts it causes cancer. Moreover, since it is fissionable, there are security problems. Anyone who obtained enough plutonium from a power plant would have the ingredients of an atomic bomb. There is a growing consensus among scientists that breeder reactors are the energy source of "last resort". That is, they should be used only if, 20 years from now, there is no better replacement for oil and natural gas.

The Fusion Process

Let's consider again the graph shown in Figure 27.6, p. 675. This time, we focus on the left side of the graph. Notice that there is a sharp decrease in mass when very small nuclei combine to give a larger nucleus. It follows that this process, called fusion, evolves a large amount of energy. A typical fusion reaction is:

$$^{2}_{1}H + ^{2}_{1}H \rightarrow ^{4}_{2}He \qquad (27.16)$$

When deuterium nuclei ($^{2}_{1}H$) combine to form an alpha particle ($^{4}_{2}He$), about 1×10^{8} kcal of energy is produced per gram of starting material. This is five times as much energy as is given off in fission. It is 10,000,000 times as large as the energy change for the combustion of a gram of petroleum.

The fusion reaction has many advantages as an energy source. In

FIGURE 27.9

Section of a Tokomak fusion chamber. Reactors of this kind may one day furnish large amounts of energy with essentially no pollution.

contrast to fission, the products are stable isotopes. This means that the radiation hazard is greatly reduced. Moreover, deuterium is much more abundant than uranium. One out of every 6000 hydrogen atoms is a deuterium atom. The oceans contain about 2×10^{23} g of deuterium. This is enough to meet our energy needs for many centuries to come.

Unfortunately, there is a catch. Fusion reactions have very high activation energies. Two deuterium nuclei, each with a +1 charge, repel each other strongly at close distances. To overcome this repulsion, the nuclei have to be given an energy of 3×10^6 kcal/mol. Energies of this order are difficult to retain long enough for fusion to take place. In any ordinary container, the nuclei would quickly lose energy by colliding with the walls. This stops the fusion process.

A possible design for a fusion reactor is shown in Figure 27.9. Here, the charged nuclei are confined in a very strong magnetic field. So far, with this technique, it has been possible to maintain fusion for 1/50 of a second. This period will have to be extended to at least one second if nuclear fusion is to be a practical energy source. The prospects are uncertain at best. Even optimists agree that fusion reactors will not be in use before the year 2000.

At the moment, the only practical fusion reactors are the sun and other stars. The sun's energy comes from a series of reactions, starting with ordinary hydrogen nuclei. The overall process is believed to be:

In the sun the temperatures and densities are large enough for fusion reactions to occur.

$$4\ {}^1_1\text{H} \rightarrow {}^4_2\text{He} + 2\ {}^0_1\text{e} \tag{27.17}$$

(${}^0_1$e is the nuclear symbol for a *positron,* a particle which has the same mass as an electron but the opposite charge). Every day, this reaction taking place on the sun produces huge quantities of energy. Of this, about 1.5×10^{18} kcal reaches the surface of the earth each day.

27.5 SOLAR ENERGY

The number just quoted, 1.5×10^{18} kcal, seems enormous. Indeed it is. The amount of solar energy received each day is 30 times the annual requirement of the earth. Solar energy might seem to be the answer to the energy crisis. There are, however, a couple of problems.

1. Sunlight is a very diffuse form of energy. Consider, for example, that the area of the earth is about 5×10^{14} m². Hence the amount of energy striking an area of one square meter in one day is:

$$\frac{1.5 \times 10^{18}\ \text{kcal}}{5 \times 10^{14}} = 3000\ \text{kcal}$$

This is only about as much energy as that obtained by burning 300 g of petroleum. To collect appreciable amounts of solar energy, large areas are necessary. This probably rules out the use of central power stations or heating plants. Very likely, each homeowner will have to collect his or her own solar energy.

2. Sunlight is not a constant source of energy. It is available only during daylight hours on sunny days. This means that a "back-up" source of energy is needed when the sun isn't shining.

At present, the largest use of solar energy is to heat water. Figure 27.10 shows a diagram of a water heater that is used in Israel, Japan,

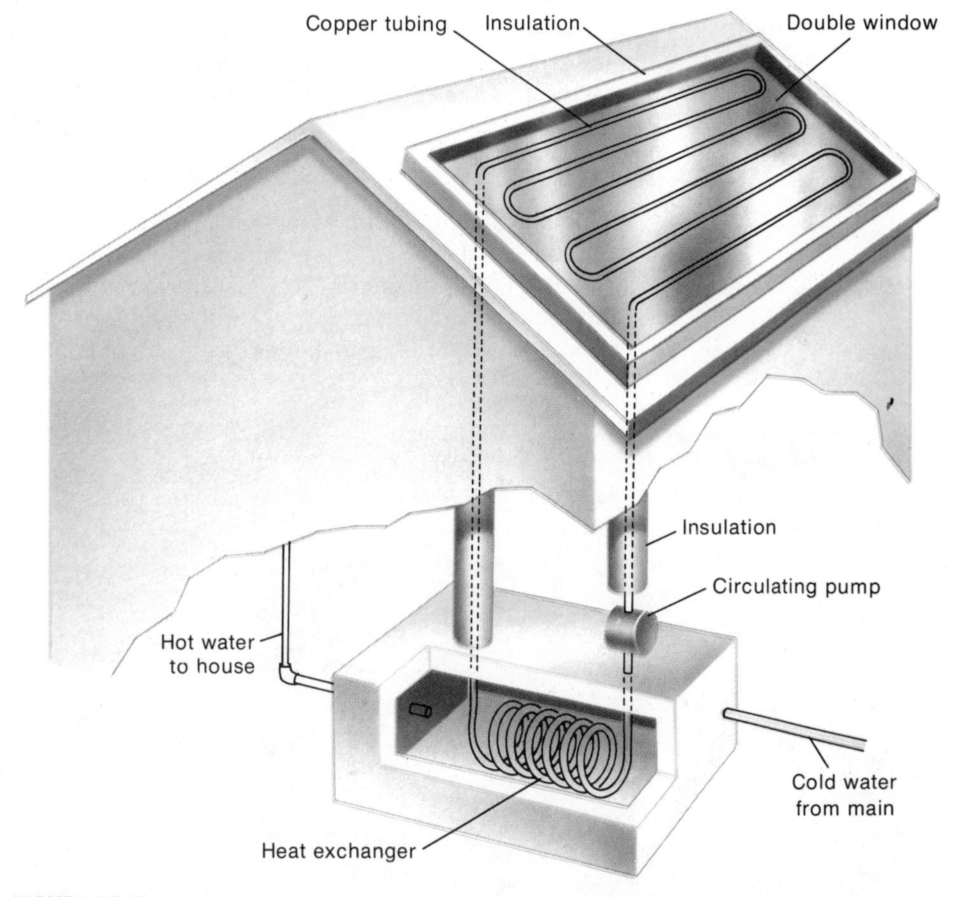

FIGURE 27.10

Diagram of a solar water heater. Water passing into the solar energy collector is warmed to a predetermined temperature by the sun and then pumped into the hot water tank in the cellar. Devices of this sort are used in several countries and are starting to appear in the U.S.

and New Zealand. The collector is a shallow metal tank. The base of the tank is painted black to absorb as much sunlight as possible. The glass cover exerts a "greenhouse" effect. That is, it allows sunlight to pass through, but prevents radiant heat from escaping. Water is circulated through the collector while the sun is shining. After a few passes, the water reaches a temperature of at least 65°C (150°F). This water is stored in a well-insulated tank. It can be drawn off as needed for bathing, cleaning, and so on.

In principle the system shown in Figure 27.10 can be adapted to heat a house. The solar-heated water is circulated through radiators as in an ordinary hot water heating system. A much larger storage tank is required. Larger collectors, perhaps covering one entire side of a roof, are needed.

At present there is much research on solar energy. A breakthrough in one or more areas could be very important in helping solve our energy problems.

Solar home heating is slowly becoming more popular. Ten years ago, only a few experimental systems were in use. Today thousands of homes and office buildings are heated wholly or in part by solar energy. By the year 2000, it is possible that sunlight may be the source of energy for at least half of all space heating. This could reduce our use of petroleum and natural gas by 10%.

The prospects for solar energy in areas other than heating are not encouraging. Ideally, we would like to convert sunlight into electrical energy. One way to do this uses a *solar cell* (Figure 27.11). Sunlight striking the cell ejects electrons from atoms.

$$X \xrightarrow{\text{sunlight}} X^+ + e^-$$

This creates a separation of charge and hence leads to a flow of current. A voltage of about 0.5 V is produced directly.

Solar cells are used in the space program, where cost is no object. They are made of extremely pure silicon or germanium. These materials cost about $60 a kilogram. Other light sensitive materials are now avail-

FIGURE 27.11

An assembly of solar cells used to convert sunlight into electrical energy. The cells are made of very pure silicon and are about 7½ cm in diameter. Department of Energy photo.

able. One of these is cadmium sulfide, CdS. It is less expensive but also less efficient. A cadmium sulfide solar cell converts less than 10% of the sunlight striking it into electrical energy. At the present time, electrical energy from any type of solar cell is much too expensive for general use. There must be a breakthrough in either cost of materials or efficiency to make these cells practical.

27.6 SUMMARY

In the United States today, petroleum and natural gas are the major sources of energy. Together they supply 75% of our energy needs. Coal, burned mostly in electrical power plants, accounts for another 22%. Energy from water power and nuclear fission together supply the remaining 3%.

In the year 2000, these figures will be different. Fossil fuels will probably continue to be the major source of energy. Very likely, coal will make a greater contribution than it does now. Virtually all of the petroleum and natural gas used in the United States will be imported. Supplies of $^{235}_{92}U$, the isotope now used in fission reactors, may well be running out. Solar energy may be a major factor in heating, but probably not for other purposes.

It is impossible to make reliable predictions about energy sources in the 21st century. However, one thing is certain. By the middle of that century, the world's deposits of petroleum and natural gas will be gone. It would seem that there are four possible replacements.

1. *Coal* The energy available from coal deposits is much greater than that from oil and gas. Probably, by the year 2000, it will be economical to produce gaseous and liquid fuels from coal. Quite possibly, you will one day use these fuels in your automobile and your home.

2. *Nuclear Fusion* We can hope that this source will be feasible by the year 2000. This is not the case at present. If a practical fusion reactor can be developed, our energy needs will be met for centuries to come. Deuterium, required for fusion, is abundant in nature.

3. *Solar Energy* By the year 2000, it may be practical to produce electrical energy from sunlight. If this happens, it will probably be through the use of solar cells. The problem is to reduce the cost and increase the efficiency of these devices.

4. *Breeder Reactors* These use the abundant isotope of uranium, $^{238}_{92}U$, as a fuel. They could meet our energy needs for at least another century. However, they pose a greater safety hazard than any other energy source.

NEW TERMS

breeder reactor
fission
fission reactor
fusion
mass-energy relation
solar energy
synthesis gas
water gas

QUESTIONS

1. About 16% of the energy we use goes for automobiles. Describe at least three ways in which the amount of energy used for this purpose could be reduced.

2. About 11% of the energy we use goes to heat homes and places of business. Suggest ways in which the use of energy for this purpose could be reduced.

3. Since 1900, the use of petroleum in the United States has increased greatly. What are some of the reasons for this change?

4. About how long would U.S. reserves of petroleum last without imports? How long are worldwide reserves likely to last?

5. What is the major use of coal?

6. What are some of the disadvantages of coal as a fuel, compared to petroleum or natural gas?

7. What substances are present in water gas? synthesis gas? How are these gaseous mixtures made?

8. Describe a method for preparing a liquid fuel from coal. Why is it important to convert coal to a liquid fuel?

9. A certain reaction absorbs energy. What is the sign of Δm for this reaction?

10. How does the change in mass in a nuclear reaction compare to that in an ordinary reaction? the change in energy?

11. Explain what is meant by nuclear fission; nuclear fusion.

12. Using Figure 27.6, explain why energy is released in fission; in fusion.

13. Which of the following undergo fission?

    a. $^{235}_{92}U$ b. $^{238}_{92}U$ c. $^{2}_{1}H$ d. $^{239}_{94}Pu$

14. In a nuclear reactor to produce electrical energy, why is it important to control the flow of neutrons?

15. Explain how fission energy is converted into electrical energy in a nuclear reactor.

16. What are some of the advantages of using nuclear energy rather than petroleum or coal to generate electricity?

17. What is the principal problem associated with fission reactors?

18. Explain the principle behind the breeder reactor.

19. Compare a breeder reactor to one using $^{235}_{92}U$ with respect to:

 a. fuel used b. safety

20. List at least two advantages of fusion as opposed to fission as an energy source. What is the main problem that must be solved for fusion to be a practical source of energy?

21. The energy of the sun comes from which one of the following sources?

 a. solar cells b. fission c. fusion d. windmills

22. What are two problems with the direct use of solar energy?

23. Solar energy is most readily used to:

 a. heat water b. run automobiles c. generate electricity

24. Explain briefly how a solar cell works. Why is so little use made of this energy-producing device?

25. How do you think the fraction of our total energy obtained from each of the following sources will change by the year 2000?

 a. petroleum b. coal c. solar energy d. water power
 e. nuclear fission f. nuclear fusion

PROBLEMS

1. *Heating Value (Example 27.1)* Using Equations 27.5 and 27.6, calculate the heating values, in kilocalories per gram, of:

 a. H_2 b. CO

2. *Heating Value (Example 27.1)* How much heat is evolved in the combustion of one mole of a gaseous fuel containing 50 mol% H_2 and 50 mol% CO?

3. *Heating Value (Example 27.1)* For the reaction:

$$2\,C_4H_{10}(g) + 13\,O_2(g) \rightarrow 8\,CO_2(g) + 10\,H_2O(l); \qquad \Delta H = -1376\text{ kcal}$$

 What is the heating value of butane, C_4H_{10}, in kcal/g?

4. *Mass-Energy Relation (Examples 27.2, 27.3)* In a certain nuclear reaction, ΔE is -6.0×10^6 kcal.

 a. Calculate Δm
 b. If the reactants weigh exactly one gram, what is the mass of the products?

5. *Mass-Energy Relation (Examples 27.2, 27.3)* Using Equation 27.9, calculate ΔE for a reaction in which Δm is:

 a. $+1.2 \times 10^{-10}$ g b. -1.6×10^{-3} g

6. *Nuclear Equations (Example 27.4)* Balance the nuclear equation:

$$^{239}_{94}Pu + ^{1}_{0}n \rightarrow ^{90}_{37}Rb + ? + 3\,^{1}_{0}n$$

7. *Nuclear Equations (Example 27.4)* A possible fusion reaction combines tritium, $^{3}_{1}H$, with $^{6}_{3}Li$ to give a single product. What is the nuclear symbol of the product?

* * * * *

8. Using Equations 27.5 and 27.6, calculate the heat evolved when one mole of a gas containing 30 mol% H_2, 30 mol% CO, and 40 mol% N_2 is burned (the nitrogen does not burn).

9. When one mole of methyl alcohol, CH_3OH, is burned, 174 kcal of heat is evolved. What is the heating value of methyl alcohol, in kcal/g?

10. For the combustion of carbon and carbon monoxide, ΔH is -94 kcal/mol and -68 kcal/mol in that order. Which fuel has the higher heating value in kcal/g?

11. A common fission reaction is:

$$^{235}_{92}U + ^{1}_{0}n \rightarrow ^{140}_{53}I + ? + 4\ ^{1}_{0}n$$

On a separate sheet of paper, complete and balance this equation.

12. A fusion reaction being studied is:

$$^{2}_{1}H + ? \rightarrow ^{4}_{2}He + ^{1}_{0}n$$

On your worksheet, complete and balance this equation.

13. In a certain nuclear reaction, starting with one gram of reactants, 0.1% of the mass is "lost". Calculate ΔE for the reaction.

14. For the nuclear reaction: $^{226}_{88}Ra \rightarrow ^{4}_{2}He + ^{222}_{86}Rn$ the molar masses are:

$^{4}_{2}He = 4.0015$ g; $^{222}_{86}Rn = 221.9703$ g; $^{226}_{88}Ra = 225.9771$

Starting with one mole of radium, calculate:

a. Δm b. ΔE

15. Referring to Table 27.1, how much energy would the world use if the value per person were equal to that in:

a. the United States b. Canada

16. Referring to Table 27.2:

a. What fraction of the energy used for transportation goes for autos?
b. What fraction of the energy used for space heating goes to homes?

17. It is estimated that oil and gas reserves in the United States amount to about 2.5×10^{17} kcal. At present, the United States uses about 1.7×10^{16} kcal of energy each year. Seventy-five percent of this comes from oil and gas. If these figures remain constant, how long would reserves last, without imports?

18. Each day the earth receives 1.5×10^{18} kcal of energy from the sun. The annual use of energy in the world is 4.9×10^{16} kcal. If all the energy from the sun falling on the earth in one day could be trapped, how long would it supply our energy needs?

19. To produce a million kilocalories of useful heat, you can burn 44 gal of heating oil at $1.00 a gallon, or (in an airtight stove) 0.40 cord of wood at $75 a cord. Which is the cheaper energy source?

*20. Reaction 27.8 is only one of several that occur when CO and H_2 react. Write a balanced equation for the reaction of CO with H_2 to give C_8H_{18} and H_2O.

*21. In the United States today, we use about 1.7×10^{16} kcal of energy per year. Each year, this number increases by about 3%. At this rate, how much energy will we be using 5 years from now? Repeat this calculation, assuming an increase of 1% per year.

*22. For the fusion reaction: $2\ {}^{2}_{1}H \rightarrow {}^{4}_{2}He$, the nuclear masses are: ${}^{2}_{1}H = 2.01355$ g; ${}^{4}_{2}He = 4.00150$ g. How much energy is evolved in the fusion of one gram of deuterium?

*23. To heat an average size house in New England requires about 2×10^{5} kcal of energy a day in winter. The amount of solar energy striking one square meter in one day is 3×10^{3} kcal. If a solar collector converts 80% of solar energy into heat, how large must its area be to supply all the heat needed for a house in New England?

Appendix 1
REVIEW OF MATHEMATICS

The mathematics that you will use in this course is relatively simple. You may be asked to:

1. Make calculations involving exponential numbers such as 6.022×10^{23}
2. Solve simple algebraic equations such as $P_2V_2 = P_1V_1$ for a quantity such as V_2.
3. Draw graphs of relationships such as $V = \text{constant} \times T$.

In this appendix, we will discuss each of these topics briefly. They are treated in more detail in "Elementary Mathematical Preparation for General Chemistry", Masterton and Slowinski, W. B. Saunders Co., 1974.

EXPONENTIAL NUMBERS

In chemistry, we often deal with very large or very small numbers. The number of atoms in one gram of carbon is:

$$50{,}150{,}000{,}000{,}000{,}000{,}000{,}000$$

At the opposite extreme, the mass of a single carbon atom is:

$$0.000\ 000\ 000\ 000\ 000\ 000\ 000\ 019\ 940\ \text{g}$$

Numbers such as these are tedious to write and awkward to work with. For this reason, we often express them in **exponential notation.** Here, the quantity is written in the form

$$C \times 10^n$$

In this expression, C is a number between 1 and 10 such as 2, 2.62, 5.30 or 8.1. The exponent n is a positive or negative integer such as 1, 2, −1, or −2. The following numbers are written in exponential notation:

$$2 \times 10^1,\ 2.62 \times 10^2,\ 5.30 \times 10^{-1},\ 8.1 \times 10^{-2}$$

These are easily translated into ordinary numbers, provided you know what the exponent means. The following rules apply.

1. **Where n is a positive integer, the expression 10^n means "mul-ply 10 by itself n times".** Thus:

$$10^1 = 10$$
$$10^2 = 10 \times 10 = 100$$
$$10^3 = 10 \times 10 \times 10 = 1000$$

2. **Where n is a negative integer, the expression 10^n means "mul-tiply $^1/_{10}$ by itself n times".** Thus:

$$10^{-1} = 0.1$$
$$10^{-2} = 0.1 \times 0.1 = 0.01$$
$$10^{-3} = 0.1 \times 0.1 \times 0.1 = 0.001$$

Applying these rules, we see that:

$$2 \times 10^1 = 2 \times 10 = 20$$
$$2.62 \times 10^2 = 2.62 \times 100 = 262$$
$$5.30 \times 10^{-1} = 5.30 \times 0.1 = 0.530$$
$$8.1 \times 10^{-2} = 8.1 \times 0.01 = 0.081$$

EXAMPLE 1

Express the following as ordinary numbers.

a. 6×10^4 b. 5.2×10^{-4}

SOLUTION

a. $10^4 = 10 \times 10 \times 10 \times 10 = 10{,}000$
Hence $6 \times 10^4 = 6 \times 10{,}000 = 60{,}000$

b. $10^{-4} = 0.1 \times 0.1 \times 0.1 \times 0.1 = 0.0001$
Hence $5.2 \times 10^{-4} = 5.2 \times 0.0001 = 0.00052$

To express an ordinary number in exponential notation, it must be written in the form $C \times 10^n$, where C is a number between 1 and 10 and n is a positive or negative integer. **To do this, first determine what C is. Then count the places that the decimal point must be moved to give C. This tells you the value of n. If the decimal point had to be moved to the left to obtain C, n is a positive integer. If it had to be moved to the right, n is a negative integer.**

Suppose you are asked to express the number 8162 in exponential notation. Here C is 8.162 (a number between 1 and 10). To obtain 8.162 from 8162, you had to move the decimal point 3 places to the left. Hence $n = +3$. That is:

$$8162 = 8.162 \times 10^3$$

Another example: How would you express 0.054 in exponential notation? Here:

$C = 5.4$
$n = -2$ (decimal point had to be moved 2 places to the right to obtain 5.4)
$0.054 = 5.4 \times 10^{-2}$

EXAMPLE 2

Express in exponential notation:

a. 62.6 b. 0.000192

SOLUTION

a. $C = 6.26$; $n = 1$ (to obtain 6.26 from 62.6, the decimal point must be moved 1 place to the *left*). $62.6 = 6.26 \times 10^1$.

b. $C = 1.92$; $n = -4$ (to obtain 1.92 from 0.000192, the decimal point must be moved 4 places to the *right*). $0.000192 = 1.92 \times 10^{-4}$.

Magnitude of Exponential Numbers

Consider the two exponential numbers:

$$6.4 \times 10^3 \text{ and } 3.1 \times 10^3$$

Which of these is the larger? Translating them into ordinary numbers:

$$6.4 \times 10^3 = 6.4 \times 1000 = 6400$$
$$3.1 \times 10^3 = 3.1 \times 1000 = 3100$$

Clearly 6.4×10^3 is larger than 3.1×10^3. This comparison illustrates the general rule:

In comparing two exponential numbers of the form $C \times 10^n$, where n is the same in both cases, the larger number is the one for which C is greater. Thus:

$$6.2 \times 10^2 > 5.4 \times 10^2 \qquad (620 > 540)$$
$$6.2 \times 10^{-2} > 5.4 \times 10^{-2} \qquad (0.062 > 0.054)$$

The comparison is a bit less obvious if the exponents differ. Consider the two exponential numbers 9×10^2 and 2×10^3. Realizing that:

$$9 \times 10^2 = 100 = 900$$
$$2 \times 10^3 = 2 \times 1000 = 2000$$

we see that 2×10^3 is larger than 9×10^2. The general rule is:

In comparing two exponential numbers of the form $C \times 10^n$, where n differs in the two cases, the larger number is the one with the larger value of n. Thus:

$$3 \times 10^2 > 4 \times 10^1 \qquad (300 > 40)$$
$$1.0 \times 10^3 > 9.9 \times 10^2 \qquad (1000 > 990)$$

This rule applies as well when one or both of the exponents are negative integers. Consider the exponential numbers:

$$1 \times 10^2 \text{ and } 4 \times 10^{-2}$$

In one case, $n = 2$; in the other, $n = -2$. Since $+2$ is larger than -2, we conclude that:

$$1 \times 10^2 > 4 \times 10^{-2} \qquad (100 > 0.04)$$

Again, compare the exponential numbers 3×10^{-2} and 3×10^{-3}. Since -2 is a larger number algebraically than -3;

$$3 \times 10^{-2} > 3 \times 10^{-3} \qquad (0.03 > 0.003)$$

EXAMPLE 3

Decide which exponential number in each of the following pairs is the larger. (Apply the two rules cited in this section.)

a. 4.2×10^{-4}, 3.6×10^{-4}

b. 3.9×10^6, 6.4×10^5

c. 4.1×10^{-12}, 1.8×10^{-10}

SOLUTION

a. n is the same; $4.2 > 3.6$. Hence: $4.2 \times 10^{-4} > 3.6 \times 10^{-4}$

b. n differs; $6 > 5$. Hence $3.9 \times 10^6 > 6.4 \times 10^5$

c. n differs; $-10 > -12$. Hence $1.8 \times 10^{-10} > 4.1 \times 10^{-12}$

Multiplying and Dividing Exponential Numbers

A major advantage of exponential notation is that it simplifies the processes of multiplication and division. To *multiply,* we *add exponents:*

$$10^1 \times 10^2 = 10^{1+2} = 10^3; \qquad 10^6 \times 10^{-4} = 10^{6+(-4)} = 10^2$$

To *divide,* we *subtract exponents:*

$$10^3/10^2 = 10^{3-2} = 10^1; \qquad 10^{-3}/10^6 = 10^{-3-6} = 10^{-9}$$

Using these principles, we arrive at the following rules.

To multiply one exponential number by another, first multiply the coefficients together. Then add exponents.

To divide one exponential number by another, first divide one coefficient by the other. Then subtract exponents.

EXAMPLE 4

Carry out the indicated operations.

a. $(5.00 \times 10^4) \times (1.60 \times 10^2)$

b. $(6.0 \times 10^{-3})/(2.4 \times 10^6)$

SOLUTION

a. For convenience, we first separate the coefficients from the exponential terms:

$$(5.00 \times 1.60) \times (10^4 \times 10^2)$$

Multiplying coefficients gives us: $5.00 \times 1.60 = 8.00$
Adding exponents gives us: $10^4 \times 10^2 = 10^6$
Hence: $(5.00 \times 10^4) \times (1.60 \times 10^2) = 8.00 \times 10^6$

b. Again, we separate the coefficients from the exponential terms:

$$\frac{6.0}{2.4} \times \frac{10^{-3}}{10^6}$$

Dividing coefficients gives us: $6.0/2.4 = 2.5$
Subtracting exponents gives us: $10^{-3}/10^6 = 10^{-9}$
Hence: $(6.0 \times 10^{-3})/(2.4 \times 10^6) = 2.5 \times 10^{-9}$

Sometimes, when exponential numbers are multiplied or divided, the answer is not in standard exponential notation. Suppose, for example, we multiply 5.0×10^4 by 6.0×10^3:

$$(5.0 \times 10^4) \times (6.0 \times 10^3) = (5.0 \times 6.0) \times (10^4 \times 10^3) = 30 \times 10^7$$

The number 30×10^7 is not in standard exponential notation. The coefficient, 30, is not a number between 1 and 10.

To express 30×10^7 in standard exponential notation, we:

—divide the coefficient by 10: $30/10 = 3.0$
—multiply the exponent by 10: $10^7 \times 10^1 = 10^8$

This leaves the value of the expression the same. That is:

$$30 \times 10^7 = 3.0 \times 10^8$$

EXAMPLE 5

Divide 4.1×10^5 by 8.2×10^3 and express your answer in standard exponential notation.

SOLUTION

Proceeding in the usual way, we have:

$$\frac{4.1}{8.2} \times \frac{10^5}{10^3} = 0.50 \times 10^2$$

To make the coefficient a number between 1 and 10, we multiply by 10: $0.50 \times 10 = 5.0$. To leave the expression unchanged, we must divide the exponent by 10: $10^2/10^1 = 10^1$
Hence:

$$\frac{4.1 \times 10^5}{8.2 \times 10^3} = 0.50 \times 10^2 = 5.0 \times 10^1$$

ALGEBRAIC EQUATIONS

The equations you will be asked to solve in this course are quite simple. They all involve a single variable (x) raised to the first power. These equations are solved by applying a basic principle of algebra which states that:

An equation remains valid if the same operation is performed on both sides.

In particular, we can:

1. *Add the same quantity to both sides.* For example, to solve the equation:

$$x - 4 = 6$$

we add 4 to each side: $x - 4 + 4 = 6 + 4$

simplifying: $x = 10$

2. *Subtract the same quantity from both sides.* To solve the equation:

$$x + 3 = 8$$

we subtract 3 from each side: $x + 3 - 3 = 8 - 3$

$$x = 5$$

3. *Multiply both sides by the same quantity.* To solve the equation:

$$\frac{x}{3} = 6$$

we multiply both sides by 3: $$\frac{3x}{3} = 6 \times 3; \; x = 18$$

4. *Divide both sides by the same quantity.* To solve the equation:

$$5x = 45$$

we divide both sides by 5: $$\frac{5x}{5} = \frac{45}{5}; \; x = 9$$

Frequently, two or more operations of this type must be carried out in succession. This is illustrated in Table 1. Examples 6 through 8 illustrate the use of these principles to solve equations that arise in chemistry.

TABLE 1 SOLUTIONS OF EQUATIONS IN ONE VARIABLE, x

Equation	Operation
1. $4x + 3 = 19$	
$4x = 16$	Subtract 3 from both sides
$x = 4$	Divide both sides by 4
2. $\frac{x}{3} - 2 = 5$	
$\frac{x}{3} = 7$	Add 2 to both sides
$x = 21$	Multiply both sides by 3
3. $5x - 2 = 3x + 6$	
$5x = 3x + 8$	Add 2 to both sides
$2x = 8$	Subtract 3x from both sides
$x = 4$	Divide both sides by 2
4. $5 = \frac{3}{x}$	
$5x = 3$	Multiply both sides by x
$x = 3/5 = 0.6$	Divide both sides by 5

EXAMPLE 6

The relationship between temperatures (Chapter 1) expressed in degrees Fahrenheit (°F) and degrees Celsius (°C) is:

$$°F = 1.8(°C) + 32°$$

a. Convert 25°C to °F

b. Convert 98.6°F (normal body temperature) to °C

SOLUTION

a. Here we can solve for °F by direct substitution.

$$\begin{aligned} °F &= 1.8(°C) + 32° \\ &= 1.8(25°) + 32° = 45° + 32° = 77° \end{aligned}$$

b. In this case, some simple algebra is involved. We might first solve the basic equation: °F = 1.8(°C) + 32° for °C:

$$1.8(°C) = °F - 32° \qquad \text{(subtract 32° from both sides)}$$

$$°C = \frac{°F - 32}{1.8} \qquad \text{(divide both sides by 1.8)}$$

Now we can substitute numbers to obtain the temperature in degrees Celsius:

$$°C = \frac{98.6° - 32°}{1.8} = \frac{66.6°}{1.8} = 37.0°$$

EXAMPLE 7

The molarity of a solution is given by the relation (Chapter 13):

$$\text{Molarity} = \frac{\text{number of moles of solute}}{\text{volume of solution in liters}}$$

Or, using symbols: $M = \frac{n}{V}$

a. Calculate the molarity of a solution containing 1.60 mol of solute in 10.0 ℓ of solution.

b. How many moles of solute are there in 2.50 ℓ of 3.20 M solution?

c. What volume of a 4.5 M solution is required to contain 0.90 mol of solute?

SOLUTION

a. By direct substitution: $M = \frac{1.60 \text{ mol}}{10.0 \ \ell} = 0.160 \frac{\text{mol}}{\ell}$

b. We first solve the equation for n, the number of moles of solute. To do this, we multiply both sides by V:

$$M = \frac{n}{V}$$

$$n = M \times V$$

Substituting: $n = 3.20 \frac{mol}{\ell} \times 2.50\ \ell = 8.00\ mol$

c. To solve the equation: $M = \frac{n}{V}$ for V, two operations are required.

$M \times V = n$ (multiply both sides by V)

$V = \frac{n}{M}$ (divide both sides by M)

Substituting: $V = \frac{0.90\ mol}{4.5 \frac{mol}{\ell}} = 0.20\ \ell$

EXAMPLE 8

The following equation relates the pressure, volume and temperature of a gas sample at two different sets of conditions (Chapter 5):

$$\frac{P_2V_2}{T_2} = \frac{P_1V_1}{T_1}$$

Solve this equation for

a. V_2 b. T_2

SOLUTION

a. Two steps are involved:

(1) Multiply both sides by T_2: $P_2V_2 = \frac{P_1V_1T_2}{T_1}$

(2) Divide both sides by P_2: $V_2 = \frac{P_1V_1T_2}{P_2T_1} = V_1 \times \frac{T_2}{T_1} \times \frac{P_1}{P_2}$

b. Several steps are required:

(1) Multiply both sides by T_2: $P_2V_2 = \frac{P_1V_1T_2}{T_1}$

(2) Divide both sides by P_1V_1: $\frac{P_2V_2}{P_1V_1} = \frac{T_2}{T_1}$

(3) Multiply both sides by T_1: $T_2 = T_1 \times \frac{P_2}{P_1} \times \frac{V_2}{V_1}$

GRAPHS

Many of the figures in this text are graphs. In several of these, only a single variable, y, is plotted. Such graphs are called *one-dimensional*. Only distances in one dimension are meaningful. An example of this type of graph is Figure 19.2, p. 481. Here the pH of various materials is shown on a scale along the vertical axis. A slightly more complex one-dimensional graph is shown in Figure 4.6, p. 88. Here heat content, H, is plotted on the vertical axis. The heat content of the "reactant", liquid water, is shown as a horizontal line at the left. That of the "product", water vapor, is shown at the right.

A more common type of graph shows the relationship between two variables, y and x. This is referred to as a *two-dimensional* graph. Distances in both vertical and horizontal dimensions have meaning. There are many such graphs in this text. A typical one is Figure 5.9, p. 117. Here volume of a gas sample is plotted on the vertical axis vs temperature on the horizontal axis. This happens to be a straight-line graph. We say that the plot is *linear*. Another two-dimensional graph is Figure 12.4, p. 291. There, the vapor pressure of water is plotted against temperature. As you can see, the graph is a curve. This implies that the relationship between vapor pressure and temperature is more complex than that between gas volume and temperature.

Sometimes a graph shows the relationship between three variables, z, y, and x. This is a *three-dimensional* graph. Distances in all three dimensions have meaning. Three-dimensional graphs are commonly used to show the shapes of electron clouds. Consider, for example, Figure 9.9, p. 227. This shows the shape of the electron clouds for the three p orbitals. One of these is concentrated along the horizontal axis. Another lies along the vertical axis. The third is directed along an axis perpendicular to the plane of the paper.

Since two-dimensional graphs are the most common type, our discussion from now on will concentrate upon them.

Constructing Graphs

A two-dimensional graph is really a collection of points, connected by a smooth curve. Each point gives a set of simultaneous values of y and x. To see how such a graph is constructed, consider the data in Table 2. This gives values of y at five different values of x.

TABLE 2

y	2	6	10	14	18
x	1	2	3	4	5

To construct a graph from this data, we need to locate each of the five points on the grid shown in Figure 1a. Let us start with the first point ($y = 2, x = 1$). We move up the vertical axis to $y = 2$. This is located

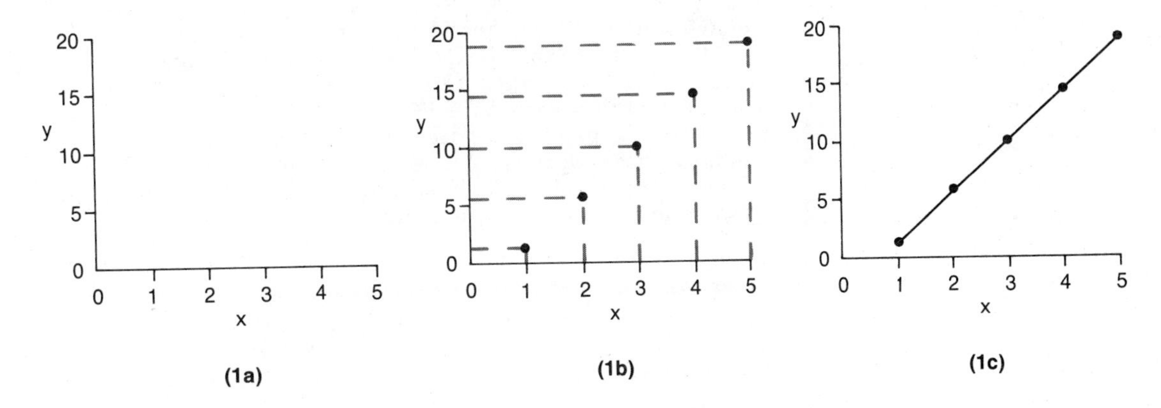

2/5, or a little less than one half, of the distance between 0 and 5. At this point, we draw a horizontal line, extending to the right. Now we locate x = 1 on the horizontal axis. At that point we draw a vertical line, extending upwards. The intersection of these two lines gives the point (y = 2, x = 1) on the graph.

This procedure is repeated for each of the points listed in Table 1. In this way, we obtain the five points shown in Figure 1b. Now we connect the points by a smooth curve. In this case, the "curve" is actually a straight line (Figure 1c).

EXAMPLE 9

Consider the following data:

y	1	3	7	12	20
x	1	2	3	4	5

Using the grid shown in Figure 2a, graph these data.

SOLUTION

First, we locate each of the points. The results are shown in Figure 2b. Note, for example, the way in which the point (y = 3, x = 2) is located. It lies at the intersection of the horizontal line

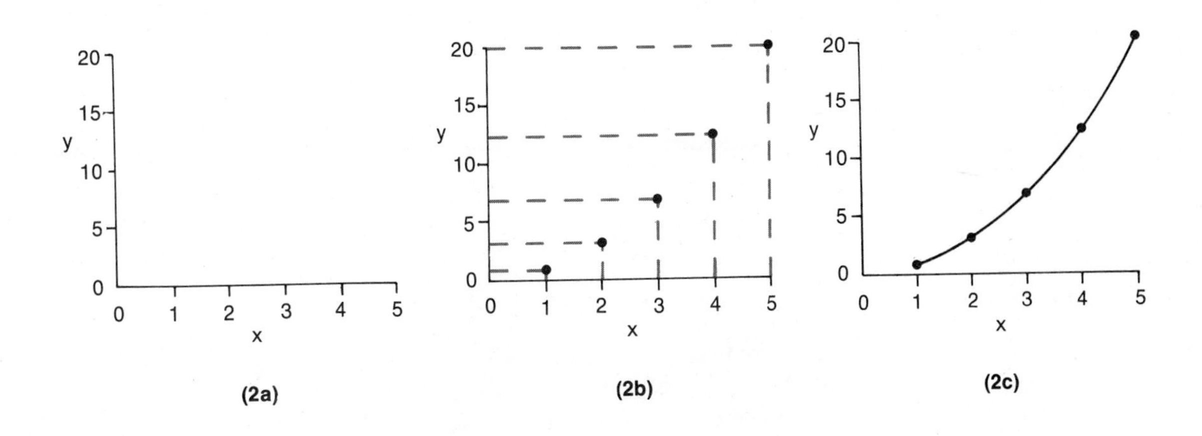

drawn from y = 3 and the vertical line x = 2. It is easy to locate x = 2 on the horizontal axis. It falls at the mark labeled "2". To find y = 3 on the vertical axis, we move up 3/5 (a little more than one half) of the distance from 0 to 5.

Once the points are located, we need to draw a smooth curve through them. If you are skillful enough, you can do this freehand. Otherwise you can use a "French curve" or other mechanical device. The results are shown in Figure 2c.

Reading Graphs

Once you understand how a graph is constructed, it is easy to interpret it. To see how this is done, consider Figure 3. This is a graph of the vapor pressure of water (mm Hg) *versus* temperature (°C). Suppose you want to know the vapor pressure of water at 10°C. To find this:

1. Draw a vertical line up from the x axis at 10°C.
2. Mark the point (A) where this line intersects the curve.
3. Draw a horizontal line across from A to the y axis. This line meets the y axis at about 9 mm Hg. We conclude that:

vapor pressure of water at 10°C = 9 mm Hg

(3)

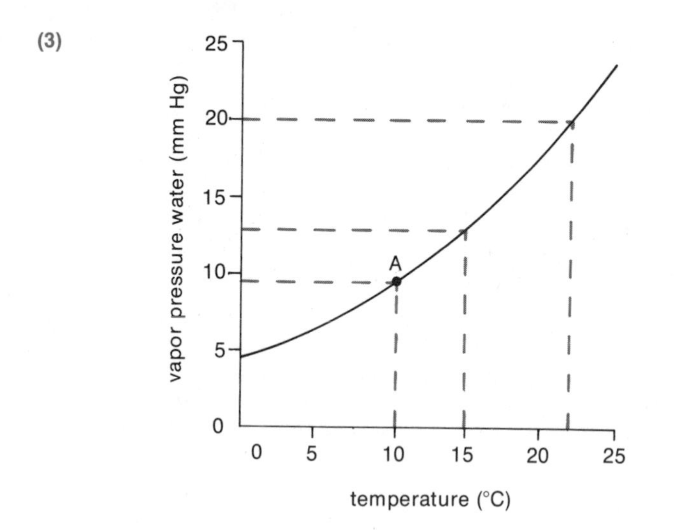

EXAMPLE 10

Using Figure 3, find:

a. the vapor pressure of water at 15°C.

b. the temperature at which the vapor pressure of water is 20 mm Hg.

SOLUTION

a. Draw a vertical line up from 15°C on the x axis. Continue this line until it meets the curve. From this point on the curve, draw a horizontal line to the y axis. You should find that it intersects the y axis at 13. Hence:

vapor pressure of water at 15°C = 13 mm Hg

b. Draw a horizontal line across from 20 mm Hg to the curve. Then draw a vertical line down to the x axis. You should find that it intersects the x axis at 22°C. This must then be the temperature at which the vapor pressure of water is 20 mm Hg.

Straight-Line Graphs

A linear or straight-line graph is common in chemistry. Three such graphs are shown in Figure 4, p. A.14. Two properties of straight-line graphs are of particular importance. These are:

1. *The y intercept.* This is the point at which the straight line intersects the y axis. It gives us the value of y when $x = 0$. The y intercept is readily found. All you have to do is to extend the line until it intersects the y axis. In Figure 4a, we see that the y intercept is 5. In other words, $y = 5$ when $x = 0$.

2. *The slope.* This is defined as the ratio $\Delta y/\Delta x$, where Δy is the change in y and Δx is the change in x. It may be found by choosing two points on the straight line. Calling these points 1 and 2:

$$\text{slope} = \frac{\Delta y}{\Delta x} = \frac{y_2 - y_1}{x_2 - x_1}$$

Looking at points 1 and 2 in Figure 4a, we see that:

$$y_2 = 15; \quad y_1 = 9$$
$$x_2 = 5; \quad x_1 = 2$$

Hence the slope is: $\frac{15 - 9}{5 - 2} = \frac{6}{3} = 2$

EXAMPLE 11

Consider the straight-line graph shown in Figure 4b. Determine the

a. y intercept b. slope

SOLUTION

a. Extending the straight line, we see that it passes through the origin ($y = 0$, $x = 0$). Hence the y intercept is 0.

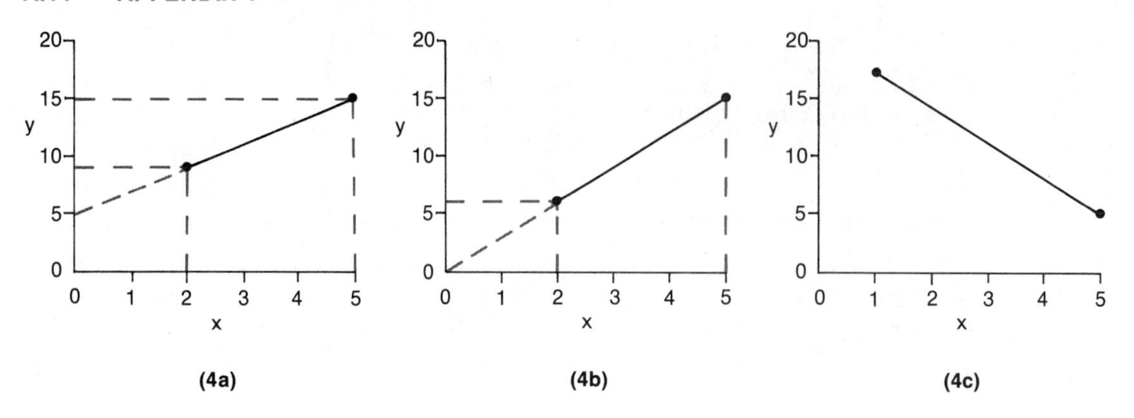

b. Using the two points at the end of the line:

$$y_2 = 15;\ y_1 = 6;\ x_2 = 5;\ x_1 = 2$$

$$\text{slope} = \frac{y_2 - y_1}{x_2 - x_1} = \frac{15 - 6}{5 - 2} = \frac{9}{3} = 3$$

Every two-dimensional graph is really a plot of an equation relating y to x. A straight-line graph is a plot of a general equation:

$$y = a + bx$$

where a and b are constants. These constants can be given a simple meaning.

1. *a is the y intercept.* At the point where the straight line intersects the y axis, $x = 0$. Thus the y intercept is the value of y when $x = 0$. Substituting $x = 0$ in the general equation above:

$$y = a$$

2. *b is the slope,* $\Delta y/\Delta x$. To see that this is the case, let us rewrite the general equation for two points, 1 and 2:

$$y_2 = a + bx_2$$

$$y_1 = a + bx_1$$

Subtracting:

$$y_2 - y_1 = bx_2 - bx_1$$

$$y_2 - y_1 = b(x_2 - x_1)$$

$$\frac{y_2 - y_1}{x_2 - x_1} = \frac{\Delta y}{\Delta x} = b$$

As we have seen, the y intercept and the slope of a straight-line graph can be determined from the graph itself. By doing this, we can determine the equation of the straight line. Consider the line shown in

Figure 4a. For that graph, we showed that:

$$y \text{ intercept} = 5; \text{ slope} = 2$$

Hence in the equation: $y = a + bx$

$$a = 5, b = 2$$

This means that the equation of the straight line is:

$$y = 5 + 2x$$

EXAMPLE 12

Using the results obtained in Example 11, determine the equation of the straight line shown in Figure 4b.

SOLUTION

In Example 11, we found that:

$$y \text{ intercept} = 0, \text{ slope} = 3$$

Hence in the equation $y = a + bx$, $a = 0$, $b = 3$. In other words, the equation is:

$$y = 3x$$

PROBLEMS

1. *(Example 1)* Express the following as ordinary numbers:

a. 2×10^2 b. 3×10^{-1} c. 4.62×10^4 d. 9.18×10^{-3}

2. *(Example 2)* Express the following numbers in exponential notation:

a. 211 b. 0.021 c. 31,620 d. 0.0003

3. *(Example 3)* Choose the largest exponential number in each of the following sets:

a. 2.0×10^5, 3.6×10^5, 1.2×10^5
b. 3.6×10^{-4}, 3.6×10^2, 3.6×10^{-1}
c. 5.9×10^{-6}, 9.1×10^{-8}
d. 4.2×10^{-5}, 1.8×10^{-3}, 2.4×10^{-7}

4. *(Example 4)* Carry out the indicated operations.

a. $(2.4 \times 10^3) \times (1.5 \times 10^2)$
b. $(3.8 \times 10^{-4}) \times (2.0 \times 10^3)$
c. $(6.6 \times 10^{-2})/(2.2 \times 10^{-7})$
d. $(7.0 \times 10^5)/(3.5 \times 10^3)$

5. *(Example 5)* Carry out the indicated operations, expressing your answers in standard exponential notation.

a. $(4.0 \times 10^3) \times (5.0 \times 10^{-2})$
b. $(3.2 \times 10^{-4})/(8.0 \times 10^9)$
c. $\dfrac{(3.2 \times 10^{-5}) \times (1.8 \times 10^3)}{(9.0 \times 10^6)}$

6. *(Example 6)* The molar volume of a gas at 1 atm, V_m (in liters) is related to the temperature in °C, t, by the equation:

$$V_m = 0.0821(t + 273°)$$

a. Calculate the molar volume of a gas at 0°C.
b. At what temperature is the molar volume 20.0 ℓ?

7. *(Example 7)* The density of a substance is given by the expression:

$$\text{density} = \text{mass/volume}$$

a. A sample of aluminum weighing 32.4 g has a volume of 12.0 cm³. What is the density of aluminum?
b. Using your answer in (a), calculate the mass of an aluminum sample that has a volume of 120 cm³.
c. What volume is occupied by 10.0 g of aluminum?

8. *(Example 8)* Solve the equation given in Example 8 for:
a. P_2 b. T_1

9. *(Example 9)* Construct a graph from the following data:

y	20	10	7	5	4
x	1	2	3	4	5

10. *(Example 10)* Using Figure 3 find:

a. the vapor pressure of water at 20°C.
b. the temperature at which the vapor pressure of water is 15 mm Hg.

11. *(Example 11)* Find the y intercept and slope of the straight line shown in Figure 4c.

12. *(Example 12)* Using the results of Problem 11, obtain the equation of the straight line shown in Figure 4c.

ANSWERS

1. a. 200 b. 0.3 c. 46,200 d. 0.00918
2. a. 2.11×10^2 b. 2.1×10^{-2} c. 3.162×10^4 d. 3×10^{-4}
3. a. 3.6×10^5 b. 3.6×10^2 c. 5.9×10^{-6} d. 1.8×10^{-3}
4. a. 3.6×10^5 b. 7.6×10^{-1} c. 3.0×10^5 d. 2.0×10^2
5. a. 2.0×10^2 b. 4.0×10^{-14} c. 6.4×10^{-9}
6. a. 22.4 ℓ b. −29°C
7. a. 2.70 g/cm³ b. 324 g c. 3.70 cm³
8. a. $P_2 = P_1 \times \frac{V_1}{V_2} \times \frac{T_2}{T_1}$ b. $T_1 = T_2 \times \frac{V_1}{V_2} \times \frac{P_1}{P_2}$
9. graph is a hyperbola
10. a. 18 mm Hg b. 18°C
11. y intercept = 20, slope = −3
12. y = 20 − 3x

Appendix 2
ANSWERS TO SELECTED PROBLEMS

CHAPTER 1

1. 0.880 g/cm^3 **2.** 1.14 cm^3 **3.** a. 234°F b. 20°C
4. 325 km **5.** 1.54 in^3 **6.** a. 799 g/ℓ b. 1.66 lb/qt
7. a. 4 b. 3 c. 2 **8.** 5.7×10^2 cm^2 **9.** 1.1 g
23. 1×10^9 cm^2; 4×10^{-2} $mile^2$ **24.** 160°C **25.** Li

CHAPTER 2

1. a. 6, 6, 6, 12 b. $^{11}_{5}B$, 5, 11 c. $^{25}_{12}Mg$, 12, 25 d. $^{9}_{4}Be$, 4, 5
2. b. 2, 2, 4 c. C_2H_6, 8 d. C_4H_8, 8
3. a. LiF b. Li_2S c. MgF_2 d. MgS **4.** 1.844
5. 27.0; Al **6.** 10.81
7. a. 26.04 b. 44.01 c. 34.08 **8.** 1.126; 0.5288
9. a. 26.04 g b. 44.01 g c. 110.26 g d. 158.04 g
10. a. 65.0 g b. 85.2 g c. 253 g
11. a. 0.0357 b. 0.0500 c. 0.0101 **21.** 9×10^{15}
22. 69.0%; 31.0% **23.** 9.39×10^{24} cm^3; 8.6×10^{24} g; 0.1%
24. 1.7×10^{-24} g; 3×10^6 g/cm^3

CHAPTER 3

1. 2.11 g N, 0.453 g H **2.** 60.31% Mg, 39.69% O
3. a. 81.71% C, 18.29% H b. 52.14% C, 13.13% H, 34.73% O
c. 67.62% U, 32.38% F
4. 699.5 g **5.** $CuCl_2$ **6.** C_2H_6O **7.** C_4H_{10}
8. a. $C_2H_4(g) + 3\ O_2(g) \rightarrow 2\ CO_2(g) + 2\ H_2O(l)$
b. $C_2H_4(g) + 2\ O_2(g) \rightarrow 2\ CO(g) + 2\ H_2O(l)$
c. $2\ Fe(s) + 3\ Cl_2(g) \rightarrow 2\ FeCl_3(s)$
d. $6\ Li(s) + N_2(g) \rightarrow 2\ Li_3N(s)$
9. a. 4.52 b. 0.128 c. 1.17×10^6

10. a. 491 g b. 0.585 **11.** a. 1.88 g b. 0.533 g
28. $CrN_9H_{18}O_9$ **29.** C_2H_4O **30.** 12 g KBr
31. 1.7×10^{16} g; 1.8×10^{16} g

CHAPTER 4

1. a. 0.219 kcal b. 1.502 kcal c. 3.60×10^3 kcal
2. a. 4.1×10^{-3} ¢ b. 0.24 ¢
3.

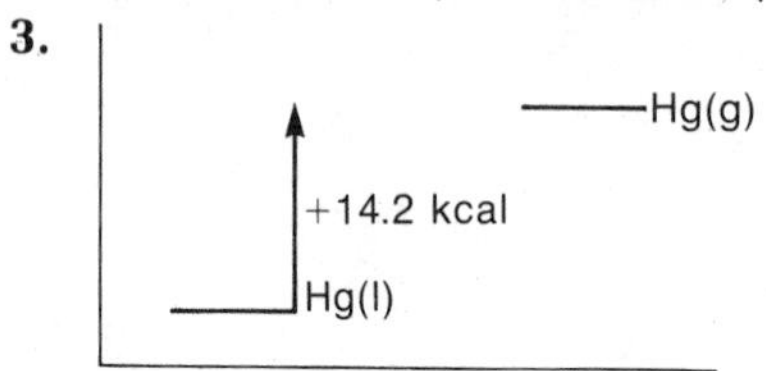

4.

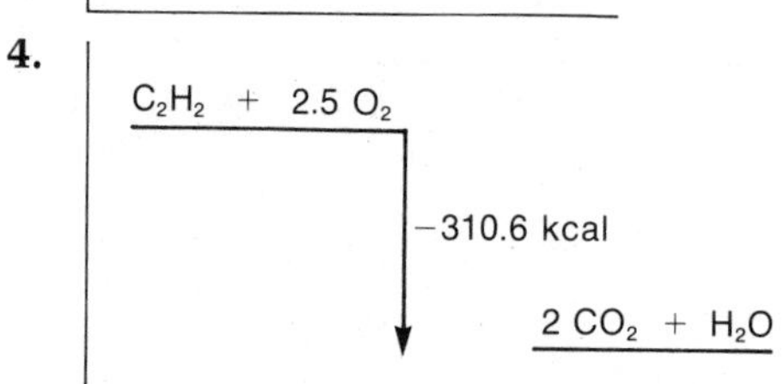

5. a. −62.4 kcal b. −0.392 kcal c. −0.491 kcal
6. a. 1.50 kcal b. 0.681 kcal **7.** 1.42 g
8. +11.16 kcal **9.** −23.5 kcal **18.** −7.3 kcal
19. 0.107 cal/g·°C **20.** 11.93 kcal/g *vs.* 12.40 kcal/g
21. 2.4×10^{-3} ¢/kcal; 2.1 ¢/kwh; use natural gas

CHAPTER 5

1. a. 353 K b. 310 K
2. a. 0.976 atm b. 14.4 lb/in² c. 98.9 kPa
3. a. 720 mm Hg b. 735 mm Hg **4.** 6.25×10^3 ℓ
5. 1.81×10^5 m³ **6.** 1.13×10^4 m³ **7.** 45.4 lb/in²
8. a. 11.2 ℓ b. 0.700 ℓ c. 0.800 ℓ **9.** 76.2
10. 33.0 ℓ; 22.0 ℓ **11.** a. 31.8 ℓ b. 6.70 ℓ
25. $C_2H_4(g) + 3\,O_2(g) \rightarrow 2\,CO_2(g) + 2\,H_2O(g)$ **26.** 1.0×10^3 g
27. a. b. c.

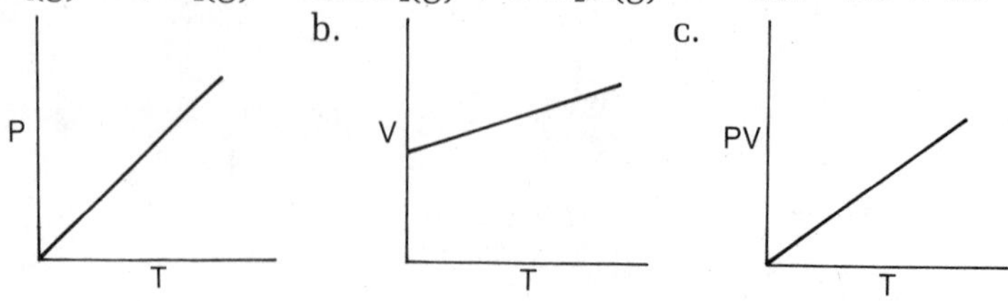

28. 45.3

CHAPTER 6

1. a. 44.44% C, 6.22% H, 49.34% O
b. 35.00% N, 5.04% H, 59.97% O
c. 20.00% C, 26.64% O, 46.64% N, 6.71% H
2. a. 0.122 mol b. 3.90 g **3.** a. 0.598 g b. 0.0144
4. 10.2 kcal **5.** a. 0.306 b. 6.85 ℓ
6. a. 0.0889 ℓ b. 0.0970 ℓ **7.** 0.179 g/ℓ
18. 1.7×10^4 g **19.** a. 12.9 g b. 98.77 c. 28.96
20. a. 1.34 g/ℓ, 2.05 g/ℓ b. 1.20 g/ℓ, 1.83 g/ℓ
21. a. too explosive b. too reactive
c. too expensive (and insoluble) d. reacts with ozone layer

CHAPTER 7

1. 2.53 g/cm³, 785°C, 1530°C, 132 kcal/mol
2. $RaCl_2$, $RaCO_3$, $Ra_3(PO_4)_2$ **3.** Ge < Ga < In
4. Ge > Ga > In **5.** Ge > Ga > In **6.** In > Ga > Ge
15. Clearly Mg and S are too high, Al and P too low.
16.

A							B
C	D	E					F
G	H	I					J
K	L	M	N	O			P
Q	R	S	T	U	V	W	X
Y	Z						

17. a. from the planet Neptune, next to Uranus in the solar system
b. Poland (Marie Curie was born in Poland)
c. France (discovered by Marguerite Perey in Paris)

CHAPTER 8

7. a. $4\ Li(s) + O_2(g) \rightarrow 2\ Li_2O(s)$
b. $2\ Li(s) + 2\ H_2O(l) \rightarrow 2\ Li^+(aq) + 2\ OH^-(aq) + H_2(g)$
8. a. $2\ KCl(l) \rightarrow 2\ K(l) + Cl_2(g)$ b. $SrCl_2(l) \rightarrow Sr(s) + Cl_2(g)$
9. a. $2\ Ba(s) + O_2(g) \rightarrow 2\ BaO(s)$
b. $Ba(s) + 2\ H_2O(l) \rightarrow Ba^{2+}(aq) + 2\ OH^-(aq) + H_2(g)$
10. a. $Mg(s) + Te(s) \rightarrow MgTe(s)$ b. $Te(s) + O_2(g) \rightarrow TeO_2(s)$
11. a. $Mg(s) + I_2(s) \rightarrow MgI_2(s)$ b. $2\ Na(s) + I_2(s) \rightarrow 2\ NaI(s)$
12. a. $2\ Rb(s) + 2\ H_2O(l) \rightarrow 2\ Rb^+(aq) + 2\ OH^-(aq) + H_2(g)$
b. $Ca(s) + 2\ H_2O(l) \rightarrow Ca^{2+}(aq) + 2\ OH^-(aq) + H_2(g)$
c. $2\ Sr(s) + O_2(g) \rightarrow 2\ SrO(s)$
13. a. $H_2(g) + Br_2(l) \rightarrow 2\ HBr(g)$
b. $Br_2(l) + 2\ I^-(aq) \rightarrow 2\ Br^-(aq) + I_2(s)$
c. $MnO_2(s) + 2\ Br^-(aq) + 4\ H^+(aq) \rightarrow Mn^{2+}(aq) + Br_2(l) + 2\ H_2O$

14. a. 0.00261 b. 0.0585 ℓ **15.** a. 0.0893 b. 8.94 g
16. 3.55 g **17.** 0.913 g
23. $C_2H_4Br_2(l) + Pb(C_2H_5)_4(l) + 16\ O_2(g) \rightarrow$
$PbBr_2(s) + 10\ CO_2(g) + 12\ H_2O(l)$
24. 11 kg
25. a. 3.818×10^{-23} g b. 1.86×10^{-8} cm c. 2.70×10^{-23} cm^3
d. 1.41 g/cm^3

CHAPTER 9

1. a. −313.6 b. −78.4 c. −34.8 d. −19.6 e. −12.5
2. a. 278.8 b. −43.6 c. 65.9
3. a. $1s^2\ 2s^2\ 2p^1$ b. $1s^2\ 2s^2\ 2p^6$ c. $1s^2\ 2s^2\ 2p^6\ 3s^2\ 3p^1$
d. $1s^2\ 2s^2\ 2p^6\ 3s^2\ 3p^5$
4. a. [Ar] $4s^2$ b. [Ar] $4s^2\ 3d^1$ c. [Ar] $4s^2\ 3d^6$
d. [Ar] $4s^2\ 3d^{10}$
5. a. $4s^2\ 4p^3$ b. $5s^1$ c. $5s^2\ 5p^2$ d. $5s^2\ 5p^5$
6. a. P b. K c. At d. Te **15.** 656 nm
16. $3 \rightarrow 1$
17. a. [Kr] $5s^2\ 4d^5$ b. [Kr] $5s^2\ 4d^{10}\ 5p^5$ c. [Xe] $6s^2\ 5d^1$
18.

Al	1, 0, 0	S	2, 1, 1
Si	1, 1, 0	Cl	2, 2, 1
P	1, 1, 1	Ar	2, 2, 2

19. a. 3 b. 6 c. 3

CHAPTER 10

1. d **2.** 35.12%; 26.52% **3.** 62.96%
4. a. CaI_2 b. K_2S c. Al_2O_3 d. SrO e. RbCl f. AlF_3
5. a. $NiCl_2$ b. NiO c. NiS d. $NiBr_2$
6. a. AgCl b. CuCl, $CuCl_2$ c. $FeCl_2$, $FeCl_3$ d. $ZnCl_2$
7. a. $Ca(s) + I_2(s) \rightarrow CaI_2(s)$
b. $2\ K(s) + S(s) \rightarrow K_2S(s)$
c. $4\ Al(s) + 3\ O_2(g) \rightarrow 2\ Al_2O_3(s)$
d. $2\ Sr(s) + O_2(g) \rightarrow 2\ SrO(s)$
e. $2\ Rb(s) + Cl_2(g) \rightarrow 2\ RbCl(s)$
f. $2\ Al(s) + 3\ F_2(g) \rightarrow 2\ AlF_3(s)$
8. a. $Ni(s) + Cl_2(g) \rightarrow NiCl_2(s)$
b. $2\ Ni(s) + O_2(g) \rightarrow 2\ NiO(s)$
c. $Ni(s) + S(s) \rightarrow NiS(s)$
d. $Ni(s) + Br_2(l) \rightarrow NiBr_2(s)$
9. a. calcium iodide b. potassium sulfide
c. aluminum oxide d. strontium oxide
e. rubidium chloride f. aluminum fluoride
10. a. iron(II) chloride b. iron(III) chloride
c. cobalt(II) bromide d. cobalt(III) fluoride
e. copper(I) nitrate
11. a. $KHCO_3$ b. $MgSO_4$ c. $Co(NO_3)_3$ **24.** −207 kcal
25. NCH_5O_3; NH_4^+, HCO_3^- **26.** $CoCl_2 \cdot 4H_2O$

CHAPTER 11

1. a. ·Ẋ· b. :Ẍ· c. ·Ẍ· d. ·Ẍ·

2. a. H—C̈l: b. P̈ (H H H) c. S̈ (H H) d. H—Si—H (H above and below)

3. a. S̈ (O, O) b. S (O, O, O) c. :S̈=S̈:

4. a. linear b. pyramidal c. bent d. tetrahedral

5. a. nonpolar b. polar c. polar d. nonpolar

6. a. polar b. polar c. polar d. nonpolar

7. a. dihydrogen oxide b. phosphorus pentachloride
c. dinitrogen trioxide d. disulfur dichloride

8.

phosphoric acid	phosphate ion	sodium phosphate
phosphorous acid	phosphite ion	sodium phosphite
hypophosphorous acid	hypophosphite ion	

23. Ṅ (O, O) C̈l (O, O)

24. a. $(:\ddot{O}—H)^-$ b. $(:\ddot{C}l—\ddot{O}:)^-$ c. $(:\ddot{O}—C—\ddot{O}:, =O:)^{2-}$

25. C_3H_4 H—C≡C—C—H (H above and below)

26. Several different structures are possible; the true structure is a tetrahedron with a P atom at each corner.

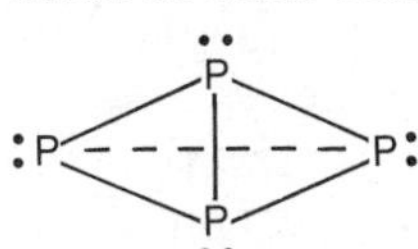

CHAPTER 12

1. 62% **2.** About 75°C **3.** 76°C

4. $d < c < b < a$ **5.** b and c **6.** a, b, d

7. a. $Ni(OH)_2(s) \rightarrow NiO(s) + H_2O(g)$
b. $2\ AgOH(s) \rightarrow Ag_2O(s) + H_2O(g)$
c. $NiCO_3(s) \rightarrow NiO(s) + CO_2(g)$
d. $Ag_2CO_3(s) \rightarrow Ag_2O(s) + CO_2(g)$

8. metallic **9.** covalent network

21. a. MgO (ionic; I_2 is molecular)
b. SiO_2 (covalent network; CO_2 is molecular)
c. Fe (metallic; ice melts at 0°C)
d. MgO (ionic bonds stronger with +2, −2 ions)

22. 14.8 kcal/mol **23.** 3.34×10^{22}; 2.69×10^{19}

CHAPTER 13

1. 8.58%, 91.42% **2.** a. 0.50 g b. 73.5 g c. 280 g

3. 2.5 M **4.** 0.050 mol; 9.7 g

5. a. 5.00 ℓ b. 0.0337 ℓ

6. a. Dissolve 170 g $AgNO_3$ to form 1.0 ℓ solution
b. Dissolve 1.5×10^2 g NH_4Br to form 0.50 ℓ solution

7. a. 34 g/100 g water; 51 g/100 g water b. 39 g water
c. 13 g KCl

8. a. B b. A c. B **9.** b **20.** −7.5°C

21. 8.6 g **22.** 2.0×10^2

CHAPTER 14

1. a. C_6H_{14} b. C_5H_{12} c. $C_{18}H_{38}$

2. a. alkane b. alkene c. alkyne d. none e. alkyne

3. a, b, and d

4.

```
                     C                     C
                     |                     |
C—C—C—C—C—C,     C—C—C—C—C,         C—C—C—C—C,

  C  C             C
  |  |             |
C—C—C—C,       C—C—C—C
                   |
                   C
```

5. c and d

6. hexane, 2-methylpentane, 3-methylpentane, 2,3-dimethylbutane, 2,2-dimethylbutane

7.

```
           H
           |
a. CH3—C—CH2—CH2—CH2—CH3
           |
          CH3

          CH3
           |
b. CH3—C—CH2—CH2—CH3
           |
          CH3
```

$$\begin{array}{ccccl} & CH_3 & & H & \\ & | & & | & \\ \text{c. } CH_3- & C & - & C & -CH_2-CH_2-CH_3 \\ & | & & | & \\ & H & & CH_2-CH_3 & \end{array}$$

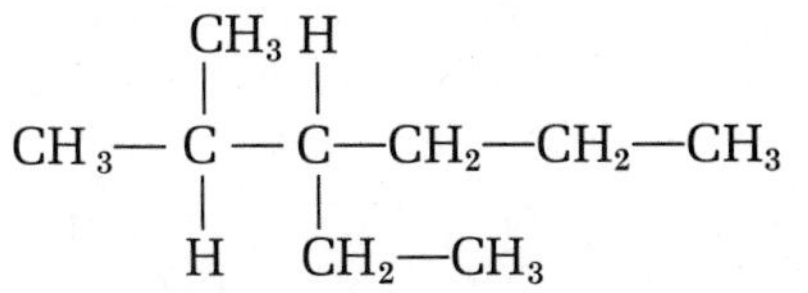

d. $CH_3-CH_2-CH_2-CH_2-CH_2-CH_2-CH_2-CH_3$

$$\begin{array}{lcl} & H & \\ & | & \\ \text{e. } CH_3-CH_2-CH_2- & C & -CH_2-CH_2-CH_3 \\ & | & \\ & CH_2-CH_3 & \end{array}$$

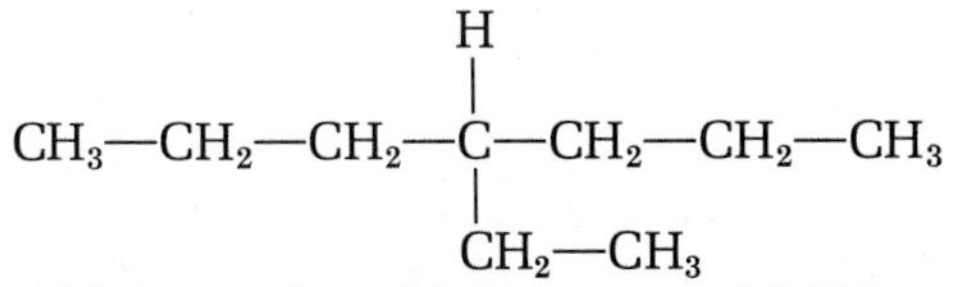

8. a. add Cl_2 b. add H_2 c. add HBr

9. a. add 1 mol H_2 b. add 1 mol HCl to (a)

17. There is one isomer with 1 Cl atom, three isomers with 2 Cl atoms, 3 isomers with 3 Cl atoms, 3 isomers with 4 Cl atoms, 1 isomer with 5 Cl atoms, 1 isomer with 6 Cl atoms, for a total of 12 isomers

18. 64 **19.** C_7H_8; CH_3

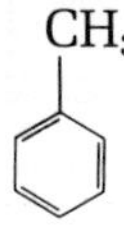

CHAPTER 15

1. $CH_3-CH_2-CH_2-CH_2-CH_2OH$,
$CH_3-CH_2-CH_2-CHOH-CH_3$,
$CH_3-CH_2-CHOH-CH_2-CH_3$

2. $CH_3-O-CH_2-CH_2-CH_2-CH_3$,
$CH_3-CH_2-O-CH_2-CH_2-CH_3$

3. $CH_3-CH_2-CH_2-COOH$,
$$\begin{array}{lcl} & H & \\ & | & \\ CH_3- & C & -COOH \\ & | & \\ & CH_3 & \end{array}$$

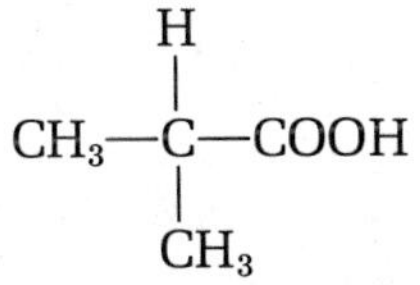

4. a. aldehyde b. acid c. ester d. ether e. alcohol

5. CH_3OH, CH_3-O-CH_3,
$$\begin{array}{lcl} CH_3- & C & -H, \\ & \| & \\ & O & \end{array} \quad \begin{array}{lcl} CH_3- & C & -CH_3, \\ & \| & \\ & O & \end{array}$$

$$\begin{array}{lcl} CH_3- & C & -OH, \\ & \| & \\ & O & \end{array} \quad \begin{array}{lcl} CH_3- & C & -O-CH_3 \\ & \| & \\ & O & \end{array}$$

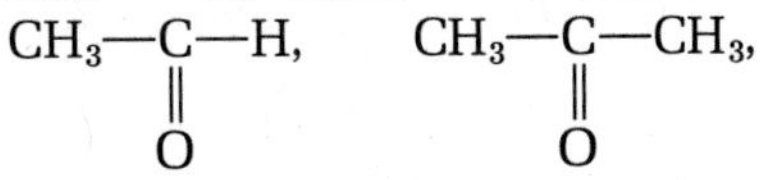

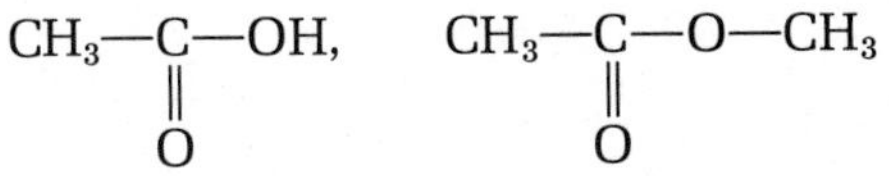

6. a. $CH_3-CH_2-CH_2OH$ or $CH_3-CHOH-CH_3$

b. $CH_3-O-CH_2-CH_3$

$$\text{c. } \begin{array}{lcl} CH_3-CH_2- & C & -H \\ & \| & \\ & O & \end{array}$$

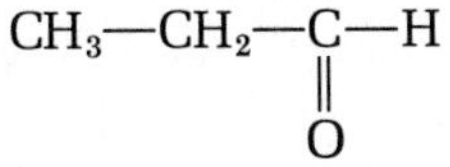

$$\text{d. } \begin{array}{lcl} CH_3- & C & -CH_3 \\ & \| & \\ & O & \end{array} \qquad \text{e. } \begin{array}{lcl} CH_3-CH_2- & C & -OH \\ & \| & \\ & O & \end{array}$$

$$\text{f. } \begin{array}{lcl} H- & C & -O-CH_2-CH_3 \\ & \| & \\ & O & \end{array} \text{ or } \begin{array}{lcl} CH_3- & C & -O-CH_3 \\ & \| & \\ & O & \end{array}$$

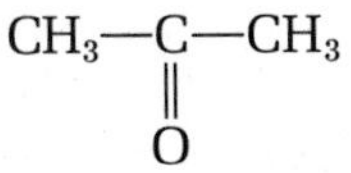

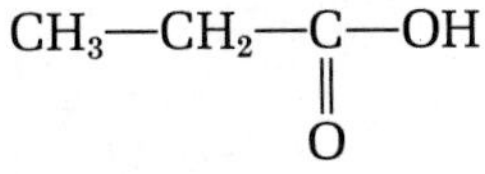

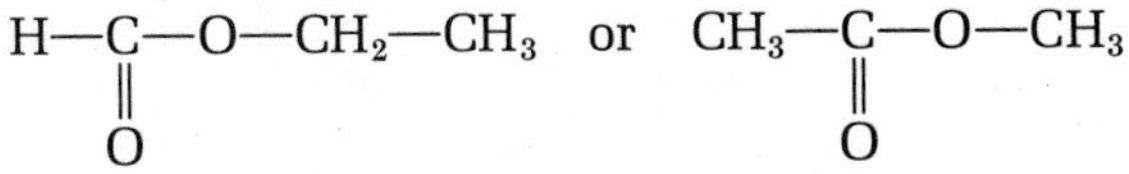

7. a. 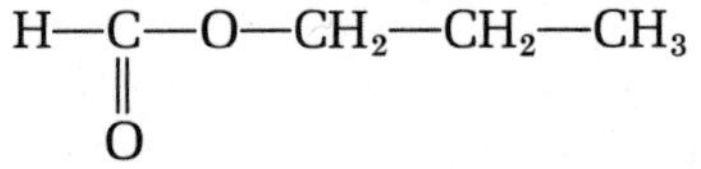
```
H—C—O—CH2—CH2—CH3
  ||
  O
```
b. 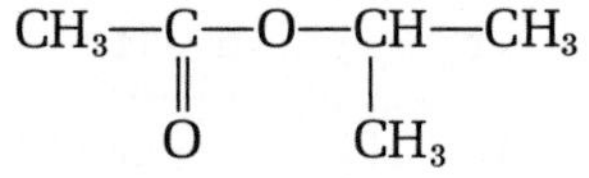
```
CH3—C—O—CH—CH3
    ||    |
    O     CH3
```

17. a. $CH_3—CH_2—OH$ (alcohol) or $CH_3—O—CH_3$ (ether)

b. 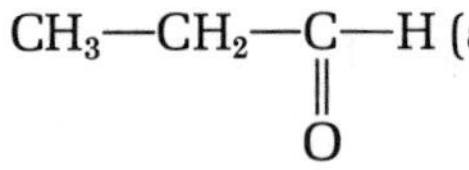
```
CH3—CH2—C—H (aldehyde)   or   CH3—C—CH3 (ketone)
        ||                        ||
        O                         O
```

c.
```
CH3—C—OH (acid)   or   H—C—O—CH3 (ester)
    ||                   ||
    O                    O
```

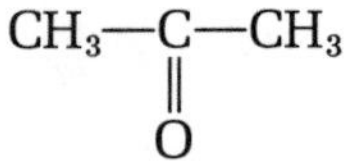

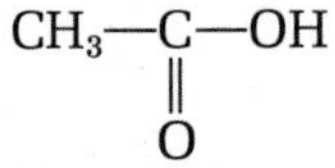

18. 2.06 g/ℓ **19.** 1.2

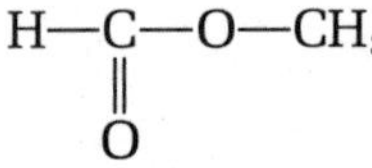

CHAPTER 16

1. a. 2.0×10^3 b. 1.0×10^3 c. 2.0×10^3

2. $CH_2{=}CHCN$

3. a. 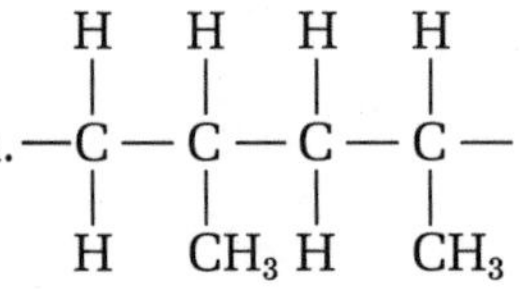
```
   H   H   H   H
   |   |   |   |
 —C — C — C — C —
   |   |   |   |
   H  CH3  H  CH3
```
b.
```
  F  F  F  F
  |  |  |  |
 —C—C—C—C—
  |  |  |  |
  F  F  F  F
```
c.
```
  H   H   H   H
  |   |   |   |
 —C — C — C — C—
  |   |   |   |
  H  C6H5 H  C6H5
```

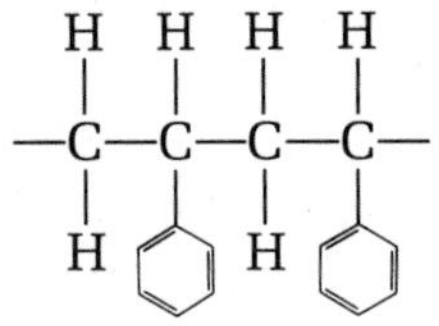

4. a. 85.63% C, 14.37% H b. 85.63% C, 14.37% H
c. 85.63% C, 14.37% H

5. 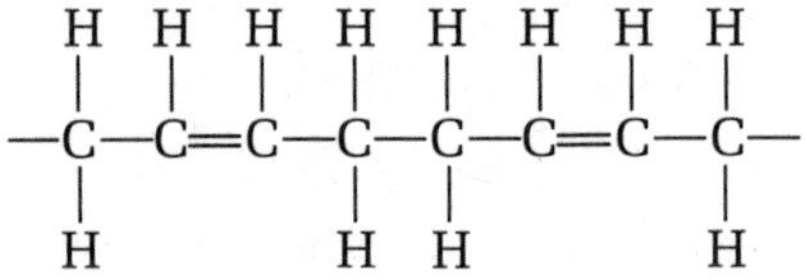
```
  H  H  H  H  H  H  H  H
  |  |  |  |  |  |  |  |
 —C—C=C—C—C—C=C—C—
  |        |  |        |
  H        H  H        H
```
6. a and c

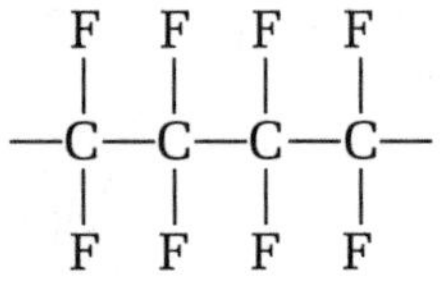

7. a. 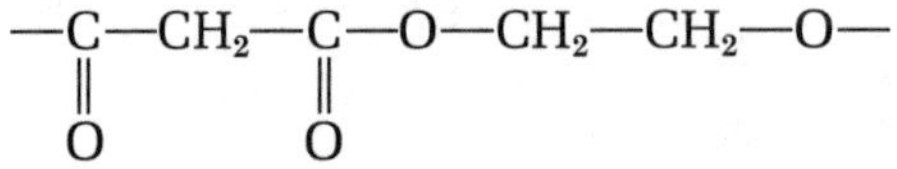
```
 —C—CH2—C—O—CH2—CH2—O—
  ||     ||
  O      O
```
c.
```
                         H
                         |
 —C—CH2—C—O—CH2—C—O—
  ||     ||              |
  O      O               CH3
```

8. 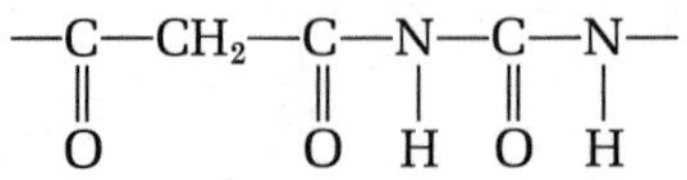
```
 —C—CH2—C—N—C—N—
  ||     || |  || |
  O      O  H  O  H
```

9. 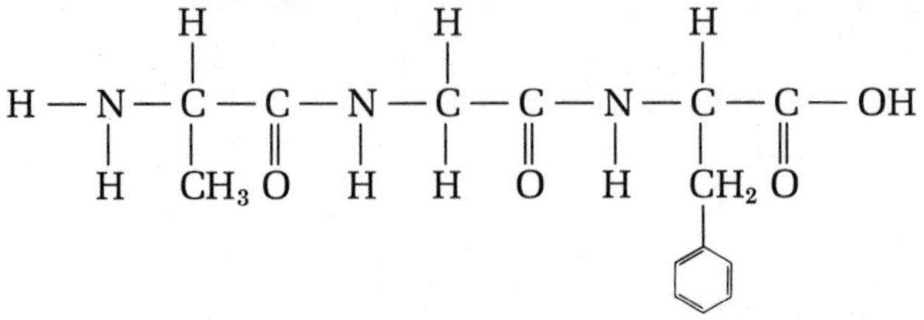
```
         H             H             H
         |             |             |
 H — N — C — C — N — C — C — N — C — C — OH
     |   |   ||  |   |   ||  |   |   ||
     H  CH3  O   H   H   O   H  CH2  O
                                 |
                                C6H5
```

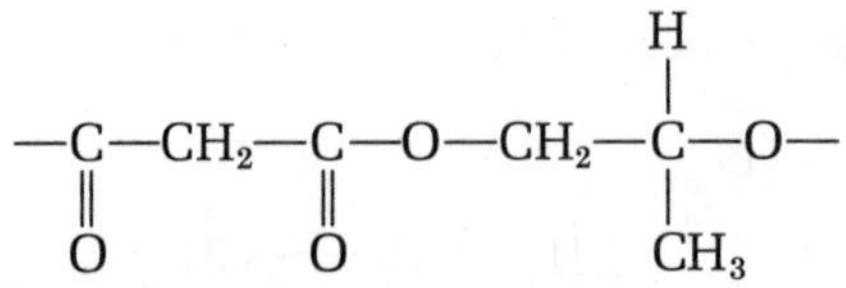

10. serine and aspartic acid **23.** 0.642 g
24. 63.68% C, 14.14% O, 12.38% N, 9.80% H **25.** 5994

CHAPTER 17

1. 30 mol/ℓ·hr **2.** 0.011 mol/ℓ·min
3. a. 0.008 mol/ℓ·min b. 0.005 mol/ℓ·min
4. a. rate increases by factor of 3
b. rate decreases to 1/5 original value
5.

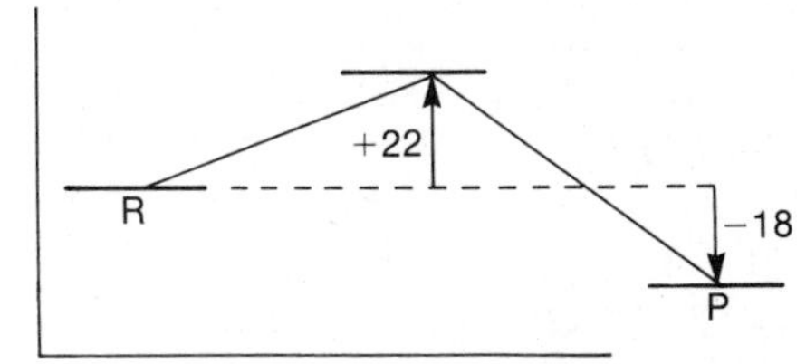

6.

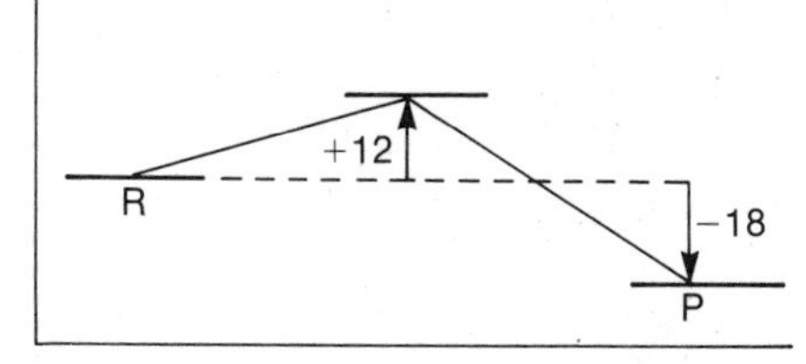

12. 3×10^{-4} mol/ℓ·s **13.** 4.5×10^{-3} mol/ℓ·min **14.** 2; 0

CHAPTER 18

1. a. $[SO_3]^2/[SO_2]^2 \times [O_2]$ b. $[HBr]^2/[H_2] \times [Br_2]$
c. $[N_2] \times [H_2]^2/[N_2H_4]$
2. a. $[H_2O]$ b. $[H_2] \times [CO_2]/[CO]$ c. $[H_2O]^3/[H_2]^3$
3. 0.77 M **4.** 0.060 M **5.** 12 **6.** 200°C
7. a. forward b. reverse c. forward d. reverse
8. a. compress b. expand c. neither
9. decrease; increase
10. a. increases rate, decreases yield b. increases rate and yield
c. increases rate
26. a. 0.025 M b. 0.025 M c. 0.025
27. a. 92.1 ℓ b. 0.0109 M c. 0.0109
28. reverse reaction occurs
29. $CuO(s) + H_2(g) \rightleftharpoons Cu(s) + H_2O(g)$
$SnO(s) + H_2(g) \rightleftharpoons Sn(s) + H_2O(g)$

CHAPTER 19

1. a. 10^{-8} M b. 5×10^{-11} M c. 2.5×10^{-3} M
2. a. 10^{-10} M b. 2×10^{-5} M c. 2.8×10^{-8} M
3. a. 8 b. 1 c. 3
4. a. 10^{-5} M b. 10^{-9} M c. 10 M
5. $HBr(aq) \rightarrow H^+(aq) + Br^-(aq)$; 0.20 M
6. $HNO_2(aq) \rightleftharpoons H^+(aq) + NO_2^-(aq)$; 0.01 M
7. a. 0.040 M b. 0.040 M c. 0.36 M d. 4.4×10^{-3}
8. $Sr(OH)_2(s) \rightarrow Sr^{2+}(aq) + 2\ OH^-(aq)$; 0.10 M
9. a, b, c: $OH^-(aq) + H^+(aq) \rightarrow H_2O$
10. a, b, c: $NH_3(aq) + H^+(aq) \rightarrow NH_4^+(aq)$
11. a. $CH_3COOH(aq) + OH^-(aq) \rightarrow CH_3COO^-(aq) + H_2O$
b. $HNO_2(aq) + OH^-(aq) \rightarrow NO_2^-(aq) + H_2O$
c. $H_2CO_3(aq) + OH^-(aq) \rightarrow HCO_3^-(aq) + H_2O$
$HCO_3^-(aq) + OH^-(aq) \rightarrow CO_3^{2-}(aq) + H_2O$
24. a. 3.7 b. 8.40 **25.** 8×10^{-3}, 7×10^{-4}
26. a. $F^-(aq) + H_2O \rightleftharpoons HF(aq) + OH^-(aq)$
b. $CH_3COO^-(aq) + H_2O \rightleftharpoons CH_3COOH(aq) + OH^-(aq)$
c. $NO_2^-(aq) + H_2O \rightleftharpoons HNO_2(aq) + OH^-(aq)$
27. a. BA: HF, H_3O^+ BB: H_2O, F^-
b. BA: H_2O, HF BB: F^-, OH^-
c. BA: HCOOH, H_2CO_3 BB: HCO_3^-, $HCOO^-$

CHAPTER 20

1. a. Ag^+, NO_3^-, Na^+, S^{2-}; Ag_2S, $NaNO_3$
b. Cu^{2+}, SO_4^{2-}, Ba^{2+}, Cl^-; $BaSO_4$, $CuCl_2$
c. Ni^{2+}, Cl^-, Ca^{2+}, OH^-; $Ni(OH)_2$, $CaCl_2$
d. Cu^{2+}, NO_3^-, Zn^{2+}, Cl^-; $Zn(NO_3)_2$, $CuCl_2$
2. a. $2\ Ag^+(aq) + S^{2-}(aq) \rightarrow Ag_2S(s)$
b. $Ba^{2+}(aq) + SO_4^{2-}(aq) \rightarrow BaSO_4(s)$
c. $Ni^{2+}(aq) + 2\ OH^-(aq) \rightarrow Ni(OH)_2(s)$
3. a. 2×10^{-10} M b. 1×10^{-12} M
4. a. 1×10^{-8} M b. 3×10^{-7} M
5. a. $[Pb^{2+}] \times [Cl^-]^2$ b. $[Cr^{3+}] \times [OH^-]^3$ c. $[As^{3+}]^2 \times [S^{2-}]^3$
d. $[Ba^{2+}] \times [CrO_4^{2-}]$
6. a. $Zn(OH)_2(s) + 2\ H^+(aq) \rightarrow Zn^{2+}(aq) + 2\ H_2O$
b. $AgOH(s) + H^+(aq) \rightarrow Ag^+(aq) + H_2O$
c. $Al(OH)_3(s) + 3\ H^+(aq) \rightarrow Al^{3+}(aq) + 3\ H_2O$
7. $BaCO_3(s) + 2\ H^+(aq) \rightarrow Ba^{2+}(aq) + CO_2(g) + H_2O$
8. a. $Cd(OH)_2(s) + 4\ NH_3(aq) \rightarrow Cd(NH_3)_4^{2+}(aq) + 2\ OH^-(aq)$
b. $CdCO_3(s) + 4\ NH_3(aq) \rightarrow Cd(NH_3)_4^{2+}(aq) + CO_3^{2-}(aq)$
c. $AgBr(s) + 2\ NH_3(aq) \rightarrow Ag(NH_3)_2^+(aq) + Br^-(aq)$
9. $Cr(OH)_3(s) + 3\ OH^-(aq) \rightarrow Cr(OH)_6^{3-}(aq)$
23. $PbCl_2(s) \rightleftharpoons Pb^{2+}(aq) + 2\ Cl^-(aq)$
$K_{sp} = [Pb^{2+}] \times [Cl^-]^2 = S(2S)^2 = 4S^3$
24. conc. Ca^{2+} × conc. $SO_4^{2-} = 2 \times 10^{-5} < K_{sp} = 3 \times 10^{-5}$. No precipitate
25. 1×10^{-5} g (1 sig. fig.)

CHAPTER 21

1. 89.99% **2.** 65.02% **3.** 0.1612 M **4.** 60.9%
5. 18.2% **6.** 3.8×10^{-5} M **17.** 0.012 M
18. 31.7% **19.** 33.6%

CHAPTER 22

1. a. $Zn^{2+}(aq) + 2\ OH^-(aq) \rightarrow Zn(OH)_2(s)$
$Zn^{2+}(aq) + S^{2-}(aq) \rightarrow ZnS(s)$
b. $Zn(OH)_2(s) + 2\ H^+(aq) \rightarrow Zn^{2+}(aq) + 2\ H_2O$
$ZnS(s) + 2\ H^+(aq) \rightarrow Zn^{2+}(aq) + H_2S(g)$
2. a. $Mg^{2+}(aq) + 2\ OH^-(aq) \rightarrow Mg(OH)_2(s)$
b. $Zn^{2+}(aq) + 2\ OH^-(aq) \rightarrow Zn(OH)_2(s)$
$Zn(OH)_2(s) + 2\ OH^-(aq) \rightarrow Zn(OH)_4^{2-}(aq)$
3. $Cu^{2+}(aq) + 2\ NH_3(aq) + 2\ H_2O \rightarrow Cu(OH)_2(s) + 2\ NH_4^+(aq)$
$Cu(OH)_2(s) + 4\ NH_3(aq) \rightarrow Cu(NH_3)_4^{2+}(aq) + 2\ OH^-(aq)$
4. Ni^{2+} **5.** Zn^{2+} **6.** a. Na^+, K^+ b. flame test
7. Ag^+ is present, Al^{3+} is absent, K^+ undetermined
8.

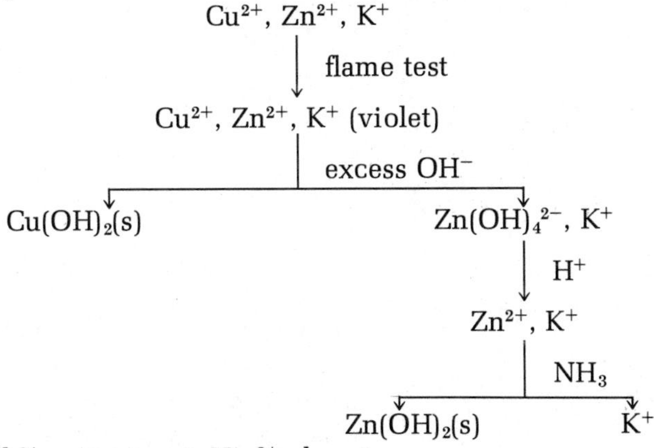

9. Ag^+, Pb^{2+} are present; Hg_2^{2+} absent
20. a. 1×10^{-22} M b. 1×10^{-5} M
21. a. H^+ b. HCl c. H^+ d. H^+, NH_3
22. a. solubility in water b. solubility in NH_3
c. add excess OH^- d. solubility in H^+
23. a. $2\ Ag^+(aq) + 2\ OH^-(aq) \rightarrow Ag_2O(s) + H_2O$
b. $Al^{3+}(aq) + 3\ S^{2-}(aq) + 3\ H_2O \rightarrow Al(OH)_3(s) + 3\ HS^-(aq)$
c. $Mg^{2+}(aq) + 2\ S^{2-}(aq) + 2\ H_2O \rightarrow Mg(OH)_2(s) + 2\ HS^-(aq)$

CHAPTER 23

1. a. +6 b. +3 c. +6 d. +6
2. a. −3, +1 b. +4, −2 c. +1, +6, −2 d. +1, +5, −2
3. a. oxid. b. oxid. c. red. d. oxid.

4. a. red. b. neither c. oxid. d. neither e. neither f. oxid.

5. a. $2\ NH_3(aq) \rightarrow N_2(g) + 6\ H^+(aq) + 6\ e^-$
b. $N_2(g) + 2\ H_2O \rightarrow 2\ NO(g) + 4\ H^+(aq) + 4\ e^-$
c. $NO_2(g) + 2\ H^+(aq) + 2\ e^- \rightarrow NO(g) + H_2O$
d. $NO(g) + 2\ H_2O \rightarrow NO_3^-(aq) + 4\ H^+(aq) + 3\ e^-$

6. $Ag(s) + NO_3^-(aq) + 2\ H^+(aq) \rightarrow Ag^+(aq) + NO_2(g) + H_2O$

7. $2\ MnO_4^-(aq) + 6\ I^-(aq) + 8\ H^+(aq) \rightarrow 2\ MnO_2(s) + 3\ I_2(s) + 4\ H_2O$

8. $CuS(s) + 2\ NO_3^-(aq) + 4\ H^+(aq) \rightarrow Cu^{2+}(aq) + S(s) + 2\ NO_2(g) + 2\ H_2O$

9. a. 0.600 b. 0.400 **10.** a. 9.27×10^{-3} mol b. 0.426 g

21. $4\ ClO_3^-(aq) \rightarrow 3\ ClO_4^-(aq) + Cl^-(aq)$

22. $Bi_2S_3(s) + 6\ NO_3^-(aq) + 12\ H^+(aq) \rightarrow$
$2\ Bi^{3+}(aq) + 3\ S(s) + 6\ NO_2(g) + 6\ H_2O$

23. a. 4.80 ℓ b. 48.0 ℓ **24.** 4.0×10^2 ℓ

CHAPTER 24

1.

Ni Ag
Ni^{2+} Ag^+

Ni anode, Ag cathode
Electrons move from Ni to Ag

2.

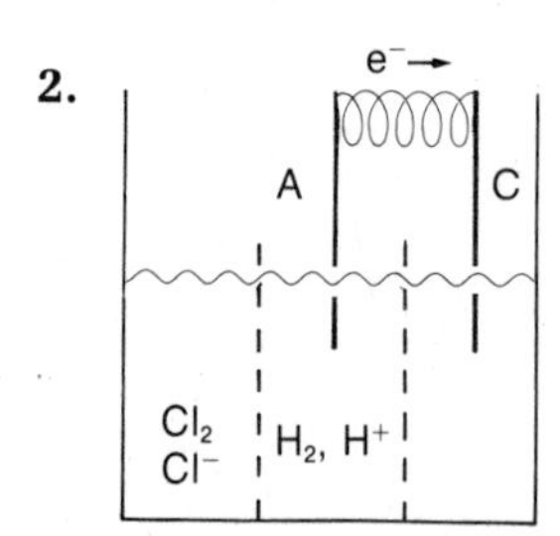

Cl^- ions move to anode (A)
H^+ ions move to cathode (C)

3. 1.05 V; 1.36 V **4.** 0.70 V **5.** $Pb > H_2 > Cu > Cl^-$

6. H^+ weakest, Au^{3+} strongest

7. All three reactions are spontaneous

8. anode: $2\ Cl^-(aq) \rightarrow Cl_2(g) + 2\ e^-$
cathode: $2\ H_2O + 2\ e^- \rightarrow H_2(g) + 2\ OH^-(aq)$
$2\ Cl^-(aq) + 2\ H_2O \rightarrow Cl_2(g) + H_2(g) + 2\ OH^-(aq)$

21. yes; $E^0 = +0.62$ V **22.** 1.90×10^{22}

23. a. $E^0 = +0.46$ V b. $E^0 = +0.03$ V
c. $Sn(s) + Sn^{4+}(aq) \rightarrow 2\ Sn^{2+}(aq)$; $E^0 = +0.29$ V

24. Increase pressure of Cl_2, increase conc. of Br^-, decrease conc. Cl^-

CHAPTER 25

1. a. 699.4 g b. 671.0 g c. 812.9 g
2. a. 0.344 g b. 0.147 g **3.** a. 75.3 g b. 13 g
4. a. $2\ AlCl_3(l) \rightarrow 2\ Al(l) + 3\ Cl_2(g)$
b. $BaCl_2(l) \rightarrow Ba(l) + Cl_2(g)$
c. $2\ KCl(l) \rightarrow 2\ K(l) + Cl_2(g)$
5. a. $2\ ZnS(s) + 3\ O_2(g) \rightarrow 2\ ZnO(s) + 2\ SO_2(g)$
b. $Cu_2S(s) + O_2(g) \rightarrow 2\ Cu(s) + SO_2(g)$
c. $2\ CoS(s) + 3\ O_2(g) \rightarrow 2\ CoO(s) + 2\ SO_2(g)$
d. $2\ As_2S_3(s) + 9\ O_2(g) \rightarrow 2\ As_2O_3(s) + 6\ SO_2(g)$
6. a. $Fe_2O_3(s) + 3\ CO(g) \rightarrow 2\ Fe(s) + 3\ CO_2(g)$
b. $ZnO(s) + CO(g) \rightarrow Zn(s) + CO_2(g)$
c. $SnO_2(s) + 2\ CO(g) \rightarrow Sn(s) + 2\ CO_2(g)$
d. $Cr_2O_3(s) + 3\ CO(g) \rightarrow 2\ Cr(s) + 3\ CO_2(g)$
15. $1.29, 0.3, 2, 0.5
16. a. $WO_3(s) + 3\ H_2(g) \rightarrow W(s) + 3\ H_2O(l)$ b. 0.366 ℓ

CHAPTER 26

1. a. 0.21 mol b. 0.42 mol c. 27 g
2. a. 5×10^{-9} g b. 5×10^{-6} g
3. a. 0.010 g b. 1.0×10^4
4. a. ${}^{235}_{92}U \rightarrow {}^{4}_{2}He + {}^{231}_{90}Th$ b. ${}^{232}_{90}Th \rightarrow {}^{4}_{2}He + {}^{228}_{88}Ra$
5. a. ${}^{137}_{55}Cs \rightarrow {}^{0}_{-1}e + {}^{137}_{56}Ba$ b. ${}^{90}_{38}Sr \rightarrow {}^{0}_{-1}e + {}^{90}_{39}Y$
6. a. 1.20 g b. 0.300 g **7.** a. 1/2 b. 1/1024
15. 1.2×10^{-3} **16.** a. 2.1×10^{-3} g b. 1.6 ppm
17. 0.42 g **18.** a. 0.50 g b. 50 ppm

CHAPTER 27

1. a. 33.9 kcal/g b. 2.42 kcal/g **2.** 68.0 kcal
3. 11.84 kcal/g **4.** a. -2.8×10^{-4} g b. 0.99972 g
5. a. 2.6 kcal b. -3.4×10^7 kcal
6. ${}^{239}_{94}Pu + {}^{1}_{0}n \rightarrow {}^{90}_{37}Rb + 3\ {}^{1}_{0}n + {}^{147}_{57}La$ **7.** ${}^{9}_{4}Be$
20. $8\ CO(g) + 17\ H_2(g) \rightarrow C_8H_{18}(l) + 8\ H_2O(l)$
21. 2.0×10^{16} kcal; 1.8×10^{16} kcal
22. -1.36×10^8 kcal **23.** 80 m^2

GLOSSARY

This glossary includes all of the "new terms" listed at the end of each chapter. The terms are defined in the context used in this text. If you want more information about the meaning of a term, look for it in the Index. This will refer you to the pages where the term is used.

A

absorbance—a quantity which is directly related to the fraction of light absorbed by a solution. The absorbance is directly proportional to the concentration of colored species in solution.

abundance—the mole percent of a particular isotope in an element occurring in nature. Example: In the element chlorine, the abundance of Cl-35 is 76%. This means that 76 of every 100 Cl atoms has a mass number of 35.

acid—a substance which, upon addition to water, produces H^+ ions. Examples include HCl, HNO_3, and CH_3COOH.

acid-base indicator—a substance which changes color as $[H^+]$ or pH changes. An example is phenolphthalein, which changes from colorless to pink at about pH 9.

acid-base reaction—the chemical change that occurs when an acid and a base are mixed. Example: When solutions of HCl and NaOH are mixed, the following acid-base reaction occurs: $H^+(aq) + OH^-(aq) \rightarrow H_2O$.

acidic solution—a solution in which $[H^+]$ is greater than 10^{-7} M (pH < 7).

actinide—an element of atomic number 90 through 103. One of the two series of "inner transition elements", the actinides appear at the bottom of the Periodic Table.

activated complex—an unstable intermediate formed in the course of a reaction. It has a higher energy than either reactants or products.

activation energy—the energy, E_a, which must be absorbed by reactant molecules in forming the activated complex.

addition polymer—a polymer formed by monomer units adding to each other. Polyethylene is an addition polymer formed when ethylene molecules add to one another.

addition reaction—a reaction in which a small molecule adds to an unsaturated hydrocarbon (alkene or alkyne). Example: $C_2H_4(g) + H_2(g) \rightarrow C_2H_6(g)$.

alcohol—an organic compound containing the —OH functional group. The —OH group is bonded to a hydrocarbon group. Example: CH_3OH (methyl alcohol).

aldehyde—an organic compound containing the functional group —C(=O)—H.

Example: $CH_3—C(=O)—H$ (acetaldehyde)

alkali metal—a metal in Group 1 of the Periodic Table. Examples: Li, Na, K.

alkaline earth metal—a metal in Group 2 of the Periodic Table. Examples: Be, Mg, Ca.

alkane—a hydrocarbon in which all the carbon-carbon bonds are single bonds. Examples:

CH_4, C_2H_6.

alkene—a hydrocarbon which contains one carbon-carbon double bond. Example:

$CH_2{=}CH_2$ (ethylene).

alkyl benzene—a hydrocarbon in which an alkyl group is substituted for a hydrogen atom of benzene. Example:

CH_3—C_6H_5 (toluene)

alkyl group—a hydrocarbon group derived from an alkane by the loss of a H atom. Example: CH_3— (methyl group).

alkyne—a hydrocarbon containing one carbon-carbon triple bond. Example:

$HC{\equiv}CH$ (acetylene).

allotropy—the existence of an element in two or more different forms in the same physical state. Example: the element oxygen exists in the allotropic forms $O_2(g)$ and $O_3(g)$.

alpha particle—a helium nucleus, He^{2+} ion. Its nuclear symbol is ^4_2He.

amide group—the —C(=O)—N(H)— group found in polyamide polymers such as nylon and in the natural polymers called proteins.

amino acid—an organic compound containing both the —COOH group and the —NH_2 group. The amino acids from which proteins are built have the general structure:

R—CH(NH_2)—COOH

ammonia—the compound of molecular formula NH_3.

ammonium ion—the polyatomic cation with the formula NH_4^+.

anion—a species carrying a negative charge. Examples: Cl^-, NO_3^-, SO_4^{2-}.

anode—the electrode at which oxidation occurs. Anions move toward the anode. Example: In the Zn—Cu^{2+} cell, the reaction occurring at the zinc anode is: $Zn(s) \rightarrow Zn^{2+}(aq) + 2\ e^-$

(aq)—the symbol used in a chemical equation to indicate that a species is in water solution. Example: $Zn^{2+}(aq)$ refers to Zn^{2+} ions in water solution.

aromatic hydrocarbon—a hydrocarbon containing a benzene ring. Examples include benzene itself, , and toluene, CH_3—

atmosphere—a unit of pressure equal to the pressure exerted by a column of mercury 760 mm high. The pressure of the atmosphere at sealevel is about 1 atm.

atom—smallest particle of an element. Matter is made up of atoms in various combinations.

atomic mass—a number which gives the average mass of an atom of an element relative to C-12. Example: The atomic mass of He is 4, which means that a helium atom is about $4/12$ or $1/3$ as heavy as a C-12 atom.

atomic number—the number of protons in the nucleus of an atom of an element. Example: the atomic number of C is 6, since there are six protons in the nucleus of every carbon atom.

atomic radius—the radius of a (spherical) atom. The Cl atom has a radius of 0.099 nm.

atomic theory—the theory, first proposed by Dalton, which holds that matter consists of tiny particles called atoms. In an element, all the atoms are chemically identical. Compounds contain two or more different kinds of atoms in a fixed ratio.

average speed—a quantity which tells us how fast, on the average, a gas molecule is moving. In $O_2(g)$ at 25°C, the average speed is 482 m/s. Some molecules are moving slower than this, some more rapidly.

Avogadro's Law—a principle stating that equal volumes of gases at the same T and P contain the same number of molecules.

Avogadro's number—the number of units in a mole, 6.022×10^{23}.

B

baking soda—sodium hydrogen carbonate, $NaHCO_3$.

balanced equation—an equation for a chemical reaction in which reactants and products contain equal numbers of each kind of atom. The equation, $CH_4(g) + 2\ O_2(g) \rightarrow CO_2(g) + 2\ H_2O(l)$, is balanced. Both reactants and products contain one C, four H, and four O atoms.

barometer—an instrument used to measure atmospheric pressure. In a mercury barometer, the height of the mercury column gives the atmospheric pressure in mm Hg.

base—a substance which, upon addition to water, produces OH^- ions. Examples include NaOH, $Ca(OH)_2$, and NH_3.

basic oxygen process—a process, described on p. 633, which is used to make steel from pig iron. It uses oxygen gas under pressure to oxidize impurities in the iron.

basic solution—a solution in which $[OH^-]$ is greater than 10^{-7} M (pH > 7).

bent molecule—a molecule containing three atoms in which the bond angle is less than 180°. An example is H_2O, where the geometry of the molecule is:

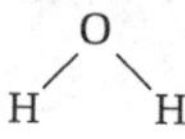

beta particle—an electron. In writing nuclear equations, the beta particle has the symbol $_{-1}^{0}e$.

blast furnace—a device, described on p. 630, which is used to convert iron ore (Fe_2O_3) to pig iron (Fe + C and other impurities).

Bohr equation—the equation which relates the energy of an electron in a hydrogen atom to the quantum number n. For one mole, the Bohr equation is:

$$E = \frac{-313.6\ \text{kcal}}{n^2}$$

where n = 1, 2, 3, ...

boiling point—the temperature at which the vapor pressure of a liquid becomes equal to the pressure above the liquid. At this point, vapor bubbles form within the liquid. The "normal" boiling point is the temperature at which the liquid boils when the pressure above it is exactly one atmosphere (760 mm Hg).

boiling point elevation—the extent to which the boiling point of a solution exceeds that of the pure solvent. Example: the normal boiling point of a 1 M sugar solution is about 100.5°C. Its boiling point elevation is 100.5°C − 100.0°C = 0.5°C.

Boyle's Law—the relation which states that the volume of a gas is inversely proportional to pressure (constant T, amount of gas). Mathematically: PV = constant, or $P_1V_1 = P_2V_2$

breeder reactor—a type of nuclear reactor which converts the common isotope of uranium, U-238, to an isotope of plutonium which is fissionable.

buret—a device used to deliver an accurately known volume of liquid (see Figure 1.5, p. 12).

C

calorie—a unit of energy equal to the amount of heat required to raise the temperature of one gram of water one degree Celsius (from 14.5 to 15.5°C).

carbohydrate—a class of organic compounds, nearly all of which have the general formula $C_x(H_2O)_y$. Examples: Glucose, $C_6H_{12}O_6$; Sucrose, $C_{12}H_{22}O_{11}$.

carbon-12 scale—scale upon which atomic masses are based. The relative mass of the $^{12}_{6}C$ atom is taken to be exactly 12.

carbonate ion—the CO_3^{2-} ion, found in such compounds as Na_2CO_3 and $CaCO_3$.

carbonyl group—the $\underset{\underset{\displaystyle O}{\|}}{-C-}$ functional group, found in organic aldehydes and ketones.

carboxyl group—the $\underset{\underset{\displaystyle O}{\|}}{-C}-OH$ functional group found in organic acids.

carboxylic acid—an organic acid containing the carboxyl group. Example: CH_3COOH (acetic acid).

catalyst—a substance which affects the rate of a reaction without being used up itself. Example: A piece of platinum foil can act as a catalyst for the combustion of methane in air.

cathode—the electrode at which reduction occurs. Cations move toward the cathode. Example: In a Zn—Cu^{2+} cell, the reaction occurring at the copper cathode is:

$$Cu^{2+}(aq) + 2\ e^- \rightarrow Cu(s).$$

cation—a species carrying a positive charge. Examples: Na^+, Fe^{3+}, NH_4^+.

Celsius scale—scale of temperatures in which 0°C is taken to be the freezing point of water and 100°C the boiling point of water at one atmosphere pressure.

centi—metric prefix meaning one hundredth. Example: 1 centimeter (1 cm) = 0.01 meter (0.01 m).

Charles' Law—a relation stating that the volume of a gas sample at constant pressure is directly proportional to its absolute temperature. That is, $V = kT$.

chemical equation—an expression which describes the nature and amounts of reactants and products taking part in a reaction. Example: $N_2(g) + 3\ H_2(g) \rightarrow 2\ NH_3(g)$. This equation tells us that one mole of nitrogen gas reacts with three moles of hydrogen gas to form two moles of ammonia gas.

chemical equilibrium—the condition that exists when a chemical system is in a state of dynamic balance. The rates of forward and reverse reactions are equal at equilibrium. Hence there is no net change in chemical composition with time.

chemical equivalence—a relation between amounts of reactants and products in a chemical reaction. In the reaction: $N_2(g) + 3\ H_2(g) \rightarrow 2\ NH_3(g)$, 1 mol of N_2, 3 mol of H_2, and 2 mol of NH_3 are chemically equivalent to one another. That is:

$$1\ mol\ N_2 \simeq 3\ mol\ H_2 \simeq 2\ mol\ NH_3.$$

chemical property—a property of a substance which is shown when it takes part in a chemical change. By studying the reaction of hydrogen with other elements, we can determine its chemical properties.

chlorate ion—the ClO_3^- ion found in such compounds as $KClO_3$.

chromate ion—the CrO_4^{2-} ion found in such compounds as K_2CrO_4.

coal gas—the gaseous mixture obtained from the destructive distillation of coal. It consists largely of methane and hydrogen.

coal tar—the viscous liquid obtained from the destructive distillation of coal.

coefficient—the number appearing before the formula of a substance in a chemical equation. Coefficients can be interpreted in terms of number of particles or number of moles. In the equation $N_2(g) + 3\ H_2(g) \rightarrow 2\ NH_3(g)$, we can say that 1 molecule of N_2 reacts with 3 molecules of H_2 to form 2 molecules of NH_3, or 1 mol of N_2 reacts with 3 mol of H_2 to form 2 mol of NH_3.

coke—the solid residue remaining from the destructive distillation of coal. It consists largely of carbon.

colligative property—a property of a solution which depends upon the concentration but not the kind of solute particle. Vapor pressure lowering, freezing point lowering, and boiling point elevation are examples of colligative properties.

colorimeter—an instrument used to determine the concentration of a colored species.

common-ion effect—an effect observed when an ion is added to a solution of a compound containing that ion. If the solution is saturated, addition of a common ion gives a precipitate. Example: Addition of 12 M HCl to a saturated solution of NaCl gives a precipitate of NaCl.

complexing agent—a reagent, such as NH_3 or NaOH, which forms a complex ion with a metal ion such as Zn^{2+}.

complex ion—a charged species consisting of a central metal cation bonded to two or more groups. The groups may be molecules as in $Zn(NH_3)_4^{2+}$ (4 NH_3 molecules bonded to Zn^{2+}). They may also be anions, as in $Zn(OH)_4^{2-}$ (4 OH^- ions bonded to Zn^{2+}).

compound—a substance containing more than one kind of atom.

concentrated solution—a solution with a relatively high ratio of solute to solvent.

concentration—a way of expressing the relative amounts of solute and solvent in a solution. A common concentration unit is molarity, M. A 6 M NaOH solution contains 6 mol of NaOH per liter of solution.

condensation polymer—a polymer formed from monomer units by splitting out a small molecule such as H_2O. Polyesters such as dacron and polyamides such as nylon are condensation polymers.

conversion factor—a ratio, numerically equal to one, by which a quantity is multiplied to obtain an equivalent quantity. Example: the conversion factor 12 in/1 ft can be used to convert a length in feet to inches.

copolymer—a polymer formed from two different kinds of monomer units. Example: the synthetic rubber SBR is a copolymer of styrene and butadiene.

covalent bond—a chemical link between two atoms, produced by shared electrons in the region between the atoms. Example: In the H_2O molecule there is a covalent bond between the O atom and each of the H atoms.

covalent network—a type of solid structure in which the atoms are linked together in a continuous framework of covalent bonds. Also referred to as "macromolecular". Diamond (C) and quartz (SiO_2) are covalent network solids.

cracking—a process in which an alkane decomposes to form an alkene. Example:

$$\underset{\text{alkane}}{C_4H_{10}(g)} \rightarrow \underset{\text{alkene}}{C_2H_4(g)} + C_2H_6(g)$$

crude mixture—a heterogeneous mixture; one in which the components are not uniformly distributed. Example: granite.

D

d orbital—any one of five orbitals within a d sublevel.

d sublevel—a sublevel with a capacity of 10 electrons which first appears in the third principal energy level.

Dalton's Law—a relation stating that the total pressure of a gas mixture is the sum of the partial pressures of its components.

density—a property of a sample equal to its mass per unit volume. Example: The density of liquid mercury is 13.5 g/cm^3.

destructive distillation—a process in which coal or wood is heated in the absence of air.

detergent—a synthetic organic compound used as a cleaning agent.

dichromate ion—the $Cr_2O_7^{2-}$ ion found in compounds such as $K_2Cr_2O_7$.

diffusion—a process by which one substance mixes with another as a result of molecular motion.

dihydrogen phosphate ion—the $H_2PO_4^-$ ion, found in compounds such as NaH_2PO_4.

dilute solution—a solution in which the ratio of solute to solvent is relatively small.

distillation—a procedure in which a liquid is vaporized in such a way that the vapor formed can be condensed and collected.

double bond—a covalent bond involving two electron pairs.

dry cell—a commercial voltaic cell, containing Zn and C electrodes. Between these electrodes is a moist paste containing NH_4Cl and MnO_2.

ductility—ability of a solid to retain strength on being forced through an opening. A ductile material can be drawn out into wire.

dynamic equilibrium—a condition where two opposing processes are occurring at equal rates. No change is observed in the system as time passes.

E

elastic collision—a molecular collision in which there is no net change in translational kinetic energy.

electrical conductivity—a term which describes the ease with which an electrical current can be passed through a sample. Metals, molten ionic compounds, and water solutions containing high concentrations of ions have high electrical conductivities.

electrode—a general name for anode or cathode.

electrolysis—the passage of a direct electric current through a liquid or solution containing ions. Chemical changes occur at the electrodes.

electrolyte—a species which exists as ions in water solution. Examples: NaCl (Na^+, Cl^- ions), HCl (H^+, Cl^- ions).

electrolytic cell—a cell in which electrolysis is carried out. An oxidation-reduction reaction is carried out within the cell. The reaction is ordinarily one that absorbs energy.

electron—the negatively charged particle found in all atoms. It carries a unit negative charge (−1) and has a very small mass.

electron cloud—the region of negative charge around the nucleus of an atom.

electron configuration—an expression giving the number of electrons in each sublevel of an atom. Example: Li has the electron configuration $1s^2\,2s^1$. This means that in the lithium atom there are two electrons in the 1s sublevel and one in the 2s sublevel.

electronegativity—a number which is directly related to the attraction that an atom shows for electrons. F, with an electronegativity of 4.0, attracts electrons more strongly than Cl, electronegativity 3.0.

electron pair—two electrons, which may either form a covalent bond between two atoms or remain unshared. In the Cl_2 molecule, which has the Lewis structure :C̤̈l—C̤̈l:, there is one electron-pair bond and a total of six unshared electron pairs.

electron pair repulsion—a principle of molecular geometry which predicts that the electron pairs around an atom will be directed in space so as to be as far apart as possible.

electron sea—a model of the particle structure of metals. According to this model, a metal consists of positive ions surrounded by a mobile "sea" of electrons.

element—a substance which contains only one kind of atom.

endothermic—describes a process which absorbs heat. For an endothermic process, ΔH is positive.

end point—the point in a titration at which an indicator changes color. Example: the end point in an acid-base titration using phenolphthalein as an indicator is at pH 9.

energy level—one of the energies permitted to an electron in an atom. We distinguish between principal energy levels and sublevels.

equilibrium—a balance between two opposing processes or forces. Once a system reaches equilibrium, it does not undergo further change unless it is disturbed.

equilibrium constant—a number which expresses the ratio of the equilibrium concentrations of products and reactants. Example: For the system $PCl_5(g) \rightleftharpoons PCl_3(g) + Cl_2(g)$, the equilibrium constant $K = [PCl_3] \times [Cl_2]/[PCl_5]$. At 250°C, K for this system is 0.050.

ester—an organic compound containing the —C(=O)—O— functional group. Example:

$$CH_3{-}\underset{\overset{\|}{O}}{C}{-}O{-}CH_3 \quad \text{(methyl acetate).}$$

ether—an organic compound which contains the —O— functional group. Example:

$$CH_3{-}CH_2{-}O{-}CH_2{-}CH_3 \quad \text{(diethyl ether).}$$

evaporation—process by which a liquid exposed to the atmosphere vaporizes.

excited state—a higher energy level, above the ground state, occupied momentarily by an electron.

exothermic—describes a process which evolves heat. For an exothermic process, ΔH is negative.

F

f orbital—any one of seven orbitals within an f sublevel.

f sublevel—a sublevel with a capacity of 14 electrons which first appears in the 4th principal energy level.

Fahrenheit scale—a temperature scale in which 32°F is taken to be the freezing point of water and 212°F the boiling point of water at one atmosphere pressure. The relationship between the Fahrenheit and Celsius scales is: °F = 1.8(°C) + 32°.

fat—an ester made from glycerol and a long-chain carboxylic acid.

filtration—a process for separating a solid-liquid mixture by passing it through a barrier with fine pores such as filter paper.

fission—a nuclear reaction in which a heavy nucleus such as U-235 splits into two fragments. A large amount of energy is evolved in the process.

fission reactor—a nuclear reactor which uses the fission process to generate electrical energy.

flame color—a characteristic color imparted to a flame by certain ions, notably those of the metals in Groups 1 and 2 of the Periodic Table.

flotation—process by which a sulfide ore is separated from rocky material. See Figure 25.8, p. 638.

flow chart—a diagram showing how a mixture of ions is separated and identified in qualitative analysis. Several typical flow charts are shown in Chapter 22.

fractional distillation—a procedure used to separate components with different boiling points from a mixture.

Frasch process—process used to extract sulfur from deep underground deposits. The sulfur is melted by superheated water. Compressed air is blown through the liquid to produce a froth which rises to the surface.

freezing point—the temperature at which solid appears when a liquid is cooled. At the freezing point, the solid and liquid are in equilibrium. The terms "freezing point" and "melting point" refer to the same temperature for a pure substance.

freezing point lowering—the extent to which the freezing point of a solution is exceeded by that of the pure solvent. Example: The freezing point of a 1 M sugar solution is about −1.9°C, as compared to 0°C for pure water. Hence, the freezing point lowering is 1.9°C.

functional group—an atom or group of atoms which is characteristic of a particular type of organic compound. Example: the —OH functional group is found in all alcohols.

fusion—a nuclear reaction in which two light nuclei combine. A typical fusion reaction is: ${}^{2}_{1}H + {}^{2}_{1}H \rightarrow {}^{4}_{2}He$. The amount of energy given off per gram of reactant is greater than in fission.

G

(g)—the symbol used after the formula of a substance in a chemical equation to indicate that it is present as a gas.

gamma ray—high energy radiation given off in nuclear reactions. The emission of gamma radiation does not change either the mass number or the atomic number.

gravimetric analysis—a type of quantitative analysis which involves measurements of only one quantity, mass.

ground state—the energy level occupied by an electron in a normal, unexcited atom. Electron configurations are ordinarily given for the ground state.

group—in the Periodic Table, the set of elements located in a single vertical column. Example: F, Cl, Br, and I are in Group 7. We also refer to groups in the qualitative analysis scheme. Example: Ag^+, Pb^{2+}, and Hg_2^{2+} are in Group I.

H

Haber process—an industrial preparation of ammonia from the elements nitrogen and hydrogen. It was developed by Fritz Haber, a German chemist.

half-equation—a chemical equation for a half-reaction, either oxidation or reduction. Examples: $Zn(s) \rightarrow Zn^{2+}(aq) + 2\ e^-$; $Cu^{2+}(aq) + 2\ e^- \rightarrow Cu(s)$.

half-life—the time for one half of a reactant, such as a radioactive isotope, to be consumed.

halide ion—a −1 ion derived from a halogen atom. The halide ions are F^-, Cl^-, Br^-, and I^-.

halogen—an element in Group 7 of the Periodic Table (fluorine, chlorine, bromine, or iodine).

hard water—water containing Ca^{2+} or Mg^{2+} ions.

heat content—a measure of the amount of energy stored in a substance or a system. This energy becomes evident in the form of heat. Physical and chemical changes are accompanied by a change in heat content. ΔH = heat content products − heat content reactants.

heat content diagram—a graph showing the relative heat contents of products and reactants. See Figures 4.5 and 4.6, Chapter 4.

heat flow—movement of energy in the form of heat. Heat flow has both direction and magnitude. When one gram of steam condenses, 540 cal of heat flows from the steam to the surroundings.

heat of combustion—the change in heat content, ΔH, for the combustion of a sample in excess oxygen. Example: The heat of combustion of $CH_4(g)$ is −212.8 kcal/mol.

heat of formation—the change in heat content, ΔH, for the formation of a compound from the elements in their stable states. Example: The heat of formation of $CO_2(g)$ from C(s) and $O_2(g)$ is −94.1 kcal/mol.

heat of fusion—the change in heat content, ΔH, for the melting of a solid. Example: The heat of fusion of ice is 80 cal/g or 1.44 kcal/mol.

heat of vaporization—the change in heat content, ΔH, for the vaporization of a liquid. Example: The heat of vaporization of water at 100°C is 540 cal/g or 9.72 kcal/mol.

Hess' Law—a relation which states that $\Delta H_3 = \Delta H_1 + \Delta H_2$, where process (3) is the sum of processes (1) and (2).

hydrate—a substance containing chemically bound water. Example: $BaCl_2 \cdot 2H_2O$

hydrated ion—a complex ion in which a simple ion is bonded to one or more water molecules. Examples: $Cu(H_2O)_4^{2+}$; H_3O^+ (hydronium ion)

hydride ion—the H^- ion found in such components as NaH.

hydrocarbon—an organic compound containing only H and C atoms.

hydrogen bond—a relatively strong type of intermolecular force. It arises from the attraction of the hydrogen atom of one molecule for a strongly electronegative atom (N, O, or F) in an adjacent molecule.

hydrogen carbonate ion—the HCO_3^- ion found in such compounds as $NaHCO_3$.

I

industrial preparation—method of preparing a substance economically on a large scale, suitable for use in the chemical industry. Example: The industrial preparation of $O_2(g)$ involves the fractionation of liquid air.

inert electrode—an electrode used in an electrical cell which takes no part in the cell reaction. Graphite and platinum are commonly used as inert electrodes.

inhibitor—a "negative" catalyst which slows down the rate of a reaction.

inner transition element—an element in either of two series: at. no. 58 through 71 (the lanthanides), at. no. 90 through 103 (the actinides). The inner transition elements are filling f sublevels (4f, 5f).

instrumental analysis—a type of quantitative analysis

in which an instrument is used to measure a physical property such as color which is related to the concentration of a species.

ion—a charged species. May be monatomic (Cl^-, Fe^{3+}) or polyatomic (CO_3^{2-}).

ionic bond—a chemical bond between ions of opposite charge.

ionic radius—the radius of a (spherical) ion. Example: the radius of the Cl^- ion is 0.181 nm.

ionization constant—the equilibrium constant for the ionization of a weak acid in water. For the weak acid HX:

$$HX(aq) \rightleftharpoons H^+(aq) + X^-(aq) \qquad K_a = \frac{[H^+] \times [X^-]}{[HX]}$$

ionization energy—ΔH for the ionization of a species. Example: the ionization energy of Li is 126 kcal/mol. That is: $Li(g) \rightarrow Li^+(g) + e^-$; ΔH = 126 kcal

isomerism—a situation where a single molecular formula represents two or more different compounds with different properties. Example: C_4H_{10} is the molecular formula of two different isomers, butane and 2-methyl propane.

isotope—an atom having the same number of protons as another atom, but with a different number of neutrons. Example: Ordinary oxygen has three isotopes. All of these have eight protons in the nucleus. One isotope has eight neutrons, another nine, and the third ten.

IUPAC name—the systematic name of an organic compound, derived according to a set of rules established by the International Union of Pure and Applied Chemistry.

J

joule—the base unit of energy in SI. 1 cal = 4.184 J

K

Kelvin scale—a temperature scale in which 0 K is taken as the lowest attainable temperature. It is related to the Celsius scale by the equation: K = °C + 273°

ketone—an organic compound containing the functional group —C(=O)—. Example: CH_3—C(=O)—CH_3

kilo—metric prefix meaning 1000. Example: 1 kg = 1000 g

kilocalorie—a unit of energy equal to 1000 cal.

kilopascal—a pressure unit used with SI. A mass of 10 g resting on a surface 1 cm² in area exerts a pressure of about 1 kPa. More exactly: 101.3 kPa = 1 atm.

kilowatt hour—a unit of energy frequently used with electrical energy. 1 kwh = 860 kcal

kinetic energy—energy of motion.

kinetic theory—a theory which explains the properties of gases in terms of the behavior of their molecules, particularly molecular motion.

L

(l)—a symbol written after the formula of a substance in a chemical equation to indicate that the substance is a pure liquid.

laboratory preparation—a method of preparing a substance suitable for use on a small scale in the chemistry laboratory.

lanthanide—an element of atomic number 58 through 71. The lanthanides are located in a separate row near the bottom of the Periodic Table.

lattice energy—ΔH for the formation of a crystal lattice from gaseous ions. Example: $Na^+(g) + Cl^-(g) \rightarrow NaCl(s)$; ΔH = lattice energy = −184 kcal

Law of Combining Volumes—a relation stating that the volumes of different gases involved in a reaction, at constant P and T, are in the same ratio as their coefficients in the balanced equation for the reaction.

Law of Conservation of Mass—a relation stating that in a chemical reaction the mass of the products equals the mass of the reactants.

Law of Constant Composition—a relation stating that a pure compound always contains the same elements in the same percentages by mass.

Le Chatelier's Principle—a relation stating that when a system at equilibrium is disturbed, it will respond in such a way as to partially counteract the change.

Lewis structure—a way of representing an atom or molecule to show the distribution of valence electrons. Examples: The Lewis structure of the Cl atom is :C̤̈l· while that of the Cl_2 molecule is :C̤̈l—C̤̈l:

limestone—calcium carbonate, $CaCO_3$.

linear molecule—a molecule in which all the atoms lie in a straight line. All the bond angles are 180°. Example: BeF_2, F—Be—F and C_2H_2, H—C≡C—H

liter—unit of volume; ℓ = 1000 cm³

luster—quality of a metal (or other substance) that gives it a shiny appearance due to reflection of light.

lye—sodium hydroxide, NaOH.

M

macromolecular—see *covalent network*

main group element—an element in any of the Groups 1 through 8 in the Periodic Table.

malleability—ability to be shaped, as by pounding with a hammer.

mass-energy relation—the equation relating the energy change in a process, ΔE, to the mass change, Δm. The equation may be written:

$$\Delta E = 2.15 \times 10^{10} \frac{kcal}{g} \times \Delta m$$

mass number—the sum of the number of protons and neutrons in an atomic nucleus. Example: The mass number of $^{37}_{17}Cl$ is 37; the nucleus of this atom contains 17 protons and 20 neutrons.

mass percentage—defined by the expression:

$$\text{mass \% A} = \frac{\text{mass A}}{\text{total mass}} \times 100.$$

mechanism—a sequence of steps that occurs during the course of a chemical reaction.

melting point—the temperature at which a liquid appears when a solid is heated. At the melting point, the solid and liquid are in equilibrium.

metallic character—having the properties of a metal, particularly the ease of loss of electrons.

metalloid—an element (B, Si, Ge, As, Sb, Te) which has

properties intermediate between those of metals and nonmetals.

milli—a metric prefix meaning one thousandth. Example: One millimeter (1 mm) = 0.001 meter (0.001 m).

millimeter of mercury—the pressure exerted by a column of mercury one millimeter high. 760 mm Hg = 1 atm

mixture—a sample of matter containing two or more substances.

molarity—a concentration unit which gives the number of moles of solute per liter of solution. Example: In 6 M HCl, there are 6 mol of HCl per liter of solution.

molar volume—the volume occupied by one mole of a substance. The molar volume of any gas at STP is approximately 22.4 ℓ.

mole—a counting unit used in chemistry. The mole represents 6.022×10^{23} items. The mass of a mole of a substance, in grams, is numerically equal to the formula mass of the substance. Examples: 1 mol of NH_3 contains 6.022×10^{23} NH_3 molecules and weighs 17.02 g. 1 mol of Cu contains 6.022×10^{23} Cu atoms and weighs 63.54 g. 1 mol of KNO_3 weighs 101.1 g.

molecular formula—an expression giving the number and kind of each atom in a molecule of a substance. Example: Since the molecular formula of hexane is C_6H_{14}, there are six C atoms and 14 H atoms in a molecule of hexane.

molecular mass—a number which gives the relative mass of a molecule on the C-12 scale. The molecular mass (MM) is obtained by adding the atomic masses of all the atoms in the molecular formula. Example: MM H_2O = 2 × AM H + AM O = 18.02. This means that the H_2O molecule is about 18/12 or 3/2 as heavy as a C-12 atom.

molecule—a group of atoms which serves as the basic building block in many substances. All gases, most common liquids, and some solids consist of molecules. Typical molecules include N_2 (nitrogen), H_2O (water), and C_6H_6 (benzene).

mole percentage—defined by the expression:

$$\text{mole \% A} = \frac{\text{no. moles A}}{\text{total no. moles}} \times 100$$

monatomic ion—a charged species derived from a single atom by the loss or gain of electrons. Examples: Na^+, Mg^{2+}, S^{2-}, Cl^-.

monohydrogen phosphate ion—the HPO_4^{2-} ion, found in compounds such as Na_2HPO_4.

monomer—the simple compound from which a polymer is derived. Example: polyethylene is derived from the monomer ethylene, C_2H_4.

N

nano—metric prefix meaning one billionth (10^{-9}). Example: One nanometer (1 nm) = 10^{-9} meter (10^{-9} m).

neutralization—the reaction: $H^+(aq) + OH^-(aq) \rightarrow H_2O$.

neutral solution—a solution of pH = 7.

neutron—one of the particles in an atomic nucleus. It has a mass number of 1 and a charge of 0.

nitrate ion—the NO_3^- ion found in compounds such as KNO_3 and $Ca(NO_3)_2$.

nitride ion—the N^{3-} ion found in compounds such as Na_3N.

noble gas—a gaseous element in Group 8 of the Periodic Table (He, Ne, Ar, Kr, Xe, Rn).

noble gas structure—the electron configuration of a noble gas. Except for He, the outer electron configuration of each noble gas is $ns^2\ np^6$, where n is the number of the period in which the noble gas is located.

nonelectrolyte—a substance that does not exist as ions in water solution. Example: ethyl alcohol, which exists in water as C_2H_5OH molecules.

nonmetal—one of the elements in the upper right-hand corner of the Periodic Table that does not show metallic properties. Example: Nitrogen gas, N_2, is a nonmetal.

nonpolar molecule—a molecule in which there are no + and − poles. Examples: H_2, Cl_2, CH_4.

nuclear equation—an equation written to represent a nuclear reaction. In such an equation, both atomic number and mass number must balance. Example: $^{226}_{88}Ra \rightarrow {}^{4}_{2}He + {}^{222}_{86}Rn$.

nucleus—the small, dense, positively charged region at the center of an atom.

O

octet rule—the principle that bonded atoms tend to be surrounded by eight valence electrons.

orbital—an electron cloud with a capacity of two electrons. An orbital is associated with a particular sublevel (s, p, d, or f).

orbital overlap—a model used to explain the stability of a covalent bond. It assumes that two orbitals, each with one electron, occupy part of the same region of space.

ore—mineral from which an element can be extracted commercially. Example: Bauxite (impure Al_2O_3) is an ore of aluminum.

organic chemistry—the branch of chemistry that deals with compounds containing carbon, hydrogen, and possibly other elements.

Ostwald process—an industrial process used to make nitric acid, HNO_3, starting with ammonia, NH_3.

oxidation—a half-reaction involving a loss of electrons or, more generally, an increase in oxidation number. Example: $Ag(s) \rightarrow Ag^+(aq) + e^-$ is an oxidation.

oxidation number—a number, assigned by a set of arbitrary rules, to an atom in a molecule or ion. Example: In the NO_3^- ion, the oxidation numbers of N and O are +5 and −2 in that order.

oxide ion—the O^{2-} ion, found in compounds such as Na_2O and CaO.

oxidizing agent—a species which accepts electrons from another species in an oxidation-reduction reaction. Example: In the reaction

$$Cl_2(g) + 2\ Br^-(aq) \rightarrow 2\ Cl^-(aq) + Br_2(l),$$

Cl_2 is the oxidizing agent.

ozone—an allotropic form of oxygen, molecular formula O_3.

P

p orbital—any one of three orbitals within a p sublevel.

p sublevel—a sublevel with a capacity of 6 electrons which first appears in the 2nd principal energy level.

partial pressure—the pressure that a component of a gas mixture would exert if it occupied the entire volume by itself (at the same temperature).

parts per million—in a liquid or solid mixture, the number of grams of a component per million grams of sample. Example: If the concentration of Pb^{2+} in water is 2 parts per million, there is 2 g of Pb^{2+} in 10^6 g of water.

percentage composition—a way of expressing the chemical composition of a compound, by giving the mass percentages of the elements present. Example: The percentage composition of water is 11.2 mass % H and 88.8 mass % O.

perchlorate ion—the ClO_4^- ion, found in compounds such as $KClO_4$.

period—a series of elements falling in the same horizontal row of the Periodic Table.

Periodic Law—a relation stating that many of the properties of elements vary in a cyclical (periodic) way with increasing atomic number.

Periodic Table—an arrangement of elements, in order of increasing atomic number, in horizontal periods of such length that elements with similar properties fall directly beneath one another. The Table is reproduced on the inside front cover of this text.

permanganate ion—the MnO_4^- ion, found in compounds such as $KMnO_4$.

pH—defined by the relation: $pH = -\log_{10}[H^+]$.

phosphate ion—the PO_4^{3-} ion, found in compounds such as Na_3PO_4.

photoelectric effect—a process in which electrons are ejected from a solid by the action of light.

physical property—a characteristic of a substance which can be measured without changing its chemical composition. Examples: density, melting point.

pig iron—the product obtained by the reduction of iron ore in the blast furnace. It consists of iron mixed with relatively large amounts of carbon and smaller amounts of other impurities.

pipet—a device used to deliver accurately a fixed volume of liquid such as 5 cm^3, 10 cm^3, ... See Figure 1.5, p. 12.

polar molecule—a molecule in which there are + and − poles. Example: The HCl molecule is polar because the bonding electrons are displaced toward the chlorine atom. The H atom acts as the positive pole, the Cl atom as the negative pole.

pollutant—a contaminant or foreign species present in a sample. Usually has an adverse effect on the quality of the sample as far as living things are concerned.

polyamide—a polymer in which there are repeating amide groups, $-\underset{\displaystyle O}{\underset{\|}{C}}-\underset{\displaystyle H}{\underset{|}{N}}-$. It is usually made from two monomers with the general structures: $HO-\underset{\displaystyle O}{\underset{\|}{C}}-R-\underset{\displaystyle O}{\underset{\|}{C}}-OH$ and $H-\underset{\displaystyle H}{\underset{|}{N}}-R'-\underset{\displaystyle H}{\underset{|}{N}}-H$.

polyatomic ion—a charged species containing more than one atom. Examples: NH_4^+, OH^-.

polyester—a polymer in which there are repeating ester groups, $-\underset{\displaystyle O}{\underset{\|}{C}}-O-$. It is usually made from two monomers with the general structures $HO-\underset{\displaystyle O}{\underset{\|}{C}}-R-\underset{\displaystyle O}{\underset{\|}{C}}-OH$ and $HO-R'-OH$.

polymer—a molecule of high molecular mass made of very small units which are bonded to each other. Example: Polyvinyl chloride, which has the structure:

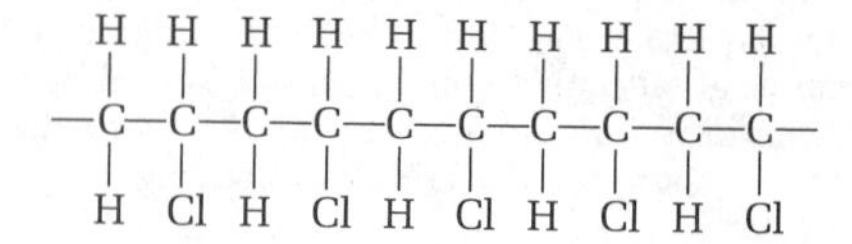

consists of a large number of vinyl chloride units, $-CH_2-CHCl-$

potential energy—energy associated with the position of a species or a system.

pound per square inch—pressure unit: 1 atm = 14.7 lb/in^2.

precipitation—a reaction in which a solid forms upon mixing two solutions. Example:

$$Ag^+(aq) + Cl^-(aq) \rightarrow AgCl(s).$$

pressure—force per unit area. Often expressed in mm Hg or in atmospheres.

principal energy level—main energy state (1st, 2nd, 3rd, . . .) of an electron in an atom. A principal level has a capacity of $2n^2$ electrons, where n is the quantum number of the level.

protein—a condensation polymer made from amino acid units found mainly in animal products.

proton—the nucleus of a hydrogen atom, the H^+ ion. A component of atomic nuclei with a charge of +1 and a mass of 1 on the C-12 scale.

pure substance—a sample of matter which consists of a single element or compound.

Q

qualitative analysis—the determination of the nature of the species present in a sample. Example: A "quarter" coin contains the two metals copper and nickel.

quantitative analysis—the determination of how much of a component is present in a sample. Example: A "quarter" coin contains 75% Cu and 25% Ni.

quantum mechanics—a science which can be applied to obtain an accurate model of the behavior of electrons in atoms.

quantum number—a number used to describe the energy levels available to an electron in an atom. The quantum number n (1, 2, 3, . . .) describes the principal energy level.

quantum theory—a theory of atomic structure which states that an electron in an atom can have only certain discrete energies.

quicklime—calcium oxide, CaO.

R

radioactive decay—the decomposition of an unstable nucleus to a more stable product.

radioactivity—a nuclear reaction in which an unstable nucleus decays by alpha, beta, or gamma emission.

rate constant—the proportionality constant in the rate equation for a reaction. Example: In the rate equation for the CO—NO_2 reaction,

$$rate = k\ (conc.\ CO) \times (conc.\ NO_2),$$

k is called the rate constant.

rate equation—the expression relating the rate of a reaction to the concentrations of reactants. Example: rate = k (conc. CO) × (conc. NO_2).

rate of reaction—the ratio: Δconc./Δt, where Δconc. is the change in concentration of a reactant or product in the time interval Δt.

recrystallization—a method used to purify a solid by the separation of crystals upon cooling a hot, saturated solution. Example: If impure KNO_3 is dissolved in the minimum amount of hot water and the solution cooled, pure KNO_3 crystallizes out.

reducing agent—a species which supplies electrons to another species in an oxidation-reduction reaction. Example: In the reaction

$$Cl_2(g) + 2\ Br^-(aq) \rightarrow 2\ Cl^-(aq) + Br_2(l),$$

Br^- is the reducing agent.

reduction—a half-reaction involving a gain of electrons or, more generally, a decrease in oxidation number. Example: $Cu^{2+}(aq) + 2\ e^- \rightarrow Cu(s)$ is a reduction.

refining—a process in which a substance is purified from small amounts of impurities. Often used as a last step in the extraction of a metal from an ore.

relative humidity—defined by the relation

$$R.H. = 100 \times P/P_0,$$

where P is the pressure of water vapor in the air and P_0 is the equilibrium vapor pressure of water at the same temperature.

resonance—a model used to explain a case where a single Lewis structure does not agree with the observed properties of a molecule. Two structures, called resonance forms, are written. These structures differ only in the distribution of valence electrons. The true structure of the molecule is believed to be intermediate between those of the resonance forms.

roasting—a process in which a sulfide ore such as ZnS or HgS is heated in air. The product may be a metal oxide (ZnO) or a metal (Hg).

rock salt—impure sodium chloride, NaCl.

S

(s)—a symbol written after the formula of a substance in a chemical equation to indicate that it is a pure solid.

s orbital—an orbital which comprises an s sublevel.

s sublevel—an energy sublevel which is found in all of the principal levels and has a capacity of 2 electrons.

saturated fat—a fat in which all the carbon-carbon bonds are single bonds.

saturated hydrocarbon—a hydrocarbon in which all the carbon-carbon bonds are single bonds. Also referred to as an alkane.

saturated solution—a solution which is in equilibrium with undissolved solute. The concentration of solute in a saturated solution is referred to as its "solubility".

significant figure—a digit which has experimental meaning. The quantities 2.26 g, 1.20×10^3 cm^3 and 0.0198 atm all contain three significant figures.

simplest formula—an expression which gives the simplest atom ratio in a compound. Example: The simplest formula of water is H_2O; that of hydrogen peroxide is HO.

skeleton structure—a preliminary structure used to obtain the Lewis structure of a molecule. In the skeleton, all the atoms are joined by single bonds.

slag—a molten waste product formed in extracting a metal from its ore. Slags often consist of silicates such as $CaSiO_3$.

slaked lime—calcium hydroxide, $Ca(OH)_2$.

soap—sodium salt of a long-chain carboxylic acid.

solar energy—energy obtained directly from the sun.

solubility product constant—equilibrium constant (K_{sp}) for the reaction by which an ionic solid dissolves in water. Example: For the reaction

$$AgCl(s) \rightleftharpoons Ag^+(aq) + Cl^-(aq),\ K_{sp} = [Ag^+] \times [Cl^-].$$

solubility rule—a statement concerning the water solubilities of members of a class of ionic compounds. Example: All nitrates are soluble in water.

solute—a component of a solution, usually one present in relatively small amounts. When a gas or solid dissolves in a liquid, the gas or solid is always referred to as the solute.

solution—a homogeneous mixture; one in which the components are distributed uniformly.

solvent—the component of a solution, usually a liquid, in which the solute is dissolved.

standard oxidation voltage (E^0_{ox})—the voltage associated with an oxidation half-reaction when all species are at standard concentrations. Example: E^0_{ox} for $Zn(s) \rightarrow Zn^{2+}(1\ M) + 2\ e^-$ is +0.76 V.

standard reduction voltage (E^0_{red})—the voltage associated with a reduction half-reaction when all species are at standard concentrations. Example: E^0_{red} for $Zn^{2+}(1\ M) + 2\ e^- \rightarrow Zn(s)$ is −0.76 V.

standard solution—a solution used to calibrate an instrument used in quantitative analysis. The standard solution is made up to a known concentration.

standard voltage (E^0)—the voltage associated with an oxidation-reduction reaction when all species are at standard concentrations. For any reaction, $E^0 = E^0_{ox} + E^0_{red}$.

STP—standard conditions of temperature and pressure (0°C and 1 atm) used in calculations involving gases.

strong acid—an acid which is completely ionized to give H^+ ions in water solution. The common strong acids are HCl, HBr, HI, HNO_3, $HClO_4$, and H_2SO_4.

strong base—a base which is completely ionized to give OH^- ions in water solution. The common strong bases are the hydroxides of the metals in Groups 1 and 2.

structural formula—an expression which shows the way in which atoms are bonded to one another in a molecule. Example: The structural formula of methyl alcohol is:

```
   H
   |
H—C—O—H
   |
   H
```

often abbreviated as CH_3OH.

structural isomers—two different compounds with the same molecular formula which differ in the way the atoms are linked together. Example: There are two structural isomers of C_4H_{10}. One of these is butane, $CH_3—CH_2—CH_2—CH_3$. The other is 2-methyl propane, $CH_3—CH—CH_3$ (with a CH_3 attached to the central CH).

subgroup—a vertical group of three transition metals in the Periodic Table. Example: The copper subgroup contains the three elements Cu, Ag, and Au.
sublevel—a division of a principal energy level. Example: The 2nd principal energy level is divided into the 2s and 2p sublevels. The 2s sublevel is slightly lower in energy than the 2p.
sublimation—a change in state in which a solid passes directly to the gas state.
sulfate ion—the SO_4^{2-} ion, found in compounds such as Na_2SO_4 and $CuSO_4$.
supersaturated solution—a solution in which the concentration of solute is greater than the equilibrium value. Such a solution is unstable in the presence of solute. If a tiny crystal of KNO_3 is added to a supersaturated solution of that compound, all of the excess solute comes out of solution.
symbol—one or two letters used to represent an element or one atom or one mole of that element. Examples: C, Ag.
synthesis gas—a gaseous mixture, made from coal, which consists principally of CO and H_2. It is used to make liquid hydrocarbons of the gasoline type.

T

temperature—a property whose magnitude determines the direction of heat flow (from high to low temperatures). Measured through its effects on the properties of matter.
tetrahedral molecule—a molecule such as CH_4 in which a central atom forms four bonds directed towards the corners of a regular tetrahedron. The bond angles are 109.5°.
tetrahedron—a figure with four sides, all of which are equilateral triangles. A tetrahedron has four corners, all equidistant from a point at the center. See Figure 11.4, p. 270.
thermionic emission—the emission of electrons from a solid, usually a metal, when it is heated.
thermochemical equation—a balanced chemical equation in which the value of ΔH is specified. Example: $CH_4(g) + 2\ O_2(g) \rightarrow CO_2(g) + 2\ H_2O(l)$; $\Delta H = -212.8$ kcal
titration—a process in which successive volumes of a solution of known concentration are added to a sample until reaction is complete.
transition element—any one of a series of 10 elements in the central region of the Periodic Table in the 4th, 5th, and 6th periods. Examples: Fe, Ag, Hg.
translational kinetic energy—energy of motion through space. A falling raindrop has translational kinetic energy, as does a molecule in motion.
triple bond—a bond consisting of three electron pairs, such as that between the carbon atoms in acetylene: H—C≡C—H.

U

uncertainty—a measure of experimental error. If an object is weighed to the nearest 0.01 g, the uncertainty in the measurement is ±0.01 g.
unsaturated fat—a fat molecule in which one or more carbon-carbon bonds is a multiple bond.
unsaturated hydrocarbon—a hydrocarbon molecule in which one or more carbon-carbon bonds is a multiple bond.
unsaturated solution—a solution in which the concentration of solute is less than the equilibrium value.
unshared electron—an electron which belongs entirely to one atom in a molecule. Example: In the :C̤̈l—C̤̈l: molecule, the electrons shown as dots are unshared.

V

valence electron—an electron in the outermost principal energy level of an atom. For the elements in Groups 1 through 8, the number of valence electrons per atom is given by the Group number.
Van der Waals forces—relatively weak intermolecular forces found in such substances as Cl_2 and ICl.
vaporization—change in state from a liquid to a gas.
vapor pressure—the pressure exerted by a vapor in equilibrium with a liquid. Example: The vapor pressure of water at 25°C is 24 mm Hg.
vapor pressure lowering—the extent to which the vapor pressure of solvent in a solution is exceeded by that of the pure solvent. Example: If the vapor pressure of water in a solution at 25°C is 20 mm Hg, the vapor pressure lowering is 24 mm Hg − 20 mm Hg = 4 mm Hg.
voltaic cell—a device in which a spontaneous oxidation-reduction reaction is used to produce electrical energy.
volumetric analysis—a type of quantitative analysis which involves at least one volume measurement.
volumetric flask—a special type of flask used to make up a solution to a desired concentration.

W

washing soda—sodium carbonate, Na_2CO_3.
water gas—a mixture of CO and H_2, produced by blowing steam through hot coke.
weak acid—an acid which is partially ionized to produce H^+ ions in water solution. Acetic acid is a weak acid because the following reaction reaches a point of equilibrium: $CH_3COOH(aq) \rightleftharpoons H^+(aq) + CH_3COO^-(aq)$
weak base—a base which is partially ionized to produce OH^- ions in water solution. Ammonia is a weak base because the following reaction reaches a point of equilibrium: $NH_3(aq) + H_2O \rightleftharpoons NH_4^+(aq) + OH^-(aq)$

INDEX

Page numbers in *italics* refer to illustrations; those followed by t refer to tables. Page numbers preceded by A and G refer to Appendix and Glossary entries.

A

D

N

O

TABLE OF ATOMIC MASSES (Based on Carbon-12)

	Symbol	Atomic No.	Atomic Mass		Symbol	Atomic No.	Atomic Mass
Actinium	Ac	89	[227]	Mercury	Hg	80	200.59
Aluminum	Al	13	26.98154	Molybdenum	Mo	42	95.94
Americium	Am	95	[243]*	Neodymium	Nd	60	144.24
Antimony	Sb	51	121.75	Neon	Ne	10	20.179
Argon	Ar	18	39.948	Neptunium	Np	93	237.0482
Arsenic	As	33	74.9216	Nickel	Ni	28	58.70
Astatine	At	85	[210]	Niobium	Nb	41	92.9064
Barium	Ba	56	137.34	Nitrogen	N	7	14.0067
Berkelium	Bk	97	[247]	Nobelium	No	102	[255]
Beryllium	Be	4	9.01218	Osmium	Os	76	190.2
Bismuth	Bi	83	208.9804	Oxygen	O	8	15.9994
Boron	B	5	10.81	Palladium	Pd	46	106.4
Bromine	Br	35	79.904	Phosphorus	P	15	30.97376
Cadmium	Cd	48	112.40	Platinum	Pt	78	195.09
Calcium	Ca	20	40.08	Plutonium	Pu	94	[244]
Californium	Cf	98	[251]	Polonium	Po	84	[210]
Carbon	C	6	12.011	Potassium	K	19	39.098
Cerium	Ce	58	140.12	Praseodymium	Pr	59	140.9077
Cesium	Cs	55	132.9054	Promethium	Pm	61	[147]
Chlorine	Cl	17	35.453	Protactinium	Pa	91	231.0359
Chromium	Cr	24	51.996	Radium	Ra	88	226.0254
Cobalt	Co	27	58.9332	Radon	Rn	86	[222]
Copper	Cu	29	63.546	Rhenium	Re	75	186.207
Curium	Cm	96	[247]	Rhodium	Rh	45	102.9055
Dysprosium	Dy	66	162.50	Rubidium	Rb	37	85.4678
Einsteinium	Es	99	[254]	Ruthenium	Ru	44	101.07
Erbium	Er	68	167.26	Samarium	Sm	62	150.4
Europium	Eu	63	151.96	Scandium	Sc	21	44.9559
Fermium	Fm	100	[257]	Selenium	Se	34	78.96
Fluorine	F	9	18.99840	Silicon	Si	14	28.086
Francium	Fr	87	[223]	Silver	Ag	47	107.868
Gadolinium	Gd	64	157.25	Sodium	Na	11	22.98977
Gallium	Ga	31	69.72	Strontium	Sr	38	87.62
Germanium	Ge	32	72.59	Sulfur	S	16	32.06
Gold	Au	79	196.9665	Tantalum	Ta	73	180.9479
Hafnium	Hf	72	178.49	Technetium	Tc	43	98.9062
Helium	He	2	4.00260	Tellurium	Te	52	127.60
Holmium	Ho	67	164.9304	Terbium	Tb	65	158.9254
Hydrogen	H	1	1.0079	Thallium	Tl	81	204.37
Indium	In	49	114.82	Thorium	Th	90	232.0381
Iodine	I	53	126.9045	Thulium	Tm	69	168.9342
Iridium	Ir	77	192.22	Tin	Sn	50	118.69
Iron	Fe	26	55.847	Titanium	Ti	22	47.90
Krypton	Kr	36	83.80	Tungsten	W	74	183.85
Lanthanum	La	57	138.9055	Uranium	U	92	238.029
Lawrencium	Lr	103	[256]	Vanadium	V	23	50.9414
Lead	Pb	82	207.2	Xenon	Xe	54	131.30
Lithium	Li	3	6.941	Ytterbium	Yb	70	173.04
Lutetium	Lu	71	174.97	Yttrium	Y	39	88.9059
Magnesium	Mg	12	24.305	Zinc	Zn	30	65.38
Manganese	Mn	25	54.9380	Zirconium	Zr	40	91.22
Mendelevium	Md	101	[258]				

*A value given in brackets denotes the mass number of the longest-lived or best-known isotope.

PERIODIC CHART OF

1	2							
1 H 1.0079								
3 Li 6.941	4 Be 9.01218							
11 Na 22.98977	12 Mg 24.305							
19 K 39.098	20 Ca 40.08	21 Sc 44.9559	22 Ti 47.90	23 V 50.9414	24 Cr 51.996	25 Mn 54.9380	26 Fe 55.847	27 Co 58.9332
37 Rb 85.4678	38 Sr 87.62	39 Y 88.9059	40 Zr 91.22	41 Nb 92.9064	42 Mo 95.94	43 Tc 98.9062	44 Ru 101.07	45 Rh 102.9055
55 Cs 132.9054	56 Ba 137.34	57 *La 138.9055	72 Hf 178.49	73 Ta 180.9479	74 W 183.85	75 Re 186.207	76 Os 190.2	77 Ir 192.22
87 Fr (223)	88 Ra 226.0254	89 ▼Ac (227)	104 § (260)	105 § (260)				

§The International Union for Pure and Applied Chemistry has not adopted official names or symbols for these elements.

* Lathanoid Series

58 Ce 140.12	59 Pr 140.9077	60 Nd 144.24	61 Pm (147)	62 Sm 150.4

▼ Actinoid Series

90 Th 232.0381	91 Pa 231.0359	92 U 238.029	93 Np 237.0482	94 Pu (244)